APPLIED FINITE MATHEMATICS

APPLIED
FINITE
MATHEMATICS

Edmond C. Tomastik

UNIVERSITY OF CONNECTICUT

Saunders College Publishing

HARCOURT BRACE COLLEGE PUBLISHERS

Philadelphia   Ft. Worth   Chicago   San Francisco   Montreal
Toronto   London   Sydney   Tokyo

Text Typeface: 10/12 Times Roman
Compositor: CRWaldman Graphic Communications
Acquisitions Editor: Jay Ricci
Developmental Editors: Anita Fallon and Anne Wightman
Managing Editor: Carol Field
Project Editor: Maureen Iannuzzi
Copy Editor: Merry Post
Manager of Art and Design: Carol Bleistine
Art Directors: Anne Muldrow and Susan Blaker
Art Assistant: Sue Kinney
Text Designer: Rebecca Lemna
Cover Designer: Jennifer Dunn
Text Artwork: Tech-Graphics
Director of EDP: Tim Frelick
Production Manager: Carol Florence
Marketing Manager: Monica Wilson
Photo Research Editor: Dena Digilio-Betz
Assistant Photo Editor: Lori Eby

On the cover: The detail on the cover is from a design adapted from a Turkish flatweave in the collection of the Textile Museum, Washington, D.C. (1978.19.1, gift of Mrs. Arthur S. Rudd). The original weave measures 160.5 cm × 229.5 cm and was purchased in 1958 in Crete. The weave is a supplementary-weft patterning.

Printed in the United States of America

Applied Finite Mathematics

ISBN: 0-03-097258-2

Library of Congress Catalog Card Number: 93-085702

9 0 1 2 3 4 5 6   016   11 10 9 8 7 6 5

In loving memory of my parents
Irene and Charles Tomastik

A Note About Custom Publishing

Courses in finite mathematics are structured in various ways, differing in length, content, and organization. To cater to these differences, Saunders College Publishing is offering *Applied Finite Mathematics* in a custom-published format. Instructors can rearrange, add, or cut chapters to produce a text that best meets their needs.

The diagram below shows the chapter dependencies in *Applied Finite Mathematics*, which instructors should consider. Beyond these dependencies, instructors, with custom publishing, are free to choose the topics they want to cover in the order they want to cover them, thereby creating a text that follows their course syllabi.

Saunders College Publishing is working hard to provide the highest quality service and product for your courses. If you have any questions about custom publishing, please contact your local Saunders sales representative.

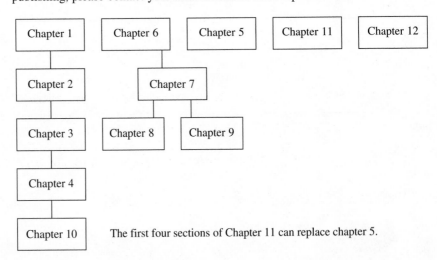

The first four sections of Chapter 11 can replace chapter 5.

Accuracy and "The Saunders Solution"

To ensure the accuracy of our math texts, Saunders is taking an additional step, "The Saunders Solution," in the proofreading process. At this time, we are the only publisher taking this extra step to ensure accuracy in our mathematics texts.

All Saunders mathematics textbooks, regardless of edition, are proofread in both the galley and pages stages of production. Instructor's Review Copies, identified as such on the book's cover, are then printed in the fall for adoption consideration only. The Instructor's Review Copy then undergoes "The Saunders Solution" accuracy check completed independently by both the author and paid accuracy reviewers. Any inconsistencies or errors in the texts are corrected in the Student Edition, which is available in the following spring.

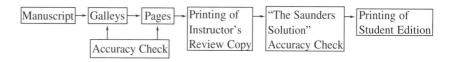

The implementation of "The Saunders Solution" ensures that all textbooks will be as clean as second printings and students will use textbooks that are as error-free and accurate as possible.

All publishers face the same concerns in dealing with the problem of errors in math texts. Only Saunders has taken an extra step to find a solution.

Preface

Applied Finite Mathematics is designed for a finite mathematics course aimed at students majoring in business, management, economics, or the life or social sciences. The text can be understood by the average student with one year of high school algebra. A wide range of topics is included, giving the instructor considerable flexibility in designing a course. Optional graphing calculator material is included.

Distinguishing Features

I have taught finite mathematics for many years, examined countless texts for adoption, used a variety of different texts in my classes, and, for a number of reasons, have found all of the current texts unsatisfactory. Thus I decided to write this text, which distinguishes itself from the others in the following ways.

First, applications truly play a central and prominent role in the text. Second, and in keeping with an emphasis on applications, I have stressed translating applied problems into mathematical equations or into an appropriate mathematical format for solution by calculators and computers. In this spirit I wrote a unique chapter, Chapter 11, on discrete dynamical systems, that teaches the skills of building appropriate mathematical models in business, finance, and science, as opposed to simply memorizing a set of formulas. The development has the considerable advantage of extending naturally to nonlinear equations and to the very frontiers of current research. Third, and again in keeping with an emphasis on applications, I have presented probability, not as an abstract mathematical subject, but as a natural phenomena that arises in business and science. By beginning with empirical probability I show how the abstract definition of probability arises, resulting in deep and practical insights into probability and statistics. Finally, I have given an easily understood geometric presentation of why and how the simplex method works, giving deep insights into this fundamentally important method.

Let me amplify on these points, one at a time. First, this text is written for *users* of mathematics. Thus, for example, a concrete applied problem is presented first as a motivation before developing a needed mathematical topic. After the mathematical topic has been developed, further applications are given so that the student understands the practical need for knowing the mathematics. This is done so consistently and thoroughly that after going through a number of chapters, the student should come to believe that mathematics is everywhere.

Second, I believe that no other skill is more important than the ability to translate a real-life problem into an appropriate mathematical format for finding the solution. Students often refer to this process as ''word problems.'' Whereas linear systems of equations, linear programming problems, and financial problems, for example, can easily be solved by computers and some calculators, no computer or calculator, now or in the foreseeable future, can translate these applied problems into the necessary mathematical language. Thus students, in their jobs, will most likely use their mathematical knowledge to translate applied problems into necessary mathematical formats for solution by computers.

To develop these needed skills many word problems, requiring the writing of one linear equation, are given in all of the sections in the first introductory chapter. This prepares the student for the many word problems that require creating systems of linear equations in the second chapter on linear systems and matrices. The word problems continue in the third chapter on linear programming, with the last section of this chapter devoted entirely to setting up linear programming problems for solution on computers. If desired, students can then use the software packages that are free to adopters of this text to solve the linear programming problems that have been written. Word problems then continue in the next chapter on the simplex method and beyond.

Chapter 11, on discrete dynamical systems, emphasizes the building of mathematical models and understanding the qualitative behavior of the solutions. Aside from developing the important skills of translating applied problems into mathematical equations, this approach has the advantage of readily extending to nonlinear equations, such as the logistic equation, and to reach the boundaries of current research with topics such as chaos. This chapter can replace the more traditionally written Chapter 5 on finance, expecially for classes less dominated by business and finance majors.

The third major area that distinguishes this text from others is the discussion found on probability. Other finite mathematics texts present probability as a *mathematical* subject with an *abstract mathematical* definition of probability given to the reader to accept on faith. If, on occasion, a text attempts to explain why this definition is used, the explanation is always confined to the equally likely outcomes case.

Since this text is not meant for mathematics majors but for business, finance, and science majors, the approach is to first present probability as a *natural* phenomena that arises in concrete problems in business, finance, and science. Thus empirical probability in a natural context is presented first. An experiment is performed N times or a survey of N people is taken. The frequency, $f(E)$, that

some particular event occurs is noted. The significance of the fraction $\dfrac{f(E)}{N}$, which
is called the empirical probability, is then easily explained. The student readily
understands that changing N will normally result in changing the empirical probability and that making N large should result in this empirical probability being
very close to some number, which is then defined as the actual probability. This,
of course, is precisely how probability arises everyday in business and science.
The basic properties of empirical probability are easily grasped, extended to probability, and then provide a basic motivation for the mathematical definition of
probability. The mathematical definition of probability then arises in a natural
context. Relating probability to relative frequencies is then used throughout the
text to give deep insights into other subjects.

For example, in statistics, the text easily demonstrates that after N trials of
an experiment with outcomes $x_1, x_2, \ldots, x_n$, the *average* outcome is just

$$x_1 \frac{f(x_1)}{N} + x_2 \frac{f(x_2)}{N} + \cdots + x_n \frac{f(x_n)}{N},$$

where x_1 is the first outcome, x_2 the second, and so on, while $f(x_1)$ is the frequency
of occurrence of the first outcome, $f(x_2)$ is the frequency of occurrence of the
second, and so on. Now when N is large, we expect the relative frequency $\dfrac{f(x_1)}{N}$
to be approximately the probability p_1 of the first outcome, $\dfrac{f(x_2)}{N}$ to be approximately the probability p_2 of the second outcome, and so on. Thus we have a
natural definition of expected value as

$$x_1 p_1 + x_2 p_2 + \cdots + x_n p_n,$$

and the student sees the somewhat mysterious expected value arising from an
everyday *average*.

Finally, a problem with current texts in finite mathematics is their presentation of the simplex method as a mechanical process with only a hint as to why
the method might work. In this text I give a geometrical presentation that my
students find easy to follow and that shows precisely how the simplex method
works. The students see precisely the roles played by the various steps, ratios,
and pivots. With these insights, the students are able to learn the additional topics
involved in the dual, the extended simplex method, and post optimal analysis.

Important Features

Style.   The text is interesting and can be understood by the average student with
a minimum of outside assistance. Material on a variety of topics is presented in
an interesting, informal, and student-friendly manner without compromising the
mathematical content and accuracy. Concepts are developed gradually, always
introduced intuitively, and culminate in a definition or result. Where possible,
general concepts are presented only after particular cases have been presented.
Scattered throughout the text, and set-off in boxes, are historical and anecdotal

comments. The historical comments are not only interesting in themselves, but also indicate that mathematics is a continually developing subject. The anecdotal comments relate the material to contemporary problems.

Applications.   The text includes many meaningful applications drawn from a variety of fields. For example, nearly every section opens by posing an interesting and relevant applied problem using familiar vocabulary, which is then solved later in the section after the appropriate mathematics have been developed. Applications are given for all the mathematics that are presented and are used to motivate the student.

Worked Examples.   About 430 worked examples, including about 130 self-help examples mentioned below, have been carefully selected to take the reader progressively from the simplest idea to the most complex. All the steps needed for the complete solutions are included.

Self-Help Exercises.   Immediately preceding each exercise set is a set of Self-Help Exercises. These 130 exercises have been very carefully selected to bridge the gap between the exposition in the chapter and the regular exercise set. By doing these exercises and checking the complete solutions provided, students will be able to test or check their comprehension of the material. This, in turn, will better prepare them to do the exercises in the regular exercise set.

Exercises.   The book contains over 2300 exercises. Each set begins with drill-type problems to build skills, and then gradually increases in difficulty. The exercise sets also include an extensive array of realistic applications from diverse disciplines. Graphing calculator exercises are included. (See below.)

Graphing Calculator Exercises.   These optional exercises appear at the end of appropriate exercise sets under the heading of Graphing Calculator Exercises, marked with the icon ▦. They are provided for those instructors who wish to supplement and complement the material by using graphing calculators.

Graphing Calculator Appendix.   Located at the end of the text, this appendix gives step-to-step explanations on how to use the TI-81. Topics covered include: basic graphing, the Gauss–Jordan method, matrix inverses and statistical calculations. Sample programs are also provided.

Student Aids

- **Boldface** is used when defining new terms.
- **Boxes** are used to highlight definitions, theorems, results, and procedures.
- **Remarks** are used to draw attention to important points that might otherwise be overlooked.
- **Warnings** alert students to common mistakes.
- **Titles** for worked examples help to identify the subject.

- **Chapter summary outlines**, at the end of each chapter, conveniently summarize all the definition, theorems, and procedures in one place.
- **Review exercises** are found at the end of each chapter.
- **Answers** to odd-numbered exercises and to all the review exercises are provided in an appendix.
- A student's **solution manual** that contains completely worked solutions to all odd-numbered exercises and to all chapter review exercises is available.
- The software package **MathPath** by George Bergman, Northern Virginia Community College, is available free to users of this text. It supplies graphical and computational support for many of the important topics in each chapter. Available for the IBM or IBM-compatible.
- **Graph 2D/3D**, a software package by George Bergeman, Northern Virginia Community College, is also available free to users. This software graphs functions in one variable and graphs surfaces of functions in two variables. It also provides computational support for solving calculus problems and investigating concepts. Available for the IBM or IBM-compatible.

Instructor Aids

- An **instructor's manual** with completely worked solutions to the even-numbered exercises is available free to adopters. A **student's solution manual** is free to adopters and contains the completely worked solutions to all odd-numbered and chapter review exercises.
- The **printed test bank** contains short-answer and multiple-choice test questions for the instructor to use in his or her own test format. Approximately 100 test questions per chapter are given. The printed test bank is available free to adopters.
- A **computerized test bank** allows instructors to quickly create, edit, and print tests or different versions of tests from the set of test questions accompanying the text. It is free to adopters and is available in IBM or Mac versions.
- The software package MathPath and Graph 2D/3D (see Student Aids) are available free to adopters.

Content Overview

Chapter 1. The first three sections contain review material on sets, cartesian coordinates, and lines. An introduction to the theory of the firm with some necessary economics background is provided to take into account the students' diverse backgrounds. The fourth (optional) section on least squares provides an example of the use of linear equations.

Chapter 2. Linear systems and the Gauss–Jordan method are covered in the first two sections of this chapter. The next three sections cover the basic material on matrices. Although many applications are included in the first five sections, the sixth section of this chapter is entirely devoted to input-output analysis, which is an application of linear systems and matrices used in economics.

Chapter 3. The first section presents linear inequalities with an emphasis on translating applied problems into an appropriate mathematical format. Basic material on geometric linear programming is included in the second section. The third section stresses translating more complicated applied problems into the linear programming format. If computers are available, all of these problems can be solved on the computer using the software available free to adopters.

Chapter 4. The first two sections present the simplex method for standard maximization problems and give considerable geometric insight into why the method works. The third section shows how to solve minimization problems by solving the dual problem, while the fourth section considers more general linear programming problems. The last section presents some post-optimal analysis.

Chapter 5. This chapter on finance does not depend on any of the other material and can be covered at any point in the course. It can also be replaced with the first four sections of Chapter 11, Discrete Dynamical Systems, in classes not dominated by business majors and for instructors less interested in covering the theory of finance and more interested in developing mathematical modeling.

Chapter 6. This chapter contains basic material on sets and counting, which is used extensively in the next chapter on probability.

Chapter 7. The basic material on probability is covered in this chapter. The second section introduces probability in a natural way by emphasizing empirical probability and the long-term behavior of the relative frequency of an event. This sets the background for the mathematical definition of probability given in the third section, motivates the definition of expected value and variance of a random variable given in the next chapter, and provides the background needed to understand the law of large numbers also given in the next chapter. Conditional probability, Bayes' theorem, and Bernoulli trials round out this chapter.

Chapter 8. This chapter contains the basic material on statistics. The expected value and variance are motivated using the notions of empirical probability and the long-term behavior of the relative frequency given in Section 7.2. Thus the discussion in Section 7.2 allows a deeper coverage of statistics than otherwise possible with other texts. The normal distribution and its approximation by the binomial distribution are covered in the fourth and fifth sections. The law of large numbers is found in Section 8.5. The last section covers the Poisson distribution.

Chapter 9. The basic material on Markov processes, covering both regular and absorbing Markov processes, is presented in this chapter.

Chapter 10. Game theory and its important connection to linear programming is presented in this chapter. This material gives the basics on the extensive interrelationship between linear programming and the celebrated theory of games developed by von Neumann and important in economic theory.

Chapter 11. This chapter is a basic introduction to the newly developing area of dynamical systems. The first four sections cover the linear theory of dynamical systems and the last two sections cover the nonlinear logistic equation, touching on the chaotic behavior of this fundamentally important equation. This chapter is less concerned with developing formulas than it is with mathematical modeling and determining the *qualitative* (as opposed to the exact) behavior of the solutions of the equations obtained in the modeling process. The first four sections can replace Chapter 5, Finance, in classes not dominated by business majors. The fifth section is an elementary introduction to the logistic equation and requires only a scientific calculator. The last section is primarily concerned with finding cycles and determining their stability. If a graphing calculator or a computer is available, a student can, in this section, explore areas of important contemporary research in mathematics.

Chapter 12. This chapter covers the basic topics in logic with an application in the last section to switching networks.

Acknowledgments

Many people at Saunders have played important roles in the production of this text. I especially want to thank Elizabeth Widdicombe, Publisher, for her generous support from the beginning to the end of this project. I also especially thank Robert Stern, my Acquisitions Editor, who signed me on and patiently guided the manuscript through many revisions, and my Executive Editor, Jay Ricci, who capably guided the project to its completion. I greatly appreciate the very important start that my first Developmental Editor, Donald Gacewicz, provided, and also greatly appreciate the considerable help given by the Developmental Editors Anne Wightman and Anita Fallon and Marketing Manager, Monica Wilson. I also thank Maureen Iannuzzi, the Project Editor, for doing such a great job and Merry Post for her excellent copyediting.

I wish to express my sincere appreciation to each of the following reviewers for their many helpful suggestions:

Paul Wayne Britt, LSU–Baton Rouge
Darrell Clevidence, Carl Sandburg College
Michael Friedberg, University of Houston
Carl T. Brezovec, University of Kentucky
Carol DeVille, LA Tech University
John Haverhals, Bradley University
Henry Mark Smith, University of New Orleans
Daniel Marks, Auburn University
Norman Martin, Northern Arizona University
Wayne Powell, Oklahoma State University

Caroline Woods, Marquette University
Harvey Yarborough, Brazosport College

A particular thanks to the following for checking the accuracy of the manuscript:

Eric Bibelnieks, Normandale Community College
Michael Friedberg, University of Houston
John Haverhals, Bradley University
William Livingston, Missouri Southern State College
David Morestad, University of North Dakota
Diane Tischer, Metropolitan Community College
Steven Winters, University of Wisconsin Oshkosh

At the University of Connecticut, there are many people that I would like to thank for playing an important role in this project. First I thank James Hurley for his continuous support, encouragement, and many suggestions. Also I thank Gerald Leibowitz and Joy Mark for using the manuscript in the classroom and providing many suggestions. I also thank Joseph McKenna for his interesting suggestions, Walter Lowrie for our discussions on linear programming, and Vince Giambalvo for his help in a variety of ways. I would like to also thank the teaching assistants Yong-Hong Chen, Yue Chen, Bill Gilmartin, Haiming Jen, Kelly Penfield, and Chunying Wang, for their considerable help in accuracy reviewing.

On a personal level, I am grateful to my wife for her love, patience, and support.

Edmond Tomastik
UNIVERSITY OF CONNECTICUT

November 1993

Contents

Applications Index

Business

Life Sciences

Social Sciences

The operation of a
factory involves
both fixed and
variable costs.

Cartesian Coordinates and Lines

In this chapter we present some very basic material that will be used throughout this text. We first give a brief introduction to sets. Introduced second is the Cartesian coordinate system, which provides a geometric way of presenting data and relationships between variables. Lines are then introduced, followed by applications of linear models.

1.1   SETS AND THE CARTESIAN COORDINATE SYSTEM

► *Sets*

► *Cartesian Coordinate System*

► *Distance Between Two Points*

► *Creating Equations*

► *Graphs*

René Descartes,
1596–1650

It is sometimes said that Descartes's work *La Geometrie* marks the turning point between medieval and modern mathematics. He demonstrated the interplay between algebra and geometry, tying together these two branches of mathematics. Due to poor health as a child, he was always permitted to remain in bed as long as he wished. He maintained this habit throughout his life and did his most productive thinking while lying in bed in the morning. Descartes was a great philosopher as well as a mathematician and felt that mathematics should be a model for other branches of study. The following is a quote from his famous *Discours sur la Méthode*: "The long chain of simple and easy reasonings by means of which geometers are accustomed to reach conclusions of their most difficult demonstrations led me to imagine that all things, to the knowledge of which man is competent, are mutually connected in the same way, and that there is nothing so far removed from us as to be beyond our reach, or so hidden that we cannot discover it, provided only we abstain from accepting the false for the true, and always preserve in our thoughts the order necessary for the deduction of one truth from another."

APPLICATION

Getting to the Church
on Time

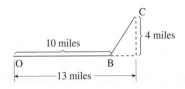

Figure 1.1

A prospective groom is located at point O in Figure 1.1 and must drive his car to his wedding taking place in 25 minutes at the church located at point C. The church at C is 13 miles east and 4 miles north of O. The only highway goes through a town located at B which is 10 miles due east of O. From B the highway goes straight to C. From O to B is a nice open stretch of highway, and he can average 50 miles per hour. The stretch from B to C has traffic lights and some congestion, and he can average only 25 miles per hour on this leg of the trip. Can he get to the church on time? See Example 5 for the answer.

Sets

The notion of a *set* forms the foundation for a great deal of mathematics. We need to introduce a few basic terms about sets and to consider some notation that is used to represent sets.

Sets

A **set** is a collection of objects, specified in a manner that enables one to determine if a given object is or is not in the collection.

EXAMPLE 1   **Determining if a Collection is a Set**

Which of the following collections are sets?

- The numbers from 1 to 10

- The past presidents of the United States
- The solutions of the equation $x^2 = 4$
- Good players who have played for the New York Yankees
- Interesting books published in 1992

Solution
The first 3 collections are sets. The last two collections are not sets since we are not given the criteria to determine whether a player is good or a book is interesting. ■

Each object in a set is called a **member** or **element** of the set. We often describe a set by listing all the objects between braces. Thus the set consisting of the first three months of the year can be written in **roster notation** as {January, February, March}.

We also use the **set-builder notation**.

Set-Builder Notation

The "set of all elements x such that x has property P" is written in set-builder notation as

$$\{x \mid x \text{ has property } P\}.$$

In this notation the vertical bar is read "such that."

E X A M P L E 2   **Using Set-Builder Notation**

Write the set {1, 2, 3, 4} in set-builder notation.

Solution   One solution is

$$\{1, 2, 3, 4\} = \{x \mid x \text{ is an integer and } 0 < x < 5\}. \quad ■$$

We also use the following notation.

$x \in A$	means	"x is an element in A."
$x \notin A$	means	"x is not an element in A."

Thus, for example,

$$3 \in \{x \mid x \text{ is an integer and } 0 < x < 5\}, \text{ and}$$

$$6 \notin \{x \mid x \text{ is an integer and } 0 < x < 5\}.$$

Equality of Sets

Two sets A and B are **equal**, written $A = B$ if, and only if, they consist of precisely the same elements.

E X A M P L E 3   **Showing Two Sets Are Equal**

a.   Determine if {1, 2, 3, 4} equals {4, 3, 2, 1}.
b.   Determine if {1, 1, 1, 1, 2, 3, 4} equals {1, 2, 3, 4}.
c.   Determine if {1, 2, 4} equals {1, 2, 3, 4}.

Solutions   a.   Since rearranging the order of the elements does not change the collection, {1, 2, 3, 4} = {4, 3, 2, 1}.
b.   Since listing the same element several times does not add any elements to the set, {1, 1, 1, 1, 2, 3, 4} = {1, 2, 3, 4}.
c.   Since 3 ∉ {1, 2, 4}, but 3 ∈ {1, 2, 3, 4}, {1, 2, 4} ≠ {1, 2, 3, 4}.   ■

Cartesian Coordinate System

It is very useful to have a geometric representation of the real numbers called the **number line**. The number line, shown in Figure 1.2, is a (straight) line with an arbitrary point selected to represent the number zero. This point is called the **origin**. A unit of length is selected, and, if the line is horizontal, the number one is placed an arbitrary distance to the right of the origin. The distance from the origin to the number one then represents the unit length. Each positive real number *x* then lies *x* units to the right of the origin, and each negative real number *x* lies −*x* units to the left of the origin. In this manner each real number corresponds to exactly one point on the number line, and each point on the number line corresponds to exactly one real number.

We will use the following convenient notation.

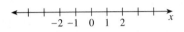

Figure 1.2

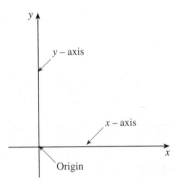

Figure 1.3

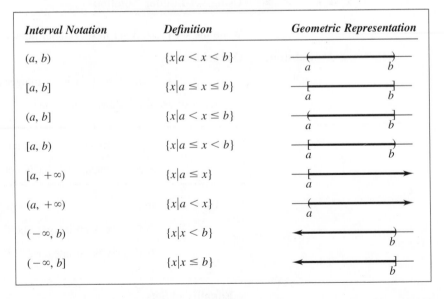

Interval Notation	Definition	Geometric Representation
(a, b)	$\{x \mid a < x < b\}$	
$[a, b]$	$\{x \mid a \leq x \leq b\}$	
$(a, b]$	$\{x \mid a < x \leq b\}$	
$[a, b)$	$\{x \mid a \leq x < b\}$	
$[a, +\infty)$	$\{x \mid a \leq x\}$	
$(a, +\infty)$	$\{x \mid a < x\}$	
$(-\infty, b)$	$\{x \mid x < b\}$	
$(-\infty, b]$	$\{x \mid x \leq b\}$	

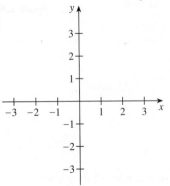

Figure 1.4

The Cartesian coordinate system (named after René Descartes) permits us to express data or relationships between two variables as geometric pictures. Representing numbers and equations as geometric pictures using the Cartesian co-

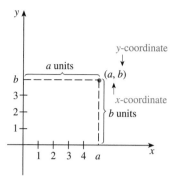

Figure 1.5

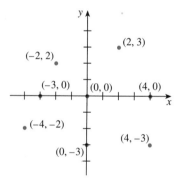

Figure 1.6

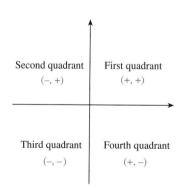

Figure 1.7

ordinate system yields deeper insight into the equations and data since we can see them graphically.

To describe a **Cartesian coordinate system** we begin with a horizontal line called the **x-axis** and a vertical line called the **y-axis**, both drawn in the plane. The point of intersection of these two lines is called the **origin**. The plane is then called the **xy-plane** or the **coordinate plane**. See Figure 1.3.

For each axis select a unit of length. The units of length for each axis need not be the same. Starting from the origin as zero, mark off the scales on each axis. For the x-axis, positive numbers are marked off to the right of the origin and negative numbers to the left. For the y-axis, positive numbers are marked off above the origin and negative numbers below the origin. See Figure 1.4.

Each point P in the xy-plane is assigned a pair of numbers (a, b) as shown in Figure 1.5. The first number a is the horizontal distance from the point P to the y-axis and is called the **x-coordinate**, and the second number b is the vertical distance to the x-axis and is called the **y-coordinate**.

Conversely, every pair of numbers (a, b) determines a point P in the xy-plane with x-coordinate equal to a and y-coordinate equal to b. Figure 1.6 indicates some examples.

We use the standard convention that $P(x, y)$ is the point P in the plane with Cartesian coordinates (x, y).

W A R N I N G .   The symbol (a, b) is used to denote both a point in the xy-plane and also an interval on the real line. The context will always indicate in which way this symbol is being used.

The x-axis and the y-axis divide the plane into four **quadrants**. The **first quadrant** is all points (x, y) with both $x > 0$ and $y > 0$. The **second quadrant** is all points (x, y) with both $x < 0$ and $y > 0$. The **third quadrant** is all points (x, y) with both $x < 0$ and $y < 0$. The **fourth quadrant** is all points (x, y) with both $x > 0$ and $y < 0$. See Figure 1.7.

Distance Between Two Points

We wish now to find the distance d between two given points $P_1(x_1, y_1)$ and $P_2(x_2, y_2)$ in the xy-plane shown in Figure 1.8.

Recall the Pythagorean theorem that says that $d^2 = a^2 + b^2$ in Figure 1.8. Then

$$d^2 = |x_2 - x_1|^2 + |y_2 - y_1|^2 = (x_2 - x_1)^2 + (y_2 - y_1)^2.$$

Taking square roots of each side gives the following.

Distance Formula

The distance $d = d(P_1, P_2)$ between the two points $P_1(x_1, y_1)$ and $P_2(x_2, y_2)$ in the xy-plane is given by

$$d(P_1, P_2) = \sqrt{(x_2 - x_1)^2 + (y_2 - y_1)^2}.$$

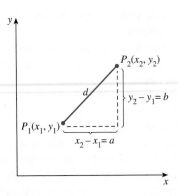

Figure 1.8

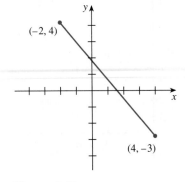

Figure 1.9

E X A M P L E   4   **Finding the Distance Between Two Points**

Find the distance between the two points $(-2, 4)$ and $(4, -3)$ shown in Figure 1.9.

Solution   We have $(x_1, y_1) = (-2, 4)$ and $(x_2, y_2) = (4, -3)$. Thus

$$d(P_1, P_2) = \sqrt{(x_2 - x_1)^2 + (y_2 - y_1)^2}$$
$$= \sqrt{[(4) - (-2)]^2 + [(-3) - (4)]^2}$$
$$= \sqrt{(6)^2 + (-7)^2} = \sqrt{85}. \ \blacksquare$$

E X A M P L E   5   **Using the Distance Between Two Points**

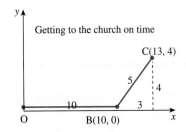

Figure 1.10

A prospective groom is located at point O in Figure 1.10 and must get, by car, to his wedding located at the church at point C in 25 minutes. The church at C is located 13 miles east and 4 miles north of O. The only highway goes through a town located at B, which is 10 miles due East of O. From O to B is a nice open stretch of highway, and he can average 50 miles per hour. The stretch from B to C has traffic lights and some congestion, and he can average only 25 miles per hour on this leg of the trip. Can he get to the church on time?

Solution   The distance from O to B is 10 miles, while the distance from B to C is $\sqrt{(13 - 10)^2 + (4 - 0)^2} = 5$ miles. Distance is the product of velocity with time, $d = v \cdot t$. Thus $t = d/v$. Then the time to travel the given route is

$$\frac{10}{50} + \frac{5}{25} = \frac{2}{5}$$

of an hour, or 24 minutes. This is less than 25 minutes, so he can make his wedding on time. $\blacksquare$

Creating Equations

In order to solve any applied problem, we must first take the problem and translate it into mathematics. Doing this requires creating equations.

E X A M P L E 6   **Creating an Equation**

A small shop makes two styles of shirts. The first style costs $7 to make, and the second $9. The shop has $252 a day to produce these shirts. If x is the number of the first style produced each day and y is the number of the second style, find an equation that x and y must satisfy.

Solution   Create a table that includes the needed information. See Table 1.1.

Since x are produced at a cost of $7 each, the cost in dollars of producing x of these shirts is $7x$. Since y are produced at a cost of $9 each, the cost in dollars of producing y of these shirts is $9y$. This is indicated in Table 1.1. Since the cost of producing the first style plus the cost of producing the second style must total $252, the equation we are seeking is

$$7x + 9y = 252. \quad \blacksquare$$

Table 1.1

Style	First	Second	Total
Cost of each	$7	$9	
Number produced	x	y	
Cost	$7x$	$9y$	252

Graphs

Given any equation in x and y (such as $y = 2x - 1$), the **graph of the equation** is the set of all points (x, y) such that x and y satisfy the equation.

> **Graph of an Equation**
>
> The **graph of an equation** is the set
>
> $$\{(x, y) | x \text{ and } y \text{ satsify the equation}\}.$$

For example the point $(0, -1)$ is on the graph of the equation $y = 2x - 1$, because when x is replaced with 0 and y replaced with -1 the equation is satisfied. That is, $(-1) = 2(0) - 1$.

E X A M P L E 7   **Determining a Graph of an Equation**

Sketch the graph of $y = 2x - 1$, that is, find the set $\{(x, y)\,|\,y = 2x - 1\}$.

Solution   We take a number of values for x, put them into the equation and solve for the corresponding values of y. For example if we take $x = -3$, then $y = 2(-3) - 1 = -7$. Table 1.2 summarizes this work

Table 1.2

x	-3	-2	-1	0	1	2	3
y	-7	-5	-3	-1	1	3	5

The points (x, y) found in Table 1.2 are graphed in Figure 1.11. We see that the graph is a straight line. ■

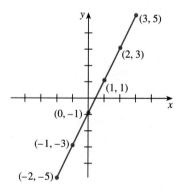

Figure 1.11

To find the graph of an equation, plot points until a pattern emerges. More complicated equations will require more points.

E X A M P L E 8   **Determining the Graph of an Equation**

Sketch the graph of $x = 2$, that is, find the set $\{(x, y)\,|\,x = 2\}$.

Solution   Points such as $(2, -1)$, $(2, 1)$, and $(2, 2)$ are all on the graph. The graph is a vertical line, as shown in Figure 1.12. ■

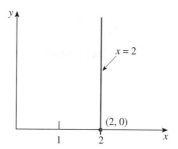

Figure 1.12

E X A M P L E   9   **Determining a Graph of an Equation**

Sketch the graph of $y = x^2$, that is, find the set $\{(x, y) | y = x^2\}$.

Solution   We take a number of values for x, put them into the equation and solve for the corresponding values of y. For example if we take $x = -2$, then $y = (-2)^2 = 4$. Table 1.3 summarizes this work.

The points (x, y) found in Table 1.3 are graphed in Figure 1.13.   ■

Table 1.3

x	-3	-2	-1	0	1	2	3
y	9	4	1	0	1	4	9

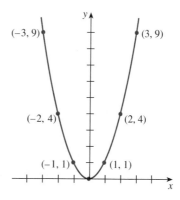

Figure 1.13

SELF-HELP EXERCISE SET 1.1

1. Plot the points $(-2, 5)$, $(-2, -4)$, and $(3, -2)$.

2. Find the distance between the two points $(-4, -2)$ and $(3, -4)$.

3. Sketch the graph of the equation $7x + 9y = 252$ found in Example 6.

EXERCISE SET 1.1

In Exercises 1 through 4 which of the statements determine sets?

1. **a.** The past presidents of Harvard University
   b. The past and future presidents of Harvard University

2. **a.** The good past presidents of Harvard University
   b. The past presidents of Harvard University who were at least 6 feet tall

3. **a.** Individuals who played for the New York Yankees in 1991
   b. All-around good New York Yankee players during 1991

4. **a.** Los Angeles Dodger players who had a batting average of at least .300 during 1991
   b. The good-looking Los Angeles Dodger players of 1991

In Exercises 5 through 8 write the sets in set-builder notation.

5. **a.** $\{1, 3, 5, 7, 9, 11, 13, 15\}$
   b. $\{3, 6, 9, 12, 15, 18, 21\}$

6. **a.** The positive integers less than 10
   b. The solutions to $(x - 2)(x - 3) = 0$

7. **a.** $\{2, 4, 6, 8, 10, 12, 14, 16\}$
   b. The odd positive integers

8. **a.** $\{5, 10, 15, 20, 25, 30\}$
   b. $\{a, b, c, d, e\}$

In Exercises 9 through 12 write the sets in roster notation.

9. **a.** $\{x | x \text{ is an integer}, 5 \le x < 12\}$
   b. The positive even integers less than or equal to 13

10. **a.** $\{x | x \text{ is a letter in the word "mathematics"}\}$
    b. $\{x | (x + 1)(x - 1) = 0\}$

11. **a.** $\{x | x \text{ is an integer}, 3 < x \le 10\}$
    b. The positive odd integers less than or equal to 13

12. **a.** $\{x | x \text{ is a letter in the word "mathematics" that appears at least twice}\}$
    b. $\{x | (x + 3)(x - 2) = 0\}$

13. If $A = \{u, v, y, z\}$, determine whether the following statements are true or false.
    a. $w \in A$, **b.** $x \notin A$, **c.** $z \in A$, **d.** $v \notin A$

14. If $A = \{u, v, y, z\}$, determine whether the following statements are true or false.
    a. $x \in A$, **b.** $y \notin A$, **c.** $y \in A$, **d.** $t \in A$

In Exercises 15 through 22 draw a number line and identify.

15. $(-1, 3)$ **16.** $[-2, 6]$

17. $(2, 5]$ **18.** $[-3, -1)$

19. $(-2, +\infty)$ **20.** $[2, +\infty)$

21. $(-\infty, 1.5]$ **22.** $(-\infty, -2)$

In Exercises 23 through 34 plot the given point in the xy-plane.

23. $(1, 2)$ **24.** $(-3, -4)$

25. $(-2, 1)$ **26.** $(2, -3)$

27. $(1, 0)$ **28.** $(-2, 0)$

29. $(0, -2)$ **30.** $(0, 4)$

31. $(-2, -1)$ **32.** $(2, 2)$

33. $(3, -4)$ **34.** $(-3, 0)$

In Exercises 35 through 46 find the set of points (x, y) in the xy-plane that satisfies the given equation or inequality.

35. $x \geq 0$ **36.** $y \geq 0$

37. $x < 0$ **38.** $y < 0$

39. $x = 2$ **40.** $x = -1$

41. $y = 3$ **42.** $y = -2$

43. $xy > 0$ **44.** $xy < 0$

45. $xy = 0$ **46.** $x^2 + y^2 = 0$

In Exercises 47 through 54 find the distance between the given points.

47. $(2, 4), (3, 2)$ **48.** $(-2, 5), (2, -3)$

49. $(-2, 5), (-3, -1)$ **50.** $(1, 3), (1, 6)$

51. $(-3, -2), (4, -3)$ **52.** $(-5, -6), (-3, -2)$

53. $(a, b), (b, a)$ **54.** $(x, y), (0, 0)$

In Exercises 55 through 64 sketch the graphs.

55. $y = -2x + 1$ **56.** $y = -2x - 2$

57. $y = 3x - 2$ **58.** $y = 2x + 3$

59. $y = 3$ **60.** $x = 4$

61. $y = 3x^2$ **62.** $y = -x^2$

63. $y = x^3$ **64.** $y = \sqrt{x}$

Applications

65. Transportation.   A truck travels a straight highway from O to A and then another straight highway from A to B. How far does the truck travel?

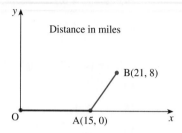

66. Transportation.   If the truck in the previous exercise travels 30 miles per hour from O to A and 60 miles per hour from A to B, how long does the trip take?

67. Navigation.   A ship leaves port and heads due east for 4 hours with a speed of 3 miles per hour, after which it turns due north and maintains this direction at a speed of 2.5 miles per hour. How far is the ship from port 6 hours after leaving?

68. Navigation.   A ship leaves port and heads due east for 1 hour at 5 miles per hour. It then turns north and heads in this direction at 4 miles per hour. How long will it take the ship to get 13 miles from port?

69. Oil Slick.   The shoreline stretches along the *x*-axis as shown with ocean consisting of the first and second quadrants. An oil spill occurs at the point (2, 4), and a coastal town is located at (5, 0) where the numbers represent miles. If the oil slick is approaching the town at a rate of one-half mile per day, how many days until the oil slick reaches the coastal town located at (5, 0)?

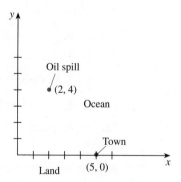

70. Transportation Charges.   A furniture store is located 5 miles east and 14 miles north of the center of town. You live 10 miles east and 2 miles north of the center of town. The store advertises free delivery within a 12-mile radius of the store. Do you qualify for free delivery?

71. **Profits.** A furniture store sells chairs at a profit of $100 each and sofas at a profit of $150 each. Let x be the number of chairs sold each week and y the number of sofas sold each week. If the profit in one week is $3000, write an equation that x and y must satisfy. Graph this equation.

72. **Revenue.** A restaurant has two specials, steak and chicken. The steak special is $13, and the chicken special is $10. Let x be the number of steak specials served and y the number of chicken specials. If the restaurant made $390 in sales on these specials, find the equation that x and y must satisfy.

73. **Nutrition.** An individual needs 800 mg of calcium daily but is unable to eat any dairy products due to the large amounts of cholesterol in such products. This individual does enjoy eating canned sardines and steamed broccoli however. Let x be the number of ounces of canned sardines consumed each day and let y be the number of cups of steamed broccoli. If there are 125 mg of calcium in each ounce of canned sardines and 190 mg in each cup of steamed broccoli, what equation must x and y satisfy for this person to satisfy the daily calcium requirement from these two sources?

(© *Spencer Grant/Photo Researchers, Inc.*)

74. **Costs.** A contractor builds ranch and split-level style homes. The ranch costs $130,000 to build and the split-level $150,000. Let x be the number of ranch-style homes built and y the number of split-level style homes. If the contractor has $1,360,000 to build these homes, find the equation that x and y must satisfy. Graph this equation.

75. **Biology.** A fisheries biologist decides to set a gill net on his next trip to a remote mountain lake. The net must stretch between points A and B shown in the figure. Since he must backpack to the lake he wants to carry no more net than necessary. He walks off the distance a from C to B and the distance b from A to C, being careful that AC is perpendicular to AB. Find the distance c from A to B in terms of a and b.

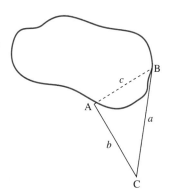

Solutions to Self-Help Exercise Set 1.1

1. The points are plotted in the following figure.

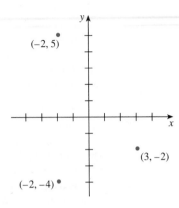

2. The distance between the two points $(-4, -2)$ and $(3, -4)$ is given by

$$\sqrt{[3 - (-4)]^2 + [-4 - (-2)]^2} = \sqrt{7^2 + (-2)^2} = \sqrt{53}$$

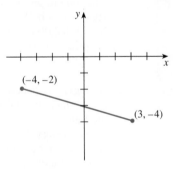

3. First solve for y and obtain $y = -\frac{7}{9}x + 28$. Now take a number of values for x, put them into the equation, and solve for the corresponding values of y. For example if we take $x = 9$, then $y = -\frac{7}{9}(9) + 28 = 21$. The following table summarizes this work.

x	0	9	18	27	36
y	28	21	14	7	0

The points (x, y) found in the table are graphed in the figure. We see that apparently the graph is a straight line.

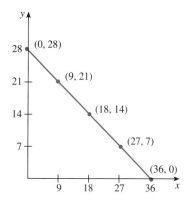

1.2   LINES

► *Slope*

► *Equations of Lines*

► *Applications*

Production of Vacuum Cleaners

A firm produces two models of vacuum cleaners, regular and deluxe. The regular model requires 8 hours of labor to assemble while the deluxe requires 10 hours of labor. Let x be the number of regular models produced and y the number of deluxe models. If the firm wishes to use 500 hours of labor to assemble these cleaners, find an equation that the two quantities x and y must satisfy. If the number of regular models to be assembled is increased by 5, how many fewer deluxe models will be assembled? See Example 8 for the answer.

Slope

In order to describe a straight line, one must first describe the "slant" or **slope** of the line. See Figure 1.14.

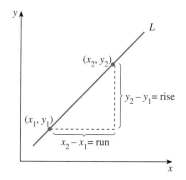

Figure 1.14

Slope of a Line

Let (x_1, y_1) and (x_2, y_2) be two points on a line L. If the line is nonvertical $(x_1 \neq x_2)$, the **slope** m of the line L is defined to be

$$m = \frac{y_2 - y_1}{x_2 - x_1}.$$

If L is vertical $(x_2 = x_1)$, then the slope is said to be **undefined**.

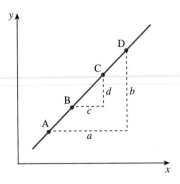

Figure 1.15

REMARK.   The term $x_2 - x_1$ is called the **run** and $y_2 - y_1$ is called the **rise**. See Figure 1.14. Thus the slope m can be thought of as

$$m = \frac{y_2 - y_1}{x_2 - x_1} = \frac{\text{rise}}{\text{run}}.$$

The slope of the line does not depend on which two points on the line are chosen to use in the above formula. To see this, notice in Figure 1.15 that the two right triangles are similar since the corresponding angles are equal. Thus the ratio of the corresponding sides must be equal. That is,

$$\frac{b}{a} = \frac{d}{c},$$

where the ratio on the left is the slope using the points A and D and the ratio on the right is the slope using the points B and C.

E X A M P L E   1   **Finding the Slope of Lines**

Find the slope (if it exists) of the line through each pair of points. Sketch the line.
(a) (1, 2), (3, 6),   (b) (−3, 1), (3, −2),   (c) (−1, 3), (2, 3),   (d) (3, 4), (3, 1)

Solutions   (a) $(x_1, y_1) = (1, 2)$ and $(x_2, y_2) = (3, 6)$. So

$$m = \frac{y_2 - y_1}{x_2 - x_1} = \frac{(6) - (2)}{(3) - (1)} = \frac{4}{2} = 2.$$

See Figure 1.16. Notice that for each unit we move to the right, the line moves up $m = 2$ units.

(b) $(x_1, y_1) = (-3, 1)$ and $(x_2, y_2) = (3, -2)$. So

$$m = \frac{y_2 - y_1}{x_2 - x_1} = \frac{(-2) - (1)}{(3) - (-3)} = \frac{-3}{6} = -\frac{1}{2}.$$

See Figure 1.17. Notice that for each unit we move to the right, the line moves down $\frac{1}{2}$ unit.

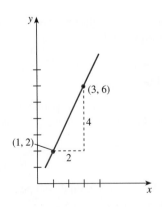

Figure 1.16

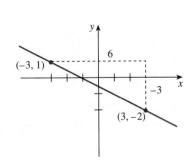

Figure 1.17

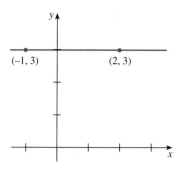

Figure 1.18

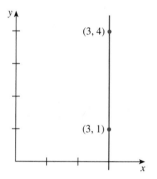

Figure 1.19

(c) $(x_1, y_1) = (-1, 3)$ and $(x_2, y_2) = (2, 3)$. So

$$m = \frac{y_2 - y_1}{x_2 - x_1} = \frac{(3) - (3)}{(2) - (-1)} = \frac{0}{3} = 0.$$

See Figure 1.18. Notice that this line is horizontal.

(d) $(x_1, y_1) = (3, 4)$ and $(x_2, y_2) = (3, 1)$. So

$$m = \frac{y_2 - y_1}{x_2 - x_1} = \frac{(1) - (4)}{(3) - (3)} = \frac{-3}{0} = \text{undefined.}$$

See Figure 1.19. Notice that this line is vertical.   ■

Consider any line L that is not vertical and let $P_1(x_1, y_1)$ be one point on the line. We noted earlier that in determining the slope of a line, *any* two points on the line can be used. We will now take a second point $P_2(x_2, y_2)$ on the line where P_2 is chosen with $x_2 = x_1 + 1$. Thus we have run $= x_2 - x_1 = 1$, and, therefore,

$$m = \frac{y_2 - y_1}{x_2 - x_1} = \frac{y_2 - y_1}{1} = y_2 - y_1 = \text{rise.} \qquad (1)$$

This says that if the run is taken as one, then the slope will equal the rise.

If now m is positive, as in Figure 1.20a, then moving over one unit results in moving *up* m units. Thus the line *rises* (moving from left to right) and the larger the value of m the steeper the rise.

If m is negative, as in Figure 1.20b, then moving over one unit results in moving *down* $|m| = -m$ units. Thus the line *falls* and the more negative the value of m, the steeper the fall.

If m is zero, as in Figure 1.20c, then moving over one unit results in moving up *no* units. Thus the line is *horizontal*.

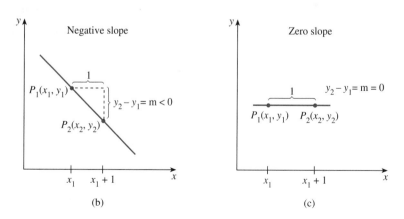

Figure 1.20

The first three parts of Example 1 give specific examples of these three general cases.

Figure 1.21 shows several lines through the same point with different slopes.

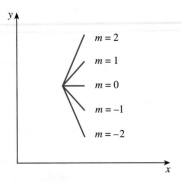

Figure 1.21

E X A M P L E 2   **Using the Slope of a Line**

Find the y-coordinate of point P if P has x-coordinate 5 and P is on the line through $(2, 7)$ with slope -2.

Solution   See Figure 1.22. Using the definition of slope, we have

$$-2 = \frac{y - 7}{5 - 2}$$

$$= \frac{y - 7}{3}$$

$$-6 = y - 7$$

$$y = 1 \quad \blacksquare$$

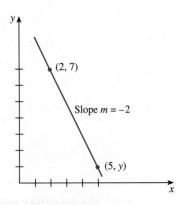

Figure 1.22

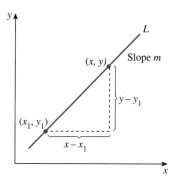

Figure 1.23

Equations of Lines

Suppose we have a line through the point (x_1, y_1) with slope m as shown in Figure 1.23. If (x, y) is any other point on the line, then we must have

$$m = \frac{y - y_1}{x - x_1},$$

or

$$y - y_1 = m(x - x_1).$$

This equation is the **point-slope equation**.

The Point-Slope Equation of a Line

The equation of the line through the point (x_1, y_1) with slope m is given by

$$y - y_1 = m(x - x_1).$$

E X A M P L E 3   **Finding the Point-Slope Equation of a Line**

Find an equation of the line through $(-2, 3)$ with slope $m = -2$. Sketch the graph.

Solution   Since $(x_1, y_1) = (-2, 3)$, the point-slope equation is

$$y - y_1 = m(x - x_1)$$
$$y - 3 = -2[x - (-2)]$$
$$y = -2x - 1 \quad \blacksquare$$

R E M A R K .   One quick way of finding the graph is to notice that when $x = 0$, $y = -2(0) - 1 = -1$. Thus $(0, -1)$ is a second point on the graph. See Figure 1.24.

A point (x, y) will be on the graph of a vertical line such as the one shown in Figure 1.25 if, and only if, $x = a$, where $(a, 0)$ is the point where the vertical line crosses the x-axis. Thus we have the following.

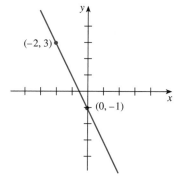

Figure 1.24

Vertical Lines

A vertical line has the equation

$$x = a$$

where $(a, 0)$ is the point at which the vertical line crosses the x-axis.

Given any line that is not vertical, then the line has a slope m and by Figure 1.26 must cross the y-axis at some point $(0, b)$. The number b is called the

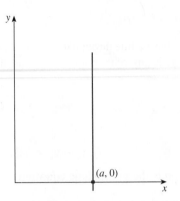

Figure 1.25   **Figure 1.26**

y-intercept. Thus we can use the point-slope equation where $(x_1, y_1) = (0, b)$ is a point on the line. This gives

$$y - b = m(x - 0)$$
$$y = mx + b$$

This is the **slope-intercept equation** of a nonvertical line.

The Slope-Intercept Equation of a Line

The equation of the line with slope m and y-intercept b is given by

$$y = mx + b.$$

E X A M P L E   4   **Finding the Slope-Intercept Equation of a Line**

Find the equation of a line with slope 3 and y-intercept -2. Draw a graph.

Solution   We have $m = 3$ and $b = -2$. Thus

$$y = mx + b = 3x - 2.$$

To draw a graph, we first notice that since the y-intercept is -2, the point $(0, -2)$ is on the graph. See Figure 1.27. Since the slope is 3, if we move over 1 unit to the right, the line will move up 3 units so that the point $(1, 1)$ is also on the line. ∎

We have already shown that every line has an equation of the general form

$$ax + by = c,$$

where a and b are not both zero. This is called a **linear equation**.

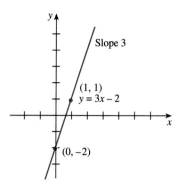

Figure 1.27

Suppose now we are given the equation $ax + by = c$, where a and b are not both zero. Then if $b \neq 0$,

$$y = -\frac{a}{b} x + \frac{c}{b},$$

which, according to the slope-intercept equation, is the equation of a line with slope $-a/b$ and y-intercept c/b. If $b = 0$, then $a \neq 0$ and $ax + by = c$ becomes $ax = c$ and thus $x = c/a$, which is the equation of a vertical line. Thus we have the following theorem.

The General Form of the Equation of a Line

The graph of a linear equation $ax + by = c$ is a straight line, and, conversely, every line is the graph of a linear equation.

We have already defined the y-intercept. We now need to define the x-intercept. Any line that is not horizontal must cross the x-axis at some point $(a, 0)$. The number a is called the **x-intercept**. To find the x-intercept of a line, simply set $y = 0$ in the equation of the line and solve for x. Similarly, to find the y-intercept of a line, set $x = 0$ in the equation of the line and solve for y. The intercepts are very useful in graphing, as the next example illustrates.

E X A M P L E   5   **Graphing an Equation of a Line**

Sketch a graph of $2x + 3y = 6$.

Solution   From the preceding theorem we know that this is a line. Perhaps the easiest way to graph this line is to find the x- and y-intercepts. Setting $x = 0$ yields $3y = 6$ or $y = 2$ as the y-intercept. Setting $y = 0$ yields $2x = 6$ or $x = 3$ as the x-intercept. See Figure 1.28. ■

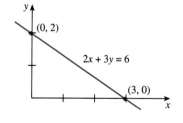

Figure 1.28

Since slope is a number that indicates the direction or slant of a line, the following theorem can be proven.

Slope and Parallel Lines

Two lines are parallel if and only if they have the same slope.

E X A M P L E   6   **Finding the Equation of a Line**

Let L_1 be the line that goes through the two points $(-3, 1)$ and $(-1, 7)$. Find the equation of the line L through the point $(2, 1)$ and parallel to L_1. See Figure 1.29.

Solution   We already have the point $(2, 1)$ on the line L. In order to use the point-slope equation, we need to now find the slope m of L. But since L is parallel to L_1, m will equal the slope of L_1. The slope of L_1 is

$$m = \frac{y_2 - y_1}{x_2 - x_1} = \frac{(7) - (1)}{(-1) - (-3)} = \frac{6}{2} = 3.$$

Now let us use the point-slope equation for L where $(2, 1)$ is a point on the line L with slope $m = 3$. Then we obtain

$$y - y_1 = m(x - x_1)$$
$$y - 1 = 3(x - 2)$$
$$y = 3x - 5. \quad \blacksquare$$

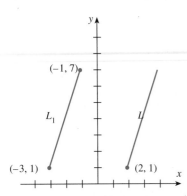

Figure 1.29

We also have the following theorem. (See Exercise 46.)

Slope and Perpendicular Lines

Two nonvertical lines with slope m_1 and m_2 are perpendicular if, and only if,

$$m_2 = -\frac{1}{m_1}.$$

E X A M P L E   7   **Finding the Equation of a Line**

Find the equation of the line L through the point $(3, -1)$ and perpendicular to the line L_1 given by $2x + 3y = 6$.

Solution   See Figure 1.30. We rewrite $2x + 3y = 6$ as

$$y = -\tfrac{2}{3}x + 2.$$

This indicates that the slope of L_1 is $-\tfrac{2}{3}$. Thus the slope m of L will be the negative reciprocal or $m = \tfrac{3}{2}$. Now use the point-slope equation for L to give

$$y - y_1 = m(x - x_1)$$
$$y - (-1) = \frac{3}{2}(x - 3)$$
$$y = \frac{3}{2}x - \frac{11}{2}. \quad \blacksquare$$

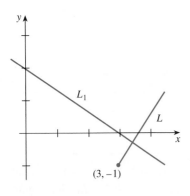

Figure 1.30

Applications

E X A M P L E 8   **Production of Vacuum Cleaners**

A firm produces two models of vacuum cleaners, regular and deluxe. The regular model requires 8 hours of labor to assemble while the deluxe requires 10 hours of labor. Let x be the number of regular models produced and y the number of deluxe models. If the firm wishes to use 500 hours of labor to assemble these cleaners, find an equation that the two quantities x and y must satisfy. If the number of the regular models that are assembled is increased by 5, how many fewer deluxe models will be assembled?

Solution   In order to find the required equation, construct a table with the given information. See Table 1.4.

Table 1.4

Type	Regular	Deluxe	Total
Work hours for each	8	10	
Number produced	x	y	
Work hours	$8x$	$10y$	500

The number of work-hours needed to assemble all of the regular models is the number of work-hours required to produce one of these models times the number produced. This gives $8x$. In a similar fashion we see that the number of work-hours needed to produce all of the deluxe models is $10y$. The number of hours allocated to assemble the regular model plus the number allocated to assemble the deluxe must equal 500. That is,

$$8x + 10y = 500.$$

This is the required equation. Solving for y yields

$$y = -\tfrac{4}{5}x + 50.$$

This is a line with slope $-\tfrac{4}{5}$. See Figure 1.31. Thus increasing x by 5 will decrease y by 4. ■

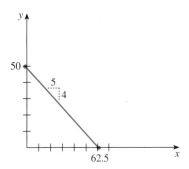

Figure 1.31

SELF-HELP EXERCISE SET 1.2

1. Find the equation of the line through the two points (3, 1) and $(-2, -3)$ and then find the y-intercept. Sketch.

2. Find the equation of the line through the point $(2, -1)$ and perpendicular to the line in the previous exercise. Sketch.

3. Sketch the graph of $2x + 3y - 12 = 0$. Find the slope of this line. Sketch.

EXERCISE SET 1.2

In Exercises 1 through 8 find the slope (if it exists) of the line through the given pair of points.

1. $(1, 3), (2, 7)$　　　　　　　　**2.** $(3, -2), (2, 5)$

3. $(-2, -2), (-3, 1)$　　　　　**4.** $(-3, 2), (5, 4)$

5. $(-1, 3), (2, 3)$　　　　　　　**6.** $(-1, 2), (-3, -4)$

7. $(-3, -1), (-3, -7)$　　　　**8.** $(-5, -2), (-3, -4)$

In Exercises 9 through 20 find the slope (if it exists) of the line given by each of the equations. Also find the x- and y-intercepts if they exist.

9. $y = 2x + 1$　　　　　　　　**10.** $y = -3x + 2$

11. $x + y = 1$　　　　　　　　**12.** $x - 2y - 3 = 0$

13. $-x + 2y - 4 = 0$　　　　**14.** $x + y = 0$

15. $x = 3$　　　　　　　　　　**16.** $x = 0$

17. $y = 4$　　　　　　　　　　**18.** $y = 0$

19. $x = 2y + 1$　　　　　　　**20.** $x = -3y + 2$

In Exercises 21 through 42 find the equation of the line given the indicated information.

21. Through $(2, 1)$ with slope -3

22. Through $(-3, 4)$ with slope 3

23. Through $(-1, -3)$ with slope 0

24. Through $(2, 3)$ with slope undefined

25. Through $(-2, 4)$ and $(-4, 6)$

26. Through $(-1, 3)$ and $(-1, 6)$

27. Through $(-2, 4)$ and $(-2, 6)$

28. Through $(-1, 4)$ and $(-3, 2)$

29. Through $(a, 0)$ and $(0, b)$, $ab \neq 0$

30. Through (a, b) and (b, a), $a \neq b$

31. Through $(-2, 3)$ and parallel to the line $2x + 5y = 3$

32. Through $(2, -3)$ and parallel to the line $2x - 3y = 4$

33. Through $(0, 0)$ and parallel to the line $y = 4$

34. Through $(0, 0)$ and parallel to the line $x = 4$

35. Through $(1, 2)$ and perpendicular to the line $y = 4$

36.   Through $(1, 2)$ and perpendicular to the x-axis

37.   Through $(-1, 3)$ and perpendicular to the line $x + 2y - 1 = 0$

38.   Through $(2, -4)$ and perpendicular to the line $x - 3y = 4$

39.   Through $(3, -1)$ and with y-intercept 4

40.   Through $(-3, -1)$ and with y-intercept 0

41. y-intercept 4 and parallel to $2x + 5y = 6$

42. y-intercept -2 and perpendicular to $x - 3y = 4$

43.   Are the two lines $2x + 3y = 6$ and $3x + 2y = 6$ perpendicular?

44.   Are the two lines $2x + 3y = 6$ and $3x - 2y = 6$ perpendicular?

45.   A person has \$2.35 in change consisting entirely of dimes and quarters. If x is the number of dimes and y is the number of quarters, write an equation that x and y must satisfy.

46.   Establish the following theorem: Two lines with slope m_1 and m_2 are perpendicular if, and only if,

$$m_2 = -\frac{1}{m_1}.$$

Give a proof in the special case shown in the figure in which the two lines intersect at the origin. (The general proof is very similar.) *Hint:* The two lines L_1 and L_2 are perpendicular if and only if the triangle AOB is a right triangle. By the Pythagorean theorem this will be the case if, and only if,

$$[d(A, B)]^2 = [d(A, O)]^2 + [d(B, O)]^2$$

Now use the distance formula for each of the three terms in the previous equation and simplify.

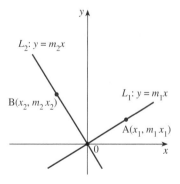

Applications

47.   Investment.   A person invests some money in a bond that yields 8% a year, and some money in a money market fund that yields 5% a year, and obtains \$576 of interest in the first year. If x is the amount invested in the bond and y

the amount invested in the money market fund, find an equation that x and y must satisfy. If the amount of money invested in bonds is increased by $100, by how much must the money invested in the money market decrease if the total interest is to stay the same?

48. Transportation Costs.   New cars are transported from Detroit to Boston and also from Detroit to New York City. It costs $90 to transport each car to Boston and $80 to transport each car to New York City. Suppose $4000 is spent on transporting these cars. If x is the number transported to Boston and y is the number transported to New York City, find an equation that x and y must satisfy. If the number of cars transported to Boston is increased by 8, how many fewer cars must be transported to New York City if the total transportation costs remained the same?

49. Scheduling.   An attorney has two types of documents to process. The first document takes 3 hours and the second takes 4 hours to process. If x is the number of the first type that she processes and y is the number of the second type, find an equation that x and y must satisfy if she works 45 hours on these documents. If she processes 8 additional of the first type of document, how many fewer of the second type of document would she process?

50. Size of Deer Antlers.   The size of a deer's antlers depends primarily on the age of the deer. Mule deer in Northern Colorado have no antlers at age 10 months, but after that the weight of the antlers increases by .12 kilogram for every 10 months of additional age. Let x be the age of these mule deer in months and y the weight of their antlers in kilograms. Assuming a linear relationship, write an equation that x and y must satisfy. What will be the weight in kilograms of antlers carried by a 40-month-old mule deer?

(© John O. Sumner / Photo Researchers, Inc.)

51. Length of Twigs.   If d is the twig diameter in inches of bitter bush, then it is estimated that the length l in inches of the twig is given by $l = 1.25 + 89.83d$. Find the slope of this line and interpret what the slope means in this instance.

Graphing Calculator Exercises

52. On your graphing calculator graph the equations $y = 3 + 2x$, $y = 3 + x$, $y = 3 - x$, $y = 3 - 2x$ in the same viewing rectangle.

53. On your graphing calculator graph $y = 1.43x$, $y = 1.43x + 1$, and $y = 1.43x + 2$ in the same viewing rectangle. Are they parallel?

Solutions to Self-Help Exercise Set 1.2

1. To find the equation of the line through the two points $(3, 1)$ and $(-2, -3)$, we first must find the slope of the line through the two points. This is

$$m = \frac{y_2 - y_1}{x_2 - x_1} = \frac{1 - (-3)}{3 - (-2)} = \frac{4}{5}.$$

We now have the slope of the line and *two* points on the line. Pick one of the points, say $(3, 1)$, and use the point-slope form of the equation of the line and obtain

$$y - y_1 = m(x - x_1)$$

$$y - 1 = \frac{4}{5}(x - 3)$$

$$y = \frac{4}{5}x - \frac{7}{5}.$$

This last equation is in the form of $y = mx + b$ with $b = -\frac{7}{5}$. Thus the y-intercept is $-\frac{7}{5}$.

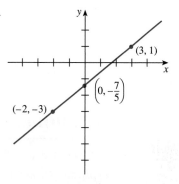

2. The equation through the point $(2, -1)$ and perpendicular to the line in the previous exercise must have a slope equal to the negative reciprocal of the slope of that line. Then the slope of the line whose equation we are seeking is $-\frac{5}{4}$. Using the point-slope form of the equation yields

$$y - y_1 = m(x - x_1)$$

$$y + 1 = -\frac{5}{4}(x - 2)$$

$$y = -\frac{5}{4}x + \frac{3}{2}.$$

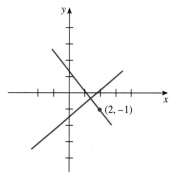

3. Perhaps the easiest way to sketch the graph of $2x + 3y - 12 = 0$ is to find the x- and y-intercepts. To find the y-intercepts, set $x = 0$ and obtain $3y = 12$ or $y = 4$. To find the x-intercept, set $y = 0$ and obtain $2x = 12$ or $x = 6$. These intercepts are plotted in the figure, and the line is drawn through them. To find the slope, write the equation in the form $y = -\frac{2}{3}x + 4$. The slope is then seen to be $m = -\frac{2}{3}$.

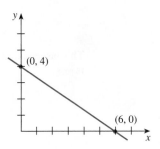

1.3   LINEAR MODELS

► *Cost, Revenue, and Profits*

► *Supply and Demand*

► *Straight-Line Depreciation*

**Augustin Cournot,
1801–1877**

*(Historical Pictures Collection/Stock
Montage, Inc.)*

The first significant work dealing with the application of mathematics to economics was Cournot's "Researches into the Mathematical Principles of the Theory of Wealth," which was published in 1836. It was Cournot who originated the supply and demand curves that are discussed in this section. Irving Fisher, a prominent economics professor at Yale and one of the first exponents of mathematical economics in the United States, wrote that Cournot's book "seemed a failure when first published. It was far in advance of the times. Its methods were too strange, its reasoning too intricate for the crude and confident notions of political economy then current."

A P P L I C A T I O N

**Finding the
Demand Equation**

Suppose that 1000 items of a certain product are sold per day by the entire industry at a price of $10 per item and that 2000 items can be sold per day by the same industry at a price of $8 per item. What is the demand equation, that is, the equation that gives the price p of each item when given the number of items x that are made and sold, if the graph of the demand equation is a straight line? See Example 3 for the answer.

Cost, Revenue, and Profits

Any manufacturing firm has two types of costs—fixed costs and variable costs. **Fixed costs** are costs that do not depend on the amount of production. These costs include real estate taxes, interest on loans, some management salaries, certain minimal maintenance, and protection of plant and equipment. **Variable costs** depend on the amount of production. They include cost of material and labor. Total cost or simply **cost** is the sum of fixed and variable costs:

$$(\text{cost}) = (\text{variable cost}) + (\text{fixed cost}).$$

Let x denote the number of units of a given product or commodity produced by a firm. In the *linear cost model* we assume that the cost m of manufacturing one item is the same no matter how many items are produced. Thus the variable cost is the number of items produced times the cost of each item:

$$(\text{variable cost}) = (\text{cost per item}) \times (\text{number of items produced})$$

$$= mx.$$

If b is the fixed cost and C is the cost, we then have the following:

$$C = \text{cost}$$

$$= (\text{variable cost}) + (\text{fixed cost})$$

$$= mx + b.$$

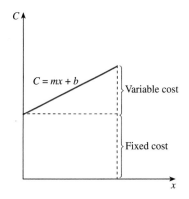

Figure 1.32

A graph is shown in Figure 1.32. Note that the y-intercept is the fixed cost and the slope is the cost per item.

In the **linear revenue model** we assume that the price p of an item sold by a firm is the same no matter how many items are sold. (This is a reasonable assumption if the number sold by the firm is small compared to the total number sold by the entire industry.) Thus the revenue is the price per item times the number of items sold. Let x be the number of items sold. (We always assume that *the number of items sold equals the number of items produced.*) Then, if we denote the revenue by R,

$$R = \text{revenue}$$

$$= (\text{price per item}) \times (\text{number sold})$$

$$= px.$$

See Figure 1.33. Notice that this straight line goes through $(0, 0)$, since nothing sold results in no revenue. The slope is the price per item.

No matter if our models of cost and revenue are linear or not, **profit** P is always revenue less cost. Thus

$$P = \text{profit}$$

$$= (\text{revenue}) - (\text{cost})$$

$$= R - C.$$

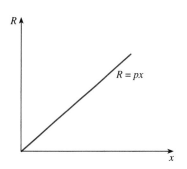

Figure 1.33

E X A M P L E 1   Finding Cost, Revenue, and Profit Equations

A firm has weekly fixed costs of \$80,000 associated with the manufacture of dresses that cost \$25 per dress to produce. The firm sells all the dresses it produces at \$75 per dress. Find the cost, revenue, and profit equations if x is the number of dresses produced per week.

Solution

a. C = (variable cost) + (fixed cost)

$$= mx + b$$

$$= 25x + 80{,}000.$$

b. R = (price per item) $\times$ (number sold)

$$= px$$

$$= 75x.$$

c. P = (revenue) $-$ (cost)

$$= R - C$$

$$= (75x) - (25x + 80{,}000)$$

$$= 50x - 80{,}000.$$

These three equations are graphed in Figure 1.34. ∎

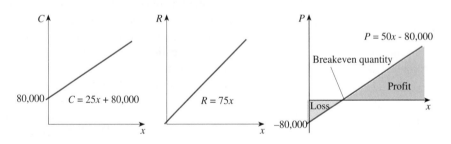

Figure 1.34

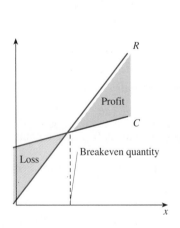

Figure 1.35

One can see in Figure 1.34c that, for smaller values of x, P is *negative*, that is, the firm has losses. For larger values of x, P turns positive and the firm has (positive) profits. The value of x at which the profit is zero is called the **break-even quantity**. Figure 1.35 shows that the break-even quantity is the value of x at which cost equals revenue. This is true since $P = R - C$, and $P = 0$ if, and only if, $R = C$.

E X A M P L E 2   **Finding the Break-Even Quantity**

Find the break-even quantity in the previous example.

Solution   Set $P = 0$ and solve for x.

$$0 = P$$

$$= 50x - 80,000$$

$$50x = 80,000$$

$$x = 1600$$

Thus the firm needs to manufacture and sell 1600 dresses for profits to be zero.   ■

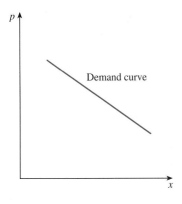

Figure 1.36

Supply and Demand

In the previous discussion we assumed that the number of items produced and sold by the given firm was small compared to the number sold by the industry. Under this assumption it was then reasonable to conclude that the price was constant and did not vary with the number x sold. But if the number of items sold by the firm represented a *large* percentage of the number sold by the entire industry, then trying to sell significantly more items could only be accomplished by *lowering* the price of each item. Thus, under these assumptions, the price p of each item would depend on the number sold. If, in addition, the price is assumed to be linear, that is, $p = mx + b$, then the graph of this equation must slope down as shown in Figure 1.36.

We assume that x is the number of a certain item produced and sold by the entire industry during a given time period and that $p = mx + b$ is the price of the item if x are sold, that is, $p = mx + b$ is the price of the xth item sold. We call $p = mx + b$ the **demand equation** and the graph the **demand curve**.

E X A M P L E 3   **Finding the Demand Equation**

Suppose that 1000 units of a certain item are sold per day by the entire industry at a price of $10 per item and that 2000 units can be sold per day by the same industry at a price of $8 per item. Find the demand equation $p = mx + b$ assuming the demand curve is linear.

Solution   Figure 1.37 shows the two points (1000, 10) and (2000, 8) that lie on the demand curve. Since we are assuming that the demand curve is a straight line, we need only find the straight line through these two points. The slope m is

$$m = \frac{8 - 10}{2000 - 1000} = -0.002.$$

Notice the slope is *negative*, indicating the line slants down moving from left to right. This is what we expect of a reasonable demand curve.

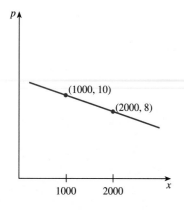

Figure 1.37

Now using the point-slope equation for a line with (1000, 10) as the point on the line, we have

$$p - 10 = m(x - 1000)$$
$$= -0.002(x - 1000)$$
$$p = -0.002x + 12. \quad \blacksquare$$

The **supply equation** $p = mx + b$ is the price p that it will take for suppliers to make available x items to the market. The graph of this equation is called the **supply curve**. Figure 1.38 shows a *linear* supply curve. A reasonable supply curve rises moving from left to right since the suppliers of an item will naturally want to sell more items if the price per item is higher.

The best known law of economics is the law of supply and demand. Figure 1.39 shows a demand equation and a supply equation that intersect. The point of intersection is the point at which supply equals demand, which is called the **equilibrium point**. The x-coordinate of the equilibrium point is called the **equilibrium quantity** and the p-coordinate is called the **equilibrium price**.

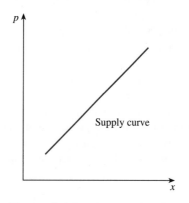

Figure 1.38

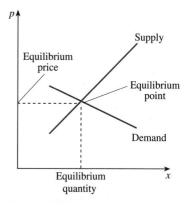

Figure 1.39

E X A M P L E   4   **Finding the Equilibrium Point**

Suppose the supply curve is given by $p = 0.008x + 5$ and the demand curve is the same as in the previous example. Find the equilibrium point.

Solution   The demand and supply equations are given by

$$\text{demand equation: } p = -0.002x + 12$$
$$\text{supply equation: } p = 0.008x + 5.$$

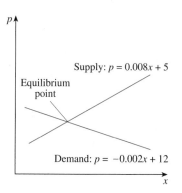

Figure 1.40

These are shown in Figure 1.40. The equilibrium point must be on *both* curves, and thus the coordinates of this point must satisfy both equations. Thus

$$-0.002x + 12 = p = 0.008x + 5$$
$$0.01x = 7$$
$$x = 700.$$

Then $x = 700$ is the equilibrium quantity, and when $x = 700$,

$$p = -0.002(700) + 12 = 10.6$$

is the equilibrium price. Thus the equilibrium point is $(700, 10.6)$. ■

Straight-Line Depreciation

Many assets such as machines or buildings have a *finite* useful life and further-more *depreciate* in value from year to year. For purposes of determining profits and taxes, various methods of depreciation can be used. In **straight-line depreciation** we assume that the value V of the asset is given by a *linear* equation in time t, say, $V = mt + b$. The slope m must be *negative* since the value of the asset *decreases* over time.

E X A M P L E 5 **Straight-Line Depreciation**

A company has purchased a new machine for $100,000 with a useful life of ten years, after which it is assumed that the scrap value of the machine is $5000. If we use straight-line depreciation, write an equation for the value V of the machine where t is measured in years. What will be the value of the machine after the first year? After the second year? After the ninth year?

Solution We assume that $V = mt + b$, where m is the slope and b is the V-intercept. We then must find both m and b. We are told that the machine is initially worth $100,000, that is, when $t = 0$, $V = 100,000$. Thus the point $(0, 100,000)$ is on the line, and thus $100,000$ is the V-intercept b. See Figure 1.41.

Since the value of the machine in 10 years will be $5000, this means that when $t = 10$, $V = 5000$. Thus $(10, 5000)$ is also on the line. From Figure 1.41, the slope can then be calculated since we now know that the two points $(0, 100,000)$ and $(10, 5000)$ are on the line. Then

$$m = \frac{5000 - 100,000}{10 - 0} = -9500.$$

Thus

$$V = -9500t + 100,000.$$

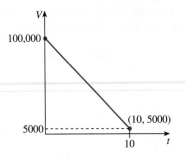

Figure 1.41

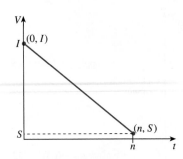

Figure 1.42

Then

$$\text{When } t = 1, \ V = -9500(1) + 100,000 = 90,500.$$

$$\text{When } t = 2, \ V = -9500(2) + 100,000 = 81,000.$$

$$\text{When } t = 9, \ V = -9500(9) + 100,000 = 14,500. \quad \blacksquare$$

In general, suppose that the initial value of the machine is I, the scrap value is S, and the useful life is n years. If we assume straight-line depreciation and $V = mt + b$ is the value after t years, then, from Figure 1.42, $(0, I)$ and (n, S) are on the graph. This is true since when $t = 0$, $V = I$, and when $t = n$, $V = S$. Thus the y-intercept is I. The slope m is

$$m = \frac{S - I}{n - 0} = -\frac{I - S}{n}.$$

Thus

$$V = mt + b = -\frac{I - S}{n} t + I.$$

SELF-HELP EXERCISE SET 1.3

1. A manufacturer has fixed costs of $10,000 per week and variable costs of $20 per item. The item is sold for $30. Let x denote the number of items manufactured and sold.
   a. Find the cost equation.   **b.** Find the revenue equation.
   c. Find the profit equation.   **d.** Find the break-even quantity.

2. A manufacturer notes that it can sell 10,000 items at $6 per item and 8,000 items at $8 per item. Find the demand curve, assuming it is a straight line.

3. Suppose the supply equation for the manufacturer in the previous exercise was $p = .002x + 4$. Find the equilibrium quantity.

EXERCISE SET 1.3

In Exercises 1 through 4 you are given the cost per item and the fixed costs over a certain time period. Assuming a linear cost model, find the cost equation, where C is cost and x is the number produced.

1. Cost per item = $3, fixed cost = $10,000

2. Cost per item = $6, fixed cost = $14,000

3. Cost per item = $0.02, fixed cost = $1000

4. Cost per item = $0.1, fixed cost = $4000

In Exercises 5 through 8 you are given the price of each item assumed to be independent of the number of items sold. Find the revenue equation, where R is revenue and x is the number sold, assuming the revenue is linear in x.

5. Price per item = $5

6. Price per item = $10

7. Price per item = $0.1

8. Price per item = $0.5

9. Using the cost equation found in Exercise 1 and the revenue equation found in Exercise 5, find the profit equation for P, assuming the number produced equals the number sold.

10. Using the cost equation found in Exercise 2 and the revenue equation found in Exercise 6, find the profit equation for P, assuming the number produced equals the number sold.

11. Using the cost equation found in Exercise 3 and the revenue equation found in Exercise 7, find the profit equation for P, assuming the number produced equals the number sold.

12. Using the cost equation found in Exercise 4 and the revenue equation found in Exercise 8, find the profit equation for P, assuming the number produced equals the number sold.

In Exercises 13 through 16 linear cost and revenue equations are given. Find the break-even quantity.

13. $C = 2x + 4, R = 4x$

14. $C = 3x + 10, R = 6x$

15. $C = 0.1x + 2, R = 0.2x$

16. $C = 0.03x + 1, R = 0.04x$

In Exercises 17 through 20 you are given a demand equation and a supply equation. Find the equilibrium point.

17. Demand: $p = -x + 6$, supply: $p = x + 3$

18. Demand: $p = -3x + 12$, supply: $p = 2x + 5$

19. Demand: $p = -10x + 25$, supply: $p = 5x + 10$

20. Demand: $p = -0.1x + 2$, supply: $p = 0.2x + 1$

Applications

21. **Straight-Line Depreciation.** A new machine that costs $50,000 has a useful life of 9 years and a scrap value of $5000. Using straight-line depreciation, find the equation for the value V in terms of t where t is in years. Find the value after one year. After 5 years.

22. **Straight-Line Depreciation.** A new building that costs $1,100,000 has a useful life of 50 years and a scrap value of $100,000. Using straight-line depreciation, find the equation for the value V in terms of t where t is in years. Find the value after one year. After two years. After 40 years.

23. **Demand.** Suppose that 500 units of a certain item are sold per day by the entire industry at a price of $20 per item and that 1500 units can be sold per day by the same industry at a price of $15 per item. Find the demand equation for p, assuming the demand curve to be a straight line.

24. **Demand.** Suppose that 10,000 units of a certain item are sold per day by the entire industry at a price of $150 per item and that 8000 units can be sold per day by the same industry at a price of $200 per item. Find the demand equation for p, assuming the demand curve to be a straight line.

25. **Machine Allocation.** A furniture manufacturer makes bookcases and small desks, each requiring use of a cutting machine. Each bookcase requires 3 hours on the machine, and each desk requires 4 hours on the machine. The machine is available 120 hours each week.
    a.   If x is the number of bookcases and y the number of desks that are manufactured each week, find the equation that x and y must satisfy if all available time is used on the machine.
    b.   Find the slope of the equation in the previous part and interpret its meaning.

26. **Labor Allocation.** A company manufacturers three-speed and ten-speed bikes. Each three-speed bike requires 6 hours of labor and each ten-speed bike requires 12 hours of labor. There are 240 hours of labor available each day.
    a.   If x is the number of three-speed bikes and y the number of ten-speed bikes that are manufactured each day, find the equation that x and y must satisfy if all available hours of labor are used.
    b.   Find the slope of the equation in the previous part and interpret its meaning.

27. **Sales Commissions.** A salesperson is offered two salary plans. The first gives a base salary of $1000 a month and a 5% commission on all sales, while the second plan gives a base salary of $1500 a month and a 4% commission.
    a.   If x is the amount of sales, and y the monthly salary, find the relationship between x and y for each plan.
    b.   How much sales are required so that each plan yields the same total salary?

28. **Inventory.** A store has 1000 units of a certain item in stock and sells 20 units a day.
    a.   If y is the number of items in stock at the end of day t, find the relationship between y and t.
    b.   How long until the inventory is exhausted?

29. **Medicine.** Two rules have been suggested by Cowling and by Friend to adjust adult drug dosage levels for young children. Let a denote the adult dosage and let t be the age (in years) of the child. The two rules are given respectively by

$$C = \frac{a}{24}(t + 1) \quad \text{and} \quad F = \frac{2}{25}at.$$

If, for a particular drug, $a = 100$ in the appropriate units, graph both functions. Find the age for which the two rules give the same dosage.

30. **Temperature.** Let C denote the temperature measured on the Celsius scale and F denote the temperature measured on the Fahrenheit scale. The freezing and boiling points of water are 0 degrees and 100 degrees, respectively, on the Celsius scale and 32 degrees and 212 degrees, respectively, on the Fahrenheit scale. If F is a linear equation in C, find the equation that relates F to C.

31. **Elections.** A candidate for the Senate of the United States obtained 41% of the vote in the previous election and spent $3 million in advertisements. The candidate's advisors estimate that each additional $1 million in advertisement expenditures will obtain an additional 6% of the vote for the next election. Write an equation that relates the advertisement expenditures in millions of dollars with the percent of the vote obtained, assuming a linear relationship. How much must be spent on advertising, according to this analysis, to obtain 50% of the vote in the next election?

32. **Weight of Whales.** It is difficult to weigh large whales, but easy to measure their length. The International Whaling Commission adopted the formula $w = 3.51l - 192$ as a reliable way of relating the length l in feet of adult blue whales with their weight w in British tons. Graph this equation. According to this formula, what must be the increase in weight of these whales for every 5-foot increase in their length?

33. **Make or Buy Decision.** A company includes a manual with each piece of software and is trying to decide whether to contract with an outside supplier or produce the manual in house. The lowest bid of any outside supplier is $.75 per manual. The company estimates that producing the manuals in house will require fixed costs of $10,000 and variable costs of $.50 per manual. Which alternative has the lower total cost if demand is 20,000 manuals?

34. **Make or Buy Decision.** Repeat the previous exercise for a demand of 50,000 manuals.

35. **Process Selection and Capacity.** A machine shop needs to drill holes in a certain plate. An inexpensive manual drill press could be purchased that will require large labor costs to operate, or an expensive automatic press can be purchased that will require small labor costs to operate. The following table summarizes the options.

Machine	Annual Fixed Costs	Variable Labor Costs	Production Rate
Manual	$1000	$16.00 per hour	10 plates per hour
Automatic	$8000	$2.00 per hour	100 plates per hour

If these are the only fixed and variable costs, determine which machine should be purchased by filling in the following table if the number produced is (a) 1000; (b) 10,000.

| Volume | Total Costs | |
	Manual	Automatic
1,000		
10,000		

36. Costs Per Unit.   If C is the total cost to produce x items, then the cost per unit is $\frac{C}{x}$. Fill in the table with the costs per unit for the indicated volumes for each of the machines in the previous exercise.

| Volume | Costs Per Unit | |
	Manual	Automatic
1,000		
10,000		
100,000		

Graphing Calculator Exercises

In Exercises 37 through 38 find the equilibrium point to two decimal places, using the zoom feature of your graphing calculator.

37.   Demand: $p = -0.07x + 5.15$, supply: $p = 0.03x + 0.65$.

38.   Demand: $p = -3.1x + 25.7$, supply: $p = 0.22x + 2.6$.

Solutions to Self-Help Exercise Set 1.3

1.   a.   A manufacturer has fixed costs of $10,000 per week and variable costs of $20 per item, and the number of items manufactured is x. The variable costs are the cost per item times the number of items, or $20x$. Thus the cost in dollars is

$$C = 20x + 10,000.$$

b.   The item is sold for $30. Thus the revenue equation is

$$R = 30x.$$

c.   Profit is always revenue less costs. Thus

$$P = R - C = 30x - (20x + 10,000) = 10x - 10,000.$$

d.   The break-even quantity is the quantity for which the profits are zero. Setting $P = 0$ then gives

$$0 = 10x - 10,000$$

$$x = 1000$$

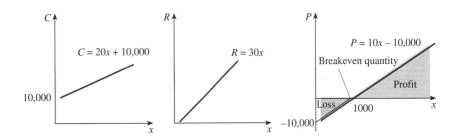

2. If a manufacturer sells 10,000 items at $6 per item and sells 8000 items at $8 per item and the demand curve is assumed to be a straight line, then the slope of this line is given by

$$m = \frac{6 - 8}{10{,}000 - 8000} = -.001$$

See the figure. Then the demand equation is given by the point-slope equation of the line as

$$p - 8 = -.001(x - 8000)$$
$$p = -.001x + 16.$$

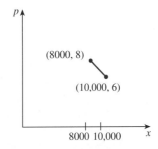

3. If the supply equation for the manufacturer in the previous exercise is $p = .002x + 4$, then to find the equilibrium quantity, we have

$$.002x + 4 = -.001x + 16$$
$$.003x = 12$$
$$x = 4000.$$

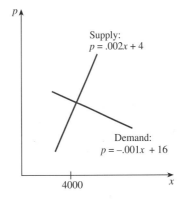

1.4   METHOD OF LEAST SQUARES*

Carl Gauss, the Missing Planet, and the Method of Least Squares

(David Eugene Smith Collection/Rare Book and Manuscript Library/Columbia University.)

Carl Gauss (1777–1855), one of the greatest mathematicians of all time, first formulated the method of least squares. He did this in 1794 at the age of 16! In 1781 Sir William Herschel discovered the planet Uranus, bringing the number of planets known at that time to seven. At that time astronomers were certain that an eighth planet existed between Mars and Jupiter, but the search for the missing planet had proved fruitless. On January 1, 1801, Giuseppe Piazzi, director of the Palermo observatory, announced the discovery of a new planet Ceres in this location. (It was subsequently realized that Ceres was actually a large asteroid.) Unfortunately, a few weeks later he lost sight of it. Gauss took up the challenge of determining the orbit from the few recorded observations. He had in his possession one remarkable mathematical tool—the method of least squares, which he had not bothered to publish. Using this method, he predicted the orbit. Astronomers around the world were astonished to find Ceres exactly where he said it would be. This incident catapulted him to fame.

We assumed in the last section that the demand equations for some products were linear. Sometimes we supposed that we were given two data points (x_1, p_1) and (x_2, p_2). That is, x_1 items were sold if the price was set at p_1, and x_2 items were sold if the price was set at p_2. If we then *assume* that the demand equation is linear, then of course there is only *one* straight line through these two points, and we can easily calculate the equation $y = ax + b$ of this line.

But suppose that we have more than two data points. Suppose, as in Table 1.5, we have five points available. The p_i are the prices in dollars for a certain product, and the x_i are the corresponding demands for the product in the number of thousands of items sold per day.

Table 1.5

x_i	1	2	3	5	9
p_i	10	9	8	7	5

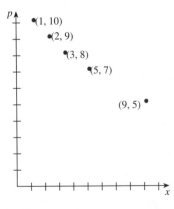

These are plotted in Figure 1.43 which is called a **scatter diagram**. If we eyeball the scatter diagram, we see clearly that the points do not lie on any straight line but seem to be scattered in a more or less linear fashion. Under such circumstances we might be justified in assuming that the demand equation was more or less a straight line. But what straight line? Any line that we draw will miss most of the points. We might then think to draw a line that is somehow closest to the

Figure 1.43

*Optional.

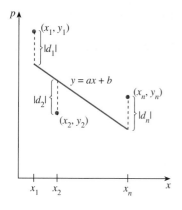

Figure 1.44

data points. To actually follow such a procedure, we need to state exactly how we are to measure this closeness.

In this section we will measure this closeness in a manner that will lead us to the **method of least squares**. To be given a nonvertical straight line is the same as to be given two numbers a and b with the equation of the straight line given as $y = ax + b$. Suppose we are given n data points (x_1, y_1), $(x_2\ y_2)$, ..., (x_n, y_n), and a line $y = ax + b$. We then define $d_1 = y_1 - (ax_1 + b)$ and note from Figure 1.44 that $|d_1|$ is just the vertical distance from the first data point (x_1, y_1) to the line $y = ax + b$. Doing the same for all the data points, we then have

$$d_1 = y_1 - (ax_1 + b)$$
$$d_2 = y_2 - (ax_2 + b)$$
$$\vdots$$
$$d_n = y_n - (ax_n + b),$$

where $|d_2|$ is the vertical distance from the second data point (x_2, y_2) to the line $y = ax + b$, and so on.

Now if all the data points were on the line $y = ax + b$, then all the distances $|d_1|, |d_2|, \ldots, |d_n|$, would be zero. Unfortunately, this will rarely be the case. We then use the sum of the squares of these distances

$$d = d_1^2 + d_2^2 + \cdots + d_n^2$$

as a measure of how close the set of data points is to the line $y = ax + b$. Notice that this number d will be different for different straight lines—large if the straight line is far removed from the data points and small if the straight line passes close to all the data points. We then seek the line or, what is the same thing, the two numbers a and b that will make this sum of squares the least. Thus the name *least squares*.

It can be shown that the numbers a and b that determine the line $y = ax + b$ closest to the set of data points must satisfy the two linear equations

$$\left.\begin{aligned}(x_1^2 + \cdots + x_n^2)a + (x_1 + \cdots + x_n)b &= x_1y_1 + \cdots + x_ny_n \\ (x_1 + \cdots + x_n)a + \qquad\qquad nb &= y_1 + \cdots + y_n\end{aligned}\right\} \quad (1)$$

We then have the following.

Method of Least Squares

The line $y = ax + b$ closest to the data points (x_1, y_1), (x_2, y_2), ..., (x_n, y_n) can be found by solving the following two linear equations for a and b:

$$(x_1^2 + \cdots + x_n^2)a + (x_1 + \cdots + x_n)b = x_1y_1 + \cdots + x_ny_n$$
$$(x_1 + \cdots + x_n)a + \qquad\qquad nb = y_1 + \cdots + y_n.$$

EXAMPLE 1   **Using the Method of Least Squares to Find a Demand Equation**

a.   Find the best fitting line through the data points in Table 1.5 and thus find a linear demand equation.

b.   Estimate the price if the demand is 6000.

Solutions   a.   Create the following table.

x_i	y_i	x_i^2	$x_i y_i$
1	10	1	10
2	9	4	18
3	8	9	24
5	7	25	35
9	5	81	45
Sum 20	39	120	132

Equation (1) then becomes

$$120a + 20b = 132$$
$$20a + 5b = 39.$$

One can solve by eliminating one of the variables. For example, to eliminate the variable b, multiply the second equation by -4 and add the result to the first, giving

first equation	$120a + 20b =$	132
-4 times second equation	$-80a - 20b =$	-156
add	$40a =$	-24

This yields $a = -0.6$. Since $a = -0.6$ and $20a + 5b = 39$,

$$b = \frac{39 - 20a}{5} = 10.2.$$

Thus the equation of the best fitting straight line that we are seeking is

$$y = -0.6x + 10.2.$$

The graph is shown in Figure 1.45.

b.   The answer to the second question is

$$p = y = (-0.6)(6) + 10.2 = 6.6,$$

that is, if 6000 items are to be sold, then the price should be $6.60.   ∎

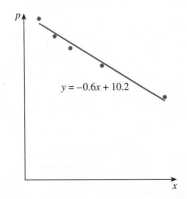

$y = -0.6x + 10.2$

Figure 1.45

SELF-HELP EXERCISE SET 1.4

1. Using the method of least squares, find the best fitting line through the three points (0, 0), (2, 2), (3, 2).

EXERCISE SET 1.4

In Exercises 1 through 8 find the best fitting straight line to the given set of data, using the method of least squares.

1. (0, 0), (1, 2), (2, 1)

2. (0, 1), (1, 2), (2, 2)

3. (0, 0), (1, 1), (2, 3), (3, 3)

4. (0, 0), (1, 2), (2, 2), (3, 0)

5. (1, 4), (2, 2), (3, 2), (4, 1)

6. (0, 0), (1, 1), (2, 2), (3, 4)

7. (0, 4), (1, 2), (2, 2), (3, 1), (4, 1)

8. (0, 0), (1, 2), (2, 1), (3, 2), (4, 4)

Applications

9. **Sales.** A firm has just introduced a new product and the sales during the first five weeks are given by the following table.

Week Number	(x_i)	1	2	3	4	5
Units Sold	(y_i)	10	16	20	26	30

Find the best fitting straight line using the method of least squares. What do you think sales will be next week?

10. **Costs.** In the last four years a firm has produced varying amounts of an item at a certain plant and incurred varying total costs given by the following table where the number of items is in thousands per year and the total cost is in millions of dollars per year.

Number Produced	(x_i)	2	3	1	3
Total Cost	(y_i)	2	4	2	3

Find the best fitting straight line using the method of least squares. What do you think costs per year would be next year if production was set at 2500 units?

11. **Sales.**   A firm has undertaken four different radio advertising campaigns, and the sales in millions of dollars versus the amount spent on advertising in millions of dollars is given in the following table.

Cost of Advertising	(x_i)	1	2.5	3.5	4
Sales in Dollars	(y_i)	3.5	5	6	6.5

Find the best fitting straight line using the method of least squares. What do you think sales would be in the next campaign if $3 million dollars were spent on radio advertising?

12. **Insect Temperature.**   An insect cannot control its body temperature; the body temperature depends on the temperature of the surrounding air. The following data have been collected for a particular insect species.

Air Temperature	(x_i)	18.2	23.0	25.6	30.4
Insect Temperature	(y_i)	19.7	24.5	26.2	31.5

Find the best fitting straight line using the method of least squares.

13. **Median Age of U.S. Population.**   The following table gives the median age of the U.S. population.

Year	(x_i)	1820	1880	1950	1990
Median Age	(y_i)	17	20	30	33

Find the best fitting straight line using the method of least squares.

 Graphing Calculator Exercises

14. **Demand.**   A firm has set different prices for its product in different cities and has obtained the following data relating the price of the product to the number of thousands sold per week at that price.

Price in Dollars	(x_i)	8	9	10	11	12	13	14	15	16	17
Sales Volume in Thousands	(y_i)	10	8	7	6	5	4.5	4	3.6	3	2

Find the best fitting straight line using the method of least squares. Graph this function. What do you think sales per week would be if the price were set at $7?

Solutions to Self-Help Exercise Set 1.4

1. Create the following table.

	x_i	y_i	x_i^2	x_iy_i
	0	0	0	0
	2	2	4	4
	3	2	9	6
Sum	5	4	13	10

This leads to the following two linear equations.

$$13a + 5b = 10$$
$$5a + 3b = 4$$

Multiplying the first equation by 3 and the second by -5 and adding them gives

Multiply first equation by 3	$39a + 15b =$	30
Multiply second equation by -5	$-25a - 15b =$	-20
Adding	$14a =$	10

or $a = \dfrac{5}{7}$. Then since $5a + 3b = 4$,

$$b = \frac{4}{3} - \frac{5}{3}a = \frac{4}{3} - \frac{5}{3} \cdot \frac{5}{7} = \frac{1}{7}.$$

The best fitting line is then $y = \dfrac{5}{7}x + \dfrac{1}{7}$.

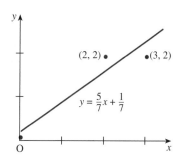

SUMMARY OUTLINE OF CHAPTER 1

- A **set** is a collection of objects, specified in a manner that enables one to determine if a given object is or is not in the collection. Each object in a set is called a **member** or **element** of the set.

- The "set of all elements x such that x has property P" is written in **set-builder notation** as $\{x|x \text{ has property } P\}$.

- Two sets A and B are **equal**, written $A = B$, if, and only if, they consist of precisely the same elements.

- Construct a **Cartesian coordinate system** with a horizontal line called the **x-axis** and a vertical line called the **y-axis** both drawn in the plane. The point of intersection of these two lines is called the **origin**. The plane is then called the **xy-plane** or the **coordinate plane**. Each point P in the xy-plane is assigned a pair of numbers (a, b). The first number a is the horizontal distance from the point P to the y-axis and is called the **x-coordinate**. The second number b is the vertical distance to the x-axis and is called the **y-coordinate**.

- The x-axis and the y-axis divide the plane into four **quadrants**. The **first quadrant** is all points (x, y) with both $x > 0$ and $y > 0$. The **second quadrant** is all points (x, y) with both $x < 0$ and $y > 0$. The **third quadrant** is all points (x, y) with both $x < 0$ and $y < 0$. The **fourth quadrant** is all points (x, y) with both $x > 0$ and $y < 0$.

- The distance $d = d(P_1, P_2)$ between the two points $P_1(x_1, y_1)$ and $P_2(x_2, y_2)$ in the xy-plane is given by

$$d(P_1, P_2) = \sqrt{(x_2 - x_1)^2 + (y_2 - y_1)^2}.$$

- The **graph of an equation** is the set $\{(x, y)|x \text{ and } y \text{ satisfy the equation}\}$.

- Let (x_1, y_1) and (x_2, y_2) be two points on a line L. If the line is nonvertical ($x_1 \neq x_2$), the **slope** m of the line L is defined to be

$$m = \frac{y_2 - y_1}{x_2 - x_1}.$$

If L is vertical ($x_2 = x_1$), then the slope is said to be **undefined**.

- **The Point-Slope Equation of a Line.**   The equation of the line through the point (x_1, y_1) with slope m is given by

$$y - y_1 = m(x - x_1).$$

- **Equation of Vertical Line.**   A vertical line has the equation $x = a$ where $(a, 0)$ is the point at which the vertical line crosses the x-axis.

- **The Slope-Intercept Equation of a Line.**   The equation of the line with slope m and y-intercept b is given by $y = mx + b$.

- **The General Form of the Equation of a Line.**   The graph of a linear equation $ax + by = c$ is a straight line, and, conversely, every line is the graph of a linear equation.

- **Slope and Parallel Lines.**   Two lines are parallel if, and only if, they have the same slope.

- **Slope and Perpendicular Lines.**   Two nonvertical lines with slope m_1 and m_2 are perpendicular if, and only if,

$$m_2 = -\frac{1}{m_1}.$$

▪ **Linear Cost, Revenue, and Profit Equations.** Let x be the number of items made and sold.

$$\textbf{variable cost} = (\text{cost per item}) \times (\text{number of items produced})$$

$$= mx$$

$$C = \textbf{cost} = (\text{variable cost}) + (\text{fixed cost})$$

$$= mx + b$$

$$R = \textbf{revenue} = (\text{price per item}) \times (\text{number sold})$$

$$= px.$$

$$P = \textbf{profit} = (\text{revenue}) - (\text{cost})$$

$$= R - C.$$

▪ The quantity at which the profit is zero is called the **break-even quantity**.

▪ Let x be the number of items made and sold and p the price of each item. A **linear demand equation**, which governs the behavior of the consumer, is of the form $p = mx + b$, where m must be negative. A **linear supply equation**, which governs the behavior of the producer, is of the form $p = mx + b$, where m must be positive.

▪ The point at which supply equals demand is called the **equilibrium point**. The x-coordinate of the equilibrium point is called the **equilibrium quantity** and the p-coordinate is called the **equilibrium price**.

▪ In **straight-line depreciation** we assume that the value V of the asset is given by a *linear* equation in time t, say, $V = mt + b$, where the slope m must be *negative*.

▪ The line $y = ax + b$ closest to the data points $(x_1, y_1), (x_2, y_2), \ldots, (x_n, y_n)$, can be found solving the following two linear equations for a and b.

$$(x_1^2 + \cdots + x_n^2)a + (x_1 + \cdots + x_n)b = x_1 y_1 + \cdots + x_n y_n$$

$$(x_1 + \cdots + x_n)a + \qquad\qquad nb = y_1 + \cdots + y_n$$

Chapter 1 Review Exercises

1. Write the set of odd positive integers in set-builder notation.

2. Write the following sets in roster notation.
   a. $\{x \mid x$ is a letter in the word Mississippi$\}$ **b.** $\{x \mid (x + 1)(x - 3) = 0\}$

In Exercises 3 through 6 plot the given point in the xy-plane.

3. $(2, 3)$ **4.** $(-2, -5)$ **5.** $(-3, 2)$ **6.** $(4, -5)$

7. Find the points (x, y) in the xy-plane that satisfy $xy \le 0$.

8. Find the distance between the two points $(2, -3)$ and $(-3, -6)$.

9. Sketch the graph of $y^2 = 9x$.

In Exercises 10 through 12 find the slope (if it exists) of the line through each pair of points.

10. $(-2, 4), (1, -3)$ **11.** $(-2, -3), (2, 3)$ **12.** $(-2, 5), (-2, -3)$

In Exercises 13 through 15 find the slope (if defined) of the lines. Find the x- and y-intercepts if they exist.

13. $y = 3x + 2$ **14.** $y = -2$ **15.** $x = 3$

16.   Find the equation of the line through the point $(-2, 5)$ with slope -2.

17.   Find the equation of the line through the two points $(-4, 2)$ and $(3, -5)$.

18.   Find the equation of the line through the point $(-3, -5)$ and parallel to the line $y = 3x - 2$.

19.   Find the equation of the line through the point $(-3, 7)$ and perpendicular to the line $y = -4x + 5$.

20.   Assuming a linear cost model, find the equation for the cost C, where x is the number produced, the cost per item is $6, and the fixed costs are $2000.

21.   Assuming a linear revenue equation, find the revenue equation for R, where x is the number sold and the price per item is $10.

22.   Assuming the cost and revenue equations in the previous two problems, find the profit equation. Also find the break-even quantity.

23.   Given that the cost equation is $C = 5x + 3000$ and the revenue equation is $R = 25x$, find the break-even quantity.

24.   Given the demand equation $p = -2x + 4000$ and the supply equation $p = x + 1000$, find the equilibrium point.

25.   **Transportation.**   A plane travels a straight path from O to A and then another straight path from A to B. How far does the plane travel? Refer to the figure.

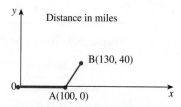

26.   **Transportation.**   If the plane in the previous exercise travels 200 miles per hour from O to A and 250 miles per hour from A to B, how long does the trip take?

27.   **Demand.**   A company notes from experience that it can sell 100,000 pens at $1 each and 120,000 of the same pens at $0.90 each. Find the demand equation, assuming it is linear.

28. Profit. It costs a publisher $2 to produce each copy of a weekly magazine. The magazine sells for $2.50 a copy, and the publisher obtains advertising revenue equal to 30% of the revenue from sales. How many copies must be sold to obtain a profit of $15,000?

29. Nutrition. A certain woman needs 15 mg of iron in her diet each day but cannot eat fish or liver. She plans to obtain all of her iron from kidney beans (4.5 mg per cup) and soybeans (5.5 mg per cup). If x is the number of cups of kidney beans and y the number of cups of soybeans she consumes each day, what linear equation must x and y satisfy? Interpret the meaning of the slope of this equation.

30. Facility Location. A company is trying to decide whether to locate a new plant in Houston or Boston. Information on the two possible locations is given below. The initial investment is in land, buildings, and equipment.

	Houston	*Boston*
Variable Cost	$.25 per item	$.22 per item
Annual Fixed Costs	$4,000,000	$4,200,000
Initial Investment	$16,000,000	$20,000,000

Suppose 10,000,000 items are produced each year.
a. Find which city has the lowest annual total costs, not counting the initial investment.
b. Which city has the lowest total cost over five years, counting the initial investment?

31. Using the method of least squares, find the best fitting line through the 4 points $(0, 0)$, $(2, 2)$, $(3, 2)$, $(4, 3)$.

Systems of equations are used in the management of portfolios. (© Jeffrey Sylvester/ FPG International Corporation.)

2

Systems of Linear Equations and Matrices

In the last chapter on several occasions we encountered a system of two linear equations in two unknowns. In this chapter we wish to find an efficient method for solving a system with any number of linear equations and any number of unknowns. For example, the following is a system of three linear equations in the four unknowns x, y, z, and u:

$$2x + 5y - 7z + 3u = 7$$
$$3x - 4y + 8z + 9u = 4$$
$$6x + 3y - 2z - 4u = 3.$$

Such a system is referred to as a *system of linear equations*.

In the first two sections of this chapter we examine the Gauss–Jordan method, which is used extensively by computers to solve systems of linear equations with many equations and unknowns. Such systems arise in a wide variety of applications. We shall examine some of these applications.

Then we shall study the basic ideas of matrix theory. We shall look at various applications of this theory, including how matrix theory is connected with solving systems of linear equations.

2.1   INTRODUCTION TO SYSTEMS OF LINEAR EQUATIONS

► *Introduction*
► *Gauss–Jordan Method*
► *Applications*

Carl Friedrich Gauss, 1777–1855
Wilhelm Jordan, 1842–1899

Most historians place Gauss among the very top mathematicians of all time. Gauss was a child prodigy born of working-class parents. Fortunately, his mother, although uneducated herself, encouraged the young Gauss to pursue his education. Gauss made fundamental contributions to algebra and number theory and also to astronomy, geodesy, and electricity. He initiated a new branch of mathematics called differential geometry. To determine the orbits of asteroids, he developed a method of solving systems of linear equations by eliminating one variable at a time. This idea was later improved by Wilhelm Jordan, a prominent professor of geodesy, who needed to solve large systems of equations that were involved in triangulation techniques needed for his major geodetic surveys of Germany and the Libyan desert. His improved method is now called the Gauss–Jordan method. It is this method that will be considered in this chapter.

APPLICATION
Allocation of Machine Time

A firm produces three products, a dress shirt, a casual shirt and a sport shirt, and uses a cutting machine, a sewing machine, and a packaging machine in the process. To produce each gross of dress shirts requires 3 hours on the cutting machine and 2 hours on the sewing machine. To produce each gross of casual shirts requires 5 hours on the cutting machine and 1 hour on the sewing machine. To produce each gross of sport shirts requires 7 hours on the cutting machine and 3 hours on the sewing machine. It takes 2 hours to package each gross of each type of shirt. If the cutting machine is available for 480 hours, the sewing machine 170 hours, and the packaging machine 200 hours, determine the number of gross of each shirt that should be produced to use all of the available time on the three machines. (A gross is a dozen dozen or 144.) See Example 3 for the answer.

Introduction

We first write the equations that are needed to solve the above problem on the allocation of machine time. As usual we let the unknowns be denoted by letters. Let x be the number of gross of dress shirts made, y the number of gross of casual shirts, and z the number of gross of sport shirts. As in the last chapter, create a table that summarizes all of the given information. See Table 2.1.

Table 2.1

	Dress Shirts	Casual Shirts	Sport Shirts	Total Available
Time on packaging machine	$2x$	$2y$	$2z$	200
Time on cutting machine	$3x$	$5y$	$7z$	480
Time on sewing machine	$2x$	y	$3z$	170

Since each gross of each style of shirt requires two hours on the packaging machine, and this machine is available for 200 hours, we must have

$$2x + 2y + 2z = 200.$$

Since the number of hours on the cutting machine is $3x + 5y + 7z$, while the total hours available is 480, we must have

$$3x + 5y + 7z = 480.$$

Looking at the time spent and available on the sewing machine gives

$$2x + y + 3z = 170.$$

The three equations together give a **system of linear equations**.

$$\left. \begin{aligned} 2x + 2y + 2z &= 200 \\ 3x + 5y + 7z &= 480 \\ 2x + y + 3z &= 170 \end{aligned} \right\} \tag{1}$$

In the previous problem there were three equations in three unknowns. In other applications one routinely encounters a system of dozens or even hundreds of linear equations with a similar number of unknowns. Such large systems cannot be solved by hand. They must be solved by a computer, and the computer needs a method to systematically solve such large systems efficiently. Such a method, called the Gauss–Jordan method, is introduced here. Our goal here is to see how this system works so that we can understand how the computer will solve such problems. This will enable us to see what types of problems can be solved by this method, to know what can be expected in the way of solutions (or if there even are solutions), and to be able to interpret our results. Another important goal will be to gain skill in converting certain applied problems into systems of linear equations.

We will solve by hand only small systems. Our goal is not so much to find the solution(s) to a system of linear equations as it is to learn how the Gauss–Jordan method works. We tried to limit the number of computations in the systems and, in particular, to avoid too many fractions.

In this section we will consider how to solve in a systematic manner a system of linear equations such as given in system (1). A **solution** of this system is a triplet of numbers (a, b, c) such that replacing x with a, replacing y with b, and replacing z with c will yield equalities. Thus the ordered pair $(30, 50, 20)$ is a solution of the above system since replacing x with 30, replacing y with 50, and replacing z with 20 yields

$$2(30) + 2(50) + 2(20) = 200$$

$$3(30) + 5(50) + 7(20) = 480$$

$$2(30) + (50) + 3(20) = 170.$$

(We will see how to obtain this solution shortly.) The set of all solutions to a system is called the **solution set**. Also two systems of equations with the same solution set are said to be **equivalent**.

Let us now consider the following system of linear equations:

$$3x + 9y = 45$$

$$2x + y = 10.$$

The graphs of each of the equations in the above system are lines. For example, the graph of the first equation $3x + 9y = 45$ is a line, say L_1. This line is graphed in Figure 2.1. The graph of the second equation $2x + y = 10$ is also a line, say L_2. This line is also graphed in Figure 2.1. The reader can easily check that $(3, 4)$ is a solution. The solution $(3, 4)$ must be the point of intersection of the two lines since the ordered pair must satisfy both equations and thus lie on both lines.

Not all systems of linear equations have solutions. Consider the following system.

$$x + y = 2$$

$$x + y = 1$$

Since we cannot have two numbers x and y that add to 2 and also add to 1, there cannot be a solution to this system. The two lines are graphed in Figure 2.2, where we see that the two lines are parallel and not equal. Thus the lines do not intersect.

There is a third possibility. The system of linear equations can have an infinite number of solutions. Consider the following:

$$x - y = 2$$

$$2x - 2y = 4.$$

Since the second equation is simply twice the first, a pair of numbers (x, y) will satisfy the second equation if, and only if, the pair satisfies the first. In other

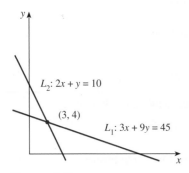

Figure 2.1

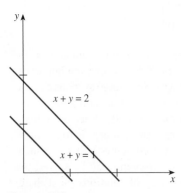

Figure 2.2

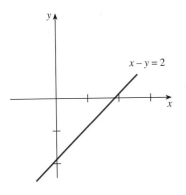

Figure 2.3

words the second equation is superfluous. Thus any point on the line $x - y = 2$ will satisfy *both* equations, and there are an infinite number of such points. Geometrically both equations represent the *same* line, as indicated in Figure 2.3. In this case the solution set is the set of all points on the line $x - y = 2$.

In this section we will consider the case when a system has exactly one solution and the case when a system has no solutions. The case when a system has an infinite number of solutions will be left until the next section.

Gauss–Jordan Method

We now will develop the **Gauss–Jordan method** to solve linear systems of equations. This method, which works for any number of equations, has the following considerable advantages.

> **Advantages of the Gauss–Jordan Method**
>
> 1. If a solution exists, this method will find it.
> 2. If no solution exists, this method will so indicate.
> 3. If an infinite number of solutions exist, this method will find all of them.
> 4. The method is readily programmable for use on computers.

The method will be illustrated with the following example.

E X A M P L E   1   **Using the Gauss–Jordan Method**

Solve

$$2x - 2y - 2z = 2$$
$$2x - 3y + z = 10$$
$$x + y - 2z = 0$$

Solution   The first step is to place a 1 as the coefficient of x in the first equation.

$$2x - 2y - 2z = 2$$
$$2x - 3y + z = 10$$
$$x + y - 2z = 0$$

STEP 1:
Want a 1 here.

To accomplish this multiply the first equation by $\frac{1}{2}$ to obtain

$$x - y - z = 1$$
$$2x - 3y + z = 10$$
$$x + y - 2z = 0.$$

STEP 2:
Want a 0 here.

Now, for the second step we wish to eliminate x from the second and third equations. To eliminate x from the second equation, add -2 times the first equation to the second equation and replace the second equation with the result.

$$
\begin{array}{lrcr}
-2 \text{ times first} & -2x + 2y + 2z & = & -2 \\
\hline
\text{second} & 2x - 3y + z & = & 10 \\
\hline
\text{new second} & -y + 3z & = & 8
\end{array}
$$

To eliminate x from the third equation add -1 times the first equation to the third equation and replace the third equation with the result.

$$
\begin{array}{lrcr}
-1 \text{ times first} & -x + y + z & = & -1 \\
\hline
\text{third} & x + y - 2z & = & 0 \\
\hline
\text{new third} & 2y - z & = & -1
\end{array}
$$

This yields

$$
\begin{array}{rcr}
x - y - z & = & 1 \\
-y + 3z & = & 8 \\
2y - z & = & -1.
\end{array}
$$

STEP 3:
Want a 1 here.

Notice that we have now eliminated x from the second and the third equations. For now we neglect the first equation and only consider the second and third equations, and repeat the application of the first two steps to these two equations. Our goal is to eliminate y from the third equation. To prepare for this, we wish to have a one as the coefficient of y in the second equation. Then for our third step we multiply the second equation by -1 and obtain

$$
\begin{array}{rcr}
x - y - z & = & 1 \\
y - 3z & = & -8 \\
2y - z & = & -1.
\end{array}
$$

STEP 4:
Want a 0 here.

Now to eliminate y from the third equation, we add -2 times the second equation to the third equation and replace the third equation with the result.

$$
\begin{array}{lrcr}
-2 \text{ times second} & -2y + 6z & = & 16 \\
\hline
\text{third} & 2y - z & = & -1 \\
\hline
\text{new third} & 5z & = & 15
\end{array}
$$

This gives

$$
\begin{array}{rcr}
x - y - z & = & 1 \\
y - 3z & = & -8 \\
5z & = & 15.
\end{array}
$$

STEP 5:
Want a 1 here.

Now we can readily solve for z by dividing the third equation by 5 and obtain

$$x - y - z = 1$$

STEP 6:
Want a 0 here.

$$y - 3z = -8$$

$$z = 3.$$

Now we wish to use the third equation to eliminate z from the first and second equations. To do this we add the third equation to the first equation and replace the result with the first equation; we also add three times the third equation to the second equation and replace the result with the second equation. This then yields

$$x - y = 4$$

STEP 7:
Want a 0 here.

$$y = 1$$

$$z = 3.$$

Finally we wish to use the second equation to eliminate y from the first equation. We accomplish this by adding the second equation to the first equation and replacing the result with the first equation giving

$$x = 5$$

$$y = 1$$

$$z = 3.$$

Thus the solution is $(5, 1, 3)$. ■

REMARK. We should check to see if this is the correct answer by replacing x with 5, y with 1, and z with 3 in the original system to obtain

$$2(5) - 2(1) - 2(3) = 2$$

$$2(5) - 3(1) + (3) = 10$$

$$(5) + (1) - 2(3) = 0.$$

This verifies the solution.

Sometimes it is necessary to interchange two equations. This is required, for example, if there is no x in the first equation but there is an x in the second equation. In such a case the two equations are interchanged. For example, in the following problem

$$3y = 6$$

$$2x + y = 1,$$

interchange the equations to obtain

$$2x + y = 1$$

$$3y = 6.$$

Then begin with Step 1 and follow the subsequent steps indicated above.

We have performed three basic operations on the system. These are called **elementary** operations; they are summarized as follows.

Elementary Operations

1. Interchange the ith equation with the jth equation ($E_i \leftrightarrow E_j$).
2. Replace the ith equation by a nonzero constant k times the ith equation ($kE_i \rightarrow E_i$).
3. Replace the ith equation with the ith equation plus k times the jth equation ($E_i + kE_j \rightarrow E_i$).

It can be proven that such elementary operations do not change the solution set to the system. Thus no solutions are introduced or lost when these operations are applied to the system.

Elementary Operations Yield Equivalent Systems

Elementary operations yield an equivalent system of equations; that is, elementary operations do not introduce or eliminate any solutions to the system.

Notice how in Example 1 we used elementary operations to go from the system

$$
\begin{aligned}
2x - 2y - 2z &= 2 \\
2x - 3y + z &= 10 \qquad \text{to the system} \\
x + y - 2z &= 0
\end{aligned}
\qquad
\begin{aligned}
x &= 5 \\
y &= 1 \\
z &= 3.
\end{aligned}
$$

Since elementary operations do not change the solution set, the solution set of the original system is the same as the solution set of the last system.

The following is an outline of our strategy for a system like (1) of three linear equations in three unknowns that has one and only one solution. We will carry out our strategy by using elementary row operations.

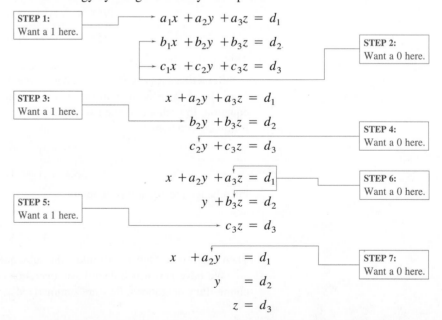

STEP 1: Want a 1 here.

$$
\begin{aligned}
a_1x + a_2y + a_3z &= d_1 \\
b_1x + b_2y + b_3z &= d_2 \\
c_1x + c_2y + c_3z &= d_3
\end{aligned}
$$

STEP 2: Want a 0 here.

STEP 3: Want a 1 here.

$$
\begin{aligned}
x + a_2y + a_3z &= d_1 \\
b_2y + b_3z &= d_2 \\
c_2y + c_3z &= d_3
\end{aligned}
$$

STEP 4: Want a 0 here.

$$
\begin{aligned}
x + a_2y + a_3z &= d_1 \\
y + b_3z &= d_2 \\
c_3z &= d_3
\end{aligned}
$$

STEP 6: Want a 0 here.

STEP 5: Want a 1 here.

$$
\begin{aligned}
x + a_2y &= d_1 \\
y &= d_2 \\
z &= d_3
\end{aligned}
$$

STEP 7: Want a 0 here.

The constants are *generic*, that is, they may change from step to step.

We continue with the following example.

E X A M P L E 2 **System With No Solution**

Solve

$$2x - 4y = 6$$
$$-3x + 6y = 1.$$

Solution

$$2x - 4y = 6 \quad \boxed{\begin{array}{l}\text{STEP 1:}\\ \text{Want a 1 here.}\end{array}} \qquad \tfrac{1}{2}E_1 \to E_1$$
$$-3x + 6y = 1$$

$$x - 2y = 3 \quad \boxed{\begin{array}{l}\text{STEP 2:}\\ \text{Want a 0 here.}\end{array}}$$
$$-3x + 6y = 1 \qquad\qquad\qquad E_2 + 3E_1 \to E_2$$

$$x - 2y = 3$$

$$0 = 10 \qquad \text{contradiction}$$

Since the second equation yields a contradiction, the system has no solution. ∎

We shall see in the next section that some systems have infinite numbers of solutions. Thus there are three possibilities in solving a system of linear equations.

The Three Possibilities in Solving a System of Linear Equations

1. There can be precisely one solution.
2. There can be no solution.
3. There can be infinitely many solutions.

Applications

The following is a typical (scaled-down) type of application in allocation theory that can be solved using the Gauss–Jordan method.

E X A M P L E 3 **Allocation of Machine Time**

A firm produces three products, a dress shirt, a casual shirt, and a sport shirt, and uses a cutting machine, a sewing machine, and a packaging machine in the process. To produce each gross of dress shirts requires 3 hours on the cutting machine and 2 hours on the sewing machine. To produce each gross of casual shirts requires 5 hours on the cutting machine and 1 hour on the sewing machine. To produce each gross of sport shirts requires 7 hours on the cutting machine and 3 hours on the sewing machine. It takes 2 hours to package each gross of each type of shirt. If the cutting machine is available for 480 hours, the sewing machine 170 hours, and the packaging machine 200 hours, determine the number of gross of each shirt that should be produced to use all of the available time on the three machines.

Solution   If x is the number of gross of dress shirts, y the number of gross of casual shirts, and z the number of gross of sport shirts produced, then we saw at the beginning of this section that x, y, and z must satisfy the following system of linear equations

$$2x + 2y + 2z = 200$$
$$3x + 5y + 7z = 480$$
$$2x + y + 3z = 170.$$

We use the Gauss–Jordan method of solution. Thus

$$2x + 2y + 2z = 200$$
$$3x + 5y + 7z = 480$$
$$2x + y + 3z = 170$$

STEP 1: Want a 1 here. $\tfrac{1}{2}E_1 \to E_1$

$$x + y + z = 100$$
$$3x + 5y + 7z = 480$$
$$2x + y + 3z = 170$$

STEP 2: Want a 0 here. $E_2 - 3E_1 \to E_2$
$E_3 - 2E_1 \to E_3$

$$x + y + z = 100$$
$$2y + 4z = 180$$
$$-y + z = -30$$

STEP 3: Want a 1 here. $\tfrac{1}{2}E_2 \to E_2$

$$x + y + z = 100$$
$$y + 2z = 90$$
$$-y + z = -30$$

STEP 4: Want a 0 here. $E_3 + E_2 \to E_3$

$$x + y + z = 100$$
$$y + 2z = 90$$
$$3z = 60$$

STEP 5: Want a 1 here. $\tfrac{1}{3}E_3 \to E_3$

$$x + y + z = 100$$
$$y + 2z = 90$$
$$z = 20$$

STEP 6: Want a 0 here. $E_1 - E_3 \to E_1$
$E_2 - 2E_3 \to E_2$

$$x + y = 80$$
$$y = 50$$
$$z = 20$$

STEP 7: Want a 0 here. $E_1 - E_2 \to E_1$

$$x = 30$$
$$y = 50$$
$$z = 20.$$

Thus the firm should produce 30 gross of dress shirts, 50 gross of casual shirts, and 20 gross of sport shirts to use all the available time on the 3 machines. ■

SELF-HELP EXERCISE SET 2.1

1. Solve the following system of equations:

$$2x + 3y = 1$$
$$6x + 11y = 1.$$

2. Find the number a such that the following system has no solution:

$$x + 2y = 3$$
$$2x + ay = 5.$$

3. A store sells 30 sweaters on a certain day. The sweaters come in three styles: A, B, and C. Style A costs $30, style B $40, and style C $50. If the store sold $1340 worth of these sweaters on that day and the number of C sold exceeded by 6 the sum of the other two styles sold, how many sweaters of each style were sold?

EXERCISE SET 2.1

In Exercises 1 through 22 use the Gauss–Jordan method to find all solutions, if any exist, of the given systems.

1. $x + 2y = 12$
    $2x + 3y = 19$

2. $x + 3y = 2$
    $3x + 4y = 1$

3. $4x - 8y = 20$
    $-x + 3y = -7$

4. $3x - 12y = 3$
    $-2x + y = -9$

5. $-2x + 8y = -6$
    $-2x + 3y = -1$

6. $-3x + 12y = -21$
    $-10x + 2y = 6$

7. $3x + 6y = 0$
    $x - y = -3$

8. $2x + 4y = 8$
    $2x - 4y = 0$

9. $2x - 4y = 8$
    $-x + 2y = 4$

10. $-3x + 6y = 10$
     $x - 2y = 4$

11. $3x - 6y = 12$
     $-x + 2y = 5$

12. $-2x + 4y = 12$
     $x - 2y = 1$

13. $-x + 3y = 7$
     $2x - 6y = 14$

14. $x - 3y = -8$
     $-2x + 6y = 16$

15. $0.1x - 0.3y = 0.4$

$\quad -2x + 6y = 4$

16. $-0.1x + 0.3y = 0.5$

$\quad 2x - 6y = 1$

17. $3x - 3y + 6z = -3$

$\quad 2x + y + 2z = 4$

$\quad 2x - 2y + 5z = -2$

18. $2x - 4y + 2z = -6$

$\quad 3x + 4y + 5z = 1$

$\quad 2x - y + z = -3$

19. $x + y + z = 10$

$\quad x - y + z = 10$

$\quad x + y - z = 0$

20. $x + y + z = 1$

$\quad 2x - y + z = 2$

$\quad 3x + 2y + 5z = 3$

21. $x + y + z = 1$

$\quad x - y + z = 2$

$\quad 3x + y + 3z = 1$

22. $x + y + z = 1$

$\quad x - y - z = 1$

$\quad 4x + 4y + 2z = 2$

23. A person has 25 coins, all of which are quarters and dimes. If the face value of the coins is \$3.25, how many of each type of coin must this person have?

24. A person has three times as many dimes as quarters. If the total face value of these coins is \$2.20, how many of each type of coin must this person have?

25. A person has 36 coins, all of which are nickels, dimes, and quarters. If there are twice as many dimes as nickels and if the face value of the coins is \$4, how many of each type of coin must this person have?

26. A person has three times as many nickels as quarters and three more dimes than nickels. If the total face value of these coins is \$2.40, how many of each type of coin must this person have?

Applications

27. **Scheduling.** An insurance company has two types of documents to process, type A and type B. The type A document needs to be examined for 2 hours by the accountant and 3 hours by the attorney, while the type B needs to be examined for 3 hours by the accountant and 1 hour by the attorney. If the accountant has 38 hours and the attorney 36 hours each week to spend working on these documents, how many documents of each type can they process?

28. **Mixture.** A small store sells mint tea at \$3.20 an ounce and peppermint tea at \$4 an ounce. The store owner decides to make a batch of 100 ounces of tea that mixes both kinds and sell the mixture for \$3.50 an ounce. How many ounces of each of the two varieties of tea should be mixed to obtain the same revenue as selling them unmixed?

29. **Investments.** An individual has a total of \$1000 in two banks. The first bank pays 8 percent a year and the second pays 10 percent a year. If the person receives \$86 of interest in the first year, how much money must have been deposited in each bank?

30. **Nutrition.** A dietitian must plan a meal for a patient using two fruits, oranges and strawberries. Each orange contains 1 gram of fiber and 75 mg of vitamin C, while each cup of strawberries contains 2 grams of fiber and 60 mg of vitamin

C. How much of each of these fruits needs to be eaten so that a total of 8 grams of fiber and 420 mg of vitamin C will be obtained?

31. **Production Scheduling.** A small plant with a cutting department and a sewing department produces two styles of sweaters, style A and style B. It takes 0.5 work-hours to cut a style A sweater and 0.6 work-hours to sew it. It takes 0.4 work-hours to cut a style B sweater and 0.3 work-hours to sew it. If the cutting department has 200 work-hours available each day and the sewing department has 186 work-hours available each day, how many sweaters of each style can be produced if both departments work at full capacity?

32. **Sales.** A store sells two types of sweaters: style A and style B. Style A costs $30 while style B costs $40. On a certain day the store sells twice as many sweaters of style A as style B. If the store sold $400 worth of these sweaters on that day, how many sweaters of each style were sold?

33. **Scheduling.** An insurance company has three types of documents to process, type A, type B, and type C. The type A document needs to be examined for 2 hours by the accountant and 3 hours by the attorney. The type B document needs to be examined for 4 hours by the accountant and 2 hours by the attorney. Finally, the last document needs to be examined for 2 hours by the accountant and 4 hours by the attorney. The secretary needs 3 hours to type each document. If the accountant has 34 hours, the attorney 35 hours, and the secretary 36 hours available to spend working on these documents, how many documents of each type can they process?

34. **Investments.** A pension fund manager has been given a total of $900,000 by a corporate client to invest in certain types of stocks and bonds over the next year. The client requires that twice as much money be in bonds as in stocks. Also stocks must be restricted to a certain class of low-risk stocks and another class of medium-risk stocks. Furthermore, the client demands an annual return of 8%. The pension fund manager assumes from historical data that the low risk stocks should return 9% annually, the medium risk stocks 11% annually, and bonds 7%. How should the pension fund manager allocate funds among the three groups to meet all the demands of the client?

35. **Investments.** Suppose in the previous problem the client demanded an annual return of 9%. What should the pension fund manager do?

36. **Production Scheduling.** A small plant with a cutting department, a sewing department, and a packaging department produces three styles of sweaters: styles A, B, and C. It takes 0.4 work-hours to cut a style A and 0.2 work-hours to sew it. It takes 0.3 work-hours to cut a style B sweater and the same to sew it. It takes 0.5 work-hours to cut a style C and 0.6 work-hours to sew it. It takes 0.1 work-hours to package each of the sweaters. If the cutting department has 110 work-hours available each day, the sewing department has 95 work-hours available each day, and the packaging department has 30 hours, how many sweaters of each style can be produced if all departments work at full capacity?

Graphing Calculator Exercises

In Exercises 37 through 40 use the zoom feature of your graphing calculator to find the solution to two decimal places to each of the following systems.

37. $3x + 4y = 12$
     $2x - 5y = 10$

38. $3x + 5y = 15$
     $3x - 4y = 12$

39. $2.1x + 4.1y = 16$
     $2.8x + 2.2y = 12$

40. $3.3x + 9.4y = 15.1$
     $0.9x + 2.2y = 4.2$

In Exercises 41 through 42 use your graphing calculator to determine if there is one solution, an infinite number of solutions, or no solution to the given system.

41. $-2x + 4y = 12$
     $x - 2y = 2$

42. $x - 4y = -8$
     $-2x + 8y = 16$

Solutions to Self-Help Exercise Set 2.1

1. Use the Gauss–Jordan method of solution and obtain

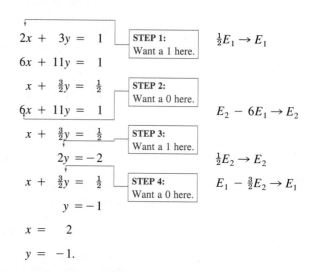

2. Use the Gauss–Jordan method of solution and obtain

$$x + 2y = 3$$
$$2x + ay = 5$$

STEP 2:
Want a 0 here.

$$E_2 - 2E_1 \rightarrow E_2$$

$$x + 2y = 3$$
$$(a - 4)y = -1.$$

Now if $a = 4$, the last line reads $0 = -1$ and is a contradiction. Thus if $a = 4$, there is no solution to the system.

3. Let x be the number of style A sweaters sold, y the number of style B, and z the number of style C. Since a total of 30 were sold, $x + y + z = 30$. The revenue in dollars from style A is $\$30x$, from style B is $\$40y$, and from style C is $\$50z$.

Since $1340 worth of the sweaters were sold, $30x + 40y + 50z = 1340$. Finally, since the number of style C exceeded by 6 the sum of the other two styles, $z = x + y + 6$, or $-x - y + z = 6$. Thus we wish to solve the system

$$
\begin{aligned}
x + y + z &= 30 \\
30x + 40y + 50z &= 1340 \\
-x - y + z &= 6.
\end{aligned}
$$

Use the Gauss–Jordan method of solution and obtain

$$
\begin{aligned}
x + y + z &= 30 \\
30x + 40y + 50z &= 1340 \qquad \boxed{\text{Want a 0 here.}} \qquad E_2 - 30E_1 \to E_2 \\
-x - y + z &= 6 \qquad\qquad\qquad\qquad\qquad\quad E_3 + E_1 \to E_3
\end{aligned}
$$

$$
\begin{aligned}
x + y + z &= 30 \\
10y + 20z &= 440 \qquad\qquad\qquad\qquad \tfrac{1}{10}E_2 \to E_2 \\
2z &= 36 \qquad \boxed{\text{Want a 1 here.}} \qquad \tfrac{1}{2}E_3 \to E_3
\end{aligned}
$$

$$
\begin{aligned}
x + y + z &= 30 \qquad \boxed{\text{Want a 0 here.}} \qquad E_1 - E_3 \to E_1 \\
y + 2z &= 44 \qquad\qquad\qquad\qquad\quad E_2 - 2E_3 \to E_2 \\
z &= 18
\end{aligned}
$$

$$
\begin{aligned}
x + y &= 12 \qquad \boxed{\text{Want a 0 here.}} \qquad E_1 - E_2 \to E_1 \\
y &= 8 \\
z &= 18
\end{aligned}
$$

$$
\begin{aligned}
x &= 4 \\
y &= 8 \\
z &= 18.
\end{aligned}
$$

Thus the store sold 4 of style A, 8 of style B, and 18 of style C.

2.2 GENERAL SYSTEMS OF LINEAR EQUATIONS

▶ *Augmented Matrix*

▶ *The Gauss–Jordan Method (Continued)*

▶ *Applications*

The Gauss–Jordan Method in Ancient China

Actually the Gauss–Jordan method was known in ancient China about 300 B.C. and was used as a practical method of solving concrete problems. The method was known as "the way of calculating by arrays." At this time in history, it had not occurred to anyone to set the unknowns in a problem equal to letters as we now do. Thus a system of linear equations such as

$$2x + 5y = 10$$
$$3x - 7y = 15$$

was written as the square array of numbers

$$\begin{array}{rrr} 2 & 5 & 10 \\ 3 & -7 & 15 \end{array}$$

However, rods replaced the numbers in the array. The number 2 was replaced with a black rod, 2 units in length, while the number 5 was replaced with a black rod, 5 units in length. A negative number was represented by a red rod. The method of eliminating one unknown at a time then proceeded just as was done in the last section, where one imagined that the first column represented the first unknown, the second column the second unknown, and the third column the number to the right of the equality. We shall now follow this system.

A P P L I C A T I O N
ıputer Purchase

A firm must purchase a total of 100 computers, some of small, some of medium, and some of large capacity. The small capacity computers cost $2000 each, the medium capacity cost $6000 each, and the large capacity cost $8000 each. If the firm plans to spend all of $400,000 on the total purchase, find the number of each type to be bought. For the answer see Example 5.

Augmented Matrix

We now introduce the augmented matrix, which will save a little work. We then extend the Gauss–Jordan method that was introduced in the last section to linear systems with any number of equations and variables. We will then consider more applications.

In the last section we solved systems of linear equations such as

$$2x + 3y = 4$$
$$5y = -2.$$

It is convenient to write this system in the abbreviated form as

$$\begin{bmatrix} 2 & 3 & 4 \\ 0 & 5 & -2 \end{bmatrix}.$$

Such a rectangular array of numbers is called a **matrix**. We will study matrices in detail in later sections. At this time we do need to know that the first line of the array is called the **first row** and the second line the **second row**, and so on if there are additional lines.

$$\begin{bmatrix} 2 & 3 & 4 \\ 0 & 5 & -2 \end{bmatrix} \begin{matrix} \leftarrow \text{Row 1} \\ \leftarrow \text{Row 2} \end{matrix}$$

It is convenient to place a vertical line to separate the last column of the matrix from the rest of the matrix.

$$\begin{bmatrix} 2 & 3 & | & 4 \\ 0 & 5 & | & -2 \end{bmatrix}$$

In this form the matrix is called the **augmented matrix** for the system. The only difference between this augmented matrix and the original system is that in the augmented matrix the symbols x, y, and $=$ have been dropped. This spares us the work of always writing these symbols for each step of the solution. Now if we wish to interchange two equations in the system, we interchange the two corresponding rows in the augmented matrix. If we wish to multiply a constant times each side of one equation, we multiply a constant times each member of the corresponding row, etc. In general we have the following.

Elementary Row Operations

1. Interchange the ith row with the jth row ($R_i \leftrightarrow R_j$).
2. Multiply each member of the ith row by a nonzero constant k ($kR_i \rightarrow R_i$).
3. Replace each element in the ith row with the corresponding element in the ith row plus k times the corresponding element in the jth row ($R_i + kR_j \rightarrow R_i$).

The Gauss–Jordan Method (Continued)

E X A M P L E   1   **System With Many Solutions**

Solve

$$-3x + 6y = 12$$
$$2x - 4y = -8.$$

Solution   We first replace this system with the augmented matrix

$$\begin{bmatrix} -3 & 6 & | & 12 \\ 2 & -4 & | & -8 \end{bmatrix}.$$

Using the Gauss–Jordan method, we have

$$\begin{bmatrix} -3 & 6 & | & 12 \\ 2 & -4 & | & -8 \end{bmatrix} \quad \boxed{\begin{array}{l}\textbf{STEP 1:} \\ \text{Want a 1 here.}\end{array}} \qquad -\tfrac{1}{3}R_1 \rightarrow R_1$$

$$\begin{bmatrix} 1 & -2 & | & -4 \\ 2 & -4 & | & -8 \end{bmatrix} \quad \boxed{\begin{array}{l}\textbf{STEP 2:} \\ \text{Want a 0 here.}\end{array}}$$

$$\qquad\qquad\qquad\qquad\qquad\qquad\qquad R_2 - 2R_1 \rightarrow R_2$$

$$\begin{bmatrix} 1 & -2 & | & -4 \\ 0 & 0 & | & 0 \end{bmatrix} \quad \text{or} \qquad \begin{array}{rcl} x - 2y &=& -4 \\ 0 &=& 0. \end{array}$$

Since the second equation is automatically satisfied, the solutions of the system will be all solutions of $x - 2y = -4$. We can solve this last equation for x or for y. We will solve for x since this is easier. If we let $y = t$, where t is any real number, and solve for x, we obtain $x = 2t - 4$. Then we have

$$x = 2t - 4$$
$$y = t,$$

and the pair $(2t - 4, t)$ is a solution for any real number t. ∎

The above variable t is called a **parameter**, and replacing t with any real number produces a **particular solution** to the system. For example, setting $t = 2$ yields the particular solution $(0, 2)$, while setting $t = 0$ yields the particular solution $(-4, 0)$. The solution $(2t - 4, t)$ in the previous problem is called the **general solution**.

Our goal is always to use elementary row operations to reduce the augmented matrix to a simple form where the solution can be read by inspection. When the augmented matrix is in this simple form, we called it the *reduced matrix*. More precisely, we have the following definition.

Reduced Matrix

A matrix is in **reduced form** if:

1. The leftmost nonzero element in each row is 1.
2. The column containing the leftmost 1 of any row has all other elements in that column equal to 0.
3. The leftmost 1 in any row is to the left of the leftmost 1 in a lower row.
4. Any row consisting of all zeros must be below any row with at least one nonzero element.

For example, all of the following matrices are in reduced form:

$$\begin{bmatrix} 1 & 0 & | & 2 \\ 0 & 1 & | & 0 \\ 0 & 0 & | & 0 \end{bmatrix}, \quad \begin{bmatrix} 1 & 4 & 0 & | & 3 \\ 0 & 0 & 1 & | & 2 \\ 0 & 0 & 0 & | & 0 \end{bmatrix}, \quad \begin{bmatrix} 1 & 0 & 3 & 0 & | & 3 \\ 0 & 1 & 2 & 0 & | & 2 \\ 0 & 0 & 0 & 1 & | & 3 \end{bmatrix}.$$

None of the following matrices are in reduced form

$$\begin{bmatrix} 1 & 0 & | & 2 \\ 0 & 2 & | & 2 \end{bmatrix}, \quad \begin{bmatrix} 1 & 2 & 3 & | & 4 \\ 0 & 0 & 1 & | & 2 \end{bmatrix}, \quad \begin{bmatrix} 0 & 1 & | & 1 \\ 1 & 0 & | & 2 \end{bmatrix}, \quad \begin{bmatrix} 1 & 0 & 0 & | & 1 \\ 0 & 0 & 0 & | & 0 \\ 0 & 1 & 0 & | & 2 \end{bmatrix}$$

since the first violates the first condition, the second violates the second condition, the third the third condition, and the fourth the fourth condition.

E X A M P L E   2   Infinite Number of Solutions

Finish the solution to the system that has the augmented matrix as follows:

$$\begin{bmatrix} 1 & 2 & 3 & | & 4 \\ 0 & 0 & 0 & | & 0 \\ 0 & 1 & 1 & | & 1 \end{bmatrix}.$$

Solution

$$\begin{bmatrix} 1 & 2 & 3 & | & 4 \\ 0 & 0 & 0 & | & 0 \\ 0 & 1 & 1 & | & 1 \end{bmatrix} \quad \boxed{\text{Need a 1 here.}} \quad R_2 \leftrightarrow R_3$$

$$\begin{bmatrix} 1 & 2 & 3 & | & 4 \\ 0 & 1 & 1 & | & 1 \\ 0 & 0 & 0 & | & 0 \end{bmatrix} \quad \boxed{\text{Need a 0 here.}} \quad R_1 - 2R_2 \to R_1$$

$$\begin{bmatrix} 1 & 0 & 1 & | & 2 \\ 0 & 1 & 1 & | & 1 \\ 0 & 0 & 0 & | & 0 \end{bmatrix} \quad \text{or} \quad \begin{matrix} x \quad\; + z = 2 \\ y + z = 1 \\ 0 = 0 \end{matrix}$$

Now we have two linear equations in three unknowns. Thus any two of these unknowns can be solved in terms of the third. The following technique indicates just how this is to be done. Consider now any nonzero row in the reduced form. The leftmost nonzero element in this row must be a one, which we refer to as a **leading one**.

$$\text{Leading one} \longrightarrow \begin{bmatrix} 1 & 0 & 1 & | & 2 \\ 0 & 1 & 1 & | & 1 \\ 0 & 0 & 0 & | & 0 \end{bmatrix}$$

The variable in the column containing the leading one is called a **leading variable**.

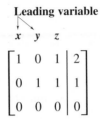

Leading variable

$$\begin{array}{ccc} x & y & z \end{array}$$

$$\begin{bmatrix} 1 & 0 & 1 & | & 2 \\ 0 & 1 & 1 & | & 1 \\ 0 & 0 & 0 & | & 0 \end{bmatrix}$$

In the above case, x and y are leading variables. The remaining variables are called **free variables**. In the above case, z is the free variable. The reduced form is such that it is always a simple matter to solve for the leading variables in terms of the free variables. In the above case, the free variable z is "free" to be any real number. We then set $z = t$, where t is any real number. Now we solve easily for the leading variables and obtain the general solution $(2 - t, 1 - t, t)$.   ∎

E X A M P L E   3   **Infinite Number of Solutions**

Given the following augmented matrix, finish the problem of finding the solution to the corresponding system:

$$\begin{array}{ccc} x & y & z \end{array}$$

$$\begin{bmatrix} 1 & 2 & 3 & | & 4 \\ 0 & 0 & 0 & | & 0 \\ 0 & 0 & 0 & | & 0 \end{bmatrix}.$$

Solution   The augmented matrix is in reduced form. Now there is one leading variable x and two free variables y and z.

Leading variable

$$\begin{array}{ccc} x & y & z \end{array}$$

Leading one $\longrightarrow$ $\begin{bmatrix} 1 & 2 & 3 & | & 4 \\ 0 & 0 & 0 & | & 0 \\ 0 & 0 & 0 & | & 0 \end{bmatrix}$

Thus both y and z are taken to be any real number. We indicate this by setting $y = s$ and $z = t$, where s and t are arbitrary numbers (that need not be equal). Thus the general solution has the form $(4 - 2s - 3t, s, t)$, where s and t are parameters. A particular solution can be obtained by setting each of the parameters equal to some number. For example, setting $s = 1$ and $t = 2$, gives $(-4, 1, 2)$ as a particular solution.   ∎

E X A M P L E 4   **Infinite Number of Solutions**

Given the following augmented matrix, finish the problem of finding the solution to the corresponding system.

$$
\begin{array}{cccccc}
x & y & z & u & v & w \\
\end{array}
$$

$$
\left[\begin{array}{cccccc|c}
1 & 2 & 0 & 4 & 5 & 0 & 4 \\
0 & 0 & 1 & 8 & 9 & 0 & 3 \\
0 & 0 & 0 & 0 & 0 & 1 & 2
\end{array}\right]
$$

Solution   The augmented matrix is in reduced form with x, z, and w the leading variables. Thus y, u, and v are the free variables.

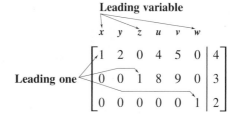

Leading variable

Leading one

$$
\left[\begin{array}{cccccc|c}
1 & 2 & 0 & 4 & 5 & 0 & 4 \\
0 & 0 & 1 & 8 & 9 & 0 & 3 \\
0 & 0 & 0 & 0 & 0 & 1 & 2
\end{array}\right]
$$

Setting $y = r$, $u = s$, and $v = t$, where r, s, and t are the parameters, the solution can then immediately be written as $x = 4 - 2r - 4s - 5t$, $y = r$, $z = 3 - 8s - 9t$, $u = s$, $v = t$, $w = 2$, or as $(4 - 2r - 4s - 5t, r, 3 - 8s - 9t, s, t, 2)$. ∎

Geometric Interpretations.   It turns out that any equation of the form $ax + by + cz = d$ (with not all of the constants a, b, and c being zero) is the equation of a plane in space. If there are 3 equations in 3 unknowns, Figure 2.4 indicates the possibilities. The three planes could intersect at a point indicating that the corresponding linear system of three equations in three unknowns has precisely one solution. The three planes could intersect in one line or in an entire plane, giving an infinite number of solutions to the corresponding system. Or there could

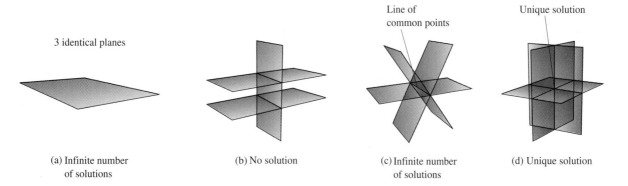

3 identical planes

Line of
common points

Unique solution

(a) Infinite number
of solutions

(b) No solution

(c) Infinite number
of solutions

(d) Unique solution

Figure 2.4

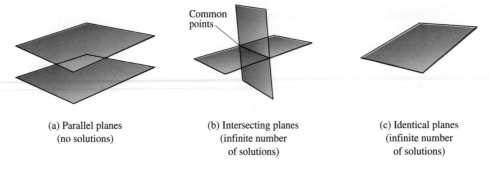

(a) Parallel planes
(no solutions)

(b) Intersecting planes
(infinite number
of solutions)

(c) Identical planes
(infinite number
of solutions)

Figure 2.5

be no point of intersection of the three planes, indicating that the corresponding system has no solution.

The geometry can give us further insights. For example, from Figure 2.5 we can see that two linear equations in three unknowns cannot have precisely one solution. If two linear equations in three unknowns had precisely one solution, this would mean that the corresponding two planes intersected in precisely one point, but this is impossible. Thus two linear equations in three unknowns has either no solution or an infinite number of solutions. This is true in general for any system with more variables than equations.

More Variables Than Equations

A system of equations that has more variables than equations has no solution or an infinite number of solutions.

By analyzing the reduced matrix one can see why this theorem must be true in general. Let us first assume the case in which there is no solution, that is, there is no row with all zeros except for the last entry. Then

$$\text{number of leading variables} = \text{number of leading ones}$$

$$\leq \text{number of equations}$$

$$< \text{number of variables}$$

Thus there is one variable left to be free, and, therefore, the system has an infinite number of solutions.

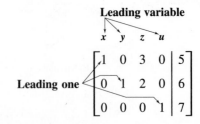

Applications

E X A M P L E   5   **Computer Purchase**
_____

A firm must purchase a total of 100 computers, some of small, some of medium, and some of large capacity. The small capacity computers cost $2000 each, the medium capacity cost $6000 each, and the large capacity cost $8000 each. If the firm plans to spend all of $400,000 on the total purchase, find the number of each type to be bought.

Solution   Let x, y, and z be the respective number of small, medium, and large capacity computers. Then the first sentence indicates that $x + y + z = 100$.

   The cost of purchasing x small computers is $2000x$ dollars, of purchasing y medium computers is $6000y$ dollars, and of purchasing z large computers is $8000z$ dollars. Since the total cost is $400,000, we have $2000x + 6000y + 8000z = 400,000$ or, in terms of thousands of dollars, $2x + 6y + 8z = 400$.

   We then have the system

$$x + y + z = 100$$
$$2x + 6y + 8z = 400.$$

Using the augmented matrix and the Gauss–Jordan method, we obtain

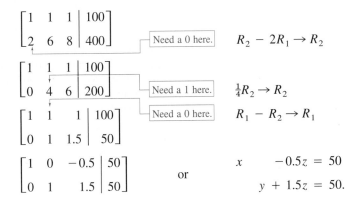

This gives

$$x = .50t + 50$$
$$y = 50 - 1.5t$$
$$z = t.$$

   First notice that t must be _even_ in order to ensure that x and y will be _integers_. Of course, x, y, and z must be nonnegative. Thus t is nonnegative. Now $0 \le x = .50t + 50$ implies $-100 \le t$. However, this gives no new information. Also $0 \le y = 50 - 1.5t$. This implies that $t \le 100/3$. Thus the firm can purchase $.50t + 50$ small computers, $50 - 1.5t$ medium computers, and t large computers, where t is any even integer with $0 \le t \le 32$. For example, the firm can take $t = 20$

giving $x = .50(20) + 50 = 60$ small capacity computers, $y = 50 - 1.5(20) = 20$ medium capacity computers, and $t = 20$ large capacity computers, spending the entire $400,000 and obtaining a total of 100 computers. ∎

SELF-HELP EXERCISE SET 2.2

1. Given the following augmented matrix, finish the problem of finding the solution to the corresponding system:

$$\begin{array}{ccc} x & y & z \end{array}$$
$$\left[\begin{array}{ccc|c} 1 & 2 & 1 & 3 \\ 0 & 0 & 1 & 2 \\ 0 & 0 & 0 & 0 \end{array}\right].$$

2. Given the following augmented matrix, finish the problem of finding the solution to the corresponding system:

$$\begin{array}{ccc} x & y & z \end{array}$$
$$\left[\begin{array}{ccc|c} 0 & 1 & 0 & 1 \\ 0 & 0 & 1 & 2 \\ 0 & 0 & 0 & 0 \end{array}\right].$$

3. A contractor has 2000 hours of labor available for three projects. The cost per work-hour for each of the three projects is $10, $12, and $14 respectively, and the total labor cost is $25,000. Find the number of work-hours that should be allocated to each project if all the available work-hours are to be used and all $25,000 spent on labor.

EXERCISE SET 2.2

In Exercises 1 through 6 determine whether each matrix is in reduced form.

1. $\left[\begin{array}{ccc|c} 1 & 0 & 2 & 1 \\ 0 & 1 & 3 & 2 \end{array}\right]$

2. $\left[\begin{array}{ccc|c} 1 & 0 & 1 & 2 \\ 0 & 0 & 1 & 3 \end{array}\right]$

3. $\left[\begin{array}{ccccc|c} 1 & 2 & 4 & 0 & 3 & 1 \\ 0 & 0 & 0 & 1 & 2 & 2 \end{array}\right]$

4. $\left[\begin{array}{ccc|c} 0 & 1 & 0 & 1 \\ 0 & 0 & 1 & 1 \\ 0 & 0 & 0 & 0 \end{array}\right]$

5. $\left[\begin{array}{ccc|c} 1 & 0 & 0 & 3 \\ 0 & 0 & 0 & 0 \\ 0 & 0 & 1 & 2 \end{array}\right]$

6. $\left[\begin{array}{ccccccc|c} 0 & 0 & 1 & 2 & 0 & 4 & 2 & 1 \\ 0 & 0 & 0 & 0 & 1 & 2 & 3 & 2 \\ 0 & 0 & 0 & 0 & 0 & 0 & 0 & 0 \end{array}\right]$

In Exercises 7 through 16 each of the matrices is in reduced form. Write the corresponding linear system and solve.

7. $\begin{array}{cc} x & y \end{array}$
$\left[\begin{array}{cc|c} 1 & 0 & 2 \\ 0 & 1 & 3 \end{array}\right]$

8. $\begin{array}{cc} x & y \end{array}$
$\left[\begin{array}{cc|c} 1 & 3 & 0 \\ 0 & 0 & 0 \end{array}\right]$

9. $\begin{array}{cc} x & y \end{array}$
$\left[\begin{array}{cc|c} 1 & 2 & 4 \\ 0 & 0 & 0 \end{array}\right]$

10. $\begin{array}{ccc} x & y & z \end{array}$
$\left[\begin{array}{ccc|c} 1 & 0 & 2 & 1 \\ 0 & 1 & 0 & 0 \end{array}\right]$

11. $\begin{array}{cccc} x & y & z & u \end{array}$
$\left[\begin{array}{cccc|c} 1 & 0 & 2 & 3 & 4 \\ 0 & 1 & 2 & 3 & 5 \end{array}\right]$

12. $\begin{array}{cccc} x & y & z & u \end{array}$
$\left[\begin{array}{cccc|c} 1 & 0 & 0 & 0 & 3 \\ 0 & 1 & 2 & 3 & 4 \end{array}\right]$

13. $\begin{array}{cccc} x & y & z & u \end{array}$
$\left[\begin{array}{cccc|c} 1 & 0 & 2 & 4 & 6 \\ 0 & 1 & 2 & 3 & 1 \\ 0 & 0 & 0 & 0 & 0 \end{array}\right]$

14. $\begin{array}{ccc} x & y & z \end{array}$
$\left[\begin{array}{ccc|c} 0 & 1 & 0 & 1 \\ 0 & 0 & 1 & 2 \\ 0 & 0 & 0 & 0 \end{array}\right]$

15. $\begin{array}{cccccc} x & y & z & u & v & w \end{array}$
$\left[\begin{array}{cccccc|c} 0 & 1 & 2 & 0 & 2 & 0 & 1 \\ 0 & 0 & 0 & 1 & 1 & 0 & 2 \\ 0 & 0 & 0 & 0 & 0 & 1 & 3 \end{array}\right]$

16. $\begin{array}{cccc} x & y & z & u \end{array}$
$\left[\begin{array}{cccc|c} 0 & 0 & 1 & 0 & 2 \\ 0 & 0 & 0 & 1 & 0 \\ 0 & 0 & 0 & 0 & 0 \end{array}\right]$

In Exercises 17 through 44 solve each of the systems by replacing the system with the augmented matrix and using the Gauss–Jordan method.

17. $\begin{aligned} x + 2y &= 5 \\ 2x - 3y &= -4 \end{aligned}$

18. $\begin{aligned} 2x + 4y &= 6 \\ 4x - y &= -6 \end{aligned}$

19. $\begin{aligned} x + y + z &= 6 \\ 2x + y + 2z &= 10 \\ 3x + 2y + z &= 10 \end{aligned}$

20. $\begin{aligned} x + 2y + z &= 6 \\ -x + 3y - 2z &= -4 \\ 2x - y - 3z &= -8 \end{aligned}$

21. $\begin{aligned} x - y + 2z &= -1 \\ 2x + y - 3z &= 6 \\ y - z &= 2 \end{aligned}$

22. $\begin{aligned} x - 2y + z &= 9 \\ 3y - 2z &= -11 \\ x + y + 4z &= 3 \end{aligned}$

23. $\begin{aligned} 2x - 4y + 8z &= 2 \\ 2x + 3y + 2z &= 3 \\ 2x - 3y + 5z &= 0 \end{aligned}$

24. $\begin{aligned} 2x - y - 3z &= 1 \\ -x - 2y + z &= -4 \\ 3x + y - 2z &= 9 \end{aligned}$

25. $\begin{aligned} -x + 2y + 3z &= 14 \\ 2x - y + 2z &= 2 \end{aligned}$

26. $\begin{aligned} 2x - 2y + 2z &= -2 \\ 3x + y - z &= 5 \end{aligned}$

27. $\begin{aligned} -2x + 6y + 4z &= 12 \\ 3x - 9y - 6z &= -18 \end{aligned}$

28. $\begin{aligned} 3x - 3y + 6z &= 15 \\ -2x + 2y - 4z &= -10 \end{aligned}$

29.
$$-x + 5y - 3z = 7$$
$$2x - 10y + 6z = -14$$

30.
$$x - y + 3z = 4$$
$$-2x + 2y - 6z = -8$$

31.
$$x + y + z = 1$$
$$x - y - z = 2$$
$$3x + y + z = 4$$

32.
$$x - y - z = 2$$
$$2x + y + z = 1$$
$$4x - y - z = 5$$

33.
$$2x + y - z = 0$$
$$3x - y + 2z = 1$$
$$x - 2y + 3z = 2$$

34.
$$x - y + z = 1$$
$$2x + 3y - 2z = 1$$
$$3x + 2y - z = 1$$

35.
$$x - 2y + 2z = 1$$
$$2x + y - z = 2$$
$$3x - y + z = 3$$

36.
$$x + y - 2z = 2$$
$$-x + 2y + z = 3$$
$$x + 4y - 3z = 7$$

37.
$$x + y + z + u = 6$$
$$y - z + 2u = 4$$
$$z + u = 3$$
$$x + 2y + 3z - u = 5$$

38.
$$x - y - z + 2u = 1$$
$$x - y + z + 3u = 9$$
$$y + 2z - u = 5$$
$$3y + z + 2u = 8$$

39.
$$x + 2y + z - u = -2$$
$$x + 2y + 2z + 2u = 9$$
$$y + z - u = -2$$
$$y - 2z + 3u = 4$$

40.
$$x + y + z + u = 6$$
$$x + 2y - z + u = 5$$
$$x + y - z + 2u = 6$$
$$2x + 2y + 2z + u = 10$$

41.
$$x + y = 4$$
$$2x - 3y = -7$$
$$3x - 4y = -9$$

42.
$$x - 3y = -7$$
$$3x + 4y = 5$$
$$2x - y = -4$$

43.
$$x + 2y = 4$$
$$2x - 3y = 5$$
$$x - 5y = 2$$

44.
$$x - 3y = 2$$
$$3x + 2y = 1$$
$$2x + 5y = 1$$

45.   Show that the system of equations

$$x + 2y + az = b$$
$$3x + 4y + cz = d$$

has a solution no matter what a, b, c, and d are.

46.   What condition must a, b, and c satisfy so that the following system has a solution?

$$x + 2y = a$$
$$3x + 4y = b$$
$$2x + 3y = c$$

47. A person has 36 coins, all of which are nickels, dimes and quarters. If the total value of the coins is $4, how many of each type of coin must this person have?

48. A person has three more dimes than nickels. If the total value of these coins is $2.40, how many of each type of coin must this person have?

Applications

49. Scheduling. An insurance company has three types of documents to process, type A, type B, and type C. The type A document needs to be examined for 2 hours by the accountant and 3 hours by the attorney. The type B document needs to be examined for 4 hours by the accountant and 2 hours by the attorney. The last document needs to be examined for 2 hours by the accountant and 4 hours by the attorney. If the accountant has 34 hours and the attorney 35 hours available to spend working on these documents, how many documents of each type can they process?

50. Investments. A pension fund manager has been given a total of $900,000 by a corporate client to invest in certain types of stocks and bonds over the next year. Stocks must be restricted to a certain class of low-risk stocks and another class of medium risk stocks. Furthermore, the client demands an annual return of 8%. The pension fund manager assumes from historical data that the low risk stocks should return 9% annually, the medium risk stocks 11% annually, and bonds 7%. How should the pension fund manager allocate funds among the three groups to meet all the demands of the client?

51. Investments. Suppose in the previous problem the client demanded an annual return of 9%. What should the pension fund manager do?

52. Production Scheduling. A small plant with a cutting department and a sewing department produces three styles of sweaters: styles A, B, and C. It takes 0.4 work-hours to cut a style A and 0.2 work-hours to sew it. It takes 0.3 work-hours to cut a style B sweater and the same to sew it. It takes 0.5 work-hours to cut a style C and 0.6 work-hours to sew it. If the cutting department has 110 work-hours available each day and the sewing department has 95 work-hours available each day, how many sweaters of each style can be produced if all departments work at full capacity?

53. Sales. A store sells a total of 300 sweaters in a certain month. The sweaters come in only three styles: styles A, B, and C. The sales of style B equaled the sum of the other two. Style A costs $30, style B costs $40, and style C $50. If the store sold $11,500 worth of these sweaters in that month, how many sweaters of each style were sold?

54. Production. Each day a firm produces 100 units of a perishable ingredient I and 200 units of another perishable ingredient II, both of which are used to manufacture 4 products, X, Y, Z, and U. The following table indicates how many units of each of the two perishable ingredients are used to manufacture each unit of the four products.

	I	II
X	0.1	0.4
Y	0.3	0.2
Z	0.6	0.4
U	0.2	0.3

Let x, y, z, u be respectively the number of units of X, Y, Z, U that are manufactured each day. If the manufacturer wishes to use all of the perishable ingredients that day, what are the options for the amounts of the four products to be manufactured?

55. **Production.**   Refer to Exercise 54. Just before production for the day is to begin, an order for 500 units of product U is received. The production manager is told to fill this order this very same day (while still using all the perishable ingredients) or be fired. Will the production manager be fired? Why or why not?

56. **Production.**   Answer the same question as in the previous problem if, instead of receiving an order for 500 units of product U, an order for 150 units of product Z is received.

57. **Nutrition.**   A dietitian must plan a meal for a patient using three fruits: oranges, strawberries, and blackberries. Each orange contains 1 gram of fiber, 75 mg of vitamin C, and 50 mg of phosphorus. Each cup of strawberries contains 2 grams of fiber, 60 mg of vitamin C, and 50 mg of phosphorus. Each cup of blackberries contains 6 grams of fiber, 30 mg of vitamin C, and 40 mg of phosphorus. How much of each of these fruits needs to be eaten so that a total of 13 grams of fiber, 375 mg of vitamin C, and 290 mg of phosphorus is obtained?

58. **Chinese Farm Problem.**   The ancient Chinese "way of calculating with arrays" can be found in Chapter 8 of the ancient text *Nine Chapters on the Mathematical Art*. The following is the first problem listed in Chapter 8. Solve it.
      There are three grades of corn. After threshing, three bundles of top grade, two bundles of medium grade, and one bundle of low grade make 39 dou (a measure of volume). Two bundles of top grade, three bundles of medium grade, and one bundle of low grade make 34 dou. The yield of one bundle of top grade, two bundles of medium grade, and three bundles of low grade make 26 dou. How many dou are contained in each bundle of each grade?

59. **Traffic Flow.**   The accompanying flow diagram indicates the traffic flow in numbers of vehicles per hour into and out of 4 intersections during rush hour. Traffic lights placed at each intersection can be timed to control the flow of traffic. Find all possible *meaningful* flow patterns. *Hint:* For each intersection write an equation that states that the total flow into the intersection must equal the total flow out. For example, the diagram indicates that

$$x_1 + x_4 = 300 + 400 = 700.$$

Now, using the diagram, find the other three equations. Solve by the Gauss–Jordan method.

(© Stockphotos Inc./The Image Bank.)

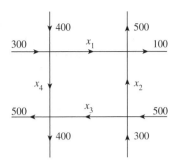

60. Traffic Flow. The accompanying flow diagram indicates the traffic flow into and out of 3 intersections during rush hour. Traffic lights placed at each intersection can be timed to control the flow of traffic. Find all possible *meaningful* flow patterns.

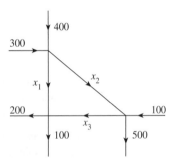

Graphing Calculator Exercises

In Exercises 61 through 62 find the solution using the matrix menu on your graphing calculator to perform the elementary row operations for the Gauss–Jordan method.

61. $1.2x + 2.1y + 3.4z = 54.2$

     $2.1x + 5.1y + 1.4z = 72.7$

     $3.7x + 1.4y + 3.8z = 62.9$

62. $2.31x + 1.04y + 3.65z = 44.554$

     $1.32x + 2.87y + 5.21z = 59.923$

     $1.66x + 1.92y + 3.41z = 42.617$

In Exercises 63 through 64 graph the equations on your graphing calculator to see if there is a solution.

63. $x + 2y = 4$

     $2x - 3y = 5$

     $x - 5y = 3$

64. $x - 3y = 2$

     $3x + 2y = 1$

     $2x + 5y = 2$

65. Demand and Supply. When the price p of pork goes up, the demand x_B for beef also goes up. In the same way, when the price q of beef goes up, the demand x_A of pork does also. Let the demand equations be given by

$$x_A = 6 - 2p + q \qquad x_B = 16 + p - 3q,$$

and let the supply equations be given as

$$p = 1 + 0.5x_A + x_B \qquad q = 1 + x_A + x_B.$$

Find the point of equilibrium, that is, the point at which all four equations are satisfied using the matrix menu to perform the elementary row operations required by the Gauss–Jordan method.

Solutions to Self-Help Exercise Set 2.2

1. Use the Gauss–Jordan method and obtain

$$\begin{bmatrix} 1 & 2 & 1 & | & 3 \\ 0 & 0 & 1 & | & 2 \\ 0 & 0 & 0 & | & 0 \end{bmatrix}$$ Need a 0 here. $R_1 - R_2 \rightarrow R_1$

$$\begin{bmatrix} 1 & 2 & 0 & | & 1 \\ 0 & 0 & 1 & | & 2 \\ 0 & 0 & 0 & | & 0 \end{bmatrix}$$ or $\begin{aligned} x + 2y &= 1 \\ z &= 2 \\ 0 &= 0. \end{aligned}$

The leading variables are x and z, while y is the free variable. Since y is taken to be arbitrary, we set $y = t$, where t is any real number. The general solution is $(1 - 2t, t, 2)$, where t is any real number.

2. The matrix is already in reduced form. The leading variables are y and z. The free variable is x. Setting x equal to the parameter t gives $z = 2$, $y = 1$, and $x = t$. The general solution is then $(t, 1, 2)$ for any t.

3. Let x, y, and z be the amount of work-hours allocated to, respectively, the first, second, and third projects. Since the total number of work-hours is 2000, then $x + y + z = 2000$. The labor cost for the first project is $\$10x$, for the second is $\$12y$, and for the third is $\$14z$. Since the total labor funds is $\$25,000$, we then have $10x + 12y + 14z = 25,000$. These two equations are then written as

$$\begin{aligned} x + \quad y + \quad z &= \quad 2000 \\ 10x + 12y + 14z &= 25,000. \end{aligned}$$

Using the augmented matrix and the Gauss–Jordan methods, we obtain

$$\begin{bmatrix} 1 & 1 & 1 & | & 2000 \\ 10 & 12 & 14 & | & 25,000 \end{bmatrix}$$ Need a 0 here. $R_2 - 10R_1 \rightarrow R_2$

$$\begin{bmatrix} 1 & 1 & 1 & | & 2000 \\ 0 & 2 & 4 & | & 5000 \end{bmatrix}$$ Need a 1 here. $\frac{1}{2}R_2 \rightarrow R_2$

$$\begin{bmatrix} 1 & 1 & 1 & | & 2000 \\ 0 & 1 & 2 & | & 2500 \end{bmatrix}$$ Need a 0 here. $R_1 - R_2 \rightarrow R_1$

$$\begin{bmatrix} 1 & 0 & -1 & | & -500 \\ 0 & 1 & 2 & | & 2500 \end{bmatrix}$$ or $\begin{aligned} x \quad - \quad z &= -500 \\ y + 2z &= \quad 2500 \end{aligned}$

This gives

$$x = t - 500$$
$$y = 2500 - 2t$$
$$z = t.$$

Since $0 \leq x = t - 500$, $t \geq 500$. Since $0 \leq y = 2500 - 2t$, $t \leq 1250$. Thus the contractor can allocate $t - 500$ work-hours to the first project, $2500 - 2t$ to the second, and t to the third, where $500 \leq t \leq 1250$. For example, taking $t = 1000$, means the contractor can allocate 500 work-hours to the first project, 500 to the second, and 1000 to the third and use all of the available work-hours with all of the $25,000 spent on labor.

2.3 INTRODUCTION TO MATRICES

▸ *Basic Definitions*

▸ *Applications*

**James Joseph Sylvester,
1814–1897**

*(David Eugene Smith Collection,
Columbia University.)*

The first mention of matrices occurred in 1850 in a paper by Sylvester. Sylvester was educated in England and became a professor at the University of Virginia in 1841. Unable to tolerate slavery, he left this position and the country after only two months. He returned to this country in 1871, accepting the chair of mathematics at Johns Hopkins University, where he was very productive, distinguishing himself as an outstanding mathematician. In 1878 he became the founding editor of the well regarded *American Journal of Mathematics.*

APPLICATION
Production

A firm makes bulldozers (B), cranes (C), and tractors (T) at two locations, New York City and Los Angeles, and the number of each produced at each site for January and February is given in tabulated form as follows.

	B	*C*	*T*
NYC	120	240	360
LA	310	0	240

January

	B	*C*	*T*
NYC	200	100	0
LA	150	200	300

February

Find the total number of each product produced at each site for the January to February period in tabulated form. See Example 4 for the answer.

Basic Definitions

Matrices were introduced in the last section. We will see that matrices are a natural way of presenting tabulated data. Certain calculations with this data correspond to certain matrix operations. These operations will be considered in this section, and some applications will be given.

As mentioned in the previous section, a matrix is a rectangular array of numbers. The matrix A

$$
\begin{array}{c}
\phantom{\text{row 1}} \quad \text{column 1} \quad \text{column 2} \quad \text{column 3} \\
\begin{array}{c} \text{row 1} \\ \text{row 2} \end{array}
\left[
\begin{array}{ccc}
2 & 3 & 5 \\
0 & -2 & 4
\end{array}
\right]
\end{array}
$$

is a matrix with 2 rows and 3 columns. The **order** of a matrix is $m \times n$ where m is the number of rows and n is the number of columns. Thus the order of the above matrix A is 2×3.

Given a matrix such as A above, the **entry** in the ith row and jth column is denoted by a_{ij}. For the above matrix A, $a_{11} = 2$, $a_{12} = 3$, $a_{13} = 5$, $a_{21} = 0$, $a_{22} = -2$, $a_{23} = 4$.

Some Matrix Terminology

A matrix A, also denoted by $(a_{ij})_{m \times n}$ or simply (a_{ij}), is a rectangular array of numbers.

The order of A is denoted by $m \times n$ where m is the number of rows and n is the number of columns.

The element in the ith row and jth column is denoted by a_{ij}.

If $A = (a_{ij})_{m \times n}$, then we have

$$
A =
\begin{array}{c}
 \\
\end{array}
\left[
\begin{array}{ccccc}
a_{11} & a_{12} & \cdots & a_{1j} & \cdots & a_{1n} \\
a_{21} & a_{22} & \cdots & a_{2j} & \cdots & a_{2n} \\
& & \cdots & & & \\
a_{i1} & a_{i2} & \cdots & a_{ij} & \cdots & a_{in} \\
& & \cdots & & & \\
a_{m1} & a_{m2} & \cdots & a_{mj} & \cdots & a_{mn}
\end{array}
\right]
\begin{array}{l}
\text{first row} \\
\text{second row} \\
\cdots \\
\text{ith row} \\
\cdots \\
\text{mth row}
\end{array}
$$

where the columns are labeled *first column*, *second column*, ..., *jth column*, ..., *nth column*.

E X A M P L E 1   **Determining the Order of Matrices**

Determine the order of

$$B = \begin{bmatrix} 1 & 2 \\ -1 & 0 \\ 4 & 8 \end{bmatrix}, \quad C = [1 \quad 2 \quad 5 \quad 0], \quad D = \begin{bmatrix} 1 \\ 3 \\ 4 \end{bmatrix}.$$

Also find $b_{22}, b_{32}, c_{12}, d_{31}$.

Solution   The order of B is 3×2.
The order of C is 1×4.
The order of D is 3×1.
Also $b_{22} = 0, b_{32} = 8, c_{12} = 2, d_{31} = 4$. ∎

There are three special types of matrices that arise often: column matrices, row matrices, and square matrices.

Row, Column, and Square Matrices

A matrix of order $1 \times n$ (one row and n columns) is called a **row matrix of dimension n**.

A matrix of order $m \times 1$ (m rows and one column) is called a **column matrix of dimension m**.

A matrix is called a **square matrix** if the number of rows equals the number of columns.

The matrix C above is a row matrix of dimension 4. The matrix D above is a column matrix of dimension 3.

Naturally we need to know when two matrices are equal. Two matrices are equal when they are absolutely identical. The following definition says just that.

Equality of Matrices

Two matrices are equal if and only if they have the same order and all corresponding entries are equal.

For example, the two matrices

$$\begin{bmatrix} 1 & 2 \\ 3 & 4 \end{bmatrix} \quad \begin{bmatrix} 1 & 2 & 0 \\ 3 & 4 & 0 \end{bmatrix}$$

are not equal since they do not have the same order.

E X A M P L E 2   **Equal Matrices**

Find x, y, and z, such that

$$\begin{bmatrix} 1 & 2 & x \\ 3 & 0 & 5 \end{bmatrix} = \begin{bmatrix} 1 & 2 & 4 \\ y & 0 & z \end{bmatrix}.$$

Solution   Since corresponding entries must equal, $x = 4$, $y = 3$, and $z = 5$. ∎

We now state three fundamental operations on matrices.

Multiplication of a Matrix by Real Numbers

If c is a real number and A is a matrix, then the matrix cA is the matrix obtained by multiplying every entry in A by c.

Addition and Subtraction of Matrices

Two matrices of the same order can be added (or subtracted) to obtain another matrix of the same order by adding (or subtracting) corresponding entries.

E X A M P L E 3   **Performing Elementary Matrix Operations**

Find the following.

(a) $2\begin{bmatrix} 3 & -1 & 0 \\ -2 & 3 & 10 \end{bmatrix} = \begin{bmatrix} 2(3) & 2(-1) & 2(0) \\ 2(-2) & 2(3) & 2(10) \end{bmatrix}$

$= \begin{bmatrix} 6 & -2 & 0 \\ -4 & 6 & 20 \end{bmatrix}$

(b) $\begin{bmatrix} 1 & 2 \\ 3 & 5 \\ -1 & 2 \end{bmatrix} + \begin{bmatrix} 1 & -2 \\ 0 & 3 \\ 2 & -4 \end{bmatrix} = \begin{bmatrix} 1+1 & 2-2 \\ 3+0 & 5+3 \\ -1+2 & 2-4 \end{bmatrix} = \begin{bmatrix} 2 & 0 \\ 3 & 8 \\ 1 & -2 \end{bmatrix}$

(c) $[1 \quad 2 \quad 3] + [2 \quad 5 \quad 6 \quad 0] =$ not defined

(d) $\begin{bmatrix} 1 & 2 \\ 3 & 0 \end{bmatrix} - \begin{bmatrix} 2 & 1 \\ 1 & 3 \end{bmatrix} = \begin{bmatrix} 1-2 & 2-1 \\ 3-1 & 0-3 \end{bmatrix} = \begin{bmatrix} -1 & 1 \\ 2 & -3 \end{bmatrix}$

(e) $2\begin{bmatrix} 2 & 1 \\ 0 & 4 \end{bmatrix} - 3\begin{bmatrix} -1 & 2 \\ 4 & 0 \end{bmatrix} = \begin{bmatrix} 4 & 2 \\ 0 & 8 \end{bmatrix} - \begin{bmatrix} -3 & 6 \\ 12 & 0 \end{bmatrix}$

$= \begin{bmatrix} 7 & -4 \\ -12 & 8 \end{bmatrix}$ ∎

The number 0 has the property that $x + 0 = 0 + x = x$ for any number x. There is a matrix, called the **zero matrix**, with an analogous property.

The Zero Matrix

The **zero matrix** of order $m \times n$ is the matrix with m rows and n columns, all of whose entries are zero.

Thus the zero matrix of order 2×3 is

$$\begin{bmatrix} 0 & 0 & 0 \\ 0 & 0 & 0 \end{bmatrix}$$

Notice that we have the following.

Properties of the Zero Matrix

If A is any matrix of order $m \times n$ and O is the zero matrix of the same order, then

$$A + O = O + A = A$$

$$A - A = O$$

The following properties of matrices can also be shown to be true.

Further Properties of Matrices

If A, B, and C are three matrices all of the same order, then

$$A + B = B + A$$

$$A + (B + C) = (A + B) + C$$

Applications

Data are often presented in matrix form. Suppose a firm makes bulldozers (B), cranes (C), and tractors (T) at two locations, New York City and Los Angeles, and the number of each produced at each location during January is given by the following table.

January

	B	C	T
NYC	120	240	360
LA	310	0	240

A **production matrix** for January can be given as

$$J = \begin{bmatrix} 120 & 240 & 360 \\ 310 & 0 & 240 \end{bmatrix},$$

where it is understood that the first row refers to NYC, the second row to LA, the first column to the number of bulldozers, etc.

E X A M P L E   4   **Adding Production Matrices**

Now suppose that the same firm as above has a February production matrix given by

$$F = \begin{bmatrix} 200 & 100 & 0 \\ 150 & 200 & 300 \end{bmatrix}.$$

Find (a) the production matrix for the January through February period and (b) the number of cranes produced in NYC during this period.

Solution   The matrix $J + F$ is the production matrix for the two-month period. Thus

(a) $J + F = \begin{bmatrix} 120 & 240 & 360 \\ 310 & 0 & 240 \end{bmatrix} + \begin{bmatrix} 200 & 100 & 0 \\ 150 & 200 & 300 \end{bmatrix}$

$= \begin{bmatrix} 320 & 340 & 360 \\ 460 & 200 & 540 \end{bmatrix}$

(b)   This is the element in the first row and second column of $J + F$, which is 340. ∎

E X A M P L E   5   **Multiplying a Number with a Production Matrix**

Suppose the production of the above firm in March increases by 10% from that of February for all products at all locations. Find (a) the production matrix M for March and (b) how many tractors were produced in LA during March.

Solution   (a)   The production matrix M for March is $1.1F$ or

$$M = 1.1F = \begin{bmatrix} 1.1(200) & 1.1(100) & 1.1(0) \\ 1.1(150) & 1.1(200) & 1.1(300) \end{bmatrix}$$

$= \begin{bmatrix} 220 & 110 & 0 \\ 165 & 220 & 330 \end{bmatrix}$

(b)   If $M = (m_{ij})$, then m_{23} is the number of tractors produced in LA in March. We see that $m_{23} = 330$. ∎

SELF-HELP EXERCISE SET 2.3

1. Suppose $A = (a_{ij})$ is a matrix of order 2×3 and $a_{ij} = i + 2j$ for all i and j. Write down the matrix.

2. Find $3 \begin{bmatrix} 3 & 2 \\ -2 & 0 \\ 5 & 1 \end{bmatrix} - 2 \begin{bmatrix} 1 & -3 \\ 0 & 2 \\ -3 & 2 \end{bmatrix}$

3. A person owns two furniture stores. The first store shows the following number of sales for the indicated two months.

Store 1

	Sofas	Love Seats	Chairs
April	30	20	40
May	30	30	50

The second store shows sales exactly 20% higher for all three types of furniture for each of the months. Use matrix operations to find the total sales in both stores.

EXERCISE SET 2.3

In Exercises 1 through 14 let

$$A = \begin{bmatrix} 3 & 1 \\ -2 & 4 \\ 0 & 3 \\ 2 & 8 \end{bmatrix}, \quad B = \begin{bmatrix} -1 & 0 & 2 & 6 \\ 3 & 1 & -2 & 4 \\ 5 & 3 & 0 & -5 \end{bmatrix}, \quad C = [-1 \ \ 2], \quad D = \begin{bmatrix} 0 \\ 1 \\ -3 \\ 4 \end{bmatrix}$$

Find the order of the following.

1. A
2. B (3×4)
3. C
4. D

If A, B, C, and D are defined as above and $A = (a_{ij})$, $B = (b_{ij})$, $C = (c_{ij})$, $D = (d_{ij})$, find the following.

5. a_{12}
6. a_{21}
7. a_{42}
8. a_{41}
9. b_{23}
10. b_{34}
11. b_{14}
12. b_{33}
13. c_{12}
14. d_{31}

15. If $A = (a_{ij})$ is a matrix of order 2×3 and $a_{ij} = 5$ for all i and j, write down A.

16. If $A = (a_{ij})$ is a matrix of order 2×3 and $a_{ij} = i + j$ for all i and j, write down A.

17. If $A = (a_{ij})$ is a matrix of order 3×3 and $a_{ij} = 1$ if $i = j$, and $a_{ij} = 0$ if $i \neq j$, write down A.

18. If $A = (a_{ij})$ is a matrix of order 3×2 and $a_{ij} = i$ for all i and j, write down A.

19. If $A = (a_{ij})$ is a matrix of order 3×3 and $a_{ij} = 0$ for all i and j, write down A.

20. If $A = (a_{ij})$ is a matrix of order 4×3 and $a_{ij} = 2i + j - 1$ for all i and j, write down A.

21. Find x, y, and z, such that

$$\begin{bmatrix} 3 & 2 \\ 4 & x \end{bmatrix} = \begin{bmatrix} 3 + y & z - 2 \\ 4 & 3 \end{bmatrix}$$

22. Find x, y, and z, such that

$$\begin{bmatrix} x & 3 \\ -1 & 0 \\ 4 & 5 \end{bmatrix} = \begin{bmatrix} 1 & 3 \\ y - 1 & 0 \\ 3 & z + x \end{bmatrix}.$$

In Exercises 23 through 42 let

$$E = \begin{bmatrix} 1 & 4 \\ 3 & 6 \\ -3 & 7 \\ 3 & -2 \end{bmatrix}, \quad F = \begin{bmatrix} 4 & 3 & -1 & -6 \\ 5 & 0 & -2 & 7 \\ 8 & 1 & -3 & 0 \end{bmatrix}, \quad G = [1 \quad 4], \quad H = \begin{bmatrix} 3 \\ -1 \\ 5 \\ 1 \end{bmatrix}$$

and let A, B, C, D be the matrices given before Exercise 1. Find the indicated quantities.

23. $A + E$	24. $B + F$	25. $C + G$
26. $D + H$	27. $A - E$	28. $B - F$
29. $C - G$	30. $D - H$	31. $2E$
32. $-3F$	33. $4G$	34. $\frac{1}{2}H$
35. $2A + E$	36. $B - 2F$	37. $2C + 3G$
38. $2D - \frac{1}{2}H$	39. $A - F$	40. $2C + D$
41. $E - E$	42. $H - H$	

In Exercises 43 through 47, let $A = (a_{ij})$, $B = (b_{ij})$, $C = (c_{ij})$, $D = (d_{ij})$, $X = (x_{ij})$ all be 2×2 matrices and let O be the 2×2 zero matrix.

43. Prove that $A + B = B + A$.

44. Prove that $A + (B + C) = (A + B) + C$.

45. Prove that $A + O = O + A = A$.

46. Prove that $(c + d)A = cA + dA$, where c and d are scalars.

47. Prove that $A - A = O$.

48. A matrix $D = (d_{ij})$ is a diagonal matrix if $d_{ij} = 0$ if $i \neq j$, that is, all entries off the "main diagonal" are zero. Suppose $A = (a_{ij})$ and $B = (b_{ij})$ are both diagonal matrices of order (4×4). Show that (a) $A + B$ is diagonal, (b) cA is diagonal where c is a real number.

49. Show that the matrix equation $X + A = B$ has the solution $X = B - A$.

50. A matrix $U = (u_{ij})$ is upper triangular if $u_{ij} = 0$ when $i > j$, that is, when all entries below the main diagonal are zero. Suppose $A = (a_{ij})$ and $B = (b_{ij})$ are both upper triangular matrices of order (4×4). Show that (a) $A + B$ is upper triangular and (b) cA is upper triangular where c is a real number.

Applications

51. **Production.** Refer to Example 5 of the text. Suppose the April production matrix was

$$A = \begin{bmatrix} 300 & 200 & 100 \\ 200 & 200 & 400 \end{bmatrix}$$

Write a production matrix for the periods from March through April as a sum of two matrices and then carry out the calculation.

52. **Production.** In the previous problem find $A - M$, where M is given in Example 5 of the text. What does $A - M$ stand for?

53. **Production.** Suppose that production of all products at all locations was reduced by 10% in May from what it was in April in the exercise before the last. Write the production matrix for May as a scalar times a matrix. Then carry out the calculation to obtain the production matrix for May.

54. **Production.** Refer to Example 5 of the text. Suppose the production for June of all products at all locations was the average of the January and February production. Write an expression using the matrices J and F for the June production matrix and then carry out the indicated calculation to obtain this matrix.

55. **Production.** A firm has a plant in NYC that has produced the following numbers of items of three products during the four quarters of the last year.

	First Quarter	Second Quarter	Third Quarter	Fourth Quarter
Product A	250	300	350	300
Product B	300	200	250	150
Product C	200	240	320	220

The following table gives the firm's production for the same period for its other production plant located in LA.

	First Quarter	Second Quarter	Third Quarter	Fourth Quarter
Product A	300	350	450	400
Product B	320	240	280	250
Product C	250	260	420	280

a.   Write a production matrix N for the NYC plant.
b.   Write a production matrix L for the LA plant.
c.   Write an expression involving the matrices N and L that gives the production matrix for the firm for the last year. Carry out the indicated calculation.
d.   Calculate $L - N$ and indicate what this matrix stands for.
e.   Next year the firm is predicting production to increase by 20% at NYC. Write the production matrix for next year for NYC as a scalar times a matrix and then carry out the indicated calculation.
f.   Write a matrix expression that gives the production matrix that represents the average production of its two plants and then carry out the indicated calculation.

Solutions to Self-Help Exercise Set 2.3

1. $a_{11} = 1 + 2(1) = 3$, $a_{12} = 1 + 2(2) = 5$, $a_{13} = 1 + 2(3) = 7$, $a_{21} = 2 + 2(1) = 4$, $a_{22} = 2 + 2(2) = 6$, $a_{23} = 2 + 2(3) = 8$. Thus $A = \begin{bmatrix} 3 & 5 & 7 \\ 4 & 6 & 8 \end{bmatrix}$

2. $3\begin{bmatrix} 3 & 2 \\ -2 & 0 \\ 5 & 1 \end{bmatrix} - 2\begin{bmatrix} 1 & -3 \\ 0 & 2 \\ -3 & 2 \end{bmatrix} = \begin{bmatrix} 9 & 6 \\ -6 & 0 \\ 15 & 3 \end{bmatrix} - \begin{bmatrix} 2 & -6 \\ 0 & 4 \\ -6 & 4 \end{bmatrix}$

$$= \begin{bmatrix} 7 & 12 \\ -6 & -4 \\ 21 & -1 \end{bmatrix}$$

3.   The indicated matrix operations are

$$\begin{bmatrix} 30 & 20 & 40 \\ 30 & 30 & 50 \end{bmatrix} + 1.2\begin{bmatrix} 30 & 20 & 40 \\ 30 & 30 & 50 \end{bmatrix} = \begin{bmatrix} 30 & 20 & 40 \\ 30 & 30 & 50 \end{bmatrix} + \begin{bmatrix} 36 & 24 & 48 \\ 36 & 36 & 60 \end{bmatrix}$$

$$= \begin{bmatrix} 66 & 44 & 88 \\ 66 & 66 & 110 \end{bmatrix}$$

2.4   MATRIX MULTIPLICATION

► *A Row Matrix Times a Column Matrix*

► *Multiplication in General*

► *Applications*

► *Systems of Equations*

The definition of multiplication of two matrices may at first seem strange. But in doing applications the reader will see how this definition becomes natural. Furthermore, this definition of multiplication plays a central role in writing systems of equations in matrix form. In the next section we will see how systems of equations with the same number of equations as unknowns can be solved using matrix operations and this definition of multiplication.

A Row Matrix Times a Column Matrix

To understand how to multiply two matrices, we first need to know how to multiply a row matrix with a column matrix of the same dimension. We therefore begin by defining such a multiplication.

Row Matrix Times Column Matrix

Let R be the row matrix $R = [r_1 \quad r_2 \quad \cdots \quad r_n]$
and let C be the column matrix

$$C = \begin{bmatrix} c_1 \\ c_2 \\ \vdots \\ c_n \end{bmatrix}$$

with the same dimension as R. Then the product RC is defined to be

$$RC = [r_1 \quad r_2 \quad \cdots \quad r_n] \begin{bmatrix} c_1 \\ c_2 \\ \vdots \\ c_n \end{bmatrix} = r_1 c_1 + r_2 c_2 + \cdots + r_n c_n.$$

REMARK.   Notice that the answer is a real number.

E X A M P L E 1   **Multiplying Row Matrices with Column Matrices**

Perform the indicated operations.

Solutions   (a) $[5 \quad \frac{1}{2}]\begin{bmatrix} 1 \\ 6 \end{bmatrix} = (5)(1) + (\frac{1}{2})(6) = 8.$

(b) $[3 \quad 0 \quad -2]\begin{bmatrix} 3 \\ 2 \\ 1 \end{bmatrix} = (3)(3) + (0)(2) + (-2)(1) = 7.$

(c) $[3 \quad 2 \quad 0]\begin{bmatrix} 3 \\ 5 \end{bmatrix} =$ undefined. ∎

Notice from (c) that if the number of elements in the row is not equal to the number of elements in the column, then the product is not defined.

The following example illustrates an application and also indicates a motivation for the above definition. The item referred to could, for example, be a television set and the ingredients could be steel, plastic, copper, glass, etc.

E X A M P L E 2   **Application**

Suppose a firm makes an item with four ingredients A, B, C, and D. Each item uses 10 units of A, 5 of B, 7 of C, and 15 of D. The cost per item of each of these four ingredients is, respectively, \$20, \$10, \$30, and \$5. (a) Write a row vector that expresses the amount of each ingredient that goes into each item. (b) Write a column matrix that represents the cost of each unit of the ingredients. (c) Express the total cost of manufacturing one item using these four ingredients by a product of a row matrix with a column matrix.

Solutions   (a) The appropriate row matrix is

$$R = [10 \quad 5 \quad 7 \quad 15].$$

(b) The appropriate column matrix is

$$C = \begin{bmatrix} 20 \\ 10 \\ 30 \\ 5 \end{bmatrix}.$$

(c) The cost of manufacturing one item using these four ingredients is

$$\text{cost of one item} = (\text{number of units of A}) \times (\text{cost of each unit of A})$$
$$+ (\text{number of units of B}) \times (\text{cost of each unit of B})$$
$$+ (\text{number of units of C}) \times (\text{cost of each unit of C})$$
$$+ (\text{number of units of D}) \times (\text{cost of each unit of D})$$

$$= (10)(20) + (5)(10) + (7)(30) + (15)(5)$$

$$= [10 \quad 5 \quad 7 \quad 15] \begin{bmatrix} 20 \\ 10 \\ 30 \\ 5 \end{bmatrix}.$$

The actual numerical answer is \$535. ■

Multiplication in General

We now turn to multiplying two general matrices and the condition that two matrices must satisfy before they can be multiplied.

We show how to multiply two general matrices by working out the specific example

$$AB = \begin{bmatrix} 3 & 1 \\ 2 & 4 \\ 5 & 0 \end{bmatrix} \begin{bmatrix} 2 & 3 \\ 4 & 6 \end{bmatrix}.$$

We obtain the product by the following rule. **Multiply the ith row of A times the jth column of B and place the resulting number in the ith row, jth column of the product.**

Notice for this process to work we must have the dimension of the ith row of A equal to the dimension of the jth column of B. In other words **the number of columns of A must equal the number of rows of B**.

We start by taking the product of the first row of A, $[3 \quad 1]$, and the first column of B, $\begin{bmatrix} 2 \\ 4 \end{bmatrix}$, and place the result in the first row and first column of the product

$$\begin{bmatrix} 3 & 1 \\ 2 & 4 \\ 5 & 0 \end{bmatrix} \begin{bmatrix} 2 & 3 \\ 4 & 6 \end{bmatrix} = \begin{bmatrix} 10 & \\ & \end{bmatrix}.$$

Now take the product of the first row of A, [3   1], and the second column of B, $\begin{bmatrix} 3 \\ 6 \end{bmatrix}$, and place the result in the first row and second column of the product

$$\begin{bmatrix} 3 & 1 \\ 2 & 4 \\ 5 & 0 \end{bmatrix} \begin{bmatrix} 2 & 3 \\ 4 & 6 \end{bmatrix} = \begin{bmatrix} 10 & 15 \\ & \\ & \end{bmatrix}.$$

Since there are no more columns of B, we move on to the second row of A. Take the product of the second row of A, [2   4], and the first column of B, $\begin{bmatrix} 2 \\ 4 \end{bmatrix}$, and place the result in the second row and first column of the product

$$\begin{bmatrix} 3 & 1 \\ 2 & 4 \\ 5 & 0 \end{bmatrix} \begin{bmatrix} 2 & 3 \\ 4 & 6 \end{bmatrix} = \begin{bmatrix} 10 & 15 \\ 20 & \\ & \end{bmatrix}.$$

Take the product of the second row of A, [2   4], and the second column of B, $\begin{bmatrix} 3 \\ 6 \end{bmatrix}$, and place the result in the second row and second column of the product

$$\begin{bmatrix} 3 & 1 \\ 2 & 4 \\ 5 & 0 \end{bmatrix} \begin{bmatrix} 2 & 3 \\ 4 & 6 \end{bmatrix} = \begin{bmatrix} 10 & 15 \\ 20 & 30 \\ & \end{bmatrix}.$$

Since there are no more columns of B, we move on to the third row of A. Take the product of the third row of A, [5   0], and the first column of B, $\begin{bmatrix} 2 \\ 4 \end{bmatrix}$, and place the result in the third row and first column of the product

$$\begin{bmatrix} 3 & 1 \\ 2 & 4 \\ 5 & 0 \end{bmatrix} \begin{bmatrix} 2 & 3 \\ 4 & 6 \end{bmatrix} = \begin{bmatrix} 10 & 15 \\ 20 & 30 \\ 10 & \end{bmatrix}.$$

Take the product of the third row of A, [5   0], and the second column of B, $\begin{bmatrix} 3 \\ 6 \end{bmatrix}$, and place the result in the third row and second column of the product

$$\begin{bmatrix} 3 & 1 \\ 2 & 4 \\ 5 & 0 \end{bmatrix} \begin{bmatrix} 2 & 3 \\ 4 & 6 \end{bmatrix} = \begin{bmatrix} 10 & 15 \\ 20 & 30 \\ 10 & 15 \end{bmatrix}.$$

Since there are no more columns of B, we move on to the next row of A. Since there are no more rows of A, we are done.

Multiplication of Matrices

Given two matrices $A = (a_{ij})_{m \times p}$ and $B = (b_{ij})_{q \times n}$, the product $AB = C = (c_{ij})$ is defined if $p = q$, that is, if the number of columns of A equals the number of rows of B. In such a case, the element in the ith row and jth column, c_{ij}, of the product matrix C is obtained by taking the product of the ith row of A and the jth column of B.

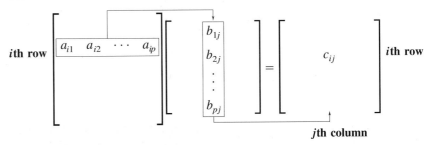

Notice that to obtain the first row of the product $C = AB$, we take the product of the first row of A and the successive columns of B until we exhaust the columns of B. This indicates that **the number of columns of the product AB must equal the number of columns of B**. Also notice that a row of B is obtained for each row of A. This indicates that **the number of rows of the product AB must equal the number of rows of A**. Thus

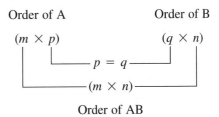

Order of AB

Order of AB

Given the two matrices $A = (a_{ij})_{m \times p}$ and $B = (b_{ij})_{p \times n}$, the matrix AB has order $m \times n$.

Applying this to the previous example we have

$$
\begin{array}{cc}
\text{Order of A} & \text{Order of B} \\
(3 \times 2) & (2 \times 2)
\end{array}
$$

$$2 = 2$$

$$(3 \times 2)$$

Order of AB

E X A M P L E 3   **Multiplying Matrices**

Perform the indicated operations.

(a)
$$\begin{bmatrix} 1 & 2 \\ 3 & 4 \end{bmatrix}\begin{bmatrix} 5 & 6 \\ 7 & 8 \end{bmatrix} = \begin{bmatrix} (1)(5) + (2)(7) & (1)(6) + (2)(8) \\ (3)(5) + (4)(7) & (3)(6) + (4)(8) \end{bmatrix}$$

$$= \begin{bmatrix} 19 & 22 \\ 43 & 50 \end{bmatrix}$$

(b)
$$\begin{bmatrix} 5 & 6 \\ 7 & 8 \end{bmatrix}\begin{bmatrix} 1 & 2 \\ 3 & 4 \end{bmatrix} = \begin{bmatrix} (5)(1) + (6)(3) & (5)(2) + (6)(4) \\ (7)(1) + (8)(3) & (7)(2) + (8)(4) \end{bmatrix}$$

$$= \begin{bmatrix} 23 & 34 \\ 31 & 46 \end{bmatrix} \quad \blacksquare$$

Notice that this example illustrates that in general AB need not equal BA even when they are both defined and of the same order.

Earlier we saw that

$$AB = \begin{bmatrix} 3 & 1 \\ 2 & 4 \\ 5 & 0 \end{bmatrix}\begin{bmatrix} 2 & 3 \\ 4 & 6 \end{bmatrix} = \begin{bmatrix} 10 & 15 \\ 20 & 30 \\ 10 & 15 \end{bmatrix}.$$

But notice that for this A and B, the product BA is not even defined, since the number of columns of B does not equal the number of rows of A.

Order of B   Order of A

(2 × 2)   (3 × 2)

∟——— 2 ≠ 3 ———⌐

BA undefined

$$BA = \begin{bmatrix} 2 & 3 \\ 4 & 6 \end{bmatrix}\begin{bmatrix} 3 & 1 \\ 2 & 4 \\ 5 & 0 \end{bmatrix} = \text{undefined}$$

At the beginning of this section we defined the product RC where R was a row matrix and C a column matrix of the same order n. The answer we obtained was a number (a 1×1 matrix). What happens if we reverse the order of the multiplication? Checking the orders we have for CR,

Order of C   Order of R

(n × 1)   (1 × n)

∟——— 1 = 1 ———⌐

∟——— (n × n) ———⌐

Order of CR

We obtain a square matrix of order $n \times n$.

E X A M P L E 4   **Multiplying Matrices**

Perform the indicated multiplication.

Solution

$$\begin{bmatrix} 1 \\ 2 \end{bmatrix} [3 \quad 4] = \begin{bmatrix} (1)(3) & (1)(4) \\ (2)(3) & (2)(4) \end{bmatrix} = \begin{bmatrix} 3 & 4 \\ 6 & 8 \end{bmatrix} \quad \blacksquare$$

We now have several examples that indicate that, in general,

$$AB \neq BA,$$

even if both matrices A and B are square.

The following, however, are true.

Properties of Matrix Multiplication

For any matrices A, B, and C,

$$A(BC) = (AB)C$$

$$A(B + C) = AB + AC,$$

whenever the indicated products are defined.

If a is any scalar, then the number 1 has the property that

$$a \cdot 1 = 1 \cdot a = a$$

There is a matrix analog to the number 1. This matrix, which must be a square matrix, is called the **identity matrix**.

The Identity Matrix

The $(n \times n)$ identity matrix, I_n, is the square matrix of order $(n \times n)$ with ones down the main diagonal and zeros elsewhere.

Thus

$$I_n = \begin{bmatrix} 1 & 0 & 0 & \cdots & 0 \\ 0 & 1 & 0 & \cdots & 0 \\ & & & \cdots & \\ 0 & 0 & 0 & \cdots & 1 \end{bmatrix} (n \text{ rows})$$

$$(n \text{ columns})$$

The identity matrix then has the following property.

> **Multiplication Property of the Identity Matrix**
>
> If A is a square matrix of order $(n \times n)$, then
>
> $$AI_n = I_nA = A.$$

For example,

$$
\begin{aligned}
AI_2 &= \begin{bmatrix} a & b \\ c & d \end{bmatrix}\begin{bmatrix} 1 & 0 \\ 0 & 1 \end{bmatrix} \\
&= \begin{bmatrix} a \cdot 1 + b \cdot 0 & a \cdot 0 + b \cdot 1 \\ c \cdot 1 + d \cdot 0 & c \cdot 0 + d \cdot 1 \end{bmatrix} \\
&= \begin{bmatrix} a & b \\ c & d \end{bmatrix} \\
&= A
\end{aligned}
$$

Applications

Matrix techniques are regularly used by firms in production planning.

We assume that a firm makes three products: P_1, P_2, and P_3. We assume that a production facility is located in St. Louis, where four ingredients, I_1, I_2, I_3, and I_4 are made that are then used to manufacture the three products. The following table and matrix gives the number of units of each ingredient needed for each product.

$$
\begin{array}{c}
 \\
P_1 \\
P_2 \\
P_3
\end{array}
\begin{array}{cccc}
I_1 & I_2 & I_3 & I_4 \\
\end{array}
$$

$$
\begin{array}{c}
P_1 \\
P_2 \\
P_3
\end{array}
\begin{bmatrix}
9 & 6 & 5 & 3 \\
5 & 5 & 8 & 4 \\
6 & 2 & 1 & 7
\end{bmatrix} = P
$$

Suppose the cost of each unit of the ingredients I_1, I_2, I_3, and I_4 is, respectively, $10, $30, $20, $40. We could summarize this in the following cost (column) matrix:

$$
C = \begin{bmatrix} 10 \\ 30 \\ 20 \\ 40 \end{bmatrix}
\begin{array}{l}
\text{cost of one unit of } I_1 \\
\text{cost of one unit of } I_2 \\
\text{cost of one unit of } I_3 \\
\text{cost of one unit of } I_4.
\end{array}
$$

The firm might naturally wish to know the total cost of the ingredients that go into each product. This is given by the (column) matrix

$$PC = \begin{array}{c} P_1 \\ P_2 \\ P_3 \end{array} \begin{bmatrix} 9 & 6 & 5 & 3 \\ 5 & 5 & 8 & 4 \\ 6 & 2 & 1 & 7 \end{bmatrix} \begin{bmatrix} 10 \\ 30 \\ 20 \\ 40 \end{bmatrix} \begin{array}{l} \text{cost of one unit of } I_1 \\ \text{cost of one unit of } I_2 \\ \text{cost of one unit of } I_3 \\ \text{cost of one unit of } I_4 \end{array}$$

$$= \begin{bmatrix} 490 \\ 520 \\ 420 \end{bmatrix} \begin{array}{l} \text{total cost of } P_1 \\ \text{total cost of } P_2 \\ \text{total cost of } P_3 \end{array}$$

where the three numbers in the last column matrix represent the total cost of ingredients in the respective products P_1, P_2, and P_3.

Suppose the firm has an order for 9000 of product P_1, 6000 of P_2, and 7000 of P_3 and then wanted to know the total number of units of each of the ingredients needed. If we expressed the amounts ordered (in thousands) as a row matrix

$$R = [9 \quad 6 \quad 7],$$

then the matrix

$$RP = [9 \quad 6 \quad 7] \begin{bmatrix} 9 & 6 & 5 & 3 \\ 5 & 5 & 8 & 4 \\ 6 & 2 & 1 & 7 \end{bmatrix} = [153 \quad 98 \quad 100 \quad 100]$$

represents, respectively, the number of units of the ingredients I_1, I_2, I_3, and I_4 needed to fill the order.

Now suppose, after the ingredients are made in the St. Louis facility, they are shipped out to the two manufacturing plants, one in LA and one in NYC, where the three products are actually produced. The following table and matrix give the number in thousands of products manufactured at each of the two plants.

	P_1	P_2	P_3
LA	2	1	1
NYC	3	1	0

$$A = \begin{bmatrix} 2 & 1 & 1 \\ 3 & 1 & 0 \end{bmatrix}$$

Suppose the firm wishes to keep track of how many units of each ingredient is going to LA and to NYC. The matrix

$$AP = \begin{bmatrix} 2 & 1 & 1 \\ 3 & 1 & 0 \end{bmatrix} \begin{bmatrix} 9 & 6 & 5 & 3 \\ 5 & 5 & 8 & 4 \\ 6 & 2 & 1 & 7 \end{bmatrix}$$

$$= \begin{array}{c} \\ LA \\ NYC \end{array} \begin{matrix} I_1 & I_2 & I_3 & I_4 \\ \begin{bmatrix} 29 & 19 & 19 & 17 \\ 32 & 23 & 23 & 13 \end{bmatrix} \end{matrix}$$

summarizes this information. Thus the element in the first row and first column, 29, is the number of thousands of units of the ingredient I_1 that are used in the LA plant.

Systems of Equations

We shall now see how to use matrices to represent systems of equations. Consider the following system of equations.

$$2x_1 + 3x_2 + 5x_3 = 2$$
$$5x_1 - 4x_2 + 3x_3 = 4$$
$$3x_1 + 2x_2 - 2x_3 = 1$$

Let

$$A = \begin{bmatrix} 2 & 3 & 5 \\ 5 & -4 & 3 \\ 3 & 2 & -2 \end{bmatrix}, \quad X = \begin{bmatrix} x_1 \\ x_2 \\ x_3 \end{bmatrix}, \quad B = \begin{bmatrix} 2 \\ 4 \\ 1 \end{bmatrix}$$

The matrix A is called the **coefficient matrix**. Then notice that

$$AX = \begin{bmatrix} 2 & 3 & 5 \\ 5 & -4 & 3 \\ 3 & 2 & -2 \end{bmatrix} \begin{bmatrix} x_1 \\ x_2 \\ x_3 \end{bmatrix}$$

$$= \begin{bmatrix} 2x_1 + 3x_2 + 5x_3 \\ 5x_1 - 4x_2 + 3x_3 \\ 3x_1 + 2x_2 - 2x_3 \end{bmatrix}$$

$$= \begin{bmatrix} 2 \\ 4 \\ 1 \end{bmatrix} = B.$$

More generally, the system of equations

$$a_{11}x_1 + a_{12}x_2 + \cdots + a_{1n}x_n = b_1$$
$$a_{21}x_1 + a_{22}x_2 + \cdots + a_{2n}x_n = b_2$$
$$\cdots$$
$$a_{m1}x_1 + a_{m2}x_2 + \cdots + a_{mn}x_n = b_m$$

can be written as

$$AX = B,$$

if we define

$$
A = \begin{bmatrix} a_{11} & a_{12} & \cdots & a_{1n} \\ a_{21} & a_{22} & \cdots & a_{2n} \\ & & \cdots & \\ a_{m1} & a_{m2} & \cdots & a_{mn} \end{bmatrix}, \quad X = \begin{bmatrix} x_1 \\ x_2 \\ \vdots \\ x_n \end{bmatrix}, \quad B = \begin{bmatrix} b_1 \\ b_2 \\ \vdots \\ b_m \end{bmatrix}.
$$

SELF-HELP EXERCISE SET 2.4

1. Perform the indicated operation.

$$
\begin{bmatrix} 1 & 2 \\ -1 & 3 \end{bmatrix} \begin{bmatrix} 3 & 0 & 4 \\ 1 & -2 & 5 \end{bmatrix}
$$

2. A firm makes two types of a product and uses three parts according to the following table.

	Part A	Part B	Part C	
Type I	3	2	1	$= P$
Type II	1	2	4	

Both types of this product are manufactured at two different factories: Factory X and Factory Y. The time in work-hours that is required to install each part is given as follows:

	Factory X	Factory Y	
Part A	3	2	
Part B	4	4	$= Q$
Part C	5	3	

Compute PQ and explain what each entry in this product matrix means.

3. Put the following system in matrix form $AX = B$, where X is a column vector with the unknowns:

$$
10x + 2y + 8z = 7
$$
$$
-3x + 5y - 7z = 5
$$
$$
-4x - 3y + 5z = 3.
$$

EXERCISE SET 2.4

In Exercises 1 through 6 perform the indicated multiplication if possible.

1. $\begin{bmatrix} 2 & 3 \end{bmatrix} \begin{bmatrix} 5 \\ 2 \end{bmatrix}$

2. $\begin{bmatrix} 1 & 4 & -2 \end{bmatrix} \begin{bmatrix} 3 \\ 2 \\ 5 \end{bmatrix}$

3. $\begin{bmatrix} -2 & 3 & 4 & 2 \end{bmatrix} \begin{bmatrix} -2 \\ -1 \\ 3 \\ 5 \end{bmatrix}$

4. $\begin{bmatrix} 2 & -3 & 0 & 4 & 2 \end{bmatrix} \begin{bmatrix} 3 \\ 4 \\ 3 \\ -3 \\ 5 \end{bmatrix}$

5. $\begin{bmatrix} 2 & 3 \end{bmatrix} \begin{bmatrix} 5 \\ 2 \\ 0 \end{bmatrix}$

6. $\begin{bmatrix} 0 & 0 & 0 \end{bmatrix} \begin{bmatrix} 0 \\ 0 \end{bmatrix}$

In Exercises 7 through 16 find the order of AB and BA when either one exists.

7. The order of A is (2×3) and the order of B is (3×4).

8. The order of A is (3×5) and the order of B is (5×7).

9. The order of A is (3×4) and the order of B is (5×3).

10. The order of A is (5×7) and the order of B is (8×5).

11. The order of A is (3×3) and the order of B is (3×5).

12. The order of A is (5×4) and the order of B is (4×4).

13. The order of A is (6×6) and the order of B is (6×6).

14. The order of A is (8×8) and the order of B is (8×8).

15. The order of A is (20×2) and the order of B is (2×30).

16. The order of A is (3×40) and the order of B is (50×3).

In Exercises 17 through 42 perform the indicated multiplications if possible.

17. $\begin{bmatrix} 2 & 5 \\ 4 & 3 \end{bmatrix} \begin{bmatrix} 2 \\ 5 \end{bmatrix}$

18. $\begin{bmatrix} 5 & -3 \\ -2 & 4 \end{bmatrix} \begin{bmatrix} -3 \\ 4 \end{bmatrix}$

19. $\begin{bmatrix} 3 & 7 \end{bmatrix} \begin{bmatrix} -2 & 5 \\ 7 & 2 \end{bmatrix}$

20. $\begin{bmatrix} 2 & 1 \end{bmatrix} \begin{bmatrix} 3 & 20 \\ -3 & 5 \end{bmatrix}$

21. $\begin{bmatrix} 0.2 & 0.1 \\ 0.4 & 0.5 \end{bmatrix} \begin{bmatrix} 2 & 3 \\ 5 & 2 \end{bmatrix}$

22. $\begin{bmatrix} -1 & -3 \\ -2 & -4 \end{bmatrix} \begin{bmatrix} 0 & 2 \\ 2 & 5 \end{bmatrix}$

23. $\begin{bmatrix} 2 & 8 & 4 \\ 3 & 0 & 2 \end{bmatrix} \begin{bmatrix} 4 \\ 2 \\ 3 \end{bmatrix}$

24. $[4 \quad 2] \begin{bmatrix} 2 & -3 & 4 \\ -1 & 3 & 5 \end{bmatrix}$

25. $\begin{bmatrix} 0.1 & 0.4 & 0.2 \\ 0.5 & 0.2 & 0.1 \end{bmatrix} \begin{bmatrix} 2 & 4 \\ -1 & 5 \\ 5 & -2 \end{bmatrix}$

26. $\begin{bmatrix} 10 & 20 \\ 30 & 40 \end{bmatrix} \begin{bmatrix} 3 & 2 & 0 \\ 1 & 3 & 2 \end{bmatrix}$

27. $\begin{bmatrix} 3 & 5 \\ 8 & 2 \end{bmatrix} \begin{bmatrix} 1 & 5 & 2 & 3 \\ 0 & 3 & 5 & 1 \end{bmatrix}$

28. $\begin{bmatrix} 4 & 2 \\ -1 & 6 \\ 3 & -2 \end{bmatrix} \begin{bmatrix} 2 \\ 3 \end{bmatrix}$

29. $[0.01 \quad 0.03] \begin{bmatrix} 1 & 3 & 2 \\ 2 & 5 & 1 \end{bmatrix}$

30. $\begin{bmatrix} -1 & -2 & -5 & -2 \\ 3 & 5 & 1 & 0 \end{bmatrix} \begin{bmatrix} 0 & 3 \\ 2 & 0 \end{bmatrix}$

31. $\begin{bmatrix} 3 & 2 & 5 \\ 1 & 3 & 2 \\ 4 & 2 & 1 \end{bmatrix} \begin{bmatrix} 2 \\ 0 \\ 4 \end{bmatrix}$

32. $\begin{bmatrix} 4 & 0 & 1 \\ 2 & -2 & 2 \\ 1 & -1 & 0 \end{bmatrix} \begin{bmatrix} 2 \\ 4 \\ 1 \end{bmatrix}$

33. $[3 \quad -2 \quad 1] \begin{bmatrix} 2 & 1 & 5 \\ 5 & 0 & -2 \\ 0 & 1 & 2 \end{bmatrix}$

34. $[-1 \quad 0 \quad 1] \begin{bmatrix} 2 & 1 & 2 \\ 1 & -1 & -1 \\ 2 & 1 & 1 \end{bmatrix}$

35. $\begin{bmatrix} 1 & 4 & 2 \\ 0 & 2 & 0 \\ 5 & 2 & 1 \end{bmatrix} \begin{bmatrix} 3 & 0 & 1 \\ 0 & 2 & 4 \\ 1 & 1 & 2 \end{bmatrix}$

36. $\begin{bmatrix} 0 & -1 & 2 \\ -1 & 0 & 2 \\ 2 & -2 & 0 \end{bmatrix} \begin{bmatrix} 1 & 5 & 2 \\ 0 & 1 & 1 \\ 2 & 4 & 2 \end{bmatrix}$

37. $\begin{bmatrix} 1 & 2 \\ 0 & 0 \\ 1 & 0 \end{bmatrix} \begin{bmatrix} 2 & 3 \\ 1 & 1 \\ 0 & 0 \end{bmatrix}$

38. $\begin{bmatrix} 0 & 0 \\ 0 & 0 \end{bmatrix} \begin{bmatrix} 1 & 1 \\ 1 & 1 \\ 1 & 1 \end{bmatrix}$

39. $\begin{bmatrix} 2 \\ 5 \end{bmatrix} [3 \quad 5]$

40. $\begin{bmatrix} 2 \\ 0 \\ -1 \end{bmatrix} [2 \quad 0 \quad -1]$

41. $\begin{bmatrix} 0 & 0 & 0 \\ 0 & 0 & 0 \end{bmatrix} \begin{bmatrix} 0 & 0 \\ 0 & 0 \end{bmatrix}$

42. $\begin{bmatrix} 1 & 1 & 1 \\ 1 & 1 & 1 \end{bmatrix} \begin{bmatrix} 1 & 1 & 1 \\ 1 & 1 & 1 \end{bmatrix}$

43. Let

$$A = \begin{bmatrix} 1 & 1 \\ 1 & 1 \end{bmatrix} \quad \text{and} \quad B = \begin{bmatrix} 1 & 1 \\ -1 & -1 \end{bmatrix}$$

Show that $AB = O$. Thus a product of two matrices may be the zero matrix without either matrix being the zero matrix.

44. Consider the four matrices

$$I_2 = \begin{bmatrix} 1 & 0 \\ 0 & 1 \end{bmatrix}, \begin{bmatrix} 1 & 0 \\ 0 & -1 \end{bmatrix}, \begin{bmatrix} -1 & 0 \\ 0 & 1 \end{bmatrix}, \begin{bmatrix} -1 & 0 \\ 0 & -1 \end{bmatrix}$$

Show that the square of each of them is I_2. Thus the identity matrix has many square roots.

45. Let

$$A = \begin{bmatrix} 1 & 1 \\ -1 & -1 \end{bmatrix}.$$

Show that $A^2 = AA = O$.

46. Let

$$A = \begin{bmatrix} a_{11} & a_{12} & a_{13} \\ a_{21} & a_{22} & a_{23} \\ a_{31} & a_{32} & a_{33} \end{bmatrix}$$

Show that $AI_3 = I_3A = A$.

47. Let

$$A = \begin{bmatrix} 1 & 2 \\ 3 & 4 \end{bmatrix}, \quad B = \begin{bmatrix} 1 & -1 \\ 1 & 2 \end{bmatrix}.$$

Show that $(A + B)^2 \neq A^2 + 2AB + B^2$.

48. Let

$$A = \begin{bmatrix} 1 & 1 \\ 2 & 1 \end{bmatrix}, \quad B = \begin{bmatrix} 2 & 2 \\ 5 & 4 \end{bmatrix}, \quad C = \begin{bmatrix} 1 & 1 \\ -1 & -1 \end{bmatrix}.$$

Show $AC = BC$, but $A \neq B$.

49. Let A, B, C be the matrices in Exercise 48. Show that

$$A(B + C) = AB + AC$$

$$A(BC) = (AB)C$$

50. A matrix $U = (u_{ij})$ is upper triangular if $u_{ij} = 0$ when $i > j$, that is, when all entries below the main diagonal are zero. Suppose $A = (a_{ij})$ and $B = (b_{ij})$ are both upper triangular matrices of order (3×3). Show that (a) A^2 is upper triangular, (b) AB is upper triangular.

51. Let

$$A = \begin{bmatrix} 0 & 1 & 2 \\ 0 & 0 & 3 \\ 0 & 0 & 0 \end{bmatrix}.$$

Show $A^3 = AAA = O$.

52. A matrix $U = (u_{ij})$ is strictly upper triangular if $u_{ij} = 0$ when $i \geq j$, that is, when all entries below and on the main diagonal are zero. If $U = (u_{ij})$ is a strictly upper triangular matrix of order (3×3), show that $U^3 = O$.

In Exercises 53 through 60, find matrices A, X, and B so that the given system of equations can be written as $AX = B$.

53. $2x_1 + 3x_2 = 5$
    $3x_1 - 5x_2 = 7$

54. $4x_1 - 5x_2 = 6$
    $2x_1 + 7x_2 = 9$

55. $2x_1 + 5x_2 + 3x_3 = 16$
    $4x_1 - 7x_2 - 2x_3 = 12$
    $5x_1 - 2x_2 + 6x_3 = 24$

56. $-3x_1 + 7x_2 + 2x_3 = 0$
    $- 7x_2 + 5x_3 = 2$
    $4x_1 + 3x_2 - 7x_3 = 4$

57. $2x_1 - 3x_2 + 3x_3 = 3$
    $5x_1 + 6x_2 - 2x_3 = 1$

58. $2x_1 - 7x_2 = 3$
    $x_1 + 8x_2 = 2$
    $3x_1 + 4x_2 = 5$

59. $2x_1 + 3x_2 = 7$

60. $3x_1 + 5x_2 - 6x_3 = 2$

Applications

61. **Spare Parts Planning.** A car rental firm is planning a maintenance program involving certain parts of various makes of compact automobiles. The following matrix indicates the number of hundreds of automobiles available for rent in the three cities in which the firm operates.

	GM	Ford	Chrysler	Toyota	
LA	1	2	2	3	
NYC	3	1	1	2	= N
SL	2	2	1	1	

Past repair records for the three cities indicate the number of each part needed per car per year. This is given in the following:

	Tires	Batteries	Plugs	
GM	4	0.5	6	
Ford	3	0.5	7	= R.
Chrysler	4	0.4	5	
Toyota	3	1.0	4	

a. Show that the quantity of each part needed in each city is given by $Q = NR$. Indicate for what each column and each row of Q stands.

b. Let the elements in the column matrix

$$C = \begin{bmatrix} 100 \\ 40 \\ 1 \end{bmatrix}$$

be, respectively, the cost per each for tires, batteries, and plugs. Determine QC and interpret what the elements stand for.

62. **Networks.** (The following application is a very scaled-down version of a technique used by the major airlines. The actual calculations are carried out by computers.) The diagram indicates the nonstop service between four cities, LA, Chicago, NYC, and Miami. For example, the arrows indicate that there is nonstop service in both directions between LA and NYC but only one-way service from Chicago to NYC. The network in the diagram can be summarized by an adjacency matrix $A = (a_{ij})$, as follows. We set $a_{ij} = 1$ if there is nonstop service from city i to city j, otherwise a_{ij} is zero. We then have

To

$$\begin{array}{c} \\ \\ \text{From} \\ \\ \end{array}\begin{array}{c} \\ \text{L} \\ \text{C} \\ \text{N} \\ \text{M} \end{array}\begin{array}{cccc} \text{L} & \text{C} & \text{N} & \text{M} \\ \left[\begin{array}{cccc} 0 & 1 & 1 & 0 \\ 0 & 0 & 1 & 0 \\ 1 & 0 & 0 & 1 \\ 0 & 0 & 1 & 0 \end{array}\right] \end{array} = A.$$

a. Compute A^2 and verify from the diagram that the element in the ith row and jth column of A^2 represents the number of one-stop routes between the city in the ith row and the city in the jth column.

b. Compute $A + A^2$ and interpret what this matrix means.

c. Compute A^3 and verify from the diagram that the element in the ith row and jth column of A^3 represents the number of two-stop routes between the city in the ith row and the city in the jth column.

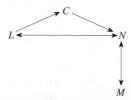

63. **Traffic Flow.** The figure shows the traffic flow through nine intersections. Traffic enters from two roads from the left and exits on four roads to the right. All roads are one-way. Given the number of cars x_1 and x_2 that enter the intersections I_1 and I_3, respectively, we wish to determine the number of cars exiting each of the four roads on the right. Notice, for example, according to the figure that 40% of the traffic that enters intersection I_1 exits due east. Thus $y_1 = 0.4x_1$. We also see that $y_2 = (0.6)(0.3)x_1 = 0.18x_1$ and $y_3 = (0.6)(0.7)x_1 + x_2$. Thus

$$y_1 = 0.4x_1$$

$$y_2 = 0.18x_1$$

$$y_3 = 0.42x_1 + x_2.$$

If

$$X = \begin{bmatrix} x_1 \\ x_2 \end{bmatrix}, \quad A = \begin{bmatrix} 0.40 & 0 \\ 0.18 & 0 \\ 0.42 & 1 \end{bmatrix}, \quad Y = \begin{bmatrix} y_1 \\ y_2 \\ y_3 \end{bmatrix},$$

then $Y = AX$.

a. Now by considering intersections I_4, I_5, I_6, find a (3×3) matrix B such that $Z = BY$, where

$$Z = \begin{bmatrix} z_1 \\ z_2 \\ z_3 \end{bmatrix}.$$

b. Now by considering intersections I_7, I_8, I_9, find a (4×3) matrix C such that $W = CZ$, where

$$W = \begin{bmatrix} w_1 \\ w_2 \\ w_3 \\ w_4 \end{bmatrix}.$$

c. Show that $W = (CBA)X$.

d. If $x_1 = 5000$ and $x_2 = 5000$, find W.

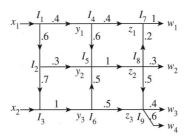

64. Social Interaction. The digraph indicates the influence in decision making that one individual has over another. For example, according to the digraph, A has influence over B and D has influence over A, B, and C. The following matrix is then formed.

$$
\begin{array}{c c}
 & \begin{array}{cccc} A & B & C & D \end{array} \\
\begin{array}{c} A \\ B \\ C \\ D \end{array} &
\begin{bmatrix}
0 & 1 & 0 & 0 \\
0 & 0 & 1 & 1 \\
0 & 0 & 0 & 1 \\
1 & 1 & 1 & 0
\end{bmatrix} = T
\end{array}
$$

A number one in the ith row and jth column indicates that the person in the ith row has direct influence over the person in the jth column, while a zero indicates no direct influence.

a. Calculate T^2 and show that a number 1 indicates that the individual in the row containing the 1 has influence over the individual in the column containing the 1 through one intermediary and that a number 2 indicates that the individual in the row containing the 2 has influence over the individual in the column containing the 2 through two different intermediaries.

b. Calculate $T + T^2$ and interpret what this matrix means.

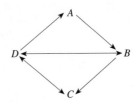

Solutions to Self-Help Exercise Set 2.4

1. $\begin{bmatrix} 1 & 2 \\ -1 & 3 \end{bmatrix}\begin{bmatrix} 3 & 0 & 4 \\ 1 & -2 & 5 \end{bmatrix}$

$= \begin{bmatrix} (1)(3) + (2)(1) & (1)(0) + (2)(-2) & (1)(4) + (2)(5) \\ (-1)(3) + (3)(1) & (-1)(0) + (3)(-2) & (-1)(4) + (3)(5) \end{bmatrix}$

$= \begin{bmatrix} 5 & -4 & 14 \\ 0 & -6 & 11 \end{bmatrix}$

2. $\begin{bmatrix} 3 & 2 & 1 \\ 1 & 2 & 4 \end{bmatrix}\begin{bmatrix} 3 & 2 \\ 4 & 4 \\ 5 & 3 \end{bmatrix}$

$= \begin{bmatrix} (3)(3) + (2)(4) + (1)(5) & (3)(2) + (2)(4) + (1)(3) \\ (1)(3) + (2)(4) + (4)(5) & (1)(2) + (2)(4) + (4)(3) \end{bmatrix}$

$\qquad\quad$ Factory X   Factory Y

$= \begin{bmatrix} 22 & 17 \\ 31 & 22 \end{bmatrix} \begin{matrix} \text{Type I} \\ \text{Type II} \end{matrix}$

3. The system is in the form $AX = B$ if

$$A = \begin{bmatrix} 10 & 2 & 8 \\ -3 & 5 & -7 \\ -4 & -3 & 5 \end{bmatrix}, \quad X = \begin{bmatrix} x \\ y \\ z \end{bmatrix}, \quad B = \begin{bmatrix} 7 \\ 5 \\ 3 \end{bmatrix}$$

2.5   INVERSE OF A SQUARE MATRIX

▸ *Definition of the Inverse of a Matrix*

▸ *Finding the Inverse*

▶ *Systems of Equations*

▶ *Applications*

Arthur Cayley,
1821–1895

(David Eugene Smith Collection,
Columbia University.)

Arthur Cayley was the first to consider the idea of an inverse to a matrix in 1855. Cayley was educated in England and, unable to find a suitable teaching position, studied and practiced law for 14 years. During this time he published about 300 papers in mathematics. He did eventually become a professor at Cambridge in 1863. Cayley became acquainted with James Joseph Sylvester, mentioned in an earlier section, and together their collaborations resulted in many outstanding results in mathematics. Cayley was an enormously prolific researcher; his collected works contain 966 papers in thirteen volumes, each volume containing about 600 pages.

APPLICATION

Allocation of Funds

An investment advisor has two mutual funds that she is recommending: a conservative bond fund with an assumed return of 7% a year and a stock fund with a projected return of 12% a year. One client has $100,000 to invest and wishes a return of 9% a year, and a second client has $50,000 and wants a 10% annual return. How should she allocate the money of each client between the bond fund and the stock fund? The answer is found in Example 5.

Definition of the Inverse of a Matrix

In this section we will define the inverse of a square matrix and will see that the inverse matrix is an extremely powerful tool in solving systems of equations where the number of unknowns is the same as the number of equations. We will find the inverse of a square matrix using the technique of Gauss–Jordan. It is this technique that is commonly used by computers to find the inverse.

The scalar equation $ax = 1$ has the solution $x = a^{-1}$ if $a \neq 0$. The number a^{-1} has the property that $aa^{-1} = a^{-1}a = 1$. Given a square ($n \times n$) matrix A, we wish to find a matrix A^{-1} called the **inverse of A** with the property that

$$AA^{-1} = A^{-1}A = I_n$$

Inverse Matrix

The $n \times n$ matrix B is said to be the **inverse** to the $n \times n$ matrix A if

$$AB = BA = I_n.$$

If the inverse matrix B exists, we write $B = A^{-1}$.

REMARK.   Do not think that the inverse matrix A^{-1} can be written as $\dfrac{1}{A}$. *We never divide matrices.*

E X A M P L E   1   Showing a Matrix is the Inverse Matrix

Let

$$A = \begin{bmatrix} 3 & 1 \\ 5 & 2 \end{bmatrix}.$$

Then show that

$$A^{-1} = \begin{bmatrix} 2 & -1 \\ -5 & 3 \end{bmatrix}.$$

Solution   The indicated matrix A^{-1} is the inverse of A since

$$AA^{-1} = \begin{bmatrix} 3 & 1 \\ 5 & 2 \end{bmatrix} \begin{bmatrix} 2 & -1 \\ -5 & 3 \end{bmatrix} = \begin{bmatrix} 1 & 0 \\ 0 & 1 \end{bmatrix} = I_2$$

and

$$A^{-1}A = \begin{bmatrix} 2 & -1 \\ -5 & 3 \end{bmatrix} \begin{bmatrix} 3 & 1 \\ 5 & 2 \end{bmatrix} = \begin{bmatrix} 1 & 0 \\ 0 & 1 \end{bmatrix} = I_2. \quad \blacksquare$$

Not every matrix has an inverse. Consider

$$B = \begin{bmatrix} 2 & 1 \\ 4 & 2 \end{bmatrix}.$$

Let us try to find a matrix X of order (2×2) such that

$$\begin{bmatrix} 1 & 0 \\ 0 & 1 \end{bmatrix} = I_2 = BX = \begin{bmatrix} 2 & 1 \\ 4 & 2 \end{bmatrix} \begin{bmatrix} a & b \\ c & d \end{bmatrix} = \begin{bmatrix} 2a + c & 2b + d \\ 4a + 2c & 4b + 2d \end{bmatrix}$$

Notice that equating elements in the first column requires that

$$2a + c = 1$$

$$4a + 2c = 0.$$

But 2 times the first of these two equations yields $4a + 2c = 2$. This is inconsistent with the second equation, and thus no solution can exist. This shows that B^{-1} does not exist.

Finding the Inverse

We shall now develop a technique that will find the inverse of a square matrix, if it exists, or indicate that the inverse does not exist.
    Consider

$$A = \begin{bmatrix} 1 & 2 \\ 3 & 7 \end{bmatrix}.$$

To find

$$A^{-1} = X = \begin{bmatrix} x_1 & y_1 \\ x_2 & y_2 \end{bmatrix},$$

we must solve the equation $AX = I_2$ or

$$AX = \begin{bmatrix} 1 & 2 \\ 3 & 7 \end{bmatrix} \begin{bmatrix} x_1 & y_1 \\ x_2 & y_2 \end{bmatrix} = \begin{bmatrix} x_1 + 2x_2 & y_1 + 2y_2 \\ 3x_1 + 7x_2 & 3y_1 + 7y_2 \end{bmatrix} = \begin{bmatrix} 1 & 0 \\ 0 & 1 \end{bmatrix}.$$

This gives the four equations

$$x_1 + 2x_2 = 1 \qquad y_1 + 2y_2 = 0$$
$$3x_1 + 7x_2 = 0 \qquad 3y_1 + 7y_2 = 1,$$

which have been written as two systems. We can then solve each system by first forming the augmented matrices

$$\begin{bmatrix} 1 & 2 & | & 1 \\ 3 & 7 & | & 0 \end{bmatrix} \quad \text{and} \quad \begin{bmatrix} 1 & 2 & | & 0 \\ 3 & 7 & | & 1 \end{bmatrix}.$$

But since both of these systems have the same coefficient matrix, they both will be solved by the Gauss–Jordan method, using precisely the same row operations. Thus we may as well solve them together. We do this by forming the augmented matrix

$$\begin{bmatrix} 1 & 2 & | & 1 & 0 \\ 3 & 7 & | & 0 & 1 \end{bmatrix}.$$

Now perform the usual row operations and obtain

$$\begin{bmatrix} 1 & 2 & | & 1 & 0 \\ 3 & 7 & | & 0 & 1 \end{bmatrix} R_2 - 3R_1 \rightarrow R_2$$

$$\begin{bmatrix} 1 & 2 & | & 1 & 0 \\ 0 & 1 & | & -3 & 1 \end{bmatrix} R_1 - 2R_2 \rightarrow R_1$$

$$\begin{bmatrix} 1 & 0 & | & 7 & -2 \\ 0 & 1 & | & -3 & 1 \end{bmatrix}.$$

This means that the solutions to the two systems are

$$x_1 = 7 \qquad y_1 = -2$$
$$x_2 = -3 \qquad y_2 = 1.$$

Thus

$$A^{-1} = \begin{bmatrix} 7 & -2 \\ -3 & 1 \end{bmatrix}.$$

In general we have the following technique.

To Find A^{-1}

1. Form the augmented matrix
$$[A|I].$$

2. Use row operations to reduce $[A|I]$ to
$$[I|B]$$

   if possible.
3. $A^{-1} = B.$

E X A M P L E   2   **Finding the Inverse of a Matrix**

Find A^{-1} if

$$A = \begin{bmatrix} 1 & 1 & 2 \\ 2 & 3 & 2 \\ 1 & 1 & 3 \end{bmatrix}.$$

Solution   Following the above technique, we have

$$\begin{bmatrix} 1 & 1 & 2 & | & 1 & 0 & 0 \\ 2 & 3 & 2 & | & 0 & 1 & 0 \\ 1 & 1 & 3 & | & 0 & 0 & 1 \end{bmatrix} \begin{matrix} \\ R_2 - 2R_1 \to R_2 \\ R_3 - R_1 \to R_3 \end{matrix}$$

$$\begin{bmatrix} 1 & 1 & 2 & | & 1 & 0 & 0 \\ 0 & 1 & -2 & | & -2 & 1 & 0 \\ 0 & 0 & 1 & | & -1 & 0 & 1 \end{bmatrix} \begin{matrix} R_1 - 2R_3 \to R_1 \\ R_2 + 2R_3 \to R_2 \\ \\ \end{matrix}$$

$$\begin{bmatrix} 1 & 1 & 0 & | & 3 & 0 & -2 \\ 0 & 1 & 0 & | & -4 & 1 & 2 \\ 0 & 0 & 1 & | & -1 & 0 & 1 \end{bmatrix} \begin{matrix} R_1 - R_2 \to R_1 \\ \\ \\ \end{matrix}$$

$$\begin{bmatrix} 1 & 0 & 0 & | & 7 & -1 & -4 \\ 0 & 1 & 0 & | & -4 & 1 & 2 \\ 0 & 0 & 1 & | & -1 & 0 & 1 \end{bmatrix}.$$

Thus

$$A^{-1} = \begin{bmatrix} 7 & -1 & -4 \\ -4 & 1 & 2 \\ -1 & 0 & 1 \end{bmatrix}. \quad \blacksquare$$

If a matrix does not have an inverse, the above technique will so indicate, as in the following.

E X A M P L E 3   **Showing That the Inverse Does Not Exist**

Find A^{-1} if

$$A = \begin{bmatrix} 1 & 2 & 3 \\ 2 & 5 & 7 \\ 2 & 4 & 6 \end{bmatrix}.$$

Solution   Forming the augmented matrix and proceeding yields

$$\begin{bmatrix} 1 & 2 & 3 & | & 1 & 0 & 0 \\ 2 & 5 & 7 & | & 0 & 1 & 0 \\ 2 & 4 & 6 & | & 0 & 0 & 1 \end{bmatrix} \begin{matrix} \\ R_2 - 2R_1 \rightarrow R_2 \\ R_3 - 2R_1 \rightarrow R_3 \end{matrix}$$

$$\begin{bmatrix} 1 & 2 & 3 & | & 1 & 0 & 0 \\ 0 & 1 & 1 & | & -2 & 1 & 0 \\ 0 & 0 & 0 & | & -2 & 0 & 1 \end{bmatrix}.$$

The last row indicates that there can be no solution. Thus A^{-1} does not exist. ∎

Systems of Equations

We shall now see how the inverse matrix can be used to solve systems of equations with the number of unknowns equal to the number of equations. Consider the system

$$a_{11}x_1 + a_{12}x_2 + \cdots + a_{1n}x_n = b_1$$
$$a_{21}x_1 + a_{22}x_2 + \cdots + a_{2n}x_n = b_2$$
$$\cdots$$
$$a_{n1}x_1 + a_{n2}x_2 + \cdots + a_{nn}x_n = b_n.$$

If

$$A = \begin{bmatrix} a_{11} & a_{12} & \cdots & a_{1n} \\ a_{21} & a_{22} & \cdots & a_{2n} \\ & & \cdots & \\ a_{n1} & a_{n2} & \cdots & a_{nn} \end{bmatrix}, X = \begin{bmatrix} x_1 \\ x_2 \\ \vdots \\ x_n \end{bmatrix}, B = \begin{bmatrix} b_1 \\ b_2 \\ \vdots \\ b_n \end{bmatrix},$$

then the system can be written as

$$AX = B.$$

Now if A^{-1} exists, multiply both sides of $AX = B$ on the left by A^{-1}, obtaining

$$A^{-1}(AX) = A^{-1}B.$$

Now the left side is

$$A^{-1}(AX) = (A^{-1}A)X = I_nX = X.$$

Thus

$$X = A^{-1}B$$

is the solution of the system.

Solution of $AX = B$

Let A be a square matrix of order n and B a column matrix of order n. If A^{-1} exists, then the linear system of equations

$$AX = B$$

has the unique solution

$$X = A^{-1}B.$$

E X A M P L E   4   **Solving Systems of Equations Using Matrices**

Solve the two systems

$$
\begin{array}{llll}
x + & y + 2z = 1 & \qquad x + & y + 2z = 2 \\
2x + & 3y + 2z = 2 & \qquad 2x + & 3y + 2z = 0 \\
x + & y + 3z = 0 & \qquad x + & y + 3z = 1
\end{array}
$$

Solution   Let

$$
A = \begin{bmatrix} 1 & 1 & 2 \\ 2 & 3 & 2 \\ 1 & 1 & 3 \end{bmatrix}, \quad
X = \begin{bmatrix} x \\ y \\ z \end{bmatrix}, \quad
B = \begin{bmatrix} 1 \\ 2 \\ 0 \end{bmatrix}, \quad
C = \begin{bmatrix} 2 \\ 0 \\ 1 \end{bmatrix}
$$

Then notice that the two systems can be written as

$$AX = B \qquad AX = C.$$

The solutions are respectively

$$X = A^{-1}B \qquad X = A^{-1}C.$$

We already found A^{-1} in Example 2. Thus the first system has the solution

$$
A^{-1}B = \begin{bmatrix} 7 & -1 & -4 \\ -4 & 1 & 2 \\ -1 & 0 & 1 \end{bmatrix}\begin{bmatrix} 1 \\ 2 \\ 0 \end{bmatrix} = \begin{bmatrix} 5 \\ -2 \\ -1 \end{bmatrix},
$$

or $x = 5$, $y = -2$, and $z = -1$.

The second system has the solution

$$A^{-1}C = \begin{bmatrix} 7 & -1 & -4 \\ -4 & 1 & 2 \\ -1 & 0 & 1 \end{bmatrix} \begin{bmatrix} 2 \\ 0 \\ 1 \end{bmatrix} = \begin{bmatrix} 10 \\ -6 \\ -1 \end{bmatrix},$$

or $x = 10$, $y = -6$, $z = -1$. ∎

Applications

EXAMPLE 5 **Investments**

An investment advisor has two mutual funds that she is recommending: a conservative bond fund with an assumed return of 7% a year and a stock fund with a projected return of 12% a year. One client has $100,000 to invest and wishes a return of 9% a year, and a second client has $50,000 and wants a 10% annual return. How should she allocate the money of each client between the bond fund and the stock fund?

Solution Let x be the amount invested in the bond fund and y the amount invested in the stock fund. If T is the total amount invested, then $x + y = T$. The return on the bond fund will be $0.07x$, and the return on the stock fund will be $0.12y$, for a total return of $0.07x + 0.12y$. If R is the return desired, then $0.07x + 0.12y = R$. We then have

$$x + y = T$$
$$0.07x + 0.12y = R.$$

If

$$A = \begin{bmatrix} 1 & 1 \\ 0.07 & 0.12 \end{bmatrix}, \quad X = \begin{bmatrix} x \\ y \end{bmatrix}, \quad B = \begin{bmatrix} T \\ R \end{bmatrix},$$

then the system can be written as $AX = B$. The solution is $X = A^{-1}B$. To find A^{-1} we use the Gauss–Jordan method.

$$[A|I] = \begin{bmatrix} 1 & 1 & | & 1 & 0 \\ 0.07 & 0.12 & | & 0 & 1 \end{bmatrix} \quad R_2 - 0.07R_1 \rightarrow R_2$$

$$\begin{bmatrix} 1 & 1 & | & 1 & 0 \\ 0 & 0.05 & | & -0.07 & 1 \end{bmatrix} \quad 20R_2 \rightarrow R_2$$

$$\begin{bmatrix} 1 & 1 & | & 1 & 0 \\ 0 & 1 & | & -1.4 & 20 \end{bmatrix} \quad R_1 - R_2 \rightarrow R_1$$

$$\begin{bmatrix} 1 & 0 & | & 2.4 & -20 \\ 0 & 1 & | & -1.4 & 20 \end{bmatrix}$$

Thus

$$A^{-1} = \begin{bmatrix} 2.4 & -20 \\ -1.4 & 20 \end{bmatrix}.$$

(a) The first client has $100,000 to invest and wishes a return of $(0.09)(\$100,000) = \$9,000$. Thus

$$X = A^{-1}B = \begin{bmatrix} 2.4 & -20 \\ -1.4 & 20 \end{bmatrix} \begin{bmatrix} 100,000 \\ 9,000 \end{bmatrix} = \begin{bmatrix} 60,000 \\ 40,000 \end{bmatrix}.$$

The first client should invest $60,000 in the bond fund and $40,000 in the stock fund.

(b) The second client has $50,000 to invest and wishes a return of $(0.10)(\$50,000) = \$5,000$. Thus

$$X = A^{-1}B = \begin{bmatrix} 2.4 & -20 \\ -1.4 & 20 \end{bmatrix} \begin{bmatrix} 50,000 \\ 5,000 \end{bmatrix} = \begin{bmatrix} 20,000 \\ 30,000 \end{bmatrix}.$$

The second client should invest $20,000 in the bond fund and $30,000 in the stock fund.   ∎

SELF-HELP EXERCISE SET 2.5

1. Find A^{-1} if

$$A = \begin{bmatrix} 1 & 0 & 0 & 1 \\ 0 & 0 & 1 & 0 \\ 0 & 0 & 0 & 1 \\ 0 & 1 & 0 & 2 \end{bmatrix}.$$

2. Solve the following system of equations using the inverse of the coefficient matrix:

$$\begin{aligned} x \quad\quad - 2z &= 1 \\ y + \quad z &= 2 \\ x + y \quad\quad &= 4. \end{aligned}$$

3. Solution A is 3% alcohol, and solution B is 7% alcohol. A lab technician needs to mix the two solutions to obtain 50 liters of a solution that is 4.2% alcohol. How many liters of each solution must be used? Use the inverse matrix to find the answer.

EXERCISE SET 2.5

In Exercises 1 through 10 determine if the given pairs of matrices are inverses to each other by showing if their product is the identity matrix.

1. $\begin{bmatrix} 1 & 2 \\ 3 & 5 \end{bmatrix}$ and $\begin{bmatrix} -5 & 2 \\ 3 & -1 \end{bmatrix}$ **2.** $\begin{bmatrix} 10 & 3 \\ 3 & 1 \end{bmatrix}$ and $\begin{bmatrix} 1 & -3 \\ -3 & 10 \end{bmatrix}$

3. $\begin{bmatrix} 4 & 7 \\ 1 & 2 \end{bmatrix}$ and $\begin{bmatrix} 2 & -7 \\ -1 & 4 \end{bmatrix}$ **4.** $\begin{bmatrix} 13 & 4 \\ 3 & 1 \end{bmatrix}$ and $\begin{bmatrix} 1 & 4 \\ -3 & 13 \end{bmatrix}$

5. $\begin{bmatrix} 1 & 2 & 0 \\ 2 & 1 & -1 \\ 3 & 1 & 1 \end{bmatrix}$ and $\frac{1}{8} \begin{bmatrix} -2 & 2 & 2 \\ 5 & -1 & -1 \\ 1 & -5 & 3 \end{bmatrix}$

6. $\begin{bmatrix} 2 & 1 & 3 \\ 3 & 1 & 3 \\ 2 & 1 & 4 \end{bmatrix}$ and $\begin{bmatrix} -1 & 1 & 0 \\ 0 & -2 & -3 \\ 1 & 0 & 1 \end{bmatrix}$

7. $\begin{bmatrix} 1 & 1 & 0 \\ 0 & 1 & 1 \\ 0 & 0 & 1 \end{bmatrix}$ and $\begin{bmatrix} 1 & -1 & 1 \\ 0 & 1 & -1 \\ 0 & 0 & 1 \end{bmatrix}$

8. $\begin{bmatrix} 1 & 2 & 3 \\ 0 & 1 & 3 \\ 0 & 0 & 1 \end{bmatrix}$ and $\begin{bmatrix} 1 & -2 & 3 \\ 0 & 1 & -3 \\ 0 & 0 & 1 \end{bmatrix}$

9. $\begin{bmatrix} 2 & 0 & 1 & 2 \\ 1 & 1 & 0 & 2 \\ 2 & -1 & 3 & 1 \\ 3 & -1 & 4 & 3 \end{bmatrix}$ and $\frac{1}{3} \begin{bmatrix} 3 & 0 & 3 & -3 \\ -3 & 5 & 5 & -3 \\ -3 & 2 & 2 & 0 \\ 0 & -1 & -4 & 3 \end{bmatrix}$

10. $\begin{bmatrix} 1 & 0 & 0 & 0 \\ 0 & 1 & 0 & 0 \\ 0 & 0 & 5 & 2 \\ 0 & 0 & 1 & 1 \end{bmatrix}$ and $\begin{bmatrix} 1 & 0 & 0 & 0 \\ 0 & 1 & 0 & 0 \\ 0 & 0 & 1 & -2 \\ 0 & 0 & 1 & 5 \end{bmatrix}$

In Exercises 11 through 26 find the inverses of the given matrices, if they exist.

11. $\begin{bmatrix} 1 & 3 \\ 3 & 10 \end{bmatrix}$ **12.** $\begin{bmatrix} 1 & 3 \\ 1 & 4 \end{bmatrix}$ **13.** $\begin{bmatrix} 5 & -1 \\ -3 & 2 \end{bmatrix}$

14. $\begin{bmatrix} 5 & -1 \\ 4 & 2 \end{bmatrix}$ **15.** $\begin{bmatrix} 4 & 2 \\ 2 & 1 \end{bmatrix}$ **16.** $\begin{bmatrix} 1 & 1 \\ 1 & 1 \end{bmatrix}$

17. $\begin{bmatrix} 1 & 2 & 1 \\ 1 & 2 & 1 \\ 2 & 0 & 1 \end{bmatrix}$ **18.** $\begin{bmatrix} 1 & 0 & 1 \\ 0 & 2 & 1 \\ 1 & 0 & 2 \end{bmatrix}$ **19.** $\begin{bmatrix} 1 & 1 & 1 \\ 0 & 2 & 2 \\ -1 & 0 & 1 \end{bmatrix}$

20. $\begin{bmatrix} 1 & 0 & 1 \\ 0 & 1 & 1 \\ 1 & 1 & 1 \end{bmatrix}$ **21.** $\begin{bmatrix} 1 & 2 & 0 \\ 1 & 1 & 2 \\ 0 & -1 & -1 \end{bmatrix}$ **22.** $\begin{bmatrix} 2 & 4 & 4 \\ 1 & 2 & 1 \\ 1 & 0 & -1 \end{bmatrix}$

23. $\begin{bmatrix} 1 & 1 & 1 \\ 1 & 2 & 4 \\ 2 & 1 & 1 \end{bmatrix}$ **24.** $\begin{bmatrix} 0 & 1 & 1 \\ 1 & 1 & 0 \\ 1 & 0 & 1 \end{bmatrix}$ **25.** $\begin{bmatrix} 1 & 1 & 1 & 1 \\ 0 & 2 & 1 & 1 \\ 1 & 0 & 2 & 1 \\ 0 & 1 & 0 & 1 \end{bmatrix}$

26. $\begin{bmatrix} 0 & 1 & 1 & 0 \\ 1 & 0 & 0 & 2 \\ 2 & 0 & 1 & 0 \\ 2 & 0 & 0 & 1 \end{bmatrix}$

In Exercises 27 through 32 solve by using the inverse of the coefficient matrix.

27. $\begin{aligned} 2x - 4y &= 10 \\ x - y &= -2 \end{aligned}$ **28.** $\begin{aligned} 2x - y &= 7 \\ x + 3y &= 7 \end{aligned}$ **29.** $\begin{aligned} x + y &= 6 \\ 2x + 3y &= 9 \end{aligned}$

30. $\begin{aligned} x + y &= 5 \\ 2x - 3y &= -5 \end{aligned}$ **31.** $\begin{aligned} 2x + 4y &= 6 \\ 3x + 5y &= 4 \end{aligned}$ **32.** $\begin{aligned} 3x - 6y &= 6 \\ x - y &= 9 \end{aligned}$

In Exercises 33 through 36 solve by using your answers to Exercises 11 through 14.

33. $\begin{aligned} x + 3y &= 2 \\ 3x + 10y &= 3 \end{aligned}$ **34.** $\begin{aligned} x + 3y &= 1 \\ x + 4y &= 6 \end{aligned}$ **35.** $\begin{aligned} 5x - y &= 4 \\ -3x + 2y &= 3 \end{aligned}$

36. $\begin{aligned} 5x - y &= -3 \\ 4x + 2y &= 5 \end{aligned}$

In Exercises 37 through 42 solve by using your answers to Exercises 21 through 26.

37. $\begin{aligned} x + 2y \quad\quad &= 1 \\ x + y + 2z &= 0 \\ - y - z &= 2 \end{aligned}$ **38.** $\begin{aligned} 2x + 4y + 4z &= 1 \\ x + 2y + z &= 1 \\ x - z &= 3 \end{aligned}$

39. $\begin{aligned} x + y + z &= 0 \\ x + 2y + 4z &= 1 \\ 2x + y + z &= 2 \end{aligned}$ **40.** $\begin{aligned} y + z &= -2 \\ x + y &= 2 \\ x \quad\quad + z &= 4 \end{aligned}$

41. $\begin{aligned} x_1 + x_2 + x_3 + x_4 &= 1 \\ 2x_2 + x_3 + x_4 &= 1 \\ x_1 \quad\quad + 2x_3 + x_4 &= 2 \\ x_2 + x_4 &= 2 \end{aligned}$ **42.** $\begin{aligned} x_2 + x_3 \quad\quad &= 2 \\ x_1 \quad\quad + 2x_4 &= -3 \\ 2x_1 \quad + x_3 \quad\quad &= 4 \\ 2x_1 \quad\quad + x_4 &= 6 \end{aligned}$

43. Let

$$A = \begin{bmatrix} a & b \\ c & d \end{bmatrix}$$

and assume that $D = ad - bc \neq 0$. Show that

$$A^{-1} = \frac{1}{D} \begin{bmatrix} d & -b \\ -c & a \end{bmatrix}.$$

44. If A^{-1} and B^{-1} exist, show that

$$(AB)^{-1} = B^{-1}A^{-1}.$$

Applications

45. Investment. Redo Example 5 with a bond fund return of 8%.

46. Investment. Redo Example 5 with a bond fund return of 8% and a stock fund return of 13%.

47. Scheduling. The material and labor costs of manufacturing two styles of chair are given as follows:

	Material	Labor
Style A	40	50
Style B	30	40

If $10,000 is allocated for both material and labor, how many of each style of chair should be manufactured to exactly use the $10,000 if
a. $4400 is allocated for material and $5600 for labor;
b. $4300 is allocated for material and $5700 for labor.

48. Mixture. A small store sells mint tea at $3.20 an ounce and peppermint tea at $4 an ounce. The store owner decides to make a batch of 100 ounces of tea that mixes both kinds and to sell the mixture for $3.50 an ounce. How many ounces of each of the two varieties of tea should be mixed to obtain the same revenue as selling them unmixed?

Graphing Calculator Exercises

In Exercises 49 through 56 use a calculator. Let

$$A = \begin{bmatrix} 1 & 4 & 5 & -3 & 6 \\ -11 & 6 & 4 & -2 & 3 \\ 4 & 16 & -1 & 8 & 1 \\ 2 & -8 & 1 & 4 & -1 \\ 1 & 9 & 10 & 4 & 8 \end{bmatrix} \quad B = \begin{bmatrix} 1 & -4 & 3 & 2 & 1 \\ 6 & 1 & -8 & 5 & 1 \\ -2 & 8 & -5 & 7 & 3 \\ 4 & -5 & -16 & 2 & 4 \\ 1 & 3 & 1 & -5 & 1 \end{bmatrix}$$

49. Does $AB = BA$?

50. Find A^{-1}.

51. Find B^{-1}.

52. Find $(AB)^{-1}$.

53. Is $(AB)^{-1} = A^{-1}B^{-1}$?

54. Is $(AB)^{-1} = B^{-1}A^{-1}$?

55. Solve $AX = C$ where $\quad C = \begin{bmatrix} 10 \\ 8 \\ 20 \\ 22 \\ 37 \end{bmatrix}$.

56. Solve $BX = C$ where C is given in the previous exercise.

Solutions to Self-Help Exercise Set 2.5

1. We have

$$\left[\begin{array}{cccc|cccc} 1 & 0 & 0 & 1 & 1 & 0 & 0 & 0 \\ 0 & 0 & 1 & 0 & 0 & 1 & 0 & 0 \\ 0 & 0 & 0 & 1 & 0 & 0 & 1 & 0 \\ 0 & 1 & 0 & 2 & 0 & 0 & 0 & 1 \end{array}\right] \begin{array}{l} \\ R_2 \leftrightarrow R_4 \\ \\ \end{array}$$

$$\left[\begin{array}{cccc|cccc} 1 & 0 & 0 & 1 & 1 & 0 & 0 & 0 \\ 0 & 1 & 0 & 2 & 0 & 0 & 0 & 1 \\ 0 & 0 & 0 & 1 & 0 & 0 & 1 & 0 \\ 0 & 0 & 1 & 0 & 0 & 1 & 0 & 0 \end{array}\right] \begin{array}{l} \\ \\ R_3 \leftrightarrow R_4 \\ \end{array}$$

$$\left[\begin{array}{cccc|cccc} 1 & 0 & 0 & 1 & 1 & 0 & 0 & 0 \\ 0 & 1 & 0 & 2 & 0 & 0 & 0 & 1 \\ 0 & 0 & 1 & 0 & 0 & 1 & 0 & 0 \\ 0 & 0 & 0 & 1 & 0 & 0 & 1 & 0 \end{array}\right] \begin{array}{l} R_1 - R_4 \rightarrow R_1 \\ R_2 - 2R_4 \rightarrow R_2 \\ \\ \end{array}$$

$$\left[\begin{array}{cccc|cccc} 1 & 0 & 0 & 0 & 1 & 0 & -1 & 0 \\ 0 & 1 & 0 & 0 & 0 & 0 & -2 & 1 \\ 0 & 0 & 1 & 0 & 0 & 1 & 0 & 0 \\ 0 & 0 & 0 & 1 & 0 & 0 & 1 & 0 \end{array}\right].$$

Thus

$$A^{-1} = \begin{bmatrix} 1 & 0 & -1 & 0 \\ 0 & 0 & -2 & 1 \\ 0 & 1 & 0 & 0 \\ 0 & 0 & 1 & 0 \end{bmatrix}.$$

2. Let

$$A = \begin{bmatrix} 1 & 0 & -2 \\ 0 & 1 & 1 \\ 1 & 1 & 0 \end{bmatrix}, X = \begin{bmatrix} x \\ y \\ z \end{bmatrix}, B = \begin{bmatrix} 1 \\ 2 \\ 4 \end{bmatrix}.$$

Then the system of equations can be written as $AX = B$. The solution is $X = A^{-1}B$. To find A^{-1}, we use the Gauss–Jordan method and obtain

$$\begin{bmatrix} 1 & 0 & -2 & | & 1 & 0 & 0 \\ 0 & 1 & 1 & | & 0 & 1 & 0 \\ 1 & 1 & 0 & | & 0 & 0 & 1 \end{bmatrix} R_3 - R_1 \rightarrow R_3$$

$$\begin{bmatrix} 1 & 0 & -2 & | & 1 & 0 & 0 \\ 0 & 1 & 1 & | & 0 & 1 & 0 \\ 0 & 1 & 2 & | & -1 & 0 & 1 \end{bmatrix} R_3 - R_2 \rightarrow R_3$$

$$\begin{bmatrix} 1 & 0 & -2 & | & 1 & 0 & 0 \\ 0 & 1 & 1 & | & 0 & 1 & 0 \\ 0 & 0 & 1 & | & -1 & -1 & 1 \end{bmatrix} \begin{matrix} R_1 + 2R_3 \rightarrow R_1 \\ R_2 - R_3 \rightarrow R_2 \end{matrix}$$

$$\begin{bmatrix} 1 & 0 & 0 & | & -1 & -2 & 2 \\ 0 & 1 & 0 & | & 1 & 2 & -1 \\ 0 & 0 & 1 & | & -1 & -1 & 1 \end{bmatrix}.$$

Thus

$$A^{-1} = \begin{bmatrix} -1 & -2 & 2 \\ 1 & 2 & -1 \\ -1 & -1 & 1 \end{bmatrix},$$

and

$$X = A^{-1}B$$

$$= \begin{bmatrix} -1 & -2 & 2 \\ 1 & 2 & -1 \\ -1 & -1 & 1 \end{bmatrix} \begin{bmatrix} 1 \\ 2 \\ 4 \end{bmatrix}$$

$$= \begin{bmatrix} (-1)(1) + (-2)(2) + (2)(4) \\ (1)(1) + (2)(2) + (-1)(4) \\ (-1)(1) + (-1)(2) + (1)(4) \end{bmatrix} = \begin{bmatrix} 3 \\ 1 \\ 1 \end{bmatrix}.$$

3. Let x and y denote, respectively, the number of liters of solution A and B to be used. Then since there must be 50 liters, we have $x + y = 50$. The amount of alcohol in liters in solution A is $.03x$, in solution B is $.07y$, and in the mixture is $.042(50) = 2.1$. Thus $.03x + .07y = 2.1$. We then have the system of equations

$$x + y = 50$$

$$.03x + .07y = 2.1.$$

This can be written in matrix form as $AX = B$ if

$$A = \begin{bmatrix} 1 & 1 \\ .03 & .07 \end{bmatrix}, \quad X = \begin{bmatrix} x \\ y \end{bmatrix}, \quad B = \begin{bmatrix} 50 \\ 2.1 \end{bmatrix}.$$

The solution is $X = A^{-1}B$. To find A^{-1} we use the Gauss–Jordan method and obtain

$$[A|I] = \begin{bmatrix} 1 & 1 & 1 & 0 \\ 0.03 & 0.07 & 0 & 1 \end{bmatrix} \begin{array}{c} \\ R_2 - 0.03R_1 \to R_2 \end{array}$$

$$\begin{bmatrix} 1 & 1 & 1 & 0 \\ 0 & 0.04 & -0.03 & 1 \end{bmatrix} 25R_2 \to R_2$$

$$\begin{bmatrix} 1 & 1 & 1 & 0 \\ 0 & 1 & -.75 & 25 \end{bmatrix} \begin{array}{c} R_1 - R_2 \to R_1 \\ \\ \end{array}$$

$$\begin{bmatrix} 1 & 0 & 1.75 & -25 \\ 0 & 1 & -.75 & 25 \end{bmatrix}$$

Thus

$$A^{-1} = \begin{bmatrix} 1.75 & -25 \\ -.75 & 25 \end{bmatrix}.$$

Then

$$X = A^{-1}B = \begin{bmatrix} 1.75 & -25 \\ -.75 & 25 \end{bmatrix} \begin{bmatrix} 50 \\ 2.1 \end{bmatrix} = \begin{bmatrix} 35 \\ 15 \end{bmatrix}.$$

Thus the technician should use 35 liters of solution A and 15 liters of solution B.

2.6   LEONTIEF INPUT–OUTPUT MODELS

► *Introduction*

► *Open Leontief Input–Output Model*

► *Closed Leontief Input–Output Model*

(© Gary Benson/Comstock, Inc.)

Introduction

There are three branches of economics that rely heavily on linear systems and matrix theory: game theory, input–output analysis, and linear programming. Game theory arrived on the scene first with the publication of a fundamental result by John von Neumann in 1928. The impact of game theory on economics, however, was considerably delayed. Game theory will be taken up in a later chapter in this text. Linear programming, which will be considered in the next chapter, appeared immediately after World War II. Input–output analysis was the second of these three branches of economics to appear. Input–output analysis was initiated by Wassily Leontief with the publication of a paper of his in 1936. In 1941 he published a full exposition of this method. For this work, he was awarded a Nobel prize in economics in 1973. Leontief divided the American economy into hundreds of sectors, such as automobiles, glass, copper, and so on, and then studied how these sectors interact with each other.

In order to effectively handle the hundreds of linear equations in the input–output analysis, one naturally needs to use computers. We will obtain a flavor for the method by considering an economy with only two or three sectors.

Consider an economy consisting of n industries, each of which produces a single product. Demand for any one of the industry's products comes from demand from each of the industries (including possibly itself) and demand from outside these industries. For example, the energy industry will need energy for its own plants, will need energy to supply other industries, and will need to supply energy to the general public.

Open Leontief Input–Output Model

In input–output analysis, we seek to determine the amount each industry must produce so that both the interindustry demand and the nonindustry demand are both met exactly, that is, so that supply and demand are exactly balanced.

The interindustry demand is summarized in an **input–output matrix** $A = (a_{ij})$. There may be as many as a hundred or even more industries in actual models. For illustrative purposes, we assume a very simple model of three industries: auto, energy, and transportation. Each of these industries acts as a supplier and a user of their products. We let a_{ij} **represent the dollar amount of the output of industry i required in producing one dollar of output in industry j.** The following matrix summarizes our simplified economy.

User

	Auto	Energy	Transportation
Auto	0.2	0.4	0.1
Energy	0.1	0.2	0.2
Transportation	0.2	0.2	0.1

$$= A$$

Thus the element $a_{11} = 0.2$ indicates that of every dollar of output *from* the automobile industry, 20% of this value is contributed by the auto industry. The element $a_{12} = 0.4$ indicates that of every dollar of output *from* the energy industry, 40% of this value is contributed by the auto industry. The element $a_{13} = 0.1$ indicates that of every dollar of output *from* the transportation industry, 10% of this value is contributed by the auto industry. The element $a_{21} = 0.1$ indicates that of every dollar of output *from* the auto industry, 10% of this value is contributed by the energy industry. The element $a_{31} = 0.2$ indicates that of every dollar of output *from* the auto industry, 20% of this value is contributed by the transportation industry. The reader can now interpret the remaining elements.

Let x_1, x_2, and x_3 represent respectively the output in dollars of auto, energy, and transportation produced. Then

$$0.2x_1 + 0.4x_2 + 0.1x_3$$

represents the interindustry demand for the first product, auto. Let d_1, d_2, and d_3 be the nonindustry demand in dollars respectively for the three products. Then since

$$\text{auto output} = \begin{pmatrix} \text{interindustry} \\ \text{demand for auto} \end{pmatrix} + \begin{pmatrix} \text{nonindustry} \\ \text{demand for auto} \end{pmatrix}$$

$$x_1 \quad = \quad 0.2x_1 + 0.4x_2 + 0.1x_3 \quad + \quad d_1$$

Continuing in this manner, we obtain the following system of three equations:

$$x_1 = 0.2x_1 + 0.4x_2 + 0.1x_3 + d_1$$
$$x_2 = 0.1x_1 + 0.2x_2 + 0.2x_3 + d_2$$
$$x_3 = 0.2x_1 + 0.2x_2 + 0.1x_3 + d_3.$$

If we define the **production column matrix** X and the **demand column matrix** D, respectively, by

$$X = \begin{bmatrix} x_1 \\ x_2 \\ x_3 \end{bmatrix}, \quad D = \begin{bmatrix} d_1 \\ d_2 \\ d_3 \end{bmatrix},$$

then these last three equations can be written in matrix form as

$$X = AX + D,$$

where

$$A = \begin{bmatrix} .2 & .4 & .1 \\ .1 & .2 & .2 \\ .2 & .2 & .1 \end{bmatrix}.$$

This can be rewritten as

$$X - AX = D$$
$$IX - AX = D$$
$$(I - A)X = D.$$

If the matrix $(I - A)$ has an inverse, then the solution is

$$X = (I - A)^{-1}D.$$

E X A M P L E 1

Finding the Production Matrix

Find the production matrix X in the previous problem if the demand matrix is

$$D = \begin{bmatrix} 474 \\ 948 \\ 474 \end{bmatrix}$$

Solution   The matrix $(I - A)$ is

$$I - A = \begin{bmatrix} 0.8 & -0.4 & -0.1 \\ -0.1 & 0.8 & -0.2 \\ -0.2 & -0.2 & 0.9 \end{bmatrix}.$$

Using the Gauss–Jordan method one obtains

$$(I - A)^{-1} = \frac{1}{474} \begin{bmatrix} 680 & 380 & 160 \\ 130 & 700 & 170 \\ 180 & 240 & 600 \end{bmatrix}.$$

Thus

$$X = (I - A)^{-1}D = \frac{1}{474} \begin{bmatrix} 680 & 380 & 160 \\ 130 & 700 & 170 \\ 180 & 240 & 600 \end{bmatrix} \begin{bmatrix} 474 \\ 948 \\ 474 \end{bmatrix} = \begin{bmatrix} 1600 \\ 1700 \\ 1260 \end{bmatrix}. \quad ■$$

If A is an input–output matrix and $(I - A)^{-1}$ has only nonnegative elements, then for any demand matrix D with nonnegative elements, $X = (I - A)^{-1}D$ will also have all nonnegative elements and thus a *meaningful* solution. This means that, in this situation, any outside (nonindustry) demand can be met. The following theorem can be proven.

When Outside Demand can be Met

If A is an input–output matrix and the sum of the elements in each row of A is less than one, then any outside demand can be met.

This was the case in Example 1.

Closed Leontief Input–Output Model

In the closed input–output model, we assume that there is *no* demand from outside the n industries under consideration. Let f_{ij} be the fraction of the total output of the jth industry purchased by the ith industry. The problem then is to find suitable prices for each of the n outputs so that the total expenditure of each industry equals its total income. Any such price structure represents an equilibrium for the economy.

Let p_i be the price charged by the ith industry for its total output. For convenience, consider again an economy with the three industries: auto, energy, transportation. As usual, we allow for the possibility that each industry consumes (and therefore buys) some of its own product. The total income of the auto industry, p_1, is then the sum of its income from the three industries or

$$p_1 = f_{11}p_1 + f_{12}p_2 + f_{13}p_3.$$

In a similar fashion, we have

$$p_2 = f_{21}p_1 + f_{22}p_2 + f_{23}p_3$$

$$p_3 = f_{31}p_1 + f_{32}p_2 + f_{33}p_3.$$

If we set $F = (f_{ij})$ and let the **price matrix** be

$$P = \begin{bmatrix} p_1 \\ p_2 \\ p_3 \end{bmatrix},$$

then these last three equations can be written in matrix form as

$$P = FP.$$

This can be rewritten as

$$P - FP = O$$

$$IP - FP = O$$

$$(I - F)P = O$$

It turns out that $(I - F)$ does *not* have an inverse. The solutions can be found using the Gauss–Jordan method.

E X A M P L E   2   **Finding the Equilibrium Prices**

Let

$$F = \begin{bmatrix} \frac{1}{2} & \frac{1}{3} & \frac{1}{4} \\ \frac{1}{3} & \frac{1}{3} & \frac{1}{4} \\ \frac{1}{6} & \frac{1}{3} & \frac{1}{2} \end{bmatrix}$$

Find the equilibrium prices.

Solution

$$I - F = \begin{bmatrix} \frac{1}{2} & -\frac{1}{3} & -\frac{1}{4} \\ -\frac{1}{3} & \frac{2}{3} & -\frac{1}{4} \\ -\frac{1}{6} & -\frac{1}{3} & \frac{1}{2} \end{bmatrix}$$

Then

$$\begin{bmatrix} \frac{1}{2} & -\frac{1}{3} & -\frac{1}{4} & | & 0 \\ -\frac{1}{3} & \frac{2}{3} & -\frac{1}{4} & | & 0 \\ -\frac{1}{6} & -\frac{1}{3} & \frac{1}{2} & | & 0 \end{bmatrix} \begin{array}{l} 2R_1 \rightarrow R_1 \\ \\ \\ \end{array}$$

$$= \begin{bmatrix} 1 & -\frac{2}{3} & -\frac{1}{2} & | & 0 \\ -\frac{1}{3} & \frac{2}{3} & -\frac{1}{4} & | & 0 \\ -\frac{1}{6} & -\frac{1}{3} & \frac{1}{2} & | & 0 \end{bmatrix} \begin{array}{l} \\ R_2 + \frac{1}{3}R_1 \rightarrow R_2 \\ R_3 + \frac{1}{6}R_1 \rightarrow R_3 \end{array}$$

$$= \begin{bmatrix} 1 & -\frac{2}{3} & -\frac{1}{2} & | & 0 \\ 0 & \frac{4}{9} & -\frac{5}{12} & | & 0 \\ 0 & -\frac{4}{9} & \frac{5}{12} & | & 0 \end{bmatrix} \begin{array}{l} \\ \frac{9}{4}R_2 \rightarrow R_2 \\ R_3 + R_2 \rightarrow R_3 \end{array}$$

$$= \begin{bmatrix} 1 & -\frac{2}{3} & -\frac{1}{2} & | & 0 \\ 0 & 1 & -\frac{15}{16} & | & 0 \\ 0 & 0 & 0 & | & 0 \end{bmatrix} \begin{array}{l} R_1 + \frac{2}{3}R_2 \rightarrow R_1 \\ \\ \end{array}$$

$$= \begin{bmatrix} 1 & 0 & -\frac{9}{8} & | & 0 \\ 0 & 1 & -\frac{15}{16} & | & 0 \\ 0 & 0 & 0 & | & 0 \end{bmatrix}$$

Thus, letting t be any real number, we have

$$p_1 = \frac{9}{8}t$$

$$p_2 = \frac{15}{16}t$$

$$p_3 = t. \quad \blacksquare$$

These are the suitable prices for each of the three outputs, that is, prices such that the total expenditure of each industry equals its total income.

Notice that the sum of the entries in each column of F is one. This *must* be the case since each of the outputs of each industry is *completely* distributed among the three industries.

SELF-HELP EXERCISE SET 2.6

1. Let the input–output matrix and the demand matrix be

$$A = \begin{bmatrix} 0.6 & 0.2 \\ 0.2 & 0.4 \end{bmatrix}, \quad D = \begin{bmatrix} 50,000,000 \\ 30,000,000 \end{bmatrix}$$

Find the production matrix.

EXERCISE SET 2.6

In Exercises 1 through 10 you are given an input–output matrix A and a demand matrix D. Assuming a Leontief open input–output model, find the production matrix.

1. $A = \begin{bmatrix} 0.3 & 0.2 \\ 0.4 & 0.3 \end{bmatrix}, \quad D = \begin{bmatrix} 10,000,000 \\ 20,000,000 \end{bmatrix}$

2. $A = \begin{bmatrix} 0.1 & 0.2 \\ 0.4 & 0.3 \end{bmatrix}, \quad D = \begin{bmatrix} 30,000,000 \\ 40,000,000 \end{bmatrix}$

3. $A = \begin{bmatrix} 0.5 & 0.2 \\ 0.4 & 0.2 \end{bmatrix}, \quad D = \begin{bmatrix} 100,000,000 \\ 200,000,000 \end{bmatrix}$

4. $A = \begin{bmatrix} 0.2 & 0.3 \\ 0.4 & 0.3 \end{bmatrix}, \quad D = \begin{bmatrix} 30,000,000 \\ 50,000,000 \end{bmatrix}$

5. $A = \begin{bmatrix} 0.5 & 0.1 \\ 0.4 & 0.1 \end{bmatrix}, \quad D = \begin{bmatrix} 200,000,000 \\ 300,000,000 \end{bmatrix}$

6. $A = \begin{bmatrix} 0.1 & 0.3 \\ 0.5 & 0.3 \end{bmatrix}, \quad D = \begin{bmatrix} 5,000,000 \\ 6,000,000 \end{bmatrix}$

7. $A = \begin{bmatrix} 0 & 0.1 & 0.2 \\ 0.4 & 0 & 0.1 \\ 0.2 & 0.3 & 0 \end{bmatrix}, \quad D = \begin{bmatrix} 20,000,000 \\ 40,000,000 \\ 10,000,000 \end{bmatrix}$

8. $A = \begin{bmatrix} 0.1 & 0.2 & 0.4 \\ 0.3 & 0.3 & 0.1 \\ 0.1 & 0.2 & 0.2 \end{bmatrix}, \quad D = \begin{bmatrix} 2,000,000 \\ 1,000,000 \\ 1,000,000 \end{bmatrix}$

9. $A = \begin{bmatrix} 0 & 0.1 & 0.1 \\ 0.2 & 0 & 0.3 \\ 0.1 & 0.3 & 0 \end{bmatrix}, \quad D = \begin{bmatrix} 30,000,000 \\ 20,000,000 \\ 20,000,000 \end{bmatrix}$

10. $A = \begin{bmatrix} 0.1 & 0.1 & 0.2 \\ 0.3 & 0.2 & 0.1 \\ 0.2 & 0.2 & 0.3 \end{bmatrix}$, $D = \begin{bmatrix} 1{,}000{,}000 \\ 2{,}000{,}000 \\ 1{,}000{,}000 \end{bmatrix}$

In Exercises 11 through 17 you are given an input–output matrix F for a closed Leontief input–output model. Find the equilibrium price matrix.

11. $F = \begin{bmatrix} 0.3 & 0.2 \\ 0.7 & 0.8 \end{bmatrix}$

12. $F = \begin{bmatrix} 0.5 & 0.4 \\ 0.5 & 0.6 \end{bmatrix}$

13. $F = \begin{bmatrix} 0.4 & 0.1 \\ 0.6 & 0.9 \end{bmatrix}$

14. $F = \begin{bmatrix} 0.2 & 0.1 \\ 0.8 & 0.9 \end{bmatrix}$

15. $F = \begin{bmatrix} 0.3 & 0.1 & 0.2 \\ 0.3 & 0.4 & 0.2 \\ 0.4 & 0.5 & 0.6 \end{bmatrix}$

16. $F = \begin{bmatrix} 0.2 & 0.1 & 0.3 \\ 0.4 & 0.3 & 0.4 \\ 0.4 & 0.6 & 0.3 \end{bmatrix}$

17. $F = \begin{bmatrix} 0.1 & 0.1 & 0.2 \\ 0.5 & 0 & 0.8 \\ 0.4 & 0.9 & 0 \end{bmatrix}$

18. Closed Leontief Input–Output Model. Suppose in the closed Leontief input–output model for a three industry system, the input–output matrix is given by $A = (a_{ij})$. We then wish to solve $(I - A)X = O$. To solve we form the augmented matrix $[I - A|O]$. In the closed model we must have

$$a_{11} + a_{21} + a_{31} = 1$$
$$a_{12} + a_{22} + a_{32} = 1$$
$$a_{13} + a_{23} + a_{33} = 1$$

Using these three equations and the operations $R_3 + R_1 \rightarrow R_3$ and $R_3 + R_2 \rightarrow R_3$ on the above augmented matrix, show that the augmented matrix becomes

$$\begin{bmatrix} 1 - a_{11} & -a_{12} & a_{13} \\ -a_{21} & 1 - a_{22} & -a_{23} \\ 0 & 0 & 0 \end{bmatrix} \begin{bmatrix} 0 \\ 0 \\ 0 \end{bmatrix}.$$

Graphing Calculator Exercise

19. Given the following input–output matrix A and demand matrix D and assuming a Leontief open input–output model, find the production matrix:

$$A = \begin{bmatrix} .20 & .20 & .15 & .10 \\ .40 & .10 & .20 & .15 \\ .02 & .03 & .01 & .12 \\ .25 & .04 & .01 & .10 \end{bmatrix}, \quad D = \begin{bmatrix} 2200 \\ 3400 \\ 2700 \\ 1600 \end{bmatrix}.$$

Solutions to Self-Help Exercise Set 2.6

1. $I - A = \begin{bmatrix} 0.4 & -0.2 \\ -0.2 & 0.6 \end{bmatrix}$

Using the Gauss–Jordan method we have

$$\begin{bmatrix} 0.4 & -0.2 & | & 1 & 0 \\ -0.2 & 0.6 & | & 0 & 1 \end{bmatrix} \frac{1}{0.4}R_1 \to R_1$$

$$\begin{bmatrix} 1 & -0.5 & | & 2.5 & 0 \\ -0.2 & 0.6 & | & 0 & 1 \end{bmatrix} R_2 + 0.2R_1 \to R_2$$

$$\begin{bmatrix} 1 & -0.5 & | & 2.5 & 0 \\ 0 & 0.5 & | & 0.5 & 1 \end{bmatrix} 2R_2 \to R_2$$

$$\begin{bmatrix} 1 & -0.5 & | & 2.5 & 0 \\ 0 & 1 & | & 1 & 2 \end{bmatrix} R_1 + 0.5R_2 \to R_1$$

$$\begin{bmatrix} 1 & 0 & | & 3 & 1 \\ 0 & 1 & | & 1 & 2 \end{bmatrix}.$$

Thus

$$(I - A)^{-1} = \begin{bmatrix} 3 & 1 \\ 1 & 2 \end{bmatrix},$$

and

$$X = (I - A)^{-1}D = \begin{bmatrix} 3 & 1 \\ 1 & 2 \end{bmatrix} \begin{bmatrix} 50,000,000 \\ 30,000,000 \end{bmatrix} = \begin{bmatrix} 180,000,000 \\ 110,000,000 \end{bmatrix}.$$

SUMMARY OUTLINE OF CHAPTER 2

- A **system of linear equations** is a set of linear equations.
- The set of all solutions to a system of linear equations is called the **solution set**.
- Two systems of equations with the same solution set are said to be **equivalent**.
- A **matrix** is a rectangular array of numbers.
- There are three **elementary operations** performed on a system of equations (matrix). Interchange two equations (rows). Replace an equation (row) with a constant times the equation (row). Replace an equation (row) with the equation (row) plus a constant times another equation (row).
- Elementary operations yield an equivalent system of equations.
- The **Gauss–Jordan** method is a procedure for systematically using elementary operations to reduce a system of linear equations to an equivalent system whose solution set can be immediately determined.

- There are three possibilities in solving a system of linear equations: There can be precisely one solution. There can be no solution. There can be infinitely many solutions.

- A matrix is in **reduced form** if:
  1. The leftmost nonzero element in each row is 1.
  2. The column containing the leftmost 1 of any row has all other elements in that column equal to 0.
  3. The leftmost 1 in any row is to the left of the leftmost 1 in a lower row.
  4. Any row consisting of all zeros must be below any row with at least one nonzero element.

- The leftmost nonzero element in a nonzero row of a reduced matrix is called a **leading one**.

- The variable in the column of a reduced matrix containing the leading one is called a **leading variable**. The remaining variables are called **free variables**.

- A **parameter** is any real number. A free variable is set equal to a parameter and is thus free to be any real number.

- The **order** of a matrix A is denoted by $m \times n$ where m is the number of rows and n is the number of columns.

- The element in the ith row and jth column of a matrix A is denoted by a_{ij}.

- A **row matrix** is a matrix with one row. A **column matrix** is a matrix with one column. A **square matrix** is a matrix with equal numbers of rows and columns.

- Two matrices are **equal** if, and only if, they have the same order and all corresponding entries are equal.

- **Multiplying a matrix by a real number.** If c is a real number and A is a matrix, then the matrix cA is the matrix obtained by mutliplying every entry in A by c.

- **Adding and subtracting matrices.** Two matrices of the same order can be added (or subtracted) to obtain another matrix of the same order by adding (or subtracting) corresponding entries.

- The **zero matrix** of order $m \times n$ is the matrix with m rows and n columns, all of whose entries are zero.

- **Some matrix properties.** Let A, B, C, and O be four matrices, all of the same order with O the zero matrix. Then

$$A + O = O + A = A$$

$$A - A = O$$

$$A + B = B + A$$

$$A + (B + C) = (A + B) + C.$$

- **Multiplying a row matrix times a column matrix.** Let R be the row matrix $R = [r_1 \quad r_2 \quad \cdots \quad r_n]$ and let C be the column matrix

$$C = \begin{bmatrix} c_1 \\ c_2 \\ \vdots \\ c_n \end{bmatrix}$$

with the same dimension as R. Then the product RC is defined to be

$$RC = [r_1 \quad r_2 \quad \cdots \quad r_n] \begin{bmatrix} c_1 \\ c_2 \\ \cdot \\ \cdot \\ \cdot \\ c_n \end{bmatrix} = r_1c_1 + r_2c_2 + \cdots + r_nc_n.$$

- **Multiplying two matrices.** Given two matrices $A = (a_{ij})_{m \times p}$ and $B = (b_{ij})_{q \times n}$, the product $AB = C = (c_{ij})$ is defined if $p = q$, that is, if the number of columns of A equals the number of rows of B. In such a case, the element in the ith row and jth column, c_{ij}, of the product matrix C is obtained by multiplying the ith row of A with the jth column of B.

- **The order of a product.** Given the two matrices $A = (a_{ij})_{m \times p}$ and $B = (b_{ij})_{p \times n}$, the matrix AB has order $m \times n$.

- The $(n \times n)$ **identity matrix**, I_n, is the square matrix of order $(n \times n)$ with ones down the main diagonal and zeros elsewhere.

- **Properties of matrix multiplication.** For any matrices A, B, and C,

$$A(BC) = (AB)C$$

$$A(B + C) = AB + AC$$

$$AI_n = I_nA = A$$

whenever the indicated products are defined.

- Given a square $(n \times n)$ matrix A, the matrix A^{-1} is called the **inverse** of A if $AA^{-1} = A^{-1}A = I_n$.

- **To Find A^{-1}.** (1) Form the augmented matrix $[A|I]$. (2) Use row operations to reduce $[A|I]$ to $[I|B]$, if possible. (3) $A^{-1} = B$.

- The system of linear equations

$$a_{11}x_1 + a_{12}x_2 + \cdots + a_{1n}x_n = b_1$$

$$a_{21}x_1 + a_{22}x_2 + \cdots + a_{2n}x_n = b_2$$

$$\cdots$$

$$a_{n1}x_1 + a_{n2}x_2 + \cdots + a_{nn}x_n = b_n$$

can be put into the **matrix form** $AX = B$ by letting

$$A = \begin{bmatrix} a_{11} & a_{12} & \cdots & a_{1n} \\ a_{21} & a_{22} & \cdots & a_{2n} \\ & & \cdots & \\ a_{n1} & a_{n2} & \cdots & a_{nn} \end{bmatrix}, \quad X = \begin{bmatrix} x_1 \\ x_2 \\ \cdot \\ \cdot \\ \cdot \\ x_n \end{bmatrix}, \quad B = \begin{bmatrix} b_1 \\ b_2 \\ \cdot \\ \cdot \\ \cdot \\ b_n \end{bmatrix}.$$

The matrix A is called the **coefficient matrix**.

- **Solution of $AX = B$.** Let A be a square matrix of order n and B a column matrix of order n. If A^{-1} exists, then the linear system of equations $AX = B$ has the unique solution $X = A^{-1}B$.

- The **input–output** matrix $A = a_{ij}$ represents the dollar amount of the output of industry i required in producing one dollar of output in industry j.

- If x_1, x_2, and x_3 represent the respective outputs in dollars of a three-industry economy and d_1, d_2, and d_3 the respective nonindustry demand in dollars for the three products produced by the three industries, then we define the **production column matrix X** and the **demand column matrix D**, respectively, by

$$X = \begin{bmatrix} x_1 \\ x_2 \\ x_3 \end{bmatrix}, \quad D = \begin{bmatrix} d_1 \\ d_2 \\ d_3 \end{bmatrix}$$

- The **production matrix X** is given by $X = (I - A)^{-1}D$.
- In the **closed Leontief input–output model** no demand is assumed from outside the industries being considered. If f_{ij} is the fraction of the total output of the jth industry purchased by the ith industry, and p_i the price charged by the ith industry for its total output, then the matrix P must satisfy the equation $(I - F)P = 0$, where $F = (f_{ij})$ is the **price matrix** and

$$P = \begin{bmatrix} p_1 \\ p_2 \\ p_3 \end{bmatrix}$$

is the equilibrium price matrix.

Chapter 2 Review Exercises

In Exercises 1 through 8 find all solutions, if any exist, of the given systems using the Gauss–Jordan method.

1. $x + 3y = 7$
$3x + 4y = 11$

2. $x + 3y = 1$
$3x + 9y = 2$

3. $x + 3y = 1$
$3x + 9y = 3$

4. $2x + 3y = 18$
$3x - y = 5$

5. $x + 3y - z = 4$
$3x - y + z = 4$

6. $x + y = 10$
$2x + 2y + 2z = 20$

7. $x + 2y - 3z = 4$
$2x - y + z = 0$
$5x - 3y + 2z = -2$

8. $x + y + z = 5$
$2x + y - z = 3$
$y + 3z = 7$

In Exercises 9 through 20 let

$$A = \begin{bmatrix} 1 & 3 \\ 2 & 1 \\ -1 & 3 \end{bmatrix}, \quad B = \begin{bmatrix} 2 & 5 \\ 1 & 0 \\ 3 & -2 \end{bmatrix}, \quad C = [1 \quad 3], \quad D = \begin{bmatrix} 2 \\ 4 \end{bmatrix}$$

$$E = \begin{bmatrix} 2 & 0 & -1 \\ 3 & 1 & 2 \end{bmatrix}, \quad F = [2 \quad 4 \quad 1], \quad G = \begin{bmatrix} 1 \\ 3 \\ -2 \end{bmatrix}$$

and find the requested quantity if defined.

9. $3A$ **10.** $-2B$ **11.** $3A - 2B$

12. CD **13.** DG **14.** AC

15. EG **16.** AF **17.** AE

18. EA **19.** AB **20.** FG

21. A matrix $D = (d_{ij})$ is a diagonal matrix if $d_{ij} = 0$ if $i \neq j$, that is, all entries off the main diagonal are zero. Suppose $A = (a_{ij})$ and $B = (b_{ij})$ are both diagonal matrices of order (3×3). Show that (a) A^2 is diagonal, (b) AB is diagonal.

In Exercises 22 through 27 find the inverse of the given matrix, if it exists.

22. $\begin{bmatrix} 5 & 1 \\ 2 & 1 \end{bmatrix}$ **23.** $\begin{bmatrix} 4 & 2 \\ 2 & 1 \end{bmatrix}$ **24.** $\begin{bmatrix} 2 & -3 \\ -3 & 1 \end{bmatrix}$

25. $\begin{bmatrix} 1 & 3 & 0 \\ 2 & 5 & 1 \\ -1 & -2 & 2 \end{bmatrix}$ **26.** $\begin{bmatrix} 1 & 4 & 0 \\ 2 & 9 & 1 \\ 5 & 22 & 2 \end{bmatrix}$ **27.** $\begin{bmatrix} 1 & 1 & 1 \\ -1 & 2 & 2 \\ 2 & 1 & 2 \end{bmatrix}$

In Exercises 28 through 31 solve the systems using the inverses found in the previous exercises.

28. $\begin{aligned} 5x + y &= 1 \\ 2x + y &= -2 \end{aligned}$ **29.** $\begin{aligned} 2x - 3y &= 14 \\ -3x + y &= 35 \end{aligned}$

30. $\begin{aligned} x + 3y &= -1 \\ 2x + 5y + z &= 3 \\ -x - 2y + 2z &= 2 \end{aligned}$ **31.** $\begin{aligned} x + y + z &= 30 \\ -x + 2y + 2z &= -9 \\ 2x + y + 2z &= 0 \end{aligned}$

In Exercises 32 through 33 you are given an input–output matrix A and a demand matrix D. Assuming a Leontief open input–output model, find the production matrix.

32. $A = \begin{bmatrix} 0.2 & 0.2 \\ 0.7 & 0.3 \end{bmatrix}, \quad D = \begin{bmatrix} 10,000,000 \\ 20,000,000 \end{bmatrix}$

33. $A = \begin{bmatrix} 0 & 0.10 & 0.25 \\ 0.50 & 0 & 0.20 \\ 0.20 & 0.30 & 0 \end{bmatrix}, \quad D = \begin{bmatrix} 10,000,000 \\ 20,000,000 \\ 30,000,000 \end{bmatrix}$

In Exercises 34 through 35 you are given an input–output matrix F for a closed Leontief input–output model. Find the equilibrium price matrix.

34. $F = \begin{bmatrix} 0.2 & 0.3 \\ 0.8 & 0.7 \end{bmatrix}$ **35.** $F = \begin{bmatrix} 0.1 & 0.1 & 0.3 \\ 0.5 & 0 & 0.7 \\ 0.4 & 0.9 & 0 \end{bmatrix}$

36. A certain 9-hole golf course has the number of par 3 holes equal to one plus the number of par 5 holes. If par at this course for 9 holes is 35, how many par 3, 4, and 5 holes are there?

37. **Sales.**   A wholesaler receives an order for a total of a dozen standard and deluxe electric sweepers. The standard sweepers cost $200 and the deluxe $300. With the order is a check for $2900, but the order neglects to specify the number of each type of sweeper. Determine how to fill the order.

38. **Sales.**   A developer sells two sizes of condominiums. One sells for $75,000 and the other for $100,000. One year the developer sold 14 condominiums for a total of $1.2 million. How many of each size of condominiums did the developer sell?

39. **Sales.**   A small restaurant sold 50 specials one day. The ham special went for $10 and the beef special for $12. The specials brought in $560. The cook was too busy to keep track of how many of each were served. Find the answer for her.

40. **Mixture.**   Three acid solutions are available. The first has 20% acid, the second 30%, and the third 40%. The three solutions need to be mixed to form 300 liters of a solution with 32% acid. If the amount of the third solution must be equal to the sum of the amounts of the first two, find the amount of each solution.

41. **Distribution.**   A firm has 3 manufacturing plants located in different parts of the country. Three major wholesalers obtain the percentages of the product from the three plants according to the following table. Also included is the demand (in numbers) from each of the wholesalers. What should be the number produced by each of the plants to meet this demand with the given percentage distributions?

	Plant 1	Plant 2	Plant 3	Demand
Wholesaler A	20%	30%	50%	105
Wholesaler B	40%	20%	40%	100
Wholesaler C	10%	30%	60%	100

Linear programming is used to determine the most profitable type of product to be made by the refinery. (Alan Pitcairn/Grant Heilman.)

Linear Programming: The Graphical Method

This chapter introduces linear programming, one of the most important practical developments in economics and business and many other areas to appear since World War II. Linear programming was originally devised to help the U.S. Air Force optimize various goals in activities such as assignments, procurement, and maintenance. Linear programming is now used in a wide variety of practical situations. For example, AT&T uses linear programming to determine the most economical telephone links between cities. Oil refineries use linear programming to determine what crude oil to buy and what to produce with it. Airlines use linear programming to determine how to maximize profits associated with assigning aircrafts to their schedules. Many companies use linear programming to schedule transportation networks, to control inventories, to plan portfolios, and so on. According to some estimates, well over $100 million in human and computer time is spent yearly on the formulation and solution of linear programming problems.

This chapter will concentrate on translating the linear programming problems into mathematical models and also to finding solutions by geometric methods. Solving

problems geometrically will give considerable insight into how linear programming works but is only practical for very small-scale problems. In the next chapter, a practical technique is introduced that can solve large systems.

3.1   GRAPHING LINEAR INEQUALITIES

- ▶ *Introduction*
- ▶ *Graphing One Linear Inequality*
- ▶ *Graphing Systems of Linear Inequalities*

George Dantzig, 1914–

(Photo by Edward W. Souza News Service, Stanford University, Stanford, CA.)

George Dantzig is considered the father of linear programming. His major accomplishments in linear programming were made during and immediately after World War II while working for the Air Force. He is often cited as a prime example of what can be accomplished with positive thinking. As a graduate student, Dantzig arrived late for a class one day, copied down what he thought was a homework assignment, and solved a famous unsolved problem in statistics.

APPLICATION

Manufacture of Wooden Boats

A small company manufactures two types of wooden boats: dinghies and skiffs. The manufacture of each boat must go through three operations: cutting (of the wood), assembly, and painting. The dinghies require 2 hours of cutting, 4 hours of assembling, and 2 hours of painting. The skiffs require 4 hours of cutting, 2 hours of assembly, and 2 hours of painting. The total work-hours available per week in the cutting section is 80, in the assembly section is 84, and in the painting section is 50. Write a system of inequalities that expresses this information.

Introduction

The information given in the above problem can be summarized in the following table.

Operation	Dinghies	Skiffs	Hours Available
Cutting	2	4	80
Assembly	4	2	84
Painting	2	2	50

Let x be the number of dinghies manufactured each week and y be the number of skiffs manufactured each week. Each of the dinghies requires 2 hours of cutting and each of the skiffs requires 4 hours of cutting. Thus the cutting section will be occupied $2x$ hours with the dinghies and $4y$ hours with the skiffs, for a total of $2x + 4y$ hours. Since this must be at most 80, we have $2x + 4y \leq 80$. Since each of the dinghies requires 4 hours to assemble and each of the skiffs requires 2 hours, the assembly section will be occupied $4x + 2y$ hours and therefore $4x + 2y \leq 84$. Finally since each of the dinghies and skiffs requires 2 hours to paint, the painting section will be occupied $2x + 2y$ hours and therefore $2x + 2y \leq 50$. Naturally we must have $x \geq 0$ and $y \geq 0$, which we refer to as the **nonnegativity constraint**. We then have the system of inequalities

$$2x + 4y \leq 80 \qquad \text{cutting constraint}$$

$$4x + 2y \leq 84 \qquad \text{assembly constraint}$$

$$2x + 2y \leq 50 \qquad \text{painting constraint}$$

$$x, y \geq 0 \qquad \text{nonnegativity constraint.}$$

We would like to find the set of points (x, y) that satisfies all of these inequalities simultaneously. This will tell us the possible numbers of each type of boat that can be made, given the labor constraints on the three operations. We will do this by looking at one inequality at a time.

Graphing One Linear Inequality

We know that if a and b are not both zero, then the graph of the linear equation

$$ax + by = c$$

is a straight line in the xy-plane. We are now interested in the graph of linear inequalities such as

$$ax + by < c, \quad ax + by \leq c, \quad ax + by > c, \quad ax + by \geq c.$$

E X A M P L E   1   **Graphing Linear Inequalities**

Graph (a) $x + y = 4$ (b) $x + y > 4$ (c) $x + y < 4$.

Solutions

(a) This can be written as $y = -x + 4$; the graph is shown in Figure 3.1a as the line L.

(b) This can be written as $y > -x + 4$. Consider now any point (x_0, y_0) that lies on the straight line $y = -x + 4$ as shown in Figure 3.1b. Then $y_0 = -x_0 + 4$. Then, as shown in Figure 3.1b, if we take any point (x_0, y) with $y > y_0 = -x_0 + 4$, the point (x_0, y) must be *above* the line L. Since this must hold for any x_0, we conclude that the graph of $y > -x + 4$ or $x + y > 4$ is the half-plane *above* the line L. See Figure 3.2a, where the line L has been drawn using a dotted line to indicate that the line is *not* on the graph.

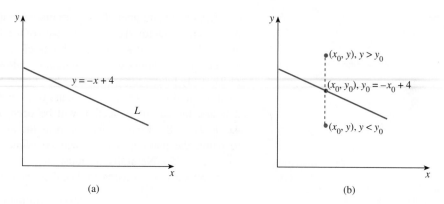

Figure 3.1

(c)   As Figure 3.1b indicates, if $y < y_0 = -x_0 + 4$, the point (x_0, y) must lie *below* the line. We conclude that the graph of $x + y < 4$ is the half-plane *below* the line L. See Figure 3.2b.   ∎

There is a simpler way to graph the linear inequalities. Since we know that the graph of $x + y < 4$ must be a half-plane, we only need to decide *which* half-plane. A practical way of deciding is to choose a convenient test point (x_0, y_0) and test to see if this point satisfies the inequality. If it does, then the half-plane containing this point must be the plane sought. If this point does not satisfy the inequality, then the half-plane containing this point is *not* the plane being sought. Thus the plane being sought is the other half-plane. Always pick the point $(0, 0)$ as a test point, if possible, since this is the easiest point to evaluate.

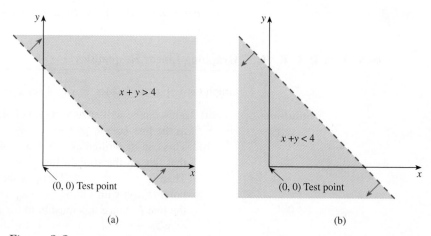

Figure 3.2

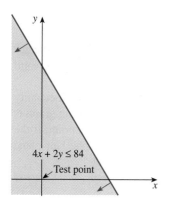

Figure 3.3

For example, in order to decide in which half-plane the graph of $x + y < 4$ lies, pick $(0, 0)$ as a test point. Since

$$0 + 0 = 0 < 4,$$

$(0, 0)$ satisfies the inequality, and thus the graph of the inequality is the half-plane containing $(0, 0)$. As Figure 3.2b indicates, $(0, 0)$ is below the line L, and thus the graph is the half-plane below the line L.

In order to decide in which half-plane the graph of $x + y > 4$ lies, pick $(0, 0)$ as a test point. Since

$$0 + 0 = 0 \not> 4,$$

$(0, 0)$ does not satisfy the inequality, and thus the graph of the inequality is the half-plane not containing $(0, 0)$. Since $(0, 0)$ is below the line L, the graph is the half-plane above the line L. See Figure 3.2a.

E X A M P L E   2   **Graphing a Linear Inequality**

Graph $4x + 2y \leq 84$, which is the assembly constraint given earlier.

Solution   We first graph $4x + 2y = 84$. This graph is the line L shown in Figure 3.3. The graph of the linear inequality $4x + 2y \leq 84$ is either the half-plane above and including L, or the half-plane below and including L. We select the test point $(0, 0)$ and evaluate the inequality at this point and obtain

$$4(0) + 2(0) = 0 \leq 84.$$

Since $(0, 0)$ does satisfy the inequality and is *below* the line L, the graph we are seeking is the half-plane below L and including L. This is indicated in Figure 3.3 using a solid line.   ■

E X A M P L E   3   **Graphing a Linear Inequality**

Graph $x < -2$.

Solution   Any point (x, y) that satisfies $x < -2$ must be to the left of the line $x = -2$, as shown in Figure 3.4. Alternately, one can pick the test point point $(0, 0)$ that is to the right of the line $x = -2$ and notice that $0 \not< -2$. This implies that $(0, 0)$ is *not* in the half-plane we are seeking. Thus the half-plane we are seeking is to the left of $x = -2$.   ■

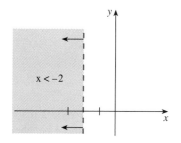

Figure 3.4

Graphing a Linear Inequality

Following are the steps required to graph the linear inequality

$$ax + by < c \text{ (or } \leq c \text{ or } > c \text{ or } \geq c).$$

1. First graph the straight line L given by $ax + by = c$, using a dotted line if the inequalities "<" or ">" are given and a solid line if the inequalities "≤" or "≥" are given. The dotted line indicates the straight line L is not part of the solution graph, while the solid line indicates that the line L is part of the solution graph.

2. Pick any convenient test point not on L and evaluate the inequality at that test point. If the inequality is true, then the solution graph is the half-plane on the same side of the line L as the test point. If the inequality is not satisfied, then the solution graph is the half-plane on that side of L that does not include the test point.

Graphing Systems of Inequalities

If we are given *several* linear inequalities, as in the wooden boat example at the beginning of this section, then we find the solution set of each of the linear inequalities individually just as previously indicated. The solution set of the system is then the region common to all the solution sets of the individual inequalities. This common region is also called the **feasible region**. In other words, any point (x, y) that satisfies all of the inequalities is in the feasible region. The feasible region is the set of all such points.

E X A M P L E   4   **Finding the Feasible Region**

A shop manufactures two styles of iron park benches. The first style requires 4 pounds of iron and 6 hours of labor while the second style requires 4 pounds of iron and 3 hours of labor. If there are 32 pounds of iron and 36 hours of labor available for the day, write a system of inequalities that expresses this information and find the feasible region.

Solution   The following table summarizes the information.

Iron Park Benches

	Style 1	*Style 2*	*Available*
Pounds of Iron	4	4	32
Hours of Labor	6	3	36

Let x be the number of style 1 benches made and y the number of style 2 made. Then $4x + 4y$ is the number of pounds of iron needed to make these benches, while the total amount available is 32 pounds. Thus $4x + 4y \le 32$. Also $6x + 3y$ is the number of hours needed to make these benches, while the total hours

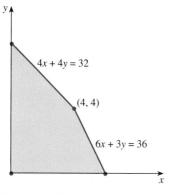

Figure 3.5

available is 36. Therefore, $6x + 3y \leq 36$. Naturally, x and y must be nonnegative. We then have

$$4x + 4y \leq 32$$

$$6x + 3y \leq 36$$

$$x, y \geq 0.$$

To determine the feasible region, first notice that since x and y are both nonnegative, the feasible region must be in the first quadrant. Now take $(0, 0)$ as a test point. Since $4(0) + 4(0) \leq 32$ and $6(0) + 3(0) \leq 36$, the point $(0, 0)$ satisfies both inequalities and thus lies below the line $4x + 4y = 32$ and also below the line $6x + 3y = 36$. Therefore the solution set for the first inequality is the half-plane below and including the line $4x + 4y = 32$, and the solution set for the second inequality is the half-plane below and including the line $6x + 3y = 36$. This is indicated in Figure 3.5. The solution set for the two inequalities is the shaded region, which is the set of points (x, y) that satisfies *both* inequalities. The corner point $(4, 4)$ is determined by solving the system

$$4x + 4y = 32$$

$$6x + 3y = 36. \quad \blacksquare$$

E X A M P L E 5   **Finding the Feasible Region**

Find the feasible region for the system of inequalities that arose in the wooden boat manufacturing problem discussed at the beginning of this section.

Solution   We have the inequalities

$$2x + 4y \leq 80 \qquad \text{cutting constraint}$$

$$4x + 2y \leq 84 \qquad \text{assembly constraint}$$

$$2x + 2y \leq 50 \qquad \text{painting constraint}$$

$$x, y \geq 0 \qquad \text{nonnegativity constraint.}$$

Since the point $(0, 0)$ satisfies each of the first 3 inequalities and lies below the three lines $2x + 4y = 80$, $4x + 2y = 84$, and $2x + 2y = 50$, the solution set for each of these inequalities is the half-plane below and including each of the lines $2x + 4y = 80$, $4x + 2y = 84$, and $2x + 2y = 50$. This is indicated in Figure 3.6. The two inequalities $x \geq 0$ and $y \geq 0$ indicate that the solution set is in the first quadrant. The corner point $(10, 15)$ shown in Figure 3.6 can be found by solving the system

$$2x + 4y = 80$$

$$2x + 2y = 50,$$

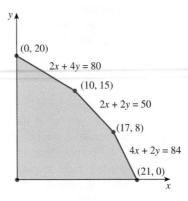

Figure 3.6

while the corner point (17, 8) shown in Figure 3.6 can be found by solving the system

$$4x + 2y = 84$$

$$2x + 2y = 50.$$

Finding the y-intercept for $2x + 4y = 80$ leads to the corner point (0, 20), while finding the x-intercept of $4x + 2y = 84$ leads to the corner (21, 0). ■

E X A M P L E 6   **Finding the Feasible Region**

A farmer can buy two types of 100-pound bags of fertilizer, type A and type B. Each 100-pound bag of type A fertilizer contains 50 pounds of nitrogen, 30 pounds of phosphoric acid, and 20 pounds of potash, whereas each 100-pound bag of type B fertilizer contains 10 pounds of nitrogen, 30 pounds of phosphoric acid, and 60 pounds of potash. The farmer requires at least 1000 pounds of nitrogen, 1800 pounds of phosphoric acid, and 2800 pounds of potash. Write a system of inequalities that expresses the given information and graph the feasible region.

Solution   We organize the information in the following table.

	Pounds Per Bag		
	Type A	Type B	Pounds Needed
Number of Bags	x	y	
Nitrogen	50	10	1000
Phosphoric acid	30	30	1800
Potash	20	60	2800

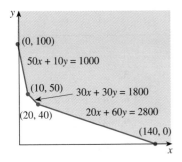

Figure 3.7

There are x bags of type A each containing 50 pounds of nitrogen and y bags of type B each containing 10 pounds of nitrogen for a total of $50x + 10y$ pounds. Since this must be at least 1000 pounds, we have $50x + 10y \geq 1000$. In a similar fashion we obtain the other two constraints together with the nonnegativity constraints $x \geq 0$ and $y \geq 0$ and have

$$50x + 10y \geq 1000 \quad \text{nitrogen constraint}$$
$$30x + 30y \geq 1800 \quad \text{phosphoric acid constraint}$$
$$20x + 60y \geq 2800 \quad \text{potash constraint}$$
$$x, y \geq \quad 0 \quad \text{nonnegativity constraint.}$$

These are then graphed in Figure 3.7. The corner point (10, 50) is found by solving the system

$$50x + 10y = 1000$$
$$30x + 30y = 1800,$$

while the corner point (20, 40) is found by solving the system

$$30x + 30y = 1800$$
$$20x + 60y = 2800.$$

The corner point (0, 100) is obtained by finding the y-intercept of $50x + 10y = 1000$, while the corner point (140, 0) is obtained by finding the x-intercept of $20x + 60y = 2800$. ■

SELF-HELP EXERCISE SET 3.1

1. Graph the feasible region.

$$3x + y \geq 5$$
$$2x + 3y \geq 8$$
$$x \quad\quad \geq 0$$
$$y \quad \geq 0$$

2. Two types of metal connectors require for their manufacture spending a certain number of minutes on a stamping machine and also on a grinding machine. The first type of connector requires 4 minutes on the stamping machine and 3 minutes on the grinding machine, while the second connector requires 5 minutes on the stamping machine and 1 minute on the grinding machine. The stamping machine is available, at most, 50 minutes each hour, and the grinding machine is available, at most, 21 minutes each hour. Furthermore, each hour a total of at least two of the connectors must be produced. Write a system of inequalities that expresses this information and find the feasible region.

EXERCISE SET 3.1

In Exercises 1 through 24 graph the solution set.

1. $x + y < 5$

2. $2x + y < 4$

3. $x + 2y \leq 1$

4. $3x + 2y \leq 6$

5. $2x + 3y \geq 12$

6. $4x + 5y \geq 20$

7. $-2x + 3y > 5$

8. $x - 2y > 8$

9. $3x - 2y < 6$

10. $-x + 2y < 8$

11. $-6x + 2y \leq 12$

12. $6x - 3y \leq 18$

13. $2x - 3y \geq -6$

14. $2x - 3y \geq -12$

15. $x > y$

16. $2x > y + 1$

17. $y \leq 3$

18. $y > 4$

19. $x > 2$

20. $x \geq 3$

21. $y \geq -2$

22. $x < -1$

23. $x \leq 0$

24. $y < 0$

In Exercises 25 through 42 graph the feasible region.

25. $x + y \leq 4$
$-x + y \leq 2$

26. $x + y \geq 4$
$-x + y \geq 2$

27. $x + y < 4$
$-x + y > 2$

28. $x + y > 4$
$-x + y < 2$

29. $x + y \leq 4$
$-x + y \geq 2$
$x, y \geq 0$

30. $x + y \leq 4$
$-x + y \geq 2$
$y \geq 0$

31. $x + y \leq 10$
$x \geq 2$
$y \geq 3$

32. $x + y \geq 10$
$y \leq 7$
$x \leq 8$

33. $x + y \geq 4$
$-x + y \geq 2$
$x \geq 0$

34. $x + y \leq 10$
$x + y \geq 4$
$x \geq 0$
$y \geq 0$

35. $-x + y \leq 4$
$x + y \geq 2$
$x \geq 0$
$y \geq 0$

36. $y \leq x$
$y \geq x - 2$
$y \leq 4$
$y \geq 0$

37. $-x + y \leq 1$
$x + y \geq 1$
$x + y \leq 4$
$y \geq 0$

38. $x + y \leq 4$
$x + y \geq 1$
$-x + y \geq -1$

39. $y \leq x$
$x + y \leq 2$

40. $x + y \leq 2$
$y \geq x$

41. $-x + y \geq 0$
$x \geq 1$
$y \leq 4$

42. $-x + y \leq 0$
$x \leq 4$
$y \geq 1$

Applications

In Exercises 43 through 50 write a system of inequalities that expresses the given information. Graph the feasible region and label all corner points.

43. **Shipping.** A manufacturer supplies its product to two distributors. Distributor A needs at least 100 of the product next month while distributor B needs at least 50. At most 200 of the product can be manufactured and delivered to the distributors.

44. **Transportation.** A plane delivers two types of cargo between two destinations. Each crate of cargo I is 5 cubic feet in volume and 100 pounds in weight. Each crate of cargo II is 5 cubic feet in volume and 200 pounds in weight. The plane has available at most 400 cubic feet and 10,000 pounds for the crates. Finally, at least twice the number of crates of I than II must be shipped.

(© Alvis Upitis/The Image Bank.)

45. **Manufacturing.** Two products, A and B, require for their manufacture spending a certain amount of hours on a stamping machine, a grinding machine, and a polishing machine. Product A requires 1 hour on the stamping machine, 2 hours on the grinding machine, and 2 hours on the polishing machine. Product B requires 2 hours on the stamping machine, 2 hours on the grinding machine, and 1 hour on the polishing machine. The stamping machine can operate a maximum of 20 hours, the grinding machine a maximum of 22 hours, and the polishing machine a maximum of 15 hours each day.

46. **Diet.** An individual needs a daily supplement of at least 500 units of vitamin C and 200 units of vitamin E and agrees to obtain this supplement by eating two foods, I and II. Each ounce of food I contains 40 units of vitamin C and 10 units of vitamin E, while each ounce of food II contains 20 units of vitamin C and also 20 units of vitamin E. The total supplement of these two foods must be at most 30 ounces.

47. **Manufacturing.** A firm manufactures tables and desks. To produce each table requires 1 hour of labor, 10 square feet of wood, and 2 quarts of finish. To produce each desk requires 3 hours of labor, 20 square feet of wood, and 1 quart of finish. Available is at most 45 hours of labor, at most 350 square feet of wood, and at most 55 quarts of finish.

48. **Mixture.** A dealer has 7600 pounds of peanuts, 5800 pounds of almonds, and 3000 pounds of cashews to be used to make two mixtures. The first mixture consists of 60% peanuts, 30% almonds, and 10% cashews. The second mixture consists of 20% peanuts, 50% almonds, and 30% cashews.

49. **Fishery.** A certain lake has smallmouth and largemouth bass and also three types of food for these fish, I, II, and III. Each month the lake can supply 800 pounds of food I, 500 pounds of food II, and 700 pounds of food III. Each month each smallmouth bass requires 1 pound of food I, 1 pound of food II, and 2 pounds of food III, and each largemouth bass requires 4 pounds of food I, 2 pounds of food II, and 1 pound of food III.

(© Grant Heilman Photography.)

50. **Scheduling.** A firm has two plants, A and B, that manufacture three products, P_1, P_2, and P_3. In one day Plant A can manufacture 30 of P_1, 30 of P_2, and 10 of P_3, whereas Plant B can manufacture 10 of P_1, 20 of P_2, and 30 of P_3. At least 240 of the first product, 390 of the second, and 410 of the third are needed.

Solutions to Self-Help Exercise Set 3.1

1. Since the point $(0, 0)$ does not satisfy either of the first 2 inequalities and lies below the 2 lines $3x + y = 5$ and $2x + 3y = 8$, the solution set for the first two inequalities is the half-plane above and including the lines $3x + y = 5$ and $2x + 3y = 8$. This is indicated in the figure. The two inequalities $x \geq 0$ and $y \geq 0$ indicate that the solution set is in the first quadrant. The feasible region is then the shaded region in the following figure.

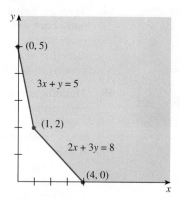

2. Most of the information can be summarized in the following table.

	First Connector	Second Connector	Minutes Available
Stamping Machine	4	5	50
Grinding Machine	3	1	21

Let x be the number of the first type of connector manufactured each hour and y be the number of the second type manufactured each hour. Each of the first type of connector requires 4 minutes on the stamping machine, and each of the second type of connector requires 5 minutes on this machine. Thus the stamping machine will be occupied $4x$ minutes with the first type of connector and $5y$ minutes with the second type of connector, for a total of $4x + 5y$ minutes. Since this must be at most 50, we have $4x + 5y \leq 50$.

Each of the first type of connector requires 3 minutes on the grinding machine, and each of the second type of connector requires 1 minute on this machine. Thus the grinding machine will be occupied $3x + y$ minutes. Thus $3x + y \leq 21$. Since at least 2 connectors must be produced, we have $x + y \geq 2$. Naturally, we must have $x \geq 0$ and $y \geq 0$. Thus we have the five inequalities

$$4x + 5y \leq 50$$

$$3x + y \leq 21$$

$$x + y \geq 2$$

$$x, y \geq 0.$$

Since $(0, 0)$ satisfies the first 2 inequalities and not the third and lies below each of the lines $4x + 5y = 50$, $3x + y = 21$, and $x + y = 2$, the solution set for the first two inequalities is the half-plane below and including the lines $4x + 5y = 50$ and $3x + y = 21$ and the half-plane above and including the line $x + y = 2$. This is indicated in the figure. The two inequalities $x \geq 0$ and $y \geq 0$ indicate that the solution set is in the first quadrant. The feasible region is then the shaded region in the following figure.

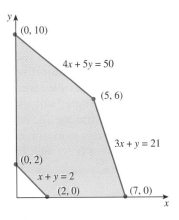

3.2   GRAPHICAL SOLUTION OF LINEAR PROGRAMMING PROBLEMS

▶ *Some Linear Programming Problems*
▶ *Problems with No Solution*
▶ *The Geometric Method*

George Dantzig   ❝*The final test of a theory is its capacity to solve the problems which originated it.*❞

APPLICATION

Maximizing Profit

Suppose the small, wooden-boat manufacturer of the last section obtained a profit of $60 for every dinghy manufactured and $80 for every skiff. How many of each type of boat should be made in order to maximize the profits? The answer is found in Example 1.

Some Linear Programming Problems

Business, economics, finance, and many other areas are replete with complex decision or allocation problems that involve the selection of specific values of many interrelated variables. Often one can focus attention on one specific objective that is to be maximized or minimized. Examples of such objectives are profit, revenue, cost, expected return on an investment, or perhaps some measure of social welfare for a government program. The variables are normally subject to constraints that take the form of a set of inequalities like the ones encountered in the previous section. These inequalities limit the possible selection of the values of the variables under consideration. In this section we will use linear programming to find the maximum or minimum of the objective function. We use the geometric approach on small-scale problems. The full power of linear programming will be taken up in the next section.

In this section we will wish to optimize (maximize or minimize) a *linear* function of two variables $z = Ax + By + C$. The function to be optimized is called the **objective function**. The variables will satisfy a system of *linear* inequalities called **constraints** that will give rise to a feasible region. Two of these inequalities will always be $x \geq 0$ and $y \geq 0$ and are called the **nonnegativity constraints**. The other constraints are called **problem constraints**. We will then attempt to maximize or minimize the function z subject to the constraint that the points (x, y) must lie in the feasible region. The following is a simple example.

E X A M P L E   1   **Geometric Method in Linear Programming**

Suppose the small wooden-boat manufacturer considered in the last section obtained a profit of $60 for every dinghy manufactured and $80 for every skiff. How many of each type of boat should be made in order to maximize the profits?

Solution   If x is the number of dinghies manufactured each week and y the number of skiffs, then the profit in dollars is $P = 60x + 80y$. Recalling the constraints that we found in the last section, we have the following linear programming problem:

Maximize	$z = 60x + 80y$	objective function
Subject to	$2x + 4y \leq 80$	cutting constraint
	$2x + 2y \leq 50$	painting constraint
	$4x + 2y \leq 84$	assembly constraint
	$x, y \geq 0$	nonnegativity constraint.

In this section we take the geometric approach to solving such problems. We first need to graph the feasible region. This was done in the previous section and is shown in Figure 3.8, where we note that the five corners have coordinates $A = (10, 15)$, $B = (17, 8)$, $C = (21, 0)$, $D = (0, 0)$, and $E = (0, 20)$. Only points (x, y) in the feasible region are feasible solutions.

First draw the straight lines $P = 60x + 80y = c$, called **isoprofit lines**, for a variety of values of c. For example, set P equal to, say 1600. The equation $60x + 80y = 1600$ is a straight line with slope $-3/4$ and is shown in Figure 3.9. Since $P = 60x + 80y$, it is clear that we can take both x and y a little larger and *still remain in the feasible region* and obtain a larger profit. If we set $P = 60x$

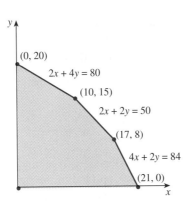

Figure 3.8

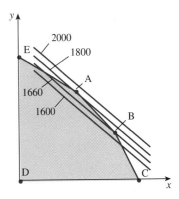

Figure 3.9

$+ 80y = 1660$, we obtain the straight line shown in Figure 3.9. (Notice the slope is also $-3/4$.) Any point along this latter straight line gives a profit of 1660, with some points in the feasible region. But we see that we can do better. Take $P = 60x + 80y = 1800$. This line also has slope $-3/4$. Notice that this line goes through the corner A, since $A = (10, 15)$ and at this point $P = 60(10) + 80(15) = 1800$. Thus only at the point A is the profit 1800 and also in the feasible region. If we take $P = 2000$, we see from Figure 3.9 that we obtain a straight line (parallel to the others) all of whose points are outside the feasible region. It thus appears that the profit is maximized at point A.

The lines for $P = 1600$, $P = 1660$, $P = 1800$, and $P = 2000$ are, as we noted, parallel. This can also be seen by setting $P = P_0$ and noting that

$$60x + 80y = P_0$$

$$y = -\frac{3}{4}x + \frac{1}{80}P_0.$$

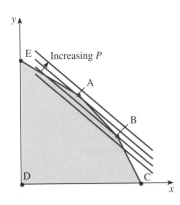

Figure 3.10

Furthermore we notice that by increasing P_0 we increase the y-intercept, and thus the line moves up (and to the right). See Figure 3.10. We then see that our strategy is to increase P_0 as much as possible while still keeping the constant profit line $P = P_0$ in contact with the feasible region. We saw that $P_0 = 1800$ meets this requirement and that the point in the feasible region at which the maximum occurs is the corner point A. (As we shall see, the optimum must always occur at a corner.)

Thus the solution to the problem is to manufacture 10 dinghies and 15 skiffs, and the weekly profit will be $1800. ∎

This problem has exactly one point at which the maximum occurs. Although this is typical, the maximum can occur at more than one point. The following is an example.

E X A M P L E   2   **Geometric Method in Linear Programming**

Redo the previous problem if the profit is given as $P = 70x + 70y$.

Solution   The feasible region is the same as in the previous example. If we set $P = 70x + 70y = P_0$, we obtain

$$y = -x + \frac{1}{70} P_0.$$

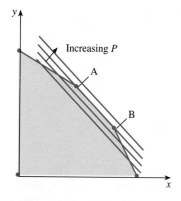

Figure 3.11

All of these lines have slope -1. Again, increasing P_0 increases the y-intercept, and thus the lines move up (and to the right), as indicated in Figure 3.11. But since the slopes of these lines are the same as the slope of the line from A to B, the maximum occurs at all the points on the line segment from A to B. But again the maximum does occur at a corner—in fact, at two adjacent corners. Thus two acceptable answers are $(10, 15)$ or $(17, 8)$, which correspond to manufacturing 10 dinghies and 15 skiffs or 17 dinghies and 8 skiffs with profits of

$$P = 70(10) + 70(15) = 1750$$

dollars per week. Other acceptable answers with the same profit are the points $(11, 14)$, $(12, 13)$, $(13, 12)$, $(14, 11)$, $(15, 10)$, and $(16, 9)$, all of which lie on the line segment between A and B and all of which will yield the same profit of \$1750. ■

We now give an example of a linear programming problem involving minimizing the objective function.

E X A M P L E   3   **A Minimizing Problem**

Solve the following linear programming problem.

$$\begin{array}{lll} \text{Minimize} & z = 3x + 2y & \text{objective function} \\ \text{Subject to} & \left.\begin{array}{r} 3x + y \geq 5 \\ x + y \geq 3 \\ x + 4y \geq 6 \end{array}\right\} & \text{problem constraints} \\ & x, y \geq 0 & \text{nonnegativity constraint} \end{array}$$

Solution   The feasible region is shown in Figure 3.12. If we set $z = z_0$, we obtain

$$3x + 2y = z_0$$

$$y = -\frac{3}{2} x + \frac{1}{2} z_0,$$

which is a straight line with slope $-3/2$. We see that decreasing z_0 brings the line $z = z_0$ closer to the origin. Furthermore, since the slope of the line segment from B to C is -1, while the slope of the line segment from A to B is -3, we

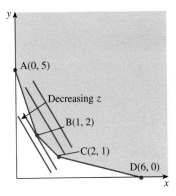

Figure 3.12

see, according to Figure 3.12, that the minimum occurs at the corner point B. At this point, $z = 3(1) + 2(2) = 7$. ∎

Problems With No Solution

We have now seen two linear programming problems with exactly one solution and one with many solutions. We now encounter two linear programming problems with no solution.

E X A M P L E 4 **Problem With No Solution**

Solve the following linear programming problem.

$$\text{Maximize} \quad P = x + y$$
$$\text{Subject to} \quad x, y \geq 0$$

Solution The feasible region is the first quadrant shown in Figure 3.13. It is clear that this problem has *no solution* since one can attain any profit $P = x + y$, no matter how large, simply by taking x and y sufficiently large. ∎

E X A M P L E 5 **Problem With No Solution**

Solve the following linear programming problem.

$$\text{Maximize} \quad z = 3x + 2y \qquad \text{objective function}$$
$$\text{Subject to} \quad x + y \leq 2 \qquad \text{problem constraint}$$
$$-x + y \geq 3 \qquad \text{problem constraint}$$
$$x, y \geq 0 \qquad \text{nonnegativity constraint}$$

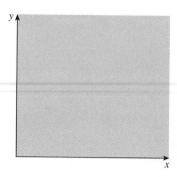

Figure 3.13

Solution   We see in Figure 3.14 that the feasible region is empty. Thus there can be no optimal solution. ∎

The Geometric Method

In all of the previous examples, the objective function was optimized at a corner point. Careful examination of other typical examples in Figure 3.15 indicates that this must be true in general. We then have the following theorem.

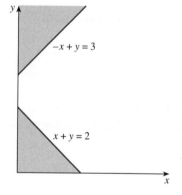

Figure 3.14

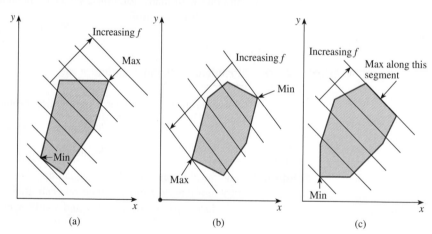

Figure 3.15

Optimal Solution of a Linear Programming Problem

If a linear programming problem has an optimal solution, then the optimum must occur at a corner of the feasible region. If the objective function is optimized at two corners of the feasible region, then any point on the line segment connecting these two corners also optimizes the objective function.

We have already seen that an objective function need not have an optimal value. We now list two important cases for which an optimal value exists.

Existence Theorem for Optimal Solutions

1. If the feasible region is nonempty, bounded, and includes its boundary, then the objective function has both a minimum and a maximum in the feasible region.
2. If the feasible region S includes its boundary, is unbounded, and the objective function $z = ax + by$ has the property that $a > 0$ and $b > 0$, then z attains a minimum value in S but has no maximum value in S.

The fact that an optimal solution, if it exists, must be at a corner gives rise to the following method of finding the optimal solution to a linear programming problem.

Method of Corners

Determine if a solution exists using the existence theorem for optimal solutions. If a solution exists then:

1. Graph the feasible region.
2. Locate all corners of the feasible region.
3. Evaluate the objective function at each of the corners.
4. The optimal solution will be the optimal value found in the previous step.

E X A M P L E   6   **The Method of Corners**

Find both the minimum and the maximum of $z = 5x + 4y$ subject to

$$-x + y \le 4$$
$$x + y \le 8$$
$$x + y \ge 4$$
$$x \le 4$$
$$x, y \ge 0.$$

Solution   The feasible region is shown in Figure 3.16. Since the feasible region is nonempty, bounded, and includes its boundary, both the maximum and the minimum exist and can be found at a corner point. There are four corners, $A = (0, 4)$, $B = (2, 6)$, $C = (4, 4)$, and $D = (4, 0)$. Evaluating z at these four corners yields

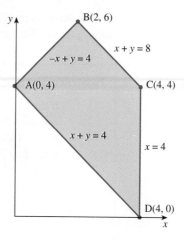

Figure 3.16

Corner	$z = 5x + 4y$
$(0, 4)$	$5(0) + 4(4) = 16 \leftarrow$ minimum
$(2, 6)$	$5(2) + 4(6) = 34$
$(4, 4)$	$5(4) + 4(4) = 36 \leftarrow$ maximum
$(4, 0)$	$5(4) + 4(0) = 20.$

Thus the maximum is $z = 36$ at the corner $C = (4, 4)$, while the minimum is $z = 16$ at the corner $A = (0, 4)$. ∎

E X A M P L E 7   **The Method of Corners**

A farmer can buy two types of 100-pound bags of fertilizer, type A and type B. Each 100-pound bag of type A fertilizer costs \$20 and contains 50 pounds of nitrogen, 30 pounds of phosphoric acid, and 20 pounds of potash, whereas each 100-pound bag of type B fertilizer costs \$30 and contains 10 pounds of nitrogen, 30 pounds of phosophoric acid, and 60 pounds of potash. The farmer requires at least 1000 pounds of nitrogen, 1800 pounds of phosphoric acid, and 2800 pounds of potash. How many bags of each type of fertilizer should be bought in order to minimize cost?

Solution   Let x be the number of bags of type A fertilizer and y the number of type B. Then we are to minimize the cost: $C = 20x + 30y$. The same constraints were given in Example 6 of the last section. Recalling our work there, we can write the linear programming problem as

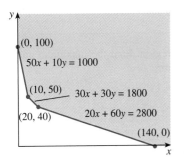

Figure 3.17

Minimize	$z = 20x + 30y$	objective function
Subject to	$50x + 10y \geq 1000$	nitrogen constraint
	$30x + 30y \geq 1800$	phosphoric acid constraint
	$20x + 60y \geq 2800$	potash constraint
	$x, y \geq \quad 0$	nonnegativity constraint.

Figure 3.17 shows the feasible region found in the last section. The region is unbounded, and the objective function $z = 20x + 30y$ has positive coefficients of x and y. Thus a minimum will exist that must be at a corner of the feasible region. We then have

Corner	$z = 20x + 30y$
(0, 100)	$20(0) + 30(100) = 3000$
(10, 50)	$20(10) + 30(50) = 1700$
(20, 40)	$20(20) + 30(40) = 1600 \leftarrow$ minimum
(140, 0)	$20(140) + 30(0) = 2800$

Thus the minimum cost is $1600, requiring purchasing 20 bags of type A and 40 bags of type B. ∎

SELF-HELP EXERCISE SET 3.2

1. Solve the following.

$$\begin{aligned} \text{Minimize} \quad & z = 4x + 6y \\ \text{Subject to} \quad & 7x + 2y \geq 20 \\ & x + 2y \geq 8 \\ & x, y \geq 0 \end{aligned}$$

2. Two types of metal connectors require for their manufacture spending a certain number of minutes on a stamping machine and on a grinding machine. The first type of connector requires 4 minutes on the stamping machine and 3 minutes on the grinding machine, while the second connector requires 5 minutes on the stamping machine and 1 minute on the grinding machine. The stamping machine is available at most 50 minutes each hour, and the grinding machine is available at most 21 minutes each hour. Furthermore, each hour a total of at least two of the connectors must be produced. If each of the first type of connector brings a profit of $0.50 and each of the second type of connector a profit of $0.60, find the number of each of these products that should be manufactured to maximize profits. (See Self-Help Exercise 2 in the last section.)

3. In the wooden-boat manufacturing problem of Example 1, suppose the profit is given by $P = ax + y$, with $a > 0$. Find all values of a so that the profit is maximized at (a) (17, 8) (b) (21, 0).

EXERCISE SET 3.2

In Exercises 1 through 6 you are given the feasible region. Find the maximum and minimum (if they exist) for the given objective function.

1. $z = x + 2y$

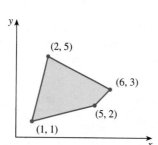

2. $z = 2x - y$

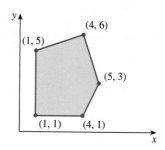

3. $z = 2x + 3y$

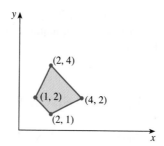

4. $z = 10x + 20y$

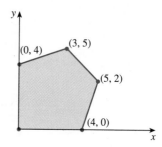

5. $z = 3x + y$

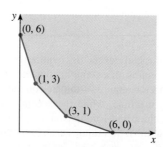

6. $z = x + y$

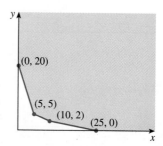

7.   Maximize $z = 20x + 30y$

     Subject to $x + 2y \le 90$

                  $x + y \le 50$

                       $x \le 30$

                 $x, y \ge 0$

8.   Maximize $z = 4x + y$

     Subject to $x + y \le 100$

                  $3x + y \le 120$

                       $y \ge 50$

                    $x, y \ge 0$

9. Maximize $z = 2x + 3y$

   Subject to $2x + y \le 100$

   $\qquad\qquad\quad y \le 40$

   $\qquad\qquad\quad x \ge 10$

   $\qquad\qquad\quad y \ge 20$

10. Maximize $z = 2x - 3y$

   Subject to $\qquad y \le x$

   $\qquad\qquad\quad y \le 30$

   $\qquad\qquad\quad y \ge 10$

   $\qquad\qquad\quad x \le 40$

   $\qquad\qquad\quad x \ge 0$

11. Minimize $z = x + y$

   Subject to $3x + y \ge 10$

   $\qquad\quad 3x + 2y \ge 17$

   $\qquad\qquad x + 2y \ge 11$

   $\qquad\qquad\quad x, y \ge 0$

12. Minimize $z = x + 3y$

   Subject to $2x + y \ge 20$

   $\qquad\qquad x + y \ge 18$

   $\qquad\qquad x + 3y \ge 41$

   $\qquad\qquad\quad x, y \ge 0$

13. Minimize $z = 2x + 3y$

   Subject to $x + y \le 10$

   $\qquad\qquad\qquad x \ge 1$

   $\qquad\qquad\qquad x \le 3$

   $\qquad\qquad\qquad y \ge 0$

14. Minimize $z = 4x + 5y$

   Subject to $\quad x + y \ge 2$

   $\qquad\qquad -x + y \le 2$

   $\qquad\qquad\quad x + y \le 6$

   $\qquad\qquad -x + y \ge -2$

15. Maximize $z = x + 10y$

   Subject to $-x + y \le 0$

   $\qquad\qquad -x + y \ge -4$

   $\qquad\qquad\qquad y \ge 1$

   $\qquad\qquad\qquad y \le 4$

16. Maximize $z = 5x + 4y$

   Subject to $-2x + y \le 0$

   $\qquad\qquad -x + y \le 0$

   $\qquad\qquad\quad x + y \le 6$

17. Minimize $z = 4x + 5y$

   Subject to $\qquad x + y \ge 5$

   $\qquad\qquad -x + y \le 1$

   $\qquad\qquad -2x + y \le 2$

   $\qquad\qquad\qquad y \ge 0$

18. Minimize $z = x + y$

   Subject to $2x + y \ge 9$

   $\qquad\qquad x + 2y \ge 9$

   $\qquad\qquad\qquad x \le 7$

   $\qquad\qquad\qquad y \le 7$

19. Maximize $z = 2x + 4y$

   Subject to $\quad x - y \le 0$

   $\qquad\qquad\quad x, y \ge 0$

20. Maximize $z = 3x + 2y$

   Subject to $\quad x - y \le 0$

   $\qquad\qquad 2x - y \ge 0$

   $\qquad\qquad\quad x, y \ge 0$

21. Minimize $z = x + 3y$

   Subject to $\quad x + y \le 0$

   $\qquad\qquad -2x + y \ge 1$

   $\qquad\qquad\quad x, y \ge 0$

22. Minimize $z = x + y$

   Subject to $x + y \le 2$

   $\qquad\qquad\quad y \ge 3$

   $\qquad\qquad\quad x \ge 0$

23. Suppose that $z = ax + y$ is the objective function to be maximized over the feasible region shown in the figure. Find all values of a such that $z = ax + y$ will be maximized at (a) A, (b) B, (c) C, (d) D.

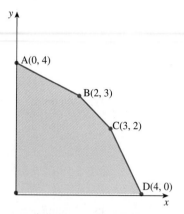

24. Suppose that $z = ax + y$ is the objective function to be minimized over the feasible region shown in the figure. Find all values of a such that $z = ax + y$ will be minimized at (a) A, (b) B, (c) C, (d) D.

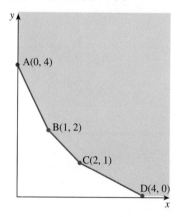

25. Suppose a linear programming problem has the feasible region shown in the figure. Determine the maximum point if the objective function is (a) $z = .20x + y$, (b) $z = .50x + y$, (c) $z = .70x + y$, (d) $z = 1.50x + y$, (e) $z = 4x + y$.

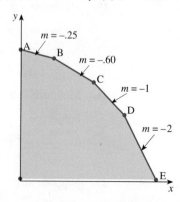

26. Suppose a linear programming problem has the feasible region shown in the figure. Determine the minimum point if the objective function is (a) $z = .20x + y$, (b) $z = .50x + y$, (c) $z = .70x + y$, (d) $z = 1.50x + y$, (e) $z = 4x + y$.

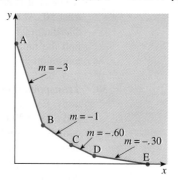

27. Suppose that $z = ax + by$ with $a > 0$ and $b > 0$ is the objective function to be maximized over the feasible region shown in the figure. Explain why the objective function must have a maximum at A.

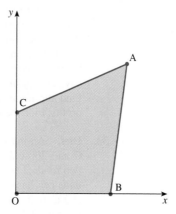

28. Suppose that $z = ax + by$ with $a > 0$ and $b > 0$ is the objective function to be minimized over the feasible region shown in the figure. Explain why the objective function must have a minimum at B.

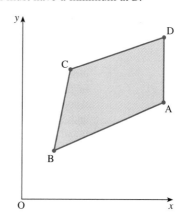

Applications

29. **Shipping.**   A manufacturer supplies its product to two distributors. Distributor A needs at least 100 of the product next month, while Distributor B needs at least 50. At most 200 of the product can be manufactured and delivered to the distributors. If it costs $30 to ship each product to Distributor A and $40 to B, find the number to be shipped to each distributor to minimize cost. Find the minimum cost. (Refer to Exercise 43 of the previous section.)

30. **Transportation.**   A plane delivers two types of cargo between two destinations. Each crate of cargo I is 5 cubic feet in volume, 100 pounds in weight, and earns $20 in revenue. Each crate of cargo II is 5 cubic feet in volume, 200 pounds in weight, and earns $30 in revenue. The plane has available at most 400 cubic feet and 10,000 pounds for the crates. Finally, at least twice the number of crates of A than B must be shipped. Find the number of crates of each cargo to ship in order to maximize revenue. Find the maximum revenue. (Refer to Exercise 44 of the previous section.)

31. **Manufacturing.**   Two products, *A* and *B*, require for their manufacture spending a certain number of hours on a stamping machine, a grinding machine, and a polishing machine. Product A requires 1 hour on the stamping machine, 2 hours on the grinding machine, and 2 hours on the polishing machine. Product B requires 2 hours on the stamping machine, 2 hours on the grinding machine, and 1 hour on the polishing machine. The stamping machine can operate 20 hours, the grinding machine 22 hours, and the polishing machine 15 hours each day. The profit on each of product A and B is, respectively, $40 and $50 each. Find the number of each product to be manufactured in order to maximize profits. Find the maximum profit. (Refer to Exercise 45 of the previous section.)

32. **Diet.**   An individual needs a daily supplement of at least 500 units of vitamin C and 200 units of vitamin E and agrees to obtain this supplement by eating two foods, I and II. Each ounce of food I contains 40 units of vitamin C and 10 units of vitamin E, while each ounce of food II contains 20 units of vitamin C and also 20 units of vitamin E. The total supplement of these two foods must be at most 30 ounces. Unfortunately, food I contains 30 units of cholesterol per ounce, and food II contains 20 units of cholesterol per ounce. Find the appropriate amounts of the two food supplements so that cholesterol is minimized. Find the minimum amount of cholesterol. (Refer to Exercise 46 of the previous section.)

33. **Manufacturing.**   A firm manufactures tables and desks. To produce each table requires 1 hour of labor, 10 square feet of wood, and 2 quarts of finish. To produce each desk requires 3 hours of labor, 20 square feet of wood, and 1 quart of finish. Available is at most 45 hours of labor, at most 350 square feet of wood, and at most 55 quarts of finish. The tables and desks yield profits of $4 and $3 respectively. Find the number of each product to be made in order to maximize profits. Find the maximum profit. (Refer to Exercise 47 of Section 3.1.)

34. **Mixture.**   A dealer has 7600 pounds of peanuts, 5800 pounds of almonds, and 3000 pounds of cashews to be used to make two mixtures. The first mixture wholesales for $2 per pound and consists of 60% peanuts, 30% almonds, and

10% cashews. The second mixture wholesales for $3 per pound and consists of 20% peanuts, 50% almonds, and 30% cashews. How many pounds of each mixture should the dealer make in order to maximize revenue? Find the maximum revenue. (Refer to Exercise 48 of Section 3.1.)

35. **Fishery.** A certain lake has smallmouth and largemouth bass and also three types of food for these fish, I, II, and III. Each month the lake can supply 800 pounds of food I, 500 pounds of food II, and 700 pounds of food III. Each month each smallmouth bass requires 1 pound of food I, 1 pound of food II, and 2 pounds of food III, and each largemouth bass requires 4 pounds of food I, 2 pounds of food II, and 1 pound of food III. What is the maximum number of these fish that the lake can support? (Refer to Exercise 49 of Section 3.1.)

36. **Scheduling.** A firm has two plants, A and B, that manufacture three products, P_1, P_2, and P_3. In one day Plant A can manufacture 30 of P_1, 30 of P_2, and 10 of P_3 and costs $30,000 a day to operate, whereas Plant B can manufacture 10 of P_1, 20 of P_2, and 30 of P_3, and costs $40,000 a day to operate. If at least 240 of the first product, 390 of the second, and 410 of the third are needed, how many days must each plant operate in order to minimize cost? Find the minimum cost. (Refer to Exercise 50 of Section 3.1.)

37. **Scheduling.** Redo the previous problem if the cost of operating Plant A is $30,000 and of Plant B is $20,000.

38. **Diet.** Redo Exercise 32 if food II contains 25 units of cholesterol and everything else is the same.

39. **Fishery.** If in Exercise 35 $2 can be obtained from harvesting the smallmouth bass and $5 for the largemouth bass, how many of each type of fish should be harvested to maximize revenue? Find the maximum revenue.

40. **Manufacturing.** A firm is planning to manufacture and sell two products with the cost and profit (in thousands of dollars) given in the following table.

	Product 1	*Product 2*
Cost	2	1
Profit	2	3

The facilities are such that a total of at most 20 of the two products and at most 15 of product 2 can be manufactured in any one day, and at most $30,000 can be allocated for the costs in any one day. With these conditions the firm can sell all of the items manufactured. Find the number of each product to manufacture in order to maximize profit. Find the maximum profit.

41. **Manufacturing.** Redo the previous exercise if the profit is given as $P = x + y$.

Graphing Calculator Exercises

In Exercises 42 through 44 solve by using the zoom feature of your graphing calculator to determine the corners of the feasible region (to two decimal places).

42. Maximize $z = 10.7x + 14.1y$

Subject to $1.22x + 3.15y \le 34.45$
$1.4x + 1.4y \le 19.6$
$1.31x + 2.46y \le 24.09$
$x, y \ge 0$

43. Minimize $z = 1.41x + 2.37y$

Subject to $1.26x + 4.17y \ge 44.31$
$2.17x + 2.17y \ge 41.23$
$1.68x + 1.03y \ge 24.12$
$x, y \ge 0$

44. **Fishery.**   A lake has rainbow trout and brown trout. Three foods A, B, C are available for these trout in the lake. Each rainbow trout requires 3.42 units of food A, 2.87 units of food B, and 2.14 units of food C each day. Each brown trout requires 5.13 units of food A, 3 units of food B, and 6 units of food C each day. If at most 2701.8 units of food A, 1797.5 units of food B, and 2515 units of food C are available daily, what is the maximum number of rainbow and brown trout that this lake can support?

Solutions to Self-Help Exercise Set 3.2

1. The feasible region is shown in the figure. The feasible region includes its boundary, is unbounded, and the objective function $z = 4x + 6y$ has positive coefficients for x and y. Thus a minimum exists and must occur at a corner point. By evaluating z at the three corners we see that the minimum of $z = 26$ occurs when $x = 2$ and $y = 3$.

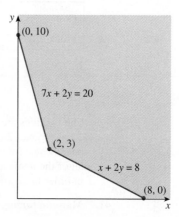

Corner	$z = 4x + 6y$
(0, 10)	$4(0) + 6(10) = 60$
(2, 3)	$4(2) + 6(3) = 26 \leftarrow$ minimum
(8, 0)	$4(8) + 6(0) = 32$

2. With x the number of the first type of connector manufactured and y the number of the second type of connector, the profit is $P = .50x + .60y$. The feasible region and the corners were found in Self-Help Exercise 2 of the last section. Evaluating P at the corners yields

Corner	$P = .50x + .60y$
(2, 0)	$.50(2) + .60(0) = 1.00$
(0, 2)	$.50(0) + .60(2) = 1.20$
(0, 10)	$.50(0) + .60(10) = 6.00$
(5, 6)	$.50(5) + .60(6) = 6.10 \leftarrow$ maximum
(7, 0)	$.50(7) + .60(0) = 3.50$

Thus the maximum profit of $6.10 will occur if 5 of the first connectors and 6 of the second are made.

3. Setting $z = ax + y = P_0$ gives $y = -ax + P_0$, which has slope $m = -a$. Notice that the slope of line AB is -1 and of BD is -2. Thus if $1 \leq a \leq 2$, the constant profit line will be as shown in the figure and B will yield the maximum. If $a \geq 2$, then the constant profit line will be steeper than the line BD and therefore D will yield the maximum.

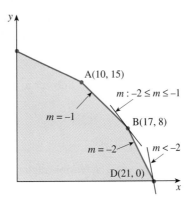

3.3   MORE LINEAR PROGRAMMING

- ▶ *Post-Optimal Analysis*
- ▶ *Portfolio Management*
- ▶ *A Transportation Problem*
- ▶ *Blending Problems*

George Dantzig   *"Linear programming models have been successful because with them many large problems can be formulated so that they are acceptable to planners and solvable on a computer."*

Post-Optimal Analysis

After an optimal solution to an applied linear programming problem has been found, further analysis is usually required. For example, analysis is needed to determine if an excess of some resource is being allocated. If this is the case, then management should consider reallocating or reducing this resource.

EXAMPLE 1   **Determining Excess Resources**

Consider the wooden-boat manufacturing problem considered in the last two sections. Recall that 10 dinghies and 15 skiffs were manufactured in order to maximize the profits subject to the three constraints on the available hours of labor for cutting, assembly, and painting. Identify any of the three operations that has an excess of needed labor.

Solution   Recall that x was the number of dinghies manufactured each week and y the number of skiffs, and the linear programming problem was

$$\text{Maximize} \qquad z = 60x + 80y$$

$$\begin{aligned}
\text{Subject to} \qquad 2x + 4y &\leq 80 \qquad \text{hours available for cutting} \\
2x + 2y &\leq 50 \qquad \text{hours available for painting} \\
4x + 2y &\leq 84 \qquad \text{hours available for assembly} \\
x, y &\geq 0 \qquad \text{nonnegativity constraint}
\end{aligned}$$

The feasible region is shown in Figure 3.18. Recall that the solution to the problem was (10, 15), that is, produce 10 dinghies and 15 skiffs. Notice from Figure 3.18 that the point (10, 15) is on the lines determined by the cutting and painting constraint. This means that all the labor hours allocated for cutting and painting are used. That is,

cutting hours used $= 2(10) + 4(15) = 80 =$ cutting hours available

painting hours used $= 2(10) + 2(15) = 50 =$ painting hours available.

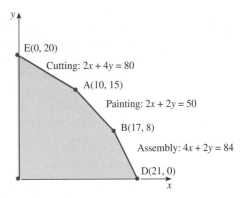

Figure 3.18

However, if 10 dinghies and 15 skiffs are manufactured, the number of labor hours of assembly is

$$4(10) + 2(15) = 70,$$

which is *less* than the 84 hours allocated. Thus we see that management can reduce the number of hours allocated to the assembly section. ■

Portfolio Management

Applied linear programming problems routinely arise with dozens and even hundreds or thousands of decision variables and problem constraints. Such problems cannot be solved by the geometric method considered in the last section, nor can they be solved by hand using the simplex method in the next chapter. These problems are solved using computers. Of course the computer must be given the problem! Thus the steps that involve translating the problem into the linear programming format must be done by humans. Undoubtedly, this represents the most important skill to acquire. Once we identify the mathematical formulation of the linear programming problem, we can usually use standard software programs to determine the solution.

Despite the existence of such software programs, we should not underestimate the need to know the geometric method and the simplex method given in the next chapter. The more we know about how and why these methods work, the more likely we will be able to avoid catastrophic errors that can occur using the software and to make a sensible analysis of the answers given.

We now consider the formulation of several linear programming problems with three or more decision variables.

E X A M P L E   2   **Allocation of Funds to Maximize Return**

An individual wishes to distribute at most $100,000 among a money market mutual fund, 30-year U.S. Treasury bonds, and a stock index fund that mimics the Standard and Poor's 500. Being basically conservative, this individual wants at all times at least $30,000 in the money market fund, at least $30,000 in treasury bonds, and the amount of money in stocks less than or equal to the total in the money market and in treasury bonds. On the other hand, some exposure to equities is desired and at least $20,000 should at all times be in stocks. Based on historical patterns, this individual assumes that the money market, bonds, and stocks will earn annual returns of 6%, 8%, and 10% (including capital gains), respectively. How should this person allocate the funds among these three investments to maximize the return?

Solution   Let us denote by x, y, and z the amount in thousands of dollars invested in, respectively, the money market, bonds, and stocks. Since the total must be at most $100,000, $x + y + z \leq 100$.

Since at least $30,000 must be in the money market and in bonds, we must have $x \geq 30$ and $y \geq 30$. Since at least $20,000 must be in stocks, $z \geq 20$. We also have the amount in stocks less than or equal to the totals in the money market and bonds. Thus $z \leq x + y$. In formulating linear programming problems, we always write the inequalities so that a constant is to the right of the inequality. We then rewrite this last inequality as $-x - y + z \leq 0$.

	Money Market	*Bonds*	*Stocks*
Variables	x	y	z
Yield	.06	.08	.10

The return R is $R = 0.06x + 0.08y + 0.10z$. This is to be maximized. The mathematical formulation in the linear programming format is then given as follows:

$$\text{Maximize} \quad R = 0.06x + 0.08y + .10z$$

$$
\begin{aligned}
\text{Subject to} \quad x + y + z &\leq 100 \\
-x - y + z &\leq 0 \\
x &\geq 30 \\
y &\geq 30 \\
z &\geq 20 \\
x, y, z &\geq 0. \quad \blacksquare
\end{aligned}
$$

A Transportation Problem

Transportation is another area where linear programming is used.

In the remainder of this section the number of decision variables will be rather large, so we will find it convenient to change the decision variables from x to x_1, y to x_2, and so on.

E X A M P L E 3   **Minimizing the Cost of Transportation**

A firm has two stockpiles of an item stored in warehouses at two different locations, Denver and Santa Fe. The item must be shipped to two plants, located in Des Moines and Fort Wayne, for assembly into a final product. At the beginning of a week the stockpile at Denver has 150 items and the stockpile at Santa Fe has 180. The Des Moines plant needs at least 100 of these items, and the Fort Wayne plant needs at least 120 that week. The cost of sending each item from the stockpile at Denver to the plants in Des Moines and Fort Wayne is, respectively, $200 and $300, while the cost of sending each item from the stockpile at Santa Fe to the plants in Des Moines and Fort Wayne is, respectively, $300 and $500. Refer to Figure 3.19. Find the number of items that should be shipped from each stockpile to each plant in order to minimize cost.

Solution   We organize the information, as usual, in a table.

	Des Moines Plant	*Fort Wayne Plant*	*Supply*
Stockpile at Denver	$200	$300	150
Stockpile at Santa Fe	$300	$500	180
Demand	100	120	

Let x_1 and x_2 be the number shipped from the stockpile at Denver to the plants at Des Moines and Fort Wayne, respectively. Since the stockpile at Denver has 150 items, we then have the constraint $x_1 + x_2 \leq 150$. Now let x_3 and x_4 be the number shipped from the stockpile at Santa Fe to the plants at Des Moines and Fort Wayne, respectively. Since the stockpile at Santa Fe has 180 items, we then have the constraint $x_3 + x_4 \leq 180$. From the table we see that the Des Moines plant is demanding at least 100 items with x_1 coming from the stockpile at Denver and x_3 coming from the stockpile at Santa Fe. Thus $x_1 + x_3 \geq 100$. Also from the table we see that the Fort Wayne plant is demanding at least 120 items with x_2 coming from the stockpile at Denver and x_4 coming from the stockpile at Santa Fe. Thus $x_2 + x_4 \geq 120$.

The cost, according to the table, is $C = 200x_1 + 300x_2 + 300x_3 + 500x_4$. The linear programming problem can now be stated as follows:

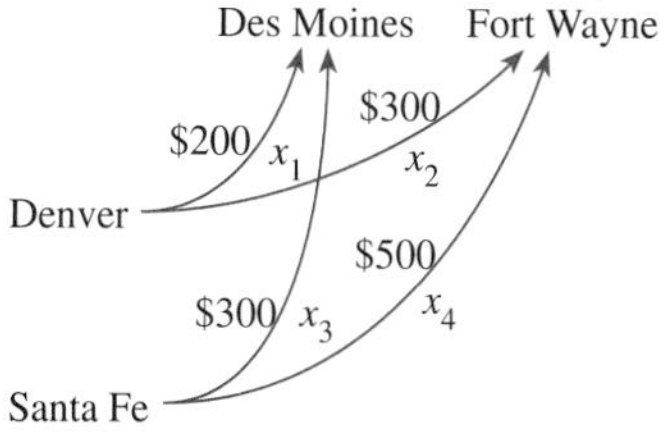

Figure 3.19

$$\text{Minimize} \quad C = 200x_1 + 300x_2 + 300x_3 + 500x_4$$

$$
\begin{aligned}
\text{Subject to} \quad x_1 + x_2 \quad\quad &\leq 150 \\
x_3 + x_4 &\leq 180 \\
x_1 \quad\quad + x_3 \quad\quad &\geq 100 \\
x_2 \quad\quad + x_4 &\geq 120 \\
x_1, x_2, x_3, x_4 &\geq 0. \quad \blacksquare
\end{aligned}
$$

Blending Problems

Problems that involve the blending of various ingredients in some desired proportion to produce a final product are often amenable to linear programming. The following gives some examples of where blending occurs.

- Various types of metal alloys need to be blended to produce various types of steels.
- Various types of chemicals need to be blended to produce other chemicals.
- Various livestock feeds need to be blended to produce low-cost feed mixtures.
- Various types of crude oil must be blended to produce different types of gasoline or heating oil.

E X A M P L E   4   Oil Refinery

An oil refinery blends three types of crude oil, crude A, crude B, and crude C, to make regular and premium gasoline. Crude A has an octane rating of 90, a sulfur content of 1%, costs $20 a barrel, and there are 50,000 barrels on hand. Crude B has an octane rating of 96, a sulfur content of 2%, costs $24 a barrel, and there are 30,000 barrels on hand. Crude C has an octane rating of 98, a sulfur content of 3%, costs $25 a barrel, and there are 20,000 barrels on hand. The octane rating for regular gasoline must be at least 92 and for premium 95. The sulfur content of regular must be at most 1.5% and for premium 2.1%. Regular gasoline sells for $26 a barrel and premium for $32. If the refinery has an order for 35,000 barrels of regular gasoline and 25,000 barrels of premium, how much of each type of crude should be used in each type of gasoline to maximize the profit?

Solution   The information is organized in the following tables.

Available Crude Oil

Crude	Octane	Sulfur	Cost/barrel	Supply (barrels)
A	90	1%	$20	50,000
B	96	2%	$24	30,000
C	98	3%	$25	20,000

Needed Gasoline

Gasoline	Octane	Sulfur	Price/barrel	Needed (barrels)
Regular	92	1.5%	$26	35,000
Premium	95	2.1%	$32	25,000

Let

$$x_1 = \text{barrels of crude A used to produce regular gasoline}$$

$$x_2 = \text{barrels of crude A used to produce premium gasoline}$$

$$x_3 = \text{barrels of crude B used to produce regular gasoline}$$

$$x_4 = \text{barrels of crude B used to produce premium gasoline}$$

$$x_5 = \text{barrels of crude C used to produce regular gasoline}$$

$$x_6 = \text{barrels of crude C used to produce premium gasoline.}$$

The total number of barrels of crude A used is $x_1 + x_2$, while the amount available is 50,000. Thus

$$x_1 + x_2 \leq 50{,}000.$$

Using the corresponding numbers for crude B and C gives

$$x_3 + x_4 \leq 30{,}000$$

$$x_5 + x_6 \leq 20{,}000.$$

The total number of barrels of regular gasoline produced is $x_1 + x_3 + x_5$, while the amount needed is 35,000. Thus

$$x_1 + x_3 + x_5 \geq 35{,}000.$$

Using the corresponding numbers for premium gasoline gives

$$x_2 + x_4 + x_6 \geq 25{,}000.$$

The sulfur constraint for regular gasoline can be given as

$$\frac{\text{total sulfur in regular}}{\text{number of barrels in regular mixture}} \leq 0.015$$

or

$$\frac{0.01x_1 + 0.02x_3 + .03x_5}{x_1 + x_3 + x_5} \leq 0.015.$$

This is not a linear inequality. But we can obtain such an inequality by multiplying both sides by $x_1 + x_3 + x_5$. Doing this gives

$$0.01x_1 + 0.02x_3 + 0.03x_5 \leq 0.015x_1 + 0.015x_3 + 0.015x_5$$

$$-0.005x_1 + 0.005x_3 + .015x_5 \leq 0.$$

For premium gasoline we have

$$\frac{0.01x_2 + 0.02x_4 + .03x_6}{x_2 + x_4 + x_6} \leq 0.021$$

$$0.01x_2 + 0.02x_4 + 0.03x_6 \leq 0.021x_2 + 0.021x_4 + 0.021x_6$$

$$-0.011x_2 - 0.001x_4 + .009x_6 \leq 0.$$

The octane levels of different crudes will blend linearly. Therefore, since $\dfrac{x_1}{x_1 + x_3 + x_5}$ is the fraction of crude A used in regular gasoline, $\dfrac{x_3}{x_1 + x_3 + x_5}$ is the fraction of crude B, and $\dfrac{x_5}{x_1 + x_3 + x_5}$ is the fraction of crude C, the octane rating for regular gasoline is

$$90\,\frac{x_1}{x_1 + x_3 + x_5} + 96\,\frac{x_3}{x_1 + x_3 + x_5} + 98\,\frac{x_5}{x_1 + x_3 + x_5}.$$

Since this must be at least 92, we have

$$90\,\frac{x_1}{x_1 + x_3 + x_5} + 96\,\frac{x_3}{x_1 + x_3 + x_5} + 98\,\frac{x_5}{x_1 + x_3 + x_5} \geq 92.$$

Again this is not a linear inequality. But multiplying both sides by $x_1 + x_3 + x_5$ gives

$$90x_1 + 96x_3 + 98x_5 \geq 92(x_1 + x_3 + x_5)$$

$$-2x_1 + 4x_3 + 6x_5 \geq 0.$$

For premium gasoline we then have

$$90\,\frac{x_2}{x_2 + x_4 + x_6} + 96\,\frac{x_4}{x_2 + x_4 + x_6} + 98\,\frac{x_6}{x_2 + x_4 + x_6} \geq 95$$

$$90x_2 + 96x_4 + 98x_6 \geq 95(x_2 + x_4 + x_6)$$

$$-5x_2 + x_4 + 3x_6 \geq 0.$$

Now we will determine the profit function. The cost of the crude is

$$C = 20(x_1 + x_2) + 24(x_3 + x_4) + 25(x_5 + x_6),$$

while the revenue is

$$R = 26(x_1 + x_3 + x_5) + 32(x_2 + x_4 + x_6).$$

Thus the profit is

$$P = R - C$$

$$= [26(x_1 + x_3 + x_5) + 32(x_2 + x_4 + x_6)]$$
$$\quad - [20(x_1 + x_2) + 24(x_3 + x_4) + 25(x_5 + x_6)]$$

$$= 6x_1 + 12x_2 + 2x_3 + 8x_4 + x_5 + 7x_6.$$

We then have the following linear programming problem to solve.

Maximize $P = 6x_1 + 12x_2 + 2x_3 + 8x_4 + x_5 + 7x_6$

$$
\begin{array}{llll}
\text{Subject to} & x_1 + x_2 & \leq 50,000 \\
& x_3 + x_4 & \leq 30,000 \\
& x_5 + x_6 \leq 20,000 \\
& x_1 + x_3 + x_5 & \geq 35,000 \\
& x_2 + x_4 + x_6 \geq 25,000 \\
& -.005x_1 + .005x_3 + .015x_5 & \leq 0 \\
& -.011x_2 - .001x_4 + .009x_6 \leq 0 \\
& -2x_1 + 4x_3 + 6x_5 & \geq 0 \\
& -5x_2 + x_4 + 3x_6 \geq 0 \\
& x_1, x_2, x_3, x_4, x_5, x_6 \geq 0
\end{array}
$$

■

SELF-HELP EXERCISE SET 3.3

1. **Diet.** A dietician has three foods, A, B, and C, to use to prepare a meal. The table indicates the constraints and the costs. If the total number of ounces must be at most 18, formulate a linear programming problem that will determine the number of ounces of each of the foods to use in order to minimize the cost.

	Units Per Ounce			
	A	B	C	Constraints in Units
Vitamin C	100	0	120	at least 500
Vitamin E	0	70	50	at least 200
Vitamin A	1000	2000	0	at most 25,000
Calories	200	150	75	at most 3000
Cost per ounce	$.60	$.40	$.70	

EXERCISE SET 3.3

In Exercises 1 through 20 translate into a mathematical formulation of a linear programming problem.

1. **Investments.** An individual has at most $100,000 to invest in four vehicles: a money market fund, a bond fund, a conservative stock fund, and a speculative stock fund. These four investments are assumed to have annual returns (including capital gains) of 6%, 8%, 10%, and 13% respectively. The investor wants at least $10,000 in the conservative stock fund, at least $10,000 and at most $40,000 in the money market fund, the amount in speculative stocks less than or equal to the amount in the money market, and finally the amount in conser-

vative stocks less than or equal to the amount in bonds. Find the appropriate allocations of funds among these four investments to maximize the return.

2. **Investments.**   Redo Example 2 of the text if the investor wants at least 40% invested in stocks with everything else remaining the same.

3. **Transportation.**   Redo Example 3 of the text using the following table.

	Plant A	Plant B	Supply
Stockpile I	$20	$40	500
Stockpile II	$50	$20	100
Demand	150	250	

4. **Transportation.**   A firm has an item on stockpile at two locations I and II. This item needs to be shipped in various quantities to three plants: A, B, and C. The following table gives the transportation cost of sending each item from the two stockpile locations to the three plants, the number of items at each stockpile, and the demanded number of this item for each of the three plants.

	Plant A	Plant B	Plant C	Supply
Stockpile I	2	1	3	600
Stockpile II	1	3	2	650
Demand	150	300	350	

Find the numbers of items to be shipped from each of the stockpiles to each of the plants in order to minimize transportation cost.

5. **Diet.**   Three foods, A, B, and C, are being used by a dietician at a hospital to obtain certain minimum supplements of three vitamins, I, II, and III. The following table gives the number of units of each vitamin in each ounce of each food, the daily minimum requirements of each vitamin, and the cost in dollars per ounce of each of the foods.

	Food A	Food B	Food C	Requirements
Vitamin I	30	30	10	500
Vitamin II	20	20	10	329
Vitamin III	35	50	20	425
Cost/oz	4	3	1	

At most 20 ounces of the three foods are to be given. How much of each of the foods should be given if the cost is to be minimized?

6. **Investments.** An investment made for a retirement plan must pay particular attention to risk. Suppose for such an investment no more than $10,000 is placed in 3 investments with annual returns of 8%, 9%, and 10% respectively. The investor requires an annual return of at least $900 on the investment and wishes to minimize the risk. The second investment is twice as risky as the first, and the third investment is 4 times as risky as the first. How should the money be allocated to minimize the risk? *Hint:* Let risk $= x_1 + 2x_2 + 4x_3$.

7. **Plant Nutrition.** A farmer can buy three types of 100-pound bags of fertilizer, type A, type B, and type C. Each 100-pound bag of type A fertilizer costs $20 and contains 40 pounds of nitrogen, 30 pounds of phosphoric acid, and 10 pounds of potash. Each 100-pound bag of type B fertilizer costs $30 and contains 20 pounds of nitrogen, 20 pounds of phosphoric acid, and 60 pounds of potash. Each 100-pound bag of type C fertilizer costs $20 and contains no nitrogen, 30 pounds of phosphoric acid, and 40 pounds of potash. The farmer requires 4400 pounds of nitrogen, 1800 pounds of phosphoric acid, and 2800 pounds of potash. How many bags of each type of fertilizer should she buy in order to minimize cost?

8. **Cost.** An airline has three types of airplanes and has contracted with a tour group to provide accommodations for a minimum of each of 80 first class, 50 tourist, and 60 economy-class passengers. The first plane costs $4000 for the trip and can accommodate 40 first-class, 10 tourist, and 10 economy-class passengers; the second plane costs $5000 for the trip and can accommodate 10 first-class, 10 tourist, and 20 economy-class passengers; while the third plane costs $6000 for the trip and can accommodate 10 first-class, 12 tourist, and 24 economy-class passengers. How many of each type of airplane should be used to minimize the operating cost?

9. **Costs.** An oil company has three refineries and must refine and deliver at least 40,000 gallons of medium-grade gasoline and 24,000 gallons of high-grade gasoline to its stations. The first plant can refine 8000 gallons of medium-grade gasoline and 3000 gallons of high-grade gasoline and costs $40,000 a day to operate. The second plant can refine 4000 gallons of medium-grade gasoline and 6000 gallons of high-grade gasoline and costs $30,000 a day to operate. The third plant can refine 4500 gallons of medium-grade gasoline and 6500 gallons of high-grade gasoline and costs $45,000 a day to operate. How many days should each plant operate to minimize the cost?

10. **Production Scheduling.** A company has 3 plants that produce 3 different sizes of their product. The North plant can produce 2000 items in the small size, 1000 in the medium size, and 1000 in the large size per day. The Center plant can produce 1000 items in the small size, 1000 in the medium size, and 2000 in the large size per day. The South plant can produce 2000 items in the small size, 2000 in the medium size, and 1000 in the large size per day. The company needs to produce at least 9000 of the small items, 8000 of the medium items, and 9000 of the large items on a given day. The cost of producing the small item is $2, the medium item $3, and the large item $2. Find the number of items each plant should produce in order to minimize the cost.

11. **Production Scheduling.** A mining company has three mines that produce three grades of ore. The North mine can produce 2 tons of low-grade ore, 1 ton of medium-grade ore, and 1 ton of high-grade ore each day. The Center mine can produce 1 ton of low-grade, 1 ton of medium-grade, and 4 tons of high-

grade ore. The South mine can produce 1 ton of low-grade, 3 tons of medium-grade, and 3 tons of high-grade ore. At a particular time the company needs at least 21 tons of low-grade ore, 19 tons of medium-grade ore, and at least 25 tons of high-grade ore. It costs $2000 per day to operate the North mine, $3000 a day for the Center mine, and $4000 for the South mine. Find the number of days each mine should be operated to minimize cost.

12. **Purchasing.**   A store decides to have a sale on pink and red azaleas. Two nurseries, East and West, will supply these plants. East nursery charges $4 for a pink azalea and $6 for a red one. West nursery charges $7 for a pink azalea and $5 for a red one. East nursery can supply at most 200 azaleas of either color, while West nursery can supply at most 400 of these azaleas. The store needs at least 250 pink azaleas and at least 300 red azaleas. How many azaleas of each color should the store order from each nursery in order to minimize the cost?

13. **School Rezoning[1].**   A small city has two high schools, NW school and NE school, located in two predominately white districts and two high schools, SW school and SE school, located in two predominately minority districts. A decision is made to bus at least 400 minority students from SW and at least 300 minority students from SE to the two schools in the predominately white districts. NW can accommodate at most 200 additional students and NE at most 600. The weekly cost per student of busing from SW to NW is $4, from SE to NW $2, from SW to NE $2, and from SE to NE is $4. Determine the numbers of students that should be bused from each of the schools in the southern districts to each of the schools in the northern districts in order to minimize the cost.

14. **Oil Refinery.**   An oil refinery blends three types of crude oil, crude A, crude B, and crude C, to make regular, premium, and super gasoline. Crude A has an octane rating of 90, costs $20 a barrel, and there are 50,000 barrels on hand. Crude B has an octane rating of 96, costs $24 a barrel, and there are 30,000 barrels on hand. Crude C has an octane rating of 98, costs $26 a barrel, and there are 20,000 barrels on hand. The octane rating for regular gasoline must be at least 92, for premium at least 96, and for super at least 97. Regular gasoline sells for $26 a barrel, premium for $30, and super for $32. If the refinery has an order for 30,000 barrels of regular gasoline, 20,000 barrels of premium, and 28,000 barrels of super, how much of each type of crude should be used in each type of gasoline to maximize the profit?

15. **Mixture.**   You are going to produce two types of candy, both of which consist solely of sugar, nuts, and chocolate. You have in stock 200 oz of sugar, 30 oz of nuts, and 60 oz of chocolate. The mixture used to make the first candy must contain at least 25% nuts. The mixture used to make the second candy must contain at least 15% nuts and at least 10% chocolate. Each ounce of the first candy can be sold for 50 cents, and each ounce of the second candy for 65 cents. Determine the ounces of each ingredient needed to make each candy in order to maximize revenue.

16. **Farming.**   A farmer has two fields in which he grows wheat and corn. Because of different conditions, there are differences in the yield and costs of growing

[1]Hickman and Taylor, "School Rezoning to Achieve Racial Balance: A Linear Programming Approach," *J. Socio-Econ. Planning Sci.* vol. 3 (1969–1970), pp. 127–134.

these crops in these two fields. The yields and cost are shown in the table. Each field has 100 acres available for cultivation; 15,000 bushels of wheat and 9000 bushels of corn must be grown. Determine a planting plan that will minimize the cost.

	Field 1	Field 2	*Need*
Corn yield/acre	150 bushels	200 bushels	9,000
Cost/acre of corn	$180	$210	
Wheat yield/acre	40 bushels	50 bushels	15,000
Cost/acre of wheat	$140	$120	

17. **Diet.** A dietician has four foods, A, B, C, and D, to use to prepare a meal. The table indicates the constraints and the costs. If the total number of ounces must be at most 18, determine the number of ounces of each of the foods to use in order to minimize the cost.

	Units Per Ounce				*Constraints in Units*
	A	*B*	*C*	*D*	
Iron	1	2	0	0	at least 10
Calcium	100	50	0	10	at least 800
Vitamin C	100	0	200	20	at least 500
Vitamin E	0	0	50	70	at least 200
Vitamin A	0	2000	0	5000	at most 25,000
Calories	200	150	75	250	at most 3000
Cholesterol	5	20	30	70	at most 130
Cost per ounce	$.50	$.40	$.60	$.75	

18. **Diet.** Redo the previous exercise if the cholesterol is to be minimized and the cost is to be at most $10.

19. **Work Schedule.** A chemical plant must operate every day of the week and requires different numbers of employees on different days of the week. Each employee must work five consecutive days and then receive two days off. For example, an employee who works Tuesday to Saturday must be off on Sunday and Monday. The table indicates the required number of employees on each day. Formulate a linear programming problem that the plant managers can use to minimize the number of employees that must be used.

Day	Employees Required
Monday	82
Tuesday	87
Wednesday	77
Thursday	73
Friday	75
Saturday	42
Sunday	23

20. **Allocation.** A corporation owns three farms. The following table gives the size, usable land, and water allocation for each.

Farm	Usable Land	Water Allocation (acre feet)
1	1200	18000
2	1500	2000
3	1000	1300

The corporation grows corn, wheat, and sorghum. The following table gives the maximum acreage that can be planted in each crop, water consumption, and net return.

Crop	Maximum Crop (acres)	Water Consumption (acre feet/acre)	Net Return (dollars/acre)
Corn	1400	3	500
Wheat	1500	2	400
Sorghum	1700	1	100

Determine the number of acres of each crop to be planted on each farm to maximize net income.

21. **Manufacturing.** A firm manufactures tables and desks. To produce each table requires 1 hour of labor, 10 square feet of wood, and 2 quarts of finish. To produce each desk requires 3 hours of labor, 20 square feet of wood, and 1 quart of finish. Available is at most 45 hours of labor, at most 350 square feet of wood, and at most 55 quarts of finish. The tables and desks yield profits of $3 and $4 respectively. Find the number of each product to be made in order to maximize profits. Identify any excess resources available if the optimal number of each product is made. (Refer to Exercise 47 of Section 3.1 where the feasible region is the same.)

22. **Mixture.** A dealer has 7600 pounds of peanuts, 5800 pounds of almonds, and 3000 pounds of cashews to be used to make two mixtures. The first mixture

wholesales for $2 per pound and consists of 60% peanuts, 30% almonds, and 10% cashews. The second mixture wholesales for $4 per pound and consists of 20% peanuts, 50% almonds, and 30% cashews. How many pounds of each mixture should the dealer make in order to maximize revenue? At the optimal solution which type of nut will be in excess? (Refer to Exercise 48 of Section 3.1 where the feasible region is the same.)

23. **Fishery.** A certain lake has smallmouth and largemouth bass and also three types of food for these fish, I, II, and III. Each month the lake can supply 800 pounds of food I, 500 pounds of food II, and 700 pounds of food III. Each month each smallmouth bass requires 1 pound of food I, 1 pound of food II, and 2 pounds of food III; and each largemouth bass requires 4 pounds of food I, 2 pounds of food II, and 1 pound of food III. If $2 can be obtained from harvesting the smallmouth bass and $3 for the largemouth bass, how many of each type of fish should be harvested to maximize revenue? At the optimal solution which type of food will be in excess? (Refer to Exercise 49 of Section 3.1 where the feasible region is the same.)

24. **Scheduling.** A firm has two plants, A and B, in which are manufactured three products, P_1, P_2, and P_3. In one day Plant A can manufacture 30 of P_1, 30 of P_2, and 10 of P_3 and costs $30,000 a day to operate, whereas Plant B can manufacture 10 of P_1, 20 of P_2, and 30 of P_3, and costs $40,000 a day to operate. If at least 240 of the first product, 390 of the second, 410 of the third are needed, how many days must each plant operate in order to minimize cost? At the optimal solution, which product will be in excess of the stated minimum needed? (Refer to Exercise 36 of Section 3.2.)

25. **Fishery.** If in Exercise 23 $1 can be obtained from harvesting the smallmouth bass and $3 for the largemouth bass, how many of each type of fish should be harvested to maximize revenue? At the optimal solution which type of food will be in excess?

26. **Scheduling.** Redo Exercise 24 if the cost of operating Plant A is $30,000 and of Plant B is $20,000.

27. **Manufacturing.** In determining how many desks and chairs to manufacture in order to maximize revenue, the feasible region for the linear programming

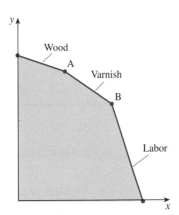

problem was determined to be the region shown in the figure where the first line comes from the constraint on the number of board-feet of wood available, the second line from the constraint on the number of gallons of varnish available, and the third line from the number of hours of labor available. Determine the excess resource if the maximum occurs at (a) *A*, (b) *B*.

28. **Mixture.** In determining how many gallons of four fruit juices to mix in order to obtain a fruit drink to maximize revenue, the feasible region for the linear programming problem was determined to be the region shown in the figure where the indicated four lines come from the constraints on the number of gallons of orange juice, grape juice, grapefruit juice, and cranberry juice. Determine the excess resource(s) if the maximum occurs at (a) *A*, (b) *B*, (c) *C*.

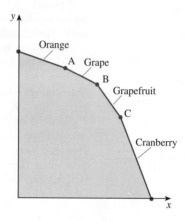

Solutions to Self-Help Exercise Set 3.3

1. Let $x_1, x_2,$ and x_3 be the respective number of ounces of foods A, B, and C. Then the linear programming problem is

Minimize $C = .60x_1 + .40x_2 + .70x_3$

Subject to

$$
\begin{array}{llll}
x_1 + & x_2 + & x_3 \le & 18 & \text{total ounces} \\
100x_1 & + 120x_3 \ge & 500 & \text{vitamin C} \\
70x_2 + & 50x_3 \ge & 200 & \text{vitamin E} \\
1000x_1 + 2000x_2 & \le 25{,}000 & \text{vitamin A} \\
200x_1 + & 150x_2 + 75x_3 \le & 3000 & \text{calories} \\
\end{array}
$$

$$x_1, x_2, x_3 \ge 0$$

SUMMARY OUTLINE OF CHAPTER 3

- A **linear inequality** is an inequality of the form

$$ax + by < c \text{ (or } \le c \text{ or } > c \text{ or } \ge c).$$

- A **system of linear inequalities** is a set of linear inequalities.

- The **feasible region** of a system of inequalities is the set of points that satisfies all the inequalities in the system.
- Graphing a Linear Inequality
  The following steps show how to graph the linear inequality

$$ax + by < c \text{ (or } \leq c \text{ or } > c \text{ or } \geq c).$$

  1. First graph the straight line L given by $ax + by = c$, using a dotted line if the inequalities "$<$" or "$>$" are given and a solid line if the inequalities "$\leq$" or "$\geq$" are given. The dotted line indicates the straight line L is not part of the solution graph, while the solid line indicates that the line L is part of the solution graph.
  2. Pick any convenient test point not on L and evaluate the inequality at that test point. If the inequality is true, then the solution graph is the half-plane on the same side of the line L as the test point. If the inequality is not satisfied, then the solution graph is the half-plane on that side of L that does not include the test point.

- **Optimal Solution of a Linear Programming Problem** If a linear programming problem has an optimal solution, then the optimum must occur at a corner of the feasible region. If the objective function is optimized at two corners of the feasible region, then any point on the line segment connecting these two corners also optimizes the objective function.

- **Existence Theorem for Optimal Solutions**

  1. If the feasible region is nonempty, bounded, and includes its boundary, then the objective function has both a minimum and a maximum in the feasible region.
  2. If the feasible region S includes its boundary, is unbounded, and the objective function $z = ax + by$ has the property that $a > 0$ and $b > 0$, then z attains a minimum value in S but has no maximum value in S.

- **Method of Corners** Determine if a solution exists using the existence theorem for optimal solutions. If a solution exists then:

  1. Graph the feasible region.
  2. Locate all corners of the feasible region.
  3. Evaluate the objective function at each of the corners.
  4. Note that the optimal solution will be the optimal value found in the previous step.

Chapter 3 Review Exercises

1. A farmer wishes to raise corn and/or soybeans. The yield per acre and the price per bushel are the same for each. However, she must fertilize the fields with at least 100 pounds of nitrogen (N) per acre and at least 150 pounds of phosphorus (F) per acre. Two materials are available to supply these nutrients: sludge, which contains 6% N and 3% P, and bone meal, which contains 4% N and 12% P. If sludge costs $15 per 100-lb sack, and bone meal costs $20 per 100-pound sack, how many sacks of each should the farmer buy to minimize the cost of fertilizer?

2. An airline has two types of airplanes and has contracted with a tour group to provide accommodations for a minimum of each of 18 first-class, 12 tourist, and

6 economy-class passengers. The first plane costs $1 per 10 yards traveled to operate and can accommodate 4 first-class, 2 tourist, and 2 economy-class passengers, while the second plane costs $1 per 10 yards to operate and can accommodate 3 first-class, 3 tourist, and 3 economy-class passengers. The airline has available at most 6 of the first type of plane and 4 of the second type. How many of each type of airplane should be used to minimize the operating cost?

3. A furniture manufacturer produces chairs and sofas. The chairs require 5 feet of wood, 5 ounces of foam rubber, and 10 square yards of material. The sofas require 20 feet of wood, 10 ounces of foam rubber, and 30 square yards of material. The manufacturer has in stock 300 feet of wood, 210 ounces of foam rubber, and 500 square yards of material. If the chairs can be sold for $200 and the sofas for $500 each, how many of each should be produced to maximize the income?

4. **a.** For the optimal values in the previous exercise determine which of the three commodities used in the manufacturing process is in excess and by how much.

   b. Redo the previous example if the chairs can be sold for $400 and everything else remains the same. For the optimal values found determine which of the three commodities used in the manufacturing process is now in excess and by how much.

5. A small shop manufactures two styles of park bench. To manufacture the first style requires one square foot of wood, 1 pound of iron, and 3 hours of labor. The second style requires 3 square feet of wood, 1 pound of iron, and 2 hours of labor. The shop has 21 square feet of wood, 9 pounds of iron, and 24 hours of available labor. If the shop makes a profit of $10 on the first style bench and $20 on the second style bench, find the number of each style of bench to manufacture that will maximize profits.

6. **a.** For the optimal values in the previous exercise determine whether iron, wood, or labor is in excess and by how much.

   b. Redo the previous example if the first style of park bench yields a profit of $20 and the second $15 and everything else remains the same. For the optimal values found determine whether iron, wood, or labor is now in excess and by how much.

7. A mining company has two mines that produce three grades of ore. The North mine can produce 3 tons of low-grade ore, 1 ton of medium-grade ore, and 1 ton of high-grade ore each day. The South mine can produce 3 tons of low-grade ore, 2 tons of medium-grade ore, and 3 tons of high-grade ore each day. It costs $3000 per day to operate the North mine and $4000 a day to operate the South mine. At a particular time the company needs at least 15 tons of low-grade ore, 8 tons of medium-grade ore, and 9 tons of high-grade ore. Find the number of days each mine should be operated in order to minimize cost and find the minimum cost.

8. A farmer has 240 acres of land on which he can raise corn and/or soybeans. Raising corn yields 120 bushels per acre, which can be sold for $2.40 per bushel, and raising soybeans yields 40 bushels per acre and sells for $6.00 per bushel. To obtain this price, the farmer must be able to store his crop in bins with a total capacity of 12,000 bushels. The farmer must limit each crop to no more than 10,800 bushels. Assuming the costs of raising corn and soybeans are the same, how much of each should the farmer plant in order to maximize his revenue? What is this maximum?

9. A health food store manager is preparing two mixtures of breakfast cereal out of a supply of 10.5 pounds of oats, 0.9 pounds of almonds, 0.66 pounds of raisins, and 50 pounds of wheat. The first mixture contains 50% oats, 4% almonds, 2% raisins, and the rest wheat and sells for $5 a pound. The second mixture contains 10% oats, 2% almonds, 2% raisins, and the rest wheat and sells for $4 a pound. How much of each mixture should be made to maximize profits? What is the maximum profit?

Linear
programming is
used in the
manufacturing
process to determine
the most profitable
product mix and to
identify excess
resources.
(© Benjamin
Mendlowitz.)

4

Linear Programming:
The Simplex Method

I n the last chapter we introduced linear programming using the geometric method. This method is practical only when the problem has a very small number of variables and constraints. For a typical problem encountered in the applications, the method is completely inadequate. Nevertheless, the geometric method gives substantial insights into the workings of the simplex method, the method considered in this chapter. The simplex method was discovered by George Dantzig in 1947.

Linear programming problems that arise in applications often involve thousands and even tens of thousands of unknowns and constraints. Thus a method is needed to solve such problems efficiently. The simplex method fills this need.

The simplex method can be readily programmed for a computer. Despite the considerable success of this method, important business and scientific problems abound that are too large in scope for this method to handle, even on the fastest known computers. Thus research continues in this area for better methods.

As we shall see, the simplex method follows the boundary of the feasible region, moving through a relatively small number of corners until the optimal corner is found. In 1979 L. G. Khachiyan discovered another way of solving linear programming problems when he constructed the "ellipsoid algorithm." The ellipsoid algorithm generates points from *outside* the feasible region that approach the optimal corner. Initially there was considerable excitement about this algorithm since it possesses certain theoretical properties that led some to conjecture that the method might be more efficient than the simplex method. Unfortunately, this turned out not to be the case in practice.

Then in 1984 Narendra Karmarkar of AT&T Bell Laboratories discovered what is now called Karmarkar's algorithm. This method generates a sequence of points in the *interior* of the feasible region that head to the optimal corner. This method is currently being evaluated to determine under what circumstances it is superior to the simplex method; it is now viewed as holding considerable promise.

4.1   SLACK VARIABLES AND THE SIMPLEX TABLEAU

▶ *The Need for an Efficient Method in Linear Programming*
▶ *Slack Variables*
▶ *Basic and Nonbasic Variables*
▶ *The Simplex Tableau and Pivoting*

The Need for an Efficient Method in Linear Programming

An important problem in linear programming involves assigning individuals to jobs. Consider a modest assignment problem of 70 jobs with 70 individuals with the conditions that every one of the 70 individuals is assigned one of the 70 jobs and every one of the 70 jobs has an individual assigned to it. The number of corners in such a problem is large beyond comprehension. To obtain some idea of just how large, imagine that large mainframe computers of the type now used were available at the time of the generally assumed beginning of the universe, 15 billion years ago. Assume, furthermore, that the earth existed at that time and was filled with such computers, all working continuously to evaluate an objective function at the corners of the feasible region for this problem. By now these computers would not have had enough time to evaluate the objective function at all the corners! (The actual number of evaluations is approximately the number 1 followed by 100 zeros.)

It is clear that for practical problems an *efficient* method is needed to find the maximum or the minimum of the objective function. George Dantzig, the father of linear programming, discovered an efficient method in 1947 to find the solution by looking at a very modest number of corners. This method, called the **simplex method**, is considered in this chapter. The simplex method can solve important practical linear programming problems in reasonable time using computers available today.

Slack Variables

We will now find it convenient to change the variables from x to x_1, y to x_2, and so on. In this section we consider linear programming problems in the following form:

$$
\begin{aligned}
\text{Maximize} \quad & z = c_1 x_1 + c_2 x_2 + \cdots + c_n x_n \\
\text{Subject to} \quad & a_{11} x_1 + a_{12} x_2 + \cdots + a_{1n} x_n \leq b_1 \\
& a_{21} x_1 + a_{22} x_2 + \cdots + a_{2n} x_n \leq b_2 \\
& \qquad\qquad\qquad \cdots \\
& a_{m1} x_1 + a_{m2} x_2 + \cdots + a_{mn} x_n \leq b_m \\
& x_1, x_2, \ldots, x_n \geq 0
\end{aligned}
\tag{1}
$$

The variables $x_1, x_2, \ldots, x_n$ are called the **decision variables**.

Consider again the small manufacturer of wooden boats considered in the last chapter. If x_1 is the number of dinghies produced, and x_2 the number of skiffs, and P is the profit, then recall that the linear programming problem was as follows:

$$
\begin{aligned}
\text{Maximize} \quad & P = 60x_1 + 80x_2 && \text{objective function} \\
\text{Subject to} \quad & 2x_1 + 4x_2 \leq 80 && \text{cutting constraint} \\
& 2x_1 + 2x_2 \leq 50 && \text{painting constraint} \\
& 4x_1 + 2x_2 \leq 84 && \text{assembly constraint} \\
& x_1, x_2 \geq 0 && \text{nonnegativity constraint.}
\end{aligned}
$$

The first constraint represents the constraint from the cutting operation, the second from the painting operation, and the third from the assembly operation. The feasible region, found in the last chapter, is shown in Figure 4.1.

Given the inequality $4 \leq 7$, we can define $s = 7 - 4 = 3$ so that $4 + s = 7$. Thus the inequality has been converted to an *equality*, where s takes up the slack and is referred to as a **slack variable**.

An important element in the simplex method is to move from a system of *inequalities* to a system of *equalities*. To accomplish this, we introduce a slack variable for each of the constraints. For example, for the first constraint define the first slack variable, s_1, so that

$$
2x_1 + 4x_2 + s_1 = 80.
$$

This can be done if $s_1 = 80 - 2x_1 - 4x_2$. Thus the inequality $2x_1 + 4x_2 \leq 80$ becomes the equality $2x_1 + 4x_2 + s_1 = 80$, with s_1 taking up the slack. Notice that the original inequality $2x_1 + 4x_2 \leq 80$ is true if and only if $s_1 \geq 0$.

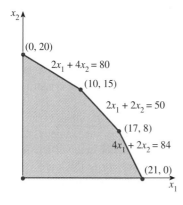

Figure 4.1

Doing the same for the other two constraints then leads to the three inequalities being converted to the following three equalities:

$$\left.\begin{aligned} 2x_1 + 4x_2 + s_1 \qquad\qquad &= 80 \\ 2x_1 + 2x_2 \qquad + s_2 \qquad &= 50 \\ 4x_1 + 2x_2 \qquad\qquad + s_3 &= 84 \end{aligned}\right\} \qquad (2)$$

with the slack variables satisfying

$$\left.\begin{aligned} s_1 &= 80 - 2x_1 - 4x_2 \\ s_2 &= 50 - 2x_1 - 2x_2 \\ s_3 &= 84 - 4x_1 - 2x_2 \end{aligned}\right\}. \qquad (3)$$

A solution of the system (2) consists of five numbers $(x_1, x_2, s_1, s_2, s_3)$.
In general we have the following.

Slack Variables

For each one of the constraints in problem (1) (not including the nonnegative constraints) one creates a slack variable. Using the slack variables s_1, s_2, $\ldots$, s_m, convert the constraints in problem (1) that involve inequalities to the following system of m linear equations:

$$\left.\begin{aligned} a_{11}x_1 + a_{12}x_2 + \cdots + a_{1n}x_n + s_1 \qquad\qquad\qquad &= b_1 \\ a_{21}x_1 + a_{22}x_2 + \cdots + a_{2n}x_n \qquad + s_2 \qquad\qquad &= b_2 \\ \cdots \\ a_{m1}x_1 + a_{m2}x_2 + \cdots + a_{mn}x_n \qquad\qquad\qquad + s_m &= b_m \end{aligned}\right\}. \qquad (4)$$

There are m equations in system (4), and, therefore, the number of variables $x_1, \ldots, x_n, s_1, \ldots, s_m$ in system (4) exceeds the number of equations. Recall from the last chapter that this means that either there is no solution or an infinite number of solutions. In the system (2), for example, one readily sees that $x_1 = 0, x_2 = 0, s_1 = 80, s_2 = 50, s_3 = 84$ is a solution and since this system has at least one solution, the system (2) has an **infinite** number of solutions. We seek the solution that will maximize the objective function subject to the constraints on the decision variables.

We now consider the importance of determining whether the slack variables are negative or not. For example, notice that $s_1 = 0$ is the same as $2x_1 + 4x_2 = 80$. See Figure 4.2a. From our knowledge of linear inequalities, if $s_1 < 0$, and thus, $2x_1 + 4x_2 > 80$, then the point (x_1, x_2) lies *above* the line $2x_1 + 4x_2 = 80$, and cannot be in the feasible region. On the other hand if $s_1 \geq 0$, then $2x_1 + 4x_2 \leq 80$ and the point (x_1, x_2) lies *below or on* the line $2x_1 + 4x_2 = 80$ and has a chance of being in the feasible region of the system of inequalities.

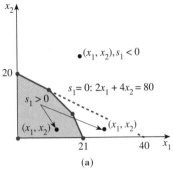

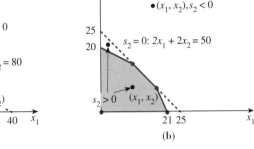

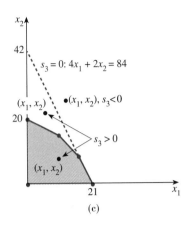

(a)   (b)   (c)

Figure 4.2

In the same way, if $s_2 < 0$, then the point (x_1, x_2) lies above the line $2x_1 + 2x_2 = 50$ and cannot be in the feasible region, while if $s_2 \geq 0$, the point (x_1, x_2) lies below or on the line $2x_1 + 2x_2 = 50$ and thus has a chance of lying in the feasible region. See Figure 4.2b.

Finally, if $s_3 < 0$, then the point (x_1, x_2) lies above the line $4x_1 + 2x_2 = 84$ and cannot be in the feasible region, while if $s_3 \geq 0$, the point (x_1, x_2) lies below or on the line $4x_1 + 2x_2 = 84$ and thus has a chance of lying in the feasible region. See Figure 4.2c.

This leads to the extremely important conclusion that, if any one of the slack variables (or, of course, any one of the decision variables x_1, x_2) is negative, then the point (x_1, x_2) *cannot be in the feasible region*. And also if all of the slack variables (and of course all of the decision variables x_1, x_2) are nonnegative, then the point (x_1, x_2) *will be in the feasible region*.

In general, we have the following.

Algebraic Criteria for Determining the Feasible Region

The point $(x_1, x_2, \ldots, x_n)$ will be in the feasible region of problem (1) if and only if all the slack variables $s_1, s_2, \ldots, s_m$ together with all the decision variables $x_1, x_2, \ldots, x_n$ are nonnegative.

For the wooden-boat manufacturing problem, this means that we have rewritten the original linear programming problem with inequality problem constraints to the following:

$$\text{Maximize} \qquad P = 60x_1 + 80x_2$$
$$\text{Subject to} \qquad 2x_1 + 4x_2 + s_1 = 80$$
$$2x_1 + 2x_2 + s_2 = 50$$
$$4x_1 + 2x_2 + s_3 = 84$$
$$x_1, x_2, s_1, s_2, s_3 \geq 0.$$

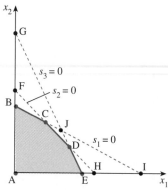

$A = (0, 0)$: $s_1 = 80$, $s_2 = 50$, $s_3 = 84$

$B = (0, 20)$: $s_1 = 0$, $s_2 = 10$, $s_3 = 44$

$C = (10, 15)$: $s_1 = 0$, $s_2 = 0$, $s_3 = 14$

$D = (17, 8)$: $s_1 = 14$, $s_2 = 0$, $s_3 = 0$

$E = (21, 0)$: $s_1 = 38$, $s_2 = 8$, $s_3 = 0$

$F = (0, 25)$: $s_1 = -20$, $s_2 = 0$, $s_3 = 34$

$G = (0, 42)$: $s_1 = -88$, $s_2 = -34$, $s_3 = 0$

$H = (25, 0)$: $s_1 = 30$, $s_2 = 0$, $s_3 = -16$

$I = (40, 0)$: $s_1 = 0$, $s_2 = -30$, $s_3 = -76$

$J = (14\frac{2}{3}, 12\frac{2}{3})$: $s_1 = 0$, $s_2 = -4\frac{2}{3}$, $s_3 = 0$

Figure 4.3

Basic and Nonbasic Variables

We saw in the last chapter that if the objective function in a linear programming problem attains a maximum or a minimum, then this maximum or minimum must occur at a corner point. Thus it is imperative that we be able to locate corner points. But we have abandoned the geometric method as inadequate for typical real applications. Thus we need an *algebraic* method that can be programmed on a computer for locating corner points. We have the first ingredient of such a method. Namely, the point $(x_1, x_2, \ldots, x_n)$ will at least be in the feasible region of problem (1) if, and only if, all the slack variables together with all the decision variables $x_1, x_2, \ldots, x_n$ are nonnegative.

Notice from Figure 4.3 that the feasible region in the wooden-boat manufacturing problem is bounded by the five straight lines $x_1 = 0$, $x_2 = 0$, $s_1 = 0$, $s_2 = 0$, $s_3 = 0$. Any corner point must be the point of intersection of two of these lines. For example, from Figure 4.3 we see that the corner point B is the intersection of $x_1 = 0$ and $s_1 = 0$, while the corner point C is the intersection of $s_1 = 0$ and $s_2 = 0$. Not all intersections of two of these five lines yield corner points. For example, the point of intersection, J, of the two lines $s_1 = 0$ and $s_3 = 0$ is *not* a corner point.

We then have the following procedure for finding the corner points. Determine the number of unknowns in the problem (1). For the wooden-boat manufacturer this is two. We then arbitrarily pick the same number, in this case two, of the variables x_1, x_2, s_1, s_2, s_3 and call them the **nonbasic variables**. Call the remaining three variables the **basic variables**. Now set the two nonbasic variables equal to zero and try to solve the three equations found in (2) for the three basic variables. Since we are dealing with a system of three equations in three unknowns, the system may have one solution, an infinite number of solutions, or no solution. If there is a solution and all five of the numbers x_1, x_2, s_1, s_2, s_3 are nonnegative, then the point (x_1, x_2) is a corner point; otherwise it is not.

In general we have the following.

Basic and Nonbasic Variables

For problem (1) arbitrarily pick n of the variables $x_1, \ldots, x_n, s_1, \ldots, s_m$, and call them the *nonbasic variables*. Call the remaining variables the *basic variables*. Set the n nonbasic variables equal to zero in the system (4) and try to solve for the basic variables. If a solution exists and if all the numbers $x_1, \ldots, x_n, s_1, \ldots, s_m$, are nonnegative, then the point $(x_1, \ldots, x_n)$ is a corner point.

For example, in the wooden boat manufacturing problem pick x_1 and s_1 as the nonbasic variables. Setting $x_1 = 0$ and $s_1 = 0$ in the system (2), yields from the first equation in (2), $2(0) + 4x_2 + 0 = 80$, or $x_2 = 20$. This in turn yields from (3) that

$$s_2 = 50 - 2x_1 - 2x_2 = 50 - 2(0) - 2(20) = 10$$

$$s_3 = 84 - 4x_1 - 2x_2 = 84 - 4(0) - 2(20) = 44$$

Thus $(0, 20, 0, 10, 44)$ is a solution of (2). Since all of these numbers are nonnegative, the point $(0, 20)$ must be a corner point. From Figure 4.3 we see that this point is B and is indeed a corner point.

For another example, pick x_1 and s_2 as the nonbasic variables. Then setting $x_1 = 0$ and $s_2 = 0$, yields, from the second equation in (2), $2(0) + 2x_2 + 0 = 50$, or $x_2 = 25$. This in turn yields from (3) that

$$s_1 = 80 - 2x_1 - 4x_2 = 80 - 2(0) - 4(25) = -20$$

$$s_3 = 84 - 4x_1 - 2x_2 = 84 - 4(0) - 2(25) = 34$$

Thus $(0, 25, -20, 0, 34)$ is a solution of (2). Since one of these numbers is negative, the point $(0, 25)$ must not be a corner point. From Figure 4.3 we see that this point is F and is indeed not a corner point.

By systematically selecting two of the five variables to be the nonbasic variables and then setting them equal to zero in system (2) and solving for the three basic variables, we can obtain the results in Table 4.1.

Table 4.1

x_1	x_2	s_1	s_2	s_3	Point (x_1, x_2)		
0	0	80	50	84	A(0, 0)	corner	$s_1, s_2, s_3 \geq 0$
0	20	0	10	44	B(0, 20)	corner	$s_1, s_2, s_3 \geq 0$
10	15	0	0	14	C(10, 15)	corner	$s_1, s_2, s_3 \geq 0$
17	8	14	0	0	D(17, 8)	corner	$s_1, s_2, s_3 \geq 0$
21	0	38	8	0	E(21, 0)	corner	$s_1, s_2, s_3 \geq 0$
0	25	-20	0	34	F(0, 25)	not corner	$s_1 < 0$
0	42	-88	-34	0	G(0, 42)	not corner	$s_1, s_2 < 0$
25	0	30	0	-16	H(25, 0)	not corner	$s_3 < 0$
40	0	0	-30	-76	I(40, 0)	not corner	$s_2, s_3 < 0$
$14\frac{2}{3}$	$12\frac{2}{3}$	0	$-4\frac{2}{3}$	0	J($14\frac{2}{3}$, $12\frac{2}{3}$)	not corner	$s_2 < 0$

Notice again from the table that the corner points have all variables nonnegative, while the points that are not corner points have at least one of the variables negative.

The Simplex Tableau and Pivoting

We now rewrite the equation $P = z = 60x_1 + 80x_2$ as

$$-60x_1 - 80x_2 + z = 0$$

and add it to the system (2) and obtain

$$\left. \begin{array}{rrrrrr} 2x_1 + 4x_2 + s_1 & & & = 80 \\ 2x_1 + 2x_2 & + s_2 & & = 50 \\ 4x_1 + 2x_2 & & + s_3 & = 84 \\ -60x_1 - 80x_2 & & & + z = 0 \end{array} \right\}. \tag{5}$$

For convenience we rewrite (5) as the following tableau.

x_1	x_2	s_1	s_2	s_3	z	
2	4	1	0	0	0	80
2	2	0	1	0	0	50
4	2	0	0	1	0	84
-60	-80	0	0	0	1	0

The simplex method will give a sequence of such tableaus. We now see how to associate a corner point with each such tableau. Recall that selecting 2 of the variables x_1, x_2, s_1, s_2, s_3 as the nonbasic variables and setting them equal to zero in the system (2) may or may not lead to a corner point. We want a procedure for selecting nonbasic variables that *will* lead to a corner point.

The key to doing this is to recognize unit and nonunit columns. A **unit column** is a column in the tableau with all entries zero except for one entry that is a one. In determining unit and nonunit columns we *ignore the last column of the tableau*. Thus in the above tableau the first and second columns are the nonunit columns and the third, fourth, fifth, and sixth columns are the unit columns.

Nonunit	*Nonunit*	*Unit*	*Unit*	*Unit*	*Unit*	
x_1	x_2	s_1	s_2	s_3	z	
2	4	1	0	0	0	80
2	2	0	1	0	0	50
4	2	0	0	1	0	84
-60	-80	0	0	0	1	0

Pick the variables associated with the nonunit columns as the nonbasic variables. This is x_1 and x_2 in the above tableau. Now set $x_1 = 0$ and $x_2 = 0$. Then we can immediately see the solution to the resulting system of equations.

$$
\begin{array}{c}
\overset{0}{}\quad\overset{0}{}
\end{array}
$$

$$
\begin{aligned}
2x_1 + 4x_2 + s_1 &= 80 \\
2x_1 + 2x_2 + s_2 &= 50 \\
4x_1 + 2x_2 + s_3 &= 84 \\
-60x_1 - 80x_2 + z &= 0
\end{aligned}
$$

We then have $s_1 = 80$, $s_2 = 50$, $s_3 = 84$, and $z = 0$. This point corresponds to the origin $(0, 0)$ with a profit of zero. The origin is definitely a corner point. We can see this from the figures or from the fact that all five of the coordinates x_1, x_2, s_1, s_2, s_3 are nonnegative.

As we shall see in the next section, the simplex method will use elementary row operations to pivot about a certain element in the tableau. For example pivoting simply replaces the original equations (2) with an equivalent set of equations in which four of the columns are unit columns and two of the columns are nonunit columns. It is the variables associated with the latter two columns that we pick as the nonbasic variables and set equal to zero. The resulting equations will then immediately yield the remaining four basic variables. If the pivot element is chosen properly, the resulting point will actually be an adjacent corner point. Furthermore, the profit at the new corner point will be at least as great as the profit at the previous corner point. Thus the simplex method will move us from corner point to adjacent corner point, always improving the profit (or at least not decreasing the profit) until the maximum is reached.

In the next section we will determine the criteria needed to determine the column that the pivot element is in and the row that the pivot element is in. In this section we merely see how to pivot. In the process we will pivot about some less optimal elements (considering what we will learn in the next section), but we will then be able to compare what happens when pivoting about an optimal element with what happens when pivoting about a less optimal element.

We first pivot about the element in the first row and second column. We will see in the next section that this is the appropriate pivot. We want the pivot element to be a 1. This requires, in this case, dividing the first row by 4. Doing this we obtain the tableau:

x_1	x_2	s_1	s_2	s_3	z		
.50	①	.25	0	0	0	20	
2	2	0	1	0	0	50	$R_2 - 2R_1 \rightarrow R_2$
4	2	0	0	1	0	84	$R_3 - 2R_1 \rightarrow R_3$
-60	-80	0	0	0	1	0	$R_4 + 80R_1 \rightarrow R_4$

Now our goal is to **obtain zeros in all the entries of the pivot column other than the pivot element**. We accomplish this by the row operations indicated to the right of the above tableau. This then leads to the new tableau:

x_1	x_2	s_1	s_2	s_3	z	
.50	1	.25	0	0	0	20
1	0	−.50	1	0	0	10
3	0	−.50	0	1	0	44
−20	0	20	0	0	1	1600

Now notice from this last tableau that the second, fourth, fifth, and sixth columns are the unit columns and the first and third columns are the nonunit columns. These latter nonunit columns determine that the nonbasic variables are x_1 and s_1. If we now set the nonbasic variables equal to zero, then we can immediately solve for the remaining variables (which are associated with the unit columns and are the basic variables).

$$
\begin{array}{rcl}
.50x_1 + x_2 + .25s_1 & = & 20 \\
x_1 - .50s_1 + s_2 & = & 10 \\
3x_1 - .50s_1 + s_3 & = & 44 \\
-20x_1 + 20s_1 + z & = & 1600
\end{array}
$$

We then obtain $x_1 = 0$, $x_2 = 20$, $s_1 = 0$, $s_2 = 10$, $s_3 = 44$. Since all of these variables are nonnegative, this point is a corner point. We recall that it is the point B in the figures. Also notice from the last tableau that setting both nonbasic variables x_1 and s_1 equal to zero yields $z = 1600$. Thus we have moved from the origin where the profit is zero to an adjacent corner point B where the profit is 1600.

We have the following rule for determining the solution from the tableaus.

Determining the Solution from the Tableau

1. For problem (1) pick n of the variables $x_1, \ldots, x_n, s_1, \ldots, s_m$, to be the nonbasic variables. The remaining m variables together with z are the basic variables.
2. a. Let the nonbasic variables be variables that correspond to non-unit columns.
   b. If there are not n nonunit columns, then we arbitrarily select some of the variables (but never z) associated with unit columns to be nonbasic variables so that there are n nonbasic variables and so that no two basic variables have identical associated columns.
3. Set all of the nonbasic variables equal to zero.
4. We can then easily find the values of the basic variables from the tableau.

In the next section we shall see how to proceed on to the maximum. For now let us pivot about some other point and see what the consequences are.

E X A M P L E   1

Pivoting

Pivot about the element in the third row and first column; determine the basic and nonbasic variables and the point that corresponds to the resulting tableau.

Solution   We first divide the third row by 4 and obtain a 1 in the pivot position.

x_1	x_2	s_1	s_2	s_3	z		
2	4	1	0	0	0	80	
2	2	0	1	0	0	50	
④	2	0	0	1	0	84	$0.25R_3 \rightarrow R_3$
-60	-80	0	0	0	1	0	

x_1	x_2	s_1	s_2	s_3	z		
2	4	1	0	0	0	80	$R_1 - 2R_3 \rightarrow R_1$
2	2	0	1	0	0	50	$R_2 - 2R_3 \rightarrow R_2$
①	.50	0	0	.25	0	21	
-60	-80	0	0	0	1	0	$R_4 + 60R_3 \rightarrow R_4$

Now the indicated row operations will place zeros in all elements in the first column except for the pivot element.

x_1	x_2	s_1	s_2	s_3	z	
0	3	1	0	$-.50$	0	38
0	1	0	1	$-.50$	0	8
1	.50	0	0	.25	0	21
0	-50	0	0	15	1	1260

This completes the pivoting.

We now see what solution this tableau corresponds to. The second and fifth columns are the nonunit columns. Thus the nonbasic variables are x_2 and s_3. The first, third, fourth, and sixth columns are the unit columns. Thus the basic variables are x_1, s_1, s_2, and z. Now we set both of the nonbasic variables x_2 and s_3 equal to zero. This immediately yields $x_1 = 21$, $x_2 = 0$, $s_1 = 38$, $s_2 = 8$, $s_3 = 0$. Since all of these variables are nonnegative, this is a corner point. In fact this is the point E in Figure 4.3. ∎

Notice that E is the other corner adjacent to the origin. From the last tableau we see that the profit is only 1260. This is less than the profit of 1600 obtained at the point B. It is for this reason that this is the less optimal point, and the pivot element used here is the less optimal element. We will have much more to say about this point in the next section.

In the next section we will see that certain ratios play an important role in determining which is the pivot row. Let us examine one of these ratios and see what the ratio means. In the last example we pivoted about the element 4 in the third row and first column. The only operation that was done to the third row was division by 4. All the other elements in the first column became zero. Looking only at the third row and first column, we ended up with

x_1	x_2	s_1	s_2	s_3	z	
0						
0						
1	.50	0	0	.25	0	$\frac{84}{4}$
0						

It is important to notice that x_1 then will equal the ratio $\frac{84}{4}$.

E X A M P L E   2   **Pivoting and Finding Basic and Nonbasic Variables**

Pivot about the element in the second row and second column, find the basic and nonbasic variables and the point that corresponds to the tableau.

Solution   Start by dividing the second row by 2 and then perform row operations to obtain zeros in the second column except for the pivot element.

x_1	x_2	s_1	s_2	s_3	z		
2	4	1	0	0	0	80	
2	②	0	1	0	0	50	$0.50R_2 \rightarrow R_2$
4	2	0	0	1	0	84	
-60	-80	0	0	0	1	0	

x_1	x_2	s_1	s_2	s_3	z		
2	4	1	0	0	0	80	$R_1 - 4R_2 \rightarrow R_1$
1	①	0	.50	0	0	25	
4	2	0	0	1	0	84	$R_3 - 2R_2 \rightarrow R_3$
-60	-80	0	0	0	1	0	$R_4 + 80R_2 \rightarrow R_4$

x_1	x_2	s_1	s_2	s_3	z	
-2	0	1	-2	0	0	-20
1	1	0	.50	0	0	25
2	0	0	-1	1	0	34
20	0	0	40	0	1	2000

Taking the nonbasic variables to be x_1 and s_2, since the first and fourth columns are nonunit columns, and setting these two variables equal to zero then yields $x_1 = 0$, $x_2 = 25$, $s_1 = -20$, $s_2 = 0$, $s_3 = 34$. Notice that one of these variables is negative, and therefore the point (x_1, x_2) cannot be a corner point. From the figures, we see that this is the point F. ■

Once again let us consider a certain ratio. In the last example we pivoted about the element 2 in the second row and second column. The only operation performed on the second row was division by 2. The other elements in the second column became zero after the pivot. Thus the second row and second column became

x_1	x_2	s_1	s_2	s_3	z	
	0					
1	1	0	.50	0	0	$\frac{50}{2}$
	0					
	0					

Notice that the value of x_2 becomes the ratio $\frac{50}{2}$.

SELF-HELP EXERCISE SET 4.1

1. Consider the following problem:

$$\begin{array}{ll} \text{Maximize} & z = 20x_1 + 35x_2 \\ \text{Subject to} & 2x_1 + 3x_2 \le 20 \\ & 2x_1 + 5x_2 \le 10 \\ & x_1, x_2 \ge 0 \end{array}$$

Introduce slack variables, converting the set of inequalities to a set of equalities. Also set up the corresponding tableau.

2. For the wooden-boat manufacturing problem considered in this section, take x_1 and s_3 as the nonbasic variables. Set the nonbasic variables equal to zero and solve equation (2) for the remaining variables.

3. For the wooden-boat manufacturing problem considered in this section, pivot about the element in the third row and second column. Determine if the point corresponding to the resulting tableau is a corner point of the feasible region.

EXERCISE SET 4.1

1. Consider the following linear programming problem:

$$\text{Maximize} \quad z = 3x_1 + 2x_2$$
$$\text{Subject to} \quad x_1 + x_2 \leq 6$$
$$2x_1 + x_2 \leq 8$$
$$x_1, x_2 \geq 0.$$

   a. Introduce two slack variables, s_1 and s_2 and convert the set of inequalities to a set of equalities.
   b. Set up the corresponding tableau.
   c. Graph the feasible region and locate all corner points.
   d. Successively set two of the variables x_1, x_2, s_1, s_2 equal to zero and solve the equations found in part (a) for the remaining variables. Verify that a point is a corner point if, and only if, none of these variables are negative.

2. Consider the following linear programming problem:

$$\text{Maximize} \quad z = 10x_1 + 20x_2$$
$$\text{Subject to} \quad x_1 + 2x_2 \leq 20$$
$$4x_1 + x_2 \leq 24$$
$$x_1, x_2 \geq 0$$

   Answer all the questions posed in the previous exercise.

3. Consider the following linear programming problem:

$$\text{Maximize} \quad z = 20x_1 + 40x_2$$
$$\text{Subject to} \quad x_1 + 3x_2 \leq 30$$
$$x_1 + x_2 \leq 12$$
$$4x_1 + x_2 \leq 24$$
$$x_1, x_2 \geq 0$$

   Answer all the questions posed in the previous exercise. (There will be three slack variables.)

4. Verify Table 4.1 in the text.

In Exercises 5 through 8 pivot about the element indicated for the tableau.

x_1	x_2	s_1	s_2	z	
2	4	1	0	0	8
5	1	0	1	0	20
-4	-2	0	0	1	0

5. First row and first column

6. First row and second column

7. Second row and first column

8. Second row and second column

In Exercises 9 through 14 pivot about the element indicated for the tableau.

x_1	x_2	s_1	s_2	s_3	z	
2	4	1	0	0	0	8
5	1	0	1	0	0	25
3	6	0	0	1	0	15
-3	-2	0	0	0	1	0

9. First row and first column

10. First row and second column

11. Second row and first column

12. Second row and second column

13. Third row and first column

14. Third row and second column

In Exercises 15 through 24 find the basic and nonbasic variables according to the rules given in the text. Set the nonbasic variables equal to zero and solve. Determine if the point is a corner point.

15.

x_1	x_2	s_1	s_2	z	
2	4	1	0	0	10
5	1	0	1	0	20
-4	-2	0	0	1	0

16.

x_1	x_2	s_1	s_2	z	
3	5	1	0	0	15
2	2	0	1	0	8
-8	-5	0	0	1	0

17.

x_1	x_2	s_1	s_2	z	
2	1	5	0	0	5
5	0	5	1	0	10
-7	0	-3	0	1	0

18.

x_1	x_2	s_1	s_2	z	
2	0	1	2	0	-2
-4	1	0	5	0	4
9	0	0	0	1	0

19.

x_1	x_2	s_1	s_2	z	
0	4	1	2	0	-2
1	3	0	3	0	4
0	-5	0	-2	1	0

20.

x_1	x_2	s_1	s_2	z	
1	10	1	0	0	50
0	10	30	1	0	20
0	-5	0	0	1	0

21.

x_1	x_2	s_1	s_2	s_3	z	
-5	0	1	0	2	0	21
-1	0	0	1	1	0	30
2	1	0	0	8	0	2
-10	0	0	0	40	1	30

22.

x_1	x_2	s_1	s_2	s_3	z	
1	0	0	1	1	0	5
3	0	1	-2	0	0	-3
7	1	0	1	0	0	6
-3	0	0	-4	0	1	10

23.

x_1	x_2	s_1	s_2	s_3	z	
0	1	1	3	0	0	0
1	8	0	6	0	0	2
0	1	0	5	1	0	-5
0	0	0	-2	0	1	8

24.

x_1	x_2	s_1	s_2	s_3	z	
0	1	0	1	5	0	10
0	5	1	0	2	0	4
1	2	0	0	1	0	1
0	-1	0	0	-2	1	10

Solutions to Self-Help Exercise Set 4.1

1. Introduce two slack variables, rewrite the equation for the objective function, and obtain

$$2x_1 + 3x_2 + s_1 \qquad\quad = 20$$
$$2x_1 + 5x_2 \qquad s_2 \quad = 10$$
$$-20x_1 - 35x_2 \qquad\qquad z = 0$$

The tableau is then

x_1	x_2	s_1	s_2	z	
2	3	1	0	0	20
2	5	0	1	0	10
-20	-35	0	0	1	0

2. Setting $x_1 = 0$ and $s_3 = 0$ in the third equation in (2) yields $4(0) + 2x_2 = 84$, or $x_2 = 42$. Now use equation (3) to solve for s_1 and s_2 and obtain

$$s_1 = 80 - 2x_1 - 4x_2 = 80 - 2(0) - 4(42) = -88$$
$$s_2 = 50 - 2x_1 - 2x_2 = 50 - 2(0) - 2(42) = -34.$$

This yields the point $(0, 42, -88, -34, 0)$, which is the point G in the figures. Since at least one of these coordinates is negative, the point $(0, 42)$ is not feasible and not a corner point.

3.

x_1	x_2	s_1	s_2	s_3	z		
2	4	1	0	0	0	80	
2	2	0	1	0	0	50	
4	②	0	0	1	0	84	$0.50R_3 \rightarrow R_3$
-60	-80	0	0	0	1	0	

x_1	x_2	s_1	s_2	s_3	z		
2	4	1	0	0	0	80	$R_1 - 4R_3 \rightarrow R_1$
2	2	0	1	0	0	50	$R_2 - 2R_3 \rightarrow R_2$
2	①	0	0	.50	0	42	
-60	-80	0	0	0	1	0	$R_4 + 80R_3 \rightarrow R_4$

x_1	x_2	s_1	s_2	s_3	z	
-6	0	1	0	-2	0	-88
-2	0	0	1	-1	0	-34
2	1	0	0	.50	0	42
100	0	0	0	40	1	3360

The first and fifth columns are the nonunit columns. Thus take the nonbasic variables to be x_1 and s_3 and set these two variables equal to zero. This yields $x_1 = 0$, $x_2 = 42$, $s_1 = -88$, $s_2 = -34$, $s_3 = 0$. Notice that at least one of these variables is negative, and, therefore, the point (x_1, x_2) cannot be a corner point. From the figures, we see that this point is the point G.

4.2 THE SIMPLEX METHOD: STANDARD MAXIMIZATION PROBLEMS

▶ *Standard Maximum Problems*
▶ *The Simplex Method*
▶ *Post-Optimal Analysis*

Dantzig and the Simplex Method

(Photo by Edward W. Souza News Service, Stanford University, Stanford, CA.)

As mentioned at the beginning of the last section, linear programming problems of even moderate size can have an unimaginably large number of corners to evaluate. In 1947 George Dantzig discovered the simplex algorithm for solving linear programming problems, which only requires evaluation of a relatively small number of corners to find the solution. Indeed Dantzig reported that most of the time the simplex method solved linear programming problems in m equations in only $2m$ or $3m$ steps. The simplex method goes from corner to adjacent corner until the solution is found. In three dimensions the feasible region can be visualized as a cut and finished diamond with faces, edges, and corners. When Dantzig first formulated the method, he had little faith in it. He assumed that the method would spend an extremely large amount of time wandering from corner to corner and take forever to find the solution. In fact after formulating the method while with the Air Force, he left the lab he was at and spent some time visiting some top mathematicians and economists to find a more efficient method. When he returned to the lab in June 1948, his group informed him that the method was working with extreme efficiency on all the test problems. He was astounded. In fact, only in the last few years have people begun obtaining insight into why the simplex algorithm works so well.

Standard Maximum Problems

We now develop the simplex method. This method is an algebraic method that can be readily programmed for use on computers. The method systematically and efficiently moves from one corner point of the feasible region to an adjacent corner point, always improving the objective function (or at least not making it worse) until the corner is found that maximizes the objective function.

In this section the simplex method is applied only to the so-called *standard maximum problems*. Subsequent sections will consider the simplex method for other problems. We have the following definition.

Standard Maximum Problem

A linear programming problem is said to be a **standard maximum problem** if the following are satisfied.

1. The decision variables are constrained to be nonnegative. This is called the *nonnegative condition*.
2. All of the problem constraint inequalities are $\leq$.
3. The constants in the problem constraints to the right of the inequality $\leq$ are never negative.
4. The objective function is to be maximized.

E X A M P L E   1   Showing a Problem is a Standard Maximization Problem

Show that the problem of the wooden-boat manufacturer of the last section is a standard maximization problem.

Solution   If x_1 is the number of dinghies produced, x_2 the number of skiffs, and z is the profit, then recall that the linear programming problem was

$$\text{Maximize} \quad z = 60x_1 + 80x_2 \qquad \text{objective function}$$

$$\text{Subject to} \quad \left. \begin{array}{r} 2x_1 + 4x_2 \leq 80 \\ 2x_1 + 2x_2 \leq 50 \\ 4x_1 + 2x_2 \leq 84 \end{array} \right\} \qquad \text{problem constraints}$$

$$x_1, x_2 \geq 0 \qquad \text{nonnegativity constraint.}$$

It is then apparent that this problem satisfies all four conditions given above and is thus a standard maximization problem. ■

The General Algebraic Form of the Standard Maximization Problem

$$\text{Maximize} \qquad z = c_1 x_1 + c_2 x_2 + \cdots + c_n x_n$$

$$\text{Subject to} \qquad a_{11} x_1 + a_{12} x_2 + \cdots + a_{1n} x_n \leq b_1$$

$$a_{21} x_1 + a_{22} x_2 + \cdots + a_{2n} x_n \leq b_2 \tag{1}$$

$$\cdots$$

$$a_{m1} x_1 + a_{m2} x_2 + \cdots + a_{mn} x_n \leq b_m$$

$$x_1, x_2, \ldots, x_n \geq 0$$

where $b_1, b_2, \ldots, b_m$ are all nonnegative.

The Simplex Method

The initial phases of the simplex method have already been completed in the previous section. First introduce a slack variable for every constraint and then form a simplex tableau.

For the problem of the wooden-boat manufacturer considered in the last section, this is

$$2x_1 + 4x_2 + s_1 \qquad\qquad\qquad = 80$$
$$2x_1 + 2x_2 \qquad + s_2 \qquad\qquad = 50$$
$$4x_1 + 2x_2 \qquad\qquad + s_3 \qquad = 84$$
$$-60x_1 - 80x_2 \qquad\qquad\qquad + z = 0$$

and

x_1	x_2	s_1	s_2	s_3	z	
2	4	1	0	0	0	80
2	2	0	1	0	0	50
4	2	0	0	1	0	84
-60	-80	0	0	0	1	0

From what we learned in the last section, the nonbasic variables in the last tableau are taken to be x_1 and x_2. Setting $x_1 = 0$ and $x_2 = 0$ means we are starting at the point $(0, 0)$, which is a corner point. After setting $x_1 = 0$ and $x_2 = 0$, notice that s_1, s_2, and s_3 must be nonnegative since, in a standard maximization problem, the constants to the right of the constraint inequalities $\leq$ must be nonnegative. In this case these constants are, respectively, 80, 50, and 84. Thus we do indeed begin with a corner point of the feasible region.

Initial Step of Simplex Method

Set up the initial tableau as indicated in the previous section. The initial tableau corresponds to setting all the variables $x_1, x_2, \ldots, x_n$ equal to zero and starting at the origin. The origin must be a feasible point in a standard maximization problem.

Now we must decide to which adjacent corner we will proceed. We must decide between setting $x_1 = 0$ and increasing x_2 or setting $x_2 = 0$ and increasing x_1. We wish to proceed to the adjacent corner that has the largest profit. The profit function is

$$z = 60x_1 + 80x_2.$$

Since 80 is larger than 60, it is clear that increasing x_2 by each unit will result in increasing the profit more than increasing x_1 by each unit. Thus we choose to increase x_2 and set $x_1 = 0$. In the tableau the profit equation appears as

$$-60x_1 - 80x_2 + z = 0.$$

Thus the correct choice is to *pick the variable with the largest negative value in the last row of the simplex tableau.*

x_1	x_2	s_1	s_2	s_3	z	
2	4	1	0	0	0	80
2	2	0	1	0	0	50
4	2	0	0	1	0	84
-60	-80	0	0	0	1	0

$\uparrow$
pivot column

We have now established a general procedure for determining the pivot column.

Finding the Pivot Column

Select as the pivot column the column containing the largest negative value in the last row of the simplex tableau, ignoring the last entry. (If there is a tie, pick any one.)

Now that we have decided to increase x_2 while holding $x_1 = 0$, we must decide just how much to increase x_2. Naturally we wish to increase x_2 so that a corner point will be reached but do not wish to increase x_2 any further since the point will not be feasible. In algebraic terms, we wish to increase x_2 as much as possible without any of the slack variables becoming negative.

Recall that the slack variables are

$$s_1 = 80 - 2x_1 - 4x_2$$

$$s_2 = 50 - 2x_1 - 2x_2$$

$$s_3 = 84 - 4x_1 - 2x_2.$$

With $x_1 = 0$, this becomes

$$s_1 = 80 - 2(0) - 4x_2 = 80 - 4x_2$$

$$s_2 = 50 - 2(0) - 2x_2 = 50 - 2x_2$$

$$s_3 = 84 - 4(0) - 2x_2 = 84 - 2x_2.$$

Since the three slack variables must be nonnegative, the last set of equations becomes

$$80 - 4x_2 \geq 0$$

$$50 - 2x_2 \geq 0$$

$$84 - 2x_2 \geq 0.$$

Solving in terms of x_2, this becomes

$$\tfrac{80}{4} \geq x_2$$

$$\tfrac{50}{2} \geq x_2$$

$$\tfrac{84}{2} \geq x_2.$$

All three of these last equations will be satisfied if we use the smallest of the three ratios

$$\frac{80}{4} \quad \frac{50}{2} \quad \frac{84}{2}$$

This is $\frac{80}{4} = 20$.

If we then pivot about the first row, the row with the smallest of the ratios, we know from the last section that the new value for x_2 will be this ratio $\frac{80}{4} = 20$. Since we picked the smallest of the three ratios, we know that all the slack variables must be nonnegative. Thus we have a feasible point.

Recall from the last section that we pivoted this tableau about the first row and second column and obtained

x_1	x_2	s_1	s_2	s_3	z	
.50	1	.25	0	0	0	20
1	0	$-.50$	1	0	0	10
3	0	$-.50$	0	1	0	44
-20	0	20	0	0	1	1600

Notice that setting $x_1 = 0$ and $s_1 = 0$ immediately yields $x_2 = 20$, which is the first ratio. This gives the corner point we are seeking.

Had we pivoted about the second or third row of the second column as we did in Example 2 and Self-Help Exercise 3 of the last section, we would have increased x_2 too much and gone out of the feasible region. In fact x_2 becomes the second ratio in the first case and the third ratio in the second case.

In the problem under consideration, all the ratios are positive. Consider now the possibility that a ratio is negative. For example, suppose the third equation above is replaced with

$$4x_1 - 2x_2 + s_3 = 84.$$

(The only way the ratio can be negative is to have the coefficient of x_2 be negative, since in a standard maximization problem the number to the right of the inequality $\leq$, in this case the number 84, must be nonnegative.)

Setting $x_1 = 0$ as before then gives $s_3 = 84 + 2x_2$. Since s_3 must be nonnegative, this requires that $84 + 2x_2 \geq 0$, or $84 \geq -2x_2$. Dividing by -2 changes the sign of the inequality and yields

$$\frac{84}{-2} \leq x_2.$$

But this is true for any nonnegative x_2. Since this does not represent any restriction on x_2, *negative ratios can be ignored.*

Using similar arguments, it can be shown that a ratio with the denominator equal to zero does not impose any restriction on the variable. Thus *ratios obtained by dividing by zero can be ignored.* Also it can be shown that *zero ratios with division by a negative number can be ignored.*

Thus, when considering the ratios, we need only consider the nonnegative ratios, ignoring the negative ratios, the ratios that come from dividing by zero, and the zero ratios that are divided by a negative number.

We summarize these findings in the following rules.

Finding the Pivot Row

1. For each row, except for the last, divide the entry in the last column by the corresponding entry in the pivot column.
2. Choose as the pivot row the row with the smallest (nonnegative) ratio, ignoring zero ratios that are divided by a negative number.

Now that the pivot element has been found, perform the pivot operation in the manner shown in the last section. Now find the new pivot element and perform another pivot operation, and so on.

We now need a stopping procedure. If any element in the last row is negative (not including the last entry), we proceed since the value of z can be increased some more. For example, from the last row of the last tableau we have that $-20x_1 + 20s_1 + z = 1600$ or $z = 20x_1 - 20s_1 + 1600$. Clearly z can be increased some more by increasing x_1.

If there are no negative numbers in the last row (not including the last entry), we stop. In the last tableau there is one negative entry in the last row. Since there is precisely one negative entry, the pivot column is then just the column with this entry. Thus the pivot column is now the first column.

In order to find the pivot row, we proceed with finding the ratios.

x_1	x_2	s_1	s_2	s_3	z		ratios
.50	1	.25	0	0	0	20	$\frac{20}{.50} = 40$
①	0	−.50	1	0	0	10	$\frac{10}{1} = 10 \leftarrow$ smallest ratio
3	0	−.50	0	1	0	44	$\frac{44}{3} = 14\frac{2}{3}$
−20	0	20	0	0	1	1600	

Since the second row yields the smallest of these ratios, the pivot row will be the second row. Thus the pivot element is the element in the second row and first column.

Then perform the elementary row operations $R_1 - .50R_2 \to R_1$, $R_3 - 3R_2 \to R_3$, $R_4 + 20R_2 \to R_4$, and obtain the new tableau

x_1	x_2	s_1	s_2	s_3	z	
0	1	.50	−.50	0	0	15
1	0	−.50	1	0	0	10
0	0	1	−3	1	0	14
0	0	10	20	0	1	1800

Since there are no more negative elements in the last row, we stop. Since the third and fourth columns are the nonunit columns, the nonbasic variables are s_1 and s_2. Setting the nonbasic variables s_1 and s_2 equal to zero then immediately yields $x_1 = 10$, $x_2 = 15$, and $s_3 = 14$. All the slack variables are nonnegative, and the maximum then occurs at the point $(x_1, x_2) = (10, 15)$. The actual maximum can be read from the last row and column of the tableau as 1800.

Notice how the simplex method started at the corner $(0, 0)$, then moved to $(0, 20)$, and then finally to $(10, 15)$, each time moving to an adjacent corner and each time improving the maximum value of the objective function. Refer to Figure 4.4.

We can now summarize the steps in the simplex method.

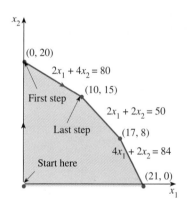

Figure 4.4

Simplex Method for Problem (1)

1. Introduce a slack variable for each of the constraints and create the initial tableau.
2. If none of the entries in the last row (ignoring the last entry) are negative, then we stop. The solution obtained from this tableau is the maximum. To obtain this solution:
   a. Pick n of the variables to be the nonbasic variables. These variables correspond to the nonunit columns. Set all of these nonbasic variables equal to zero. If there are not n nonunit columns, then pick variables (but never z) associated with unit columns

to be nonbasic variables in such a manner that no two basic variables have identical associated columns.

b. The values of the basic variables can then be found immediately. (The basic variables always correspond to unit columns.)

3. If one of the entries in the last row (except for the last entry) is negative, then construct a new tableau choosing the pivot element as follows:

a. Locate the most negative element (other than the last entry) in the last row of the tableau. The column with this entry will be the pivot column. If there is a tie, pick any one of these columns.

b. Divide each of the elements in the last column (except for the bottom most) by the corresponding elements in the pivot column. Choose as the pivot row the row that has the smallest of the nonnegative ratios, ignoring zero ratios that are divided by a negative number. If this is impossible to do, then stop. There is no solution.

4. Obtain a new simplex tableau by doing the following.

a. Divide each entry in the pivot row by the pivot element, obtaining a 1 in the pivot position.

b. Use elementary row operations with the pivot row to obtain zeros in all elements in the pivot column other than the pivot position.

5. Go to Step 2.

E X A M P L E   2   **Using the Simplex Method**

$$\text{Maximize} \quad z = 3x_1 + 2x_2$$
$$\text{Subject to} \quad -x_1 + x_2 \leq 2$$
$$2x_1 + x_2 \leq 8$$
$$x_1, x_2 \geq 0$$

Solution   Introducing two slack variables and rewriting the objective function yields

$$-x_1 + x_2 + s_1 \qquad\qquad = 2$$
$$2x_1 + x_2 \qquad + s_2 \qquad = 8$$
$$-3x_1 - 2x_2 \qquad\qquad + z = 0$$

The initial tableau then becomes

x_1	x_2	s_1	s_2	z		ratio
-1	1	1	0	0	2	
②	1	0	1	0	8	$\frac{8}{2}$ ← smallest ratio
-3	-2	0	0	1	0	

The most negative element in the last row is -3, which indicates that the pivot column is the first column. There is only one positive ratio, $\frac{8}{2}$, which is in the second row. Thus the second row is the pivot row. The pivot is the element in the second row and the first column. Dividing the second row by 2 yields

x_1	x_2	s_1	s_2	z	
-1	1	1	0	0	2
①	$\frac{1}{2}$	0	$\frac{1}{2}$	0	4
-3	-2	0	0	1	0

Performing the row operations $R_1 + R_2 \rightarrow R_1$ and $R_3 + 3R_2 \rightarrow R_3$ yields the new tableau

x_1	x_2	s_1	s_2	z		ratios
0	③⁄②	1	$\frac{1}{2}$	0	6	$\dfrac{6}{\frac{3}{2}} = 4 \leftarrow$ smallest ratio
1	$\frac{1}{2}$	0	$\frac{1}{2}$	0	4	$\dfrac{4}{\frac{1}{2}} = 8$
0	$-\frac{1}{2}$	0	$\frac{3}{2}$	1	12	

There is only one negative element in the last row, so the second column will be the pivot column. The two ratios indicate that the first row is the new pivot row. Dividing the first row by $\frac{3}{2}$ and performing the indicated row operations then yields

x_1	x_2	s_1	s_2	z		
0	①	$\frac{2}{3}$	$\frac{1}{3}$	0	4	
1	$\frac{1}{2}$	0	$\frac{1}{2}$	0	4	$R_2 - \frac{1}{2}R_1 \rightarrow R_2$
0	$-\frac{1}{2}$	0	$\frac{3}{2}$	1	12	$R_3 + \frac{1}{2}R_1 \rightarrow R_3$

x_1	x_2	s_1	s_2	z	
0	1	$\frac{2}{3}$	$\frac{1}{3}$	0	4
1	0	$-\frac{1}{3}$	$\frac{1}{3}$	0	2
0	0	$\frac{1}{3}$	$\frac{5}{3}$	1	14

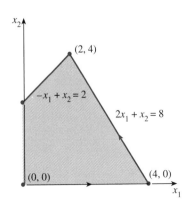

Figure 4.5

Since the last row contains no negative elements, we stop. The solution is $x_1 = 2$ and $x_2 = 4$, with the maximum value of 14. Refer to Figure 4.5. ∎

E X A M P L E 3   **Using the Simplex Method**

$$\text{Maximize} \qquad z = 10x_1 + 20x_2 + 10x_3$$

$$\text{Subject to} \qquad \begin{aligned} x_1 + x_2 + x_3 &\le 16 \\ -2x_1 + x_2 + 2x_3 &\le 20 \\ x_1, x_2, x_3 &\ge 0 \end{aligned}$$

Solution   Introducing two slack variables yields

$$\begin{aligned} x_1 + x_2 + x_3 + s_1 \qquad\qquad &= 16 \\ -2x_1 + x_2 + 2x_3 \qquad + s_2 \qquad &= 20 \\ -10x_1 - 20x_2 - 10x_3 \qquad\qquad + z &= 0 \end{aligned}$$

The initial tableau then becomes

x_1	x_2	x_3	s_1	s_2	z		ratios
1	①	1	1	0	0	16	$\frac{16}{1}$ ← smallest ratio
-2	1	2	0	1	0	20	$\frac{20}{1}$
-10	-20	-10	0	0	1	0	

$$R_2 - R_1 \rightarrow R_2$$
$$R_3 + 20R_1 \rightarrow R_3$$

The most negative element in the last row is -20, which indicates that the pivot column is the second. The smallest ratio is 16, which indicates that the pivot row is the first. Following the indicated elementary row operations we obtain the tableau

x_1	x_2	x_3	s_1	s_2	z	
1	1	1	1	0	0	16
-3	0	1	-1	1	0	4
10	0	10	20	0	1	320

Since there are no negative elements in the last row, the solution is $x_1 = 0$, $x_2 = 16$, $x_3 = 0$, with maximum equal to 320. ■

E X A M P L E 4   **An Example with No Maximum**

$$\text{Maximize} \qquad z = 10x_1 + 20x_2$$

$$\text{Subject to} \qquad \begin{aligned} 4x_1 - 2x_2 &\le 12 \\ x_1 - x_2 &\le 1 \\ x_1, x_2 &\ge 0 \end{aligned}$$

Solution   The initial tableau is

x_1	x_2	s_1	s_2	z	
4	-2	1	0	0	12
1	-1	0	1	0	1
-10	-20	0	0	1	0

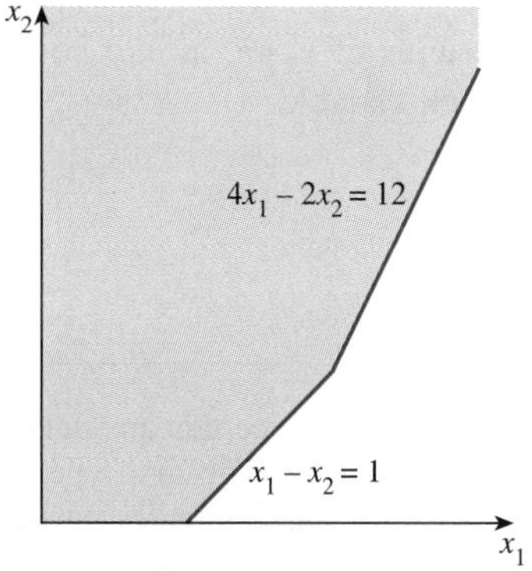

Since -20 is the most negative element in the last row, the pivot column is the second column. But all the ratios are negative! Since we cannot pick a pivot row, the simplex method stops. There is no solution. ∎

Recall from our previous analysis that having all ratios negative in the previous example indicates that x_2 can be increased without bound. Therefore, since $z = 10x_1 + 20x_2$, there can be no maximum.

The graph of the feasible region is shown in Figure 4.6. Notice the feasible region is unbounded. Since the objective function has positive numbers in front of the decision variables, no maximum can exist.

Figure 4.6

Post-Optimal Analysis

After an optimal solution to an applied linear programming problem has been found, further analysis is usually required. For example, an analysis is needed to determine if an excess of some resource is being allocated. If this is the case, then management should consider reallocating or reducing this resource.

E X A M P L E 5   **Determining Excess Resources**

For the wooden-boat manufacturing problem considered earlier, identify any of the three operations that has an excess of needed labor and determine the amount.

Solution   To answer this question we need to examine the slack variables for the final tableau obtained when using the simplex method. Recall that $s_1 = 0$, $s_2 = 0$, and $s_3 = 14$. Recall also that s_1 and s_2 are associated with the cutting constraint and the painting constraint, respectively. Thus there is no slack in the cutting and painting departments. However, $s_3 = 14$, and since this slack variable is associated with the assembly constraint, there is a slack of 14 hours per week in the assembly department. Management should consider eliminating or reallocating these hours. ∎

In general we can make the following statement.

Economic Interpretation of Slack Variables

The value of a slack variable in the final tableau indicates how much unused capacity is available for the associated constraint.

REMARK.   Recall that the assembly constraint was

$$4x_1 + 2x_2 \le 84.$$

It would have simplified calculations if we had divided this inequality by two *first* and then added a slack variable obtaining

$$2x_1 + x_2 + s_3' = 42.$$

But now we must be very careful to note that the units of s_3' are in *half* hours when we give an economic interpretation to this slack variable.

SELF-HELP EXERCISE SET 4.2

1. Two types of metal connectors require for their manufacture a certain amount of minutes on a stamping machine and on a grinding machine. The first type of connector requires 4 minutes on the stamping machine and 3 minutes on the grinding machine, while the second connector requires 5 minutes on the stamping machine and 1 minute on the grinding machine. The stamping machine is available at most 50 minutes each hour, and the grinding machine is available at most 21 minutes each hour. If each of the first type connector brings a profit of $0.50 and each of the second type of connector a profit of $0.60, use the simplex method to find the number of each of these products that should be manufactured to maximize profits. (See Self-Help Exercise 2 of Section 3.1 and Self-Help Exercise 2 of 3.2, where one of the constraints has been eliminated.)

EXERCISE SET 4.2

In Exercises 1 through 10 use the simplex method to obtain the next tableau. If the next tableau should be the final tableau, then indicate the solution.

1.

x_1	x_2	s_1	s_2	z	
2	4	1	0	0	8
5	1	0	1	0	25
-4	-2	0	0	1	0

2.

x_1	x_2	s_1	s_2	z	
3	5	1	0	0	15
2	2	0	1	0	8
-8	-5	0	0	1	0

3.

x_1	x_2	s_1	s_2	z	
2	1	5	0	0	5
5	0	5	1	0	10
-7	0	-3	0	1	0

4.

x_1	x_2	s_1	s_2	z	
2	0	1	2	0	2
-4	1	0	5	0	4
9	0	0	0	1	0

5.

x_1	x_2	s_1	s_2	z	
0	-4	1	2	0	2
1	3	0	3	0	3
0	-5	0	-2	1	0

6.

x_1	x_2	s_1	s_2	z	
1	10	1	0	0	50
0	10	30	1	0	20
0	-5	0	0	1	0

7.

x_1	x_2	s_1	s_2	s_3	z	
-5	0	1	0	2	0	21
-1	0	0	1	1	0	30
2	1	0	0	8	0	2
-10	0	0	0	40	1	30

8.

x_1	x_2	s_1	s_2	s_3	z	
1	0	0	1	1	0	5
3	0	1	-2	0	0	3
7	1	0	1	0	0	6
-3	0	0	-4	0	1	10

9.

x_1	x_2	s_1	s_2	s_3	z	
0	1	1	-3	0	0	0
0	8	0	6	1	0	2
1	1	0	0	0	0	2
0	2	0	-2	0	1	8

10.

x_1	x_2	s_1	s_2	s_3	z	
0	1	0	1	5	0	10
0	5	1	0	2	0	4
1	2	0	0	1	0	1
0	-1	0	0	-2	1	10

In Exercises 11 through 32 add the nonnegativity constraint and use the simplex method to solve. Where there are only 2 variables, graph the feasible region and note which corners the simplex method moves through.

11. Maximize $z = 2x_1 + 3x_2$

    Subject to $x_1 + 2x_2 \le 24$
                  $5x_1 + x_2 \le 30$

12. Maximize $z = 10x_1 + 20x_2$

    Subject to $x_1 + 3x_2 \le 36$
                  $5x_1 + x_2 \le 40$

13. Maximize $z = 3x_1 + 2x_2$

    Subject to $-x_1 + 2x_2 \le 1$
                  $x_1 - x_2 \le 1$

14. Maximize $z = 2x_1 + x_2$

    Subject to $x_1 + x_2 \le 8$
                  $2x_1 - x_2 \le 4$

15. Maximize $z = 2x_1 + x_2$

   Subject to $x_1 + x_2 \leq 8$

   $x_1 \leq 2$

16. Maximize $z = x_1 + 2x_2$

   Subject to $x_1 + x_2 \leq 8$

   $x_2 \leq 4$

17. Maximize $z = x_1 + 2x_2$

   Subject to $x_1 + x_2 \leq 6$

   $-x_1 + 2x_2 \leq 0$

   $x_2 \leq 4$

18. Maximize $z = 2x_1 + x_2$

   Subject to $-2x_1 + x_2 \leq 0$

   $x_1 + x_2 \leq 6$

   $x_1 \leq 4$

19. Maximize $z = x_1 + 2x_2$

   Subject to $x_1 + 3x_2 \leq 36$

   $x_1 + x_2 \leq 16$

   $2x_1 + x_2 \leq 24$

20. Maximize $z = 3x_1 + 2x_2$

   Subject to $x_1 + 2x_2 \leq 28$

   $x_1 + x_2 \leq 16$

   $2x_1 + x_2 \leq 24$

21. Maximize $z = x_1 + 2x_2 + x_3$

   Subject to $x_1 + x_2 + x_3 \leq 12$

   $2x_1 + x_2 + 4x_3 \leq 24$

22. Maximize $z = 2x_1 + x_2 + x_3$

   Subject to $x_1 + x_2 + x_3 \leq 4$

   $x_1 + 2x_2 + 3x_3 \leq 6$

23. Maximize $z = x_1 + x_2 + 2x_3$

   Subject to $x_1 + x_2 + x_3 \leq 4$

   $x_1 - 2x_2 + 2x_3 \leq 10$

   $2x_1 - x_2 + 3x_3 \leq 18$

24. Maximize $z = x_1 + 3x_2 + x_3$

   Subject to $x_1 + x_2 + 2x_3 \leq 2$

   $2x_1 - x_2 + 5x_3 \leq 3$

   $3x_1 + 2x_2 - 7x_3 \leq 6$

25. Maximize $z = 10x_1 + 20x_2 + 5x_3$

   Subject to $x_1 + 3x_3 \leq 1$

   $2x_1 + x_2 + x_3 \leq 2$

   $x_1 - x_2 + 2x_3 \leq 0$

   $3x_1 - 2x_2 + 4x_3 \leq 5$

26. Maximize $z = x_1 + 2x_2 + 3x_3$

   Subject to $3x_1 - x_2 \leq 5$

   $2x_1 + x_2 - 4x_3 \leq 7$

   $x_1 + 2x_2 + x_3 \leq 1$

   $x_1 + 3x_2 + 2x_3 \leq 6$

27. Maximize $\quad z = 2x_1 + 4x_2 + x_3$

    Subject to $\quad x_1 + x_2 + 2x_3 \le 200$
    $$10x_1 + 8x_2 + 5x_3 \le 2000$$
    $$2x_1 + x_2 \qquad \le 100$$

28. Maximize $\quad z = x_1 + 3x_2 + 4x_3$

    Subject to $2x_1 + x_2 + 2x_3 \le 4$
    $$x_1 + 2x_2 + 2x_3 \le 10$$
    $$3x_1 + 2x_2 + 2x_3 \le 12$$

29. Maximize $z = x_1 + 2x_2 + 3x_3$

    Subject to $2x_1 \qquad + x_3 \le 2$
    $$x_1 + 2x_2 + x_3 \le 4$$
    $$x_1 + x_2 + x_3 \le 8$$

30. Maximize $\quad z = x_1 + 2x_2 + x_3$

    Subject to $x_1 - x_2 + 2x_3 \le 3$
    $$x_1 + x_2 \qquad \le 1$$
    $$2x_1 \qquad + x_3 \le 4$$

31. Maximize $\quad z = x_1 + 2x_2$

    Subject to $\quad -x_1 + 2x_2 \le 4$
    $$-3x_1 + x_2 \le 12$$

32. Maximize $z = 3x_1 + 2x_2$

    Subject to $-x_1 + 3x_2 \le 2$
    $$x_1 - 4x_2 \le 2$$

Applications

33. **Manufacturing.**   Three types of connectors, A, B, and C, require for their manufacture a certain number of minutes on each of three machines: a stamping machine, a grinding machine, and a polishing machine. Connector A requires 1 minute on the stamping machine, 2 minutes on the grinding machine, and 3 minutes on the polishing machine. Connector B requires 2 minutes on the stamping machine, 1 minute on the grinding machine, and 1 minute on the polishing machine. Connector C requires 2 minutes on the stamping machine, 2 minutes on the grinding machine, and 2 minutes on the polishing machine. The stamping machine can operate 12 minutes each hour, the grinding machine 8 minutes each hour, and the polishing machine 16 minutes each hour. The profit on each of the connectors A, B, and C is, respectively, $35, $40, and $50 each. Find the number of each of the connectors to be manufactured in order to maximize profits. Find the maximum profit.

34. **Manufacturing.**   A firm manufactures three products: tables, desks, and bookcases. To produce each table requires 1 hour of labor, 10 square feet of wood, and 2 quarts of finish. To produce each desk requires 3 hours of labor, 35 square feet of wood, and 1 quart of finish. To produce each bookcase requires 0.50 hours of labor, 15 square feet of wood, and 1 quart of finish. Available is at most 25 hours of labor, at most 350 square feet of wood, and at most 55 quarts of finish. The table yields a profit of $4, the desk $3, and the bookcase $3. Find the number of each product to be made in order to maximize profits. Find the maximum profit.

35. **Mixture.** A dealer has 8 pounds of cashews, 24 pounds of almonds, and 36 pounds of peanuts to be used to make three mixtures. Mixture A sells for $2 per pound and consists of 20% cashews, 20% almonds, and 60% peanuts. Mixture B sells for $4 per pound and contains 20% cashews, 40% almonds, and 40% peanuts. Mixture C sells for $3 per pound and contains 10% cashews, 30% almonds, and 60% peanuts. How many pounds of each mixture should the dealer make in order to maximize revenue? Find the maximum revenue.

36. **Fishery.** A certain lake has three species of trout, rainbow, brown, and brook, and also three types of food for these fish, I, II, and III. Each month the lake can supply 800 pounds of food I, 500 pounds of food II, and 1100 pounds of food III. Each month each rainbow trout requires 1 pound of food I, 1 pound of food II, and 2 pounds of food III. Each brown trout requires 4 pounds of food I, 2 pounds of food II, and 1 pound of food III. Each brook trout requires 0.50 pounds of each of the foods. What is the maximum number of these trout that the lake can support and the number of each type of trout?

37. **Manufacturing.** A furniture manufacturer produces chairs, sofas, and love seats. The chairs require 5 feet of wood, 1 pound of foam rubber, and 10 square yards of material. The sofas require 35 feet of wood, 2 pounds of foam rubber, and 20 square yards of material. The love seats require 9 feet of wood, 0.2 pounds of foam rubber, and 10 square yards of material. The manufacturer has in stock 405 feet of wood, 25 pounds of foam rubber, and 410 square yards of material. If the chairs yield a profit of $300, the sofas $200, and the love seats $220 each, how many of each should be produced to maximize the income? Find the maximum income.

38. **Fishery.** If in Exercise 36, $3 can be obtained from harvesting the rainbow trout, $2 for the brown trout, and $2 for the brook trout, how many of each trout should be harvested to maximize revenue? Find the maximum revenue.

39. **Profits.** A company makes 3 styles of sleds: Classic, Sport, and Deluxe. The Classic uses 1 board-foot of wood, 0.5 pounds of iron, and 1.25 hours of labor. The Sport uses 4 board-feet of wood, 2 pounds of iron, and 1 hour of labor. The Deluxe uses 1 board-foot of wood, 1.5 pounds of iron, and .25 hours of labor. There are 56 board-feet of wood, 44 pounds of iron, and 34 hours of labor available. The profit on the Classic is $3, on the Sport $8, and on the Deluxe $4. Find the number of each type sled to make in order to maximize the profits and find the maximum profits.

40. **Profits.** A firm manufactures 3 products, each of which uses 3 ingredients. The first product requires 2 units of the first ingredient, 1 unit of the second, and 1 unit of the third. The second product requires 1 unit of the first ingredient, 1 unit of the second, and 3 units of the third. The third product requires 1 unit of the first ingredient, 3 units of the second, and 3 units of the third. There are only 20 units of the first ingredient, 30 units of the second ingredient, and 40 units of the third ingredient available. The profit on each of the first product is $26, on the second is $24, and on the third is $30. Find the number of each product that should be made in order to maximize profits and find the maximum profits.

41. Give an economic interpretation of the slack variables associated with the optimal solutions found in Exercises 35, 37, and 39 and determine in each case which resources are in excess.

42. Give an economic interpretation of the slack variables associated with the optimal solutions found in Exercises 36, 38, and 40 and determine in each case which resources are in excess.

Graphing Calculator Exercises

Do Exercises 43 through 44 using the row operations on your graphing calculator.

43. Maximize $z = 90x_1 + 210x_2 + 52x_3$

Subject to $11x_1 + 12x_2 + 21x_3 \leq 2100$
$12x_1 + 9x_2 + 6x_3 \leq 1700$
$8x_1 + 4x_2 \qquad \leq 400$
$x_1, x_2, x_3 \geq 0$

44. Maximize $z = 81x_1 + 89x_2 + 153x_3$

Subject to $22x_1 + 9x_2 + 43x_3 \leq 3500$
$19x_1 + 48x_2 + 102x_3 \leq 9000$
$42x_1 + 28x_2 + 8x_3 \leq 4900$
$x_1, x_2, x_3 \geq 0$

Solutions to Self-Help Exercise Set 4.2

1. If x_1 is the number of the first type connector and x_2 the number of the second type connector, then the problem is to maximize the objective function $z = 50x_1 + 60x_2$ subject to the constraints

$$4x_1 + 5x_2 \leq 50$$
$$3x_1 + x_2 \leq 21$$
$$x_1, x_2 \geq 0$$

The initial tableau is

x_1	x_2	s_1	s_2	z		
4	⑤	1	0	0	50	$\dfrac{50}{5} = 10$
3	1	0	1	0	21	$\dfrac{21}{1} = 21$
-50	-60	0	0	1	0	

Since -60 is the most negative element in the last row, the pivot column is the second column. The above ratios indicate that the pivot row is the first row. Dividing the first row by 5 yields

x_1	x_2	s_1	s_2	z		
$\frac{4}{5}$	①	$\frac{1}{5}$	0	0	10	
3	1	0	1	0	21	$R_2 - R_1 \rightarrow R_2$
-50	-60	0	0	1	0	$R_3 + 60R_1 \rightarrow R_3$

Performing the indicated row operations then yields

x_1	x_2	s_1	s_2	z		
$\frac{4}{5}$	1	$\frac{1}{5}$	0	0	10	$\dfrac{10}{\frac{4}{5}} = 12.5$
⑪⁄₅	0	$-\frac{1}{5}$	1	0	11	$\dfrac{11}{\frac{11}{5}} = 5$
-2	0	12	0	1	600	

Notice that this tableau corresponds to the point $(0, 10)$.

The last row contains only one negative element, thus the pivot column is the first column. The two ratios indicate that the pivot row is the second row. Dividing the second row by $\frac{11}{5}$, then yields the next tableau

x_1	x_2	s_1	s_2	z		
$\frac{4}{5}$	1	$\frac{1}{5}$	0	0	10	$R_1 - \frac{4}{5}R_2 \rightarrow R_1$
①	0	$-\frac{1}{11}$	$\frac{5}{11}$	0	5	
-2	0	12	0	1	600	$R_3 + 2R_2 \rightarrow R_3$

Performing the indicated row operations then gives

x_1	x_2	s_1	s_2	z	
0	1	$\frac{3}{11}$	$-\frac{4}{11}$	0	6
1	0	$-\frac{1}{11}$	$\frac{5}{11}$	0	5
0	0	$11\frac{9}{11}$	$\frac{10}{11}$	1	610

This yields $x_1 = 5$ and $x_2 = 6$. This solution was also found in the previous chapter by the geometric method. Notice the simplex method started at the corner $(0, 0)$, moved to $(0, 10)$, and then finally to $(5, 6)$. Since in the final tableau the two slack variables are zero, there is no slack in the operation of the two machines.

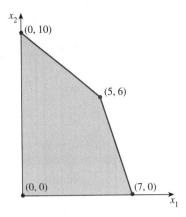

4.3 MINIMIZATION PROBLEMS AND THE DUAL

▶ *The Dual Problem*

▶ *Economic Interpretation of Dual Problems*

▶ *Matrix Formulation*

In the last two sections we considered maximization problems with the inequality ≤ in all the problem constraints. In this section we will consider minimization problems with the inequality ≥ in all the problem constraints. To solve these minimization problems we will need to consider an associated problem called the *dual*.

Associated with any linear program problem, referred to as the **primal problem**, is another linear programming problem, called the **dual**. The dual problem is important for several reasons. First, the solution of the dual problem can be used to find the solution of the primal problem. Secondly, the dual problem gives an interesting economic interpretation that adds additional insights into the primal problem.

The Dual Problem

We first consider a typical minimization problem.

EXAMPLE 1 **Dual of a Minimization Problem**

An animal rancher requires at least 2 units per day of nutrient A, 4 units of nutrient B, and 3 units of nutrient C for his animals. The rancher has two different mixtures of feeds available, feed 1 and 2. If each bag of feed 1 costs $12 and yields 1 unit of each of the nutrients and each bag of feed 2 costs $24 and yields 2 units of nutrient A, 1 unit of B, and 4 units of C, find the linear programming problem needed to find the number of bags of each feed that should be used to minimize the cost.

Solution   The information can be summarized in the following table.

	Feed 1	Feed 2	Requirements
Nutrient A	1	2	2
Nutrient B	1	1	4
Nutrient C	1	4	3
Costs	12	24	

If y_1 is the number of bags of nutrient 1 and y_2 is the number of bags of nutrient 2, then the linear programming problem that we must solve is

Minimize $w = 12y_1 + 24y_2$ objective function

Subject to
$$y_1 + 2y_2 \geq 2$$
$$y_1 + y_2 \geq 4$$
$$y_1 + 4y_2 \geq 3$$
problem constraints

$$y_1, y_2 \geq 0$$ nonnegativity constraint ∎

We shall see how to solve this problem using the simplex method. However, to do this we must first formulate and solve an associated problem called the dual problem. We will refer to the original problem as the **primal problem**.

If the primal problem is a minimization problem with all problem constraints containing the inequality $\geq$, then the dual problem is a maximization problem with all problem constraints containing the inequality $\leq$. The number of decision variables in the dual problem equals the number of problem constraints in the primal problem, while the number of problem constraints in the dual problem equals the number of decision variables in the primal problem. Following are further details for determining the dual problem for Example 1.

Primal Problem   **Dual Problem**

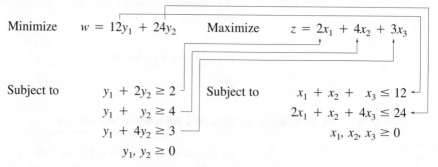

Minimize $w = 12y_1 + 24y_2$ Maximize $z = 2x_1 + 4x_2 + 3x_3$

Subject to $y_1 + 2y_2 \geq 2$ Subject to $x_1 + x_2 + x_3 \leq 12$
$y_1 + y_2 \geq 4$ $2x_1 + x_2 + 4x_3 \leq 24$
$y_1 + 4y_2 \geq 3$ $x_1, x_2, x_3 \geq 0$
$y_1, y_2 \geq 0$

First column becomes first row.

coefficient matrix $\begin{bmatrix} 1 & 2 \\ 1 & 1 \\ 1 & 4 \end{bmatrix}$ coefficient matrix $\begin{bmatrix} 1 & 1 & 1 \\ 2 & 1 & 4 \end{bmatrix}$

Second column becomes second row.

Formation of the Dual Problem

Given a minimization problem with $\geq$ problem constraints, the following rules apply.

1.   The dual problem is a maximization problem with $\leq$ problem constraints.
2.   The coefficients of the decision variables in the primal problem become the constants to the right of the inequality $\leq$ in the problem constraints of the dual.
3.   The constants to the right of the inequality $\geq$ in the problem constraints of the primal problem become the coefficients of the decision variables in the dual.
4.   The coefficients of the ith decision variable in the problem constraints for the primal problem become the coefficients of the decision variables in the ith problem constraint of the dual.

More precisely, we have the following.

The Dual Problem

If the primal problem has the form

$$\left.\begin{array}{rl} \text{Maximize} & z = c_1x_1 + c_2x_2 + \cdots + c_nx_n \\ \text{Subject to} & a_{11}x_1 + a_{12}x_2 + \cdots + a_{1n}x_n \leq b_1 \\ & a_{21}x_1 + a_{22}x_2 + \cdots + a_{2n}x_n \leq b_2 \\ & \cdots \\ & a_{m1}x_1 + a_{m2}x_2 + \cdots + a_{mn}x_n \leq b_m \\ & x_1, x_2, \ldots, x_n \geq 0 \end{array}\right\} \quad (1)$$

then the dual problem is

$$\left.\begin{array}{rl} \text{Minimize} & w = b_1y_1 + b_2y_2 + \cdots + b_my_m \\ \text{Subject to} & a_{11}y_1 + a_{21}y_2 + \cdots + a_{m1}y_m \geq c_1 \\ & a_{12}y_1 + a_{22}y_2 + \cdots + a_{m2}y_m \geq c_2 \\ & \cdots \\ & a_{1n}y_1 + a_{2n}y_2 + \cdots + a_{mn}y_m \geq c_n \\ & y_1, y_2, \ldots, y_m \geq 0 \end{array}\right\} \quad (2)$$

Also if the primal problem is (2), then the dual is (1).

R E M A R K .   In this section we will reserve $x_1, x_2, \ldots$ for the decision variables in the maximization problems and $y_1, y_2, \ldots$ for the decision variables in the minimization problems.

E X A M P L E 2   **Dual of a Maximization Problem**

Find the dual problem for the wooden-boat manufacturing problem considered in this and the last chapter. Recall that the problem is

$$\text{Maximize} \quad z = 60x_1 + 80x_2$$
$$\text{Subject to} \quad 2x_1 + 4x_2 \leq 80$$
$$2x_1 + 2x_2 \leq 50$$
$$4x_1 + 2x_2 \leq 84$$
$$x_1, x_2 \geq 0$$

Solution   The dual must be a minimization problem with $\geq$ constraints and is

$$\text{Minimize} \quad w = 80y_1 + 50y_2 + 84y_3$$
$$\text{Subject to} \quad 2y_1 + 2y_2 + 4y_3 \geq 60$$
$$4y_1 + 2y_2 + 2y_3 \geq 80$$
$$y_1, y_2, y_3 \geq 0 \quad \blacksquare$$

To solve a minimization problem such as found in Example 1, first form the dual and then use the following theorem.

Fundamental Theorem of Duality

The objective function z in the linear programming problem (1) assumes a maximum value if, and only if, the dual linear programming problem (2) assumes a minimum value. Furthermore,

1. The maximum value of z in (1) equals the minimum value of w in (2).
2. The optimal solution to the primal problem can be found under the slack variables in the last row of the final tableau associated with the dual problem.

We will now use this theorem to solve the problem in Example 1.

E X A M P L E 3   **Using the Fundamental Theorem of Duality**

Use the fundamental theorem of duality to solve the primal problem given in Example 1.

Solution   The dual problem was found in Example 1. The initial tableau for this dual problem is

x_1	x_2	x_3	s_1	s_2	z		ratios
1	①	1	1	0	0	12	$\frac{12}{1}$ ← smallest ratio
2	1	4	0	1	0	24	$\frac{24}{1}$
-2	-4	-3	0	0	1	0	

↑
pivot column

Pivoting about the first row and second column yields

x_1	x_2	x_3	s_1	s_2	z	
1	1	1	1	0	0	12
1	0	3	-1	1	0	12
2	0	1	4	0	1	48

This is the final tableau. Now consider the last row.

x_1	x_2	x_3	s_1	s_2	z	
2	0	1	4	0	1	48

Looking under the slack variables in the last row, we see a 4 and a 0. Thus the solution to the primal problem is (4, 0). The minimum of the primal problem is the same as the maximum of the dual problem, which is found in the last entry of the last row to be 48.

Thus 4 bags of feed 1 and no bags of feed 2 should be purchased in order to obtain the minimum cost of $48 while meeting the nutritional requirement given in the constraints. ∎

The following problem was formulated in Example 3 of Section 3.3.

E X A M P L E 4 **Using the Dual to Solve a Minimization Problem**

A firm has two stockpiles of an item stored in warehouses at two different locations, Denver and Santa Fe. The item must be shipped to two plants, located in Des Moines and Fort Wayne, for assembly into a final product. At the beginning of a week the stockpile at Denver has 150 items and the stockpile at Santa Fe has 180. The Des Moines plant needs at least 100 of these items, and the Fort Wayne plant needs at least 120 that week. The cost of sending each item from the stockpile at Denver to the plants in Des Moines and Fort Wayne is, respectively, $200 and $300, while the cost of sending each item from the stockpile at Santa Fe to

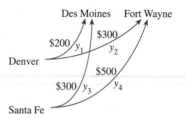

Figure 4.7

the plants in Des Moines and Fort Wayne is, respectively, $300 and $500. Refer to Figure 4.7. Find the number of items that should be shipped from each stockpile to each plant in order to minimize cost.

Solution   The following table summarizes the cost of shipping the items.

	Des Moines Plant	*Fort Wayne Plant*	*Supply*
Stockpile at Denver	$200	$300	150
Stockpile at Santa Fe	$300	$500	180
Demand	100	120	

Let y_1 and y_2 be the number of items shipped from the stockpile in Denver to the plants in Des Moines and Fort Wayne, respectively. Let y_3 and y_4 be the number of items shipped from the stockpile in Santa Fe to the plants in Des Moines and Fort Wayne, respectively.

Then the problem is

$$\text{Minimize} \quad w = 200y_1 + 300y_2 + 300y_3 + 500y_4$$

$$\begin{aligned}
\text{Subject to} \quad y_1 + y_2 \quad\quad\quad &\leq 150 \\
y_3 + y_4 &\leq 180 \\
y_1 \quad\quad + y_3 \quad\quad &\geq 100 \\
y_2 \quad\quad + y_4 &\geq 120 \\
y_1,\, y_2,\, y_3,\, y_4 &\geq \quad 0
\end{aligned}$$

The first two problem constraints have $\leq$ inequalities. We correct this by multiplying both of these inequalities by -1 and obtain

$$\text{Minimize} \quad w = 200y_1 + 300y_2 + 300y_3 + 500y_4$$

$$\begin{aligned}
\text{Subject to} \quad -y_1 - y_2 \quad\quad\quad &\geq -150 \\
- y_3 - y_4 &\geq -180 \\
y_1 \quad\quad + y_3 \quad\quad &\geq \quad 100 \\
y_2 \quad\quad + y_4 &\geq \quad 120 \\
y_1,\, y_2,\, y_3,\, y_4 &\geq \quad\quad 0
\end{aligned}$$

At first there seems to be a serious difficulty with this formulation of the problem since the constants to the right of the first two inequalities are *negative*. But we will solve this problem by first solving the dual problem, which does not have this difficulty. The dual problem is

$$\text{Maximize} \quad z = -150x_1 - 180x_2 + 100x_3 + 120x_4$$

$$\text{Subject to} \qquad \begin{array}{rcl} -x_1 \quad\quad + x_3 \quad\quad & \leq & 200 \\ -x_1 \quad\quad\quad\quad + x_4 & \leq & 300 \\ -x_2 + x_3 \quad\quad & \leq & 300 \\ -x_2 \quad\quad + x_4 & \leq & 500 \\ x_1, x_2, x_3, x_4 & \geq & 0 \end{array}$$

As we see, this problem is a *standard maximization problem*. The initial tableau is

x_1	x_2	x_3	x_4	s_1	s_2	s_3	s_4	z		ratios
-1	0	1	0	1	0	0	0	0	200	
-1	0	0	①	0	1	0	0	0	300	$\frac{300}{1} \leftarrow$ smallest ratio
0	-1	1	0	0	0	1	0	0	300	
0	-1	0	1	0	0	0	1	0	500	$\frac{500}{1}$
150	180	-100	-120	0	0	0	0	1	0	

$$\uparrow$$
$$\text{pivot column}$$

Pivoting about the second row and fourth column yields

x_1	x_2	x_3	x_4	s_1	s_2	s_3	s_4	z		ratios
-1	0	①	0	1	0	0	0	0	200	$\frac{200}{1} \leftarrow$ smallest ratio
-1	0	0	1	0	1	0	0	0	300	
0	-1	1	0	0	0	1	0	0	300	$\frac{300}{1}$
1	-1	0	0	0	-1	0	1	0	200	
30	180	-100	0	0	120	0	0	1	$36{,}000$	

$$\uparrow$$
$$\text{pivot column}$$

Pivoting about the first row and third column yields

x_1	x_2	x_3	x_4	s_1	s_2	s_3	s_4	z		ratios
-1	0	1	0	1	0	0	0	0	200	
-1	0	0	1	0	1	0	0	0	300	
①	-1	0	0	-1	0	1	0	0	100	$\frac{100}{1}$ ← smallest ratio
1	-1	0	0	0	-1	0	1	0	200	$\frac{200}{1}$
-70	180	0	0	100	120	0	0	1	56,000	

↑
pivot column

Pivoting about the third row and first column yields

x_1	x_2	x_3	x_4	s_1	s_2	s_3	s_4	z	
0	-1	1	0	0	0	1	0	0	300
0	-1	0	1	-1	1	1	0	0	400
1	-1	0	0	-1	0	1	0	0	100
0	0	0	0	1	-1	-1	1	0	100
0	110	0	0	30	120	70	0	1	63,000

This is the final tableau. Since the maximum of this problem is 63,000, the minimum of the primal problem is also 63,000. We find the solution to the primal problem under the slack variables in the last row to be (30, 120, 70, 0). ∎

E X A M P L E 5   No Solution

Solve

$$\text{Minimize} \qquad w = 2y_1 + 4y_2$$
$$\text{Subject to} \qquad y_1 - y_2 \geq 1$$
$$-2y_1 + y_2 \geq 3$$
$$y_1, y_2 \geq 0$$

Solution   The dual is

$$\text{Maximize} \qquad z = x_1 + 3x_2$$
$$\text{Subject to} \qquad x_1 - 2x_2 \leq 2$$
$$-x_1 + x_2 \leq 4$$
$$x_1, x_2 \geq 0$$

The initial tableau is

x_1	x_2	s_1	s_2	z		ratio
1	−2	1	0	0	2	
−1	①	0	1	0	4	$\frac{4}{1}$ ← smallest ratio
−1	−3	0	0	1	0	

Pivoting about the second row, second column then gives

x_1	x_2	s_1	s_2	z	
−1	0	1	2	0	10
−1	1	0	1	0	4
−4	0	0	3	1	12

The pivot column is the first, however, there are only negative ratios. Thus, according to the simplex method, there is no solution. ∎

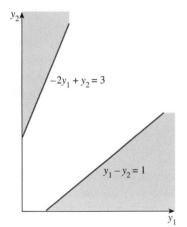

Figure 4.8

Figure 4.8 indicates that the feasible region for the primal problem in the last example is *empty*. Thus there can be no solution.

In the last section we noted that dividing a problem constraint by a constant will still lead to the correct answer, but that there will be a change in the economic interpretation of the corresponding slack variable. However, serious problems can arise when dividing a problem constraint by a constant in a dual problem! Suppose, for example, that we decide to solve a primal problem by solving the dual problem. *We must never divide any problem constraint in the dual problem by a constant since this will lead to an incorrect answer to the primal problem.* The difficulty arises because dividing a problem constraint by a constant in the dual problem will change the corresponding slack variable. Using the fundamental theorem of duality will then lead to an incorrect answer to the primal problem.

WARNING. When using the dual problem to solve a primal problem, never divide a problem constraint by a constant in the dual problem, since this will lead to an incorrect answer to the primal problem.

To illustrate this point, you are invited to divide the second problem constraint in the dual problem in Self-Help Exercise 2 by 2. The last row of the final tableau will yield slack variables of 2 and 6, which in turn would give the solution (2, 6) to the primal problem. This is not correct, the correct solution is (2, 3).

We can, however, divide a *primal* problem constraint by a constant without causing any difficulties.

Economic Interpretation of Dual Problems

Let us now consider the economic interpretation of the dual problem to the problem of the wooden-boat builder of Example 2. First recall the table that gives the allocation of hours per week among the three operations.

Operation	Dinghies	Skiffs	Hours Available
Cutting	2	4	80
Painting	2	2	50
Assembly	4	2	84
Profit	60	80	

Suppose an entrepreneur wishes to purchase all of the resources of the wooden-boat company. Then the entrepreneur must determine the price to be paid for each of the resources. Let us define

y_1 = price paid for 1 hour of labor in cutting

y_2 = price paid for 1 hour of labor in painting

y_3 = price paid for 1 hour of labor in assembly.

Economists refer to these prices as **shadow prices**. (Shadow prices will be discussed in more detail in Section 4.5.) We will now show that the resource prices, y_1, y_2, y_3, should be determined by solving the dual problem. The total cost for all of these resources (per week) is $80y_1 + 50y_2 + 84y_3$. Since the entrepreneur wishes to minimize this cost, the problem is to

$$\text{minimize} \quad 80y_1 + 50y_2 + 84y_3.$$

This is precisely the objective function for the dual.

Now let us look at the constraints on the resource prices. Naturally the resource prices must be set high enough to induce the wooden-boat manufacturer to sell each one of the resources. For example, the boat manufacturer, in order to be induced to sell the production of dinghies, will demand that the cost of the labor, $2y_1 + 2y_2 + 4y_3$, be at least $60, or the boat manufacturer will simply continue to build the dinghies and make more profits. Thus for the entrepreneur to hope to buy the production of dinghies,

$$2y_1 + 2y_2 + 4y_3 \geq 60.$$

Similarly, to induce the boat manufacturer to sell the production of skiffs,

$$4y_1 + 2y_2 + 2y_3 \geq 80.$$

These are then the two constraints in the dual problem.

We now turn to the economic interpretation of a minimization problem.

E X A M P L E   6   **Economic Interpretation of the Dual**

Give an economic interpretation of the dual to the linear programming problem in Example 1.

Solution   The information can be summarized in the following table.

	Feed 1	Feed 2	Requirements
Nutrient A	1	2	2
Nutrient B	1	1	4
Nutrient C	1	4	3
Costs	12	24	

The primal and dual problems are

<table>
<tr><td colspan="2" align="center">**Primal**</td><td colspan="2" align="center">**Dual**</td></tr>
<tr><td>Minimize</td><td>$w = 12y_1 + 24y_2$</td><td>Maximize</td><td>$z = 2x_1 + 4x_2 + 3x_3$</td></tr>
<tr><td>Subject to</td><td>$y_1 + 2y_2 \geq 2$</td><td>Subject to</td><td>$x_1 + x_2 + x_3 \leq 12$</td></tr>
<tr><td></td><td>$y_1 + y_2 \geq 4$</td><td></td><td>$2x_1 + x_2 + 4x_3 \leq 24$</td></tr>
<tr><td></td><td>$y_1 + 4y_2 \geq 3$</td><td></td><td>$x_1, x_2, x_3 \geq 0$</td></tr>
<tr><td></td><td>$y_1, y_2 \geq 0$</td><td></td><td></td></tr>
</table>

To interpret the dual of this diet problem, suppose you are a salesman, selling nutrients A, B, and C. You want to ensure that the rancher will obtain the required amount of the three nutrients from you. You must determine

$$x_1 = \text{price per unit of nutrient A to charge rancher}$$

$$x_2 = \text{price per unit of nutrient B to charge rancher}$$

$$x_3 = \text{price per unit of nutrient C to charge rancher.}$$

You wish to maximize your revenue from selling the rancher the required nutrients for the animals. You will receive $2x_1 + 4x_2 + 3x_3$ dollars in revenue from the rancher. Thus you wish to

$$\text{maximize} \quad 2x_1 + 4x_2 + 3x_3.$$

This is precisely the objective function for the dual problem.

You must set the prices of the nutrients low enough to induce the rancher to buy all the needed nutrients from you. Your price for feed 1 is $x_1 + x_2 + x_3$. This must be at most $12. Thus

$$x_1 + x_2 + x_3 \leq 12.$$

Similarly, considering your price for feed 2,

$$2x_1 + x_2 + 4x_3 \leq 24.$$

These are the two constraints in the dual problem.   ∎

Matrix Formulations*

Problem (1) can be placed in a convenient matrix form no matter what the signs of the a_{ij}'s and the b_i's. If we define

$$A = \begin{bmatrix} a_{11} & a_{12} & \cdots & a_{1n} \\ a_{21} & a_{22} & \cdots & a_{2n} \\ & & \cdots & \\ a_{m1} & a_{m2} & \cdots & a_{mn} \end{bmatrix}, \quad X = \begin{bmatrix} x_1 \\ x_2 \\ \vdots \\ x_n \end{bmatrix}, \quad Y = \begin{bmatrix} y_1 \\ y_2 \\ \vdots \\ y_m \end{bmatrix},$$

$$B = \begin{bmatrix} b_1 \\ b_2 \\ \vdots \\ b_m \end{bmatrix}, \quad C = [c_1 \quad c_2 \quad \cdots \quad c_n],$$

then (1) can be written as follows.

Matrix Form of (1)

$$\text{Maximize} \quad z = CX$$
$$\text{Subject to} \quad AX \leq B$$
$$X \geq 0$$

We now introduce a convenient matrix operation. Given a matrix such as

$$\begin{bmatrix} 1 & 3 & 4 \\ 2 & 5 & 9 \end{bmatrix},$$

we can create another matrix whose rows are the columns of A. We call this matrix the **transpose** and denote the transpose by A^T. Thus if A is given above,

$$A^T = \begin{bmatrix} 1 & 2 \\ 3 & 5 \\ 4 & 9 \end{bmatrix}.$$

*Optional

Transpose of a Matrix

The transpose of a matrix A, denoted by A^T, is the matrix obtained by interchanging the rows and columns of A, with the first row of A becoming the first column of A^T, the second row of A becoming the second column of A^T, and so on.

An examination of problems (1) and (2) indicates that the dual problem (2) can be written using the transpose notation as

---

Matrix Form of (2)

$$\text{Minimize} \quad w = B^T Y$$
$$\text{Subject to} \quad A^T Y \geq C^T$$
$$Y \geq 0$$

The definition of the dual can then be given in matrix form. We have the following.

Matrix Form of a Dual Problem

If the primal problem is

$$\left.\begin{array}{rl} \text{Maximize} & z = CX \\ \text{Subject to} & AX \leq B \\ & X \geq 0 \end{array}\right\} \tag{1}$$

Then the dual problem is

$$\left.\begin{array}{rl} \text{Minimize} & w = B^T Y \\ \text{Subject to} & A^T Y \geq C^T \\ & Y \geq 0 \end{array}\right\} \tag{2}$$

Also if (2) is the primal problem, then (1) is the dual problem.

E X A M P L E   7   **Finding the Matrix Formulation of a Linear Programming Problem**

Give the matrix formulation of the linear programming problem in the last chapter involving the maximization of profits in a wooden-boat builder.

$$\text{Maximize} \quad z = 60x_1 + 80x_2$$
$$\text{Subject to} \quad 2x_1 + 4x_2 \leq 80$$
$$2x_1 + 2x_2 \leq 50$$
$$4x_1 + 2x_2 \leq 84$$
$$x_1, x_2 \geq 0$$

Solution   If

$$A = \begin{bmatrix} 2 & 4 \\ 2 & 2 \\ 4 & 2 \end{bmatrix}, \quad X = \begin{bmatrix} x_1 \\ x_2 \end{bmatrix}, \quad B = \begin{bmatrix} 80 \\ 50 \\ 84 \end{bmatrix}, \quad C = \begin{bmatrix} 60 & 80 \end{bmatrix},$$

then the problem can be written as

$$\begin{aligned}
\text{Maximize} \quad & z = CX \\
\text{Subject to} \quad & AX \le B \\
& X \ge 0 \quad \blacksquare
\end{aligned}$$

E X A M P L E 8   **Finding the Dual Using the Matrix Formulation**

Give the matrix formulation of the dual problem to the linear programming problem in Example 7.

Solution   The matrix formulation of the dual problem is

$$\begin{aligned}
\text{Minimize} \quad & w = B^T Y \\
\text{Subject to} \quad & A^T Y \ge C^T \\
& Y \ge 0,
\end{aligned}$$

where

$$B^T = \begin{bmatrix} 80 \\ 50 \\ 84 \end{bmatrix}^T = \begin{bmatrix} 80 & 50 & 84 \end{bmatrix}, \quad A^T = \begin{bmatrix} 2 & 4 \\ 2 & 2 \\ 4 & 2 \end{bmatrix}^T = \begin{bmatrix} 2 & 2 & 4 \\ 4 & 2 & 2 \end{bmatrix},$$

$$C^T = \begin{bmatrix} 60 & 80 \end{bmatrix}^T = \begin{bmatrix} 60 \\ 80 \end{bmatrix}. \quad \blacksquare$$

SELF-HELP EXERCISE SET 4.3

1. Rambler Motors has two plants that produce two distinctive vehicles, the Rambler and the Convertible. The first plant can make three Ramblers and one convertible each month and costs $30,000 per month to operate. The second plant can make two Ramblers and two Convertibles each month and costs $40,000 per month to operate. If at least 12 Ramblers and at least 8 Convertibles are needed, how many months should each plant operate to minimize costs? For convenience let the cost function be in tens of thousands of dollars.

2. Give the economic interpretation of the dual in the previous exercise.

EXERCISE SET 4.3

In Exercises 1 through 18 add the nonnegativity constraint and solve by using the dual. In some exercises you will need to multiply an inequality by -1 first.

1. Minimize $w = 3y_1 + 4y_2$

Subject to $\quad 2y_1 + y_2 \geq 3$

$\qquad\qquad y_1 + y_2 \leq 2$

2. Minimize $w = 4y_1 + 5y_2$

Subject to $3y_1 + y_2 \geq 4$

$\qquad\qquad y_1 + 2y_2 \geq 3$

3. Minimize $w = 3y_1 + 4y_2$

Subject to $3y_1 + 2y_2 \geq 11$

$\qquad\qquad\quad y_1 \leq 3$

$\qquad\qquad\quad y_2 \leq 4$

4. Minimize $w = y_1 + 2y_2$

Subject to $\quad y_1 + y_2 \geq 2$

$\qquad\qquad y_1 + y_2 \leq 4$

5. Minimize $w = 10y_1 + 20y_2$

Subject to $\quad y_1 + 2y_2 \geq 7$

$\qquad\qquad\quad y_1 \leq 3$

$\qquad\qquad\quad y_1 \geq 1$

$\qquad\qquad\quad y_2 \leq 4$

6. Minimize $w = y_1 + 2y_2$

Subject to $\quad y_1 + y_2 \geq 2$

$\qquad\qquad\quad y_1 \geq 1$

$\qquad\qquad\quad y_2 \geq 1$

7. Minimize $w = 2y_1 + 3y_2$

Subject to $4y_1 + y_2 \geq 8$

$\qquad\qquad y_1 + y_2 \geq 5$

$\qquad\qquad y_1 + 2y_2 \geq 6$

8. Minimize $w = 20y_1 + 30y_2$

Subject to $\qquad 3y_1 + y_2 \geq 8$

$\qquad\qquad\quad y_1 + y_2 \geq 4$

$\qquad\qquad\qquad\quad y_2 \geq 1$

9. Minimize $w = 3y_1 + 4y_2$

Subject to $3y_1 + 2y_2 \geq 14$

$\qquad\qquad\quad y_1 \geq 2$

$\qquad\qquad\quad y_2 \geq 1$

10. Minimize $w = y_1 + 2y_2$

Subject to $-y_1 + y_2 \geq 0$

$\qquad\qquad y_1 + y_2 \geq 4$

$\qquad\qquad\qquad y_2 \geq 1$

11. Minimize $w = 2y_1 + 2y_2 + y_3$

Subject to $\quad y_1 + y_2 + y_3 \leq 8$

$\qquad\qquad\quad y_2 + y_3 \geq 1$

$\qquad\qquad y_1 + y_2 \qquad \geq 1$

12. Minimize $w = y_1 + 2y_2 + y_3$

Subject to $y_1 + 2y_2 + 2y_3 \geq 4$

$\qquad\quad y_1 + 2y_2 + y_3 \leq 24$

$\qquad\quad y_1 + y_2 + y_3 \geq 1$

13. Minimize $w = 16y_1 + 20y_2 + 9y_3$

Subject to $\quad y_1 + 4y_2 + y_3 \geq 66$

$\qquad\qquad 4y_1 + 2y_2 + y_3 \geq 72$

$\qquad\qquad 4y_1 + 10y_2 + 5y_3 \geq 200$

14. Minimize $w = y_1 + 2y_2 + 2y_3$

Subject to $y_1 + y_2 + y_3 \leq 4$

$y_1 \geq 1$

$y_1 + y_2 + y_3 \geq 1$

15. Minimize $w = 2y_1 + y_2 + y_3$

Subject to $y_1 + 2y_2 + y_3 \leq 4$

$y_1 + y_2 \geq 1$

$y_3 \geq 2$

16. Minimize $w = 7y_1 + 8y_2$

Subject to $y_1 + y_2 \geq 10$

$4y_1 + 5y_2 \leq 80$

$3y_1 + 4y_2 \geq 36$

17. Minimize $w = 2y_1 + 3y_2$

Subject to $y_1 - 2y_2 \geq 2$

$-4y_1 + y_2 \geq 1$

18. Minimize $w = 7y_1 + 8y_2$

Subject to $y_1 - y_2 \leq 1$

$y_1 - 2y_2 \geq 2$

19. Write the linear programming problems in Exercises 7 through 10 in matrix form. Then write the duals in matrix form.

20. Write the linear programming problems in Exercises 11 through 14 in matrix form by multiplying equations by -1 if necessary. Then write the duals in matrix form.

Applications

21. Nutrition. A dietitian must arrange a meal for a patient using two fruits— oranges and strawberries. Each orange costs 75 cents and contains 1 gram of fiber and 75 mg of vitamin C, while each cup of strawberries costs 75 cents and contains 2 grams of fiber and 60 mg of vitamin C. How many of each of these fruits needs to be eaten to minimize the cost and to ensure that a total of at least 8 grams of fiber and 420 mg of vitamin C will be consumed? What is the minimum cost?

22. Nutrition. Suppose in the previous exercise the cost of each orange is 25 cents. Now answer the same questions.

23. Costs. An oil company has two refineries and must refine and deliver at least 40,000 gallons of medium-grade gasoline and 24,000 gallons of high-grade gas- oline to its stations. The first plant can refine 8000 gallons of medium-grade gasoline and 3000 gallons of high-grade gasoline and costs $40,000 a day to operate. The second plant can refine 4000 gallons of medium-grade gasoline and 6000 gallons of high-grade gasoline and costs $35,000 a day to operate. How many days should each plant operate to minimize the cost and what is the minimum cost? *Hint:* Minimize cost in thousands of dollars and divide each constraint by an appropriate constant.

24. Cost. An airline has two types of airplanes and has contracted with a tour group to provide accommodations for a minimum of each of 80 first-class, 50 tourist, and 60 economy-class passengers. The first plane costs $4000 for the trip and can accommodate 40 first-class, 10 tourist, and 10 economy-class pas- sengers, while the second plane costs $5000 for the trip and can accommodate

10 first-class, 10 tourist, and 20 economy-class passengers. How many of each type of airplane should be used to minimize the operating cost? *Hint:* Divide each of the constraints by 10 to simplify and minimize cost in thousands of dollars.

25. **Investments.** An investment made for a retirement plan must pay particular attention to risk. Suppose for such an investment no more than $10,250 is to be placed in three investments with estimated annual returns of 8%, 9%, and 10% respectively. The investor requires an annual return of at least $900 on the investment and wishes to minimize the risk. The second investment is twice as risky as the first, and the third investment is 4 times as risky as the first. How should the money be allocated to minimize the risk? *Hint:* Let risk = $y_1 + 2y_2 + 4y_3$.

26. **Diet.** An individual needs a daily supplement of at least 500 units of vitamin C and 200 units of vitamin E and agrees to obtain this supplement by eating two foods, I and II. Each ounce of food I contains 40 units of vitamin C and 10 units of vitamin E, while each ounce of food II contains 20 units of vitamin C and also 20 units of vitamin E. The total amount of these two foods must be at most 25 ounces. Unfortunately, food I contains 30 units of cholesterol per ounce, and food II contains 20 units of cholesterol per ounce. Find the appropriate amounts of the two food supplements so that cholesterol is minimized.

27. **Plant Nutrition.** A farmer can buy three types of 100-pound bags of fertilizer, type A, type B, and type C. Each 100-pound bag of type A fertilizer costs $20 and contains 40 pounds of nitrogen, 30 pounds of phosphoric acid, and 10 pounds of potash. Each 100-pound bag of type B fertilizer costs $30 and contains 20 pounds of nitrogen, 20 pounds of phosphoric acid, and 55 pounds of potash. Each 100-pound bag of type C fertilizer costs $20 and contains no nitrogen, 30 pounds of phosophoric acid, and 40 pounds of potash. The farmer requires 4000 pounds of nitrogen, 2000 pounds of phosphoric acid, and 2000 pounds of potash. How many bags of each type of fertilizer should be bought in order to minimize cost and what will be the minimum cost?

28. **Production Scheduling.** A company has 3 plants that produce 3 different sizes of their product. The North plant can produce 2000 items in the small size, 1000 in the medium size, and 1000 in the large size per day. The Center plant can produce 1000 items in the small size, 1000 in the medium size, and 2000 in the large size per day. The South plant can produce 2000 items in the small size, 2000 in the medium size, and 1000 in the large size per day. The company needs to produce at least 9000 of the small items, 8000 of the medium items, and 9000 of the large items on a given day. The cost per item of producing the small item is $2, the medium item $3, and the large item $2. Find the number of days each plant should operate in order to minimize the cost and find the minimum cost. *Hint:* Find the minimum cost in thousands of dollars and divide the constraints by an appropriate constant.

29. **Production Scheduling.** A mining company has three mines that produce three grades of ore. The North mine can produce 2 tons of low-grade ore, 1 ton of medium-grade ore, and 1 ton of high-grade ore each day. The Center mine can produce 1 ton of low-grade, 1 ton of medium-grade, and 4 tons of high-grade ore. The South mine can produce 1 ton of low-grade, 3 tons of medium-grade, and 3 tons of high-grade ore. At a particular time the company needs at

(© Grant Heilman, Inc.)

least 21 tons of low-grade ore, at least 19 tons of medium-grade ore, and at least 25 tons of high-grade ore. It costs $2000 per day to operate the North mine for a day, $3000 a day for the Center mine, and $4000 for the South mine. Find the number of days each mine should operate to minimize cost and find the minimum cost. *Hint:* Minimize cost in thousands of dollars.

30. **Transportation.**   Redo Example 4 of the text using the following table.

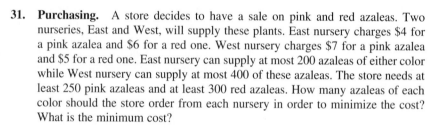

	Plant A	*Plant B*	*Supply*
Stockpile I	$20	$40	500
Stockpile II	$50	$20	100
Demand	150	250	

(© Mike Chuang/FPG International Corporation.)

31. **Purchasing.**   A store decides to have a sale on pink and red azaleas. Two nurseries, East and West, will supply these plants. East nursery charges $4 for a pink azalea and $6 for a red one. West nursery charges $7 for a pink azalea and $5 for a red one. East nursery can supply at most 200 azaleas of either color while West nursery can supply at most 400 of these azaleas. The store needs at least 250 pink azaleas and at least 300 red azaleas. How many azaleas of each color should the store order from each nursery in order to minimize the cost? What is the minimum cost?

32. **Transportation.**   Redo Example 4 of the text using the following table. Notice there are now three plants.

	Plant A	*Plant B*	*Plant C*	*Supply*
Stockpile I	$20	$30	$40	500
Stockpile II	$100	$40	$50	300
Demand	150	200	100	

33. **School Rezoning.**   A small city has two high schools, NW School and NE School, located in two predominately white districts and two high schools, SW School and SE School, located in two predominately minority districts. A decision is made to bus at least 400 minority students from SW and at least 300 minority students from SE to the two schools in the predominately white districts. NW can accommodate at most 200 additional students and NE at most 600. The weekly cost per student of busing from SW to NW is $4, from SW to NE $2, from SE to NW $2, and from SE to NE is $4. Determine the number of students that should be bused from each of the schools in the southern districts to each of the schools in the northern districts in order to minimize cost. What is the minimum cost per week?

34. **School Rezoning.**   Redo the previous exercise if the total miles bused needs to be minimized, where the distance from SW to NW is 4, from SE to NW is 2, from SW to NE is 3, and from SE to NE is 5.

35. **Profits.** Give an economic interpretation to the dual of the following problem. A firm manufactures 3 products, each of which uses 3 ingredients. The first product requires 2 units of the first ingredient, 1 unit of the second, and 1 unit of the third. The second product requires 1 unit of the first ingredient, 1 unit of the second, and 3 units of the third. The third product requires 1 unit of the first ingredient, 4 units of the second, and 3 units of the third. There are only 20 units of the first ingredient, 30 units of the second ingredient, and 40 units of the third ingredient available. The profit on each of the first product is $26, on the second is $24, and on the third is $30. Find the number of each product that should be made in order to maximize profits.

36. **Profits.** Give an economic interpretation to the dual of the following problem. A company makes 3 styles of sleds—Classic, Sport, and Deluxe. The Classic uses 2 board-feet of wood, 1 pound of iron, and 2 hours of labor. The Sport uses 1 board-foot of wood, 1 pound of iron, and 2 hours of labor. The Deluxe uses 1 board-foot of wood, 2 pounds of iron, and 1 hour of labor. There are 20 board-feet of wood, 30 pounds of iron, and 20 hours of labor available. The profit on the Classic is $12, on the Sport $11, and on the Deluxe $12. Find the number of each type of sled to make in order to maximize the profits.

37. **Nutrition.** State the dual problem to Exercise 21 and give an economic interpretation to the dual problem.

38. **Cost.** State the dual problem to Exercise 24 and give an economic interpretation to the dual problem.

39. **Plant Nutrition.** State the dual problem to Exercise 27 and give an economic interpretation to the dual problem.

40. **Diet.** State the dual problem to Exercise 26 and give an economic interpretation to the dual problem.

41. **Production Scheduling.** State the dual problem to Exercise 29 and give an economic interpretation to the dual problem.

42. **Production Scheduling.** State the dual problem to Exericse 28 and give an economic interpretation to the dual problem.

 ## Graphing Calculator Exercises

Solve the next two exercises by solving the dual problem.

43. Minimize $z = 79x_1 + 51x_2 + 118x_3$

Subject to
$$9x_1 + 11x_2 + 12x_3 \geq 380$$
$$29x_1 + 11x_2 + 28x_3 \geq 800$$
$$27x_1 + 61x_2 + 82x_3 \geq 2200$$
$$x_1, x_2, x_3 \geq 0$$

44. Minimize $z = 32x_1 + 16x_2 + 30x_3$

Subject to
$$48x_1 + 29x_2 + 41x_3 \geq 460$$
$$49x_1 + 58x_2 + 77x_3 \geq 1300$$
$$21x_1 + 6x_2 + 15x_3 \geq 310$$
$$x_1, x_2, x_3 \geq 0$$

Solutions to Self-Help Exercise Set 4.3

1. Let y_1 and y_2 be the respective number of months that the first and second plants operate and let the cost C be in tens of thousands of dollars. Then the problem is

$$\text{Minimize} \qquad C = 3y_1 + 4y_2$$
$$\text{Subject to} \qquad 3y_1 + 2y_2 \geq 12$$
$$y_1 + 2y_2 \geq 8$$
$$y_1, y_2 \geq 0$$

The dual of this problem is then

$$\text{Maximize} \qquad z = 12x_1 + 8x_2$$
$$\text{Subject to} \qquad 3x_1 + x_2 \leq 3$$
$$2x_1 + 2x_2 \leq 4$$
$$x_1, x_2 \geq 0$$

The initial tableau for the dual problem given above is

x_1	x_2	s_1	s_2	z		
③	1	1	0	0	3	$\frac{3}{3} = 1$
2	2	0	1	0	4	$\frac{4}{2} = 2$
-12	-8	0	0	1	0	

Pivoting about the first row and first column yields

x_1	x_2	s_1	s_2	z		
1	$\frac{1}{3}$	$\frac{1}{3}$	0	0	1	$\frac{1}{\frac{1}{3}} = 3$
0	④⁄₃	$-\frac{2}{3}$	1	0	2	$\frac{2}{\frac{4}{3}} = 1.5$
0	-4	4	0	1	12	

Pivoting about the second row and second column gives

x_1	x_2	s_1	s_2	z	
1	0	$\frac{1}{2}$	$-\frac{1}{4}$	0	$\frac{1}{2}$
0	1	$-\frac{1}{2}$	$\frac{3}{4}$	0	$\frac{3}{2}$
0	0	2	3	1	18

This is the final tableau. Looking under the slack variables in the last row, we see a 2 and a 3. Thus the solution of the primal problem is (2, 3). This means that the first plant should operate for 2 months and the second for 3 months. The minimum of the primal problem is the same as the maximum of the dual problem, which is found in the last entry of the last row to be 18. Thus, considering the units, the cost is $18 \times 10,000 = 180,000$ dollars.

2. You have your own manufacturing plants that can produce these automobiles. You offer to sell to Rambling Motors a Rambler for x_1 tens of thousands of dollars and a Convertible for x_2 tens of thousands of dollars. Now 12 Ramblers and 8 Convertibles will yield revenue in tens of thousands of dollars equal to $12x_1 + 8x_2$. It is this revenue you wish to maximize. In the first plant Rambling Motors makes 3 Ramblers and 1 Convertible a month at a cost of $30,000. In the second plant Rambling Motors makes 2 Ramblers and 2 Convertibles a month at a cost of $40,000. Thus to get Rambling to buy the vehicles from you, you need to set prices x_1 and x_2 so that you can sell Rambling 3 Ramblers and 1 Convertible for *at most* $30,000 and also sell 2 Ramblers and 2 Convertibles for *at most* $40,000. That is,

$$3x_1 + x_2 \le 3$$
$$2x_1 + 2x_2 \le 4$$

Thus your problem is

$$\text{Maximize} \quad R = 12x_1 + 8x_2$$
$$\text{Subject to} \quad 3x_1 + x_2 \le 3$$
$$2x_1 + 2x_2 \le 4$$
$$x_1, x_2 \ge 0$$

4.4 THE SIMPLEX METHOD: NONSTANDARD AND MINIMIZATION PROBLEMS

▶ *Nonstandard Maximum Problem*

▶ *Minimization Problems*

Tjalling Koopmans

(UPI/Bettmann.)

Dantzig visited the well known mathematical economist Koopmans in June 1947 to talk to him about linear programming. Koopmans quickly realized that a good part of economics could be translated into the linear programming format. He then led a group of economists who developed the theory of allocation of resources and its relation to linear programming. For this work he was awarded the Nobel prize in economics in 1975. At that time he expressed regret that Dantzig was not sharing the honor with him.

In this section we will expand the simplex method to include maximization problems in which the constants to the right of the inequality $\leq$ may be negative. We will also see how this technique can be used to solve minimization problems.

There are two methods for handling these problems: the big-M method, developed by Charnes and Cooper in 1961 and Crown's method, developed by Crown in 1982. We will use Crown's method since this method does not use the artificial variables of the other method and normally requires fewer tableaus.

Nonstandard Maximum Problem

In a standard maximization problem the constraints are in the form using the inequality $\leq$ and the constants to the right of the inequalities $\leq$ are positive. In the last chapter we encountered maximization problems with inequalities of the form

$$x_1 + x_2 \geq 2.$$

We shall now multiply such an inequality by -1, yielding

$$-x_1 - x_2 \leq -2.$$

This yields the inequality $\leq$, but now a *negative* constant appears to the right of the inequality $\leq$.

We shall see how to pivot about certain elements until the constants to the right of all the inequalities $\leq$ become nonnegative. The problem is then in standard form. We then proceed as in the previous section.

Consider now the following example.

E X A M P L E   1   **Simplex Method on Nonstandard Problem**

Two types of metal connectors require for their manufacture spending a certain amount of minutes on a stamping machine and on a grinding machine. The first type of connector requires 4 minutes on the stamping machine and 3 minutes on the grinding machine, while the second connector requires 5 minutes on the stamping machine and 1 minute on the grinding machine. The stamping machine is available at most 50 minutes each hour and the grinding machine is available at most 21 minutes each hour. Furthermore, each hour a total of at least two of the connectors must be produced. If each of the first type connector brings a profit of $0.50 and each of the second type of connector a profit of $0.60, use the simplex method to find the number of each of these products that should be manufactured to maximize profits.

Solution   This problem was considered in the last chapter in Self-Help Exercise 3.2. If x_1 and x_2 are, respectively, the number of the first type connector and the second type connector manufactured, then the problem is

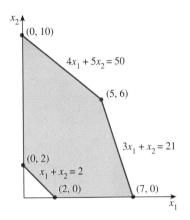

Figure 4.9

Maximize $z = 50x_1 + 60x_2$

Subject to $4x_1 + 5x_2 \leq 50$

$3x_1 + x_2 \leq 21$

$x_1 + x_2 \geq 2$

$x_1, x_2 \geq 0$

For convenience z is given in cents. Figure 4.9 shows the feasible region. The first step is to multiply the third constraint by -1. This then yields

$$4x_1 + 5x_2 \leq 50$$

$$3x_1 + x_2 \leq 21$$

$$-x_1 - x_2 \leq -2$$

$$x_1, x_2 \geq 0$$

The initial tableau is then

x_1	x_2	s_1	s_2	s_3	z		ratios
4	5	1	0	0	0	50	$\frac{50}{4} = 12.5$
③	1	0	1	0	0	21	$\frac{21}{3} = 7 \leftarrow$ smallest of the $\frac{+}{+}$ ratios
-1	-1	0	0	1	0	-2	
-50	-60	0	0	0	1	0	

The -2 in the last column indicates that this tableau is not in standard form. From Figure 4.9 we see that $(0, 0)$ is not feasible.

We now consider the third row that contains the -2 in the last entry. We now **pick the most negative element in the third row**. Since in this case there is a tie, we will pick the first -1 and **use the column this element is in as the pivot column**.

Now form all the ratios as usual. First ignore all ratios that are of the form $\frac{\text{negative number}}{\text{negative number}}$, and pick the smallest nonnegative one according to the usual rules, picking the first to break any ties. The row this element is in then determines the pivot row. If this procedure should fail, that is, if there are no such ratios to be considered, then take the **largest** of the ratios of the form $\left(\frac{\text{negative}}{\text{negative}}\right)$ and use the row this element is in as the pivot row. (There must always be such a ratio.)

According to this rule, the second row is the pivot row. Dividing the second row by 3, we obtain

x_1	x_2	s_1	s_2	s_3	z		
4	5	1	0	0	0	50	$R_1 - 4R_2 \rightarrow R_1$
①	$\frac{1}{3}$	0	$\frac{1}{3}$	0	0	7	
-1	-1	0	0	1	0	-2	$R_3 + R_2 \rightarrow R_3$
-50	-60	0	0	0	1	0	$R_4 + 50R_2 \rightarrow R_4$

Performing the indicated row operations gives

x_1	x_2	s_1	s_2	s_3	z		ratios
0	⑪⁄₃	1	$-\frac{4}{3}$	0	0	22	$\frac{22}{\frac{11}{3}} = 6 \leftarrow$ smallest of the $\frac{+}{+}$ ratios
1	$\frac{1}{3}$	0	$\frac{1}{3}$	0	0	7	$\frac{7}{\frac{1}{3}} = 21$
0	$-\frac{2}{3}$	0	$\frac{1}{3}$	1	0	5	
0	$-\frac{130}{3}$	0	$\frac{50}{3}$	0	1	350	

Notice that the problem has now become a standard problem. The point corresponding to this last tableau is (7, 0). The pivot column is the second and the above ratios indicate that the first row is the pivot row. Dividing the first row by $\frac{11}{3}$ yields

x_1	x_2	s_1	s_2	s_3	z		
0	①	$\frac{3}{11}$	$-\frac{4}{11}$	0	0	6	
1	$\frac{1}{3}$	0	$\frac{1}{3}$	0	0	7	$R_2 - \frac{1}{3}R_1 \rightarrow R_2$
0	$-\frac{2}{3}$	0	$\frac{1}{3}$	1	0	5	$R_3 + \frac{2}{3}R_1 \rightarrow R_3$
0	$-\frac{130}{3}$	0	$\frac{50}{3}$	0	1	350	$R_4 + \frac{130}{3}R_1 \rightarrow R_4$

The above row operations yield

x_1	x_2	s_1	s_2	s_3	z	
0	1	$\frac{3}{11}$	$-\frac{4}{11}$	0	0	6
1	0	$-\frac{1}{11}$	$\frac{5}{11}$	0	0	5
0	0	$\frac{2}{11}$	$\frac{1}{11}$	1	0	9
0	0	$\frac{130}{11}$	$\frac{10}{11}$	0	1	610

This is the final tableau and gives the point $(5, 6)$ as the point at which the objective function attains the maximum of 610. ∎

We can think of the solution as a two-stage process. In the first stage we pivot about appropriate elements until the tableau is in standard form. Then in the second stage we proceed as in Section 4.2.

We now state the rules for applying the simplex method in the case when some constants to the right of the inequalities $\leq$ are negative.

Extended Simplex Method

If some constant to the right of an inequality $\leq$ is negative, then go to Stage I; otherwise, go to Stage II.

Stage I   Select the first negative element in the last column (ignoring the entry in the last row).

1. Consider the row this element is in. Pick the most negative element in this row (ignoring the last entry). The column containing this element will be the pivot column. If there is no such negative entry, then stop. There is no solution.

2. Form all the ratios as usual. First ignore all ratios that are of the form $\left(\dfrac{\text{negative}}{\text{negative}}\right)$ and pick the smallest nonnegative one according to the usual rules, picking the first to break any ties. The row this element is in then determines the pivot row. If this procedure should fail, that is, there are no such ratios to be considered, then take the **largest** ratio of the form $\left(\dfrac{\text{negative}}{\text{negative}}\right)$, and use the row this element is in as the pivot row. (There must always be such a ratio.)

3. Pivot about the pivot element and repeat as necessary until the tableau is in standard form.

Stage II   Proceed as in Section 4.2.

Minimization Problems

We can readily convert a minimization problem into a maximization problem by observing that the minimum of an objective function such as $z = 10 - 2x_1 + 3x_2$ occurs at the same point as the maximum of the negative of this function $-z = -10 + 2x_1 - 3x_2$. Furthermore, the minimum of the original objective function $z = 10 - 2x_1 + 3x_2$ is the maximum of the negative of this function $-z = -10 + 2x_1 - 3x_2$.

Thus, when asked to minimize an objective function, we instead maximize the negative of this function.

We now look at an important type of problem: the transportation problem. This problem was also done in the last section using the dual. We now see how to solve this problem without using the dual.

E X A M P L E   2   **Extended Simplex Method on Nonstandard Problem**

A firm has two stockpiles of an item stored in warehouses at two different locations, Denver and Santa Fe. The item must be shipped to two plants, located in Des Moines and Fort Wayne, for assembly into a final product. At the beginning of a week the stockpile at Denver has 150 items and the stockpile at Santa Fe has 180. The Des Moines plant needs at least 100 of these items and the Fort Wayne plant needs at least 120 that week. The cost of sending each item from the stockpile at Denver to the plants in Des Moines and Fort Wayne is, respectively, $200 and $300, while the cost of sending each item from the stockpile at Santa Fe to the plants in Des Moines and Fort Wayne is, respectively, $300 and $500. Find the number of items that should be shipped from each stockpile to each plant in order to minimize cost.

Solution   The following table summarizes the cost of shipping the items.

	Des Moines Plant	Fort Wayne Plant	Supply
Stockpile at Denver	$200	$300	150
Stockpile at Santa Fe	$300	$500	180
Demand	100	120	

Let x_1 and x_2 be the number of items shipped from the stockpile in Denver to the plants in Des Moines and Fort Wayne, respectively. Let x_3 and x_4 be the number of items shipped from the stockpile in Santa Fe to the plants in Des Moines and Fort Wayne, respectively.

Then the problem is

$$\text{Minimize} \quad C = 200x_1 + 300x_2 + 300x_2 + 500x_4$$

$$\text{Subject to} \qquad x_1 + x_2 \qquad\qquad \leq 150$$
$$x_3 + x_4 \leq 180$$
$$x_1 \qquad + x_3 \qquad \geq 100$$
$$x_2 \qquad + x_4 \geq 120$$
$$x_1, x_2, x_3, x_4 \geq 0$$

This can then be considered as the following equivalent problem.

$$\text{Maximize} \quad z = -C = -200x_1 - 300x_2 - 300x_3 - 500x_4$$

$$\text{Subject to} \qquad x_1 + x_2 \qquad\qquad \leq \quad 150$$
$$x_3 + x_4 \leq \quad 180$$
$$-x_1 \qquad -x_3 \qquad \leq -100$$
$$-x_2 \qquad -x_4 \leq -120$$
$$x_1, x_2, x_3, x_4 \geq \quad 0$$

The initial tableau is

x_1	x_2	x_3	x_4	s_1	s_2	s_3	s_4	z		ratio
①	1	0	0	1	0	0	0	0	150	$\frac{150}{1}$ ← smallest $\frac{+}{-}$ ratio
0	0	1	1	0	1	0	0	0	180	
−1	0	−1	0	0	0	1	0	0	−100	
0	−1	0	−1	0	0	0	1	0	−120	
200	300	300	500	0	0	0	0	1	0	

Since the third row has a −100 as the last entry and is the first such negative encountered in the last column, we first consider the third row. There are two elements with a −1. We arbitrarily pick the first one and thus select the first column as the pivot column. There is only one ratio to consider. This indicates pivoting about the first row and first column. We then obtain

x_1	x_2	x_3	x_4	s_1	s_2	s_3	s_4	z		ratios
1	1	0	0	1	0	0	0	0	150	$\frac{150}{1}$
0	0	1	1	0	1	0	0	0	180	
0	①	−1	0	1	0	1	0	0	50	$\frac{50}{1}$ ← smallest $\frac{+}{-}$ ratio
0	−1	0	−1	0	0	0	1	0	−120	
0	100	300	500	−200	0	0	0	1	−30,000	

The −120 in the last entry of the fourth column results in selecting the fourth row for consideration. The first −1 is in the second column. Thus the second column becomes the pivot column. The two ratios indicate that the third row is the pivot row. Pivoting about the 1 in the third row and second column yields

x_1	x_2	x_3	x_4	s_1	s_2	s_3	s_4	z		ratios
1	0	①	0	0	0	−1	0	0	100	$\frac{100}{1}$ ← smallest $\frac{+}{-}$ ratio
0	0	1	1	0	1	0	0	0	180	$\frac{180}{1}$
0	1	−1	0	1	0	1	0	0	50	
0	0	−1	−1	1	0	1	1	0	−70	
0	0	400	500	−300	0	−100	0	1	−35,000	

The next pivot is in the first row and third column. Pivoting gives

x_1	x_2	x_3	x_4	s_1	s_2	s_3	s_4	z		ratios
1	0	1	0	0	0	-1	0	0	100	$\frac{100}{1}$
-1	0	0	1	0	1	1	0	0	80	
1	1	0	0	1	0	0	0	0	150	$\frac{150}{1}$
①	0	0	-1	1	0	0	1	0	30	$\frac{30}{1} \leftarrow$ smallest $\frac{+}{+}$ ratio
-400	0	0	500	-300	0	300	0	1	$-75{,}000$	

The tableau is now in standard form. We continue as in Section 4.2 and pivot about the fourth row and first column, obtaining

x_1	x_2	x_3	x_4	s_1	s_2	s_3	s_4	z	
0	0	1	1	-1	0	-1	-1	0	70
0	0	0	0	1	1	1	1	0	110
0	1	0	1	0	0	0	-1	0	120
1	0	0	-1	1	0	0	1	0	30
0	0	0	100	100	0	300	400	1	$-63{,}000$

This is the final tableau. The solution is the point (30, 120, 70, 0), and the minimum is the negative of the entry in the last row and last column, or $63,000. Thus ship 30 items from the stockpile in Denver to the plant in Des Moines, 120 items from the stockpile in Denver to the plant in Fort Wayne, 70 items from the stockpile in Santa Fe to the plant in Des Moines, and no items from the stockpile in Sante Fe to the plant in Fort Wayne. This will result in a minimum transportation cost of $63,000. ∎

The same technique can readily be used to solve a transportation problem with any number of stockpiles and any number of plants.

E X A M P L E 3   **Simplex Method on Minimization Problem**

An animal rancher requires at least 12 units per day of nutrient A and at least 8 units per day of nutrient B for his animals. The rancher has two different mixtures of feeds available, feed 1 and 2. If each bag of feed 1 costs $3 and yields 3 units of nutrient A and 1 unit of B and each bag of feed 2 costs $4 and yields 2 units of nutrient A and 2 units of B, find the number of bags of each feed that should be used to minimize the cost.

Solution   If x_1 and x_2 are respectively the number of bags of feed 1 and feed 2, then the problem is

$$\text{Minimize} \qquad C = 3x_1 + 4x_2$$
$$\text{Subject to} \qquad 3x_1 + 2x_2 \geq 12$$
$$x_1 + 2x_2 \geq 8$$
$$x_1, x_2 \geq 0$$

We convert the problem to

$$\text{Maximize} \qquad w = -C = -3x_1 - 4x_2$$
$$\text{Subject to} \qquad -3x_1 - 2x_2 \leq -12$$
$$-x_1 - 2x_2 \leq -8$$
$$x_1, x_2 \geq 0$$

The initial tableau is

x_1	x_2	s_1	s_2	w		ratios
-3	-2	1	0	0	-12	$\frac{-12}{-3} = 4$
$\enclose{circle}{-1}$	-2	0	1	0	-8	$\frac{-8}{-1} = 8 \leftarrow$ largest $\frac{(-)}{(-)}$ ratio. No $\frac{+}{+}$ ratio.
3	4	0	0	1	0	

We begin with the first row. The most negative element in this row is in the first column. Thus the pivot column is the first column. Now there are only ratios of the form $\left(\dfrac{\text{negative}}{\text{negative}} \right)$. Picking the **largest** of these gives the second row as the pivot row. Pivoting gives

x_1	x_2	s_1	s_2	w		ratio
0	$\enclose{circle}{4}$	1	-3	0	12	$\frac{12}{4} = 3 \leftarrow$ smallest $\frac{+}{+}$ ratio
1	2	0	-1	0	8	$\frac{8}{2} = 4$
0	-2	0	3	1	-24	

This is in standard form. Pivoting about the first row and second column gives

x_1	x_2	s_1	s_2	w	
0	1	$\frac{1}{4}$	$-\frac{3}{4}$	0	3
1	0	$-\frac{1}{2}$	$\frac{1}{2}$	0	2
0	0	$\frac{1}{2}$	$\frac{3}{2}$	1	-18

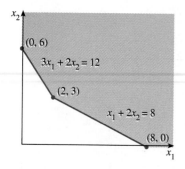

Figure 4.10

This is the final tableau. The minimum occurs at the point $(2, 3)$ with minimum equal to $-(-18) = 18$. Thus the rancher should buy 2 bags of feed 1 and 3 bags of feed 2. The minimum cost is \$18.

Figure 4.10 shows the feasible region. ∎

SELF-HELP EXERCISE SET 4.4

1. Minimize $z = 2x_1 + 3x_2$ subject to the constraints

$$x_1 + x_2 \geq 3$$
$$x_1 \leq 2$$
$$x_2 \leq 2$$
$$x_1, x_2 \geq 0$$

EXERCISE SET 4.4

In all of the following exercises use the extended simplex method to solve. In Exercises 1 through 26 assume the additional constraint that all the variables are nonnegative.

1. Maximize $z = 2x_1 + x_2$

Subject to $x_1 + x_2 \leq 4$
            $-x_1 + 2x_2 \geq 2$

2. Maximize $z = x_1 + 2x_2$

Subject to $x_1 + x_2 \leq 3$
            $x_1 \geq 1$

3. Maximize $z = x_1 + 2x_2$

Subject to $x_1 + 3x_2 \geq 12$
         $x_1 + x_2 \leq 6$
            $x_2 \leq 4$

4. Maximize $z = 3x_1 + 4x_2$

Subject to $-x_1 + x_2 \leq -1$
         $x_1 + x_2 \leq 3$

5. Maximize $z = 2x_1 - 3x_2$

Subject to $-x_1 + x_2 \leq 0$
         $x_1 + x_2 \leq 8$
           $x_1 \geq 2$

6. Maximize $z = 2x_1 + 3x_2$

Subject to $x_1 + x_2 \leq 3$
           $x_2 \leq 2$
           $x_2 \geq 1$

7. Maximize $z = 2x_1 + 3x_2$

Subject to $-x_1 + x_2 \leq 1$
       $-x_1 + x_2 \geq -1$
         $x_1 + x_2 \leq 5$
            $x_1 \geq 1$
            $x_2 \geq 1$

8. Maximize $z = 2x_1 + 3x_2$

Subject to $x_1 + x_2 \geq 2$
         $x_1 + x_2 \leq 4$
           $x_2 \leq 2$

9. Maximize $z = 2x_1 + 3x_2 + x_3$

   Subject to
   $$x_1 + x_2 + x_3 \geq 660$$
   $$4x_1 + 5x_2 + 3x_3 \leq 2600$$
   $$x_1 + 3x_2 + 2x_3 \leq 1360$$
   $$8x_1 + 5x_2 + 10x_3 \leq 7300$$

10. Maximize $z = x_1 + 5x_2 + 3x_3$

    Subject to
    $$3x_1 + x_2 + x_3 \geq 300$$
    $$5x_1 + 16x_2 + 8x_3 \geq 120$$
    $$3x_1 + 6x_2 + 2x_3 \leq 450$$

11. Minimize $z = 3x_1 + 4x_2$

    Subject to
    $$2x_1 + x_2 \geq 3$$
    $$x_1 + x_2 \leq 2$$

12. Minimize $z = 4x_1 + 5x_2$

    Subject to
    $$3x_1 + x_2 \geq 4$$
    $$x_1 + 2x_2 \geq 3$$

13. Minimize $z = 3x_1 + 4x_2$

    Subject to
    $$3x_1 + 2x_2 \geq 11$$
    $$x_1 \leq 3$$
    $$x_2 \leq 4$$

14. Minimize $z = x_1 + 2x_2$

    Subject to
    $$x_1 + x_2 \geq 2$$
    $$x_1 + x_2 \leq 4$$

15. Minimize $z = 10x_1 + 20x_2$

    Subject to
    $$x_1 + 2x_2 \geq 7$$
    $$x_1 \leq 3$$
    $$x_1 \geq 1$$
    $$x_2 \leq 4$$

16. Minimize $z = x_1 + 2x_2$

    Subject to
    $$x_1 + x_2 \geq 2$$
    $$x_1 \geq 1$$
    $$x_2 \geq 1$$

17. Minimize $z = 2x_1 + 3x_2$

    Subject to
    $$4x_1 + x_2 \geq 8$$
    $$x_1 + x_2 \geq 5$$
    $$x_1 + 2x_2 \geq 6$$

18. Minimize $z = 20x_1 + 30x_2$

    Subject to
    $$3x_1 + x_2 \geq 8$$
    $$x_1 + x_2 \geq 4$$
    $$x_2 \geq 1$$

19. Minimize $z = 3x_1 + 4x_2$

    Subject to
    $$3x_1 + 2x_2 \geq 14$$
    $$x_1 \geq 2$$
    $$x_2 \geq 1$$

20. Minimize $z = x_1 + 2x_2$

    Subject to
    $$-x_1 + x_2 \geq 0$$
    $$x_1 + x_2 \geq 4$$
    $$x_2 \geq 1$$

21. Minimize $z = 2x_1 + 2x_2 + x_3$

    Subject to
    $$x_1 + x_2 + x_3 \leq 8$$
    $$x_2 + x_3 \geq 1$$
    $$x_1 + x_2 \geq 1$$

22. Minimize $z = x_1 + 2x_2 + x_3$

Subject to $x_1 + 2x_2 + 2x_3 \geq 4$

$\qquad x_1 + 2x_2 + x_3 \leq 24$

$\qquad x_1 + x_2 + x_3 \geq 1$

23. Minimize $z = 16x_1 + 20x_2 + 9x_3$

Subject to $x_1 + 4x_2 + x_3 \geq 66$

$\qquad 4x_1 + 2x_2 + x_3 \geq 72$

$\qquad 4x_1 + 10x_2 + 5x_3 \geq 200$

24. Minimize $z = x_1 + 2x_2 + 2x_3$

Subject to $x_1 + x_2 + x_3 \leq 4$

$\qquad\qquad\quad x_1 \geq 1$

$\qquad x_1 + x_2 + x_3 \geq 1$

25. Minimize $z = 2x_1 + x_2 + x_3$

Subject to $x_1 + 2x_2 + x_3 \leq 4$

$\qquad x_1 + x_2 \geq 1$

$\qquad\qquad\quad x_3 \geq 2$

26. Minimize $z = 7x_1 + 8x_2$

Subject to $x_1 + x_2 \geq 10$

$\qquad 4x_1 + 5x_2 \leq 80$

$\qquad 3x_1 + 4x_2 \geq 36$

Applications

27. **Transportation.** Redo Example 2 of the text using the following table.

	Plant A	Plant B	Supply
Stockpile I	$20	$40	500
Stockpile II	$50	$20	100
Demand	150	250	

28. **Transportation.** Redo Example 2 of the text using the following table. Notice there are now three plants.

	Plant A	Plant B	Plant C	Supply
Stockpile I	$20	$30	$40	500
Stockpile II	$100	$40	$50	300
Demand	150	200	100	

29. **Investments.** In Example 2 of Section 3.3 the following linear programming problem was formulated (but not solved): Maximize $R = .06x + .08y + .10z$, subject to $x + y + z \leq 100$, $-x - y + z \leq 0$, $x \geq 30$, $y \geq 30$, $z \geq 20$. Solve this problem.

30. **Investments.** An investment made for a retirement plan must pay particular attention to risk. Suppose for such an investment no more than $10,000 is placed in 3 investments with annual returns of 8%, 9%, and 10% respectively. The investor requires an annual return of at least $900 on the investment and wishes to minimize the risk. The second investment is twice as risky as the first, and the third investment is 4 times as risky as the first. How should the money be allocated to minimize the risk? *Hint:* Let risk $= x_1 + 2x_2 + 4x_3$.

31. **Plant Nutrition.** A farmer can buy three types of 100-pound bags of fertilizer, type A, type B, and type C. Each 100-pound bag of type A fertilizer costs $20 and contains 40 pounds of nitrogen, 30 pounds of phosphoric acid, and 10 pounds of potash. Each 100-pound bag of type B fertilizer costs $30 and contains 20 pounds of nitrogen, 20 pounds of phosphoric acid, and 55 pounds of potash. Each 100-pound bag of type C fertilizer costs $20 and contains no nitrogen, 30 pounds of phosophoric acid, and 40 pounds of potash. The farmer requires 4000 pounds of nitrogen, 2000 pounds of phosphoric acid, and 2000 pounds of potash. How many bags of each type of fertilizer should the farmer buy in order to minimize cost? Find the minimum cost.

(© Guido A. Rossi/The Image Bank.)

32. **Cost.** An airline has two types of airplanes and has contracted with a tour group to provide accommodations for a minimum each of 80 first-class, 50 tourist, and 60 economy-class passengers. The first plane costs $4000 for the trip and can accommodate 40 first-class, 10 tourist, and 10 economy-class passengers, while the second plane costs $5000 for the trip and can accommodate 10 first-class, 10 tourist, and 20 economy-class passengers. How many of each type of airplane should be used to minimize the operating cost? Find the minimum cost. *Hint:* Divide each of the constraints by 10 to simplify and minimize cost in thousands of dollars.

33. **Costs.** An oil company has two refineries and must refine and deliver at least 40,000 gallons of medium-grade gasoline and 24,000 gallons of high-grade gasoline to its stations. The first plant can refine 8000 gallons of medium-grade gasoline and 3000 gallons of high-grade gasoline and costs $40,000 a day to operate. The second plant can refine 4000 gallons of medium-grade gasoline and 6000 gallons of high-grade gasoline and costs $30,000 a day to operate. How many days should each plant operate to minimize the cost? Find the minimum cost. *Hint:* Minimize cost in thousands of dollars and divide each constraint by an appropriate constant.

34. **Production Scheduling.** A company has 3 plants that produce 3 different sizes of their product. The North plant can produce 2000 items in the small size, 1000 in the medium size, and 1000 in the large size per day. The Center plant can produce 1000 items in the small size, 1000 in the medium size, and 2000 in the large size per day. The South plant can produce 2000 items in the small size, 2000 in the medium size, and 1000 in the large size per day. The company needs to produce at least 9000 of the small items, 8000 of the medium items, and 9000 of the large items on a given day. The cost per item of producing the small item is $2, the medium item $3, and the large item $2. Find the number of days each plant should operate in order to minimize the cost; find the minimum cost. *Hint:* Find the minimum cost in thousands of dollars and divide the constraints by an appropriate constant.

35. **Production Scheduling.** A mining company has three mines that produce three grades of ore. The North mine can produce 2 tons of low-grade ore, 1 ton of medium-grade ore, and 1 ton of high-grade ore each day. The Center mine can produce 1 ton of low-grade, 1 ton of medium-grade, and 4 tons of high-grade ore. The South mine can produce 1 ton of low-grade, 3 tons of medium-grade, and 3 tons of high-grade ore at least. At a particular time the company needs at least 21 tons of low-grade ore, at least 19 tons of medium-grade ore, and at least 25 tons of high-grade ore. It costs \$2000 per day to operate the North mine for a day, \$3000 a day for the Center mine, and \$4000 for the South mine. Find the number of days each mine should operate to minimize cost; find the minimum cost. *Hint:* Take cost to be in thousands of dollars.

Graphing Calculator Exercises

36. Minimize $z = 16x_1 + 19x_2 + 10x_3$

Subject to $41x_1 + 19x_2 + 11x_3 \geq 680$

$51x_1 + 59x_2 + 19x_3 \geq 1400$

$42x_1 + 58x_2 + 27x_3 \geq 1380$

$x_1, x_2, x_3 \geq 0$

37. Maximize $z = 51x_1 + 49x_2 + 87x_3$

Subject to $21x_1 + 9x_2 + 38x_3 \leq 870$

$12x_1 + 31x_2 + 21x_3 \leq 1100$

$28x_1 + 9x_2 + 29x_3 \leq 910$

$x_1, x_2, x_3 \geq 0$

Solutions to Self-Help Exercise Set 4.4

1. Convert the problem to:

Maximize $z = -2x_1 - 3x_2$

Subject to $-x_1 - x_2 \leq -3$

$x_1 \leq 2$

$x_2 \leq 2$

$x_1, x_2 \geq 0$

The initial tableau is

x_1	x_2	s_1	s_2	s_3	z		
-1	-1	1	0	0	0	-3	
①	0	0	1	0	0	2	$\frac{2}{1}$
0	1	0	0	1	0	2	
2	3	0	0	0	1	0	

Pivoting about the second row and first column gives

x_1	x_2	s_1	s_2	s_3	z		
0	-1	1	1	0	0	-1	
1	0	0	1	0	0	2	
0	①	0	0	1	0	2	$\frac{2}{1}$
0	3	0	-2	0	1	-4	

Pivoting about the third row and second column gives

x_1	x_2	s_1	s_2	s_3	z		
0	0	1	1	①	0	1	$\frac{1}{1}$
1	0	0	1	0	0	2	
0	1	0	0	1	0	2	$\frac{2}{1}$
0	0	0	-2	-3	1	-10	

The tableau is now in standard form and corresponds to the point $(2, 2)$. See the figure. Pivoting about the first row and fifth column yields

x_1	x_2	s_1	s_2	s_3	z	
0	0	1	1	1	0	1
1	0	0	1	0	0	2
0	1	-1	-1	0	0	1
0	0	3	1	0	1	-7

This is the final tableau. The minimum occurs at the point $(2, 1)$. The minimum is $-(-7) = 7$.

4.5   POST-OPTIMAL ANALYSIS*

▶ *Introduction*

▶ *Changes in the* b_i's

▶ *Shadow Prices*

▶ *Complementary Slackness*

*Optional

Suppose the management of the manufacturer of wooden boats considered earlier in this chapter increased the cutting hours by 2, increased the painting hours by 4 and decreased the assembly hours by 1. Can you find the new optimal solution and value without doing the simplex from the beginning? For the answer see Example 1.

Introduction

When an optimal solution has been found to a linear programming problem, usually the work has only begun. The assumption has been made that the constants in the linear programming problem are *known*. Usually these constants are only *estimates* based on what the research team believes conditions will be during the production process. Thus further analysis is normally needed when these "constants" are changed. For this reason the constants are usually referred to as *parameters* and can be changed within certain bounds. The constants, b_i, in particular, are very often set as a result of a company policy decision as to how much resource to allocate to the production process. A further analysis must be done to determine the consequences of changes in this allocation.

For all of these reasons it is important to perform a **sensitivity analysis** to determine the effect on the optimal solution and the maximum or minimum of any changes in the constants. Sensitivity analysis would be very expensive if each change in a constant required redoing the simplex method from the beginning. Fortunately, theoretical analysis has revealed certain relationships that can reduce the computations to a minimum and do not require the simplex method to be performed again.

Changes in the b_i's

In this section we shall restrict the discussion to the standard maximization problem

$$\text{Maximize} \qquad z = CX$$

$$\text{Subject to} \qquad AX \leq B$$

$$X \geq 0$$

where all the components of the column matrix B are nonnegative.

We will only look at the consequences of changing the b_i's, that is, the components of the column matrix B. The analysis of the changes in the other constants involves more time than is available here.

We first look at the wooden-boat manufacturer again. Recall that the problem was

$$\text{Maximize} \quad P = 60x_1 + 80x_2$$

$$\text{Subject to} \quad 2x_1 + 4x_2 \leq 80$$

$$2x_1 + 2x_2 \leq 50$$

$$4x_1 + 2x_2 \leq 84$$

$$x_1, x_2 \geq 0.$$

Thus the column matrix B is

$$B = \begin{bmatrix} 80 \\ 50 \\ 84 \end{bmatrix}$$

where the first component, 80, represents the constraint from the number of hours in the cutting operation, 50 represents the constraint from the number of hours in the painting operation, and 84 represents the constraint from the number of hours in the assembly operation.

The initial simplex tableau is

x_1	x_2	s_1	s_2	s_3	z	
2	4	1	0	0	0	80
2	2	0	1	0	0	50
4	2	0	0	1	0	84
-60	-80	0	0	0	1	0

Suppose now we consider changing the first constraint to $80 + h_1$, changing the second constraint to $50 + h_2$, and the third to $84 + h_3$, where h_1, h_2, and h_3 could be positive or negative. We wish to analyze how changes in each of the components h_1, h_2, and h_3 will affect the optimal value of the objective function and the value of the optimal solution.

The simplex tableau remains the same except for the last column. The last column now becomes

$$\begin{bmatrix} 80 + h_1 \\ 50 + h_2 \\ 84 + h_3 \\ \hline 0 \end{bmatrix} = \begin{bmatrix} B + H \\ \hline 0 \end{bmatrix}$$

where

$$H = \begin{bmatrix} h_1 \\ h_2 \\ h_3 \end{bmatrix}.$$

If we now carry out the simplex method for this new last column, only the last column will change for any of the intermediate tableaus (and thus also for the last tableau). Therefore we simply follow the changes in the last column of the simplex method. Recalling these elementary row operations we obtain

$80 + h_1$	$\frac{1}{4}R_1 \rightarrow R_1$
$50 + h_2$	
$84 + h_3$	
0	

$\rightarrow$

$20 + \frac{1}{4}h_1$	
$50 + h_2$	$R_2 - 2R_1 \rightarrow R_2$
$84 + h_3$	$R_3 - 2R_1 \rightarrow R_3$
0	$R_4 + 80R_1 \rightarrow R_4$

$\rightarrow$

$20 + \frac{1}{4}h_1$	$R_1 - \frac{1}{2}R_2 \rightarrow R_1$
$10 - \frac{1}{2}h_1 + h_2$	
$44 - \frac{1}{2}h_1 + h_3$	$R_3 - 3R_2 \rightarrow R_3$
$1600 + 20h_1$	$R_4 + 20R_2 \rightarrow R_4$

$\rightarrow$

$15 + \frac{1}{2}h_1 - \frac{1}{2}h_2$
$10 - \frac{1}{2}h_1 + h_2$
$14 + h_1 - 3h_2 + h_3$
$1800 + 10h_1 + 20h_2$

This is the last column of the final tableau. The optimal solution is $1800 + 10h_1 + 20h_2$ and the optimal point is

$$(10 - \tfrac{1}{2}h_1 + h_2, \ 15 + \tfrac{1}{2}h_1 - \tfrac{1}{2}h_2) \tag{1}$$

This latter statement is true *if all the elements in the last column (except for the last) remain nonnegative.* If values of h_1, h_2, and h_3 are chosen that make one of these elements negative, then the solution is not feasible. Thus in order for (1) to be the optimal solution, we must have

$$15 + \tfrac{1}{2}h_1 - \tfrac{1}{2}h_2 \geq 0$$

$$10 - \tfrac{1}{2}h_1 + h_2 \geq 0 \tag{2}$$

$$14 + h_1 - 3h_2 + h_3 \geq 0$$

It is of considerable interest to notice that when we increase the number of hours allocated to the assembly process, that is, for any positive h_3, *no change* occurs in the maximum. Thus more than enough hours have already been allocated to the assembly process. Perhaps management might consider changing this resource allocation.

Consider the following specific example.

E X A M P L E   1

Suppose management increased the cutting hours by 2, increased the painting hours by 4, and decreased the assembly hours by 1. Find the new optimal solution and value.

Solution   We see that (2) is still satisfied and the new maximum profit becomes

$$1800 + 10h_1 + 20h_2 = 1800 + 10(2) + 20(4) = 1900$$

while the new optimal solution becomes

$$(10 - \tfrac{1}{2}(2) + (4), 15 + \tfrac{1}{2}(2) - \tfrac{1}{2}(4)) = (13, 14)$$

Thus the new maximum profit is $1900, and the new solution is to make 13 dinghies and 14 skiffs. ■

Naturally management would have to determine if the additional number of dinghies manufactured could actually be sold.

To repeat the above procedure for a large simplex tableau would represent an unacceptable cost. Fortunately, there is a simple way of seeing the result.

First recall that the final tableau is

x_1	x_2	s_1	s_2	s_3	z	
0	1	.50	$-.50$	0	0	15
1	0	$-.50$	1	0	0	10
0	0	1	-3	1	0	14
0	0	10	20	0	1	1800

If we let D be the square matrix sitting under the slack variables and above the last line in the final tableau, and let E be the row matrix under the slack variables in the last row of the final tableau, then it can be shown that the change in the last column of the final tableau can be written as $\boxed{\begin{array}{c} DH \\ \hline EH \end{array}}$

We notice above that

$$D = \begin{bmatrix} \tfrac{1}{2} & -\tfrac{1}{2} & 0 \\ -\tfrac{1}{2} & 1 & 0 \\ 1 & -3 & 1 \end{bmatrix}, \quad E = \begin{bmatrix} 10 & 20 & 0 \end{bmatrix}$$

Then

$$DH = \begin{bmatrix} \tfrac{1}{2} & -\tfrac{1}{2} & 0 \\ -\tfrac{1}{2} & 1 & 0 \\ 1 & -3 & 1 \end{bmatrix} \begin{bmatrix} h_1 \\ h_2 \\ h_3 \end{bmatrix} = \begin{bmatrix} \tfrac{1}{2}h_1 - \tfrac{1}{2}h_2 \\ -\tfrac{1}{2}h_1 + h_2 \\ h_1 - 3h_2 + h_3 \end{bmatrix}$$

and

$$EH = [10 \quad 20 \quad 0]\begin{bmatrix} h_1 \\ h_2 \\ h_3 \end{bmatrix} = 10h_1 + 20h_2$$

We now give the following theorem, which has been verified in the above case.

Sensitivity Theorem

If the linear programming problem is a standard maximization problem of the form

$$\text{Maximize} \qquad z = CX$$
$$\text{Subject to} \qquad AX \le B$$
$$X \ge 0,$$

then changing the matrix B to $B + H$ results in changing only the last column of the final tableau by adding the quantity $\dfrac{DH}{EH}$ to the last column of the old final tableau, where D is the square matrix sitting under the slack variables and above the last line in the final tableau, and E is the row matrix under the slack variables in the last row of the final tableau. This tableau is an acceptable final tableau as long as all the elements in the last column of the new tableau (except the last element) are nonnegative.

Shadow Prices

It is no accident that in the wooden-boat manufacturing problem the row matrix E, which corresponds to the elements under the slack variables in the last row of the final tableau, is the **solution to the dual problem and also the shadow prices**. Thus if Y is the column matrix that gives the solution to the dual problem, then $E = Y^T$. This is always the case in a standard maximization problem.

Shadow prices are defined in terms of marginal analysis. Thus we have the following definition.

Shadow Prices

The ith shadow price is the amount by which the optimal z-value is improved if the right-hand side of the ith constraint is increased by 1.

E X A M P L E 2

Find the shadow prices in the wooden-boat manufacturing problem.

Solution   In the wooden-boat manufacturing problem, the optimal value is given by

$$z_{opt} = 1800 + EH = 1800 + 10h_1 + 20h_2$$

Thus increasing the cutting hours by 1 and leaving the other constraints unchanged is the same as setting $h_1 = 1$, $h_2 = 0$, and $h_3 = 0$, obtaining

$$z_{opt} = 1800 + 10(1) + 20(0) = 1800 + 10.$$

The increase, 10, is the first shadow price. Also, increasing the painting hours by 1 and leaving the other constraints unchanged is the same as setting $h_1 = 0$, $h_2 = 1$, and $h_3 = 0$, obtaining

$$z_{opt} = 10(1) = 1800 + 10(0) + 20(1) = 1800 + 20.$$

The increase, 20, is the second shadow price. Finally, increasing the assembly hours by 1 and leaving the other constraints unchanged is the same as setting $h_1 = 0$, $h_2 = 0$, and $h_3 = 1$, obtaining

$$z_{opt} = 10(1) = 1800 + 10(0) + 20(0) = 1800.$$

The increase, 0, is the third shadow price.   ■

We now state the connection between shadow prices and the solution to the dual problem.

Shadow Price Theorem

Given a standard maximization problem

$$\begin{aligned} \text{Maximize} \quad & z = CX \\ \text{Subject to} \quad & AX \leq B \\ & X \geq 0 \end{aligned}$$

Let Y be the solution of the dual problem

$$\begin{aligned} \text{Minimize} \quad & w = B^T Y \\ \text{Subject to} \quad & A^T Y \geq C^T \\ & Y \geq 0 \end{aligned}$$

Then the ith component of Y, y_i, is the ith shadow price.

Complementary Slackness

We state one form of the complementary slackness theorem.

Complementary Slackness Theorem

1. If the ith primal slack variable is > 0, then the ith dual variable equals zero.
2. If the ith dual variable is > 0, then the ith primal slack variable is zero.

To interpret the first statement in the complementary slackness theorem in the wooden-boat manufacturing problem, recall that $s_3 > 0$. Thus according to (1) above, we must have $y_3 = 0$, that is, the third shadow price is zero. This makes sense. Since there is positive slack in the (third) assembly constraint, an extra hour of assembly labor is worthless.

To interpret the second part (2), recall that in the same problem, $y_1 > 0$ and $y_2 > 0$. Thus according to (2), we must have $s_1 = 0$ and $s_2 = 0$. This is indeed the case. This also is reasonable since a positive second and third shadow price means that an extra hour of cutting and an extra hour of painting have some value. This can only happen if all of the hours for cutting and painting are being used; that is, there is no slack in the cutting and painting constraints.

SELF-HELP EXERCISE SET 4.5

In determining how many chairs (x_1), sofas (x_2), and love-seats (x_3) to manufacture to maximize revenue in a standard maximization problem, suppose the final tableau of the corresponding linear programming problem is

x_1	x_2	x_3	s_1	s_2	s_3	z	
0	1	2	1	0	2	0	10
0	0	1	0	1	-1	0	12
1	0	1	1	0	1	0	8
0	0	5	2	0	1	1	15

The first constraint refers to the number of yards of fabric available, the second constraint to the number of board-feet of wood available, and the third constraint to the number of hours of available labor.

1. Determine the final tableau if the number of yards of fabric were changed by h_1, the number of board-feet of lumber by h_2, and the number of hours of labor by h_3.

2. Determine the constraints on h_1, h_2, and h_3 in order to assure that the tableau found in (1) is the final tableau.

3. Find the new optimal solution and the new maximum of the objective function if the number of yards of fabric is increased by 1, the number of board-feet of lumber is increased by 2, and the number of work-hours is increased by 3.

4. What are the shadow prices, and what do they mean?

EXERCISE SET 4.5

1. To determine how many desks (x_1) and chairs (x_2) to manufacture to maximize revenue in a standard maximization problem, suppose the final tableau of the corresponding linear programming problem is

x_1	x_2	s_1	s_2	s_3	z	
1	0	0	-2	0	0	20
0	0	1	-1	-1	0	30
0	1	0	-1	3	0	25
0	0	0	4	3	1	40

The first constraint refers to the number of gallons of varnish available, the second constraint to the number of board-feet of wood available, and the third constraint to the number of hours of available labor.

a. Determine the new tableau if the number of gallons of varnish available were changed by h_1, the number of board-feet of lumber available by h_2, and the number of hours of labor available by h_3.

b. Determine the constraints on h_1, h_2, and h_3 to ensure that the new tableau found in (a) is the final tableau.

c. Assuming that h_1, h_2, and h_3 satisfy the constraints in (b), find the new optimal solution.

d. Assuming that h_1, h_2, and h_3 satisfy the constraints in (b), find the new maximum.

e. How much can the second constraint be increased while keeping the first and third constraint unchanged without changing the optimal solution or the maximum value of the objective function?

f. Find the new optimal solution and the new maximum of the objective function if the number of gallons of varnish is increased by 2, the number of board-feet of lumber is increased by 3, and the number of work-hours is decreased by 2.

g. What are the shadow prices, and what do they mean?

2. In determining how many pounds of fudge 1 (x_1) and fudge 2 (x_2) to make to maximize revenue in a standard maximization problem, suppose the final tableau of the corresponding linear programming problem is

x_1	x_2	s_1	s_2	s_3	z	
0	1	−3	4	0	0	30
1	0	−1	2	0	0	20
0	0	−1	1	1	0	25
0	0	5	3	0	1	50

The first constraint refers to the number of pounds of chocolate available, the second constraint to the number of pounds of marshmallow available, and the third constraint to the number of pounds of nuts available.

a. Determine the final tableau if the number of pounds of chocolate were changed by h_1, the number of pounds of marshmallow by h_2, and the number of pounds of nuts by h_3.

b. Determine the constraints on h_1, h_2, and h_3 to ensure that the new tableau found in (a) is the final tableau.

c. Assuming that h_1, h_2, and h_3 satisfy the constraints in (b), find the new optimal solution.

d. Assuming that h_1, h_2, and h_3 satisfy the constraints in (b), find the new maximum.

e. How much can the third constraint be decreased while keeping the first and second constraint unchanged, without changing the optimal solution or the maximum value of the objective function?

f. Find the new optimal solution and the new maximum of the objective function if the number of pounds of chocolate is increased by 3, the number of pounds of marshmallow is increased by 2, and the number of pounds of nuts is decreased by 1.

g. What are the shadow prices, and what do they mean?

3. In determining how many gallons of 4 fruit juices to mix to obtain fruit drink 1 (x_1), fruit drink 2 (x_2), and fruit drink 3 (x_3) in order to maximize revenue in a standard maximization problem, suppose the final tableau of the corresponding linear programming problem is

x_1	x_2	x_3	s_1	s_2	s_3	s_4	z	
0	0	0	1	4	1	5	0	20
1	0	0	3	2	0	1	0	30
0	1	0	2	1	0	2	0	25
0	0	1	1	3	0	6	0	40
0	0	0	2	4	0	5	1	90

The first constraint refers to the number of gallons of orange juice available, the second constraint to the number of gallons of grape juice available, the third constraint to the number of gallons of grapefruit juice available, and the fourth constraint to the number of gallons of cranberry juice available.

a. Determine the new tableau if the number of gallons of orange juice, grape juice, grapefruit juice, and cranberry juice available were changed respectively by h_1, h_2, h_3, and h_4.

b. Determine the constraints on h_1, h_2, h_3, and h_4 in order to assure that the new tableau found in (a) is the final tableau.

c. Assuming that h_1, h_2, h_3, and h_4 satisfy the constraints in (b), find the new optimal solution.

d. Assuming that h_1, h_2, h_3, and h_4 satisfy the constraints in (b), find the new maximum.

e. How much can the third constraint be decreased while keeping the other constraints unchanged, without changing the optimal solution or the maximum value of the objective function?

f. Find the new optimal solution and the new maximum of the objective function if the number of gallons of the 4 fruit juices is increased by 2, increased by 3, decreased by 4, and increased by 1, respectively.

g. What are the shadow prices, and what do they mean?

4. Redo Exercise 2 with the given tableau replaced with

x_1	x_2	s_1	s_2	s_3	z	
0	0	2	1	1	0	50
1	0	3	0	2	0	40
0	1	1	0	1	0	45
0	0	2	0	4	1	60

except that (e) is changed to the following.

e. How much can the second constraint be decreased while keeping the first and third constraint unchanged, without changing the optimal solution or the maximum value of the objective function?

If, when changing the column matrix from B to $B + H$, one of the elements in the last column of the new tableau (not including the last entry) becomes negative, then one can use the new tableau as an intermediate tableau and proceed as in Section 4.4 to the final tableau. This, in general, will change the optimal solution and the optimal maximum of the objective function. Do this for the next four exercises to obtain the new solution and optimal value of z for the indicated changes.

5. In the tableau in Exercise 1 decrease the third constraint by 9 without changing the other constraints.

6. In the tableau in Exercise 2 increase the first constraint by 7 without changing the other constraints.

7. In the tableau in Exercise 1 increase the second constraint by 11 without changing the other constraints.

8. In the tableau in Exercise 2 decrease the third constraint by 26 without changing the other constraints.

Solutions to Self-Help Exercise Set 4.5

1. We notice above that

$$D = \begin{bmatrix} 1 & 0 & 2 \\ 0 & 1 & -1 \\ 1 & 0 & 1 \end{bmatrix}, \quad E = \begin{bmatrix} 2 & 0 & 1 \end{bmatrix}$$

Then

$$DH = \begin{bmatrix} 1 & 0 & 2 \\ 0 & 1 & -1 \\ 1 & 0 & 1 \end{bmatrix}\begin{bmatrix} h_1 \\ h_2 \\ h_3 \end{bmatrix} = \begin{bmatrix} h_1 + 2h_3 \\ h_2 - h_3 \\ h_1 + h_3 \end{bmatrix}$$

and

$$EH = \begin{bmatrix} 2 & 0 & 1 \end{bmatrix}\begin{bmatrix} h_1 \\ h_2 \\ h_3 \end{bmatrix} = 2h_1 + h_3$$

Thus the new final tableau is the same as the old one, except the last column is

$10 + h_1 + 2h_3$
$12 + h_2 - h_3$
$8 + h_1 + h_3$
$15 + 2h_1 + h_3$

2. The constraints are

$$10 + h_1 + 2h_3 \geq 0$$
$$12 + h_2 - h_3 \geq 0$$
$$8 + h_1 + h_3 \geq 0$$

3. Setting $h_1 = 1, h_2 = 2, h_3 = 3$, in the above constraints gives

$$10 + h_1 + 2h_3 = 10 + (1) + 2(3) = 17 \geq 0$$
$$12 + h_2 - h_3 = 12 + (2) - (3) = 11 \geq 0$$
$$8 + h_1 + h_3 = 8 + (1) + (3) = 12 \geq 0$$

Since all of these are satisfied, the new tableau will be the final tableau. The last entry in the last column is

$$15 + EH = 15 + 2h_1 + h_3 = 15 + 2(1) + (3) = 20,$$

so the final tableau is

x_1	x_2	x_3	s_1	s_2	s_3	z	
0	1	2	1	0	2	0	17
0	0	1	0	1	-1	0	11
1	0	1	1	0	1	0	12
0	0	5	2	0	1	1	20

The new optimal solution is (12, 17, 0) with maximum 20.

4. The shadow prices are the elements of the row matrix E, and thus are 2, 0, 1. Thus increasing the amount of available fabric by 1 square yard will result in an increase of profit of \$2; increasing the number of board-feet available by 1 will leave the profit unchanged; and increasing the number of work-hours by 1 will result in profits increasing by \$1.

SUMMARY OUTLINE OF CHAPTER 4

- Given a problem constraint in a linear programming problem such as $a_1x_1 + a_2x_2 + a_3x_3 \leq b_1$, a **slack variable** s_1 is defined so that $a_1x_1 + a_2x_2 + a_3x_3 + s_1 = b_1$.

- **The General Algebraic Form of the Standard Maximization Problem**

$$\text{Maximize} \qquad z = c_1x_1 + c_2x_2 + \cdots + c_nx_n$$

$$\text{Subject to} \qquad a_{11}x_1 + a_{12}x_2 + \cdots + a_{1n}x_n \leq b_1$$

$$a_{21}x_1 + a_{22}x_2 + \cdots + a_{2n}x_n \leq b_2$$

$$\cdots$$

$$a_{m1}x_1 + a_{m2}x_2 + \cdots + a_{mn}x_n \leq b_m$$

$$x_1, x_2, \ldots, x_n \geq 0$$

- **Algebraic Criteria for Determining the Feasible Region.** Given a linear programming problem with n decision variables and m problem constraints, the point $(x_1, x_2, \ldots, x_n)$ will be in the feasible region of a standard maximization problem if, and only if, all the slack variables $s_1, s_2, \ldots, s_m$, together with all the decision variables $x_1, x_2, \ldots, x_n$, are nonnegative.

- **Basic and Nonbasic Variables.** Arbitrarily pick n of the variables $x_1, \ldots, x_n$, $s_1, \ldots, s_m$, and call them the *nonbasic variables*. Call the remaining variables the *basic variables*. Set the n nonbasic variables equal to zero and try to solve for the basic variables. If a solution exists and if all the numbers $x_1, \ldots, x_n$, $s_1, \ldots, s_m$, are nonnegative, then the point $(x_1, \ldots, x_n)$ is a corner point.

- **Determining the Solution from the Tableau**
  1. Pick n of the variables $x_1, \ldots, x_n, s_1, \ldots, s_m$, to be the nonbasic variables. The remaining m variables together with z are the basic variables.
  2. (a) Let the nonbasic variables be variables that correspond to nonunit columns. (b) If there are not n nonunit columns, then we arbitrarily select some of the variables (but never z) associated with unit columns to be nonbasic

variables, so that there are n nonbasic variables and so that no two basic variables have identical associated columns.

3. Set all of the nonbasic variables equal to zero.
4. We can then easily find the values of the basic variables from the tableau.

- **Simplex Method.**   See pages 207–208.

- **Dual Problem.**   See page 220.

- **Crown Method.**   See page 243.

Chapter 4 Review Exercises

In Exercises 1 through 10 add the nonnegativity constraint and solve.

1.   Maximize $z = x_1 + x_2$

Subject to $x_1 + 4x_2 \leq 800$
$$x_1 + 2x_2 \leq 500$$
$$2x_1 + x_2 \leq 1100$$

2.   Maximize $z = 2x_1 + 4x_2$

Subject to $x_1 + x_2 \leq 200$
$$10x_1 + 8x_2 \leq 2000$$
$$2x_1 + x_2 \leq 100$$

3.   Maximize $z = x_1 + 2x_2$

Subject to $x_1 - x_2 \leq 3$
$$x_1 + x_2 \leq 1$$
$$x_1 \leq 2$$

4.   Maximize $z = 2x_1 - 5x_2$

Subject to $x_1 + 3x_2 \leq 35$
$$2x_1 - 4x_2 \geq 40$$

5.   Maximize $z = 3x_1 + 8x_2$

Subject to $x_1 - x_2 \leq 6$
$$2x_1 + 6x_2 \leq 21$$
$$x_1 - 2x_2 \geq 6$$

6.   Maximize $z = 4x_1 + 3x_2$

Subject to $x_1 + 3x_2 \geq 12$
$$2x_1 + x_2 \geq 8$$
$$5x_1 + 2x_2 \leq 30$$
$$3x_1 + 4x_2 \leq 28$$

7.   Minimize $z = 2x_1 + 3x_2$

Subject to $2x_1 + x_2 \geq 15$
$$x_1 + x_2 \leq 14$$
$$x_1 + 2x_2 \leq 15$$

8.   Minimize $z = 2x_1 + 3x_2$

Subject to $2x_1 + x_2 \geq 31$
$$x_1 + x_2 \geq 29$$
$$x_1 + 4x_2 \geq 35$$

9.   Minimize $z = 2x_1 + 3x_2$

Subject to $4x_1 + 2x_2 \geq 440$
$$3x_1 + 2x_2 \leq 180$$
$$x_1 + 6x_2 \leq 280$$

10.   Minimize $z = 4x_1 + 5x_2$

Subject to $4x_1 + x_2 \geq 8$
$$x_1 + x_2 \geq 5$$
$$x_1 + 2x_2 \geq 6$$

11.   **Farming.**   A farm has 2 fields to grow wheat and corn, the first field has 400 acres and the second 100 acres. In both fields the corn yield is 125 bushels per acre and the wheat yield is 40 bushels per acre. The cost per acre of growing corn is \$90 for the first field and \$120 for the second field. The cost per acre of growing wheat is \$120 for the first field and \$90 for the second field. If 13,750 bushels of corn and 8000 bushels of wheat must be grown, how many acres of each crop should be planted in each field in order to minimize cost? Find the minimum cost. Use the extended simplex method.

12. **Farming.** Solve the previous problem by using the dual.

13. **Manufacturing.** In determining how many leather purses (x_1) and belts (x_2) to manufacture to maximize revenue in a standard maximization problem, suppose the final tableau of the corresponding linear programming problem is

x_1	x_2	s_1	s_2	s_3	z	
0	1	1	0	1	0	50
1	0	-1	0	-2	0	60
0	0	1	1	3	0	40
0	0	2	0	4	1	70

The first constraint refers to the number of square feet of leather available, the second constraint to the number of ounces of brass available, and the third constraint to the number of hours of available labor.

a. Find the new solution if the number of hours of labor available were changed by $+5$.

b. What are the shadow prices and what do they mean?

How can we determine the monthly car payment? (Chester Higgins, Jr./Photo Researchers, Inc.)

Mathematics of Finance

In the first two sections we consider what happens to an initial lump sum of money deposited into an account that is earning interest at some given rate and that is compounding in some given way. In the next two sections, we consider periodic payments into an account and payment schedules for loans.

5.1   SIMPLE INTEREST AND DISCOUNT

► *Simple Interest*
► *Present Value*
► *Discounts*

APPLICATION

Simple Interest Rates

A person deposits $1000 into an account earning simple interest. What simple interest rate is being charged if the amount at the end of 8 months is $1060? See Example 3 for the answer.

APPLICATION

Discounted Notes

A borrower signs a note and agrees to pay $20,000 to a bank in 9 months at 10% simple discount. What is the discount and what does the borrower actually receive from the bank? See Example 5 for the answer.

Simple Interest

If you lend someone a sum of money, the sum of money is called the **principal** or **present value**. You then charge for the use of this money. This charge is called **interest**. Thus the future amount F of money at any time is given as the principal P plus the interest I or

$$\text{future amount} = \text{principal} + \text{interest}$$

$$F = P + I$$

Normally, the interest paid is given in terms of some **interest rate** expressed as a percent. To calculate the interest, this percent must be converted to a decimal. For example, if the account pays 5% interest per year, and the principal (present value) at the beginning of the year is $1000, then the interest rate is $r = 0.05$, and the interest is

$$I = Pr = \$1000(.05) = \$50$$

The **simple interest** after three years in the previous example is just 3 times $50 or $150. The simple interest for half a year is one half of $50 or $25. In general, we have the following.

Simple Interest

Suppose a sum of money P, called the principal or present value, is invested for t years at an annual *simple* interest rate of r, where r is given as a decimal. Then the interest I at the end of t years is given by

$$I = Prt.$$

The future value F at the end of t years is

$$F = P + I = P + Prt = P(1 + rt).$$

EXAMPLE 1   **Monthly Interest on Credit Card Account**

Suppose you have borrowed $1000 on a credit card that charges simple interest at an annual rate of 18%. What is your interest for the first month?

Solution   Here $r = 0.18$, $P = \$1000$, and $t = \frac{1}{12}$. Thus the interest for the first month is

$$I = Prt = \$1000(0.18)\frac{1}{12} = \$1000(0.015) = \$15. \quad \blacksquare$$

EXAMPLE 2   **The Future Value in an Account Paying Simple Interest**

An account with an initial amount of $1000 earns simple interest of 9% annually. How much is in the account after 4 years? After t years?

Solutions

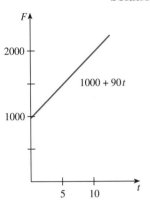

Figure 5.1

(a) Here $P = \$1000$, $i = 0.09$, and $t = 4$. Thus the amount in this account after 4 years is

$$F = P(1 + rt) = \$1000[1 + (.09)(4)] = \$1360.$$

(b) Here $P = \$1000$, $i = 0.09$, and $t = t$. Thus the dollar amount in this account after t years is

$$F = P(1 + rt) = 1000[1 + (.09)(t)] = 1000 + 90t. \quad \blacksquare$$

REMARK. We saw that the amount in the account in Example 2 after t years was $1000 + 90t$. The equation $y = 1000 + 90t$ is the equation of a straight line with y-intercept 1000 and slope 90. See Figure 5.1.

EXAMPLE 3   **Determining the Annual Rate**

A person deposits $1000 into an account earning simple interest. What simple interest rate is being charged if the amount F at the end of 8 months is $1060?

Solution   We substitute $P = 1000$, $F = 1060$, and $t = \frac{8}{12} = \frac{2}{3}$ into the equation $F = P + Prt$ and obtain

$$F = P + Prt$$

$$1060 = 1000 + 1000r\,\frac{2}{3}$$

$$60 = \frac{2000}{3}\,r$$

$$\frac{3}{2000}\,60 = r$$

$$r = .09,$$

which is an annual rate of 9%. ■

Present Value

Very often we find ourselves needing a certain amount of money at a specific time in the future. We will now consider how to determine the present amount or **present value** of money needed to deposit into an account earning simple interest in order to attain this future need. Suppose we need the future amount F at the end of t years where money is deposited at a simple annual rate of r, expressed as a decimal. Then to find the present amount needed to be deposited into this account, we solve for P in the equation

$$F = P(1 + rt),$$

and obtain

$$P = \frac{F}{1 + rt}.$$

We then have the following.

Present Value

The **present value** needed to deposit into an account earning a simple annual rate of r, expressed as a decimal, in order to have a future amount F after t years is

$$P = \frac{F}{1 + rt}.$$

When asked how to amass $1 million the easy way, the banker quipped "start with $950,000."

E X A M P L E   4   Present Value

How much should be placed into an account paying simple interest of 8% so that after 6 months the future value of the account will be $1,000,000?

Solution   We have $F = \$1,000,000$, $r = .08$, and $t = \frac{6}{12} = \frac{1}{2}$. Thus

$$P = \frac{F}{1 + rt}$$

$$= \frac{\$1,000,000}{1 + (.08)\frac{1}{2}}$$

$$= \$961,538.46 \quad \blacksquare$$

Thus you can have $1 million in this account in 6 months if you deposit $961,538.46 now.

Discounts

Often when a bank loans money for some specific length of time, the bank will deduct interest when the loan is made. Such a loan or note is said to be **discounted**. The interest discounted from the loan is referred to as the **simple discount** or just **discount**. The amount that the borrower actually receives is called the **proceeds**, and the amount to be repaid is called the **maturity value**. Keep in mind that the bank gives the borrower the maturity value *less* the discount.

We can use the simple interest formula to obtain the discount for a simple discounted loan.

Discount and Proceeds

The discount D on a discounted loan of M dollars at a simple annual interest rate of r for t years is

$$D = Mrt,$$

where

$$D = \text{discount} \qquad \text{(interest paid at time of loan)}$$

$$M = \text{maturity value} \qquad \text{(amount borrowed)}$$

$$r = \text{discount rate} \qquad \text{(annual simple interest rate)}$$

$$t = \text{length of loan.}$$

The **proceeds** P of the loan is the actual amount the borrower receives when the loan is made and is given by

$$P = M - D.$$

E X A M P L E   5   **Determining Discounts and Proceeds**

A borrower signs a note and agrees to pay $20,000 to a bank in 9 months at 10% simple discount. What is the discount and what does the borrower actually receive from the bank?

Solution   We have $M = \$20{,}000$, $r = .10$, and $t = \frac{9}{12} = \frac{3}{4}$. The discount is

$$D = Mrt = \$20{,}000(.10)\,\frac{3}{4} = \$1500.$$

The proceeds are

$$P = M - D = \$20{,}000 - \$1500 = \$18{,}500.$$

Thus the borrower will receive $18,500 at the time of the loan and pay the bank $20,000 in 9 months. ∎

E X A M P L E   6   **Discounted Loan with Given Proceeds**

Suppose the borrower in the previous example wanted to receive $20,000. What should the maturity value be?

Solution   We need to solve for the maturity value M in the equation

$$P = M - D$$
$$= M - Mrt$$
$$= M(1 - rt).$$

We have the proceeds $P = \$20,000$, $r = .10$, and $t = \frac{3}{4} = .75$. Thus

$$M = \frac{P}{1 - rt}$$

$$= \frac{\$20,000}{1 - (.10)(.75)}$$

$$= \$21,621.62.$$

Thus the borrower will receive the proceeds of $20,000 at the time of the loan and must pay the bank $21,621.62 after 9 months. ∎

In Example 5 the bank quoted an annual interest rate of 10%. But what is the *actual* interest rate? The rate of interest actually paid is called the **effective rate**.

E X A M P L E   7   **Effective Rate of Interest**

What is the effective rate of interest for the borrower in Example 5?

Solution   The borrower must pay the bank $20,000 in 9 months after receiving $18,500. The discount or interest was $1500. We can calculate the effective interest rate r_{eff} by using the formula $D = I = Pr_{\text{eff}}t$ and solving for r_{eff}, where $D = I = \$1500$, $P = \$18,500$, and $t = \frac{3}{4} = 0.75$. We have

$$D = Pr_{\text{eff}}t$$

$$r_{\text{eff}} = \frac{D}{Pt}$$

$$= \frac{1500}{18,500(.75)}$$

$$= .1081.$$

Thus the effective annual interest is 10.81%, whereas the advertised discount rate r was 10%. ∎

We can calculate the effective interest rate using only the discount rate r and the length t of the loan. We know from Example 7 that

$$r_{\text{eff}} = \frac{D}{Pt}$$

$$= \frac{Mrt}{(M - Mrt)t}$$

$$= \frac{r}{1 - rt}.$$

Effective Rate of a Discounted Loan

The effective interest rate given on a discounted loan of length t years with a discount rate of r is

$$r_{eff} = \frac{r}{1 - rt}.$$

E X A M P L E 8 **Effective Rate on a Discounted Loan**

Use the above formula to find the effective rate of the discounted loan in Example 5.

Solution We have $r = .10$ and $t = .75$. Thus

$$r_{eff} = \frac{r}{1 - rt} = \frac{.10}{1 - .10(.75)} = .1081, \text{ or } 10.81\%. \quad \blacksquare$$

SELF-HELP EXERCISE SET 5.1

1. A bank borrows $100,000 for 3 months at a simple interest rate of 6% per year. How much must be repaid at the end of the 3 months?

2. The United States government borrows substantial sums of money by issuing treasury bills (T-bills). T-bills do not specify an interest rate, but rather are sold at public auction. A bank wishes to purchase a 6-month $1 million T-bill. For such a T-bill the bank will receive $1 million at the end of 6 months. If the bank wishes to earn 5% simple discount interest on this T-bill, what should the bank bid?

3. If the bank obtains a 6-month T-bill at a discount rate of 5%, what is the effective yield?

EXERCISE SET 5.1

In Exercises 1 through 4 an amount of P dollars is borrowed for the given length of time at an annual interest rate of r. Find the simple interest that is owed.

1. $P = \$1000$, $r = 8.0\%$, 4 months

2. $P = \$2000$, $r = 6.0\%$, 3 months

3. $P = \$6000$, $r = 4.0\%$, 2 years

4. $P = \$5000$, $r = 7.0\%$, 3 years

In Exercises 5 through 8 an amount of P dollars is borrowed for the given length of time at an annual simple interest rate of r. Find the amount due at the end of the given length of time.

5. $P = \$2000$, $r = 4.0\%$,   7 months

6. $P = \$5000$, $r = 5.0\%$,   4 months

7. $P = \$1000$, $r = 6.0\%$,   2 years

8. $P = \$8000$, $r = 7.0\%$,   3 years

In Exercises 9 through 12 an amount of P dollars is borrowed for the given length of time with the amount F due at the end of the given length of time. Find the annual simple interest rate r.

9. $P = \$2000$, $F = \$2100$,   7 months

10. $P = \$5000$, $F = \$5200$,   4 months

11. $P = \$1000$, $F = \$1070$,   2 years

12. $P = \$8000$, $F = \$9200$,   3 years

In Exercises 13 through 16 find the present amount needed to attain a future amount of F dollars in the given time using an annual simple interest rate of r.

13. $F = \$2000$, $r = 4.0\%$,   3 months

14. $F = \$5000$, $r = 5.0\%$,   8 months

15. $F = \$6000$, $r = 6.0\%$,   2 years

16. $F = \$8000$, $r = 7.0\%$,   3 years

In Exercises 17 through 20 find the effective yield on a discount loan with the given discount rate r and the time.

17. $r = 5\%$,   5 months

18. $r = 6\%$,   9 months

19. $r = 7\%$,   3 months

20. $r = 8\%$,   10 months

In Exercises 21 through 24 find the simple discount and the proceeds for the simple discounted loans.

21. $M = \$1000$,   5 months, $r = .04$

22. $M = \$3000$,   8 months, $r = .05$

23. $M = \$5000$,   20 months, $r = .06$

24. $M = \$7000$,   2 years, $r = .08$

Applications

25. **Future Value.**   A principal of $5000 earns 8% per year simple interest. How long will it take for the future value to become $6000?

26. **Future Value.**   A principal of $2000 earns 6% per year simple interest. How long will it take for the future value to become $2300?

27. **Doubling Time.**   A principal earns 8% per year simple interest. How long will it take for the future value to double?

28. **Tripling Time.** A principal earns 6% per year simple interest. How long will it take for the future value to triple?

29. **Simple Discount.** A borrower signs a note and agrees to pay $50,000 to a bank in 5 months at 8% simple discount. What is the discount and what does the borrower actually receive from the bank?

30. **Simple Discount.** A borrower signs a note and agrees to pay $40,000 to a bank in 7 months at 6% simple discount. What is the discount and what does the borrower actually receive from the bank?

Solutions to Self-Help Exercise Set 5.1

1. Since $P = \$100,000$, $r = .06$, and $t = \frac{3}{12} = .25$,
$$F = P(1 + rt) = \$100,000[1 + (.06)(.25)] = \$101,500.$$

2. Since $M = \$1,000,000$, $r = .05$, and $t = \frac{6}{12} = .50$,
$$P = M - D = M - Mrt = M(1 - rt)$$
$$= \$1,000,000[1 - (.05)(.50)] = \$975,000.$$

3. Since $r = .05$ and $t = .50$,
$$r_{eff} = \frac{r}{1 - rt} = \frac{.05}{1 - .05(.50)} = .0513,$$
or 5.13%.

5.2   COMPOUND INTEREST

▶ *Compound Interest*
▶ *Effective Yield*
▶ *Present Value*

One bank advertises an annual rate of 9.1% compounded semiannually. A second bank advertises an annual rate of 9% compounded daily. In which bank would you deposit your money? The answer is given in Example 5.

How much money must grandparents set aside at the birth of their grandchild if they wish to have $20,000 when the grandchild reaches eighteen? They can earn 9% compounded quarterly. The answer is given in Example 6.

Compound Interest

The most common type of interest is **compound interest**. We now discuss this type of interest.

If the principal is invested for a certain fraction t of a year at an annual *simple* interest rate of r, where r is given as a decimal, then as we have seen in the last section, the amount at the end of the period is

$$\text{principal} + \text{interest} = P + Pi = P(1 + i), \tag{1}$$

where $i = rt$. The interest $i = rt$ is called the **interest per period**.

If, for example, the annual simple interest on a credit card is 18% and the time period is one month, then $i = .18(\frac{1}{12}) = .015$ is the interest per month.

A useful way of looking at formula (1) is to think of it as the *new principal* equals the *old principal* times $(1 + i)$, or

$$(\text{new principal}) = (\text{old principal})(1 + i). \tag{2}$$

If the interest and principal are left in the account for more than one period and interest is calculated not only on the principal but also on the previous interest earned, we say that the interest is being **compounded**.

Consider now what happens during the second period of the compounding. At the start of the second period the *old principal* is $P(1 + i)$. Thus the amount of money in the account at the end of the second period of time (the *new principal*) is according to (2)

$$(\text{new principal}) = (\text{old principal})(1 + i) = [P(1 + i)](1 + i) = P(1 + i)^2.$$

If this compounding continues for a third period of time, the amount of money in the account at the end of the third period of time (the *new principal*) is

$$(\text{new principal}) = (\text{old principal})(1 + i) = [P(1 + i)^2](1 + i) = P(1 + i)^3.$$

Continuing in this manner, the future amount of money F in the account at the end of n periods will be

$$F = P(1 + i)^n.$$

E X A M P L E   1   **Finding Compound Interest**

Suppose $1000 is deposited into an account with an annual yield of 8% compounded quarterly. Find the amount in the account at the end of five years, ten years, twenty years, thirty years, and forty years.

Solutions   We have $P = \$1000$, while the interest per quarter is $i = .08(\frac{1}{4}) = 0.02$. Thus $F = \$1000(1 + .02)^n$. Now using either the y^x key on a calculator or Table B.2 in Appendix B we can obtain

Years	Periods ($n = 4t$)	Future Value
5	20	$\$1000(1 + .02)^{20} = \$1485.95$
10	40	$\$1000(1 + .02)^{40} = \$2208.04$
20	80	$\$1000(1 + .02)^{80} = \$4875.44$
30	120	$\$1000(1 + .02)^{120} = \$10,765.16$
40	160	$\$1000(1 + .02)^{160} = \$23,769.91$

Also see Figure 5.2. ∎

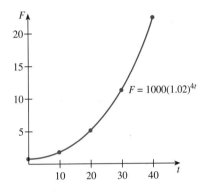

Figure 5.2

Notice that during the first ten years the account grows by about $1200, but during the last ten years grows by about $13,000. In fact, each year the account grows by more than in the previous year. Recall that for simple compounding, the account grows by the *same* amount every year.

R E M A R K .   We will use the letter i to designate the interest rate for any period, whether annual or not. We will reserve the letter r to always designate an annual rate.

Suppose r is the annual interest rate expressed as a decimal, and interest is compounded m times a year. Then the interest rate per time period is $i = r/m$. If the compounding goes on for t years, then there are $n = mt$ time periods and the amount F after t years is

$$F = P(1 + i)^n = P\left(1 + \frac{r}{m}\right)^{mt}.$$

Compound Interest

Suppose a principal P earns interest at the annual rate of r, expressed as a decimal, and interest is compounded m times a year. Then the amount F after t years is

$$F = P(1 + i)^n = P\left(1 + \frac{r}{m}\right)^{mt},$$

where $n = mt$ is the number of time periods and $i = \dfrac{r}{m}$ is the interest per period.

E X A M P L E 2   **Finding Compounded Interest**

Suppose $1000 is deposited into an account that yields 9% annually. Find the amount in the account at the end of the fifth year if the compounding is (a) quarterly, (b) monthly.

Solutions   (a)   Here $P = \$1000$, $r = .09$, $m = 4$, and $t = 5$. Using the y^x key on your calculator gives

$$F = P\left(1 + \frac{r}{m}\right)^{mt} = \$1000\left(1 + \frac{.09}{4}\right)^{4(5)} = \$1560.51.$$

(b)   Here $P = \$1000$, $r = .09$, $m = 12$, and $t = 5$. Using the y^x key on your calculator gives

$$F = P\left(1 + \frac{r}{m}\right)^{mt} = \$1000\left(1 + \frac{.09}{12}\right)^{12(5)} = \$1565.68. \quad \blacksquare$$

R E M A R K .   Notice that the amount at the end of the time period is larger if the compounding is done more often.

E X A M P L E 3   **Calculating Interest Over a Long Period of Time**

Suppose that when the Canarsie Indians sold Manhattan for $24 in 1626, the money was deposited into a Dutch guilder account that yielded an annual rate of 6% compounded quarterly. How much would be in this account in 1994?

Solution   We have

$$F = P\left(1 + \frac{r}{m}\right)^{mt}$$

$$= \$24\left(1 + \frac{.06}{4}\right)^{368(4)}$$

$$\approx \$79 \text{ billion.} \quad \blacksquare$$

The following table indicates what the value of this account would have been at some intermediate times.

Year		Future Value
1626		$24
1650	$24 $(1 + \frac{.06}{4})^{4(24)} \approx$	$100
1700	$24 $(1 + \frac{.06}{4})^{4(74)} \approx$	$2,000
1750	$24 $(1 + \frac{.06}{4})^{4(124)} \approx$	$39,000
1800	$24 $(1 + \frac{.06}{4})^{4(174)} \approx$	$760,000
1850	$24 $(1 + \frac{.06}{4})^{4(224)} \approx$	$15,000,000
1900	$24 $(1 + \frac{.06}{4})^{4(274)} \approx$	$293,000,000
1950	$24 $(1 + \frac{.06}{4})^{4(324)} \approx$	$6,000,000,000
1994	$24 $(1 + \frac{.06}{4})^{4(368)} \approx$	$79,000,000,000

REMARK. It is, of course, very unlikely that any investment could have survived through the upheavals of wars and financial crises that occurred during this 368-year span of time. Nonetheless, it is examples like this one that inspire some to use the phrase ''the wonders of compounding.''

If an account earns interest at a rate of i, expressed as a decimal, per compounding period, how long will it take for the principal of P to grow to some future value F? To answer this question we solve for the number n of time periods in the equation

$$F = P(1 + i)^n,$$

by first taking common logarithms of each side.

$$\log F = \log P(1 + i)^n$$
$$= \log P + \log(1 + i)^n$$
$$= \log P + n \log(1 + i)$$
$$n \log(1 + i) = \log F - \log P$$
$$n \log(1 + i) = \log F/P$$
$$n = \frac{\log F/P}{\log(1 + i)}$$

An important case concerns the time for an account to double. For this case we must have $F = 2P$. This yields

$$n = \frac{\log F/P}{\log(1 + i)} = \frac{\log 2P/P}{\log(1 + i)} = \frac{\log 2}{\log(1 + i)}.$$

See Figure 5.3.

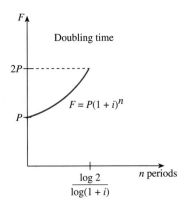

Figure 5.3

We have the following.

The Time for Money to Grow from P to F

The number of periods n it takes for a principal of P to grow to a future value of F in an account that earns interest compounded at a rate of i per period, expressed as a decimal, is

$$n = \frac{\log F/P}{\log(1 + i)}.$$

The number of periods for the principal to double is given by

$$n = \frac{\log 2}{\log(1 + i)}.$$

E X A M P L E   4   **Doubling Time**

Determine the time it takes for an account earning interest at an annual rate of 4% compounded annually to double.

Solution   Here $i = .04$. Since the time periods are measured in years, the number of years for this account to double is

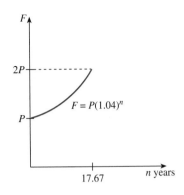

Figure 5.4

$$n = \frac{\log 2}{\log(1 + i)}$$

$$= \frac{\log 2}{\log(1 + .04)}$$

$$= 17.67 \qquad \text{using a calculator}$$

or almost exactly 17 years and 8 months. Refer to Figure 5.4. ∎

The following table gives the doubling times for various annual interest rates for accounts that are compounding annually.

Interest Rate	Doubling Time in Years
4%	17.67
5%	14.21
6%	11.90
7%	10.24
8%	9.01
9%	8.04
10%	7.27
15%	4.96
20%	3.80

Effective Yield

If $1000 is invested at an annual rate of 9% compounded monthly, then at the end of a year there is

$$F = \$1000 \left(1 + \frac{.09}{12}\right)^{12} = \$1093.81$$

in the account. This is the same amount obtainable if the same principal of $1000 was invested for one year at an *annual* rate of 9.381% (or 0.09381, expressed as a decimal). We call the rate 9.381% the **effective annual yield**. The 9% annual rate is often referred to as the **nominal rate**.

Suppose r is the annual interest rate expressed as a decimal and interest is compounded m times a year. If the compounding goes on for one year, then the amount F after one year is

$$F = P \left(1 + \frac{r}{m}\right)^m.$$

If we let r_{eff} be the effective annual yield, then r_{eff} must satisfy

$$P\left(1 + \frac{r}{m}\right)^m = P(1 + r_{eff}).$$

Solving for r_{eff} we obtain

$$r_{eff} = \left(1 + \frac{r}{m}\right)^m - 1.$$

Effective Yield

Suppose a sum of money is invested at an annual rate of r expressed as a decimal and is compounded m times a year. The effective yield r_{eff} is

$$r_{eff} = \left(1 + \frac{r}{m}\right)^m - 1.$$

E X A M P L E 5   **Comparing Investments**

One bank advertises a nominal rate of 9.1% compounded semiannually. A second bank advertises a nominal rate of 9% compounded daily. What are the effective yields? In which bank would you deposit your money?

Solutions   For the first bank $r = .091$ and $m = 2$. Then

$$r_{eff} = \left(1 + \frac{.091}{2}\right)^2 - 1 \approx .0931,$$

or as a percent, 9.31%.

For the second bank $r = .09$ and $m = 365$. Then

$$r_{eff} = \left(1 + \frac{.09}{365}\right)^{365} - 1 \approx .0942,$$

or as a percent, 9.42%.

Despite the higher nominal rate given by the first bank, the effective yield for the second bank is higher than the first. Thus money deposited in the second bank will grow faster than money deposited in the first bank. ∎

Present Value

If we have an account initially with P and earning interest at an annual rate of r expressed as a decimal and interest is compounded m times a year, then the amount F in the account after t years is

$$F = P\left(1 + \frac{r}{m}\right)^{mt}.$$

If we wish to know how many dollars P to set aside now in this account so that we will have a future amount of F dollars after t years, we simply solve the above expression for P. Thus

$$P = \frac{F}{\left(1 + \dfrac{r}{m}\right)^{mt}}.$$

This is called the **present value**.

Present Value

Suppose an account earns an annual rate of r expressed as a decimal and is compounded m times a year. Then the amount P, called the **present value**, needed presently in this account so that a future amount of F will be attained in t years is given by

$$P = \frac{F}{\left(1 + \dfrac{r}{m}\right)^{mt}}.$$

E X A M P L E   6   **Finding the Present Value of a Future Balance**

How much money must grandparents set aside at the birth of their grandchild if they wish to have $20,000 when the grandchild turns eighteen? They can earn 9% compounded quarterly.

Solution   Here $r = 0.09$, $m = 4$, $t = 18$, and $F = \$20{,}000$. Thus

$$P = \frac{F}{\left(1 + \dfrac{r}{m}\right)^{mt}}$$

$$= \frac{\$20{,}000}{\left(1 + \dfrac{.09}{4}\right)^{4(18)}} \approx \$4029.69 \quad \blacksquare$$

SELF-HELP EXERCISE SET 5.2

1. An account with $1000 earns interest at an annual rate of 8% compounded monthly. Find the amount in this account after 10 years.

2. Find the effective yield if the annual rate is 8% and the compounding is weekly.

3. How much money should be deposited in a bank account earning the annual interest rate of 8% compounded quarterly in order to have $10,000 in the account at the end of 10 years?

EXERCISE SET 5.2

In Exercises 1 through 10 find how much is in the accounts after the given years where P is the initial principal and r is the annual rate given as a percent with the indicated compounding.

1. $P = \$1000$, $r = 10\%$, compounded annually for 1 year

2. $P = \$2000$, $r = 9\%$, compounded annually for 1 year

3. After 1 year where $P = \$1000$, $r = 8\%$, compounded (a) annually, (b) quarterly, (c) monthly, (d) weekly, (e) daily

4. After 1 year where $P = \$1000$, $r = 10\%$, compounded (a) annually, (b) quarterly, (c) monthly, (d) weekly, (e) daily

5. After 40 years where $P = \$1000$, $r = 8\%$, compounded (a) annually, (b) quarterly, (c) monthly, (d) weekly, (e) daily

6. After 40 years where $P = \$1000$, $r = 10\%$, compounded (a) annually, (b) quarterly, (c) monthly, (d) weekly, (e) daily,

7. After 40 years where $P = \$1000$, compounded annually, r equal to (a) 3%, (b) 5%, (c) 7%, (d) 9%, (e) 12%, (f) 15%

8. After 40 years where $P = \$1000$, compounded annually, r equal to (a) 4%, (b) 6%, (c) 8%, (d) 10%, (e) 20%, (f) 25%

9. $P = \$1000$, $r = 9\%$, compounded annually, after (a) 5, (b) 10, (c) 15, (d) 30 years

10. $P = \$1000$, $r = 7\%$, compounded annually, after (a) 5, (b) 10, (c) 15, (d) 30 years

In Exercises 11 through 12 find the effective yield given the annual rate r and the indicated compounding.

11. $r = 8\%$, compounded (a) semiannually, (b) quarterly, (c) monthly, (d) weekly, (e) daily

12. $r = 10\%$, compounded (a) semiannually, (b) quarterly, (c) monthly, (d) weekly, (e) daily

In Exercises 13 through 16 find the present value of the given amounts F with the indicated annual rate of return r, the number of years t, and the indicated compounding

13. $F = \$10,000$, $r = 9\%$, $t = 20$, compounded (a) annually, (b) monthly, (c) weekly

14. $F = \$10,000$, $r = 10\%$, $t = 20$, compounded (a) annually, (b) quarterly, (c) daily

15. $F = \$100,000$, $r = 9\%$, $t = 40$, compounded (a) annually, (b) monthly, (c) weekly

16. $F = \$100,000$, $r = 10\%$, $t = 40$, compounded (a) annually, (b) quarterly, (c) daily

17. Your rich uncle has just given you your high school graduation present of $1,000,000. The present is in the form of a 40-year bond with an annual interest rate of 9% compounded annually. The bond says it will be worth $1,000,000 in 40 years. What is this $1,000,000 gift worth at the present time?

18. Redo Exercise 17 if the annual interest rate is 6%.

19. Your second rich uncle gives you a high school graduation present of $2,000,000. The present is in the form of a 50-year bond with an annual interest rate of 9% compounded annually. The bond says it will be worth $2,000,000 in 50 years. What is this $2,000,000 gift worth now? Compare your answer to Exercise 17.

20. Redo Exercise 19 with an annual interest rate of 6%.

21. In Example 3 find the amount in 1994 if the annual interest was 7% compounded quarterly. Compare your answer to that of Example 3 in the text.

22. In Example 3 find the amount in 1994 if the annual interest was 5% compounded quarterly. Compare your answer to that of Example 3 in the text.

23. An account grew from $1000 to $1100 in one year. The interest was compounded monthly. What was the nominal interest rate?

24. An account grew from $1000 to $1150 in one year. The interest was compounded quarterly. What was the nominal interest rate?

25. How long does it take an account earning interest at an annual rate of 6% and compounding monthly to grow from $1000 to $1500? Redo if the compounding is weekly.

26. How long does it take an account earning interest at an annual rate of 6% and compounding quarterly to grow from $1000 to $1600? Redo if the compounding is weekly.

27. How long does it take an account to double in value if the account earns interest at an annual rate of 6.5% and the compounding is (a) monthly, (b) daily?

28. How long does it take an account to double in value if the account earns interest at an annual rate of 5.5% and the compounding is (a) monthly, (b) daily?

Applications

29. **Real Estate Appreciation.** The United States paid about 4 cents an acre for the Louisiana Purchase in 1803. Suppose the value of this property grew at an annual rate of 5.5% compounded annually. What would an acre be worth in 1994? Does this seem realistic?

30. **Real Estate Appreciation.** Redo the previous problem using a rate of 6% instead of 5.5%. Compare your answer with the answer to the previous problem.

31. **Coin Appreciation.** What constant annual rate when compounded annually was necessary for the value of a United States half-penny to go from 0.50 cents in 1800 to $2000 in 1994?

(Courtesy of Coin Magazine.)

32. **Comparing Rates at Banks.** One bank advertises a nominal rate of 6.5%

compounded quarterly. A second bank advertises a nominal rate of 6.6% compounded daily. What are the effective annual yields? In which bank would you deposit your money?

33. **Comparing Rates at Banks.**   One bank advertises a nominal rate of 8.1% compounded semiannually. A second bank advertises a nominal rate of 8% compounded weekly. What are the effective yields? In which bank would you deposit your money?

34. **Saving for Machinery.**   How much money should a company deposit in an account with a nominal rate of 8% compounded quarterly in order to have $100,000 for a certain piece of machinery in 5 years?

35. **Saving for Machinery.**   Repeat the previous exercise with an annual rate of 7% and monthly compounding.

Solutions to Self-Help Exercise Set 5.2

1.   Here $P = \$1000$, $r = 0.08$, and $t = 10$, $m = 12$.
     Thus

$$F = P \left(1 + \frac{r}{m} \right)^{mt} = \$1000 \left(1 + \frac{.08}{12} \right)^{12(10)} = \$2219.64.$$

2.   If the annual rate is 8%, the effective yield is given by

$$r_{eff} = \left(1 + \frac{r}{m} \right)^{m} - 1 = \left(1 + \frac{.08}{52} \right)^{52} - 1 \approx .0832, \text{ or } 8.32\%.$$

3.   The present value of $10,000 if the annual interest rate is 8% compounded quarterly for 10 years is

$$P = \frac{F}{(1 + \frac{r}{m})^{mt}} = \frac{10,000}{(1 + \frac{.08}{4})^{4(10)}} = 4528.90.$$

Thus a person must deposit $4528.90 in an account earning 8% compounded quarterly so that there will be $10,000 in the account after 10 years.

5.3   ANNUITIES AND SINKING FUNDS

▶ *Annuities*

▶ *Sinking Funds*

APPLICATION

**Future Balance
in an Annuity**

An individual is trying to save money for a down payment on a house to be purchased in 5 years. She can deposit $100 at the end of each month into an account that pays interest at an annual rate of 9% compounded monthly. How much is in this account after 5 years? (The answer can be found in Example 1.)

A P P L I C A T I O N

A Sinking Fund

A corporation wishes to set up a fund in order to have the money necessary to replace a current machine. It is estimated that the machine will need to be replaced in 10 years and will cost $100,000. How much per quarter should be deposited into an account with an annual interest rate of 8% compounded quarterly to meet this future obligation? (The answer can be found in Example 2.)

Annuities

We have previously studied lump sum payments. We now wish to study periodic payments.

An **annuity** is a sequence of equal payments made at equal time periods. An **ordinary annuity** is one in which the payments are made at the *end* of the time periods of compounding. The **term** of an annuity is the time from the beginning of the first period to the end of the last period. The total amount in the account, including interest, at the end of the term of an annuity is called the **future value of the annuity**. Examples of annuities are regular deposits into a savings account, monthly home mortgage payments, and monthly insurance payments.

We begin with an example of regular deposits into a savings account. Suppose we make deposits of $1000 at the end of each year into a savings account that earns 8% per year. We want to determine how much is in the account at the end of the sixth year. Notice that the first payment is made at the end of the first year and we want the total in the account immediately after the last deposit is made.

Figure 5.5 shows what happens to each deposit. For example, the first deposit of $1000 is in the account for 5 years and becomes $1000(1.08)^5$ at the end of the sixth year. The second deposit of $1000 is in the account for 4 years and becomes $1000(1.08)^4$ at the end of the sixth year, and so on. The amount in the account at the end of the sixth year is the sum of these six numbers or

$$S = 1000 + 1000(1.08) + 1000(1.08)^2 + 1000(1.08)^3 \tag{1}$$
$$+ 1000(1.08)^4 + 1000(1.08)^5.$$

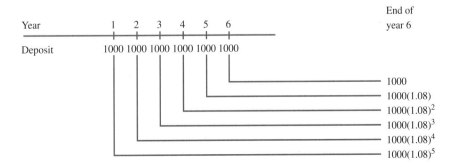

Figure 5.5

It is possible to greatly simplify this quantity. First notice that this quantity is in the form of

$$S = a + ab + ab^2 + \cdots + ab^{n-2} + ab^{n-1} \tag{2}$$

where $a = 1000$, $b = 1.08$, and $n = 6$.

We now see how to find a simplified expression for any sum in the form of equation (2). If $b = 1$, then $S = na$. If $b \neq 1$, then multiply S by b and obtain

$$bS = ab + ab^2 + ab^3 + \cdots + ab^{n-1} + ab^n.$$

If we now subtract S from bS, notice that all but two of the terms cancel and we have

$$bS - S = ab^n - a$$

$$(b - 1)S = a(b^n - 1)$$

$$S = a\frac{b^n - 1}{b - 1}.$$

Thus we have the following.

Summation Formula

For any positive integer n and any real numbers a and b, with $b \neq 1$,

$$a + ab + ab^2 + ab^3 + \cdots + ab^{n-1} = a\frac{b^n - 1}{b - 1}. \tag{3}$$

Using formula (3) with $a = 1000$, $b = 1.08$, and $n = 6$, the sum of the six terms in (1) is

$$S = a\frac{b^n - 1}{b - 1} = 1000\frac{(1.08)^6 - 1}{1.08 - 1} \approx 7335.93.$$

Thus there will be $7335.93 in this account at the end of the sixth year.

Now more generally, suppose that R dollars is deposited into an account at the end of each of n periods with an interest rate of i per period. We wish to know the amount at the end of the nth period. Figure 5.6 indicates what each of the deposits become by the end of the nth period. The value of the account at the end of the nth period is the sum of these n values, denoted by S, and is given by

$$S = R + R(1 + i) + R(1 + i)^2 + \cdots + R(1 + i)^{n-1}.$$

We can now use the summation formula with $a = R$ and $b = (1 + i)$ and obtain

$$S = a\frac{b^n - 1}{b - 1} = R\frac{[(1 + i)^n - 1]}{1 + i - 1} = R\frac{[(1 + i)^n - 1]}{i}.$$

It is common to write

$$s_{n\rceil i} = \frac{[(1 + i)^n - 1]}{i}.$$

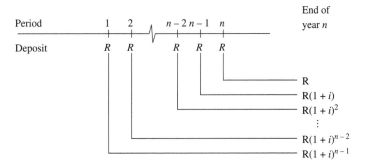

Figure 5.6

The symbol $s_{n\rceil i}$ is read "s angle n at i" and can be found in tables. It is also common to use the symbol FV (future value) for S and to use the symbol PMT (payment) for R. We thus have the following.

Future Value of an Ordinary Annuity

The future value FV of an ordinary annuity of n payments of PMT dollars paid at the end of each period into an account that earns interest at the rate of i per period is

$$FV = PMT\frac{[(1+i)^n - 1]}{i} = PMTs_{n\rceil i}. \qquad (4)$$

R E M A R K . When the periodic payment, PMT, is \$1, the future value FT is just $s_{n\rceil i}$. Thus we have that $s_{n\rceil i}$ is the future value of an annuity after n payments of \$1 have been made, where i is the interest rate per period and the compounding is per the same period.

E X A M P L E 1   **Calculating the Future Value of an Ordinary Annuity**

An individual is trying to save money for a downpayment on a house to be purchased in 5 years. She can deposit \$100 at the end of each month into an account that pays interest at an annual rate of 9% compounded monthly. How much is in this account after 5 years? Also find the amount of interest that has been earned.

Solution   To use the above formula for FV, we need to find PMT, i and n. The monthly payment is \$100, so $PMT = 100$. The interest rate per month is $0.09/12 = 0.0075$. The number of periods is $n = 12(5) = 60$. Thus

$$FV = PMT\frac{[(1+i)^n - 1]}{i} \quad \text{or} \quad PMT\ s_{n\rceil i}$$

$$= \$100\frac{[(1.0075)^{60} - 1]}{0.0075} \quad \text{or} \quad \$100\ s_{60\rceil 0.0075}$$

$$\approx \$100(75.4241) \qquad \begin{array}{l} \text{Use calculator or} \\ \text{Table B.2 in Appendix B.} \end{array}$$

$$= \$7542.41. \qquad \blacksquare$$

To find the interest, subtract the total deposits from *FV*. Total deposits are equal to the number of payments times the amount of each deposit or 60($100). Thus

$$\text{interest} = FV - nPMT = \$7542.41 - \$60(100) = \$1542.41.$$

Sinking Funds

Often an individual or corporation knows that at some future date a certain amount *FV* of money will be needed. Any account that is established for accumulating funds to meet a future need is called a **sinking fund**. The question is how much money *PMT* should be put away periodically so that the deposits plus interest will be *FV* at the needed point in time. In such a case, *FV*, *n*, and *i* are known and one wishes to calculate the periodic payment *PMT*.

To find *PMT* start with the formula

$$FV = PMT \frac{[(1 + i)^n - 1]}{i} = PMT s_{n\rceil i},$$

and solve for *PMT*. Thus

$$PMT = FV \frac{i}{[(1 + i)^n - 1]} = \frac{FV}{s_{n\rceil i}}.$$

We have the following.

Sinking Fund Payment

The periodic payment *PMT* that is required to accumulate the sum *FV* over *n* periods of time with interest at the rate of *i* per period is

$$PMT = FV \frac{i}{[(1 + i)^n - 1]} = \frac{FV}{s_{n\rceil i}}. \tag{5}$$

E X A M P L E   2   **Calculating a Sinking Fund Payment**

A corporation wishes to set up a sinking fund in order to have the funds necessary to replace a current machine. The machine will need to be replaced in 10 years and cost $100,000. How much should be deposited at the end of each quarter into an account with an annual interest rate of 8% compounded quarterly to meet this future obligation? What will be the total amount of the payments and what will be the interest earned?

Solution   We have $FV = 100,000$. Also $i = \frac{0.08}{4} = 0.02$. The number of periods is $n = 4(10) = 40$. Thus

$$PMT = FV \frac{i}{[(1 + i)^n - 1]} \quad \text{or} \quad \frac{FV}{s_{\overline{n}|i}}$$

$$= \$100,000 \frac{0.02}{[(1.02)^{40} - 1]} \quad \text{or} \quad \frac{\$100,000}{s_{\overline{40}|0.02}}$$

$$\approx \$100,000(0.0165557) \quad \text{or} \quad \frac{\$100,000}{60.40198} \quad \begin{array}{l} \text{Use a calculator or} \\ \text{Table B.2 in Appendix B.} \end{array}$$

$$= \$1655.57.$$

Thus if \$1655.57 is placed into this sinking fund at the end of every quarter, there will be \$100,000 in the fund at the end of 10 years.

     Since there are 40 payments, the total amount of payments is \$1655.57(40) = \$66,222.80. The total interest earned is then \$100,000 less the payments or \$100,000 − \$66,222.80 = \$33,777.20. ∎

E X A M P L E   3   **Equity in a Sinking Fund**

Just 4 years after making the sinking fund payment in Example 2 the corporation decides to use the accumulated money (**equity**) for another purpose. Determine the equity.

Solution   We need to find the future value of an annuity with a quarterly payment of \$1655.57 and an annual interest rate of 8% compounded quarterly at the end of 4 years. The number of periods is $n = 4(4) = 16$. We use formula (4).

$$FV = PMT \frac{[(1 + i)^n - 1]}{i} \quad \text{or} \quad PMT \, s_{\overline{n}|i}$$

$$= \$1655.57 \frac{[(1.02)^{16} - 1]}{0.02} \quad \text{or} \quad \$1655.57 \, s_{\overline{16}|0.02}$$

$$\approx \$1655.57(18.6393) \quad \begin{array}{l} \text{Use calculator or} \\ \text{Table B.2 in Appendix B.} \end{array}$$

$$= \$30,858.64$$

Thus after 4 years this fund has grown to \$30,858.64. ∎

     The following table indicates the equity in this sinking fund for selected times. The values can be found in the same way as in Example 3. The total interest is found as in Example 2, that is, by subtracting the total payments from the future value.

End of Year	Number of $1655.57 Payments	Total Payments	Interest Earned (Equity less total payments)	Equity
2	8	$13,244.56	$965.15	$14,209.71
4	16	26,489.12	4369.52	30,858.64
6	24	39,733.68	10,631.84	50,365.52
8	32	52,978.24	20,242.70	73,220.94
10	40	66,222.80	33,777.20	100,000.00

The above corporation with a sinking fund will need to know the interest per quarter since this interest contributes to profits. To determine the interest for any period, we note that the value of the sinking fund increases in any one period due to one payment into the fund and the interest earned during that period. Thus we have

$$\text{Interest in } m\text{th period} = \begin{array}{l} \text{value at end of } m\text{th period} \\ \text{less value at end of previous period} \\ \text{less payment} \end{array}$$

or

$$\text{Interest in } m\text{th period} = PMT\, s_{m\rceil i} - PMT\, s_{m-1\rceil i} - PMT$$
$$= PMT(s_{m\rceil i} - s_{m-1\rceil i}) - PMT.$$

Interest per Period in a Sinking Fund

The interest earned during the mth period of a sinking fund with payments of PMT and earning interest at a rate of i per period is

$$\text{interest in } m\text{th period} = PMT(s_{m\rceil i} - s_{m-1\rceil i}) - PMT.$$
$$= PMT(s_{m\rceil i} - s_{m-1\rceil i} - 1)$$

E X A M P L E   4   **Interest per Period in a Sinking Fund**

Find the interest earned by the sinking fund in Example 2 during the first quarter of the fifth year.

Solution   The first quarter of the fifth year is the 17th period. Taking $m = 17$ in the above formula and using Table B.2 in Appendix B gives

$$\begin{array}{l} \text{interest in} \\ \text{17th period} \end{array} = \$1655.57(s_{17\rceil 0.02} - s_{16\rceil 0.02}) - \$1655.57$$

$$= \$1655.57(20.012071 - 18.639285) - \$1655.57 = \$617.17$$

Thus $617.17 of interest was earned on this sinking fund during the first quarter of the fifth year.

SELF-HELP EXERCISE SET 5.3

1. At the end of every six months an individual places $1000 into an account earning an annual rate of 10% compounded semiannually. Find the amount in the account at the end of 15 years.

2. A person will need $20,000 to start a small business in 5 years. How much should be deposited at the end of every 3 months into an account paying an annual rate of 9% compounded quarterly to meet this goal?

EXERCISE SET 5.3

In Exercises 1 through 6 find the future values of each of the ordinary annuities at the given annual rate r compounded as indicated. The payments are made to coincide with the periods of compounding.

  1. $PMT = 1200$, $r = 0.08$, compounded annually for 10 years

  2. $PMT = 600$, $r = 0.08$, compounded semiannually for 10 years

  3. $PMT = 300$, $r = 0.08$, compounded quarterly for 10 years

  4. $PMT = 100$, $r = 0.08$, compounded monthly for 10 years

  5. $PMT = 100$, $r = 0.12$, compounded monthly for 20 years

  6. $PMT = 100$, $r = 0.12$, compounded weekly for 40 years

In Exercise 7 through 10 find the future value of each of the annuities at the *end* of the given nth year where the payments are made at the *beginning* of each year and r is the annual interest rate compounded annually. Such an annuity is called an **annuity due**.

  7. $PMT = \$1000$, $r = 0.08$, 10 years

  8. $PMT = \$1000$, $r = 0.10$, 20 years

  9. $PMT = \$1000$, $r = 0.15$, 40 years

  10. $PMT = \$500$, $r = 0.12$, 30 years

In Exercises 11 through 16 find the periodic payment for each sinking fund that is needed to accumulate the given sum under the given conditions.

  11. $FV = \$20,000$, $r = 0.09$, compounded annually for 10 years

12. $FV = \$20,000$, $r = 0.09$, compounded monthly for 20 years

13. $FV = \$10,000$, $r = 0.10$, compounded quarterly for 10 years

14. $FV = \$1,000,000$, $r = 0.09$, compounded weekly for 40 years

15. $FV = \$1,000,000$, $r = 0.15$, compounded weekly for 40 years

16. $FV = \$1,000,000$, $r = 0.12$, compounded monthly for 40 years

17. In formula (4) in the text solve for n and obtain

$$n = \frac{\log\left(\frac{iFV}{PMT} + 1\right)}{\log(1 + i)}.$$

18. Use the formula in Exercise 17 to find how many years it would take to accumulate $1 million with an annual payment of $500 and an interest rate of 15% per year compounded annually.

19. Use the formula in Exercise 17 to find how many years it would take to accumulate $1 million with an annual payment of $5000 and an interest rate of 9% per year compounded annually.

20. Show that the interest in mth period

$$PMT(s_{\overline{m}|i} - s_{\overline{m-1}|i} - 1) = PMT[(1 + i)^{m-1} - 1]$$

Applications

21. **Retirement.** An individual earns an extra $2000 each year and places this money at the end of each year into an Individual Retirement Account (IRA) in which both the original earnings and the interest in the account are not subject to taxation. If the account has an annual interest rate of 9% compounded annually, how much is in the account at the end of 40 years?

22. **Retirement.** Repeat the previous problem if the annual interest rate is 12% per year.

23. **Retirement.** Suppose the personal income tax is set at $33\frac{1}{3}\%$. An individual earns an extra $2000 each year and pays income taxes on these earnings and at the end of each year places the remaining funds $2000($\frac{2}{3}$) into a regular savings account in which the interest in the account is subject to the personal income tax mentioned before. If the account earns an annual interest rate of 9% compounded annually and the individual pays income taxes owed on the interest out of these funds, how much is in the account at the end of 40 years? Compare your answer to Exercise 21.

24. **Retirement.** Repeat the previous problem if the annual interest rate is 12% per year. Compare your answer to Exercise 22.

25. **Education Fund.** New parents wish to save for their newborn's education and wish to have $50,000 at the end of 18 years. How much should they place at the end of each year into a savings account that earns an annual rate of 9% compounded annually? How much interest would they earn over the life of this account? Determine the equity in this fund after 10 years. How much interest was earned during the tenth year?

26. **Education Fund.** Repeat the previous problem if the annual interest rate is 7%.

27. **House Down Payment.** A couple will need $20,000 at the end of 5 years for a down payment on a house. How much should they place at the end of each month into a savings account earning an annual rate of 9% compounded monthly to meet this goal? Determine the equity in this fund at the end of each year.

28. **House Down Payment.** Repeat the previous problem if the annual interest rate is 7%.

29. **Equipment.** A corporation creates a sinking fund in order to have $1 million to replace some machinery in 10 years. How much should be placed into this account at the end of each quarter if the annual interest rate is 10% compounded quarterly? Determine the equity in this fund at the end of every two years. How much interest was earned over the life of this fund? Determine the interest earned during the second quarter of the fifth year.

30. **Equipment.** Repeat the previous problem if the annual interest rate is 6%.

Solutions to Self-Help Exercise Set 5.3

1. The amount in an account at the end of 15 years into which $1000 has been placed every 6 months and that pays 10% compounded semiannually is given by

$$FV = PMT \frac{[(1 + i)^n - 1]}{i},$$

where $PMT = \$1000$ and $i = \frac{.10}{2} = .05$ and $n = 2(15) = 30$. Thus

$$FV = \$1000 \frac{[(1.05)^{30} - 1]}{0.05} = \$66,438.85.$$

2. The periodic payment PMT that is required to accumulate the sum of $FV = \$20,000$ over the $n = 4(5) = 20$ quarters of time with interest at the rate of $i = \frac{.09}{4} = .0225$ per quarter is

$$PMT = FV \frac{i}{[(1 + i)^n - 1]} = \$20,000 \frac{.0225}{[(1.0225)^{20} - 1]} = \$802.84.$$

5.4 PRESENT VALUE OF ANNUITIES AND AMORTIZATION

► *Present Value of Annuities*

► *Amortization*

► *Amortization Schedule*

APPLICATION
**Finding the Actual Value of
$1,000,000 Lotto Prize**

Suppose you run a lottery for your state government and you have just had a "$1 million winner." The winner is immediately given $50,000 and you must make the arrangements that will ensure that this winner will receive

$50,000 at the end of each of the next 19 years. You notice that short, intermediate, and long-term interest rates are all at 8%. How much money will it presently cost your department to ensure that these 20 payments of $50,000 each will be made? In other words, how much has the "$1 million winner" *really* won? (The answer to this question is given below.)

APPLICATION

Finding Car Payments

You wish to borrow $12,000 from the bank to purchase a car. Interest is 1% a month, and there are to be 48 equal monthly payments with the first to begin in one month. What must the payments be so that the loan will be paid off after 48 months? (The answer is given in Example 4.)

Present Value of Annuities

We now answer the question posed in the first paragraph above.

You know that if an amount equal to P earns interest at 8% per year, then the amount after n years is given by $F = P(1.08)^n$. Thus

$$P = \frac{F}{(1.08)^n} = F(1.08)^{-n}$$

is the present amount of money (present value) needed to attain the value F after n years if compounded annually at 8%.

After making the initial payment of $50,000, we will now see how much money must be placed into an account earning 8% per year so that a payment of $50,000 can be made at the end of each of the next 19 years. If P_1 is the present amount needed to be set aside now in order to make the $50,000 payment at the end of the first year, then, according to the above formula,

$$P_1 = 50,000(1.08)^{-1}.$$

The present amount of money P_2 needed to be set aside now in order to make the $50,000 payment at the end of the second year is

$$P_2 = 50,000(1.08)^{-2}.$$

And in general the present amount of money P_n needed to be set aside now to make the $50,000 payment at the end of the nth year is

$$P_n = 50,000(1.08)^{-n}.$$

See Figure 5.7. The sum of these 19 payments, denoted by P, is then the amount needed.

$$P = P_1 + P_2 + \cdots + P_{19}$$

$$P = 50,000(1.08)^{-1} + 50,000(1.08)^{-2} + \cdots + 50,000(1.08)^{-19} \quad (1)$$

Recall the summation formula mentioned in the last section

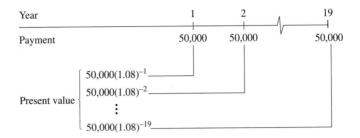

Figure 5.7

$$a + ab + ab^2 + ab^3 + \cdots + ab^{n-1} = a\,\frac{b^n - 1}{b - 1}.$$

The sum of the 19 numbers in (1) can be found using this last formula with $a = 50{,}000(1.08)^{-1}$, $b = (1.08)^{-1}$, and $n = 19$, to obtain

$$P = a\,\frac{b^{19} - 1}{b - 1}$$

$$= 50{,}000(1.08)^{-1}\,\frac{(1.08)^{-19} - 1}{(1.08)^{-1} - 1}$$

$$= \frac{50{,}000}{(1.08)}\,\frac{1 - (1.08)^{-19}}{1 - (1.08)^{-1}}$$

$$= 50{,}000\,\frac{1 - (1.08)^{-19}}{1.08 - 1}$$

$$= 50{,}000\,\frac{1 - (1.08)^{-19}}{0.08}.$$

Thus we have

$$P = 50{,}000\,\frac{1 - (1.08)^{-19}}{0.08}. \tag{2}$$

If we carry out the calculation, we obtain

$$P \approx 50{,}000(9.6036) \approx 480{,}180.$$

So you need to put away into this 8% account only $480,180 to pay out the remaining $950,000 in 19 payments of $50,000 each. Thus the total cost to your department for this "$1 million" winner is just $480,180 + $50,000 = $530,180.

In general, if R is the periodic payment, i the rate of interest per period, and n the number of periods, we can proceed as above and in analogy with formula (2) we obtain

$$P = R\,\frac{[1 - (1 + i)^{-n}]}{i}.$$

It is the custom to use *PV* for the present value, *PMT* for the periodic payment, and to use the notation

$$a_{n\rceil i} = \frac{[1 - (1 + i)^{-n}]}{i}.$$

We then have the following.

Present Value of an Ordinary Annuity

The present value *PV* of an ordinary annuity of *n* payments of *PMT* dollars each made at the end of the period with interest compounded at the rate of *i* per period is

$$PV = PMT \frac{[1 - (1 + i)^{-n}]}{i} = PMTa_{n\rceil i}. \tag{3}$$

REMARK.   If the payment PMT is $1, then the present value is just $a_{n\rceil i}$. Thus $a_{n\rceil i}$ is the value of an account needed to provide for *n* payments (or withdrawals) of $1 each when interest is paid at a rate of *i* per period and the compounding is per the same period.

E X A M P L E   1   **Calculating the Present Value of an Ordinary Annuity**

You have announced to your company that you will retire in one year. Your pension plan requires the company to pay you $25,000 in a lump sum at the end of one year and every year thereafter until your demise. The company makes the assumption that you will live to receive 15 payments. Interest rates are 7% per year compounded annually. What amount of money should the company set aside now to ensure that they can meet their pension obligations to you?

Solution   We use formula (3) with *PMT* = 25,000, *i* = 0.07, and *n* = 15, and obtain

$$PV = PMT \frac{[1 - (1 + i)^{-n}]}{i} \quad \text{or} \quad PMT \, a_{n\rceil i}$$

$$= \$25{,}000 \frac{[1 - (1.07)^{-15}]}{0.07} \quad \text{or} \quad \$25{,}000 \, a_{15\rceil 0.07}$$

$$\approx \$25{,}000(9.107914) \qquad \text{Use a calculator or}$$
$$\text{Table B.2 in Appendix B.}$$

$$= \$227{,}698$$

Thus if the company now puts away $227,698 into this account, they can make payments of $25,000 each at the end of each of the next 15 years (with no money left in the account after the last payment). ∎

EXAMPLE 2   **Comparing Investments**

You represent a bank that has $1 million in cash to invest. An oil company seeking money to hunt for oil approaches you wishing to borrow $1 million using one of their producing oil wells as collateral. They have the long-term oil contracts in hand and can guarantee you a payment of $100,000 at the end of each of the next 20 years. Interest rates available to the bank on the open market are at 8% per year compounded annually. Is this a good deal?

Solution   You ask yourself how much would be needed to invest at current interest rates to guarantee a payment of $100,000 at the end of each of the next 20 years. Use formula (3) with $PMT = 100,000$, $i = 0.08$, and $n = 20$, and obtain

$$PV = PMT \frac{[1 - (1 + i)^{-n}]}{i} \qquad \text{or} \quad PMT \, a_{\overline{n}|i}$$

$$= \$100,000 \frac{[1 - (1.08)^{-20}]}{0.08} \qquad \text{or} \quad \$100,000 \, a_{\overline{20}|0.08}$$

$$\approx \$100,000(9.818147) \qquad\qquad \begin{array}{l}\text{Use a calculator or}\\ \text{Table B.2 in Appendix B.}\end{array}$$

$$= \$981,814.74$$

Thus only about $982,000 is needed at current interest rates to obtain the same cash payments as the oil company is offering for $1 million. This is not a good deal. ▪

EXAMPLE 3   **Capital Expenditure Analysis**

A corporation wishes to increase productivity and thus save money by purchasing new machinery. The corporation can buy a machine for $100,000 that will save $30,000 annually and has a useful life of 5 years or a second machine for $160,000 that will save $35,000 annually and has a useful life of 7 years. If the annual interest rate is 8% compounded annually, which machine should the corporation purchase? Assume that the annual savings occurs at the end of each year.

Solution   The first machine saves $30,000 annually for 5 years. The present value of an annuity with an annual payment of $30,000 for 5 years with an annual interest rate of 8% compounded annually is

$$PV = PMT \, a_{\overline{5}|0.08}$$

$$= \$30,000(3.992710) \qquad \text{From Table B.2 in Appendix B}$$

$$= \$119,781.30$$

This means that an annual savings of $30,000 for 5 years is equivalent to a lump sum of $119,781.30 now for a savings of

$$\$119,781.30 - \$100,000 = \$19,781.30.$$

The second machine saves \$35,000 annually for 7 years. The present value of an annuity with an annual payment of \$35,000 for 7 years with an annual interest rate of 8% compounded annually is

$$PV = PMT\, a_{7\rceil 0.08}$$

$$= \$35,000(5.206370) \qquad \text{From Table B.2 in Appendix B}$$

$$= \$182,222.95$$

This means that an annual savings of \$35,000 for 7 years is equivalent to a lump sum of \$182,222.95 now for a savings of

$$\$182,222.95 - \$160,000 = \$22,222.95.$$

The second machine offers the larger savings and should be bought. ■

Amortization

Now look again at Example 2 but this time from the perspective of the oil company, that is, from the perspective of the borrower. If the oil company wishes to borrow an amount PV at an interest rate of i with equal payments at the end of each year with interest charged on the unpaid balance so that the entire debt will be repaid after n years, what will be the payments PMT? The very same formula used by the bank that relates all of these quantities must apply, namely,

$$PV = PMT\,\frac{[1 - (1 + i)^{-n}]}{i},$$

Now we are interested in solving for PMT. Solving for PMT, we obtain

$$PMT = PV\,\frac{i}{[1 - (1 + i)^{-n}]}.$$

This process of paying the debt is called **amortizing** (or killing) the debt. We then have

---

Amortization Formula

The periodic payment PMT to be made at the end of each period on a loan of PV dollars that is to be amortized over n periods with interest at the rate of i per period is

$$PMT = PV\,\frac{i}{[1 - (1 + i)^{-n}]} = \frac{PV}{a_{n\rceil i}}. \qquad (4)$$

E X A M P L E   4   **Calculating Payments on a Loan**

You wish to borrow \$12,000 from the bank to purchase a car. The bank charges interest at an annual rate of 12%. There are to be 48 equal monthly payments with the first to begin in one month. What must the payments be so that the loan will be paid off after 48 months? Find the total interest paid on this loan.

Solution   The rate per month is the annual rate divided by 12 or $i = \frac{.12}{12} = .01$. Using formula (4) with $PV = 12,000$, $i = .01$, $n = 48$, we obtain

$$PMT = PV \frac{i}{[1 - (1 + i)^{-n}]} \qquad \text{or} \qquad \frac{PV}{a_{\overline{n}|i}}$$

$$= \$12,000 \frac{0.01}{[1 - (1.01)^{-48}]} \qquad \text{or} \qquad \frac{\$12,000}{a_{\overline{48}|0.01}}$$

$$\approx \$12,000(0.0263338) \qquad \text{or} \qquad \frac{\$12,000}{37.973959} \quad \begin{array}{l} \text{Use calculator} \\ \text{or Table B.2} \\ \text{in Appendix B.} \end{array}$$

$$= \$316.01$$

Thus you will have to make 48 monthly payments of $316.01 in order to pay off this loan at the given interest rate.

The total interest paid is the total payments less the loan, or

$$\$316.01(48) - \$12,000 = \$15,168.48 - \$12,000 = \$3168.48. \quad \blacksquare$$

Amortization Schedule

We now look at what happens if you wish to make a lump-sum payment to pay off a loan. For example, it is routine for an individual to sell a house and buy another. This requires paying off the old mortgage. How much needs to be paid? To answer this we look closely at how much is owed at the end of each payment period. The next example indicates how to create such a schedule.

E X A M P L E   5   **Calculating an Amortization Schedule**

You agree to sell a small piece of property and grant a loan of $9000 to the buyer with annual interest at 10% compounded annually, with payments of equal amounts made at the end of each of the next 6 years. Construct a table that gives for each period the interest, payment toward principal, and the outstanding balance.

Solution   To find the payments we use formula (4) with $PV = 9000$, $i = 0.10$, $n = 6$ and obtain

$$PMT = PV \frac{i}{[1 - (1 + i)^{-n}]}$$

$$= \$9000 \frac{0.10}{[1 - (1.10)^{-6}]}$$

$$\approx \$9000(0.2296074) \qquad \text{Use calculator}$$

$$= \$2066.47$$

The interest paid for the first year is $9000(0.10) = 900. Thus the payment on the principal is

$$\$2066.47 - \$900 = \$1166.47.$$

The remaining principal is then

$$\$9000 - \$1166.47 = \$7833.53.$$

The interest paid for the second year is $\$7833.53(0.10) = \$783.35$. Thus the payment on the principal is

$$\$2066.47 - \$783.35 = \$1283.12.$$

The remaining principal is then

$$\$7833.53 - \$1283.12 = \$6550.41.$$

The following table summarizes the remaining calculations.

End of Period	Repayment Made	Interest Charged	Payment Toward Principal	Outstanding Principal
0				9000
1	2066.47	900.00	1166.47	7833.53
2	2066.47	783.35	1283.12	6550.41
3	2066.47	655.04	1411.43	5138.98
4	2066.47	513.90	1552.57	3586.41
5	2066.47	358.64	1707.83	1878.58
6	2066.44	187.86	1878.61	0000.00

The last payment is the one that adjusts for any roundoff errors. ■

E X A M P L E   6   **Calculating the Outstanding Principal**

In the loan in Example 4, suppose the individual wishes to pay off the loan after 20 payments. What is the outstanding principal or, alternatively, the **equity**?

Solution   One way of finding the outstanding principal or unpaid balance is to make a table as in the previous example, but there is a less tedious way. The bank views the loan as an annuity that they purchased from the borrower. Thus the unpaid balance of a loan with $(48 - 20) = 28$ remaining payments is the present value of the annuity that can be found using formula (3). We have the payments $PMT = \$316.01$, $i = 0.01$, and $n = 28$. Thus the present value is

$$PV = PMT \frac{[1 - (1 + i)^{-n}]}{i} \quad \text{or} \quad PMT\, a_{\overline{n}|i}$$

$$= \$316.01 \frac{[1 - (1.01)^{-28}]}{0.01} \quad \text{or} \quad \$316.01 \, a_{\overline{28}|0.01}$$

$$\approx \$316.01(24.316443) \qquad \begin{array}{l}\text{Use calculator or Table B.2}\\ \text{in Appendix B.}\end{array}$$

$$= \$7684.24$$

This is the amount the bank will require to pay off the loan at this point. ■

In the previous example let $n = 48$ be the total number of payments and $m = 20$ be the number of payments made, then notice that

$$PV = PMT \, a_{\overline{28}|0.01} = PMT \, a_{\overline{n-m}|0.01}.$$

In general we have the following.

Outstanding Principal

Let n be the total number of payments required to pay off an amortized loan and let m be the number of payments of PMT made; then the outstanding principal P is given by

$$P = PMT \, a_{\overline{n-m}|i}.$$

We can also use this formula to determine the amount of any payment that is applied to the principal. To see how to do this notice that

$$\text{the amount owed after } m - 1 \text{ payments} = PMT \, a_{\overline{n-m+1}|i}$$

$$\text{the amount owed after } m \text{ payments} = PMT \, a_{\overline{n-m}|i}$$

The difference of these two is the amount of the mth payment that is applied to the principal.

Payment Toward Principal

For an amortized loan requiring n equal payments of PMT dollars each and interest at the rate of i per period, the amount of the mth payment that is applied toward the principal is

$$\text{payment toward principal} = PMT(a_{\overline{n-m+1}|i} - a_{\overline{n-m}|i}).$$

E X A M P L E 7   **Finding the Payment Toward Principal**

For the amortized loan in Example 5 find the amount of the second payment R made toward the principal.

Solution   Since each payment is \$2066.47, $n = 6$, $m = 2$, and $i = .10$,

$$R = PMT(a_{n-m+1\rceil i} - a_{n-m\rceil i})$$

$$= \$2066.47(a_{5\rceil 0.10} - a_{4\rceil 0.10})$$

$$= \$2066.47(3.790787 - 3.169865) \qquad \text{Use a calculator.}$$

$$= \$1283.12$$

This agrees with what we found in Example 5.   ∎

SELF-HELP EXERCISE SET 5.4

1. Refer to Example 2 of the text. The very next day the Federal Reserve chairman announces a concerted attempt to control inflation. The financial markets respond immediately with a 1% drop in interest rates. Do you want the deal now?

2. You borrow $100,000 to purchase a house. The bank charges interest of 9% per year (on the unpaid balance). Payments will be made every month for 30 years.
    a. What are the monthly payments to ensure that the loan is paid in full by the end of 30 years?
    b. What is the total interest paid on this mortgage?
    c. What are the amounts of the 10th and 300th payments made toward the principal?

EXERCISE SET 5.4

For Exercises 1 through 6 find the amount needed to deposit into an account today that will yield a typical pension payment of $25,000 at the end of each of the next 20 years for the given annual interest rates.

1. 4%   **2.** 5%   **3.** 6%   **4.** 7%   **5.** 8%   **6.** 9%

In Exercises 7 through 10 find the annual payment needed to amortize a $10,000 loan in 6 payments if the interest is compounded annually at the given rate. Then find the total interest paid on each loan.

7. 10%   **8.** 12%   **9.** 15%   **10.** 18%

In Exercises 11 through 14 find the monthly payment needed to amortize the following typical $100,000 mortgage loans amortized over 30 years at the given annual interest rate compounded monthly. Then find the total interest paid on each loan.

11. 7%   **12.** 8%   **13.** 9%   **14.** 10%

In Exercises 15 through 18 find the monthly payments needed to amortize the following typical $100,000 mortgage loans amortized over 15 years at the given annual interest rate compounded monthly. Then find the total interest paid on each loan and compare with your answers to Exercises 11 through 14.

15. 7%   **16.** 8%   **17.** 9%   **18.** 10%

In Exercises 19 through 22 find the monthly payment needed to amortize the following typical $10,000 automobile loans over 4 years at the given annual interest rate compounded monthly. Then find the total interest paid on each loan.

19. 10%   **20.** 11%   **21.** 12%   **22.** 14%

In Exercises 23 through 26 find the monthly payment needed to amortize the following typical $1,000 credit card loans over 3 years at the given annual interest rate compounded monthly. Then find the total interest paid on each loan.

23. 14%   **24.** 16%   **25.** 18%   **26.** 20%

27. An individual wishes to pay off the mortgage in Exercise 11 after 120 payments. What amount is owed? What are the amounts of the 15th and 340th payments paid toward the principal?

28. An individual wishes to pay off the mortgage in Exercise 12 after 120 payments. What amount is owed? What are the amounts of the 15th and 340th payments paid toward the principal?

29. An individual wishes to pay off the mortgage in Exercise 15 after 60 payments. What amount is owed? What are the amounts of the 10th and 150th payments paid toward the principal?

30. An individual wishes to pay off the mortgage in Exercise 16 after 60 payments. What amount is owed? What are the amounts of the 10th and 150th payments paid toward the principal?

In Exercises 31 through 34 prepare an amortization schedule as in the table in Example 5 for each of the following loans.

31. The loan in Exercise 7   **32.** The loan in Exercise 8

33. The loan in Exercise 9   **34.** The loan in Exercise 10

35. From either formula (3) or (4) in the text show that

$$n = \frac{\log \dfrac{PMT}{PMT - PV(i)}}{\log(1 + i)}$$

Suppose you wished to take out a $10,000 loan at 1% a month and could make a monthly payment of $200. Use this formula to determine the number of payments needed to amortize this loan.

36. Show that the payment toward principal

$$PMT\,(a_{n-m+1\rceil i} - a_{n-m\rceil i}) = PMT\,(1 + i)^{m - n - 1}$$

Applications

37. Comparing Present Values. Two oil wells are for sale. The first will yield payments of $10,000 at the end of each of the next 10 years, while the second will yield payments of $6500 at the end of each of the next 20 years. Interest rates are assumed to hold steady at 8% per year over the next 20 years. Which has the higher present value?

38. Comparing Present Values. Redo the previous exercise if interest rates were 6% over the next 20 years.

(© John Spragens, Jr./Photo Researchers, Inc.)

39. Comparing Present Values. You are being offered a "half a million dollar" retirement package to be given in $50,000 payments at the end of each of the next 10 years. You are also given the option of accepting a $350,000 lump sum payment now. Interest rates are at 9% per year. Which looks better to you? Why?

40. Comparing Present Values. You are being offered a "million dollar" retirement package to be given in $50,000 payments at the end of each of the next 20 years. You are also given the option of accepting a $500,000 lump-sum payment now. Interest rates are at 9% per year. Which looks better to you? Why?

41. Capital Expenditure. A corporation wishes to increase productivity and thus save money by purchasing new machinery. The corporation can buy a machine for $95,000 that will save $30,000 annually and has a useful life of 5 years or a second machine for $160,000 that will save $35,000 annually and has a useful life of 7 years. Assuming that the annual savings occurs on the last day of each year, which machine should the corporation purchase if the annual interest rate is (a) 8% compounded annually, (b) 6% compounded annually?

42. Leasing. A corporation can either lease a machine with a useful life of 6 years for $20,000 per year paid at the end of the year or buy it for $100,000. Which should the corporation do if annual interest rates are (a) 7% compounded annually (b) 5% compounded annually?

43. Car Loan. You purchase a new car for $12,000 and make a $2000 down payment. You then owe $10,000 for the car which you intend to pay off in 3 years. Initially, you intended to finance your car loan through your credit union that charges 9% compounded monthly for new-car loans. The salesman tells you, however, that GM is having a promotion and you can finance your car through GMAC for 4% interest compounded monthly or you can receive a rebate of $800 that can be applied to the purchase price. Should you finance the full $10,000 with GMAC for the lower interest rate or take the rebate and finance $9200 with your credit union? Explain your answer.

Solutions to Self-Help Exercise Set 5.4

1. Do the same calculation as in Example 2 of the text but now use $i = 0.07$ and obtain

$$PV = PMT \frac{[1 - (1 + i)^{-n}]}{i} = 100,000 \frac{[1 - (1.07)^{-20}]}{0.07} \approx 1,059,400$$

Now it will take about $1,059,400 at current interest rates to obtain the same cash payments as the oil company is offering for less than $1 million. The deal is now a good one (without consideration of the riskiness of the deal).

2. a. The monthly payment of PMT on a loan of $PV = \$100,000$ that is to be amortized over $n = 12(30) = 360$ periods with interest at the rate of $i = .09/12 = .0075$ per month is, using formula (4),

$$PMT = PV \frac{i}{[1 + (1 + i)^{-n}]} = 100,000 \frac{.0075}{[1 - (1.0075)^{-360}]} = 804.62$$

or $804.62 as the monthly payment.

b. The interest is the total payments less the principal or

$$804.62(360) - 100,000 = 189,664.14.$$

c. The amount of the 10th payment toward the principal is

$$R = \$804.62(a_{\overline{360-10+1}|0.0075} - a_{\overline{360-10}|0.0075})$$

$$= \$804.62(123.65224 - 123.57963)$$

$$= \$58.42$$

The amount of the 300th payment toward the principal is

$$R = \$804.62(a_{\overline{360-300+1}|0.0075} - a_{\overline{360-300}|0.0075})$$

$$= \$804.62(48.807319 - 48.173374)$$

$$= \$510.08$$

SUMMARY OUTLINE OF CHAPTER 5

- **Simple Interest.** Suppose a sum of money P, called the principal or present value, is invested for t years at an annual *simple* interest rate of r, where r is given as a decimal. Then the interest I at the end of t years is given by

$$I = Prt.$$

The future value F at the end of t years is

$$F = P + I = P + Prt = P(1 + rt).$$

- **Present Value.** The **present value** needed to deposit into an account earning a simple annual rate of r, expressed as a decimal, in order to have a future amount F after t years is

$$P = \frac{F}{1 + rt}.$$

- **Discount and Proceeds.** The discount D on a discounted loan of M dollars at a simple annual interest rate of r, expressed as a decimal, for t years is

$$D = Mrt,$$

where D is the discount (interest paid at time of loan), M is the maturity value (amount borrowed), r is the discount rate (annual simple interest rate), and t is the length of the loan. The **proceeds** P of the loan is the actual amount the borrower receives when the loan is made and is given by $P = M - D$.

- **Effective Rate of a Discounted Loan.** The effective interest rate given on a discounted loan of length t with a discount rate of r, expressed as a decimal, is

$$r_{eff} = \frac{r}{1 - rt}$$

- **Compound Interest.** Suppose a principal P earns interest at the annual rate of r, expressed as a decimal, and interest is compounded m times a year. Then the amount F after t years is

$$F = P(1 + i)^n = P\left(1 + \frac{r}{m}\right)^{mt},$$

where $n = mt$ is the number of time periods and $i = \frac{r}{m}$ is the interest per period.

- **The Time for Money to Grow from P to F.**   The number of periods n it takes for a principal of P to grow to a future value of F in an account that earns interest per period at a rate of i, expressed as a decimal, is

$$n = \frac{\log F/P}{\log(1 + i)}.$$

The number of periods for the principal to double is given by

$$n = \frac{\log 2}{\log(1 + i)}.$$

- **Effective Yield.**   Suppose a sum of money is invested at an annual rate of r expressed as a decimal and is compounded m times a year. The effective yield r_{eff} is

$$r_{eff} = \left(1 + \frac{r}{m}\right)^m - 1.$$

- **Present Value.**   Suppose an account earns an annual rate of r expressed as a decimal and is compounded m times a year. Then the amount P, called the **present value,** needed presently in this account so that a future amount of F will be attained in t years is given by

$$P = \frac{F}{\left(1 + \frac{r}{m}\right)^{mt}}.$$

- An **annuity** is a sequence of equal payments made at equal time periods.

- An **ordinary annuity** is one in which the payments are made at the *end* of the time periods and the periods of compounding are the same time periods.

- The **term** of an annuity is the time from the beginning of the first period to the end of the last period.

- The total amount in the account, including interest, at the end of the term of an annuity is called the **future value of the annuity**.

- **Summation Formula.**   For any positive integer n and any real numbers a and b with $b \neq 1$

$$a + ab + ab^2 + ab^3 + \cdots + ab^{n-1} = a\frac{b^n - 1}{b - 1}.$$

- **Future Value of an Ordinary Annuity.**   The future value FV of an ordinary annuity of n payments of PMT dollars paid at the end of each period into an account that earns interest at the rate of i per period is

$$FV = PMT\frac{[(1 + i)^n - 1]}{i} = PMT\, s_{n\rceil i}$$

- **Sinking Fund Payment.**   The periodic payment PMT that is required to accumulate the sum FV over n periods of time with interest at the rate of i per period is

$$PMT = FV\frac{i}{[(1 + i)^n - 1]} = \frac{FV}{s_{n\rceil i}}$$

- **Interest per Period in a Sinking Fund.** The interest earned during the mth period of a sinking fund with payments of PMT and earning interest at a rate of i per period is

$$\text{Interest in } m\text{th period} = PMT(s_{\overline{m}|i} - s_{\overline{m-1}|i}) - PMT.$$

- **Present Value of an Ordinary Annuity.** The present value PV of an ordinary annuity of n payments of PMT dollars each made at the end of the period with interest compounded at the rate of i per period is

$$PV = PMT\,\frac{[1 - (1 + i)^{-n}]}{i} = PMT\,a_{\overline{n}|i}$$

- **Amortization Formula.** The periodic payment PMT to be made at the end of each period on a loan of PV dollars that is to be amortized over n periods with interest at the rate of i per period is

$$PMT = PV\,\frac{i}{[1 - (1 + i)^{-n}]} = \frac{PV}{a_{\overline{n}|i}}$$

- **Outstanding Principal.** Let n be the total number of payments required to pay off an amortized loan and let m be the number of payments of PMT made; then the outstanding principal P is given by

$$P = PMT\,a_{\overline{n-m}|i}.$$

- **Payment Toward Principal.** For an amortized loan requiring n equal payments of PMT dollars each and interest at the rate of i per period, the amount of the mth payment that is applied toward the principal is

$$\text{payment toward principal} = PMT(a_{\overline{n-m+1}|i} - a_{\overline{n-m}|i}).$$

Chapter 5 Review Exercises

In Exercises 1 through 2 find how much is in the following accounts after the given t years where P is the initial principal, r is the annual rate, and the compounding is as indicated.

1. $t = 5$, $P = \$1000$, $r = 8\%$, compounded (a) annually, (b) quarterly, (c) monthly, (d) weekly, (e) daily

2. $t = 5$, P $= \$1000$, $r = 10\%$, compounded (a) annually, (b) quarterly, (c) monthly, (d) weekly, (e) daily

In Exercises 3 through 4 find the effective yield given the annual rate r and the indicated compounding.

3. $r = 9\%$, compounded (a) semiannually, (b) quarterly, (c) monthly, (d) weekly, (e) daily

4. $r = 11\%$, compounded (a) semiannually, (b) quarterly, (c) monthly, (d) weekly, (e) daily

In Exercises 5 through 6 find the present value of the given amounts A with the indicated annual rate of return r, the number of years t, and the indicated compounding.

5. $A = \$10,000$, $r = 8\%$, $t = 10$, compounded (a) annually, (b) monthly, (c) weekly

6. $A = \$10,000, r = 9\%, t = 10$, compounded (a) annually, (b) quarterly, (c) daily

7.   You have just won a $1,000,000 lottery. Your entire prize will be given to you in 20 years in a lump sum. If current interest rates are 8% compounded annually, what is this $1,000,000 prize worth at the present?

8.   How much money should you deposit now in an account earning 9% annually in order to have $10,000 in 5 years?

9.   What annual rate is required for an account that is compounded annually to double in 8 years?

10.   If an account yields an annual return of r where r is expressed as a decimal and if T is the time it takes for this account to triple, show that

$$T = \frac{\log 3}{\log(1 + r)}.$$

11.   **Savings.**   A self-employed individual places $3000 a year into a Keogh account in which taxes are not paid on the interest. If interest rates remain at 9%, how much will be in this account after 30 years?

12.   **Sinking Fund.**   A large corporation creates a sinking fund in order to have a $2 million cash bonus for the president of the company on his retirement in 10 years. How much should be placed into this account at the end of each quarter if the interest rate is 8% compounded quarterly? Over the life of this sinking fund how much interest accumulates? How much interest is earned in the third quarter of the fifth year?

13.   **Lottery Prize.**   The state lottery has just had a $2 million winner and needs to pay the recipient $100,000 now and the same amount at the end of each of the next 19 years. The state does not wish to be involved in the administration of this disbursement and wishes to give this obligation together with a lump sum payment to an insurance company to handle the payments. You are an investment advisor for an insurance company. Interest rates are at 8%. What is the very least that you would accept for this obligation?

14.   **Amortization.**   You plan to borrow $100,000 to buy a house. Interest rates are at 9% a year compounded monthly. What will be your monthly payments if the length of the mortgage is 25 years? 15 years? What will be the total interest paid in each case? In each case how much of the tenth payment will go toward the principal? The 120th?

How many different poker hands are there? (Dennis Drenner.)

chapter six

6

Sets and Counting

The first section gives the fundamental ideas of set theory, which we will then use throughout the remainder of the text. In particular these ideas are used in the remaining sections of this chapter to develop means to count the number of elements in various sets. We shall see in the next chapter that this is fundamental to finding probability.

6.1   SETS

► *Introduction*

► *Set Operations*

► *Additional Rules and Laws for Sets*

► *Applications*

George Boole was born into a lower class family in Lincoln, England, and had only a common school education. He was largely self-taught and managed to become an elementary school teacher. Up to this time any rule of algebra such as $a(x + y) = ax + ay$ was understood to apply only to numbers and magnitudes. Boole developed an "algebra" of sets where the elements of the sets could be not just numbers but *anything*. This then laid down the foundations for a fundamental way of thinking. Bertrand Russell, a great mathematician and philosopher of the 20th century, said that the greatest discovery of the 19th century was the nature of pure mathematics, which he asserted was discovered by George Boole. Boole's pamphlet "The Mathematical Analysis of Logic" maintained that the essential character of mathematics lies in its form rather than in its content. Thus mathematics is not merely the science of measurement and number but any study consisting of symbols and precise rules of operation. Boole founded not only a new algebra of sets but also a formal logic.

Introduction

Chapter 1 gave a brief introduction to sets. It would be beneficial to review that material before reading this section. This section discusses operations on sets and laws governing these set operations. These are fundamental notions that will be used throughout the remainder of this text. In the next two chapters we will see that probability and statistics are based on counting the elements in sets and manipulating set operations. Thus we first need to understand clearly the notion of sets and their operations.

Subsets

If every element of a set A is also an element of another set B, we say that **A is a subset of B** and write $A \subset B$. If A is not a subset of B, we write $A \not\subset B$.

Thus $\{1, 2, 4\} \subset \{1, 2, 3, 4\}$, but $\{1, 2, 3, 4\} \not\subset \{1, 2, 4\}$.

Empty Set

The **empty set**, written as ϕ, is the set with no elements.

The empty set can be used to conveniently indicate that an equation has no solution. For example

$$\{x \mid x \text{ is real and } x^2 = -1\} = \phi.$$

By the definition of subset, given any set A, we must have $\phi \subset A$ and $A \subset A$.

E X A M P L E 1 **Finding Subsets**

Find all the subsets of $\{a, b, c\}$.

Solution The subsets are

$$\phi, \{a\}, \{b\}, \{c\}, \{a, b\}, \{a, c\}, \{b, c\}, \{a, b, c\}. \quad \blacksquare$$

The empty set is the set with no elements. At the other extreme is the **universal set**. This set is the set of all elements being considered and is denoted by U. If, for example, we are to take a national survey of voter satisfaction with the President, the universal set is the set of all voters in this country. If the survey is to determine the effects of smoking on pregnant women, the universal set is the set of all pregnant women. The context of the problem under discussion will determine the universal set for that problem. The universal set must contain every element under discussion.

A **Venn diagram** is a way of visualizing sets. For example, given a universal set U and two subsets A and B, Figure 6.1a is a Venn diagram that visualizes the concept that $A \subset U$, while Figure 6.1b is a Venn diagram that visualizes the concept $B \subset A$.

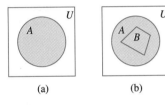

(a)　　　　　(b)

Figure 6.1

Set Operations

The first set operation we consider is the complement.

A^c

Figure 6.2

Complement

Given a universal set U and a set $A \subset U$, the **complement of A**, written A^c, is the set of all elements that are in U but not in A, that is,

$$A^c = \{x \mid x \in U, x \notin A\}.$$

A Venn diagram visualizing A^c is shown in Figure 6.2.

E X A M P L E 2 **Finding Complements of Sets**

Let $U = \{1, 2, 3, 4, 5, 6, 7, 8, 9\}$, $A = \{1, 3, 5, 7, 9\}$, $B = \{1, 2, 3, 4, 5\}$. Find A^c, B^c, U^c, ϕ^c, $(A^c)^c$.

Solution   We have

$$A^c = \{2, 4, 6, 8\}$$

$$B^c = \{6, 7, 8, 9\}$$

$$U^c = \phi$$

$$\phi^c = U$$

$$(A^c)^c = \{2, 4, 6, 8\}^c$$

$$= \{1, 3, 5, 7, 9\} = A. \quad \blacksquare$$

If U is a universal set, we must always have

$$U^c = \phi, \qquad \phi^c = U.$$

If A is any subset of a universal set U, then

$$(A^c)^c = A.$$

This can be seen using the Venn diagram in Figure 6.2, since the complement of A^c is all elements in U but not in A^c. This can be seen in Figure 6.2 to be just the set A.

Set Union

The **union** of two sets A and B, written $A \cup B$, is the set of all elements that belong to A, or to B, or to both. Thus

$$A \cup B = \{x | x \in A \text{ or } x \in B \text{ or both}\}.$$

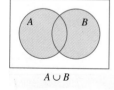

$A \cup B$

Figure 6.3

R E M A R K .   Our convention will be to drop the phrase ''or both'' but still maintain the same meaning. Note very carefully that this gives a particular definition to the word ''or.'' Thus we will normally write

$$A \cup B = \{x | x \in A \text{ or } x \in B\}.$$

We can also say that $A \cup B$ is all elements in A together with all elements in B. A Venn diagram is shown in Figure 6.3.

E X A M P L E   3   **Finding the Union of Two Sets**

Find $\{1, 2, 3, 4\} \cup \{1, 4, 5, 6\}$.

Solution   We add to the first set any elements in the second set that are not already there. Thus

$$\{1, 2, 3, 4\} \cup \{1, 4, 5, 6\} = \{1, 2, 3, 4, 5, 6\}. \quad \blacksquare$$

From Figure 6.2, we can see that if U is a universal set and $A \subset U$, then

$$A \cup A^c = U.$$

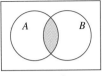

$A \cap B$

Figure 6.4

Set Intersection

The **intersection** of two sets A and B, written $A \cap B$, is the set of all elements that belong to both the set A and to the set B. Thus,

$$A \cap B = \{x \mid x \in A \text{ and } x \in B\}$$

A Venn diagram is shown in Figure 6.4.

E X A M P L E   4   **Find the Intersection of Two Sets**

Find $\{a, b, c, d\} \cap \{a, c, e\}$ and $\{a, b\} \cap \{c, d\}$.

Solutions

(a)   Only a and c are in both of the sets. Thus

$$\{a, b, c, d\} \cap \{a, c, e\} = \{a, c\}.$$

(b)   The two sets $\{a, b\}$ and $\{c, d\}$ have no elements in common. Thus

$$\{a, b\} \cap \{c, d\} = \phi. \quad \blacksquare$$

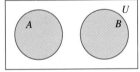

A and B disjoint

Figure 6.5

Disjoint Sets

Two sets A and B are **disjoint** if they have no elements in common, that is, if $A \cap B = \phi$. See Figure 6.5.

An examination of Figure 6.2 or referring to the definition of A^c indicates that for any set A, A and A^c are disjoint. That is,

$$A \cap A^c = \phi.$$

Additional Rules and Laws for Sets

The following rules can be established.

Rules for Set Operations

Let U be a universal set and let A, B, and C be any subsets of U. Then

$A \cup B = B \cup A$	Commutative law for union
$A \cap B = B \cap A$	Commutative law for intersection
$A \cup (B \cup C) = (A \cup B) \cup C$	Associative law for union
$A \cap (B \cap C) = (A \cap B) \cap C$	Associative law for intersection
$A \cup (B \cap C) = (A \cup B) \cap (A \cup C)$	Distributive law for union
$A \cap (B \cup C) = (A \cap B) \cup (A \cap C)$	Distributive law for intersection

$$(A \cup B)^c = A^c \cap B^c \qquad \text{De Morgan law}$$

$$(A \cap B)^c = A^c \cup B^c \qquad \text{De Morgan law}$$

E X A M P L E   5   **Establishing a De Morgan Law**

Use a Venn diagram to show that

$$(A \cup B)^c = A^c \cap B^c.$$

Solution   We first consider the right side of this equation. Figure 6.6 shows a Venn diagram of A^c and B^c and $A^c \cap B^c$. We then notice from Figure 6.3 that this is $(A \cup B)^c$.   ■

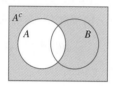

Figure 6.6

E X A M P L E   6   **Establishing the Distributive Law for Union**

Use a Venn diagram to show that

$$A \cup (B \cap C) = (A \cup B) \cap (A \cup C).$$

Solution   Consider first the left side of this equation. In Figure 6.7a the sets A, $B \cap C$, and the union of these two are shown. Now for the right side of the equation refer to Figure 6.7b, where the sets $A \cup B$, $A \cup C$, and the intersection of these two sets are shown. We see that we end up with the same set in both cases.   ■

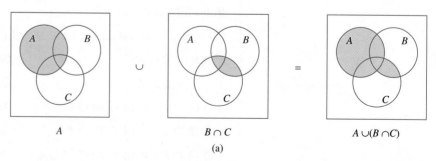

(a)

Figure 6.7

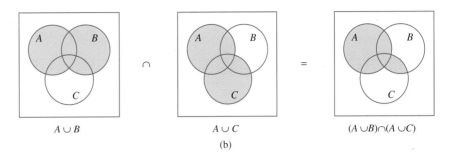

$A \cup B$ $\cap$ $A \cup C$ $=$ $(A \cup B) \cap (A \cup C)$

(b)

Figure 6.7 (cont.)

Applications

E X A M P L E   7   **Using Set Operations to Write Expressions**

Let U be the universal set consisting of the set of all students taking classes at the University of Hawaii and

$$B = \{x \in U \,|\, x \text{ is currently taking a business course}\}$$

$$E = \{x \in U \,|\, x \text{ is currently taking an English course}\}$$

$$M = \{x \in U \,|\, x \text{ is currently taking a math course}\}.$$

Write an expression using set operations for each of the following:

  (a)   The set of students at the University of Hawaii taking a course in at least one of the above three fields.

  (b)   The set of all students at the University of Hawaii taking both an English course and a math course but not a business course.

  (c)   The set of all students at the University of Hawaii taking a course in exactly one of the three fields above.

Solutions   (a)   This is $B \cup E \cup M$. See Figure 6.8.

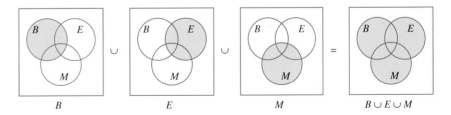

B $\cup$ E $\cup$ M $=$ $B \cup E \cup M$

Figure 6.8

  (b)   This can be described as the set of students taking an English course (E) and also (intersection) a math course (M) and also (intersection) not a business course (B^c) or

$$E \cap M \cap B^c.$$

This is the set of points in the universal set that are in both E and M but not in B and is shown in Figure 6.9.

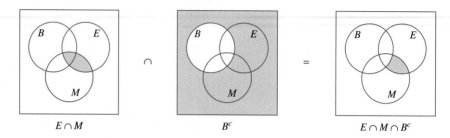

Figure 6.9

    (c) We describe this set as the set of students taking business but not taking English or math ($B \cap E^c \cap M^c$) together with (union) the set of students taking English but not business or math ($E \cap B^c \cap M^c$) together with (union) the set of students taking math but not business or English ($M \cap B^c \cap E^c$) or

$$(B \cap E^c \cap M^c) \cup (E \cap B^c \cap M^c) \cup (M \cap B^c \cap E^c).$$

This is the union of the three sets shown in Figure 6.10. The first, $B \cap E^c \cap M^c$, consists of those points in B that are outside E and also outside M. The second set $E \cap B^c \cap M^c$ consists of those points in E that are outside B and M. The third set $M \cap B^c \cap E^c$ is the set of points in M that are outside B and E. The union of these three sets is then shown on the right in Figure 6.10. ∎

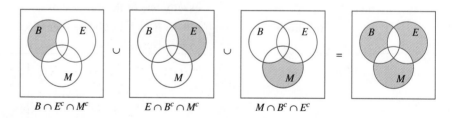

Figure 6.10

SELF-HELP EXERCISE SET 6.1

1. Let $U = \{1, 2, 3, 4, 5, 6, 7\}$, $A = \{1, 2, 3, 4\}$, $B = \{3, 4, 5\}$, $C = \{2, 3, 4, 5, 6\}$. Find the following.

   a. $A \cup B$,   **b.** $A \cap B$,   **c.** A^c,

   d. $(A \cup B) \cap C$,   **e.** $(A \cap B) \cup C$,   **f.** $A^c \cup B \cup C$.

2. Let U denote the set of all corporations in this country and P those that made profits during the last year, D those that paid a dividend during the last year, and L those that increased their labor force during the last year. Describe the following using the three sets P, D, L, and set operations.
   a. Corporations in this country that had profits and also paid a dividend last year
   b. Corporations in this country that either had profits or paid a dividend last year
   c. Corporations in this country that did not have profits last year
   d. Corporations in this country that had profits, paid a dividend, and did not increase their labor force last year
   e. Corporations in this country that had profits or paid a dividend, and did not increase their labor force last year

EXERCISE SET 6.1

In Exercises 1 through 4 determine whether the statements are true or false.

1. a. $\phi \in A$,   b. $A \in A$

2. a. $0 = \phi$,   b. $\{x, y\} \in \{x, y, z\}$

3. a. $\{x | 0 < x < -1\} = \phi$,   b. $\{x | 0 < x < -1\} = 0$

4. a. $\{x | x(x - 1) = 0\} = \{0, 1\}$,   b. $\{x | x^2 + 1 < 0\} = \phi$

5. If $A = \{u, v, y, z\}$, determine whether the following statements are true or false.
   a. $w \in A$,   b. $x \notin A$,
   c. $\{u, x\} \subset A$,   d. $\{y, z, v, u\} = A$

6. If $A = \{u, v, y, z\}$, determine whether the following statements are true or false.
   a. $x \notin A$ b. $\{v, w\} \notin A$
   c. $\{x, w\} \not\subset A$ d. $\phi \subset A$

7. List all the subsets of (a) $\{3\}$, (b) $\{3, 4\}$.

8. List all the subsets of (a) ϕ, (b) $\{3, 4, 5\}$.

9. Use Venn diagrams to indicate the following.
   a. $A \subset U, B \subset U, A \subset B^c$ b. $A \subset U, B \subset U, B \subset A^c$

10. Use Venn diagrams to indicate the following.
    a. $A \subset U, B \subset U, C \subset U, C \subset (A \cup B)^c$
    b. $A \subset U, B \subset U, C \subset U, C \subset A \cap B$

11. On the accompanying figure, indicate where the following sets are.
    a. $A \cap B^c$ b. $A \cap B$
    c. $A^c \cap B$ d. $A^c \cap B^c$

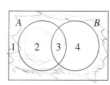

12. Can you find any two sets among the list of four sets given in Exercise 11 that are not disjoint?

On the accompanying figure indicate where the following sets are.

13. $A \cup B^c$ **14.** $A^c \cup B^c$

15. $(A \cup B)^c$ **16.** $(A \cap B)^c$

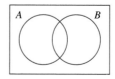

17. Match up the following sets to the appropriate region in the accompanying Venn diagram.

    a. $A \cap B \cap C$,   **b.** $A \cap B^c \cap C^c$ **c.** $A \cap B \cap C^c$,

    d. $B \cap A^c \cap C^c$,   **e.** $A^c \cap B^c \cap C^c$ **f.** $A \cap C \cap B^c$,

    g. $B \cap C \cap A^c$,   **h.** $C \cap A^c \cap B^c$

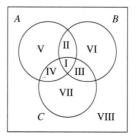

On the accompanying figure indicate where the following sets are.

18. $A \cup (B \cap C)$

19. $(A \cup B) \cap C^c$ **20.** $A \cap B \cap C^c$

21. $(A \cap B)^c \cap C$ **22.** $A^c \cap B^c \cap C^c$

23. $(A \cup B)^c \cap C$ **24.** $(A \cup B \cup C)^c \cap A$

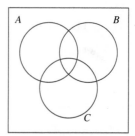

In Exercises 25 through 32 let $U = \{1, 2, 3, 4, 5, 6, 7, 8, 9, 10\}$, $A = \{1, 2, 3, 4, 5, 6\}$, $B = \{4, 5, 6, 7, 8\}$, $C = \{5, 6, 7, 8, 9, 10\}$ and find the indicated sets.

25. a. $A \cap B$ **b.** $A \cup B$

26. a. A^c **b.** $A^c \cap B$

27. a. $A \cap B^c$ **b.** $A^c \cap B^c$

28. a. $A^c \cup B^c$ **b.** $(A^c \cup B^c)^c$

29. a. $A \cap B \cap C$ **b.** $(A \cap B \cap C)^c$

30. a. $A \cap (B \cup C)$ **b.** $A \cap (B^c \cup C)$

31. a. $A^c \cap B^c \cap C^c$ **b.** $(A \cup B \cup C)^c$

32. a. $A^c \cap B^c \cap C$ **b.** $A^c \cap B \cap C^c$

Let U be the set of all residents of your state and let

$A = \{x \in U \mid x \text{ owns an automobile}\}$
$H = \{x \in U \mid x \text{ owns a house}\}.$

In Exercises 33 through 36 describe each of the sets in words.

33. a. A^c,   **b.** $A \cup H$,   **c.** $A \cup H^c$

34. a. H^c,   **b.** $A \cap H$,   **c.** $A^c \cap H$

35. a. $A \cap H^c$,   **b.** $A^c \cap H^c$,   **c.** $A^c \cup H^c$

36. a. $(A \cap H)^c$,   **b.** $(A \cup H)^c$,   **c.** $(A^c \cap H^c)^c$

In Exercises 37 through 40 let U, A, and H be as in the previous four problems, let $P = \{x \in U \mid x \text{ owns a piano}\}$, and describe each of the sets in words.

37. a. $A \cap H \cap P$ **b.** $A \cup H \cup P$ **c.** $(A \cap H) \cup P$

38. a. $(A \cup H) \cap P$ **b.** $(A \cup H) \cap P^c$ **c.** $A \cap H \cap P^c$

39. a. $(A \cap H)^c \cap P$ **b.** $A^c \cap H^c \cap P^c$ **c.** $(A \cup H)^c \cap P$

40. a. $(A \cup H \cup P)^c \cap A$ **b.** $(A \cup H \cup P)^c$ **c.** $(A \cap H \cap P)^c$

In Exercises 41 through 48 let U be the set of major league baseball players and let

$N = \{x \in U \mid x \text{ plays for the New York Yankees}\}$

$S = \{x \in U \mid x \text{ plays for the San Francisco Giants}\}$

$F = \{x \in U \mid x \text{ is an outfielder}\}$

$H = \{x \in U \mid x \text{ has hit 20 homers in one season}\}.$

Write the set that represents the following descriptions.

41. a. Outfielders for the New York Yankees
   b. New York Yankees who have never hit 20 homers in a season

42. **a.** San Francisco Giants who have hit 20 homers in a season
    b. San Francisco Giants who do not play outfield

43. **a.** Major league ball players who play for the New York Yankees or the San Francisco Giants
    b. Major league ball players who play for neither the New York Yankees nor the San Francisco Giants

44. **a.** San Francisco Giants who have never hit 20 homers in a season
    b. Major league ball players who have never hit 20 homers in a season

45. **a.** New York Yankees or San Francisco Giants who have hit 20 homers in a season
    b. Outfielders for the New York Yankees who have never hit 20 homers in a season

46. **a.** Outfielders for the New York Yankees or San Francisco Giants
    b. Outfielders for the New York Yankees who have hit 20 homers in a season

47. **a.** Major league outfielders who have hit 20 homers in a season and do not play for the New York Yankees or the San Francisco Giants
    b. Major league outfielders who have never hit 20 homers in a season and do not play for the New York Yankees or the San Francisco Giants

48. **a.** Major league players who do not play outfield, who have hit 20 homers in a season, and do not play for the New York Yankees or the San Francisco Giants
    b. Major league players who play outfield, who have never hit 20 homers in a season, and do not play for the New York Yankees or the San Francisco Giants

In Exercises 49 through 54 let $U = \{1, 2, 3, 4, 5, 6, 7, 8, 9, 10\}$, $A = \{1, 2, 3, 4, 5\}$, $B = \{4, 5, 6, 7\}$, $C = \{5, 6, 7, 8, 9, 10\}$. Verify that the identities are true for these sets.

49. $A \cup (B \cup C) = (A \cup B) \cup C$

50. $A \cap (B \cap C) = (A \cap B) \cap C$

51. $A \cup (B \cap C) = (A \cup B) \cap (A \cup C)$

52. $A \cap (B \cup C) = (A \cap B) \cup (A \cap C)$

53. $(A \cup B)^c = A^c \cap B^c$

54. $(A \cap B)^c = A^c \cup B^c$

Solutions to Self-Help Exercise Set 6.1

1. **a.** $A \cup B$ is the elements in A or B or both. Thus $A \cup B = \{1, 2, 3, 4, 5\}$.
   b. $A \cap B$ is the elements in both A and B. Thus $A \cap B = \{3, 4\}$.
   c. A^c is the elements not in A (but in U). Thus $A^c = \{5, 6, 7\}$.
   d. $(A \cup B) \cap C$ is those elements in $A \cup B$ and also in C. From (a) we have

$$(A \cup B) \cap C = \{1, 2, 3, 4, 5\} \cap \{2, 3, 4, 5, 6\} = \{2, 3, 4, 5\}.$$

   e. $(A \cap B) \cup C$ is those elements in $A \cap B$ or in C. Thus from (b)

$$(A \cap B) \cup C = \{3, 4\} \cup \{2, 3, 4, 5, 6\} = \{2, 3, 4, 5, 6\}.$$

f. $A^c \cup B \cup C$ is elements in B, or in C, or not in A. Thus

$$A^c \cup B \cup C = \{2, 3, 4, 5, 6, 7\}.$$

2. a. Corporations in this country that had profits and also paid a dividend last year is represented by $P \cap D$.

b. Corporations in this country that either had profits or paid a dividend last year is represented by $P \cup D$.

c. Corporations in this country that did not have profits last year is represented by P^c.

d. Corporations in this country that had profits, paid a dividend, and did not increase their labor force last year is represented by $P \cap D \cap L^c$.

e. Corporations in this country that had profits or paid a dividend, and did not increase their labor force last year is represented by $(P \cup D) \cap L^c$.

6.2   THE NUMBER OF ELEMENTS IN A SET

▶ *Counting*

▶ *Applications*

Augustus De Morgan, 1806–1871

(David H. Smith Collection/Columbia University.)

APPLICATION

Counting

It was De Morgan who got George Boole interested in set theory and formal logic and then made significant advances upon Boole's epochal work. He discovered the De Morgan laws mentioned in the last section. Boole and De Morgan are together considered the founders of the algebra of sets and of mathematical logic. De Morgan was a champion of religious and intellectual toleration and on several occasions resigned his professorships in protest of the abridgements of academic freedom of others.

In a survey of 120 adults, 55 said they had an egg for breakfast that morning, 40 said they had juice for breakfast, and 70 said they had an egg or juice for breakfast. How many had both an egg and juice for breakfast? See Example 1 for the answer.

Counting

This section shows the relationship between the number of elements in $A \cup B$ and the number of elements in A, B, and $A \cap B$. This is our first counting principle. The examples and exercises in this section give some applications. In other applications in Chapter 7 we will count the number of elements in various sets to find probability.

Figure 6.11

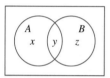

Figure 6.12

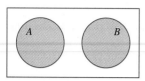

Figure 6.13

> **The Notation $n(A)$**
>
> If A is a set with a finite number of elements, we denote the number of elements in A by $n(A)$.

There are two results that are rather apparent. First, since the empty set ϕ has no elements, $n(\phi) = 0$. For the second refer to Figure 6.11 where the two sets A and B are disjoint.

> **The Number in the Union of Disjoint Sets**
>
> If the sets A and B are disjoint, then
> $$n(A \cup B) = n(A) + n(B).$$

A consequence of the last result is the following. In Figure 6.12, we are given a universal set U and a set $A \subset U$. Then since $A \cap A^c = \phi$ and $U = A \cup A^c$,

$$n(U) = n(A \cup A^c) = n(A) + n(A^c).$$

Now consider the more general case shown in Figure 6.13. We assume that x is the number in the set A that are not also in B, z is the number in the set B that are not also in A, and finally, y is the number in both A and B. Then

$$n(A \cup B) = x + y + z$$
$$= (x + y) + (y + z) - y$$
$$= n(A) + n(B) - n(A \cap B).$$

Alternatively, we can see that the total $n(A) + n(B)$ counts the number in the intersection $n(A \cap B)$ twice. Thus to obtain the number in the union $n(A \cup B)$, we must subtract $n(A \cap B)$ from $n(A) + n(B)$.

> **The Number in the Union of Two Sets**
>
> For any finite sets A and B,
> $$n(A \cup B) = n(A) + n(B) - n(A \cap B). \tag{1}$$

Applications

E X A M P L E   1   **An Application of Counting**

In a survey of 120 adults, 55 said they had an egg for breakfast that morning, 40 said they had juice for breakfast, and 70 said they had an egg or juice for breakfast. How many had both an egg and juice for breakfast? How many had an egg but no juice for breakfast? How many had neither an egg nor juice for breakfast?

Solutions   Let U be the set of adults surveyed, E the subset that had an egg for breakfast, and J the subset that had juice for breakfast. A Venn diagram is shown in Figure 6.14. From the survey, we have that

$$n(E) = 55, \qquad n(J) = 40, \qquad n(E \cup J) = 70.$$

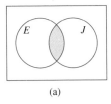 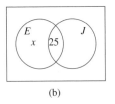 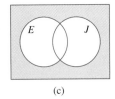

(a)   (b)   (c)

Figure 6.14

(a)   The number that had an egg and juice for breakfast is given by $n(E \cap J)$ and is shown as the shaded region in Figure 6.14a. Since $n(E \cup J) = n(E) + n(J) - n(E \cap J)$, we have that

$$n(E \cap J) = n(E) + n(J) - n(E \cup J)$$
$$= 55 + 40 - 70$$
$$= 25.$$

(b)   We first place the number 25, just found, in the $E \cap J$ area in the Venn diagram in Figure 6.14b. If we let x be the number that had an egg but no juice, then according to Figure 6.14b,

$$x + 25 = n(E) = 55$$
$$x = 30.$$

(c)   We wish to find $n((E \cup J)^c)$. This is shown as the shaded region in Figure 6.14c. The unshaded region is $E \cup J$. We then have that

$$n(E \cup J) + n((E \cup J)^c) = n(U)$$
$$n((E \cup J)^c) = n(U) - n(E \cup J)$$
$$= 120 - 70$$
$$= 50. \quad \blacksquare$$

E X A M P L E   2   **An Application of Counting**

In a survey of 200 people that had just returned from a trip to Europe, the following information was gathered.

- 142 visited England
- 95 visited France

- 65 visited Germany
- 70 visited both England and France
- 50 visited both England and Germany
- 30 visited both France and Germany
- 20 visited all three of these countries

   (a)   How many went to England but not France or Germany?
   (b)   How many went to exactly one of these three countries?
   (c)   How many went to none of these three countries?

Solutions   Let U be the set of 200 people that were surveyed and let

- $E = \{x \in U \,|\, x$ visited England$\}$
- $F = \{x \in U \,|\, x$ visited France$\}$
- $G = \{x \in U \,|\, x$ visited Germany$\}$.

We first note that the last piece of information from the survey indicates that

$$n(E \cap F \cap G) = 20.$$

Place this in the Venn diagram shown in Figure 6.15a. Recall that 70 visited both England and France, that is, $n(E \cap F) = 70$. If a is number that visited England and France but not Germany, then, according to Figure 6.15a, $20 + a = n(E \cap F) = 70$. Thus $a = 50$. In the same way, if b is the number that visited England and Germany but not France, then $20 + b = n(E \cap G) = 50$. Thus $b = 30$. Also if c is the number that visited France and Germany but not England, then $20 + c = n(G \cap F) = 30$. Thus $c = 10$. All of this information is then shown in Figure 6.15b.

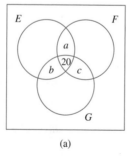
(a)

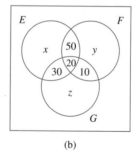
(b)

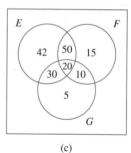
(c)

Figure 6.15

   (a)   Let x denote the number that visited England but not France or Germany. Then, according to Figure 6.15b, $20 + 30 + 50 + x = n(E) = 142$. Thus $x = 42$, that is, the number that visited England but not France or Germany is 42.

   (b)   Since $n(F) = 95$, the number that visited France but not England or Germany is given from Figure 6.15b by $95 - (50 + 20 + 10) = 15$.

Since $n(G) = 65$, the number that visited Germany but not England or France is, according to Figure 6.15b, given by $65 - (30 + 20 + 10) = 5$. Thus, according to Figure 6.15c, the number who visited just one of the three countries is

$$42 + 15 + 5 = 62.$$

(c) According to Figure 6.15c, the number that visited none of these three countries is given by

$$200 - (42 + 15 + 5 + 50 + 30 + 10 + 20) = 200 - 172$$
$$= 28. \quad \blacksquare$$

SELF-HELP EXERCISE SET 6.2

1. Given that $n(A \cup B) = 100$, $n(A \cap B^c) = 50$, and $n(A \cap B) = 20$, find $n(A^c \cap B)$.

2. The registrar reported that among 2000 students, 700 did not register for a math or English course, while 400 registered for both of these two courses. How many registered for exactly one of these courses?

EXERCISE SET 6.2

1. If $n(A) = 100$, $n(B) = 75$, $n(A \cap B) = 40$, what is $n(A \cup B)$?

2. If $n(A) = 200$, $n(B) = 100$, $n(A \cup B) = 250$, what is $n(A \cap B)$?

3. If $n(A) = 100$, $n(A \cap B) = 20$, $n(A \cup B) = 150$, what is $n(B)$?

4. If $n(B) = 100$, $n(A \cup B) = 175$, $n(A \cap B) = 40$, what is $n(A)$?

5. If $n(A) = 100$ and $n(A \cap B) = 40$, what is $n(A \cap B^c)$?

6. If $n(U) = 200$ and $n(A \cup B) = 150$, what is $n(A^c \cap B^c)$?

7. If $n(A \cup B) = 500$, $n(A \cap B^c) = 200$, $n(A^c \cap B) = 150$, what is $n(A \cap B)$?

8. If $n(A \cap B) = 50$, $n(A \cap B^c) = 200$, $n(A^c \cap B) = 150$, what is $n(A \cup B)$?

9. In a survey of 1200 households, 950 said they had aspirin in the house, 350 said they had acetaminophen, and 200 said they had both aspirin and acetaminophen.
   a. How many in the survey had at least one of the two medications?
   b. How many in the survey had aspirin but not acetaminophen?
   c. How many in the survey had neither aspirin nor acetaminophen?

10. In a survey of 1000 households, 600 said they received the morning paper but not the evening paper, 300 said they received both papers, and 100 said they received neither paper.
    a. How many received the evening paper but not the morning paper?
    b. How many received at least one of the papers?

11. The registrar reported that among 1300 students, 700 students did not register for either a math or English course, 400 registered for an English course, and 300 registered for both types of courses.
    a. How many registered for an English course but not a math course?
    b. How many registered for a math course?

12. In a survey of 500 people, a pet food manufacturer found that 200 owned a dog but not a cat, 150 owned a cat but not a dog, and 100 owned neither a dog nor a cat.
    a. How many owned both a dog and a cat?
    b. How many owned a dog?

13. If $n(A \cap B) = 150$ and $n(A \cap B \cap C) = 40$, what is $n(A \cap B \cap C^c)$?

14. If $n(A \cap C) = 100$ and $n(A \cap B \cap C) = 60$, what is $n(A \cap B^c \cap C)$?

15. If $n(A) = 200$, $n(A \cap B \cap C) = 40$, $n(A \cap B \cap C^c) = 20$, $n(A \cap B^c \cap C) = 50$, what is $n(A \cap B^c \cap C^c)$?

16. If $n(B) = 200$, $n(A \cap B \cap C) = 40$, $n(A \cap B \cap C^c) = 20$, $n(A^c \cap B \cap C) = 50$, what is $n(A^c \cap B \cap C^c)$?

Given $n(U) = 100$, $n(A) = 40$, $n(B) = 37$, $n(C) = 35$, $n(A \cap B) = 25$, $n(A \cap C) = 22$, $n(B \cap C) = 24$, and $n(A \cap B \cap C^c) = 10$, find the following.

17. $n(A \cap B \cap C)$ 18. $n(A^c \cap B \cap C)$ 19. $n(A \cap B^c \cap C)$

20. $n(A \cap B^c \cap C^c)$ 21. $n(A^c \cap B \cap C^c)$ 22. $n(A^c \cap B^c \cap C)$

23. $n(A \cup B \cup C)$ 24. $n((A \cup B \cup C)^c)$

25. Using a Venn diagram show that

$$n(A \cup B \cup C) = n(A) + n(B) + n(C) - n(A \cap B)$$
$$- n(A \cap C) - n(B \cap C) + n(A \cap B \cap C).$$

26. Give a proof of the formula in Exercise 25. *Hint:* set $B \cup C = D$ and use formula (1) of the text on $n(A \cup D)$. Now use formula (1) two more times, recalling from the last section that $A \cap (B \cup C) = (A \cap B) \cup (A \cap C)$.

Applications

27. **Sales.** A survey by a fast-food chain of 1000 adults found that in the past month 500 had been to Burger King, 700 to McDonald's, 400 to Wendy's, 300 to Burger King and McDonald's, 250 to McDonald's and Wendy's, 220 to Burger King and Wendy's, and 100 to all three. How many went to
    a. Wendy's but not the other two?
    b. Only one of them?
    c. None of these three?

28. **Investments.** A survey of 600 adults over age 50 found that 200 owned stocks or real estate but no bonds, 220 owned real estate or bonds but no stock, 60 owned real estate but no stocks or bonds, and 130 owned both stocks and bonds. How many owned none of the three?

29. **Entertainment.** A survey of 500 adults found that 190 played golf, 200 skied, 95 played tennis, 100 played golf but did not ski or play tennis, 120 skied but did not play golf or tennis, 30 played golf and skied but did not play tennis, and 40 did all three.
    a. How many played golf and tennis but did not ski?
    b. How many played tennis but did not play golf or ski?
    c. How many participated in at least one of the three above sports?

30. **Transportation.** A survey of 600 adults found that during the last year, 100 traveled by plane but not by train, 150 traveled by train but not by plane, 120 traveled by bus but not by train or plane, 100 traveled by both bus and plane, 40 traveled by all three, and 360 traveled by plane or train. How many did not travel by any of these three modes of transportation?

31. **Education.** In a survey of 250 business executives, 40 said they did not read *Money*, *Fortune*, or *Business Week*, while 120 said they read exactly one of these three and 60 said they read exactly two of them. How many read all three?

32. **Sales.** A furniture store held a sale that attracted 100 people to the store. Of these, 57 did not buy anything, 9 bought both a sofa and loveseat, 8 bought both a sofa and chair, 7 bought both a loveseat and chair. There were 24 sofas, 18 loveseats, and 20 chairs sold. How many people bought all three items?

Solutions to Self-Help Exercise Set 6.2

1. The accompanying Venn diagram indicates that $n(A \cap B^c) = 50$, $n(A \cap B) = 20$, and $x = n(A^c \cap B)$. Then, according to the diagram, $50 + 20 + x = n(A \cup B) = 100$. Thus $x = 30$.

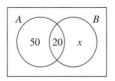

2. The number of students that registered for exactly one of the courses is the number that registered for math but not English, $n(M \cap E^c)$, plus the number that registered for English but not math, $n(M^c \cap E)$. Denote this sum by x. Then, according to the accompanying Venn diagram, $x + 400 + 700 = 2000$. Thus $x = 900$. That is, 900 students registered for exactly one of math or English.

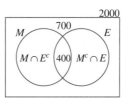

6.3   THE MULTIPLICATION PRINCIPLE AND PERMUTATIONS

▶ *Multiplication Principle*

▶ *Factorials*

▶ *Permutations*

APPLICATION
Number of License Plates

A certain state uses license plates with three letters followed by three numbers with no repeats of letters or numbers. How many such license plates can be made? See Example 10 for the answer.

Multiplication Principle

Some of the basic counting techniques that will be used in the study of probability will now be considered. The "multiplication principle" given here is fundamental to all of the counting methods that will follow.

Before stating the multiplication principle, we consider the following example.

E X A M P L E   1   **A Simple Counting Problem**

A manufacturer makes four flavors of yogurt, and each flavor comes in two sizes. Blueberry comes in small and medium sizes, cherry comes in medium and large, strawberry comes in large and extra large, and vanilla comes in small and large. How many different possibilities are there?

Solution   We can think of the procedure as a sequence of two choices. The first is to pick a flavor and the second is to pick a size. We can denote the choice of picking cherry in the large size as the ordered pair (cherry, large). All possible outcomes can be visualized in the **tree diagram** in Figure 6.16. Counting all the possibilities gives 8. ■

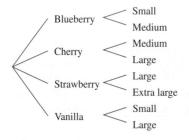

Figure 6.16

If we examine more closely why we obtained the answer 8, we can notice from Figure 6.16 that we had four groups of two each since, no matter what the flavor, there were the same number of choices for the size. This gives $4 \times 2 = 8$, where there were 4 choices for flavors, and for each flavor there were 2 choices for the size.

E X A M P L E   2   **A Simple Counting Problem**

A committee of three individuals, A, B, and C, must pick one of them to be chairman and a different one to be secretary. How many ways can this be done?

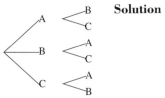

Solution

The tree diagram in Figure 6.17 breaks this problem into an operation of first picking a chairman and then picking a secretary. Notice that there are three choices for chairman and, no matter who is picked for chairman, there are always two choices for secretary. Thus there are three groups of two each or $3 \times 2 = 6$ possible outcomes. ∎

Figure 6.17

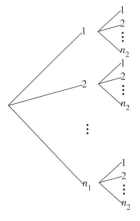

Now suppose there is an operation that consists of making a sequence of two choices with n_1 possible outcomes for the first choice and no matter what the first choice there are always n_2 possible outcomes for the second choice. Then Figure 6.18 indicates a tree diagram. From this figure we see that there are n_1 groups, each with n_2 elements. Thus there is a total of $n_1 \cdot n_2$ possible outcomes. This is the **multiplication principle**.

> **Multiplication Principle**
>
> Suppose there is an operation that consists of making a sequence of two choices with n_1 possible outcomes for the first choice and, no matter what the first choice, there are always n_2 possible outcomes for the second choice. Then there are $n_1 \cdot n_2$ possible ways in which the sequence of two choices can be made.

Figure 6.18

E X A M P L E   3   **Using the Multiplication Principle**

Ten horses are running in the first race at Aqueduct. You wish to buy a Perfecta ticket, which requires picking the first and second finishers in order. How many possible tickets are there?

Solution

You must make a choice: (first, second). Think of this as a sequence of two blanks (__, __) where we must fill in the blanks. There are 10 choices for first. No matter what horse you pick for first, there are always 9 left for second. This can be summarized as ($\underline{10}$, $\underline{9}$). Thus, by the multiplication principle, there are $10 \times 9 = 90$ possible tickets. ∎

We now can see in the same way the general multiplication principle.

> **General Multiplication Principle**
>
> Suppose there is an operation that consists of making a sequence of k choices with n_1 possible outcomes for the first choice and, no matter what the first choice, there are always n_2 possible outcomes for the second choice, and, no matter what the first two choices, there are always n_3 possible outcomes for the third choice, and so on. Then there are $n_1 \cdot n_2 \cdots n_k$ possible ways in which the sequence of k choices can be made.

For example, in Figure 6.19, there are $n_1 \cdot n_2$ elements in the second column, as we noted in Figure 6.18. Emanating from each one of these elements there are n_3 "branches" in the third column. Thus there are $n_1 \cdot n_2 \cdot n_3$ elements in the third column.

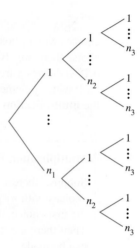

Figure 6.19

E X A M P L E 4   **Using the General Multiplication Principle**

Ten horses are running in the first race at Aqueduct. You wish to buy a Trifecta ticket that requires picking the first three finishers in order. How many possible tickets are there?

Solution   You must make a choice: (first, second, third). Thus we have a sequence of three blanks (__, __, __), and we must fill in the blanks. There are 10 choices for first. No matter what horse you pick for first there are always 9 left for second, and no matter which horses you pick for the first and second choices, there are always 8 left for third. This can be summarized as (10, 9, 8). Thus by the general multiplication principle there are $10 \times 9 \times 8 = 720$ possible tickets. ∎

E X A M P L E 5   **Using the General Multiplication Principle**

In the seventh race at Aqueduct, which is running 10 horses, you have learned through impeccable sources that the overwhelming favorite will be held back from winning. You wish to be assured of having a winning Trifecta ticket. How many must you buy?

Solution   Again you must make a choice: (first, second, third). This time, from your inside knowledge, you know that there are only 9 choices for first, and, given any of these 9 choices, there remain 9 horses to pick for second. No matter what the first and second pick is, there are always 8 remaining choices for third. This can be summarized as (9, 9, 8). Thus, by the multiplication principle, there are $9 \times 9 \times 8 = 648$ possible choices. ∎

E X A M P L E 6   **Using the General Multiplication Principle**

How many three letter words that all begin with consonants and have exactly one vowel can be made using the first seven letters of the alphabet where using a letter twice is permitted but having two consonants next to each other is not?

Solution   Since two consonants cannot be next to each other, one must make the choice (consonant, vowel, consonant). There are 5 choices for the first consonant (b, c, d, f, or g). No matter what the first choice there are 2 choices for vowels (a and e), and no matter what the first two choices, there remain the same 5 consonants to pick from. This can be summarized as (5, 2, 5). Thus the multiplication principle gives $5 \times 2 \times 5 = 50$ possible words. ∎

Factorials

We will be encountering expressions such as $6 \cdot 5 \cdot 4 \cdot 3 \cdot 2 \cdot 1$ and will need symbols to denote them. We have the following definition.

Factorial

For any natural number n

$$n! = n(n - 1)(n - 2) \cdots 3 \cdot 2 \cdot 1$$

$$0! = 1$$

E X A M P L E 7   **Calculating Some Factorials**

$$4! = 4 \cdot 3 \cdot 2 \cdot 1 = 24.$$

$$5! = 5 \cdot 4 \cdot 3 \cdot 2 \cdot 1 = 120.$$

$$6! = 6 \cdot 5 \cdot 4 \cdot 3 \cdot 2 \cdot 1 = 720. \quad ∎$$

Permutations

The set of elements {1, 2, 3, 4} can be arranged in various orders. For example, they can be listed as {1, 3, 4, 2} or perhaps {4, 3, 1, 2}. Each of these ordered arrangements is called a *permutation*. We have the following definition.

Permutation

A permutation of a set of elements is an ordered arrangement of all the elements.

By "arrangement" it is understood that elements can be used only once.

E X A M P L E   8   **Counting the Number of Permutations**

Find all permutations of the set $\{a, b, c\}$ and count the total number.

Solution   Given the set $\{a, b, c\}$, we can use a tree diagram to find all possible permutations. See Figure 6.20. There are 6 possibilities, and these can be seen also by the multiplication principle to be $3 \cdot 2 \cdot 1 = 3!$ ∎

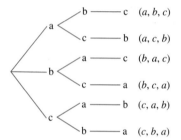

Suppose a set of n distinct objects is given and we wish to find the number of permutations that are possible. We denote this number by $P(n, n)$. There are n possibilities for the first one. No matter which one is chosen first, there always remain $(n - 1)$ for the second choice. No matter how the first two have been chosen, there remains $(n - 2)$ for the third choice, and so on. By the multiplication principle, we then have the total number of possibilities to be

$$P(n, n) = n(n - 1)(n - 2) \cdots 3 \cdot 2 \cdot 1 = n!$$

Figure 6.20   We thus have shown the following.

Number of Permutations of n Objects

The number of permutations of n distinct objects is given by

$$P(n, n) = n(n - 1)(n - 2) \cdots 3 \cdot 2 \cdot 1 = n!$$

E X A M P L E   9   **Counting the Number of Permutations**

In how many ways can a football team of eleven players arrange themselves to trot out onto the football field one at a time?

Solution   The answer is given by

$$P(11, 11) = 11! = 39,916,800.   ∎$$

There are situations in which we have a set of n distinct objects and wish to select an ordered arrangement of r of them. The total number of such arrangements is denoted by $P(n, r)$. We also refer to this as the **number of permutations of r objects taken from a set of size n**. Example 4 was such a case where you selected a first, second, and third place from a set of 10 horses running in a race. Recall from the multiplication principle that we have

$$P(10, 3) = 10 \cdot 9 \cdot 8.$$

Given a set of n distinct objects, the multiplication principle indicates that the number of ways of selecting an ordered arrangement of 3 of them is

$$P(n, 3) = n(n - 1)(n - 2),$$

of 4 of them is

$$P(n, 4) = n(n - 1)(n - 2)(n - 3),$$

and, in general, of r of them is

$$P(n, r) = \underbrace{n(n - 1)(n - 2) \cdots (n - r + 1)}_{r \text{ factors}}.$$

It is sometimes convenient to write $P(n, r)$ in a different form as follows.

$$P(n, r) = n(n - 1)(n - 2) \cdots (n - r + 1)$$

$$= n(n - 1)(n - 2) \cdots (n - r + 1) \frac{(n - r)(n - r - 1) \cdots 2 \cdot 1}{(n - r)(n - r - 1) \cdots 2 \cdot 1}$$

$$= \frac{n(n - 1)(n - 2) \cdots (n - r + 1)(n - r)(n - r - 1) \cdots 2 \cdot 1}{(n - r)(n - r - 1) \cdots 2 \cdot 1}$$

$$= \frac{n!}{(n - r)!}$$

We have established the following.

Number of Permutations of r Objects Taken from a Set of Size n

The number of permutations of r distinct objects taken from a set of size n is given by

$$P(n, r) = n(n - 1)(n - 2) \cdots (n - r + 1) = \frac{n!}{(n - r)!}.$$

Since $0! = 1$,

$$P(n, n) = \frac{n!}{(n - n)!} = \frac{n!}{1} = n!,$$

which is what we obtained earlier.

E X A M P L E   1 0   **Using the Multiplication Principle and Permutations**

A certain state uses license plates with three letters followed by three numbers with no repeats of letters or numbers. How many such license plates can be made?

Solution   Although this problem could be done directly from the multiplication principle with $k = 6$, we will instead use our knowledge of permutations. We can think of deciding on a license plate as a two-step process. First choose the three letters

(Dennis Drenner.)

and then choose the three numbers. The number of ways of arranging the three letters without repetitions and when order is important is the number of permutations of 26 letters taken 3 at a time or $P(26, 3)$. The number of ways of arranging the three numbers without repetitions and when order is important is the number of permutations of 10 digits taken 3 at a time or $P(10, 3)$. By the multiplication principle (when $k = 2$), the total number of different license plates will be

$$P(26, 3)P(10, 3) = (26 \cdot 25 \cdot 24)(10 \cdot 9 \cdot 8) = 11{,}232{,}000. \quad \blacksquare$$

SELF-HELP EXERCISE SET 6.3

1. A restaurant serves 3 soups, 4 salads, 10 main dishes, and 6 desserts. How many different meals can be served if one of each category is chosen?

2. From a group of 6 junior executives, one is to be chosen to go to the Denver office, one to go to the St. Louis office, and one to the Atlanta office. In how many ways can this be done?

3. Six junior executives and 3 senior executives are to line up for a picture. The senior executives must all be lined up together on the left and the junior executives must be all together on the right. In how many ways can this be done?

EXERCISE SET 6.3

In Exercises 1 through 8 evaluate the given expression.

1. $P(5, 3)$ **2.** $P(5, 2)$ **3.** $P(8, 5)$

4. $P(8, 3)$ **5.** $P(7, 7)$ **6.** $P(7, 1)$

7. $P(9, 8)$ **8.** $P(9, 1)$

9. A manufacturer offers 4 styles of sofas with 30 fabrics for each sofa. How many different sofas are there?

10. A manufacturer offers 3 grades of carpet with 10 colors for each grade. How many different carpets are there?

11. A restaurant offers 4 types of salads, 10 main dishes, and 5 desserts. How many different complete meals are there?

12. A card is picked from a standard deck of 52, and then a coin is flipped twice. How many possible outcomes are there?

13. A license plate has 6 digits with repetitions permitted. How many possible license plates of such type are there?

14. A license plate has 5 letters with repetitions not permitted. How many possible license plates of such type are there?

15. An automobile manufacturer offers a certain style car with two types of radios, 10 choices of exterior colors, 5 different interior colors, and 3 types of engines. How many different automobiles are offered?

16. A contractor has 4 styles of homes, each with 3 styles of garages, 4 styles of decks, and 5 styles of carpeting. How many possibilities are there?

17. A state makes license plates with three letters followed by three digits with repetitions permitted. How many of these license plates are there?

18. A state makes license plates with three letters followed by three numbers with no repetitions of letters permitted. How many of these license plates are there?

19. How many 3-letter words can be made from the first 8 letters of the alphabet if consonants cannot be next to each other and letters cannot be repeated?

20. How many 5-letter words can be made from the first 8 letters of the alphabet if consonants cannot be next to each other, vowels cannot be next to each other, and vowels cannot be repeated but consonants can be?

21. At an awards ceremony, 5 men and 4 women are to be called one at a time to receive an award. In how many ways can this be done if men and women must alternate?

22. At an awards ceremony, 5 women and 4 men are each to receive one award and are to be presented their award one at a time. Two of the awards are to be first given to two of the women, and then the remaining awards will alternate between men and women. How many ways can this be done?

23. In how many ways can the 5 members of a basketball team line up in a row for a picture?

24. In how many ways can the individuals in a foursome of golfers tee off in succession?

25. An executive is scheduling meetings with 12 people in succession. The first two meetings must be with two directors on the board, the second 4 with 4 vice-presidents, and the last 6 with 6 junior executives. How many ways can this schedule be made out?

26. An executive is scheduling trips to the company's European plants. First the 4 French plants will be visited, followed by the 3 Italian plants, and then the 5 German ones. How many ways can this schedule be made out?

27. The starting 9 players on the school baseball team and the starting 5 players on the basketball team are to line up for a picture with all members of the baseball team together on the left. How many ways can this be done?

28. The 7 starting offensive linemen and 4 starting offensive backs of the New York Giants are to line up for a picture with the 7 linemen in the middle. How many ways can this be done?

29. On a baseball team, the 3 outfielders can play any of the 3 outfield positions, and the 4 infielders can play any of the 4 infield positions. How many different arrangements of these 7 players can be made?

30. On a football team, the 7 linemen can play any of the 7 linemen positions and the 4 backs can play any of the backfield positions. How many different arrangements of these 11 players can be made?

31. A group of 12 must select a president, a vice-president, a treasurer, and a secretary. How many ways can this be done?

32. In the Superfecta, one must pick the first four finishers of a horse race in correct order. If there are 10 horses running in a race, how many different tickets are there?

33. A buyer for a furniture store selects 8 different style sofas from a group of 10 and has each style shipped on successive weeks. How many ways can this be done?

34. A chef can make 12 main courses. Every day a menu is formed by selecting 7 of the main courses and listing them in order. How many different such menus can be made?

35. Two groups are formed with 10 in the first group and 8 different people in the second. A president, vice-president, and a secretary/treasurer is to be chosen in each group. How many ways can this be done?

36. At a race track you have the opportunity to buy a ticket that requires you to pick the first and second place horses in the first two races. If the first race runs 8 horses and the second runs 10, how many different tickets are possible?

37. A picture is to be taken by lining up 4 of the 11 players from the football team on the left, then 3 of the 9 players from the baseball team in the center, and finally 2 of the 5 players from the basketball team on the right. How many ways can this arrangement be done?

38. A tourist has 8 cities in Great Britain, 6 in France, 5 in Italy, and 7 in Germany on a list she would like to visit. She decides that she will first go to Great Britain and visit 4 of the cities on her list, then on to France to visit 3 cities on the list, then on to Italy for 2 cities, and then on to Germany to visit 4 on the list. How many ways can her itinerary be made out?

Solutions to Self-Help Exercise Set 6.3

1. A restaurant that serves 3 soups, 4 salads, 10 main dishes, and 6 desserts can by the multiplication principle serve $3 \times 4 \times 10 \times 6 = 720$ different meals.

2. The number of ways 6 junior executives can be sent to 3 different offices is the same as the number of permutations of 6 objects taken 3 at a time. This is

$$P(6, 3) = 6 \times 5 \times 4 = 120$$

ways.

3. Six junior executives can line up in 6! ways on the right. For each of these ways the 3 senior executives can line up in 3! ways on the left. Thus by the multiplication principle, the two groups can line up in

$$3! \times 6! = 4320$$

ways.

6.4   COMBINATIONS

▸ *Combinations*
▸ *Counting the Number of Sequences*

APPLICATION
Connecticut Lotto Game

In the state of Connecticut's lotto game, 6 numbered ping pong balls are randomly selected without replacement from a set of 44 to determine a winning set of numbers (without regard to order). If no one picks these 6 numbers, the money wagered stays in the "pot" for the next drawing. After several consecutive games with no winners, the pot gets large and attracts a lot of attention and ticket sales. When the Connecticut lotto game started, the total number of ping pong balls was 36. After a number of years when there were very few long streaks with no winners, lotto officials changed the total number of ping pong balls from 36 to 44. Suppose a number of weeks has gone by without a winner and the pot has grown very large. You wish to organize a syndicate of investors that will purchase every possible ticket to ensure obtaining a winning ticket. How many tickets will the syndicate have to buy if there are 36 ping pong balls? If there are 44? The answer can be found in the discussion before Example 1.

Combinations

When a permutation of n distinct objects is taken r at a time, one selects r of the objects in a specific order. A **combination** of r distinct objects taken from a set of size n is merely a selection of r of the objects (without concern for order). We consider combinations in this section.

Given the set $\{a, b, c\}$ we know from the last section that there are $P(3, 2) = 3 \times 2 = 6$ ways of selecting 2 of these at a time when order is important. The 6 ways are

$$(a, b), (b, a),$$
$$(a, c), (c, a),$$
$$(b, c), (c, b).$$

If now we wish to select 2 at a time when order is not important, then (a, b) is the same as (b, a), and (a, c) is the same as (c, a), and (b, c) is the same as (c, b). Thus there are only 3 ways of selecting 2 objects at a time when order is not important: $(a, b), (a, c), (b, c)$. These are referred to as *combinations*.

Combinations

A **combination** of r distinct objects taken from a set of size n is a selection of r of the objects (without concern for order).

For example, if from a group of 4 people, we wish to select a president, a vice-president, and a secretary/treasurer, then order is important. If, on the other hand, we wished to select a committee of three people from a set of 4, then order is not important since the duties and title of each committee member is the same no matter in what order they are selected.

We know from the last section that the number of permutations of r objects taken from a set of size n is given by $P(n, r)$. We denote the number of combinations of r objects taken from a set of size n by $C(n, r)$. We wish to find a formula for $C(n, r)$.

We can see this by viewing the process of selecting all *permutations* of r objects taken from a set of size n as a sequence of two operations. In the first operation select r objects. In the second operation place these r objects in some order. That is, first select a *combination* of r distinct objects and then order them. We know from the last section that any r distinct objects can be ordered in $r!$ ways. Thus we can obtain all permutations of r objects taken from a set of size n by first selecting all combinations of r objects taken from a set of size n and then permuting each of the r combinations in $(r!)$ ways. Notice that no matter what combination we take, we then always permute by the same number, namely $(r!)$. Thus by the multiplication principle we then must have

$$P(n, r) = C(n, r)r!$$

or

$$C(n, r) = \frac{1}{r!} P(n, r) = \frac{n!}{(r!)(n - r)!}.$$

See Figure 6.21 where the distinct combinations are listed as $C_1, C_2, \ldots, C(n, r)$. We have proven the following.

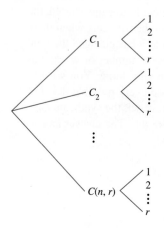

Figure 6.21

Number of Combinations of r Objects taken from a Set of Size n

The number of combinations of r distinct objects taken from a set of size n, denoted by $C(n, r)$, is given by

$$C(n, r) = \frac{n!}{(r!)(n - r)!}.$$

We now solve the problem posed at the beginning of this chapter. Since the order in which the numbers are selected is of no consequence, we are looking for $C(36, 6)$ in the first case and $C(44, 6)$ in the second case. These are

$$C(36, 6) = \frac{36!}{6!(36 - 6)!}$$

$$= \frac{36 \cdot 35 \cdot 34 \cdot 33 \cdot 32 \cdot 31 \cdot (30!)}{6!(30!)}$$

$$= \frac{36 \cdot 35 \cdot 34 \cdot 33 \cdot 32 \cdot 31}{6!}$$

$$= 1{,}947{,}792$$

and

$$C(44, 6) = \frac{44!}{6!(44 - 6)!}$$

$$= \frac{44 \cdot 43 \cdot 42 \cdot 41 \cdot 40 \cdot 39 \cdot (38!)}{6!(38!)}$$

$$= \frac{44 \cdot 43 \cdot 42 \cdot 41 \cdot 40 \cdot 39}{6!}$$

$$= 7,059,052.$$

Thus with 36 ping pong balls, the syndicate must buy about 2 million tickets. If there are 44 balls, then about 7 million tickets must be bought to guarantee a winning ticket.

(Dennis Drenner.)

> **Standard Deck of 52 Playing Cards**
>
> A standard deck of 52 playing cards has four 13-card suits: clubs ♣, diamonds ♢, hearts ♡, and spades ♠. The diamonds and hearts are red, while the clubs and spades are black. Each 13-card suit contains cards numbered from 2 to 10, a jack, a queen, a king, and an ace. The jack, queen, king, and ace can be considered respectively as number 11, 12, 13, and 14. In poker the ace can be either a 14 or a 1.

E X A M P L E 1 **Calculating the Number of Different Poker Hands**

How many different 5-card poker hands can be dealt from a standard deck of 52 cards?

Solution Since the order in which the cards in a poker hand are dealt is of no consequence, we are looking for $C(52, 5)$ which is

$$C(52, 5) = \frac{52!}{5!(52 - 5)!}$$

$$= \frac{52 \cdot 51 \cdot 50 \cdot 49 \cdot 48 \cdot (47!)}{5!(47!)}$$

$$= \frac{52 \cdot 51 \cdot 50 \cdot 49 \cdot 48}{5!}$$

$$= 2,598,960. \quad ■$$

E X A M P L E 2 **A Counting Problem in Poker**

Find the number of poker hands with 3 queens and 2 jacks.

Solution View this as a sequence of two operations. In the first operation, select 3 queens from a deck with 4 queens. In the second operation, select 2 jacks from a deck with 4 jacks. Since order is not important in either case, the first selection can be made in $C(4, 3)$ ways and the second selection can be made in $C(4, 2)$ ways. Since no matter how the first selection is made, the second selection can always be made in $C(4, 2)$ ways, the multiplication principle then indicates that the number of hands with 3 queens and 2 jacks is given by

$$C(4, 3)C(4, 2) = \left(\frac{4!}{3!(4-3)!}\right)\left(\frac{4!}{2!(4-2)!}\right)$$

$$= \left(\frac{4(3!)}{3!}\right)\left(\frac{4 \cdot 3 \cdot 2}{2 \cdot 2}\right)$$

$$= (4)(6) = 24. \quad \blacksquare$$

E X A M P L E 3   **Counting Using Both Combinations and Permutations**

In how many ways can a committee be formed with a chairman, a vice chairman, a secretary/treasurer, and 4 additional people all chosen from a group of 10 people?

Solution   One way of doing this problem is to see the process as a sequence of two selections. The first is to pick the chairman, vice chairman, and secretary/treasurer from the set of 10. Since order is important, this can be done in $P(10, 3)$ ways. The second selection is to then pick the other 4 members of the committee from the remaining 7 people. This can be done in $C(7, 4)$ ways since order is not important. Furthermore, since the second selection can always be made in $C(7, 4)$ ways no matter how the first selection was made, the multiplication principle indicates that the total number of ways of selecting the committee is

$$P(10, 3)C(7, 4) = 10 \cdot 9 \cdot 8 \frac{7 \cdot 6 \cdot 5 \cdot 4}{4 \cdot 3 \cdot 2}. \quad \blacksquare$$

Another way of viewing this is to select 4 of the committee members from the group of 10 and then select the chairman, vice chairman, and secretary/treasurer from the remaining 6. This gives

$$C(10, 4)P(6, 3) = \frac{10 \cdot 9 \cdot 8 \cdot 7}{4 \cdot 3 \cdot 2} 6 \cdot 5 \cdot 4,$$

which is the same answer as before.

Yet another way of looking at this is to select the committee of 7 from the group of 10 and then select the chairman, vice chairman, and secretary/treasurer from the committee of 7. This gives

$$C(10, 7)P(7, 3) = \frac{10 \cdot 9 \cdot 8}{3 \cdot 2} 7 \cdot 6 \cdot 5,$$

which agrees with the other answers.

E X A M P L E   4   **A Counting Problem Using Combinations**

A committee of 15 people consists of 8 males and 7 females. In how many ways can a subcommittee of 5 be formed if the subcommittee consists of

(a) Any 5 committee members?
(b) All men?
(c) At least 3 men?

Solutions

(a) Since the order of selection is of no consequence, the answer is

$$C(15, 5) = \frac{15!}{5!(15 - 5)!}$$

$$= \frac{15 \cdot 14 \cdot 13 \cdot 12 \cdot 11 \cdot (10!)}{5!(10!)}$$

$$= \frac{15 \cdot 14 \cdot 13 \cdot 12 \cdot 11}{5!}$$

$$= 3003.$$

(b) Here we must pick 5 men out of 8 possible men. This can be done in $C(8, 5)$ ways since again order is not important. Thus

$$C(8, 5) = \frac{8!}{5!(8 - 5)!}$$

$$= \frac{8 \cdot 7 \cdot 6 \cdot (5!)}{(5!)(3 \cdot 2)}$$

$$= 56.$$

(c) At least 3 men means all subcommittees with 3 men and 2 women, plus all subcommittees with 4 men and 1 women, plus all with 5 men. This is

$$C(8, 3)C(7, 2) + C(8, 4)C(7, 1) + C(8, 5)C(7, 0)$$

$$= \frac{8 \cdot 7 \cdot 6}{3 \cdot 2}\left(\frac{7 \cdot 6}{2}\right) + \frac{8 \cdot 7 \cdot 6 \cdot 5}{4 \cdot 3 \cdot 2}(7)$$

$$+ \frac{8 \cdot 7 \cdot 6 \cdot 5 \cdot 4}{5 \cdot 4 \cdot 3 \cdot 2}(1)$$

$$= (56)(21) + (70)(7) + (56)(1)$$

$$= 1722. \quad \blacksquare$$

Counting the Number of Sequences

E X A M P L E   5   **A Counting Problem Involving a Sequence**

An investor has selected a growth mutual fund from a large set of growth funds and will consider any of the next 10 years a success (S) if this mutual fund performs above average in the set of funds and a failure (F) otherwise.

(a)   How many different outcomes are possible?

(b)   How many different outcomes have exactly 6 successes?

(c)   How many different outcomes have at least 3 successes?

Solutions

(a)   An outcome consists of 10 operations in sequence. Each operation assigns a S or F. One such example is

$$\{S,\ S,\ F,\ S,\ S,\ F,\ F,\ F,\ S,\ S\}.$$

No matter what the assignments of S's and F's in any of the prior years, there are always two possibilities for the current year: S or F. Using the multiplication principle with $k = 10$ and $n_1 = n_2 = \ldots = n_{10} = 2$ yields

$$2^{10} = 1024$$

as the total number of possible outcomes.

(b)   An outcome with exactly 6 S's was given in (a). Notice that this amounts to filling in 6 years with S's and 4 years with F's. A particular outcome will be determined once we fill in 6 S's in 6 of the years. This can be done in $C(10, 6)$ ways. This is

$$C(10, 6) = \frac{10!}{6!(10 - 6)!} = \frac{10 \cdot 9 \cdot 8 \cdot 7}{4 \cdot 3 \cdot 2} = 210.$$

(c)   The answer to this is the number with exactly 3 S's plus the number with exactly 4 S's plus . . . the number with exactly 10 S's, or

$$C(10, 3) + C(10, 4) + \cdots + C(10, 10).$$

A shorter way can be given by noticing that this is just the total number less the number with at most 2 S's. This is

$$1024 - [C(10, 0) + C(10, 1) + C(10, 2)]$$
$$= 1024 - [1 + 10 + 45] = 968. \quad \blacksquare$$

SELF-HELP EXERCISE SET 6.4

1.   A quinella ticket at a race track allows one to pick the first two finishers without regard to order. How many different tickets are possible in a race with 10 horses?

2.   A company wishes to select 4 junior executives from the San Francisco office, 5 from the Dallas office, and 5 from the Miami office to bring to the New York City headquarters. In how many ways can this be done if there are 10 junior executives in San Francisco, 12 in Dallas, and 15 in Miami?

EXERCISE SET 6.4

In Exercises 1 through 13 calculate the indicated quantity.

1. $C(8, 3)$ **2.** $C(8, 4)$ **3.** $C(8, 5)$

4. $C(12, 12)$ **5.** $C(12, 1)$ **6.** $C(12, 0)$

7. $C(7, 4)$ **8.** $C(7, 3)$ **9.** $C(15, 2)$

10. $C(n, 1)$ **11.** $C(n, 0)$ **12.** $C(n, n - 1)$

13. $C(n, n)$

14. Show that $C(n, r) = C(n, (n - r))$.

15. Find all permutations of $\{a, b, c\}$ taken 2 at a time by first finding all combinations of the set taken 2 at a time and then permuting each combination. Construct a tree similar to Figure 6.21 in the text.

16. Find all permutations of $\{a, b, c, d\}$ taken 3 at a time by first finding all combinations of the set taken 3 at a time and then permuting each combination. Construct a tree similar to Figure 6.21 in the text.

17. If you have a penny, a nickel, a dime, a quarter, and a half-dollar in your pocket or purse, how many different tips can you leave using 3 coins?

18. From a list of 40 captains, 5 are to be promoted to major. In how many ways can this be done?

19. In a certain lotto game, 6 numbered ping pong balls are randomly selected without replacement from a set of balls numbered from 1 to 46 to determine a winning set of numbers (without regard to order). Find the number of possible outcomes.

20. A boxed Trifecta ticket at a horse track allows you to pick the first three finishers without regard to order. How many different tickets are possible in a race with 10 horses?

21. From a list of 20 recommended stocks from your brokerage firm, you wish to select 5 of them for purchase. In how many ways can you do this?

22. A restaurant offers 8 toppings on its pizza. In how many ways can you select 3 of them?

23. If you join a book club, you can purchase 4 books at a sharp discount from a list of 20. In how many ways can you do this?

24. A firm is considering expanding into 4 of 9 possible cities. In how many ways can this be done?

25. In how many ways can an inspector select 5 bolts from a batch of 40 for inspection?

26. A chef has 20 dinners that she can make. In how many ways can she select 6 of them for the menu for today?

27. In her last semester, a student must pick 3 mathematics courses and 2 computer science courses to graduate with a degree in mathematics with a minor in computer science. If there are 11 mathematics courses and 7 computer science courses available to take, how many different ways can this be done?

28. A committee of 12 U.S. Senators is to be formed with 7 Democrats and 5 Republicans. In how many ways can this be done if there are 53 Democratic senators and 47 Republican?

29. A chef has 20 main courses and 6 soups that he can prepare. How many different menus could he require if he always has 7 main courses and 3 soups on each menu?

30. A firm must select 4 out of a possible 10 sites on the East Coast and 3 out of a possible 8 sites on the West Coast for expansion. In how many ways can this be done?

31. A firm has 12 junior executives. Three are to be sent to Pittsburgh, one to Houston, one to Atlanta, and one to Boston. In how many ways can this be done?

32. Six prizes are to be given to six different people in a group of nine. In how many ways can a first prize, a second prize, a third prize, and three fourth prizes be given?

33. In a new group of 11 employees 4 are to be assigned to production, 1 to sales, and 1 to advertising. In how many ways can this be done?

34. A parent of 7 children wants 2 children to make dinner, 1 to dust, and 1 to vacuum. In how many ways can this be done?

35. In how many ways can the 9 member Supreme Court give a five-to-four decision upholding a lower court?

36. In how many ways can a committee of 5 reach a majority decision if there are no abstentions?

37. A coin is flipped 8 times in succession. In how many ways can exactly 5 heads occur?

38. A coin is flipped 8 times in succession. In how many ways can at least 6 heads occur?

39. A coin is flipped 8 times in succession. In how many ways can at least 2 heads occur?

40. A baseball team takes a road trip and plays 12 games. In how many ways could they win 7 and lose 5?

41. A banana split is made with 3 scoops of different flavors of ice cream, 3 different syrups, 2 different types of nuts, and with or without whipped cream. How many different banana splits can be made if there are 12 ice creams, 8 syrups, and 4 types of nuts to choose from?

42. A salesman has 10 customers in New York City, 8 in Dallas, and 6 in Denver. In how many ways can he see 4 customers in New York City, 3 in Dallas, and 4 in Denver?

43. Find the number of different poker hands that contain exactly 3 aces, while the remaining 2 cards do not form a pair.

44. Find the number of full houses in a poker hand, that is, the number of poker hands with 3 of a kind and 2 of a kind.

45. Find the number of poker hands with two pairs, that is, two different two of a kinds with the fifth card a third different kind.

46. In how many ways can a doubles game of tennis be arranged from 8 boys and 4 girls if each side must have 1 boy and 1 girl?

Solutions to Self-Help Exercise Set 6.4

1. The number of tickets is the same as the number of ways of selecting 2 objects from 10 when order is not important. This is

$$C(10, 2) = \frac{10 \cdot 9}{2} = 45.$$

2. The number of ways that junior executives can be selected from San Francisco, Dallas, and Miami, is, respectively, $C(10, 4)$, $C(12, 5)$, and $C(15, 5)$. By the multiplication principle the total number of ways this can be done is

$$C(10, 4)C(12, 5)C(15, 5) = \frac{10!}{4!(10-4)!} \frac{12!}{5!(12-5)!} \frac{15!}{5!(15-5)!}$$

$$= \frac{10 \cdot 9 \cdot 8 \cdot 7}{4 \cdot 3 \cdot 2} \frac{12 \cdot 11 \cdot 10 \cdot 9 \cdot 8}{5 \cdot 4 \cdot 3 \cdot 2} \frac{15 \cdot 14 \cdot 13 \cdot 12 \cdot 11}{5 \cdot 4 \cdot 3 \cdot 2}$$

$$= (10 \cdot 3 \cdot 7)(11 \cdot 9 \cdot 8)(7 \cdot 13 \cdot 3 \cdot 11)$$

$$= 499,458,960.$$

6.5 BINOMIAL THEOREM, PARTITIONS*

▶ *Binomial Theorem*

▶ *Pascal's Triangle*

▶ *Ordered Partitions*

APPLICATION

The Number of Options for an Executive

A study committee reports to an executive that to solve a particular problem any number of 4 different actions can be taken, including doing nothing. How many options does the executive have? See Example 3 for the answer.

APPLICATION

Counting the Ways of Dividing a Committee to Perform Tasks

In how many ways can a group of 14 people be divided into 3 committees, each assigned a different task, the first committee consisting of 8 people, the second 4 people, and the third 2 people? See Example 4 for the answer.

The Binomial Theorem

The expansion $(x + y)^2 = x^2 + 2xy + y^2$ is familiar. In the first part of this section formulas for the expansion of $(x + y)^n$ where n is any integer will be given. In the last part of this section, ordered partitions are introduced. These are generalizations of the notion of combinations that have already been studied.

*This section is optional.

We wish to develop a systematic way of writing the expansion of expressions of the form $(x + y)^n$ where n is a positive integer.

First calculating by direct multiplication we can obtain

$$(x + y)^0 = 1$$

$$(x + y)^1 = x + y$$

$$(x + y)^2 = x^2 + 2xy + y^2$$

$$(x + y)^3 = x^3 + 3x^2y + 3xy^2 + y^3$$

$$(x + y)^4 = x^4 + 4x^3y + 6x^2y^2 + 4xy^3 + y^4$$

$$(x + y)^5 = x^5 + 5x^4y + 10x^3y^2 + 10x^2y^3 + 5xy^4 + y^5$$

$$(x + y)^6 = x^6 + 6x^5y + 15x^4y^2 + 20x^3y^3 + 15x^2y^4 + 6xy^5 + y^6$$

Notice that in the expansions $(x + y)^n$ the powers of x decrease by one and the powers of y increase by one as we move to the next term. Also notice that for the expressions x^ay^b, we have $a + b = n$.

How can we predict the coefficients of such terms? Take $(x + y)^5$ as an example. Write

$$(x + y)^5 = (x + y)(x + y)(x + y)(x + y)(x + y) \qquad (1)$$

as the product of 5 factors. We can obtain an x^2y^3 term in the product by selecting y from exactly 3 of the factors on the right of (1). We can think of this as filling in the blanks of

$$\{\_\_, \_\_, \_\_, \_\_, \_\_\}$$

with exactly 3 y's. (The other blanks then must be x.) For example, $\{y, x, y, y, x\}$ indicates that we have selected y from only the first, third, and fourth factors on the right of (1). The number of ways we can select exactly 3 blanks to put a y in from the 5 possible blanks is $C(5, 3)$. This is

$$C(5, 3) = \frac{5 \cdot 4 \cdot 3}{3 \cdot 2} = 10,$$

which agrees with the coefficient of x^2y^3 in the above expansion of $(x + y)^5$.

In general, when looking at the expansion

$$(x + y)^n = (x + y)(x + y)(x + y) \cdots (x + y),$$

we will obtain an $x^{n-k}y^k$ term by selecting y from exactly k of the n factors. This can be done in $C(n, k)$ ways. Thus the coefficient of the $x^{n-k}y^k$ term must be $C(n, k)$. We have proven the following theorem.

The Binomial Theorem

The coefficient of $x^{n-k}y^k$ in the expansion of $(x + y)^n$ is $C(n, k)$.

E X A M P L E   1   **Using the Binomial Theorem**

Find the coefficient of $x^7 y^3$ in the expansion $(x + y)^{10}$.

Solution   According to the binomial theorem this must be

$$C(10, 3) = \frac{10 \cdot 9 \cdot 8}{3 \cdot 2} = 120. \quad \blacksquare$$

It is common to use the notation

$$\binom{n}{k} = C(n, k).$$

With this notation we then have

$$(x + y)^n = \binom{n}{0} x^n + \binom{n}{1} x^{n-1} y + \binom{n}{2} x^{n-2} y^2$$

$$+ \cdots +$$

$$\binom{n}{n-2} x^2 y^{n-2} + \binom{n}{n-1} xy^{n-1} + \binom{n}{n} y^n.$$

E X A M P L E   2   **Using the Binomial Theorem**

Write out the expansion of $(a - 2b)^4$.

Solution   By setting $x = a$ and $y = -2b$ in the binomial theorem and using the above notation, we have

$$(a - 2b)^4 = (a + (-2b))^4$$

$$= \binom{4}{0} a^4 + \binom{4}{1} a^3(-2b) + \binom{4}{2} a^2(-2b)^2$$

$$+ \binom{4}{3} a(-2b)^3 + \binom{4}{4} (-2b)^4$$

$$= a^4 + 4a^3(-2b) + 6a^2(-2b)^2 + 4a(-2b)^3 + (-2b)^4$$

$$= a^4 - 8a^3 b + 24a^2 b^2 - 32ab^3 + 16b^4. \quad \blacksquare$$

A consequence of the binomial theorem is the following.

The Number of Subsets of a Set

A set with n distinct elements has 2^n distinct subsets.

Before giving a proof, let us list all the subsets of $\{a, b, c\}$ by listing all subsets with 3 elements, all with 2 elements, all with 1 element, and all with no elements. We have

$$\{a, b, c\}$$

$$\{a, b\}, \{a, c\}, \{b, c\}$$

$$\{a\}, \{b\}, \{c\}$$

$$\phi.$$

There are 8, which is $8 = 2^3$.

To establish the theorem, the total number of subsets of a set with n distinct elements is the number of subsets with n elements, plus the number of subsets with $(n - 1)$ elements, plus the number with $(n - 2)$ elements, and so on. This is just

$$\binom{n}{n} + \binom{n}{n - 1} + \binom{n}{n - 2} + \cdots + \binom{n}{1} + \binom{n}{0}.$$

Now setting $x = y = 1$ in the binomial theorem gives

$$2^n = (1 + 1)^n$$

$$= \binom{n}{n} (1)^n + \binom{n}{n - 1} (1)^{n-1}(1) + \binom{n}{n - 2} (1)^{n-2}(1)^2$$

$$+ \cdots +$$

$$\binom{n}{1} (1)(1)^{n - 1} + \binom{n}{0} (1)^n$$

$$= \binom{n}{n} + \binom{n}{n - 1} + \binom{n}{n - 2} + \cdots + \binom{n}{1} + \binom{n}{0}.$$

which, as we have just seen, is the total number of subsets we are seeking.

E X A M P L E   3   **The Number of Options for an Executive**

A study committee reports to an executive that to solve a particular problem any number of 4 different actions can be taken, including doing nothing. How many options does the executive have?

Solution   The executive has the option of selecting any subset from a set with 4 elements in it. This can be done in $2^4 = 16$ ways.   ■

Pascal's Triangle

Figure 6.22 lists the coefficients from the expansions of $(x + y)^n$ given at the beginning of this section. This is called **Pascal's Triangle**, named after its discoverer, Blaise Pascal (1623–1662). Notice that there are always 1's at the two

Figure 6.22

sides and that any coefficient inside the triangle can be obtained by adding the coefficient above and to the left with the coefficient above and to the right. This is another way the coefficients can be obtained.

Ordered Partitions

We will now look at combinations from a slightly different perspective, which will permit us to generalize the notion of combinations to the notion of "ordered partitions."

If we have a set $S = \{a_1, a_2, \ldots, a_n\}$ with n elements, that is, $n(S) = n$, and divide the set into two parts or subsets with the first part having k elements and the second part having the remaining $n - k$ elements, we call this an **ordered partition of S of type $(k, n - k)$.** We can think of this as a two-step process. In the first step we select a subset of k elements. This can be done in $C(n, k)$ ways. In the second step we place the remaining $n - k$ elements in the second part, but this can be done in only one way. Thus the number of such ordered partitions is the same as the number of combinations of n distinct elements taken k at a time or

$$C(n, k) = \frac{n!}{k!(n - k)!}.$$

If we let $n_1 = k$ and $n_2 = n - k$, then this becomes

$$\frac{n!}{n_1! n_2!}$$

where

$$n_1 + n_2 = n.$$

We use the following notation

$$\binom{n}{n_1, n_2} = \frac{n!}{n_1! n_2!}.$$

Figure 6.23 gives a schematic of what we have done. We have divided the set S into two parts, the first part is called S_1 and the second part is called S_2, with $n_1 = n(S_1)$, $n_2 = n(S_2)$, and $n_1 + n_2 = n = n(S)$. As we noted this can be done in

$$\binom{n}{n_1, n_2} = \frac{n!}{n_1! n_2!}$$

ways.

Thus if we have a group of 14 people and wish to divide the group into 2 committees, with the first committee consisting of 8 people to perform some task

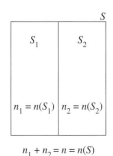

Figure 6.23

and the remaining 6 people into another committee to form another task, this can be done in

$$\binom{14}{8,\,6} = \frac{14!}{8!6!}$$

ways. This is referred to as an *ordered* partition since it is important which committee an individual is on.

E X A M P L E   4   **Counting the Ways of Dividing a Committee to Perform Tasks**

In how many ways can a group of 14 people be divided into 3 committees, each assigned a different task, the first committee consisting of 8 people, the second 4 people, and the third 2 people?

Solution   We can select 8 people for the first committee in $C(14, 8)$ ways and then can select 4 people for the second committee from among the remaining 6 people in $C(6, 4)$ ways. The remaining 2 people can then go into the third committee in only 1 way. By the multiplication principle, the total number of ways this can be done is

$$C(14, 8)C(6, 4) = \frac{14!}{8!6!}\frac{6!}{4!2!}$$

$$= \frac{14!}{8!4!2!}.$$

We leave the answer in this form in view of what we are about to do.   ■

We can think of the 3 committees in the previous problem as a division of the set of 14 people into three groups. We refer to this as an *ordered partition of type* (8, 4, 2). We use the term *ordered* since it is important which committee an individual will be on.

We now wish to generalize this notion. In Figure 6.24 we have a set S of n distinct elements and have partitioned the set into the 3 subsets shown. We call this an *ordered partition of type* (n_1, n_2, n_3). Another way of looking at this is to think of the three rectangles in Figure 6.24 initially as empty boxes. Now select n_1 elements from S and place them into S_1. Now select n_2 elements from the remaining $n - n_1$ elements in S and place them into S_2. Now place the remaining n_3 elements of S into S_3. In how many ways can this be done?

We can view this as a three-step process. First select n_1 elements from the set of n. This can be done in $C(n, n_1)$ ways. Then from among the remaining $n - n_1$ elements, select n_2 for S_2. This can be done in $C(n - n_1, n_2)$ ways. Now

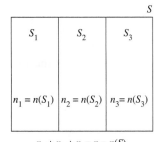

$$n_1 + n_2 + n_3 = n = n(S)$$

Figure 6.24

for the third step place the remaining n_3 elements into S_3. This can be done in only one way. The total number of ways this can be done is

$$C(n, n_1)C(n - n_1, n_2) = \frac{n!}{n_1!(n - n_1)!} \frac{(n - n_1)!}{n_2!(n - n_1 - n_2)!}$$

$$= \frac{n!}{n_1!n_2!(n - n_1 - n_2)!}$$

$$= \frac{n!}{n_1!n_2!n_3!}.$$

We use the notation

$$\binom{n}{n_1, n_2, n_3} = \frac{n!}{n_1!n_2!n_3!}.$$

We have the more general definition.

Ordered Partition of Type $(n_1, n_2, \ldots, n_k)$

Given a set S of n elements, an ordered partition of S of type $(n_1, n_2, \ldots, n_k)$ is a division of S into k subsets, $S_1, S_2, \ldots, S_k$, with order being important and where $n_1 = n(S_1)$, $n_2 = n(S_2)$, $\ldots$, $n_k = n(S_k)$. The number of such ordered partitions is denoted by

$$\binom{n}{n_1, n_2, \ldots, n_k}.$$

See Figure 6.25. Notice that since none of the sets $S_1, S_2, \ldots, S_k$, can overlap, we must have

$$n_1 + n_2 + \cdots + n_k = n.$$

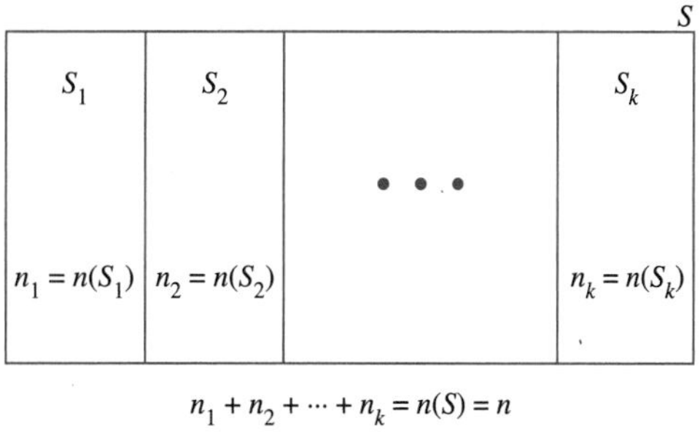

$$n_1 + n_2 + \cdots + n_k = n(S) = n$$

Figure 6.25

The following is then true.

Number of Ordered Partitions of Type $(n_1, n_2, \ldots, n_k)$

$$\binom{n}{n_1, n_2, \ldots, n_k} = \frac{n!}{n_1!n_2! \cdots n_k!}$$

We have already established this for $k = 3$. The general proof is similar.

E X A M P L E   5   **The Number of Ordered Partitions of Stocks**

A brokerage firm follows 20 stocks for its clients. It divides the stocks into 4 groups: buy, hold, sell, short. Find the number of ways the firm can place 5 stocks into the buy group, 10 into the hold group, 3 into the sell group, and 2 into the short group.

Solution   From a set with 20 elements, we are to find the number of ordered partitions of type (5, 10, 3, 2). This is

$$\binom{20}{5, 10, 3, 2} = \frac{20!}{5!10!3!2!} = 465,585,120. \quad \blacksquare$$

SELF-HELP EXERCISES 6.5

1. Find the coefficient of x^4y^5 in the expansion of $(x + y)^9$.

2. A company hires 10 junior executives. Five are to go to the Boston office, 3 to the Chicago office, and 2 to the San Diego office. In how many ways can this be done?

EXERCISE SET 6.5

In Exercises 1 through 10 expand using the binomial theorem.

1. $(a - b)^5$ 2. $(2a + b)^4$ 3. $(2x + 3y)^5$

4. $(3x - 2y)^4$ 5. $(1 - x)^6$ 6. $(2 + x)^6$

7. $(2 - x^2)^4$ 8. $(1 + 2x^2)^6$ 9. $(s^2 + t^2)^6$

10. $(s^2 - 1)^5$

In Exercises 11 through 18 determine the first 3 and last 3 terms in the expansion of each of the terms.

11. $(a - b)^{10}$ 12. $(a + b)^{12}$ 13. $(x + y)^{11}$

14. $(x - y)^8$ **15.** $(1 - z)^{12}$ **16.** $(1 + x)^{10}$

17. $(1 - x^3)^{12}$ **18.** $(x^2 - 1)^{10}$

In Exercises 19 through 26 evaluate the given term.

19. $\begin{pmatrix} 7 \\ 2, 2, 3 \end{pmatrix}$ **20.** $\begin{pmatrix} 8 \\ 2, 3, 3 \end{pmatrix}$ **21.** $\begin{pmatrix} 9 \\ 2, 3, 4 \end{pmatrix}$

22. $\begin{pmatrix} 8 \\ 2, 2, 2, 2 \end{pmatrix}$ **23.** $\begin{pmatrix} 10 \\ 2, 2, 2, 2, 2 \end{pmatrix}$ **24.** $\begin{pmatrix} 12 \\ 2, 3, 4, 3 \end{pmatrix}$

25. $\begin{pmatrix} 8 \\ 2, 2, 2, 1, 1 \end{pmatrix}$ **26.** $\begin{pmatrix} 10 \\ 2, 3, 2, 3 \end{pmatrix}$

27. Determine the next row in Pascal's triangle shown in Figure 6.22.

28. Find the coefficient of $x^4 y^4$ in the expansion of $(x + y)^8$ using Figure 6.22.

29. A manufacturing company buys a certain component from 3 different vendors. In how many ways can the company order 8 components with 4 from the first vendor and 2 each from the other vendors?

30. A mutual fund has 20 stocks in its portfolio. On a given day 3 stocks move up, 15 stay the same, and 2 move down. In how many ways could this happen?

31. An advertising firm has 12 potential clients and 3 different salesmen. In how many ways can it divide the potential clients equally among the 3 salesmen?

32. In how many ways can a class of 10 students be assigned 1 A, 2 B's, 4 C's, 2 D's, and 1 F?

33. Two scholarships of $10,000 each, 3 of $5000 each, and 5 of $2000 each are to be awarded to 10 finalists. In how many ways can this be done?

34. The 12 directors of a company are to be divided equally into 3 separate committees to study sales, recent products, and labor relations. In how many ways can this be done?

35. Find the number of arrangements of each of the following words that can be distinguished. (a) $a_1 a_2 b$, (b) aab, (c) $a_1 a_2 b_1 b_2$, (d) $aab_1 b_2$, (e) $aabb$

36. Find the number of arrangements of each of the following words that can be distinguished. (a) $a_1 a_2 b_1 b_2 b_3$, (b) $aab_1 b_2 b_3$, (c) $aabbb$

37. Find the number of arrangements of each of the following words that can be distinguished. (a) Mississippi, (b) Tennessee

38. Suppose a word has n symbols made from k distinct elements with n_1 of the first element, n_2 of the second element, ..., n_k of the kth element. If $n_1 + n_2 + \cdots + n_k = n$, show that the number of distinguishable arrangements of the n symbol word is

$$\begin{pmatrix} n \\ n_1, n_2, \ldots, n_k \end{pmatrix}.$$

Verify that this works for the previous exercise.

Solutions to Self-Help Exercise Set 6.5

1. By the binomial theorem the coefficient of $x^4 y^5$ in the expansion of $(x + y)^9$ is

$$C(9, 5) = \frac{9!}{5!(9 - 5)!} = \frac{9 \cdot 8 \cdot 7 \cdot 6 \cdot 5}{5 \cdot 4 \cdot 3 \cdot 2} = 126.$$

2. The 10 junior executives are partitioned into 3 groups of size 5, 3, and 2, where which group one is in is important. The number of ways this can be done is

$$\binom{10}{5, 3, 2} = \frac{10!}{5!3!2!} = \frac{10 \cdot 9 \cdot 8 \cdot 7 \cdot 6 \cdot 5!}{5!(3 \cdot 2)(2)} = 2520.$$

SUMMARY OUTLINE OF CHAPTER 6

- If every element of a set A is also an element of another set B, we say that **A is a subset of B** and write $A \subset B$. If A is not a subset of B, we write $A \not\subset B$.

- The **empty set**, written as ϕ, is the set with no elements.

- Given a universal set U and a set $A \subset U$, the **complement of A**, written A^c, is the set of all elements that are in U but not in A, that is,

$$A^c = \{x | x \in U, x \notin A\}.$$

- The **union** of two sets A and B, written $A \cup B$, is the set of all elements that belong to A, or to B, or to both. Thus

$$A \cup B = \{x | x \in A \text{ or } x \in B \text{ or both}\}.$$

- The **intersection** of two sets A and B, written $A \cap B$, is the set of all elements that belong to both the set A and to the set B.

- **Rules for Set Operations**

$A \cup B = B \cup A$	Commutative law for union
$A \cap B = B \cap A$	Commutative law for intersection
$A \cup (B \cup C) = (A \cup B) \cup C$	Associative law for union
$A \cap (B \cap C) = (A \cap B) \cap C$	Associative law for intersection
$A \cup (B \cap C) = (A \cup B) \cap (A \cup C)$	Distributive law for union
$A \cap (B \cup C) = (A \cap B) \cup (A \cap C)$	Distributive law for intersection
$(A \cup B)^c = A^c \cap B^c$	De Morgan law
$(A \cap B)^c = A^c \cup B^c$	De Morgan law

- If A is a set with a finite number of elements, we denote the number of elements in A by $n(A)$.

- If the sets A and B are disjoint, then

$$n(A \cup B) = n(A) + n(B)$$

- For any finite sets A and B,

$$n(A \cup B) = n(A) + n(B) - n(A \cap B)$$

- **General Multiplication Principle.** Suppose there is an operation that consists of making a sequence of k choices with n_1 possible outcomes for the first choice, and, no matter what the first choice, there are always n_2 possible outcomes for the second choice, and, no matter what the first two choices, there are always n_3 possible outcomes for the third choice, and so on. Then there are $n_1 \cdot n_2 \cdots n_k$ possible ways in which the sequence of k choices can be made.

- For any natural number n

$$n! = n(n - 1)(n - 2) \cdots 3 \cdot 2 \cdot 1$$

$$0! = 1.$$

- A **permutation** of a set of elements is an ordered arrangement of all the elements.

- A **permutation of r objects taken from a set of size n** is a selection of r of the objects with order being important.

- The number of permutations of n distinct objects is given by

$$P(n, n) = n(n - 1)(n - 2) \cdots 3 \cdot 2 \cdot 1 = n!$$

- The number of permutations of r distinct objects taken from a set of size n is given by

$$P(n, r) = n(n - 1)(n - 2) \cdots (n - r + 1) = \frac{n!}{(n - r)!}.$$

- A **combination** of r distinct objects taken from a set of size n is a selection of r of the objects (without concern for order).

- The number of combinations of r distinct objects taken from a set of size n, denoted by $C(n, r)$, is given by

$$C(n, r) = \frac{n!}{(r!)(n - r)!}.$$

- **The Binomial Theorem.** The coefficient of $x^{n-k}y^k$ in the expansion of $(x + y)^n$ is $C(n, k)$.

- Given a set S of n elements, an ordered partition of S of type $(n_1, n_2, \ldots, n_k)$ is a division of S into k subsets, $S_1, S_2, \ldots, S_k$, with order being important and where $n_1 = n(S_1)$, $n_2 = n(S_2)$, $\ldots$, $n_k = n(S_k)$. The number of such ordered partitions is denoted by

$$\binom{n}{n_1, n_2, \ldots, n_k}.$$

- **Number of Ordered Partition of Type** $(n_1, n_2, \ldots, n_k)$.

$$\binom{n}{n_1, n_2, \ldots, n_k} = \frac{n!}{n_1! n_2! \cdots n_k!}$$

Chapter 6 Review Exercises

1. Determine which of the following are sets.
   a. Current members of the board of Chase Manhattan Bank.
   b. Past and present board members of Chase Manhattan Bank that have done an outstanding job.
   c. Current members of the board of Chase Manhattan Bank who are over 10 feet tall.

2. Write in set-builder notation: $\{5, 10, 15, 20, 25, 30, 35, 40\}$.

3. Write in roster notation: $\{x \mid x^3 - 2x = 0\}$.

4. List all the subsets of $\{A, B, C\}$.

5. On a Venn diagram indicate where the following sets are.

$$A \cap B \cap C \qquad A^c \cap B \cap C \qquad (A \cup B)^c \cap C$$

6. Let $U = \{1, 2, 3, 4, 5, 6\}$, $A = \{1, 2, 3\}$, $B = \{2, 3, 4\}$, $C = \{4, 5\}$. Find the following sets.

$$A \cup B \qquad A \cap B \qquad B^c$$

$$A \cap B \cap C \qquad (A \cup B) \cap C \qquad A \cap B^c \cap C$$

7. Let U be the set of all your current instructors and let

$$H = \{x \in U \mid x \text{ is at least 6 feet tall}\}$$

$$M = \{x \in U \mid x \text{ is a male}\}$$

$$W = \{x \in U \mid x \text{ weighs more than 180 pounds}\}.$$

Describe each of the following sets in words.

(a) H^c (b) $H \cup M$ (c) $M^c \cap W^c$

(d) $H \cap M \cap W$ (e) $H^c \cap M \cap W$ (f) $(H \cap M^c) \cup W$

8. Using the set H, M, and W in the previous exercise and set operations, write the set that represents the following statements.
   a. My current female instructors
   b. My current female instructors who weigh at most 180 pounds
   c. My current male instructors who are at least 6 feet tall or else weigh more than 180 pounds

9. For the sets given in Exercise 6, verify that

$$A \cup (B \cap C) = (A \cup B) \cap (A \cup C).$$

10. Use a Venn diagram to show that $(A \cap B)^c = A^c \cup B^c$.

11. If $n(A) = 100$, $n(B) = 40$, $n(A \cap B) = 20$, find $n(A \cup B)$.

12. If $n(A) = 40$ and $n(A \cap B^c) = 30$, find $n(A \cap B)$.

13. In a consumer advertising survey of 100 men, it was found that 20 watched the first game of the last World Series, 15 watched the first game of the last World Series and also watched the Super Bowl, while 30 did not watch either. How many watched the last Super Bowl but not the first game of the last World Series?

14. A consumer survey of 100 women with young daughters found that

    57 had a Barbie doll
    68 had a teddy bear
    11 had a toy piano

45 had a Barbie doll and a teddy bear
8 had a teddy bear and a toy piano
7 had a Barbie doll and a toy piano
5 had all three.

    a. How many had a Barbie doll and a teddy bear but not a toy piano?
    b. How many had exactly 2 of these toys?
    c. How many had none of these toys?

15. Find $P(10, 4)$ and $C(10, 4)$.

16. If instead of a social security *number*, we had a social security *word* where letters could be repeated, what would be the length of the words needed to have at least 250 million different words?

17. Before 5 labor leaders and 6 management personnel begin negotiating a new labor contract, they decided to first take a picture with the 5 labor leaders together on the left. In how many different ways can such a picture be taken?

18. In the previous exercise suppose a picture is to be taken of 3 of the labor leaders and 3 of the management personnel with the labor leaders grouped together on the right. In how many ways can this be done?

19. Refer to Exercise 17. Suppose a committee of 3 labor leaders and 4 management personnel is formed to study the issue of pensions. In how many ways can this be done?

20. Refer to Exercise 17. In a straw vote on a proposal from the labor leaders, the 6 management personnel each cast a vote with no abstentions. In how many ways can these 6 individuals come to a majority decision?

21. An investor decides that her investment year is a success if her portfolio of stocks beats the S&P 500. In how many ways can her next 10 years have exactly 7 successes?

22. Expand $(2 - x)^5$ using the binomial theorem.

23. Find the last 3 terms in the expansion of $(1 - x)^9$.

24. Find $\begin{pmatrix} 10 \\ 4, 4, 2 \end{pmatrix}$

25. Find $\begin{pmatrix} 9 \\ 3, 2, 2, 2 \end{pmatrix}$

26. In how many ways can a laboratory divide 12 scientists into 4 groups of equal size in order to perform 4 different experiments?

Probability problems in business are very similar to probability problems in gambling. (Michael Engler/The Image Bank.)

Probability

Probability is now used widely throughout business and the social and physical sciences. Probability deals with events that are random but yet display a certain statistical regularity. Thus, although it is not possible to predict what will happen next, it is possible to predict what will happen in the long run.

For example, we know that the outcomes of tossing a die or flipping a coin cannot be predicted, and thus the outcomes do not occur with a deterministic regularity. These are examples of *random phenomena*. After many repetitions or trials of either of these experiments, we know that we would observe a regularity of sorts. Namely, any one number on the die would show about as often as any other number, and heads would come up about as often as tails. We now wish to study this type of regularity in random phenomena.

There is a funny thing about probability. Everyone has an intuitive idea of what it is, but when it comes to actually determining probabilities, mistakes are very common. For a tricky subject such as probability we use a formal approach, carefully setting out the "spaces" on which we operate in the first section. This will help a great deal to keep the mistakes to a minimum. The second section continues with an intuitive definition of probability widely used in science, business, and everyday life. The third section presents the mathematical definition of probability, which is then shown to be consistent with the intuitive definition. Many useful properties of probability are then established using the mathematical definition of probability. The chapter continues with a number of important notions and applications in probability.

7.1   THE BASIS OF PROBABILITY

▶ *Preliminaries*
▶ *Events*

APPLICATION

Popular Television Game Show Problem

A contestant is shown 3 closed doors and is correctly informed that behind one of the doors is a car and behind each of the other 2 doors is a goat. The contestant is then asked to select a door. After the contestant picks a door, the game show host, who knows where the goats are, then chooses one of the 2 doors not selected by the contestant that has a goat behind it and opens this door for the contestant to see that there is a goat behind this door. Now the host gives the contestant the opportunity to change the selection of the door. If the car is behind the door selected, the car is won. Should the contestant switch doors or not? Or does it not matter?

The correct solution to this problem was given in the "Ask Marilyn" column of *Parade* magazine. The solution caused enormous controversy with over 10,000 letters written to *Parade*, with the vast majority disagreeing with the solution given. Many who disagreed with the correct solution were well educated and well trained individuals. A discussion of the controversy appeared in *The New York Times*, July 21, 1991. At the end of the chapter is given a completely elementary solution.

Preliminaries

We first take the time to define the basic notions that underlie probability theory. The reward will be a better understanding of the subject and the ability to obtain correct answers to probability problems. In the next section, the intuitive notion of probability is discussed.

We begin the preliminaries by stating the following definitions.

Experiments and Outcomes

An **experiment** is an activity that has observable results.
The results of the experiment are called **outcomes**.

The following are some examples of experiments. Flip a coin and observe whether it falls "heads" or "tails." Throw a die (a small cube marked on each face with from one to six dots) and observe the number of dots on the top face. Select a transistor from a bin and observe whether or not it is defective.

The following are some additional terms that are needed.

Sample Spaces and Trials

A **sample space** of an experiment is the set of all possible outcomes of the experiment. Each repetition of an experiment is called a **trial**.

For the experiment of throwing a die and observing the number of dots on the top face the sample space is the set

$$S = \{1, 2, 3, 4, 5, 6\}.$$

In the experiment of flipping a coin and observing whether it falls heads or tails, the sample space is $S = \{\text{heads, tails}\}$.

E X A M P L E   1   **Determining a Sample Space**

An experiment consists of noting whether the price of the stock of the Ford Corporation rose, fell, or remained unchanged on the most recent day of trading. What is the sample space for this experiment?

Solution   There are three possible outcomes depending on whether the price rose, fell, or remained unchanged. Thus the sample space S is

$$S = \{\text{rose, fell, unchanged}\}. \quad \blacksquare$$

E X A M P L E   2   **Determining a Sample Space**

Two dice, identical except that one is white and the other is red, are tossed and the number of dots on the top face of each is observed. What is the sample space for this experiment?

Solution   The outcomes can be considered order pairs. For example (2, 3) will mean 2 dots on the top face of the white die and 3 dots on the top face of the red die. The sample space S is

(Dennis Drenner.)

$$S = \{(1, 1), (1, 2), (1, 3), (1, 4), (1, 5), (1, 6),$$
$$(2, 1), (2, 2), (2, 3), (2, 4), (2, 5), (2, 6),$$
$$(3, 1), (3, 2), (3, 3), (3, 4), (3, 5), (3, 6),$$
$$(4, 1), (4, 2), (4, 3), (4, 4), (4, 5), (4, 6),$$
$$(5, 1), (5, 2), (5, 3), (5, 4), (5, 5), (5, 6),$$
$$(6, 1), (6, 2), (6, 3), (6, 4), (6, 5), (6, 6)\}. \blacksquare$$

If the experiment of tossing 2 dice consists of just observing the total number of dots on the top faces of the two dice, then the sample space would be

$$S = \{2, 3, 4, 5, 6, 7, 8, 9, 10, 11, 12\}.$$

In short, the sample space depends on the precise statement of the experiment.

E X A M P L E   3   **Determining the Sample Space**

A coin is flipped twice to observe whether heads or tails shows; order is important. What is the sample space for this experiment?

Solution   The sample space S consists of the 4 outcomes

$$S = \{(H, H), (H, T), (T, H), (T, T)\}. \blacksquare$$

Events

> **Events and Elementary Events**
>
> Given a sample space S for an experiment, an **event** is any subset E of S. An **elementary event** is an event with a single outcome.

E X A M P L E   4   **Determining an Event and an Elementary Event**

In Example 3 find the event: "At least one head comes up." Find the elementary event: "Two tails come up."

Solutions   (a)   "At least one head comes up." $= \{(H, H), (H, T), (T, H)\}$
(b)   "Two tails come up." $= \{(T, T)\} \blacksquare$

We can use union, intersection, and complement to describe events.

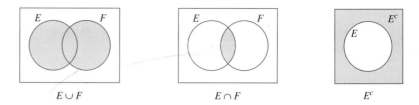

Figure 7.1

Union of Two Events

If E and F are two events, then $E \cup F$ is the union of the two events and consists of the set of outcomes that are in E or F or both.

Thus the event $E \cup F$ is the event that "E or F occurs." Refer to Figure 7.1.

Intersection of Two Events

If E and F are two events, then $E \cap F$ is the intersection of the two events and consists of the set of outcomes that are in both E and F.

Thus the event $E \cap F$ is the event that "E and F both occur."

Complement of an Event

If E is an event, then E^c is the complement of E and consists of the set of outcomes that are not in E.

Thus the event E^c is the event that "E does not occur."

E X A M P L E   5   **Determining Union, Intersection, and Complement**

Consider the sample space given in Example 2. Let E consist of those outcomes for which the number of dots on the top faces of both dice is an even number. Let F be the event that the sum of the number of dots on the top faces of the two dice is 6. Let G be the event that the sum of the number of dots on the top faces of the two dice is less than 11. (a) List the elements of E and F. (b) Find $E \cup F$. (c) Find $E \cap F$. (d) Find G^c.

Solutions

(a) $E = \{(2, 2), (2, 4), (2, 6), (4, 2), (4, 4), (4, 6), (6, 2), (6, 4), (6, 6)\}$
     $F = \{(1, 5), (2, 4), (3, 3), (4, 2), (5, 1)\}$

(b) $E \cup F = \{(2, 2), (2, 4), (2, 6), (4, 2), (4, 4), (4, 6), (6, 2), (6, 4), (6, 6),$
                                                                    $(1, 5), (3, 3), (5, 1)\}$

(c) $E \cap F = \{(2, 4), (4, 2)\}$.

(d) $G^c = \{(5, 6), (6, 5), (6, 6)\}$ ■

If S is a sample space, $\phi \subset S$, and thus ϕ is an event. We call the event ϕ the **impossible event** since the event ϕ means that no outcome has occurred, whereas, in any experiment some outcome *must* occur.

The Impossible Event

The empty set, ϕ, is called the **impossible event**.

For example, if H is the event that a head shows on flipping a coin and T is the event that a tail shows, then $H \cap T = \phi$. The event $H \cap T$ means that both heads and tails show, which is impossible.

Since $S \subset S$, S is itself an event. We call S the **certainty event** since any outcome of the experiment must be in S.

The Certainty Event

Let S be a sample space. The event S is called the **certainty event**.

We also have the following definition.

Mutually Exclusive Events

Two events E and F are said to be **mutually exclusive** or **disjoint** if

$$E \cap F = \phi.$$

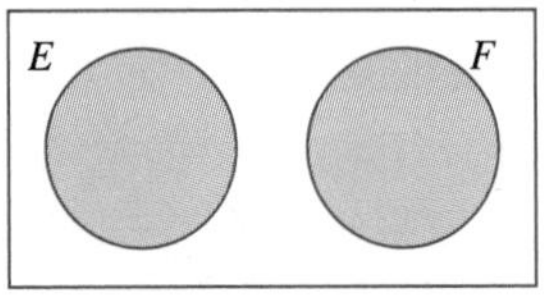

E F

Figure 7.2

See Figure 7.2.

E X A M P L E   6   **Determining if Sets are Mutually Exclusive**

Let a card be chosen from a standard deck of 52 cards. Let E be the event consisting of drawing a 3. Let F be the event of drawing a heart. Let G be the event of drawing a jack. Are E and F mutually exclusive? Are E and G?

Solution   Since $E \cap F = \{3\heartsuit\} \neq \phi$

and $E \cap G = \phi$,

E and F are not mutually exclusive while E and G are.   ■

SELF-HELP EXERCISE SET 7.1

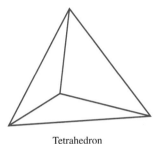

Tetrahedron

1. Two tetrahedrons (4 sided), each with equal sides numbered from 1 to 4, are identical except that one is red and the other white. If the two tetrahedrons are tossed and the number on the bottom face of each is observed, what is the sample space for this experiment?

2. Consider the sample space given in the previous exercise. Let E consist of those outcomes for which both (tetrahedron) dice show an odd number. Let F be the event that the sum of the two numbers on these dice is 5. Let G be the event that the sum of the two numbers is less than 7. (a) List the elements of E and F. (b) Find $E \cap F$. (c) Find $E \cup F$. (d) Find G^c.

EXERCISE SET 7.1

1. Let $S = \{a, b, c\}$ be a sample space. Find all the events.

2. Let the sample space be $S = \{a, b, c, d\}$. How many events are there?

3. A coin is flipped 3 times, and heads or tails is observed after each flip. What is the sample space? Indicate the outcomes in the event "at least 2 heads are observed."

4. A coin is flipped, and it is noted whether heads or tails show. A die is tossed, and the number on the top face is noted. What is the sample space of this experiment?

5. A coin is flipped three times. If heads show, one is written down. If tails show, zero is written down. What is the sample space for this experiment? Indicate the outcomes if "one is observed at least twice."

6. Two tetrahedrons (4 sided), each with equal sides numbered from 1 to 4, are identical except that one is red and the other white. If the two tetrahedrons are tossed and the number on the bottom face of each is observed, indicate the outcomes in the event "the sum of the numbers is 4."

7. An urn holds 10 identical balls except that 1 is white, 4 are black, and 5 are red. An experiment consists of selecting a ball from the urn and observing its color. What is the sample space for this experiment? Indicate the outcomes in the event "the ball is not white."

8. For the urn in Exercise 7, an experiment consists of selecting 2 balls in succession without replacement and observing the color of each of the balls. What is the sample space of this experiment? Indicate the outcomes of the event "no ball is white."

9. An executive must select for promotion 3 people from a group of 5, $\{A, B, C, D, F\}$. What is the sample space? Indicate the outcomes of the event "A is selected."

10. An experiment consists of selecting a letter from the word CONNECTICUT

and observing it. What is the sample space? Indicate the outcomes of the event "a vowel is selected."

11. An inspector selects 10 transistors from the production line and notes how many are defective.
  a. Determine the sample space.
  b. Find the outcomes in the set corresponding to the event E "at least 6 are defective."
  c. Find the outcomes in the set corresponding to the event F "at most 4 are defective."
  d. Find and describe the sets $E \cup F, E \cap F, E^c, E \cap F^c, E^c \cap F^c$.
  e. Find all pairs of sets among the nonempty ones listed in (d) that are mutually exclusive.

12. A survey indicates first whether a person is in the lower income group (L), middle income group (M), or upper income group (U), and second which of these groups the father of the person is in.
  a. Determine the sample space using the letters L, M, and U.
  b. Find the outcomes in the set corresponding to the event E "the person is in the lower income group."
  c. Find the outcomes in the set corresponding to the event F "the person's father is in the higher income group."
  d. Find and describe the sets $E \cup F, E \cap F, E^c, E \cap F^c, E^c \cap F^c$.
  e. Find all pairs of sets listed in (d) that are mutually exclusive.

13. A corporate president decides that for each of the next 3 fiscal years success (S) will be declared if the earnings per share of the company go up at least 10% that year and failure (F) will occur if less than 10%.
  a. Determine the sample space using the letters S and F.
  b. Find the outcomes in the set corresponding to the event E "at least 2 of the next 3 years is a success."
  c. Find the outcomes in the set corresponding to the event G "the first year is a success."
  d. Find and describe the sets $E \cup G, E \cap G, G^c, E^c \cap G, (E \cup G)^c$.
  e. Find all pairs of sets listed in (d) that are mutually exclusive.

14. Let E be the event that the life of a certain lightbulb is at least 100 hours and F that the life is at most 200 hours. Describe the sets $E \cap F, F^c, E^c \cap F, (E \cup F)^c$.

In Exercises 15 through 20 S is a sample space and $E, F,$ and G are 3 events. Use the symbols $\cup, \cap, {}^c$ to describe the given events.

15. E but not F **16.** F but not E

17. Not F and not E **18.** Not F or not E

19. E and F but not G **20.** Not F, nor E, nor G

21. Let S be the 26 letters of the alphabet, E be the 5 vowels $\{a, e, i, o, u\}$, F the remaining 21 letters, and G the first five letters of the alphabet. Find the events $E \cup F \cup G, E^c \cap F^c \cap G^c, E \cap F \cap G, E \cup F^c \cup G$.

22. Let S be a sample space consisting of all the integers from 1 to 20 inclusive, E the first 10 of these, and F the last five of these. Find $E \cap F, E^c \cap F, (E \cup F)^c, E^c \cap F^c$.

Solutions to Self-Help Exercise Set 7.1

1. Consider the outcomes as ordered pairs, with the number on the bottom of the red one the first number and the number on the bottom of the white one the second number. The sample space is

$$S = \{(1, 1), (1, 2), (1, 3), (1, 4),$$
$$(2, 1), (2, 2), (2, 3), (2, 4),$$
$$(3, 1), (3, 2), (3, 3), (3, 4),$$
$$(4, 1), (4, 2), (4, 3), (4, 4)\}.$$

2. **a.** $E = \{(1, 1), (1, 3), (3, 1), (3, 3)\}$
    $F = \{(1, 4), (2, 3), (3, 2), (4, 1)\}$
    b. $E \cap F = \phi$
    c. $E \cup F = \{(1, 1), (1, 3), (3, 1), (3, 3), (1, 4), (2, 3), (3, 2), (4, 1)\}$
    d. $G^c = \{(3, 4), (4, 3), (4, 4)\}$

7.2 EMPIRICAL AND INTUITIVE PROBABILITY

▶ *Empirical Probability*
▶ *Intuitive Probability*
▶ *Basic Properties of Probability*

Blaise Pascal
1623–1662

(David H. Smith Collection/Columbia University.)

The Beginnings of Mathematical Probability

In 1654 the famous mathematician Blaise Pascal had a friend, Chevalier de Méré, a member of the French nobility and a gambler, who wanted to adjust gambling stakes so that he would be assured of winning if he played long enough. This gambler raised questions with Pascal such as the following: In eight throws of a die a player attempts to throw a one, but after three unsuccessful trials the game is interrupted. How should he be indemnified? Pascal wrote to the leading mathematician of that day, Pierre de Fermat (1601–1665), about these problems, and their resulting correspondence represents the beginnings of the modern theory of mathematical probability.

Empirical Probability

A very important type of problem that arises every day in business and science is to find a *practical* way to estimate the likelihood of certain events. For example, a food company may seek a practical method of estimating the likelihood that a new type of candy will be enjoyed by consumers. The most obvious procedure for the company to follow is to select a consumer, have the consumer taste the candy, and then record the result. This should be repeated many times and the final totals tabulated to give the fraction of tested consumers who enjoy the candy. This fraction is then a practical estimate of the likelihood that all consumers will enjoy this candy.

In mathematical terms, let L be the event that the consumer enjoys the candy and N the number of consumers that have been tested. If $f(L)$ is the number or frequency of times a consumer in this group enjoys the candy, then the fraction of consumers who enjoy the candy is given by $\dfrac{f(L)}{N}$. We call this fraction the **relative frequency**.

For a general event E in a sample space S we have the following.

Relative Frequency

If an experiment is repeated N times and an event E occurs $f(E)$ times, then the fraction

$$\frac{f(E)}{N}$$

is called the **relative frequency** of the event.

The food company will take N as large as time and money will permit. The resulting relative frequency $\dfrac{f(L)}{N}$ is then a *practical* estimate of the likelihood that a randomly chosen consumer will enjoy this candy. We refer to such a relative frequency as the **empirical probability**.

In general we have the following.

Empirical Probability

Let S be a sample space of an experiment and E an event in S. If the experiment has actually been performed N times, the relative frequency $\dfrac{f(E)}{N}$ is called the **empirical probability** of E and is denoted by $p(E)$.

The Beginnings of Empirical Probability

Empirical probability began with the emergence of insurance companies. Insurance seems to have been originally used to protect merchant vessels and was in use even in Roman times. The first marine insurance companies began in Italy and Holland in the 14th century and spread to other countries by the 16th century. In fact, the famous Lloyd's of London was founded in the late 1600s. The first life insurance seems to have been written in the late 16th century in Europe. All of these insurance companies naturally needed to

> know with what relative frequencies certain events would occur. These relative frequencies were determined by collecting data over long periods of time.

The London merchant John Graunt (1620–1674) with the publication of *Natural and Political Observations Made upon the Bills of Mortality* in 1662 seems to have been the first person to have gathered data on mortality rates and determined relative frequencies from them. The data were extremely difficult to obtain. His then-famous London Life Table is reproduced below, showing the number of survivors through certain ages per 100 people.

London Life Table

Age	0	6	16	26	36	46	56	66	76
Survivors	100	64	40	25	16	10	6	3	1

E X A M P L E   1   **Finding Empirical Probability**

Using the London Life Table, find the empirical probability of a randomly chosen person living in London in the first half of the 17th century surviving until age 46.

Solution   In the London Life Table $N = 100$. If E is the event "survive to age 46," then according to the table $f(E) = 10$. Thus the empirical probability of people living in London at that time surviving until age 46 was

$$p(E) = \frac{f(E)}{N} = \frac{10}{100} = 0.1. \quad \blacksquare$$

E X A M P L E   2   **Finding Empirical Probability**

In a survey of 1000 people selected at random throughout the country, 110 said that they planned to purchase an automobile, 30 said they planned to buy a house, and 200 planned to buy a major appliance during the next year. What is the empirical probability that a person during the next year in this country will (a) buy an automobile, (b) buy a house, or (c) buy a major appliance?

Solutions   The empirical probabilities are calculated from the given data using the relative frequencies. Thus the empirical probabilities are the following respective relative frequencies.

(a) $\dfrac{110}{1000} = .11$ (b) $\dfrac{30}{1000} = .03$ (c) $\dfrac{200}{1000} = .2$ $\blacksquare$

E X A M P L E   3   **Finding Empirical Probability**

An inspector finds 4 defective parts from among 100 removed from an assembly line. What is the empirical probability that a part on the assembly line is defective?

Solution   The relative frequency and thus the empirical probability is $4/100 = 0.04$. Thus based on this data alone, we expect about 4% of the parts to be defective. ∎

Consider now a poorly made die purchased at a discount store. Dice are made by drilling holes in the sides and then backfilling. Cheap dice are, of course, not carefully backfilled. So when a lot of holes are made in a face, such as for the side with 6, and they are not carefully backfilled, that side will not be quite as heavy as the others. Thus a 6 will tend to come up more often on the top. Even a die taken from a craps table in Las Vegas, where the dice are of very high quality, will have some tiny imbalance.

E X A M P L E   4   **Finding Empirical Probability**

A die with 6 sides numbered from 1 to 6, such as used in the game of craps, is suspected to be somewhat lopsided. A laboratory has tossed this die 1000 times and obtained the results shown in the table. Find the empirical probabilities of the elementary outcomes and also find the empirical probability that an even number will occur.

Outcome	1	2	3	4	5	6
Number Observed	161	179	148	177	210	125

Solutions   Dividing each of the numbers by 1000 gives the relative frequencies and thus the empirical probabilities. Thus

$$p(1) = .161, \qquad p(2) = .179, \qquad p(3) = .148,$$
$$p(4) = .177, \qquad p(5) = .210, \qquad p(6) = .125.$$

For convenience we write $p(1)$ instead of the more precise $p(\{1\})$.

To find the probability of an even number, denoted by $p(\{2, 4, 6\})$, we count the number of times a 2, 4, or 6 has occurred, which we denote by $f(\{2, 4, 6\})$ and divide by the number of trials $N = 1000$. Since the number of times an even number occurs is just the sum of the number of occurrences of each of the even numbers, or

$$f(\{2, 4, 6\}) = f(2) + f(4) + f(6),$$

we have

$$p(\{2, 4, 6\}) = \frac{f(\{2, 4, 6\})}{N}$$

$$= \frac{f(2) + f(4) + f(6)}{N}$$

$$= \frac{f(2)}{N} + \frac{f(4)}{N} + \frac{f(6)}{N}$$

$$= p(2) + p(4) + p(6)$$

$$= .179 + .177 + .125 = .481. \quad \blacksquare$$

It is very important to note that the empirical probability of an event $E = \{e_1, e_2, \ldots, e_r\}$ is simply obtained as the sum of the empirical probabilities of all of the individual outcomes in the event. This follows simply because, in repeated trials of an experiment, the total number of times an event E occurs just equals the sum of the number of occurrences of each of the outcomes that comprise the event, that is,

$$f(\{e_1, e_2, \ldots, e_r\}) = f(e_1) + f(e_2) + \cdots + f(e_r).$$

This implies the following for relative frequencies:

$$\frac{f(\{e_1, e_2, \ldots, e_r\})}{N} = \frac{f(e_1)}{N} + \frac{f(e_2)}{N} + \cdots + \frac{f(e_r)}{N}.$$

Since the empirical probability is defined by

$$p(E) = \frac{f(E)}{N},$$

this last equation can also be written as

$$p(\{e_1, e_2, \ldots, e_r\}) = p(e_1) + p(e_2) + \cdots + p(e_r).$$

We thus have the following.

Finding the Empirical Probability of an Event

Suppose an experiment has been repeated N times. Let $E = \{e_1, e_2, \ldots, e_r\}$ be any event, and $p(E)$ the empirical probability. Then

$$p(\{e_1, e_2, \ldots, e_r\}) = p(e_1) + p(e_2) + \cdots + p(e_r). \qquad (1)$$

Frederick Mosteller and the Dice Experiment

Frederick Mosteller has been president of the American Association for the Advancement of Science, the Institute of Mathematical Statistics, and the American Statistical Association. He once decided that "It would be nice to see if the actual outcome of a real person tossing real dice would match up with the theory." He then engaged Willard H. Longcor to buy some dice, toss them, and keep careful records of the outcomes. Mr. Longcor then tossed the dice on his floor at home so that the dice would bounce on the floor and then up against the wall and then land back on the floor. After doing this several thousand times, his wife became troubled by the noise. He then placed a rug on the floor and on the wall, and then proceeded to quietly toss his dice *millions* of times, keeping careful records of the outcomes. In fact, he was so careful and responsible about his task, that he threw away his data on the first 100,000 tosses, since he had a nagging worry that he might have made some mistake keeping perfect track.

E X A M P L E   5   **Finding Empirical Probabilities**

In the previous example, find the empirical probability that the number on the die is at least 4.

Solution   We have

$$p(\{4, 5, 6\}) = p(4) + p(5) + p(6) = .177 + .210 + .125 = .512. \quad \blacksquare$$

Intuitive Probability

We have just given a practical way to assign a number to an event that indicates the approximate likelihood that the event will occur. We now see how to assign a number to an event that gives the *exact* likelihood that the event will occur. This number is assigned on the basis of the percentage of times the event E will occur over the long-term when the experiment is repeated over and over again. More specifically, the number is assigned on the basis of the long-term behavior of the relative frequency $\dfrac{f(E)}{N}$.

Consider a *theoretical* example. We flip a fair coin, that is, a coin that shows no preference for heads or tails. We naturally expect heads to show as often as tails. Thus we expect heads to show one-half the time, and if H is the event that a head shows, we expect that as the number N of times the coin is flipped becomes larger and larger, the relative frequencies $\dfrac{f(H)}{N}$ should become closer and closer to exactly .50. The number .50 is called the **intuitive probability** or just the **probability** of H, and is denoted by $p(H)$.

To obtain a specific example, the author flipped a fair coin 100,000 times. More precisely, the author used a random number generator on a computer to simulate flipping a fair coin.

Figure 7.3 shows in graphical form what happened to the relative frequencies. Figure 7.3(a) shows the relative frequencies for the first 10,000 "tosses" in increments of 100. Figure 7.3(b) shows the relative frequencies in increments of 5000 tosses. First notice that since we are dealing with a *random* phenomenon, we should not be surprised to see the relative frequencies bounce *randomly*. Very importantly, however, we notice that as N becomes larger and larger, the relative frequency, although bouncing around, seems to be approaching the number $p(H)$ = .50, the probability of H. This is typical of random phenomena in general.

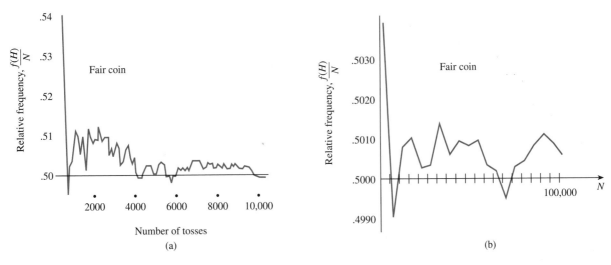

Figure 7.3

For a fair coin we know, on the basis of a theoretical analysis, that as N becomes larger, the relative frequencies $\dfrac{f(H)}{N}$ must approach the number $p(H) =$ .50. But if a coin were not known ahead of time to be fair, then how could we determine $p(H)$? Unfortunately, we cannot determine $p(H)$ *exactly*. However, if we flipped the coin a very large number N of times, the empirical probability $\dfrac{f(H)}{N}$ should be approximately equal to $p(H)$. Thus in the situation in which we have no theoretical analysis that can give us the exact probability, we must repeat the experiment a large number of times and approximate the actual probability with the empirical probability.

For a general event E in a sample space S, we have the following (intuitive) definition.

The Probability of an Event E

Let S be the sample space of an experiment, let E be an event in S, and let N be the number of times the experiment has been repeated. The **probability of E**, denoted by $p(E)$, is the number that the relative frequency $\dfrac{f(E)}{N}$ approaches as N becomes larger and larger.

R E M A R K .   We also refer to probability defined in this way as **intuitive probability**.

In Section 8.5 we shall delve deeper into the long-term behavior of the relative frequency. In that section we shall see that the relative frequency $\dfrac{f(E)}{N}$ can be made as close to $p(E)$ as we like with any degree of certainty (short of absolute certainty) by taking N sufficiently large, and we shall find formulas for N toward this end.

Suppose we want to flip a fair coin enough times to be reasonably certain that the relative frequency is within .01 of $\frac{1}{2}$. In Section 8.5 we show that if a fair coin is flipped $N = 10,000$ times, then 95 times out of 100 the relative frequency of heads will be within 0.01 of $\frac{1}{2}$. We also show in the same section that if an experiment consists of flipping a fair coin $N = 1,000,000$ times, then 95 times out of 100 the relative frequency of heads will be within 0.001 of $\frac{1}{2}$.

If we wish even greater accuracy with even more certainty, then there will exist a number N that will meet our increased demands. It is in this sense that we expect the relative frequency to approach the exact probability as N becomes large.

Some Basic Properties of Probability

We now consider some basic properties of probability. The definition immediately implies several things about the numbers $p(E)$. First, if E is an event in a sample space S and the experiment is repeated N times, then the number of occurrences $f(E)$ of the event E can be at most N and, of course, at least 0. That is, $0 \le f(E) \le N$. This implies that

$$0 \le \frac{f(E)}{N} \le 1.$$

Since for large N, $\dfrac{f(E)}{N} \approx p(E)$, we then must have $0 \le p(E) \le 1$. Second, since some outcome must occur on every one of the N trials, for every N we have $f(S) = N$, and $\dfrac{f(S)}{N} = \dfrac{N}{N} = 1$. Thus we must have $p(S) = 1$. Finally, since some outcome must occur on every trial, no outcome is impossible and never happens. Therefore for every N we have $f(\phi) = 0$, and $\dfrac{f(\phi)}{N} = \dfrac{0}{N} = 0$. Thus we must have $p(\phi) = 0$.

Properties of Probability

Let E be an event in a sample space S. Then

Property 1 $0 \le p(E) \le 1$.

Property 2 $p(S) = 1$.

Property 3 $p(\phi) = 0$.

We are now prepared to consider another important property of probability. We will now see that equation (1) is not only satisfied by empirical probability, but also by probability in general. We then have the following.

Property 4 Given any event $E = \{e_1, e_2, \ldots, e_r\}$, where $e_1, e_2, \ldots, e_r$, are all distinct, then

$$p(\{e_1, e_2, \ldots, e_r\}) = p(e_1) + p(e_2) + \cdots + p(e_r).$$

We have already indicated that Property 4 holds for empirical probability. To see why the fourth property is true in general, suppose the experiment is repeated N times. Then in terms of relative frequencies

$$\frac{f(\{e_1, e_2, \ldots, e_r\})}{N} = \frac{f(e_1)}{N} + \frac{f(e_2)}{N} + \cdots + \frac{f(e_r)}{N}.$$

Now, for very large N, the left-hand side of this equation should be very close to $p(\{e_1, e_2, \ldots, e_r\})$ while the right-hand side should be very close to $p(e_1) + p(e_2) + \cdots + p(e_r)$. Thus $p(\{e_1, e_2, \ldots, e_r\})$ should equal $p(e_1) + p(e_2) + \cdots + p(e_r)$. This is then Property 4.

DISCUSSION QUESTIONS

Answers to the following discussion questions can be found after the exercise set.

1. You buy a new die and toss it 1000 times. A 1 comes up 165 times. Is it true that the probability of a 1 showing on this die is .165?

2. A fair coin is to be tossed 100 times. Naturally you expect tails to come up 50 times. After 60 tosses, heads has come up 40 times. Is it true that now heads is likely to come up less often than tails during the next 40 tosses?

3. You are playing a game at a casino and have correctly calculated the probability of winning any one game to be 0.48. You have played for some time and have won 60% of the time. You are on a roll. Should you keep playing?

4. You are watching a roulette game at a casino and notice that red has been coming up unusually often. (Red should come up as often as black.) Is it true that, according to "the law of averages," black is likely to come up unusually often in the next number of games to "even things up"?

5. You buy a die and toss it 1000 times and notice that a 1 came up 165 times. You decide that the probability of a 1 on this die is .165. Your friend takes this die and tosses it 1000 times and notes that a 1 came up 170 times. He concludes that the probability of a one is .17. Who is correct?

6. People who frequent casinos and play lotteries are gamblers, but those who run the casinos and lotteries are not. Do you agree? Why or why not?

SELF-HELP EXERCISE SET 7.2

A retail store that sells sweaters notes the following number of sweaters of each size that were sold last year.

small	medium	large	extra large
10	40	30	20

1. What is the empirical probability that a customer that buys a sweater will buy one of size small? medium? large? extra large?

2. Of size medium or large?

3. Or size larger than small?

EXERCISE SET 7.2

1. Suppose the probability of an event E is $p(E) = .2317$. What is the expected frequency of occurrence $f(E)$ of the event E if (a) $N = 100$, (b) $N = 1000$, (c) $N = 10,000$?

2. Suppose the probability of an event E is $p(E) = \frac{1}{9}$. What is the expected frequency of occurrence $f(E)$ of the event E if (a) $N = 100$, (b) $N = 900$, (c) $N = 1000$?

3. Using the London Life Table given in the text, find the empirical probability that a person living in London at that time would survive to age 16.

4. Repeat the previous exercise, replacing 16 with 76.

5. A somewhat lopsided die is tossed 1000 times with 1 showing on the top face 150 times. What is the empirical probability that a 1 will show?

6. A coin is flipped 10,000 times with heads showing 5050 times. What is the empirical probability that heads will show?

7. The speed of 500 vehicles on a highway with a speed limit of 55 mph was observed, with 400 going between 55 and 65 mph, 60 going less than 55 mph, and 40 going over 65 mph. What is the empirical probability that a vehicle chosen at random on this highway will be going (a) under 55 mph, (b) between 55 and 65 mph, (c) over 65 mph?

8. In a survey of 1000 randomly selected consumers, 50 said they bought brand A cereal, 60 said they bought brand B, and 80 said they bought brand C. What is the empirical probability that a consumer will purchase (a) brand A cereal, (b) brand B, (c) brand C?

9. Weather records indicate that during the past 10 years there has been a measurable snowfall in a certain city on a total of 62 days in January. What is the empirical probability that a day in January in this city will have a measurable snowfall?

10. A salesman has noticed that his records indicate that he sold at least one major appliance on 80 different days during the last year. If he worked 240 days last year, what is the empirical probability that he sells at least one major appliance on a working day?

11. A large dose of a suspected carcinogen has been given to 500 white rats in a laboratory experiment. During the next year, 280 rats get cancer. What is the empirical probability that a rat chosen randomly from this group of 500 will get cancer?

12. A new brand of sausage is tested on 200 randomly selected customers in grocery stores with 40 saying they like the product, the others saying they do not. What is the empirical probability that a consumer will like this brand of sausage?

13. Over a number of years the grade distribution in a mathematics course was observed to be

A	B	C	D	F
25	35	80	40	20

What is the empirical probability that a randomly selected student taking this course will receive a grade of A? B? C? D? F?

14. A store sells 4 different brands of VCRs. During the past year the following number of sales of each of the brands were found.

Brand A	Brand B	Brand C	Brand D
20	60	100	70

What is the empirical probability that a randomly selected customer who buys a VCR at this store will pick brand A? brand B? brand C? brand D?

15. A somewhat lopsided die is tossed 1000 times with the following results.

1	2	3	4	5	6
150	200	140	250	160	100

What is the empirical probability that each of the numbers will occur?

16. A retail store that sells sneakers notes the following number of sneakers of each size that were sold last year.

7	8	9	10	11	12
20	40	60	30	40	10

What is the empirical probability that a customer who buys a pair of sneakers will buy size 7? 8? 9? 10? 11? 12?

17. What is the empirical probability that one of the students in Exercise 13 will make a grade higher than a C?

18. What is the empirical probability that one of the students in Exercise 13 will make a grade of C or lower?

19. What is the empirical probability that one of the people in Exercise 14 will not select brand D?

20. What is the empirical probability that one of the people in Exercise 14 will not select brand A or B?

21. What is the empirical probability that an even number shows in Exercise 15?

22. What is the empirical probability that a number bigger than 4 shows in Exercise 15?

23. What is the empirical probability that a customer in Exercise 16 buys a pair of sneakers of size 7 or 12?

24. What is the empirical probability that a customer in Exercise 16 buys a pair of sneakers size 10 or larger?

25. If E is any event in a sample space S, give an intuitive argument based on relative frequencies that indicates that $p(E^c) = 1 - p(E)$. *Hint:* Show $f(E) + f(E^c) = N$.

26. Let E and F be two mutually disjoint events in a sample space S. Give an intuitive argument based on relative frequencies that indicates that $p(E \cup F) = p(E) + p(F)$. *Hint:* Use the fact that $f(E) + f(F) = f(E \cup F)$.

27. Let the sample space $S = \{s_1, s_2, \ldots, s_n\}$ and let $p(s_1) = p_1$, $p(s_2) = p_2$, $\ldots, p(s_n) = p_n$. Then give an intuitive argument that indicates that $p_1 + p_2 + \cdots + p_n = 1$. *Hint:* Follow the discussion of the proof of Property 4 and use Property 2.

28. Let E and F be two events in a sample space S. Give an intuitive argument based on relative frequencies that indicates that $p(E \cup F) = p(E) + p(F) - p(E \cap F)$. *Hint:* Use the fact that $f(E \cup F) = f(E) + f(F) - f(E \cap F)$.

Answers to Discussion Questions

1. You do not know what the actual probability is. You do know that the empirical probability is $\frac{f(H)}{N} = \frac{165}{1000} = .165$. This represents the best guess for the actual probability. But if you tossed the coin more times, the relative frequency and the new empirical probability would most likely change.

2. Since the coin is assumed to be fair, the probability of heads on any one toss is $\frac{1}{2}$. No matter how many heads or tails have occurred in the past, the coin remains a fair coin. Therefore, the probability must remain the same. Whether heads has come up often in the past or seldom in the past, the probability of heads on the

next toss must remain at $\frac{1}{2}$ and not change. Thus we still expect heads to come up as often as tails during the last 40 tosses.

3. The probabilities in the game are constant and do not change just because you are on a winning streak. Thus no matter what has happened to you in the past, the probability of winning any one game remains constant at .48. Thus if you continue to play, you should expect to win 48% of the time in the future. You have been lucky to have won 60% of the time up until now.

4. Since there are just as many red slots as black slots in a roulette game, the probability of a red must always equal the probability of a black. Just because an unusual number of reds has come up cannot affect this fundamental physical fact. The physical setup of the game and the physical laws of the universe do not change from one game to another. Thus no matter what has happened in the past, the probabilities for red and black remain the same.

5. After reading the first discussion problem above, we know that it is, in fact, impossible to determine with certainty the actual probability precisely. Since the die has been tossed a total of 2000 times and a one has come up 335 times, our best guess at the actual probability is $\frac{335}{2000} = .1675$.

6. Casinos and lotteries make money based on the fact that when the "experiment" of playing a casino game or picking a number is repeated a very large number of times, the relative frequency will very likely be very close to the actual probabilities. These probabilities have been carefully calculated so that the casino can be assured the customer will take back about 90 cents of every dollar bet, while lotteries generally give about 50 cents back for every dollar bet. Thus any individual player who plays for an extended number of times can count on the likelihood that the relative frequencies will be close to the actual probabilities, thus assuring losses for the player.

Solutions to Self-Help Exercise Set 7.2

1. There were $10 + 40 + 30 + 20 = 100$ sold. Let the sample space be denoted by $\{s, m, l, x\}$ with the obvious designations. To find the empirical probability of each outcome, count the frequency of each outcome and divide by 100. The probabilities are as follows:

$$p(s) = \frac{f(s)}{N} = \frac{10}{100} = .10 \qquad p(m) = \frac{f(m)}{N} = \frac{40}{100} = .40$$

$$p(l) = \frac{f(l)}{N} = \frac{30}{100} = .30 \qquad p(x) = \frac{f(x)}{N} = \frac{20}{100} = .20$$

2. $p(\{m, l\}) = \dfrac{f(\{m, l\})}{N} = \dfrac{40 + 30}{100} = .70$

3. $p(\{m, l, x\}) = \dfrac{f(\{m, l, x\})}{N} = \dfrac{40 + 30 + 20}{100} = .90.$

7.3   MATHEMATICAL PROBABILITY

▶ *Mathematical Definition of Probability*

▶ *Additional Properties of Probability*

▶ *Odds*

Further Historical Developments in Probability

Neither Pascal nor Fermat published their initial findings on probability. Christian Huygens (1629–1695) became acquainted with the work of Pascal and Fermat and subsequently published in 1657 the first tract on probability: *On Reasoning in Games of Dice*. This little pamphlet remained the only published work on probability for the next 50 years. James Bernoulli (1654–1705) published the first substantial tract on probability when his *Art of Reasoning* appeared 7 years after his death. This expanded considerably on Huygens' work. The next major milestone in probability occurred with the publication in 1718 of Abraham De Moivre's work *Doctrine of Chance: A Method of Calculating the Probability of Events in Play*. Before 1770, probability was almost entirely restricted to the study of gambling and actuarial problems, although some applications in errors of observation, population, and certain political and social phenomena had been touched on. It was Pierre Simon Laplace (1749–1827) who broadened the mathematical treatment of probability beyond games of chance to many areas of scientific research. The theory of probability undoubtedly owes more to Laplace than to any other individual.

APPLICATION
Finding the Probability of Obtaining a Contract

Three companies are competing for a contract. It is known that the first two companies have equal probability of obtaining the contract and that the third company has only one-half the probability of the first one. What is the probability that the first company will get the contract? See Example 4 for the answer.

Mathematical Definition of Probability

In the last section we took the point of view that probability is part of the study of natural and human phenomena in the same way that physics or economics is. Probability was discussed only in the context of events in business or science. Subjects such as "randomness" and the long-term behavior of the relative frequency of an event were deemed intuitive and within the experience of the reader.

We take an alternate point of view of probability in this section. We will now regard probability as a part of *mathematics*. As such, we need to establish abstract definitions and construct a set of axioms that govern the theory, placing the theory on a firm foundation. Using the logic of mathematics we can then derive other properties that probability must satisfy. Such a mathematical theory is said to be a **model** of the phenomenon of probability. One important advantage of this approach is that all terms are clearly defined and "experience" is not a prerequisite for understanding the development. The axioms that describe the phenomenon of probability must be carefully chosen so that the resulting theory actually describes the phenomenon that we are studying. In other words, the theory must not contradict our intuitive notions of probability gained in the last section. How can we then choose the axioms that will govern the theory? Naturally we must look to the intuitive notions gained in the last section. The four properties of probability that we learned of in the last section play a fundamental role.

We have interpreted probability in terms of the long-term behavior of relative frequencies and found this very useful in applications. This concept is intuitive, however, and was not made mathematically precise in the last section. We could make this intuitive notion mathematically precise, but the process is not worth the effort since a more fruitful but still mathematically precise approach is available. Naturally any definition of probability must be consistent with the intuitive interpretation given in the last section so that the results can be used to model real applications.

We first need to have a set S called the sample space. To define a probability, we must associate with each event E, that is with each subset of S, a real number $p(E)$ called the probability of E.

We choose as axioms the four properties of probability given in the last section. These four, in effect, capture the essence of what we intuitively think of as probability.

We then have the following definition.

Definition of Probability

Let S be a sample space. Assign to each event E in S a number $p(E)$ so that the following four properties are satisfied.

Property 1 $0 \leq p(E) \leq 1$ for any event E.

Property 2 $p(S) = 1$.

Property 3 $p(\phi) = 0$.

Property 4 Given any event $E = \{e_1, e_2, \ldots, e_r\}$, where $e_1, e_2, \ldots, e_r$, are all distinct, then

$$p(\{e_1, e_2, \ldots, e_r\}) = p(e_1) + p(e_2) + \cdots + p(e_r).$$

Then for any event E, we called $p(E)$ the **probability** of E and say that p is a probability on S.

Just how to assign the probabilities to each outcome must be determined by the information on hand. The following example indicates what is possible in a simple situation.

E X A M P L E 1   **Assigning Probabilities**

Let a sample space by $S = \{H, T\}$ and define

$$p(H) = p_1 \geq 0, \quad p(T) = p_2 \geq 0, \quad p(S) = 1, \quad p(\phi) = 0,$$

and $p_1 + p_2 = 1$. Show that p is a probability on S.

Solution   This p certainly satisfies Properties 1, 2, and 3. For Property 4, notice that

$$p(\{H, T\}) = p(S) = 1,$$

while

$$p(H) + p(T) = p_1 + p_2 = 1.$$

This shows that p is a probability on S. ∎

This probability could model a fair or unfair coin (depending on whether p_1 was .5 or was not) where $p(H) = p_1$. But this could also model any situation with 2 outcomes and where the probability of one outcome is p_1. In a 2 outcome experiment, we often refer to one outcome as "success" and the other as "failure." Then the sample space is {success, failure}. The probability of "success" is p_1 and the probability of "failure" is p_2, where $p_1 + p_2 = 1$. Also we are often interested in the probability of a specific event E occurring or not occurring. In such a situation we have the two events: E and E^c. As we will very shortly see, $p(E) + p(E^c)$ must be 1.

Additional Properties of Probability

The four defining properties of probability have a number of consequences. We now give two.

Property 5

Let the sample space be $S = \{s_1, s_2, \ldots, s_n\}$ and let $p(s_1) = p_1, p(s_2) = p_2, \ldots, p(s_n) = p_n$. Then

$$p_1 + p_2 + \cdots + p_n = 1.$$

Property 5 follows by replacing S with E in Property 4 and using Property 2. Thus

$$1 = p(S) = p(s_1) + p(s_2) + \cdots + p(s_n) = p_1 + p_2 + \cdots + p_n.$$

Property 6

Let G and H be any two events. Then

$$p(G \cup H) = p(G) + p(H) \quad \text{if} \quad G \cap H = \phi.$$

To establish Property 6, let $G = \{g_1, \ldots, g_k\}$ and $H = \{h_1, \ldots, h_j\}$. Since $G \cap H = \phi$, all the elements in the set

$$\{g_1, \ldots, g_k, h_1, \ldots, h_j\}$$

are distinct. Thus from Property 4,

$$p(G \cup H) = p(\{g_1, \ldots, g_k, h_1, \ldots, h_j\})$$
$$= p(g_1) + \cdots + p(g_k) + p(h_1) + \cdots + p(h_j)$$
$$= [p(g_1) + \cdots + p(g_k)] + [p(h_1) + \cdots + p(h_j)]$$
$$= p(G) + p(H).$$

This establishes Property 6.

In the last section we had mentioned that probabilities can often be determined by theoretical means. We decided by an intuitive argument based on interpreting probabilities in terms of relative frequencies that if in a coin-flipping experiment heads is as likely as tails, then $p(H) = p(T) = .5$. We now give another reason for this, this time based on the above mathematical definition of probability.

E X A M P L E 2   **Finding the Probability of an Event**

Suppose in a coin-flipping experiment we assume that the probability of heads is the same as the probability of tails. What are $p(H)$ and $p(T)$?

Solution   We have $p(H) = p(T)$. By Property 6

$$1 = p(H) + p(T) = 2p(H).$$

Thus $p(H) = \frac{1}{2} = p(T)$.   ■

In the next two examples, the usefulness of Property 6 is shown.

E X A M P L E 3   **Finding the Probability of an Event**

The probability that any of the first 5 numbers of a loaded die will come up is the same while the probability that a 6 comes up is .25. What is the probability that a 1 will come up?

Solution   We are given that

$$p(1) = p(2) = p(3) = p(4) = p(5), \qquad p(6) = .25.$$

From Property 6

$$1 = p(1) + p(2) + p(3) + p(4) + p(5) + p(6)$$
$$= 5p(1) + .25$$
$$5p(1) = .75$$
$$p(1) = .15.   ■$$

E X A M P L E   4   **Finding the Probability of an Event**

Three companies are competing for a contract. It is known that the first two companies have equal probability of obtaining the contract and that the third company has only one-half the probability of the first one. What is the probability that the first company will get the contract?

Solution   Let $p(1)$, $p(2)$, $p(3)$, be respectively the probability of the first, second, and third company obtaining the contract. Then we have that $p(2) = p(1)$ and $p(3) = p(1)/2$. Thus from Property 6

$$1 = p(1) + p(2) + p(3) = p(1) + p(1) + \frac{1}{2}p(1) = \frac{5}{2}p(1).$$

Thus $p(1) = .40$ ■

The first 6 properties have a number of further consequences. We now give some of them.

Further Properties of Probability

Let p be a probability on a sample space S and E and F two events in S. Then the following are true.

Property 7 $p(E^c) = 1 - p(E)$.

Property 8 $p(E \cup F) = p(E) + p(F) - p(E \cap F)$.

Property 9 $p(F) \le p(E)$ if $F \subset E$.

Property 10   Suppose $E_1, E_2, \ldots, E_k$ are pairwise mutually disjoint, that is $E_i \cap E_j = \phi$ if $i \ne j$. Then

$$p(E_1 \cup E_2 \cup \cdots \cup E_k) = p(E_1) + p(E_2) + \cdots + p(E_k).$$

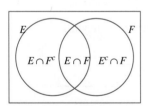

Figure 7.4

The proofs of Property 7, Property 9, and Property 10 are left as exercises. Property 10 represents an extension of Property 6 to any number of mutually disjoint sets.

To establish Property 8 notice from Figure 7.4 that since $E \cap F^c$ and $E \cap F$ are mutually exclusive and $E = (E \cap F^c) \cup (E \cap F)$, Property 6 implies that

$$p(E) = p(E \cap F^c) + p(E \cap F).$$

Thus

$$p(E \cap F^c) = p(E) - p(E \cap F).$$

We can see from Figure 7.5 that we can write $E \cup F$ as the union of the two sets $E \cap F^c$ and F and these two sets are mutually exclusive. Thus by Property 6 and the above

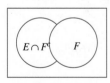

Figure 7.5

$$p(E \cup F) = p((E \cap F^c) \cup F)$$
$$= p(E \cap F^c) + p(F)$$
$$= p(E) - p(E \cap F) + p(F).$$

This gives Property 8.

E X A M P L E 5 **Finding the Probability of an Event**

A salesman makes two stops when in Pittsburgh. The first stop yields a sale 10% of the time, the second stop 15% of the time, and both stops yield a sale 4% of the time. What proportion of the time does a trip to Pittsburgh result in no sale?

Solution Let E be the event a sale is made at the first stop and F the event that a sale is made at the second stop. Then we are given that

$$p(E) = .10, \qquad p(F) = .15, \qquad p(E \cap F) = .04.$$

Since $p(E \cap F) = .04$, we place .04 in the region $E \cap F$ in Figure 7.6. Now since $p(E) = .10$, we can see from Figure 7.6 that $p(E \cap F^c) = .10 - .04 = .06$. In a similar fashion we have $p(E^c \cap F) = .15 - .04 = .11$. Thus we readily see from Figure 7.6 that

$$p(E \cup F) = .06 + .04 + .11 = .21.$$

Then by Property 7

$$p((E \cup F)^c) = 1 - p(E \cup F) = 1 - .21 = .79.$$

Thus no sale is made in Pittsburgh 79% of the time. ∎

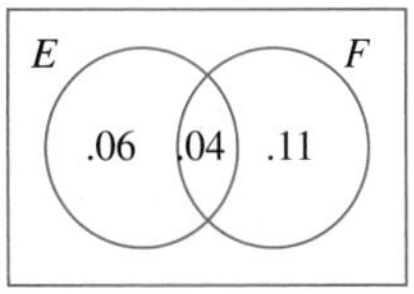

R E M A R K .   We could have obtained $p(E \cup F)$ directly from Property 8 as follows.

$$P(E \cup F) = p(E) + p(F) - p(E \cap F) = .10 + .15 - .04 = .21.$$

E   F

.06 .04 .11

Figure 7.6

Odds*

One can interpret probabilities in terms of **odds** in a bet. Suppose in a sample space S we are given an event E with probability $p = p(E)$. Suppose we engage in a bet in which we agree to give a dollars if E does not occur and receive b dollars if E does occur. Let us decide what the relationship should be between a and b so that the bet will be fair. In a large number of such bets we expect to win b dollars a fraction p of the time and lose a dollars a fraction $1 - p$ of the time. On average we expect to obtain $bp - a(1 - p)$ dollars. If the wager is to be fair, this should be zero. Thus $bp = a(1 - p)$ or

_____

*Optional

$$\frac{a}{b} = \frac{p}{1 - p}.$$

This ratio is called the **odds** and is normally written as the ratio of two integers.

Odds of an Event

Let $p = p(E)$ be the probability of E, then the **odds for E** is defined to be

$$\text{Odds for } E = \frac{p}{1 - p} \qquad \text{if } p \neq 1.$$

If the odds are given as $\dfrac{a}{b}$, we say that **the odds are a to b** and also write $a : b$.

E X A M P L E   6   **Determining the Odds of an Event**

You believe that a horse has a probability of $\frac{1}{4}$ of winning a race. What are the odds of this horse winning? What are the odds of this horse losing? What profit should a winning $2 bet return for the bet to be fair?

Solutions   (a)   Since $p = \frac{1}{4}$, the odds for winning are

$$\frac{p}{1 - p} = \frac{\frac{1}{4}}{\frac{3}{4}} = \frac{1}{3},$$

that is, $1 : 3$.

(b)   Since the probability of winning is $\frac{1}{4}$, the probability of losing is $1 - \frac{1}{4} = \frac{3}{4}$. Then

$$\text{odds for losing} = \frac{\frac{3}{4}}{1 - \frac{3}{4}} = \frac{3}{1},$$

or $3 : 1$.

(c)   Since the fraction $\frac{1}{3}$ can also be written as $\frac{2}{6}$ with odds $2 : 6$, a fair $2 bet should return a profit of $6 for a winning bet.   ∎

Notice that making this same bet many times, we expect to win $6 one-fourth of the time and lose $2 three-fourths of the time. Our average winnings would be

$$6(\tfrac{1}{4}) - 2(\tfrac{3}{4}) = 0$$

dollars.

The way in which probability can be interpreted in terms of odds is as follows. Suppose that the odds for an event E occurring is given as $\dfrac{a}{b}$. If $p = p(E)$, then

$$\frac{a}{b} = \frac{p}{1 - p}$$

$$a(1 - p) = bp$$

$$a = bp + ap = p(a + b)$$

$$p = \frac{a}{a + b}.$$

Obtaining Probability from Odds

Suppose that the odds for an event E occurring is given as $\frac{a}{b}$. If $p = p(E)$, then

$$p = \frac{a}{a + b}.$$

SELF-HELP EXERCISE SET 7.3

1. If $S = \{a, b, c\}$ with $p(a) = p(b) = p(c)$, find $p(a)$.

2. A company has bids on two contracts. They believe that the probability of obtaining the first contract is .4 and of obtaining the second contract is .3, while the probability of obtaining both contracts is .1. (a) Find the probability that they will obtain exactly one of the contracts. (b) Find the probability that they will obtain neither of the contracts.

3. What are the odds that the company in the previous exercise will obtain both of the contracts?

EXERCISE SET 7.3

In all the following, S is assumed to be a sample space and p a probability function on S.

1. Let $S = \{a, b, c\}$ with $p(a) = .1$, $p(b) = .4$, $p(c) = .5$. Let $E = \{a, b\}$ and $F = \{b, c\}$. Find $p(E)$ and $p(F)$.

2. Let $S = \{a, b, c, d, e, f\}$ with $p(a) = .1$, $p(b) = .2$, $p(c) = .25$, $p(d) = .15$, $p(e) = .12$, $p(f) = .18$. Let $E = \{a, b, c\}$ and $F = \{c, d, e, f\}$ and find $p(E)$ and $p(F)$.

3. Let $S = \{a, b, c, d, e, f\}$ with $p(b) = .2$, $p(c) = .25$, $p(d) = .15$, $p(e) = .12$, $p(f) = .1$. Let $E = \{a, b, c\}$ and $F = \{c, d, e, f\}$. Find $p(a)$, $p(E)$, and $p(F)$.

4. Let $S = \{a, b, c, d, e, f\}$ with $p(b) = .3$, $p(c) = .15$, $p(d) = .05$, $p(e) = .2$, $p(f) = .13$. Let $E = \{a, b, c\}$ and $F = \{c, d, e, f\}$. Find $p(a)$, $p(E)$, and $p(F)$.

5. If $S = \{a, b, c, d\}$ with $p(a) = p(b) = p(c) = p(d)$, find $p(a)$.

6. If $S = \{a, b, c\}$ with $p(a) = p(b)$ and $p(c) = .4$, find $p(a)$.

7. If $S = \{a, b, c, d, e, f\}$ with $p(a) = p(b) = p(c) = p(d) = p(e) = p(f)$, find $p(a)$.

8. If $S = \{a, b, c\}$ with $p(a) = 2p(b) = 3p(c)$, find $p(a)$.

9. If $S = \{a, b, c, d, e, f\}$ with $p(a) = p(b) = p(c), p(d) = p(e) = p(f) = .1$, find $p(a)$.

10. If $S = \{a, b, c, d, e\ f\}$ and if $p(a) = p(b) = p(c), p(d) = p(e) = p(f), p(d) = 2p(a)$, find $p(a)$.

11. If E and F are two mutually disjoint events in S with $p(E) = .2$ and $p(F) = .4$, find $p(E \cup F)$, $p(E^c)$, and $p(E \cap F)$.

12. Why is it not possible for E and F to be two mutually disjoint events in S with $p(E) = .5$ and $p(F) = .7$?

13. If E and F are two mutually disjoint events in S with $p(E) = .4$ and $p(F) = .3$, find $p(E \cup F)$, $p(F^c)$, $p(E \cap F)$, $p((E \cup F)^c)$, $p((E \cap F)^c)$.

14. Why is it not possible for $S = \{a, b, c\}$ with $p(a) = .3, p(b) = .4, p(c) = .5$?

15. Let E and F be two events in S with $p(E) = .3, p(F) = .5, p(E \cap F) = .2$. Find $p(E \cup F)$ and $p(E \cap F^c)$.

16. Let E and F be two events in S with $p(E) = .3, p(F) = .5, p(E \cap F) = .2$. Find $p(E^c \cap F)$ and $p(E^c \cap F^c)$.

17. Let E and F be two events in S with $p(E) = .3, p(F) = .5, p(E \cup F) = .6$. Find $p(E \cap F)$ and $p(E \cap F^c)$.

18. Why is it not possible to have E and F two events in S with $p(E) = .3$ and $p(E \cap F) = .5$?

19. Let E and F be two events in S with $p(E) = .5, p(F) = .7$. Just how small could $p(E \cap F)$ possibly be?

20. Let E and F be two events in S with $p(E) = .3, p(F) = .4$. Just how large could $p(E \cup F)$ possibly be?

In Exercises 21 through 24 let E, F, and G be events in S with $p(E) = .55, p(F) = .4, p(G) = .45, p(E \cap F) = .3, p(E \cap G) = .2, p(F \cap G) = .15, p(E \cap F \cap G) = .1$.

21. Find $p(E \cap F \cap G^c)$, $p(E \cap F^c \cap G)$, $p(E \cap F^c \cap G^c)$.

22. Using the results of the previous exercise, find $p(E^c \cap F \cap G)$, $p(E^c \cap F \cap G^c)$, $p(E^c \cap F^c \cap G)$.

23. Using the results of the previous two exercises, find $p(E \cup F \cup G)$.

24. Using the result of the previous exercise, find $p(E^c \cup F^c \cup G^c)$.

25. For the loaded die in Example 3 of the text, what are the odds that (a) a 2 will occur, (b) a 6 will occur?

26. For the loaded die in Example 3 of the text, what are the odds that (a) a 3 will occur, (b) a 1 will occur?

27. In Example 4 of the text, what are the odds that (a) the first company will receive the contract, (b) the third one will?

28. In Example 5 of the text, what are the odds that the salesman will make a sale on (a) the first stop, (b) on the second stop, (c) on both stops?

29. It is known that the odds that E will occur are $1:3$ and that the odds that F will occur are $1:2$, and that both E and F cannot occur simultaneously. What are the odds that E or F will occur?

30. Show that if the odds for E are $a:b$, then the odds for E not occurring are $b:a$.

31. If the odds for the Giants winning the World Series are $1:4$, what is the probability the Giants will win the Series?

32. If the odds for a successful marriage are $1:2$, what is the probability for a successful marriage?

33. Establish Property 7: For any event E, $p(E^c) = 1 - p(E)$. *Hint:* Notice that E and E^c are mutually exclusive and $E \cup E^c = S$. Now use Property 2 and Property 6.

34. Establish Property 9: If E and F are two events with $F \subset E$, then $p(F) \le p(E)$. *Hint:* If $F \subset E$, the two sets F and $E \cap F^c$ are mutually exclusive and have union equal to E. Use Property 6 and then Property 1.

35. Establish Property 10 for 3 pairwise mutually disjoint sets.

36. If E, F, and G are events in a sample space S, show that

$$p(E \cup F \cup G) = p(E) + p(F) + p(G) - p(E \cap F) - \\ p(E \cap G) - P(F \cap G) + p(E \cap F \cap G).$$

Hint: Let $H = F \cup G$ and use Property 8.

37. If E and F are two events in a sample space S, the event $(E \cap F^c) \cup (E^c \cap F)$ is the event that either E or F occur but not both, that is, that exactly one of the events E or F occurs. Show that

$$p((E \cap F^c) \cup (E^c \cap F)) = p(E) + p(F) - 2p(E \cap F).$$

38. For any two events E and F on a sample space S, show

$$p(E \cap F) \le p(E) \le p(E \cup F).$$

39. For any two events E and F on a sample space S, show

$$p(E^c \cap F) = p(F) - p(E \cap F).$$

Applications

40. **Bidding on Contracts.** An aerospace firm has 3 bids on government contracts and knows that the contracts are most likely to be divided up among a number of companies. The firm decides that the probability of obtaining exactly one

contract is .6, of exactly two contracts is .15, and of exactly three contracts is .04. What is the probability that the firm will obtain at least one contract? No contracts?

41. **Quality Control.**   An inspection of computers manufactured at a plant reveals that 2% of the monitors are defective, 3% of the keyboards are defective and 1% of the computers have both defects. (a) Find the probability that a computer at this plant has at least one of these defects. (b) Find the probability that a computer at this plant has none of these defects.

42. **Medicine.**   A new medication produces headaches in 5% of the users, upset stomach in 15%, and both in 2%. (a) Find the probability that at least one of these side effects is produced. (b) Find the probability that neither of these side effects is produced.

43. **Manufacturing.**   A manufactured item is guaranteed for one year and has three critical parts. It has been decided that during the first year the probability of failure of the first part is .03, of the second part .02, the third part .01, both the first and second .005, both the first and third .004, both the second and third .003, and all three parts .001.
   a.   What is the probability that exactly one of these parts will fail in the first year?
   b.   What is the probability that at least one of these parts will fail in the first year?
   c.   What is the probability that none of these parts will fail in the first year?

44. **Marketing.**   A survey of business executives found that 40% read *Business Week*, 50% read *Fortune*, 40% read *Money*, 17% read both *Business Week* and *Fortune*, 15% read both *Business Week* and *Money*, 14% read both *Fortune* and *Money*, and 8% read all three of these magazines.
   a.   What is the probability that one of these executives reads exactly one of these three magazines?
   b.   What is the probability that one of these executives reads at least one of these three magazines?
   c.   What is the probability that one of these executives reads none of these three magazines?

45. **Advertising.**   A firm advertises three different products, A, B, C, on television. From past experience, it expects 1.5% of listeners to buy exactly one of the products, 1% to buy exactly two of the products, 1.2% to buy A, .4% to buy both A and B, .3% to buy both A and C, and .6% to buy A but not the other two.
   a.   Find the probability that a listener will buy only B or only C.
   b.   Find the probability that a listener will buy all three.
   c.   Find the probability that a listener will buy both B and C.
   d.   Find the probability that a listener will buy none of the three.

46. **Sales.**   A salesman always makes a sale at one of three stops in Atlanta and 30% of the time makes a sale at only the first stop, 15% at only the second stop, 20% at only the third stop, and 35% of the time at exactly two of the stops. Find the probability that the salesman makes a sale at all three stops in Atlanta.

Solutions to Self-Help Exercise Set 7.3

1. If $S = \{a, b, c\}$ and $p(a) = p(b) = p(c)$, then from Property 6

$$1 = p(a) + p(b) + p(c) = 3p(a).$$

Thus $p(a) = \frac{1}{3}$.

2. **a.** Let E be the event that the company obtains the first contract and let F be the event that the company obtains the second contract. The event that the company obtains the first contract but not the second is $E \cap F^c$, while the event that the company obtains the second contract but not the first is $E^c \cap F$. These two sets are mutually exclusive, so the probability that the company receives exactly one of the contracts is

$$p(E \cap F^c) + p(E^c \cap F).$$

Now $p(E) = .40$, $p(F) = .30$, and since $E \cap F$ is the event that the company receives both contracts, $p(E \cap F) = .10$.

Notice on the accompanying diagram that $E \cap F^c$ and $E \cap F$ are mutually disjoint and that $(E \cap F^c) \cup (E \cap F) = E$. Thus

$$p(E \cap F^c) + p(E \cap F) = p(E)$$

$$p(E \cap F^c) + .10 = .40$$

$$p(E \cap F^c) = .30.$$

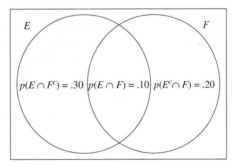

Also notice on the accompanying diagram that $E^c \cap F$ and $E \cap F$ are mutually disjoint and that $(E^c \cap F) \cup (E \cap F) = F$. Thus

$$p(E^c \cap F) + p(E \cap F) = p(F)$$

$$p(E^c \cap F) + .10 = .30$$

$$p(E^c \cap F) = .20.$$

Thus the probability that the company will receive exactly one of the contracts is

$$p(E \cap F^c) + p(E^c \cap F) = .30 + .20 = .50.$$

2. b. The event that the company obtains neither contracts is given by $(E \cup F)^c$. From the diagram

$$p(E \cup F) = .30 + .10 + .20 = .60.$$

Thus

$$p((E \cup F)^c) = 1 - p(E \cup F) = 1 - .60 = .40.$$

The probability that the company receives neither contract is .40.

3. From the formula for odds, we have

$$\text{odds for } (E \cap F) = \frac{.10}{1 - .10} = \frac{1}{9}.$$

Thus the odds are 1 to 9 that the company will obtain both contracts.

7.4   SPACES WITH EQUALLY LIKELY OUTCOMES

▶ *Counting Techniques in Probability*

▶ *A Nonintuitive Example*

A bin contains 15 identical (to the eye) transistors except that 6 are defective and 9 are not. What is the probability of selecting 5 transistors from this bin with 2 defective and 3 not defective? See Example 3 for the answer.

Counting Techniques in Probability

In this section we consider sample spaces in which the probability of any one elementary event is the same as any other. We shall see that for such spaces finding the probability reduces to counting the numbers in sets. To do this we will rely on the counting techniques that were given in the last chapter.

In this section we assume that the space S is finite and has n distinct elements with $S = \{s_1, s_2, \ldots, s_n\}$. We furthermore assume that the space S is a **uniform sample space**, that is, the probability of each of the elementary events is the same. Thus

$$p(s_1) = p(s_2) = \cdots = p(s_n).$$

If we let p be this common probability, then from Property 5 of the last section we have

$$1 = p(s_1) + p(s_2) + \cdots + p(s_n) = np.$$

Thus

$$p = \frac{1}{n}.$$

Now give any event $E = \{e_1, e_2, \ldots, e_r\}$, we have, using Property 4 of the last section, that

$$p(E) = p(\{e_1, e_2, \ldots, e_r\})$$

$$= p(e_1) + p(e_2) + \cdots + p(e_r)$$

$$= r\frac{1}{n}$$

$$= \frac{n(E)}{n(S)}$$

Thus we have the following.

Probability of an Event in the Uniform Sample Space

If S is a finite uniform sample space and E is any event, then

$$p(E) = \frac{\text{Number of elements in E}}{\text{Number of elements in S}} = \frac{n(E)}{n(S)}.$$

E X A M P L E   1   **Calculating Probability in a Uniform Sample Space**

A bin contains 15 identical (to the eye) transistors except that 6 are defective and 9 are not. What is the probability that a transistor selected at random is defective?

Solution   We can do an equivalent problem in which the 15 transistors are labeled from 1 to 15 in a manner that the human eye cannot detect, with the defective ones numbered from 1 to 6. Then the sample space is $S = \{1, 2, \ldots, 15\}$. Since the transistors are identical to the eye, the probability of selecting any one is the same as selecting any other. Since the event "the transistor is defective" is given by $E = \{1, 2, 3, 4, 5, 6\}$, we have

$$p(E) = \frac{n(E)}{n(S)} = \frac{6}{15} = .4. \quad \blacksquare$$

E X A M P L E   2   **Calculating Probability Using Counting Techniques**

Suppose one transistor is selected from the bin in Example 1 and then another is selected without replacing the first. What is the probability that both transistors are defective?

Solution   In order to solve this problem, we need to first know the sample space. The sample space S consists of some subset of the ordered pairs (x, y) with x and y integers from 1 to 15. We again label the transistors as before. Since order is not important, there are

$$C(15, 2) = \frac{15 \cdot 14}{2 \cdot 1} = 105$$

ways of selecting 2 transistors from the bin of 15. Thus there are 105 elementary events in the sample space.

Now we must decide what is the probability of each of these 105 elementary events. One might consider whether some particular selection, such as $\{2, 4\}$, is any more or less likely than another selection, say $\{3, 7\}$. Considering the physical set-up, there does not appear to be any reason to argue that one is any more or less likely than the other. Thus it would appear that the probability of each of the 105 elementary events is the same. In any such situation, we will always *assume* that this is the case, if not explicitly stated.

Then, since we are assuming the uniform probability, we count the number of ways of selecting 2 defective transistors from the 6 in the bin and divide by 105, which is the number of ways of selecting 2 transistors from the bin of 15. If E is the event that both transistors are defective, we have

$$p(E) = \frac{n(E)}{n(S)} = \frac{C(6, 2)}{C(15, 2)} = \frac{\frac{6 \cdot 5}{2 \cdot 1}}{105} = \frac{15}{105} = \frac{1}{7}. \quad \blacksquare$$

E X A M P L E   3   **Calculating Probability Using Counting Techniques**

What is the probability of selecting 5 transistors from the bin in Example 1 with 2 defective and 3 not defective?

Solution   The number of ways of selecting 5 from the 15 is

$$C(15, 5) = \frac{15 \cdot 14 \cdot 13 \cdot 12 \cdot 11}{5 \cdot 4 \cdot 3 \cdot 2 \cdot 1} = 7 \cdot 13 \cdot 3 \cdot 11 = 3003.$$

This is the number in the sample space S.

The number of ways of selecting 2 defectives from 6 is $C(6, 2)$, while the number of ways of selecting 3 nondefective ones from 9 is $C(9, 3)$. Thus by the multiplication principle, the number of ways of doing both is

$$C(6, 2) \cdot C(9, 3) = \frac{6 \cdot 5}{2 \cdot 1} \frac{9 \cdot 8 \cdot 7}{3 \cdot 2 \cdot 1} = 1260.$$

If E is the probability of selecting 2 defective transistors and 3 nondefective ones, then

$$p(E) = \frac{n(E)}{n(S)} = \frac{1260}{3003} \approx .420. \quad \blacksquare$$

E X A M P L E   4   **Using Counting**

Two fair dice are tossed. Assuming that the probability of every elementary event shown in Example 2 of Section 7.1 is equally likely, find the probability that a sum shows that is equal to (a) 3, (b) 7.

Solutions   We first comment that the assumption that any elementary event is just as likely as any other seems reasonable since it does not seem possible to argue based on the physical set-up how one outcome, such as (2, 4), can be any more or less likely than any other outcome, such as (3, 6).

The total number of outcomes is $n(S) = 36$.

(a)   The event "the sum is 3" is given by $E = \{(1, 2), (2, 1)\}$. Thus

$$p(E) = \frac{n(E)}{n(S)} = \frac{2}{36} = \frac{1}{18}.$$

(b)   The event "the sum is 7" is given by

$$F = \{(1, 6), (2, 5), (3, 4), (4, 3), (5, 2), (6, 1)\}.$$

Thus

$$p(F) = \frac{n(F)}{n(S)} = \frac{6}{36} = \frac{1}{6}. \quad \blacksquare$$

E X A M P L E   5   **Using Counting Techniques**

A fair coin is tossed 6 times. Assuming that any outcome is as likely as any other, find the probability of obtaining exactly 3 heads.

Solution   Again the physical set-up implies the reasonableness of the assumption that each of the elementary events is equally likely.

By the multiplication principle of the last chapter, there are $2^6 = 64$ possible outcomes. The number of ways of obtaining exactly 3 heads is the number of ways of selecting 3 slots from among 6 to place the heads. This is

$$C(6, 3) = \frac{6 \cdot 5 \cdot 4}{3 \cdot 2 \cdot 1} = 20.$$

Thus if E is the event that exactly 3 heads occur, then

$$p(E) = \frac{n(E)}{n(S)} = \frac{20}{64} = \frac{5}{16}. \quad \blacksquare$$

E X A M P L E   6   **The Probability of a Poker Hand**

Find the probability of drawing a flush, but not a straight flush, in a poker game, assuming that any 5-card hand is just as likely as any other.

Solution   A deck in poker has 52 cards. Thus there are

$$C(52, 5) = \frac{52 \cdot 51 \cdot 50 \cdot 49 \cdot 48}{5 \cdot 4 \cdot 3 \cdot 2 \cdot 1} = 2{,}598{,}960$$

possible hands.

A flush consists of 5 cards in a single suit. There are 4 suits, each of 13 cards. Thus the number of ways of obtaining a flush in a particular suit is

$$C(13, 5) = \frac{13 \cdot 12 \cdot 11 \cdot 10 \cdot 9}{5 \cdot 4 \cdot 3 \cdot 2 \cdot 1} = 1287.$$

A straight consists of 5 cards in sequence. There are 10 such straight flushes in each suit: $\{1, 2, 3, 4, 5\}, \ldots, \{10, J, Q, K, A\}$. Thus the number of flushes that are not straights is $4(1287 - 10) = 5108$. Thus if E is the event of drawing a flush, but not a straight flush,

$$p(E) = \frac{n(E)}{n(S)} = \frac{5108}{2{,}598{,}960} \approx .0020. \quad \blacksquare$$

A Nonintuitive Example

We now calculate the probability of a certain event and obtain a very surprising (nonintuitive) result.

E X A M P L E   7   **The Birthday Problem**

Suppose there are 50 people in a room with you. What is the probability that at least two of these people will have the same birthday?

Solution   Suppose more generally there are n people in the room. Let E be the event that at least two of these people will have the same birthday. We will ignore leap years and assume that every one of the 365 days of the year is just as likely to be a birthday as any other. It is much easier to first find $p(E^c)$, the probability that no two have the same birthday.

First notice that there are 365 possible birthdays for each individual. Thus, by the multiplication principle, there are 365^n possible birthdays for the n individuals. This is the total number in the sample space. To find $p(E^c)$ notice that there are 365 possible birthdays for the first individual and since the second individual cannot have the same birthday as the first, there are 364 possible birthdays for the second, and then 363 for the third, and so on. We have

$$p(E) = 1 - p(E^c) = 1 - \frac{(365)(364) \cdots (365 - n + 1)}{365^n}.$$

The table gives this for several values of n. Notice the very surprising result that for n equal to only 23 the probability is about .51.

n	15	20	23	30	50
$p(E)$	.25	.41	.51	.71	.97

With 50 people in the room the probability is approximately .97 that 2 or more of these people will have the same birthday.   ■

SELF-HELP EXERCISE SET 7.4

1. Two tetrahedrons (4 sided), each with equal sides, are identical except that one is red and the other white. Both have the four sides numbered from 1 to 4. The two tetrahedrons are tossed, and the number on the bottom face of each is observed. Let E consist of those outcomes for which both tetrahedrons show an odd number. Let F be the event that the sum of the two numbers on these tetrahedrons is 5. Let G be the event that the sum of the two numbers is less than 7. (a) Find $p(E)$. (b) Find $p(F)$. (c) Find $p(E \cap F)$. (d) Find $p(E \cup F)$. (e) Find $p(G^c)$. (Refer to Self-Help Exercises 1 and 2 of Section 7.1.)

2. A lotto game consists of picking (in any order) the correct 6 numbers drawn from 1 to 42 without replacement. (a) What is the probability of any one pick winning? (b) What is the probability of a winning pick being all consecutive numbers?

EXERCISE SET 7.4

In all the following exercises assume that all elementary events in the same sample space are equally likely.

1. A fair coin is flipped 3 times. What is the probability of obtaining exactly 2 heads? At least 1 head?

2. A family has 3 children. Assuming a boy is as likely as a girl to have been born, what is the probability that 2 are boys and 1 is a girl? That at least one is a boy?

3. A fair coin is flipped and a fair die is tossed. What is the probability of obtaining a head and a 3?

4. A fair coin is flipped twice and a fair die is tossed. What is the probability of obtaining 2 heads and a 3?

5. A pair of fair dice are tossed. What is the probability of obtaining a sum of 2? 4? 8?

6. A pair of fair dice are tossed. What is the probability of obtaining a sum of 5? 6? 11?

In Exercises 7 through 8, let an urn have 10 balls, identical except that 4 are white and 6 are red.

7. If 3 are selected randomly without replacement, what is the probability that 2 are white and 1 is red? At least 2 are white?

8. If 5 are selected randomly without replacement, what is the probability that 3 are white and 2 are red? At least 3 are white?

In Exercises 9 through 12, let an urn have 21 identical balls except that 6 are white, 7 are red, and 8 are blue.

9. What is the probability that there is one of each color if 3 are selected randomly without replacement?

10. What is the probability that all are white if 3 are selected randomly without replacement?

11. What is the probability that 3 are white, 2 are red, and 1 is blue if 6 are selected randomly without replacement?

12. What is the probability that at least 5 are white if 6 are selected randomly without replacement?

In Exercises 13 through 18 a 2-card hand is drawn from a standard deck of 52 cards. Find the probability that the hand contains the given cards.

13. Two kings

14. Two spades

15. A pair

16. Two of the same suit

17. Two consecutive cards

18. No face card

In Exercises 19 through 26 find the probability of obtaining each of the given in a 5-card poker hand. (*Hint:* the probabilities increase.)

19. Royal flush: ace, king, queen, jack, ten in the same suit

20. Straight flush: five cards in sequence in the same suit but not a royal flush

21. Four of a kind: four queens, four sevens, etc.

22. Full house: three of a kind together with a pair

23. Straight: five cards in sequence not all in the same suit

24. Three of a kind

25. Two pairs

26. One pair

27. Assume that the probability of an individual being born in any month is the same and that there are n individuals in a room. Find the probability that at least two individuals have their birthdays in the same month when $n = 2, 3, 4, 5$.

28. Suppose n different letters have been written with n corresponding addressed envelopes, and the letters are inserted *randomly* into the envelopes. What is the probability that no letter gets into its correct envelope for $n = 2, 3, 4, 5$?

29. What is the probability that at least two members of the 434-member United States House of Representatives have their birthdays on the same day?

30. What is the probability that at least 2 of the 100 Senators of the U.S. Congress have the same birthday?

Applications

31. Quality Control. A bin has 4 defective transistors and 6 nondefective ones. If 2 are picked randomly from the bin, what is the probability that both are defective?

32. Stock Selection. Among a group of 20 stocks, suppose that 10 stocks will perform above average and the other 10 below average. If you pick 3 stocks from this group, what is the probability that all 3 will be above average in performance?

33. Mutual Funds. Suppose in any year a certain mutual fund is just as likely to perform above average as not. Find the probability that this fund will perform above average in at least 8 of the next 10 years.

34. Committees. A committee of 3 is to be selected at random from a group of 3 senior and 4 junior executives. What is the probability that the committee will have more senior than junior executives?

35. Testing. A company places a dozen of the same product in one box. Before sealing, 3 of the product are tested. If any of the 3 is defective, the entire box will be rejected. Suppose a box has 2 defective products. What is the probability the box will be rejected?

36. Awarding of Contracts. Suppose that there are 3 corporations competing for 4 different government contracts. If the contracts are awarded randomly, what is the probability that each corporation will get a contract?

Solutions to Self-Help Exercise Set 7.4

1. Referring to Solutions to Self-Help Exercise Set 7.1, Exercises 1 and 2, we see that $n(S) = 16$, $n(E) = 4$, $n(F) = 4$, $n(E \cap F) = 0$, $n(E \cup F) = 8$, $n(G^c) = 3$. Furthermore there is no reason to assume that any of the elementary events in S are any more or less likely than any other. So we assume the uniform probability. Thus

$$p(E) = \frac{n(E)}{n(S)} = \frac{4}{16} = \frac{1}{4}$$

$$p(F) = \frac{n(F)}{n(S)} = \frac{4}{16} = \frac{1}{4}$$

$$p(E \cap F) = \frac{n(E \cap F)}{n(S)} = \frac{0}{16} = 0$$

$$p(E \cup F) = \frac{n(E \cup F)}{n(S)} = \frac{8}{16} = \frac{1}{2}$$

$$p(G^c) = \frac{n(G^c)}{n(S)} = \frac{3}{16}.$$

2. a. The number in the sample space S is the number of ways of selecting 6 objects (without replacement) from a set of 42 where order is not important. This is

$$C(42, 6) = \frac{42 \cdot 41 \cdot 40 \cdot 39 \cdot 38 \cdot 37}{6 \cdot 5 \cdot 4 \cdot 3 \cdot 2 \cdot 1} = 5{,}245{,}786.$$

Thus the probability of any one pick is

$$\frac{1}{5{,}245{,}786}.$$

2. b. The picks in which the numbers are consecutive are

$$\{1, 2, 3, 4, 5, 6\}, \ldots, \{37, 38, 39, 40, 41, 42\}.$$

There are 37 such selections. Thus the probability of any one of these being the winning number is

$$\frac{37}{5{,}245{,}786}.$$

7.5   CONDITIONAL PROBABILITY

▶ *Conditional Probability*

▶ *Product Rule*

▶ *Probability Trees*

▶ *Independent Events*

APPLICATION

Locating Defective Parts

A company makes the components for a product at a central location. These components are shipped to 3 plants, 1, 2, and 3, for assembly into a final product. The percentages of the product assembled by the 3 plants are, respectively, 50%, 20%, and 30%. The percentages of defective products coming from these 3 plants are, respectively, 1%, 2%, and 3%. What is the probability of randomly choosing a product made by this company that is defective from Plant 1? See Example 5 for the answer.

Conditional Probability

The probability of an event is often affected by the occurrences of other events. For example, there is a certain probability that an individual will die of lung cancer. But if a person smokes heavily, then the probability that this person will die of lung cancer is higher. That is, the probability has changed with the additional information. In this section we study such *conditional probabilities*. This important idea is further considered in the next section.

Given two events E and F, we call the probability that E will occur given that F has occurred the **conditional probability** and write $p(E|F)$.

Conditional probability normally arises in situations where the old probability is "updated" based on new information. Suppose, for example, a new family moves in next door. The real estate agent mentions that this new family has two children. Based on this information, you can calculate the probability that both children are boys. Now a neighbor mentions that they met the oldest child from the new family and this child is a boy. Now the probability that both children from the new family are boys has changed given this new information.

E X A M P L E   1   **Finding Conditional Probability**

A new family has moved in next door and is known to have two children. (a) Find the probability that both children are boys. (b) Find the probability that both children are boys given that the oldest is a boy. Assume that a boy is as likely as a girl.

Solutions
(a) With no information at all about either child, the sample space is

$$S = \{BB, BG, GB, GG\}$$

where the first element in each ordered pair refers to the youngest child and the second element to the oldest. We know from the last section that it is reasonable to assign uniform probability to this sample space of 4 outcomes and if $E = \{BB\}$, $p(E) = \frac{1}{4}$.

(b) Given the information that the oldest child is a boy, the new sample space becomes $S_2 = \{BB, GB\}$. There is no reason to assign anything other than the uniform probability to this new sample space with 2 outcomes. Doing this we then obtain the probability of the event $\{BB\}$ in the new sample space is $\frac{1}{2}$. That is, if $F = \{BB, GB\}$, $p(E|F) = \frac{1}{2}$. ∎

E X A M P L E   2   **Finding Conditional Probability**

A card is drawn randomly from a deck of 52 cards. (a) What is the probability that this card is an ace? (b) What is the probability that this card is an ace given that the card is known to be a 10 or higher?

Solutions

(a)   The sample space consists of 52 cards and has the uniform probability. Thus, from the last section, if $E = \{ace\}$, $p(E) = \frac{4}{52}$.

(b)   With the new information, the new sample space S_2 is now the space consisting of the 20 cards that are 10 or higher. Since there is no reason to assume that any of these is any more or less likely than any other, we assume the uniform probability on S_2. Letting F be the event in S consisting of all cards 10 or higher (which of course is also S_2), then the probability of obtaining an ace, given that the card is a 10 or higher, is

$$p(E|F) = \frac{4}{20}. \quad \blacksquare$$

We have seen in the previous two examples that finding the conditional probability requires identifying the *new* sample space and determining the *new* probability. The procedure used illustrated these important points but is not adequate in general. We now see how to define conditional probability in a more precise and useful manner.

To do this we return to our intuitive notions of probability. Suppose an experiment yields a sample space S, and E and F are two events in S with $p(E \cap F) = .12$ and $p(F) = .36$ as shown in Figure 7.7. If $N = 100$ is the number of times the experiment has been repeated, then we expect the frequency of occurrence of $E \cap F$ and F to be $f(E \cap F) = 12$ and $f(F) = 36$. The conditional probability $p(E|F)$ is the expected fraction of those experiments in which F occurred that E also occurred. Therefore we expect

$$p(E|F) = \frac{f(E \cap F)}{f(F)} = \frac{12}{36} = \frac{1}{3}.$$

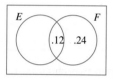

Figure 7.7

Another instructive way of viewing this is to divide the numerator and the denominator of the last fraction by $N = 100$ and obtain

$$p(E|F) = \frac{f(E \cap F)/100}{f(F)/100} = \frac{.12}{.36} = \frac{p(E \cap F)}{p(F)}.$$

More generally, suppose we are given a sample space S and any two events E and F with $p(F) > 0$. If N is the number of times the experiment is repeated and N is very large, then we expect that

$$\frac{f(F)}{N} \approx p(F) \quad \text{and} \quad \frac{f(E \cap F)}{N} \approx p(E \cap F).$$

On the other hand, the conditional probability $p(E|F)$ is the expected fraction of those experiments in which F occurred that E also occurred. We therefore expect

$$p(E|F) \approx \frac{f(E \cap F)}{f(F)}.$$

We now divide the numerator and denominator of the last fraction by N and obtain

$$p(E|F) \approx \frac{f(E \cap F)}{f(F)} = \frac{f(E \cap F)/N}{f(F)/N} \approx \frac{p(E \cap F)}{p(F)}.$$

This motivates the following definition.

Conditional Probability

Let E and F be two events in a sample space S. The conditional probability that E occurs given that F has occurred is defined to be

$$p(E|F) = \frac{p(E \cap F)}{p(F)} \quad \text{if} \quad p(F) > 0.$$

REMARK. It is worthwhile to notice that the events E, F, and $E \cap F$ are all in the original sample space S and $p(E)$, $p(F)$, and $p(E \cap F)$ are all probabilities defined on S. However $p(E|F)$ is a probability defined on the new sample space $S' = F$. See Figure 7.8.

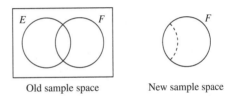

Old sample space   New sample space

Figure 7.8

E X A M P L E   3

Calculating a Conditional Probability

A new family has moved in next door and is known to have two children. Find the probability that both children are boys given that at least one is a boy. Assume that a boy is as likely as a girl.

Solution   Let E be the event that both children are a boy and F the event that at least one is a boy. Then

$$S = \{BB, BG, GB, GG\},$$

$E = \{BB\}$, $F = \{BB, BG, GB\}$, $E \cap F = \{BB\}$, and

$$p(E|F) = \frac{p(E \cap F)}{p(F)} = \frac{\frac{1}{4}}{\frac{3}{4}} = \frac{1}{3}. \quad \blacksquare$$

Product Rule

We will now see how to write $p(E \cap F)$ in terms of a product of two probabilities. From the definition of conditional probability, we have

$$p(E|F) = \frac{p(E \cap F)}{p(F)}, \qquad p(F|E) = \frac{p(F \cap E)}{p(E)},$$

if $p(E) > 0$ and $p(F) > 0$. Solving for $p(E \cap F)$ and $p(F \cap E)$, we obtain

$$p(E \cap F) = p(F)p(E|F), \qquad p(F \cap E) = p(E)p(F|E).$$

Since $E \cap F = F \cap E$, it follows that

$$p(E \cap F) = p(F)p(E|F) = p(E)p(F|E).$$

This is called the product rule.

Product Rule

If E and F are two events in a sample space S with $p(E) > 0$ and $p(F) > 0$, then

$$p(E \cap F) = p(F)p(E|F) = p(E)p(F|E).$$

E X A M P L E 4   **Using the Product Rule**

Two bins contain transistors. The first has 5 defective and 15 nondefective while the second has 3 defective and 17 nondefective ones. If the probability of picking either bin is the same, what is the probability of picking the first bin and a good transistor?

Solution   The sample space is $S = \{1D, 1N, 2D, 2N\}$ where the number refers to picking the first or second bin and the letter refers to picking a defective (D) or nondefective (N) transistor.

If E is the event "pick the first bin" and F is the event "pick a nondefective transistor," then $E = \{1D, 1N\}$ and $F = \{1N, 2N\}$. The probability of picking a nondefective transistor given that the first bin has been picked is the conditional probability $p(F|E) = \frac{15}{20}$. The event "picking the first bin and a nondefective transistor" is $E \cap F$. From the product rule

$$p(E \cap F) = p(E)p(F|E) = \frac{1}{2}\frac{15}{20} = \frac{3}{8}. \quad \blacksquare$$

Probability Trees

We shall now consider a finite sequence of experiments in which the outcomes and associated probabilities of each experiment depend on the outcomes of the preceding experiments. Such a finite sequence of experiments is called a **finite stochastic process**. Stochastic processes can be effectively described by probability trees that we now consider. The following example should be studied carefully since we will return to it at the beginning of the next section.

E X A M P L E 5   **Using a Probability Tree**

A company makes the components for a product at a central location. These components are shipped to 3 plants, 1, 2, and 3, for assembly into a final product. The percentages of the product assembled by the 3 plants are, respectively, 50%,

20%, and 30%. The percentages of defective products coming from these 3 plants are, respectively, 1%, 2%, and 3%. (a) What is the probability of randomly choosing a product made by this company that is defective from Plant 1? 2? 3? (b) What is the probability of randomly choosing a product made by this company that is defective?

Before proceeding, some notational conventions will be discussed that will be used in this problem and others. The sample space is

$$S = \{1N,\ 1D,\ 2N,\ 2D,\ 3N,\ 3D\},$$

where for example "$1D$" means selecting Plant 1 and a defective product from this plant. Thus the event "Plant 1 has been selected" is given by $\{1N,\ 1D\}$. However, this is rather cumbersome and we follow the usual convention and just call this event "1." The event "a defective product has been selected" is given by $\{1D,\ 2D,\ 3D\}$. Again this is rather inconvenient, and we just refer to this event as "D." Then the event "select a defective product from Plant 1" is the intersection of the previous two events, expressed in our shorthand notation as

$$1 \cap D = \{1N,\ 1D\} \cap \{1D,\ 2D,\ 3D\} = \{1D\}.$$

Solutions   Using these conventions, the probability of the part being assembled in Plant 1 is $p(1) = .5$, in Plant 2 is $p(2) = .2$, and in Plant 3 is $p(3) = .3$. We are also given the conditional probabilities, $p(D|1) = .01$, $p(D|2) = .02$, $p(D|3) = .03$. This is all shown in the tree diagram in Figure 7.9.

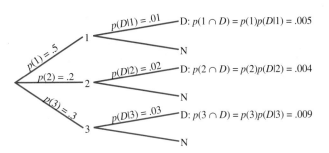

Figure 7.9

(a)   Use the product rule and obtain

$$p(1D) = p(1 \cap D) = p(1)p(D|1) = (.5)(.01) = .005$$

$$p(2D) = p(2 \cap D) = p(2)p(D|2) = (.2)(.02) = .004$$

$$p(3D) = p(3 \cap D) = p(3)p(D|3) = (.3)(.03) = .009.$$

(b)   Using (a) gives

$$p(D) = p(\{1D,\ 2D,\ 3D\}) = p(1D) + p(2D) + p(3D)$$

$$= .005 + .004 + .009 = .018. \quad \blacksquare$$

E X A M P L E   6   **Using a Probability Tree**

A box contains 3 defective (D) parts and 4 nondefective (N) ones. One randomly selects a part (without replacement) until a nondefective one is obtained. What is the probability that the number of parts selected is (a) one, (b) two, (c) three?

Solutions

(a)   There are 4 nondefective parts in a box of 7. So the probability of selecting a nondefective part is $\frac{4}{7}$.

   We note for further reference that the probability of selecting a defective part is $\frac{3}{7}$.

(b)   The only way it can take 2 selections to obtain a nondefective part is for the first selection to be defective. In this case there are 2 defective and 4 nondefective parts left in the box. Thus the probability of selecting a nondefective part given that a defective part was chosen first is $\frac{4}{6}$. From the tree in Figure 7.10, the probability of the branch DN is $\frac{3}{7}\frac{4}{6} = \frac{2}{7}$.

   We note for further reference that the probability of selecting a defective part on the second selection given that a defective part was selected first is $\frac{2}{6}$. See Figure 7.10.

(c)   The only way it can take 3 selections to obtain a nondefective part is for the first two selections to be defective. In this case there are 1 defective and 4 nondefective parts left in the box. Thus the probability of selecting a nondefective part given that a defective part was chosen the first and second time is $\frac{4}{5}$. From the tree in Figure 7.10, the probability of the branch DDN is $\frac{3}{7}\frac{2}{6}\frac{4}{5} = \frac{4}{35}$.   ∎

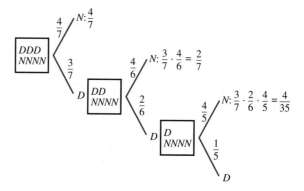

Figure 7.10

Independent Events

Intuitively, two events E and F are **independent** if the outcome of one does not affect the outcome of the other. For example, the probability of obtaining a head on a second flip of a coin is independent from what happened on the first flip. This is intuitively clear since the coin cannot have any memory of what happened on the first flip. Indeed the laws of physics determine the probability of heads occurring. Thus for the probability of heads to be different on the second flip the laws of physics must be different on the second flip. On the other hand, the probability of selecting an ace on the second draw (without replacement) clearly depends on what happens on the first draw. Thus drawing an ace on the second draw without replacement is not independent from drawing an ace on the first draw.

Intuitively, two events E and F are independent if

$$p(E|F) = p(E) \quad \text{and} \quad p(F|E) = p(F),$$

that is, the probability of E given that F has occurred is just the probability of E, and, similarly, the probability of F given that E has occurred is just the probability of F.

Independent Events

Two events E and F are said to be independent if

$$p(E|F) = p(E) \quad \text{and} \quad p(F|E) = p(F)$$

We shall now obtain a result that is more convenient to apply when attempting to determine if two events are independent.

If then two events E and F are independent, the previous comments together with the product rule indicate that

$$p(E \cap F) = p(E|F)p(F) = p(E)p(F).$$

Now consider the case that $p(E) > 0$ and $p(F) > 0$. Assume that $p(E \cap F) = p(E)p(F)$, then

$$p(E|F) = \frac{p(E \cap F)}{p(F)} = \frac{p(E)p(F)}{p(F)} = p(E)$$

$$p(F|E) = \frac{p(E \cap F)}{p(E)} = \frac{p(E)p(F)}{p(E)} = p(F).$$

This discussion then yields the following theorem.

Independent Events Theorem

Let E and F be two events with $p(E) > 0$ and $p(F) > 0$. Then E and F are independent if, and only if,

$$p(E \cap F) = p(E)p(F).$$

Although at times one is certain whether or not two events are independent, often one can only tell by doing the calculations.

E X A M P L E   7   **Determining if Two Events are Independent**

A study of 1000 men over 65 indicated that 250 smoked and 50 of these smokers had some signs of heart disease, while 100 of the nonsmokers showed some signs of heart disease. Let E be the event "smokes" and H be the event "has signs of heart disease." Are these two events independent?

Solution   The Venn diagram is given in Figure 7.11. From this diagram $p(H) = .15$, $p(E) = .25$, and $p(H \cap E) = .05$. Thus

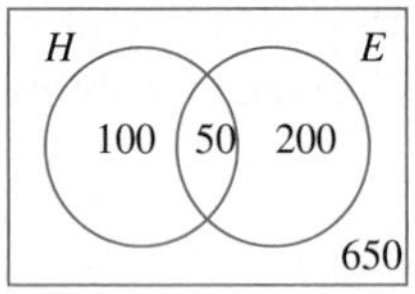

$$p(H)p(E) = (.15)(.25)$$
$$= .0375$$
$$\neq .05$$
$$= p(H \cap E),$$

Figure 7.11

and the two events are not independent.   ■

E X A M P L E   8   **Determining if Two Events are Independent**

Answer the previous question if a study of 1000 men over 65 indicated that 500 smoked and 100 of these smokers have some sign of heart disease, while 100 of the nonsmokers showed some signs of heart disease.

Solution   From the Venn diagram in Figure 7.12, we have

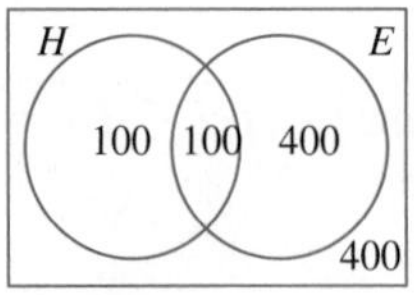

$$p(H)p(E) = (.20)(.50)$$
$$= .10$$
$$= p(H \cap E).$$

Figure 7.12

Thus these events are independent.   ■

W A R N I N G .   Notice that saying two events are independent is not the same as saying that they are mutually exclusive. The sets E and H in both previous examples were not mutually exclusive, but in one case the sets were independent and in the other case they were not.

The notion of independence can be extended to any number of finite sets.

Independent Set of Events

A set of events $\{E_1, E_2, \ldots, E_n\}$ is said to be independent if, for any k of these events, the probability of the intersection of these k events is the product of the probabilities of each of the k events. This must hold for any $k = 2$, $3, \ldots, n$.

For example, for $\{E, F, G\}$ to be independent all of the following must be true:

$$p(E \cap F) = p(E)p(F), \quad p(E \cap G) = p(E)p(G), \quad p(F \cap G) = p(F)p(G),$$

$$p(E \cap F \cap G) = p(E)p(F)p(G).$$

It is intuitively clear that if two events E and F are independent, then so also are E and F^c, E^c and F, E^c and F^c. (See Exercises 38 and 39.) Similar statements are true about a set of events.

E X A M P L E   9   **Independent Events and Safety**

An aircraft has a system of three computers, each independently able to exercise control of the flight. The computers are considered 99.9% reliable during a routine flight. What is the probability of having a failure of the control system during a routine flight?

Solution   Let the events E_i, $i = 1, 2, 3$ be the three events given by the reliable performance of respectively the first, second, and third computer. Since the set of events $\{E_1, E_2, E_3\}$ is independent, so is the set of events $\{E_1^c, E_2^c, E_3^c\}$. The system will fail only if all three computers fail. Thus the probability of failure of the system is given by

$$p(E_1^c \cap E_2^c \cap E_3^c) = p(E_1^c)p(E_2^c)p(E_3^c) = (.001)^3,$$

which, of course, is an extremely small number.   ∎

SELF-HELP EXERCISE SET 7.5

1. Three companies A, B, and C are competing for a contract. The probabilities that they receive the contract are, respectively, $p(A) = \frac{1}{6}$, $p(B) = \frac{1}{3}$, and $p(C) = \frac{1}{2}$. What is the probability that A will receive the contract if C pulls out of the bidding?

2. Two bins contain transistors. The first has 4 defective and 15 nondefective, while the second has 3 defective and 22 nondefective ones. If the probability of picking either bin is the same, what is the probability of picking the second bin and a defective transistor?

3. Success is said to breed success. Suppose you are in a best-of-3-game tennis match with an evenly matched opponent. However, if you win a game, your probability of winning the next increases from $\frac{1}{2}$ to $\frac{2}{3}$. Suppose, however, that if you lose, the probability of winning the next match remains the same. (Success does not breed success for your opponent.) What is your probability of winning the match? *Hint:* Draw a tree.

4. A family has 3 children. Let E be the event "at most one boy" and F the event "at least one boy and at least one girl." Are E and F independent if a boy is as likely as a girl? *Hint:* Write down every element in the sample space S and the events E, F, $E \cap F$, and find the appropriate probabilities by counting.

EXERCISE SET 7.5

In Exercises 1 through 6 refer to the accompanying Venn diagram to find the conditional probabilities.

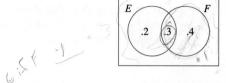

1. **a.** $p(E|F)$,   **b.** $p(F|E)$

2. **a.** $p(E^c|F)$,   **b.** $p(F^c|E)$

3. **a.** $p(E|F^c)$,   **b.** $p(F|E^c)$

4. **a.** $p(E^c|F^c)$,   **b.** $p(E \cup F|F)$

5. **a.** $p(F|E \cap F)$,   **b.** $p(F^c|F)$

6. **a.** $p(E^c \cap F|F)$,   **b.** $p(E \cap F^c|F)$

In Exercises 7 through 12, let $p(E) = .4$, $p(F) = .6$, and $p(E \cap F) = .2$. Draw a Venn diagram and find the conditional probabilities.

7. **a.** $p(E^c|F)$,   **b.** $p(F^c|E)$

8. **a.** $p(E|F)$,   **b.** $p(F|E)$

9. **a.** $p(F^c|E^c)$,   **b.** $p(E \cup F|E)$

10. **a.** $p(E|F^c)$,   **b.** $p(F|E^c)$

11. **a.** $p(E^c \cap F|E)$,   **b.** $p(E \cap F^c|E)$

12. **a.** $p(F|E \cap F)$,   **b.** $p(E^c|E)$

In Exercises 13 through 20, determine if the given events E and F are independent.

13. $p(E) = .3$, $p(F) = .5$, $p(E \cap F) = .2$

14. $p(E) = .5$, $p(F) = .7$, $p(E \cap F) = .3$

15. $p(E) = .2$, $p(F) = .5$, $p(E \cap F) = .1$

16. $p(E) = .4$, $p(F) = .5$, $p(E \cap F) = .2$

17. $p(E) = .4$, $p(F) = .3$, $p(E \cup F) = .6$

18. $p(E \cap F^c) = .3$, $p(E \cap F) = .2$, $p(E^c \cap F) = .2$

19. $p(E \cap F^c) = .3$, $p(E \cap F) = .3$, $p(E^c \cap F) = .2$

20. $p(E) = .2$, $p(F) = .5$, $p(E \cup F) = .6$

21. A pair of fair dice is tossed. What is the probability that a sum of seven has been tossed if it is known that at least one of the numbers is a 3?

22. A single fair die is tossed. What is the probability that a 3 occurs on the top if it is known that the number is a prime?

23. A fair coin is flipped 3 times. What is the probability that heads occurs 3 times if it is known that heads occurs at least once?

24. A fair coin is flipped 4 times. What is the probability that heads occurs 3 times if it is known that heads occurs at least twice?

25. Three cards are randomly drawn without replacement from a standard deck of 52 cards.
    a. What is the probability of drawing an ace on the third draw?
    b. What is the probability of drawing an ace on the third draw given that at least one ace was drawn on the first 2 draws?

26. Three balls are randomly drawn from an urn that contains 4 white and 6 red balls.
    a. What is the probability of drawing a red ball on the third draw?
    b. What is the probability of drawing a red ball on the third draw given that at least one red ball was drawn on the first 2 draws?

27. From the tree diagram find (a) $p(A \cap E)$, (b) $p(A)$, (c) $p(A|E)$.

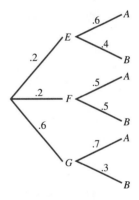

28. From the tree diagram find (a) $p(A \cap E)$, (b) $p(A)$, (c) $p(A|E)$.

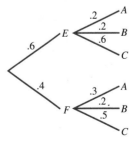

29. An urn contains 5 white, 3 red, and 2 blue balls. Two balls are randomly drawn. What is the probability that one is white and one is red if the balls are drawn (a) without replacement? (b) with replacement after each draw?

30. An urn contains 4 white and 6 red balls. Two balls are randomly drawn. If the first one is white, the ball is replaced. If the first one is red, the ball is not replaced. What is the probability of drawing at least one white ball?

31. In a family of 4 children, let E be the event "at most one boy" and F the event "at least one girl and at least one boy." If a boy is as likely as a girl, are these two events independent?

32. A fair coin is flipped three times. Let E be the event "at most one head" and F the event "at least one head and at least one tail." Are these two events independent?

33. The two events E and F are independent with $p(E) = .3$, and $p(F) = .5$. Find $p(E \cup F)$.

34. The two events E and F are independent with $p(E) = .4$, and $p(F) = .6$. Find $p(E \cup F)$.

35. The three events E, F, and G are independent with $p(E) = .2$, $p(F) = .3$, $p(G) = .5$. What is $p(E \cup F \cup G)$?

36. The three events E, F, and G are independent with $p(E) = .3$, $p(F) = .4$, $p(G) = .6$. What is $p(E^c \cup F^c \cup G^c)$?

37. Given the probabilities shown in the accompanying Venn diagram, show that the events E and F are independent if, and only if,

$$p_1 p_3 = p_2 p_4.$$

What must the Venn diagram look like if the sets are mutually disjoint?

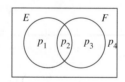

38. Show that if E and F are independent, then so are E and F^c.

39. Show that if E and F are independent, then so are E^c and F^c.

40. Show that two events are indepenent if they are mutually exclusive and the probability of one of them is zero.

41. Show that two events E and F are not independent if they are mutually exclusive and both have nonzero probability.

42. Show that if E and F are independent events, then

$$p(E \cup F) = 1 - p(E^c)p(F^c).$$

43. If $p(F) > 0$, then show that

$$p(E^c | F) = 1 - p(E | F).$$

44. If E, F, G are three events and $p(G) > 0$, show that

$$p(E \cup F | G) = p(E | G) + p(F | G) - p(E \cap F | G).$$

45. If E and F are two events with $F \subset E$, then show that $p(E|F) = 1$.

46. If E and F are two events with $E \cap F = \phi$, then show that $p(E|F) = 0$.

47. If E and F are two events, show that

$$p(E|F) + p(E^c|F) = 1.$$

Applications

48. Manufacturing. Two machines turn out all the products in a factory, with the first machine producing 40% of the product and the second 60%. The first machine produces defective products 2% of the time and the second machine 4% of the time.
  a. What is the probability that a defective product is produced at this factory given that it was made on the first machine?
  b. What is the probability that a defective product is produced at this factory?
  c. Given a defective product, what is the probability it was produced on the first machine?

49. Manufacturing. A plant has 3 assembly lines with the first line producing 50% of the product and the second 30%. The first line produces defective products 1% of the time, the second line 2% of the time, and the third 3% of the time.
  a. What is the probability that a defective product is produced at this plant given that it was made on the second assembly line?
  b. What is the probability that a defective product is produced at this plant?
  c. Given a defective product, what is the probability it was produced on the second assembly line?

50. Advertising. A television ad for a company's product has been seen by 20% of the population. Of those who see the ad, 10% then buy the product. Of those who do not see the ad, 2% buy the product. Find the probability that a person buys the product.

51. Suppliers. A manufacturer buys 40% of a certain part from one supplier and the rest from a second supplier. It notes that 2% of the parts from the first supplier are defective, and 3% are defective from the second supplier. Given that a part is defective, what is the probability it came from the first supplier?

52. Psychology and Sales. A door-to-door salesman expects to makes a sale 10% of the time when starting the day. But making a sale increases his enthusiasm so much that the probability of a sale to the next customer is .2. If he makes no sale, the probability for a sale stays at .1. What is the probability that he will make at least 2 sales with his first 3 visits?

53. Quality Control. A box contains 2 defective (D) parts and 5 nondefective (N) ones. You randomly select a part (without replacement) until you get a nondefective part. What is the probability that the number of parts selected is (a) one, (b) two, (c) three?

54. Medicine. The probability of residents of a certain town contracting cancer is .01. Let x be the percent of residents that work for a certain chemical plant and suppose that the probability of both working for this plant and of contracting cancer is .001. What must x be for the two events "gets cancer" and "works for the chemical plant" to be independent?

55. **Medicine.**   In a study of 250 men over 65, 100 smoked, 60 of the smokers had some signs of heart disease, and 90 of the nonsmokers showed some signs of heart disease. Let E be the event "smokes" and H be the event "has signs of heart disease." Are these two events independent?

56. **Sales.**   A company sells machine tools to two firms in a certain city. In 40% of the years it makes a sale to the first firm, 30% of the years to the second firm, and 10% to both. Are the two events "a sale to the first firm" and "a sale to the second firm" independent?

57. **Stocks.**   A firm checks the last 200 days on which its stock has traded. On 100 of these occasions the stock has risen in price with a broad-based market index also rising on 70 of these particular days. The same market index has risen on 90 of the 200 trading days. Are the movement of the firm's stock and the movement of the market index independent?

58. **Contracts.**   A firm has bids on two contracts. It is known that the awarding of these two contracts are independent events. If the probability of receiving these contracts are .3 and .4 respectively, what is the probability of not receiving either?

59. **Reliability.**   A firm is making a very expensive optical lens to be used in an earth satellite. To be assured that the lens has been ground correctly, three independent tests using entirely different techniques are used. The probability is .99 that any of one of these tests will detect a defect in the lens. What is the probability that the lens has a defect even though none of the three tests so indicates?

(Photo courtesy of NASA.)

Solutions to Self-Help Exercise Set 7.5

1. If E is the event that A obtains the contract and F the event that either A or B obtain the contract, then $E = \{A\}$, $F = \{A, B\}$, and $E \cap F = \{A\}$, and the conditional probability that A will receive the contract if C pulls out of the bidding is

$$p(E|F) = \frac{p(E \cap F)}{p(F)} = \frac{\frac{1}{6}}{\frac{1}{2}} = \frac{1}{3}.$$

2. The sample space is $S = \{1D, 1N, 2D, 2N\}$ where the number refers to picking the first or second bin and the letter refers to picking a defective (D) or nondefective (N) transistor. If E is picking the first bin and F is picking a defective transistor, then

$$p(E \cap F) = p(E)p(F|E) = \frac{1}{2} \frac{3}{25} = \frac{3}{50}.$$

3. The appropriate tree is given. The probability of winning the match is then

$$\frac{1}{3} + \frac{1}{12} + \frac{1}{6} = \frac{7}{12}.$$

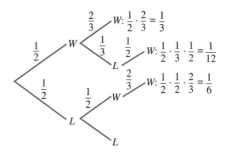

4. The elements in the spaces S, E, F, and $E \cap F$ are

$$S = \{BBB, BBG, BGB, BGG, GBB, GBG, GGB, GGG\}$$

$$E = \{BGG, GBG, GGB, GGG\}$$

$$F = \{BBG, BGB, BGG, GBB, GBG, GGB\}$$

$$E \cap F = \{BGG, GBG, GGB\}.$$

Thus counting elements gives

$$p(E \cap F) = \frac{3}{8} \qquad p(E)p(F) = \frac{4}{8}\frac{6}{8} = \frac{3}{8}.$$

Since these two numbers are the same, the two events are independent.

7.6 BAYES' THEOREM

▶ *Bayes' Theorem*

▶ *Applications*

APPLICATION

Finding a Probability of a Defective Part

Recall Example 5 of the last section. A company makes the components for a product at a central location. These components are shipped to 3 plants, 1, 2, and 3, for assembly into a final product. The percentages of the product assembled by the 3 plants are, respectively, 50%, 20% and 30%. The percentages of defective products coming from these 3 plants are, respectively, 1%, 2%, and 3%. Given a defective product, what is the probability it was assembled at Plant 1? At Plant 2? At Plant 3? See the discussion immediately below for the answers to these questions.

Bayes' Theorem

We have been concerned with finding the probability of an event that will occur in the future. We now look at calculating probabilities after the events have occurred.

We begin this section by answering the question posed above. Refer to Figure 7.13. Recall from the last section that we had

$$p(1D) = p(1 \cap D) = p(1)p(D|1) = (.5)(.01) = .005$$

$$p(2D) = p(2 \cap D) = p(2)p(D|2) = (.2)(.02) = .004$$

$$p(3D) = p(3 \cap D) = p(3)p(D|3) = (.3)(.03) = .009,$$

and thus

$$p(D) = p(\{1D, 2D, 3D\}) = p(1D) + p(2D) + p(3D)$$

$$= .005 + .004 + .009$$

$$= .018.$$

Using this information and the definition of conditional probability, we then have

$$p(1|D) = \frac{p(1 \cap D)}{p(D)} = \frac{.005}{.005 + .004 + .009} = \frac{5}{18}.$$

We now notice that the numerator of this fraction is the probability of the branch $1D$, while the denominator is the sum of all the probabilities of all branches that end in D. We also have

$$p(2|D) = \frac{p(2 \cap D)}{p(D)} = \frac{.004}{.018} = \frac{4}{18}.$$

Now notice that the numerator of this fraction is the probability of the branch $2D$, while the denominator is the sum of all the probabilities of all branches that end in D. A similar statement can be made for

$$p(3|D) = \frac{p(3 \cap D)}{p(D)} = \frac{.009}{.018} = \frac{9}{18}.$$

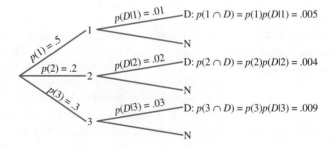

Figure 7.13

We will see that this is an example of **Bayes' theorem**. This theorem was discovered by the Presbyterian minister Thomas Bayes (1702–1763). We will now indicate how to establish this result and the exact conditions needed.

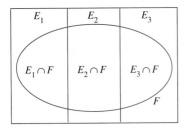

Figure 7.14

We suppose that we are given a sample space S and three mutually exclusive events E_1, E_2, E_3, with $E_1 \cup E_2 \cup E_3 = S$ as indicated in Figure 7.14. Notice that the three events divide the space S into 3 partitions. Given another event F, the tree diagram of possibilities is shown in Figure 7.15. We now wish to show that the probability $p(E_1|F)$ is the fraction with the numerator given by the probability of the branch E_1F while the denominator is the sum of all the probabilities of all branches that end in F.

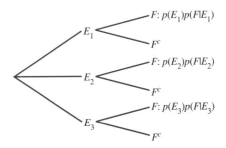

Figure 7.15

To see this first notice from Figure 7.14 that

$$F = (E_1 \cap F) \cup (E_2 \cap F) \cup (E_3 \cap F)$$

and

$$p(F) = p(E_1 \cap F) + p(E_2 \cap F) + p(E_3 \cap F).$$

Using the product rule, this can be written as

$$p(F) = p(E_1)p(F|E_1) + p(E_2)p(F|E_2) + p(E_3)p(F|E_3).$$

On the other hand notice that

$$p(F|E_1) = \frac{p(E_1 \cap F)}{p(E_1)} \quad \text{implies} \quad p(E_1 \cap F) = p(E_1)p(F|E_1).$$

Using this then gives

$$p(E_1|F) = \frac{p(E_1 \cap F)}{p(F)} = \frac{p(E_1)p(F|E_1)}{p(F)}$$

$$= \frac{p(E_1)p(F|E_1)}{p(E_1)p(F|E_1) + p(E_2)p(F|E_2) + p(E_3)p(F|E_3)}.$$

Similar statements can be made about the other probabilities. For example, $p(E_2|F)$ is the fraction with the numerator given by the probability of the branch E_2F, while the denominator is the sum of all the probabilities of all branches that end in F. This gives Bayes' theorem, which is true for any number of events E_i.

Bayes' Theorem

Let $E_1, E_2, \ldots, E_n$, be mutually exclusive events in a sample space S with $E_1 \cup E_2 \cup \cdots \cup E_n = S$. If F is any event in S, then for any $i = 1, 2, \ldots, n$,

$$p(E_i|F) = \frac{\text{probability of branch } E_iF}{\text{sum of all probabilities of all branches that end in } F}$$

$$= \frac{p(E_i)p(F|E_i)}{p(E_1)p(F|E_1) + p(E_2)p(F|E_2) + \cdots + p(E_n)p(F|E_n)}$$

Applications

A very interesting example occurs in medical tests for disease. We begin with a test that gives every appearance of being excellent, but an important consequence may be disappointing.

E X A M P L E 1   **A Medical Application of Bayes' Theorem**

The standard tine test for tuberculosis attempts to identify carriers, that is, people who have been infected by the tuberculin bacteria. The probability of a false negative is .08, that is, the probability of the tine test giving a negative reading to a carrier is $p(-|C) = .08$. The probability of a false positive is .04, that is, the probability of the tine test giving a positive indication when a person is a noncarrier is $p(+|N) = .04$. The probability of a random person in the United States having tuberculosis is .0075. Find the probability that a person is a carrier given that the tine test gives a positive indication.

Solution   The probability we are seeking is $p(C|+)$. Figure 7.16 shows the appropriate tree diagram where C is the event "is a carrier," N the event "is a noncarrier," $+$ the event "test yields positive result." Then Bayes' theorem can be used and $p(C|+)$ is the probability of branch $C+$ divided by the sum of all probabilities that end in $+$. Then

$$p(C|+) = \frac{p(C)p(+|C)}{p(C)p(+|C) + p(N)p(+|N)}$$

$$= \frac{(.0075)(.92)}{(.0075)(.92) + (.9925)(.04)}$$

$$\approx .15.$$

This says that only 15% of people with positive tine test results actually carry TB. ∎

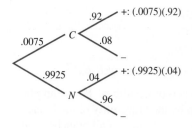

Figure 7.16

This number is surprisingly low. Does this indicate that the test is of little value? As Self-Help Exercise 1 will show, a person whose tine test is negative has a probability of .999 of not having tuberculosis. Such an individual can feel safe. The individuals whose tine test is positive are probably okay also but will need to undergo further tests, such as a chest x-ray.

In some areas of the United States the probability of being a carrier can be as high as .10. The following example examines the tine test under these conditions.

E X A M P L E 2   **A Medical Application of Bayes' Theorem**

Find $p(C|+)$ again when $p(C) = .10$.

Solution   See Figure 7.17. Using Bayes' Theorem exactly as before, we obtain

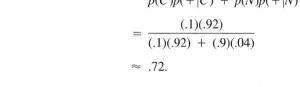

$$p(C|+) = \frac{p(C)p(+|C)}{p(C)p(+|C) + p(N)p(+|N)}$$

$$= \frac{(.1)(.92)}{(.1)(.92) + (.9)(.04)}$$

$$\approx .72.$$

Thus 72% of these individuals who have a positive tine test result will be carriers. ∎

Figure 7.17

Thus in the first example, when the probability of being a carrier is low, the tine test is useful for determining those who do not have TB. In the second example, when the probability of being a carrier is much higher, the tine test is useful for determining those who are carriers, although, naturally, these latter individuals will undergo further testing.

E X A M P L E 3   **An Application of Bayes' Theorem**

Suppose there are only 4 economic theories that can be used to predict expansions and contractions in the economy. By polling economists on their beliefs on which theory is correct, the probability that each of the theories is correct has been determined as follows:

$$p(E_1) = .40, \qquad p(E_2) = .25, \qquad p(E_3) = .30, \qquad p(E_4) = .05.$$

The economists who support each theory then use the theory to predict the likelihood of a recession (R) in the next year. These are as follows:

$$p(R|E_1) = .01, \qquad p(R|E_2) = .02, \qquad p(R|E_3) = .03, \qquad p(R|E_4) = .90.$$

Now suppose a recession actually occurs in the next year. How would the probabilities of the correctness of the fourth and first theories be changed?

Solution   We first note that the fourth theory E_4 has initially a low probability of being correct. Also notice that this theory is in sharp disagreement with the other 3 on whether there will be a recession in the next year.

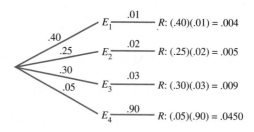

Figure 7.18

Bayes' theorem gives $p(E_4|R)$ as the probability of the branch E_4R divided by the sum of all the probabilities of all the branches that end in R. See Figure 7.18. Similarly for the others. Thus

$$p(E_4|R) = \frac{p(E_4)p(R|E_4)}{p(E_1)p(R|E_1) + p(E_2)p(R|E_2) + p(E_3)p(R|E_3) + p(E_4)p(R|E_4)}$$

$$= \frac{(.05)(.9)}{(.4)(.01) + (.25)(.02) + (.3)(.03) + (.05)(.9)}$$

$$= \frac{.045}{.063} \approx .71,$$

and

$$p(E_1|R) = \frac{p(E_1)p(R|E_1)}{p(E_1)p(R|E_1) + p(E_2)p(R|E_2) + p(E_3)p(R|E_3) + p(E_4)p(R|E_4)}$$

$$= \frac{(.4)(.01)}{(.4)(.01) + (.25)(.02) + (.3)(.03) + (.05)(.9)}$$

$$= \frac{.004}{.063} \approx .06.$$

Thus, given that the recession did occur in the next year, the probability that E_4 is correct has jumped dramatically, while the probability that E_1 is true has plunged. ■

Although this is an artificial example, probabilities are indeed reevaluated in this way based on new information.

SELF-HELP EXERCISE SET 7.6

1. Referring to Example 1 of the text, find the probability that an individual in the United States is not a carrier given that the tine test is negative.

2. Suppose a new tine test had the remarkable property that $p(+|C) = 1$ but $P(+|N)$ is still the same .04. Now find the probability that an individual in the United States is a carrier given a positive reading on this new tine test.

EXERCISE SET 7.6

1. Given the tree diagram, find $p(E_1|F)$ and $p(E_1|F^c)$.

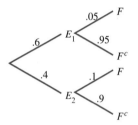

2. Given the same tree diagram as in the previous exercise, find $p(E_2|F)$ and $p(E_2|F^c)$.

3. Given the tree diagram, find $p(E_1|F)$ and $p(E_1|F^c)$.

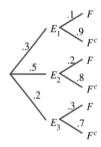

4. Given the same tree diagram as in the previous exercise, find $p(E_3|F)$ and $p(E_3|F^c)$.

5. Two urns each contain 10 balls. The first contains 2 white and 8 red, the second 7 white and 3 red. An urn is selected, and a ball is randomly drawn from this urn. If this ball is white and if the probability of selecting the first urn is twice the probability of selecting the second, find the probability that the first urn was selected.

6. In the previous exercise, find the probability that the second urn was selected.

7. Three urns each contain 10 balls. The first contains 2 white and 8 red balls, the second 5 white and 5 red, and the third all 10 white. An urn is selected, and a ball is randomly drawn from this urn. If this ball is white and if the probability of selecting any urn is the same, find the probability that the first urn was selected.

8. In the previous exercise, find the probability that the third urn was selected.

9. Using the urns in Exercise 5, an urn is selected with the given probabilities and two balls are drawn from this urn (without replacement); both are white. What is the probability that the urn selected was the (a) first one, (b) the second one?

10. Using the urns in Exercise 5, an urn is selected with the given probabilities and one ball is drawn, replaced, and then another drawn. Suppose both are white. What is the probability that the urn selected was the (a) first one, (b) the second one?

11. Using the urns in Exercise 7, an urn is selected with the given probabilities and two balls are drawn from this urn (without replacement); both are white. What is the probability that the urn selected was the (a) first one, (b) the second one?

12. Using the urns in Exercise 7, an urn is selected with the given probabilities and one ball is drawn, replaced, and then another drawn. Suppose both are white. What is the probability that the urn selected was the (a) first one, (b) the second one?

Applications

13. **Medicine.** Do Example 2 of the text if all the information remains the same except that the tine test has the remarkable property that $p(+|C) = 1$. Compare your answer to the one in Example 2.

14. **Medicine.** Using the information in Example 2 of the text, find $p(N|-)$, where $-$ is the event "test shows negative."

15. **Economics.** Using the information in Example 3 of the text, find $p(E_1|R^c)$ and $p(E_4|R^c)$. Compare your answer with $p(E_1)$ and $p(E_4)$.

16. **Manufacturing.** For the example discussed in the beginning of this section (p. 421), find $p(E_1|F^c)$.

17. **Quality Control.** One of two bins is selected at random, one as likely to be selected as the other, and from the bin selected a transistor is chosen at random. The transistor is tested and found to be defective. It is known that the first bin contains 2 defective and 4 nondefective transistors, while the second bin contains 5 defective and 1 nondefective transistors. Find the probability that the second bin was selected.

18. **Quality Control.** Suppose in the previous exercise there is a third bin with 5 transistors, all of which are defective. Now one of the 3 bins is selected at random, one as likely to be selected as any other, and from this bin a transistor is chosen at random. If the transistor is defective, find the probability it came from the third bin.

19. **Quality Control.** A typical box of 100 transistors contains only 1 defective one. It is realized that among the last 10 boxes, one box has 10 defective transistors. An inspector picks one box at random, and the first transistor selected is found to be defective. What is the probability that this box is the bad one?

20. **Quality Control.** A typical box of 100 transistors contains only 1 defective one. It is realized that among the last 10 boxes, one box has 10 defective transistors. An inspector picks one box at random, inspects two transistors from this box on a machine and discovers that one of them is defective and one is not. What is the probability that this box is the bad one?

21. **Manufacturing.** A manufacturing firm has 4 machines that produce the same component. Using the table, given that a component is defective find the probability that the defective component was produced by (a) machine 1, (b) machine 2.

Machine	Percentage of Components Produced	Percentage of Defective Components
1	20	1
2	30	2
3	40	3
4	10	4

22. **Social Sciences.** A man claimed not to be the father of a certain child. On the basis of evidence presented, the court felt that this man was twice as likely to be the father as not, and, hardly satisfied with these odds, required the man to take a blood test. The mother of the child had a different blood type than the child; therefore the blood type of the child was completely determined by the father. If the man's blood type was different than the child, then he could not be the father. The blood type of the child occurred in only 10% of the population. The blood tests indicated that the man had the same blood type as the child. What is the probability that the man is the father?

23. **Medical Diagnosis.** A physician examines a patient and, on the basis of the symptoms, determines that he may have one of four diseases; the probability of each is given in the table. She orders a blood test, which indicates that the blood is perfectly normal. Data are available on the percentage of patients with each disease whose blood tests are normal. On the basis of the normal blood test, find all probabilities that the patient has each disease.

Diseases	Probability of Disease	Percentage of Normal Blood With this Disease
1	.1	60
2	.2	20
3	.3	20
4	.4	10

Solutions to Self-Help Exercise Set 7.6

1. Bayes' theorem can be used if $n = 2$, $E_1 = N$, $E_2 = C$, and $F = -$. Then

$$p(N|-) = \frac{p(N)p(-|N)}{p(N)p(-|N) + p(C)p(-|C)}$$

$$= \frac{(.9925)(.96)}{(.9925)(.96) + (.0075)(.08)}$$

$$\approx .999.$$

2. Using Bayes' theorem we obtain

$$p(C|+) = \frac{p(C)p(+|C)}{p(C)p(+|C) + p(N)p(+|N)}$$

$$= \frac{(.0075)(1)}{(.0075)(1) + (.9925)(.04)}$$

$$\approx .16.$$

This is about the same as before.

7.7   BERNOULLI TRIALS

▶ *Bernoulli Trials*

▶ *Applications*

A P P L I C A T I O N

Finding Probabilities of Nondefective Microchips

A computer manufacturer uses 8 microchips in each of its computers. It knows that 1% of these chips are defective. What is the probability that at least 6 are good? See Example 3 for the answer.

Bernoulli Trials

In this section we consider the simplest possible experiments: those with just two outcomes.

We refer to experiments in which there are just 2 outcomes as **Bernoulli trials**. Some examples are as follows. Flip a coin and see if heads or tails turns up. Test a transistor to see if it is defective or not. Examine a patient to see if a particular disease is present or not. Take a free throw in basketball and make the basket or not.

We commonly refer to the two outcomes of a Bernoulli trial as "success" (S) or "failure" (F). We agree always to write p for the probability of "success" and q for the probability of "failure." Naturally $q = 1 - p$.

In this section we are actually not so much interested in performing an experiment with two outcomes once, but rather many times. We refer to this as a **repeated Bernoulli trial**. We make a very fundamental assumption: The successive Bernoulli trials are independent of one another. Thus, for example, flipping a coin 10 times is a repeated Bernoulli trial. Tossing a die 20 times and seeing if an even number or an odd number occurs each time is another example.

Given a Bernoulli trial repeated n times, we are interested in determining the probability that a specific number of successes occurs. If k is the number of successes, then we denote $b(k : n, p)$ the probability of exactly k successes in n repeated Bernoulli trials where the probability of success is p.

E X A M P L E   1 **Finding the Probability of an Outcome in a Repeated Bernoulli Trial**

Suppose a basketball player makes on average two free throws of every three attempted and that success and failure on any one free throw does not depend on the outcomes of the other shots. If the player shoots 10 free throws, find the probability of making exactly 6 of them, that is, find $b(6 : 10, \frac{2}{3})$.

Solution We let S designate ''making the basket'' and F ''not making the basket.'' A typical sequence of exactly 6 successes looks like

$$SSFSFSSFFS.$$

Now the product rule indicates that the probability of this occurring is

$$\left(\frac{2}{3}\right)\left(\frac{2}{3}\right)\left(\frac{1}{3}\right)\left(\frac{2}{3}\right)\left(\frac{1}{3}\right)\left(\frac{2}{3}\right)\left(\frac{2}{3}\right)\left(\frac{1}{3}\right)\left(\frac{1}{3}\right)\left(\frac{2}{3}\right)$$

or

$$\left(\frac{2}{3}\right)^6\left(\frac{1}{3}\right)^4.$$

(Frank A. Cezus/FPG International Corporation.)

We might think of the process of obtaining exactly 6 successes as lining up 10 boxes in a row and picking exactly 6 of them in which to place an S. Thus the previous sequence of successes and failures would be

Since every such sequence must contain exactly 6 S's and 4 F's, the probability of any one of these occurring is always $(\frac{2}{3})^6(\frac{1}{3})^4$. To find $b(6 : 10, \frac{2}{3})$, we then must count the number of ways there can be exactly 6 successes. But this is the number of ways of selecting 6 of the boxes to place an S inside. This can be done in $C(10, 6)$ ways. Thus

$$b\left(6 : 10, \frac{2}{3}\right) = C(10, 6)\left(\frac{2}{3}\right)^6\left(\frac{1}{3}\right)^4$$

$$= \frac{10 \cdot 9 \cdot 8 \cdot 7 \cdot 6 \cdot 5}{6 \cdot 5 \cdot 4 \cdot 3 \cdot 2 \cdot 1}\left(\frac{2}{3}\right)^6\left(\frac{1}{3}\right)^4$$

$$= 210\left(\frac{2}{3}\right)^6\left(\frac{1}{3}\right)^4$$

$$\approx .228. \quad \blacksquare$$

In general, to find the probability of exactly k successes in n repeated Bernoulli trials, first notice that any particular sequence with exactly k successes must have exactly $n - k$ failures. Thus by the product rule, the probability of any one of these occurring is

$$p^k q^{n-k}.$$

The number of ways of obtaining exactly k successes in n trials is the number of ways of selecting k objects (the boxes with S's inside in the above discussion) from a total of n. This is $C(n, k)$. Thus

$$\boxed{b(k : n, p) = C(n, k)p^k q^{n-k}.}$$

Applications

Ken Griffey, Jr. played baseball for the Seattle Mariners during the 1990 season and was the youngest big league player that year although 1990 was Griffey, Jr.'s second season. His batting average that year was .300, that is, the probability of his getting a hit on any official time at bat[1] was .300.

E X A M P L E   2   **Finding the Probability of an Outcome in a Repeated Bernoulli Trial**

What was the probability of Griffey, Jr. getting at least 2 hits in a game in 1990 if we assume that he came to bat 4 official times in that game and if we assume coming to bat each time is an independent trial.

Solution   If we designate success as a hit, then $p = .300$ and $q = 1 - p = .700$. We then are looking for

$$b(2 : 4, .300) + b(3 : 4, .300) + b(4 : 4, .300).$$

We can save some calculations if we notice that this is one less the probability of at most 1 hit, which is

$$1 - [b(0 : 4, .300) + b(1 : 4, .300)],$$

or

$$1 - [C(4, 0)(.300)^0(.700)^4 + C(4, 1)(.300)^1(.700)^3] = 1 - [(.7)^4 + 4(.3)(.7)^3]$$

$$= .3483.$$

So about 35% of the time Griffey, Jr. should have obtained at least 2 hits out of 4 times-at-bat. ■

---

[1]An official time at bat does not include a time when the batter walked, hit a sacrifice fly or bunt, or was hit by a pitched ball.

E X A M P L E 3   **Finding Probabilities of Outcomes in a Repeated Bernoulli Trial**

A computer manufacturer uses 8 microchips in each of its computers. It knows that 1% of these chips is defective. What is the probability that (a) all 8 chips are good? (b) at least 6 are good?

Solutions   If we let S be the event "not defective," then $p = .99$ and $q = .01$.

(a) The answer is

$$b(8 : 8, .99) = C(8, 8)(.99)^8(.01)^0$$
$$= (.99)^8 \approx .92274.$$

(b) The answer is the sum of the three probabilities

$$b(8 : 8, .99) + b(7 : 8, .99) + b(6 : 8, .99).$$

This is

$$.92274 + C(8, 7)(.99)^7(.01)^1 + C(8, 6)(.99)^6(.01)^2$$
$$\approx .92274 + 8(.99)^7(.01) + 28(.99)^6(.01)^2$$
$$= .92274 + .07457 + .00264 = .99995. \quad \blacksquare$$

SELF-HELP EXERCISE SET 7.7

1. If a repeated Bernoulli trial is performed 6 times, find the probability of obtaining 2 successes and 4 failures if $p = .20$.

2. A retail store sells 2 brands of TVs, with the first brand comprising 60% of these sales. What is the probability that the next 5 sales of TVs will consist of at most 1 of the first brand?

EXERCISE SET 7.7

For Exercises 1 through 6, a repeated Bernoulli trial is performed. Find the probability of obtaining the indicated number of successes and failures for the indicated value of p.

1. 4 S's, 1 F, $p = .2$

2. 4 S's, 3 F's, $p = .3$

3. 3 S's, 4 F's, $p = .5$

4. 4 S's, 4 F's, $p = .5$

5. 4 S's, 4 F's, $p = .25$

6. 2 S's, 2F's, $p = .1$

Evaluate the term in Exercises 7 through 12.

7. $b(3 : 6, .5)$

8. $b(4 : 6, .5)$

9. $b(4 : 7, .1)$

10. $b(3 : 4, .2)$

11. $b(3 : 5, .1)$

12. $b(3 : 8, .2)$

In Exercises 13 through 18 flip a fair coin 10 times. Find the probability of getting the following outcomes.

13. Exactly 8 heads

14. Exactly 3 heads

15. At least 8 heads

16. At least 7 heads

17. At most 1 head

18. At most 2 heads

In Exercises 19 through 23 an event E has probability $p = p(E) = .6$ in some sample space. Suppose the experiment that yields this sample space is repeated 7 times and the outcomes are independent. Find the probability of getting the following outcomes.

19. E exactly 6 times

20. E exactly 3 times

21. E at least 6 times

22. E at least 5 times

23. E at most 2 times

24. Show

$$b(k : n, p) = b(n - k : n, 1 - p).$$

Ty Cobb has the highest lifetime batting average of any big league baseball player with a remarkable average of .367. Assume in Exercises 25 through 28 that Cobb came to bat officially 4 times in every game played.

25. What would be Cobb's probability of getting at least one hit in a game?

26. What would be Cobb's probability of getting at least 3 hits in a game?

27. What would be Cobb's probability of getting at least one hit in 10 successive games? (Use the result in Exercise 25.)

28. What would be Cobb's probability of getting at least one hit in 20 successive games? (Use the result in Exercise 25.)

Babe Ruth holds the record for the highest lifetime percent (8.5%) of home runs per time-at-bat. (Ralph Kiner comes in a distant second at 7.1%.) Assume in Exercise 29 through 32 that Ruth came to bat officially 4 times in every game played.

29. What would be Ruth's probability of getting at least two home runs in a game?

30. What would be Ruth's probability of getting at least one home run in a game?

31. What would be Ruth's probability of getting at least two home runs in 3 successive games? (Use the result in Exercise 29.)

32. What would be Ruth's probability of getting 4 home runs in a game?

Applications

Oil Drilling.   An oil company estimates that only 1 oil well in 20 will yield commercial quantities of oil. Assume that successively drilled wells represent independent events. If 12 wells are drilled, find the probability of obtaining a commercially successful well for the following number of times.

33. Exactly 1

34. None

35. At most 2

36. Exactly 4

Personnel.   A company finds that one out of five workers it hires turns out to be unsatisfactory. Assume that the satisfactory performance of any hired worker is independent of that of any other hired workers. If the company hires 20 people, what is the probability that the following number of people will turn out satisfactory?

37.   Exactly 10   **38.**   At most 2

39.   At least 18   **40.**   Exactly 20

Medicine.   A certain type of heart surgery in a certain hospital results in mortality in 5% of the cases. Assume that the death of a person undergoing this surgery is independent from the death of any others who have undergone this same surgery. If 20 people have this heart surgery at this hospital, find the probability that the following number of people will not survive the operation.

41.   Exactly 2   **42.**   At most 2

43.   At most 3   **44.**   Exactly 10

Solutions to Self-Help Exercise Set 7.7

1.   In a repeated Bernoulli trial the probability of obtaining 2 successes and 4 failures if $p = .20$ is

$$b(k : n, p) = C(n, k)p^k q^{n-k} = C(6, 2)(.20)^2(.80)^4 = \frac{6 \cdot 5}{2 \cdot 1}(.20)^2(.80)^4 \approx .246.$$

2.   If a retail store sells 2 brands of TVs, with the first brand comprising 60% of these sales, then the probability that the next 5 sales of TVs will consist of at most 1 of the first brand is the probability that the next 5 sales will consist of exactly 0 of the first brand plus exactly 1. This is

$$C(5, 0)(.60)^0(.40)^5 + C(5, 1)(.60)^1(.40)^4 = (.40)^5 + 5(.60)(.40)^4$$

$$= .01024 + .0768 = .08704.$$

SUMMARY OUTLINE OF CHAPTER 7

- An **experiment** is an activity that has observable results. The results of the experiment are called **outcomes**.

- A **sample space** of an experiment is the set of all possible outcomes of the experiment. Each repetition of an experiment is called a **trial**.

- Given a sample space S for an experiment, an **event** is any subset E of S. An **elementary event** is an event with a single outcome.

- If E and F are two events, then $E \cup F$ is the **union** of the two events and consists of the set of outcomes that are in E or F or both.

- If E and F are two events, then $E \cap F$ is the **intersection** of the two events and consists of the set of outcomes that are in both E and F.

- If E is an event, then E^c is the **complement** of E and consists of the set of outcomes that are not in E.

- The empty set, ϕ, is called the **impossible event**.

- Let S be a sample space. The event S is called the **certainty event**.

- Two events E and F are said to be **mutually exclusive** or **disjoint** if $E \cap F = \phi$.

- If an experiment is repeated N times and an event E occurs $f(E)$ times, then the fraction $\dfrac{f(E)}{N}$ is called the **relative frequency** of the event.

- Let S be the sample space of an experiment and E an event in S. If the experiment has actually been performed N times, the relative frequency $\dfrac{f(E)}{N}$ is called the **empirical probability** of E.

- Let S be the sample space of an experiment, let E be an event in S, and let N be the number of times the experiment has been repeated. The **intuitive probability of E**, denoted by $p(E)$, is the number that the relative frequency $\dfrac{f(E)}{N}$ approaches as N becomes larger and larger.

- **Definition of Probability.**   Let S be a sample space. Assign to each event E in S a number $p(E)$ so that the following four properties are satisfied.

  Property 1. $0 \le p(E) \le 1$ for any event E.

  Property 2. $p(S) = 1$.

  Property 3. $p(\phi) = 0$.

  Property 4.   Given any event $E = \{e_1, e_2, \ldots, e_r\}$, where $e_1, e_2, \ldots, e_r$, are all distinct, then

  $$p(\{e_1, e_2, \ldots, e_r\}) = p(e_1) + p(e_2) + \cdots + p(e_r).$$

  Then for any event E, we called $p(E)$ the **probability** of E and say that p is a probability on S.

- **Further Properties of Probability.**   Let p be a probability on a sample space $S = \{s_1, s_2, \ldots, s_n\}$ and E and F two events in S. Then the following are true.

  Property 5. $p(s_1) + p(s_2) + \cdots + p(s_n) = 1$.

  Property 6. $p(G \cup H) = p(G) + p(H)$ if $G \cap H = \phi$.

  Property 7. $p(E^c) = 1 - p(E)$.

  Property 8. $p(E \cup F) = p(E) + p(F) - p(E \cap F)$.

  Property 9. $p(F) \le p(E)$ if $F \subset E$.

  Property 10.   Suppose $E_1, E_2, \ldots, E_k$ are pairwise mutually disjoint, that is, $E_i \cap E_j = \phi$ if $i \ne j$. Then

  $$p(E_1 \cup E_2 \cup \cdots \cup E_k) = p(E_1) + p(E_2) + \cdots + p(E_k).$$

- Let $p = p(E)$ be the probability of E, then the **odds for E** is defined to be

$$\text{Odds for } E = \frac{p}{1 - p}   \quad \text{if } p \ne 1.$$

- Suppose that the odds for an event E occurring is given as $\dfrac{a}{b}$. If $p = p(E)$, then

$$p = \frac{a}{a + b}.$$

- If S is a finite uniform sample space and E is any event, then

$$p(E) = \frac{\text{Number of elements in } E}{\text{Number of elements in } S} = \frac{n(E)}{n(S)}.$$

- Let E and F be two events in a sample space S. The **conditional probability** that E occurs given that F has occurred is defined to be

$$p(E|F) = \frac{p(E \cap F)}{p(F)} \quad \text{if} \quad p(F) > 0.$$

- If E and F are two events in a sample space S with $p(E) > 0$ and $p(F) > 0$, then

$$p(E \cap F) = p(F)p(E|F) = p(E)p(F|E).$$

- Two events E and F are said to be **independent** if

$$p(E|F) = p(E) \quad \text{and} \quad p(F|E) = p(F).$$

- Let E and F be two events with $p(E) > 0$ and $p(F) > 0$. Then E and F are independent if, and only if, $p(E \cap F) = p(E)p(F)$.
  A set of events $\{E_1, E_2, \ldots, E_n\}$ is said to be independent if, for any k of these events, the probability of the intersection of these k events is the product of the probabilities of each of the k events. This must hold for any $k = 2, 3, \ldots, n$.

- **Bayes' Theorem**   Let $E_1, E_2, \ldots, E_n$, be mutually exclusive events in a sample space S with $E_1 \cup E_2 \cup \cdots \cup E_n = S$. If F is any event in S, then for any $i = 1, 2, \ldots, n$,

$$p(E_i|F) = \frac{\text{probability of branch } E_iF}{\text{sum of all probabilities of all branches that end in } F}$$

$$= \frac{p(E_i)p(F|E_i)}{p(E_1)p(F|E_1) + p(E_2)p(F|E_2) + \cdots + p(E_n)p(F|E_n)}.$$

- The probability of exactly k successes in n repeated Bernoulli trials where the probability of success is p and failure is q is given by $b(k : n, p) = C(n, k)p^k q^{n-k}$.

Chapter 7 Review Exercises

1. During a recent four-round golf tournament, the following frequencies of scores were recorded on a par 5 hole.

Score	3	4	5	6	7	8
Frequency	4	62	157	22	4	1

  a. Find the probability that each of the scores was made by a random player.
  b. Find the probability that a score of par or lower was recorded.
  c. Find the probability that a score of less than par was recorded.

2. An urn has 10 white, 5 red, and 15 blue balls. A ball is drawn at random. What is the probability that the ball will be (a) red? (b) red or white? (c) not white?

3. If E and F are mutually disjoint sets in a sample space S with $p(E) = .25$ and $p(F) = .35$, find (a) $p(E \cup F)$, (b) $p(E \cap F)$, and (c) $p(E^c)$.

4. If E and F are two events in the sample space S with $p(E) = .20$, $p(F) = .40$, and $p(E \cap F) = .05$, find (a) $p(E \cup F)$, (b) $p(E^c \cap F)$, and (c) $p((E \cup F)^c)$.

5. Consider the sample space $S = \{a, b, c, d\}$ and suppose that $p(a) = p(b)$, $p(c) = p(d)$, and $p(d) = 2p(a)$. Find $p(b)$.

6. If the odds for a company obtaining a certain contract are 3 to 1, what is the probability that the company will receive the contract?

7. A furniture manufacturer notes that 6% of its reclining chairs have a defect in the upholstery, 4% a defect in the reclining mechanism, and 1% have both defects. (a) Find the probability that a recliner has at least one of these defects. (b) Find the probability that a recliner has none of these defects.

8. A survey of homeowners indicated that during the last year: 22% had planted some vegetables, 30% some flowers, 10% some trees, 9% vegetables and flowers, 7% vegetables and trees, 5% flowers and trees, and 4% all three of these. (a) Find the probability that a homeowner planted vegetables but not flowers. (b) Find the probability that exactly 2 of the items were planted. (c) Find the probability that none of these three items were planted.

9. For the urn in Exercise 2 what is the probability that there is one of each color if 3 are selected randomly without replacement? With replacement?

10. For the urn in Exercise 2, what is the probability that 3 are white, 4 are red, and 2 are blue, if 9 are selected randomly without replacement? With replacement?

11. Let $p(E) = .3$, $p(F) = .5$, and $p(E \cap F) = .2$. Draw a Venn diagram and find the indicated conditional probabilities: (a) $p(E|F)$, (b) $p(E^c|F)$, and (c) $p(F^c|E^c)$.

12. If $p(E) = .5$, $p(F) = .6$, and $p(E \cap F) = .4$, determine if E and F are independent events.

13. **Reliability.** A spacecraft has 3 batteries that can operate all systems independently. If the probability that any battery will fail is .05, what is the probability that all three will fail?

14. **Basketball.** A basketball player sinks a free-throw 80% of the time. If she sinks one, the probability of sinking the next goes to .90. If she misses, the probability of sinking the next goes to .70. Find the probability that she will sink exactly 2 out of 3 free-throws.

15. **Manufacturing.** A manufacturing firm has 5 machines that produce the same component. Using the table, find the probability that a defective component was produced by (a) machine 1, (b) machine 4.

Machine	Percentage of Components Produced	Percentage of Defective Components
1	20	1
2	30	2
3	30	3
4	10	4
5	10	10

16. **Assembly Line.** A machine on an assembly line is malfunctioning randomly and produces defective parts 30% of the time. What is the probability that this machine will produce exactly 3 defective parts among the next 6?

17. **Teaching Methods.** An instructor finds that only 55% of her college algebra students pass the course. She then tries a new teaching method and finds that 85% of the 20 students in the first class with the new method pass. Assuming that the probability at this school of passing college algebra is .55 and that any one student passing is independent of any other student passing, what is the probability that at least 85% of a college algebra class of 20 will pass this course? Is the instructor justified in claiming that the new method is superior to the old method?

18. **Drug Testing.** A company tests its employees for drug usage with a test that gives a positive reading 95% of the time when administered to a drug user and gives a negative reading 95% of the time when administered to a nondrug user. If 5% of the employees are drug users, find the probability that an employee is a nondrug user given that this person had a positive reading on the test. (The answer is shocking and illustrates the care that must be exercised in using such tests in determining guilt.)

Solution to Game Show Problem

The solution given here was selected since it relies on the commonly held intuitive notions of probability.

A contestant is shown 3 closed doors and is correctly informed that behind one of the doors is a car and behind each of the other 2 doors is a goat. The contestant is then asked to select a door. After the contestant picks a door, the game show host then chooses one of the 2 doors not selected that has a goat behind it and opens this door for the contestant to see the goat. Now the host gives the contestant the opportunity to change the selection of the door. If the car is behind the door selected, the contestant wins the car. Should the contestant switch doors or not, or does it matter?

Solution Our approach to the solution of this problem is to first assume that the contestant is allowed to play the game a large number of times, say 300. For the sake of definiteness, assume that *the third door is always initially selected.* (We will see that this is unimportant.)

Naturally the game show will place the car behind a door in a random fashion. Thus, from our intuitive understanding of probability, the car should be behind the first door about 100 times, behind the second door about 100 times, and behind the third door about 100 times.

Let us now assume that *the car is behind the first door exactly 100 times, behind the second door exactly 100 times, and behind the third door exactly 100 times.* Then if the contestant does not switch doors, she will win 100 out of 300 times, or one-third of the time.

Now let us see what happens if the contestant always switches. First consider the 100 times that the car is behind the first door: (car, goat, goat). In every one of these 100 cases, the third door has been selected and the car is behind the first door, and thus in every one of these cases, the game show has no choice but to reveal the goat behind the second door. The contestant then switches the selection to the first door and wins every time in these 100 cases.

Consider now the 100 cases when the car is behind the second door: (goat, car, goat). This is very similar to the previous situation. Since the game show is forced to reveal the goat behind the first door, the contestant switches to the second door and wins every one of these 100 cases.

Finally consider the 100 cases when the car is behind the third door: (goat, goat, car). Since in every one of these 100 cases the contestant switches to either the first or second door, every one of these 100 cases will be lost.

If the contestant always initially selects another door besides the third one, the analysis is the same and leads to the same conclusion. The reader is invited to do this analysis.

Thus for the 300 cases, switching results in winning 200 times and losing 100 times. That is, the probability of winning by switching is $\frac{2}{3}$, and of losing by switching is $\frac{1}{3}$.

The importance of this example is to realize that, when confronted with a confusing problem, it never hurts to return to the basic intuitive notation of probability.

The cost of disaster insurance is based on the likelihood of the disaster occurring. (National Center for Atmospheric Research/National Science Foundation.)

8

Statistics and Probability Distributions

T he application of statistics to business and the sciences can be divided into two broad areas. In the first area, large masses of data need to be described by graphs or by certain descriptive numbers such as the "average," "standard deviation," and so on. This is referred to as *descriptive statistics*. The second area is involved in making inferences about the data based on sampling. Since this chapter is a short introduction to the area of statistics, we will just touch on the second area.

Fundamental to a statistical analysis is knowledge of how the underlying probabilities are distributed. We will discuss a number of important probability distributions and their applications. These include the binomial, the Poisson, and the normal distributions.

8.1   PROBABILITY DISTRIBUTIONS

►   *Models and the Need for Statistics*

►   *Random Variables*

►   *Probability Distributions*

A P P L I C A T I O N

Picking a Mutual Fund

> Suppose you wish to select a mutual fund from a list of 64 funds, all of which involve similar risks. You notice that one single fund has performed above the average of these 64 funds in every one of the last 6 years. Is this a clear indication that the managers of this fund have the ability to keep their fund above average in performance? See the discussion immediately below for the answer.

Models and the Need for Statistics

In this section we wish to see how to present numerical data in a manner that will permit making interpretations and comparisons. We begin with an example from finance mentioned above.

A very important problem in the theory of finance is to determine if it is possible for investment advisors or theories to predict, with some degree of accuracy, the movements in the various financial markets. One very highly touted way of investing is to pick an investment advisor on the basis of past performance. This certainly appears reasonable.

Unfortunately, a great deal of research has tended to indicate that the future performance of an advisor is *independent* of the past performance. Certainly the research in this area indicates that if there is a connection between past and future performance, it is a weak and subtle one.

The data in Table 8.1 refer to the past 6-year annual performance records of 64 mutual funds considered to be growth funds. Each of the 64 funds were ranked according to whether their percent increase in asset value (including capital gains and dividends) for any one year was above the average for that year of the group of 64 or not. The number of years a fund was above average was then counted. The table gives the number of funds that had an above-average performance for a given number of years during the 6-year time period. For example, the table indicates that 14 funds had an above-average performance in 4 of the 6 years studied and 22 had an above-average performance in 3 of the 6 years. It is interesting to notice that one fund had an above-average performance in every single year of the 6-year period and one fund did not perform above average for any of these years. Would it make sense *based solely on this data* to invest in the one

Table 8.1

No. of Yrs. Above Avg.	0	1	2	3	4	5	6
Frequency	1	4	17	22	14	5	1
Empirical Probability	$\frac{1}{64}$	$\frac{4}{64}$	$\frac{17}{64}$	$\frac{22}{64}$	$\frac{14}{64}$	$\frac{5}{64}$	$\frac{1}{64}$

fund with above-average performance all 6 years and to avoid the one that never had an above-average performance during this period?

For later use, the last line of Table 8.1 gives the empirical probabilities (the relative frequencies).

Table 8.1 contains fictitious data. However, it does mirror actual data.[1] In the actual studies performance was adjusted to take into account the risk each fund was assuming. Studies indicate that higher rates of return can, in general, be obtained by taking higher risks. Precisely how to define and measure risk will be taken up in a subsequent section. For the 64 funds in Table 8.1 we suppose they all assumed the same risk. This enormously simplifies the analysis and is the reason the actual published data is not used here.

Let us assume that Table 8.1 contains actual data. Let us now consider a model that we think might explain why the data looks as it does. In our model suppose we have 64 funds and let us assume that the performance of a fund in this group in any year is *independent* from its performance in the previous years and furthermore assume that the probability of the performance of any fund being above average is .50. Our model is saying that a fund is determined to be above average or not in any year the same way we determine heads or tails by flipping a fair coin. Then we can view the performance of the funds in any year as a Bernoulli trial with $p = .50$ and the performance over the 6-year period of time as a repeated Bernoulli trial with $n = 6$. Using the formulas in the last section, the probability that a fund (under the stated assumptions in the model) has an above average performance exactly k of the 6 years is given by

$$b(k : 6, .50) = C(6, k)(.5)^k(.5)^{6-k} = C(6, k)(.5)^6.$$

By recalling Pascal's triangle, the binomial coefficients are 1, 6, 15, 20, 15, 6, 1. From this and the fact that $(.5)^6 = \frac{1}{64}$ we can create Table 8.2, which gives the probability for every value of k in the model.

What is striking is the similarity of Table 8.2 to Table 8.1. This can be further demonstrated by presenting the data in the two tables in graphical form. The numbers 0, 1, 2, 3, 4, 5, 6, represent the possible outcomes. These numbers are located on the horizontal axis as in Figure 8.1. Above each number is drawn a rectangle with base equal to one unit and height equal to the probability of that number. For example, above the number 5 is a rectangle with height $\frac{5}{64}$ in the first case and $\frac{6}{64}$ in the second case. Such graphs are called **histograms**; they give a vivid description of how the probability is distributed. As a consequence, comparison can be made more easily between the two histograms than between the two tables of probability.

Table 8.2

No. of Yrs. Above Avg.	0	1	2	3	4	5	6
Theoretical Prob. in Model	$\frac{1}{64}$	$\frac{6}{64}$	$\frac{15}{64}$	$\frac{20}{64}$	$\frac{15}{64}$	$\frac{6}{64}$	$\frac{1}{64}$

---

[1]Cf. Michael C. Jensen, "The Performance of Mutual Funds in the Period 1945–64," *Journal of Finance*, vol. 23 no. 2 (May 1968), pp. 389–416.

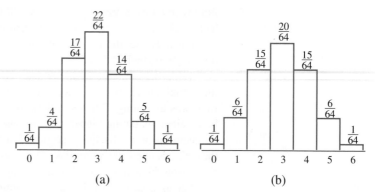

Figure 8.1

Given the remarkable similarity between the *theoretical* model we created and the actual data in Table 8.1, we might be very inclined to believe that the assumptions underlying the model are correct. That is, if the assumptions of the model are correct, then the data shown in Table 8.2 can be viewed as a prediction of what should be happening. Since the predictions indicated in Table 8.2 are so close to the actual data given in Table 8.1, we would tend to think that the assumptions of the model are correct.

These assumptions state that determining whether a fund is above average in any year is the same as flipping a fair coin and interpreting the occurrence of (say) "heads" with "above average performance." The model is then the same as having 64 people each flip a fair coin 6 times in succession and declaring their performance to be above average if a head occurs. In such a model we naturally expect, for example, one person to obtain heads all 6 times. Would you conclude that this person is a talented "head flipper" and then wager large sums of money on this person's alleged ability to flip heads?

Thus in our model we fully expect one fund to be above average in performance every single one of the 6 years. Now we see the possibility that the single fund in Table 8.1 that had the above-average performance in every one of the 6 years may not be a result at all of the brilliance of the management but an inevitable result of chance.

It is apparent that some measurements of the data in the two tables are needed that would measure the similarity of the two sets of data. This would be especially helpful in cases where the similarity is not so obvious. We will do this in subsequent sections.

Critical to the above analysis was the need to find the probability of every outcome in each of the cases given in the two tables. We then needed to see how the probability was *distributed* in each case. Furthermore, we needed to be able to compare the manner in which the probability was distributed in Table 8.1 with the manner it was distributed in Table 8.2. In the next two sections we will develop ways of making such comparisons. This will enable us to state in what ways two sets of data are similar and in what ways they are not. This will be especially helpful in cases where the similarity is not so obvious as it is when comparing Table 8.1 with Table 8.2.

Random Variables

As we already noted, the outcomes in the above experiments were any of the numbers 0, 1, 2, 3, 4, 5, or 6. Of course, outcomes of experiments are not always real numbers. The outcomes of experiments can be ''heads,'' ''above average,'' or ''defective''—none of which are real numbers. It is often useful to assign a real number to each outcome of an experiment. For example, the outcome of the experiment ''your grade in math'' is a *letter*. But for purposes that you are well aware of, each letter is assigned a real number. Usually 4 is assigned to an A, 3 to a B, 2 to a C, 1 to a D, and 0 to an F. Usually there is some rational basis on which the assignment of a real number to an outcome is made. For example, since a grade of *A* is ''better'' than a grade of *B*, it makes sense to assign a *higher* number to the grade A than to B. Also the assignment of the numbers to each letter is done in such a manner that the numerical difference between two successive letters will be equal. Assigning zero to F makes sense since no credit is given for a grade of F. Finally, using the numbers 0, 1, 2, 3, and 4 makes it easy to calculate the ''average'' grade and for this number to have a readily understood meaning.

When numbers are assigned to the outcomes of experiments according to some rule, the rule is referred to as a **random variable**. As we have seen in assigning numbers to letter grades, the assignments of numbers to the outcomes of experiments are normally done in a manner that is reasonable and, most important, in a manner that permits these numbers to be used for interpretation and comparison.

Random Variables

A **random variable** is a rule that assigns precisely one real number to each outcome of an experiment.

REMARK. Unless otherwise specified, when the outcomes of an experiment are themselves numbers, the random variable is the rule that simply assigns each number to itself.

There are three types of random variables, as we now indicate.

Finite Discrete, Infinite Discrete, and Continuous Random Variables

1. A random variable is **finite discrete** if it assumes only a finite number of values.
2. A random variable is **infinite discrete** if it takes on an infinite number of values that can be listed in a sequence, so that there is a first one, a second one, a third one, and so on.
3. A random variable is said to be **continuous** if it can take any of the infinite number of values in some interval of real numbers.

If a random variable denotes the number of years one of the funds mentioned at the beginning of this section has an above-average performance in a 6-year period, then the random value can take the finite number of values in the set $\{0, 1, 2, \ldots, 6\}$, and thus is finite discrete. If the random variable denotes the number of flips it takes to obtain a head, then the random value can take any of the infinitely many values in the set $\{1, 2, \ldots\}$ and thus is infinite discrete. If a random variable denotes the heights in feet of adult men in this country, then the random variable can take on any value in the interval $[2, 8]$ and thus is continuous.

Probability Distributions

Suppose now that the random variable associated with the outcomes of an experiment are the finite set of real numbers $\{x_1, x_2, \ldots, x_n\}$. We have usually denoted the probability of the outcome x_k by $p(x_k)$. We now denote this same probability by

$$P(X = x_k).$$

We will use this latter notation since it is widely used in statistics and has some advantages, as we shall soon see. The **probability distribution of the random variable** X is a listing of all the probabilities associated with all possible values of the random variable. Such a listing is often given in a table.

Probability Distribution of the Random Variable X

Suppose the random variable X can take the values $x_1, \ldots, x_n$. The probability distribution of the random variable X is a listing of all the probabilities associated with all possible values of the random variable, that is, $p_1, \ldots, p_n$, where $p_1 = p(x_1), \ldots, p_n = p(x_n)$.

Using the new notation, we can rewrite Table 8.2 as follows.

x	0	1	2	3	4	5	6
$P(X = x)$	$\frac{1}{64}$	$\frac{6}{64}$	$\frac{15}{64}$	$\frac{20}{64}$	$\frac{15}{64}$	$\frac{6}{64}$	$\frac{1}{64}$

This particular distribution is referred to as a **binomial distribution** since the listed probabilities are the terms in the binomial expansion of $(p + q)^6 = (\frac{1}{2} + \frac{1}{2})^6$. In general, if $p + q = 1$,

$$1 = (p + q)^n = C(n, n)p^n q^0 + C(n, n - 1)p^{n-1}q + \cdots + C(n, 0)p^0 q^n.$$

Recalling the material on Bernoulli trials this can be written as

$$b(n : n, p) + b(n - 1 : n, p) + \cdots + b(0 : n, p).$$

Binomial Distribution

Given a sequence of n Bernoulli trials with the probability of success p and the probability of failure q, the **binomial distribution** is given by

$$P(X = k) = b(k : n, p) = C(n, k)p^k q^{n-k}.$$

E X A M P L E 1   **Finding a Probability Distribution and Corresponding Histogram**

Suppose a pair of fair dice is tossed. Let X denote the random variable that gives the sum of the top faces. Find the probability distribution. Draw a histogram.

Solution   The random variable can take any of the values $2, 3, \ldots, 12$. Using the new notation and recalling the work done in Example 2 of Section 7.1, we have

$$P(X = 2) = p(\{(1, 1)\}) = \frac{1}{36}$$

$$P(X = 3) = p(\{(1, 2), (2, 1)\}) = \frac{2}{36},$$

and so forth. This then yields the probability distribution given in the following table.

x	2	3	4	5	6	7	8	9	10	11	12
$P(X = x)$	$\frac{1}{36}$	$\frac{2}{36}$	$\frac{3}{36}$	$\frac{4}{36}$	$\frac{5}{36}$	$\frac{6}{36}$	$\frac{5}{36}$	$\frac{4}{36}$	$\frac{3}{36}$	$\frac{2}{36}$	$\frac{1}{36}$

The histogram is drawn in Figure 8.2. ∎

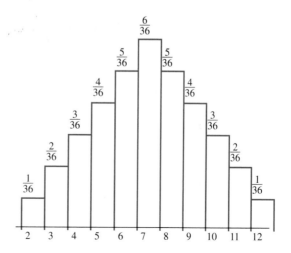

Figure 8.2

E X A M P L E   2   **Finding a Probability Distribution**

Suppose a pair of fair dice is tossed. Let X be the random variable given by 1 if the sum of the top two faces is a prime number and 0 otherwise. Find the probability distribution.

Solution   Since the primes are 2, 3, 5, 7, and 11, the table in the previous example indicates that

$$P(X = 1) = \frac{1 + 2 + 4 + 6 + 2}{36} = \frac{15}{36}.$$

Thus

Event	Prime	Not a Prime
Random Variable x	1	0
$P(X = x)$	$\frac{15}{36}$	$\frac{21}{36}$

E X A M P L E   3   **Finding a Probability Distribution and Corresponding Histogram**

Ted Williams had a batting average of .406 in 1941. During that year he had 112 singles, 33 doubles, 3 triples, and 37 home runs in 456 (official) times-at-bat. Let the random variable X be 0 for no hit, 1 for a single, 2 for a double, 3 for a triple, and 4 for a home run. Find a probability distribution for Ted Williams obtaining a 0, 1, 2, 3, or 4 for each time-at-bat during 1941. Draw a histogram.

Solution   The probability of hitting a home run is $37/456 = .081$, for a triple is $3/456 = .007$, for a double is $33/456 = .072$, for a single is $112/456 = .246$. For no hit, we have $[456 - (37 + 3 + 33 + 112)]/456 = 271/456 = .594$.

Event	No Hit	Single	Double	Triple	Home Run
Random Variable x	0	1	2	3	4
Frequency	271	112	33	3	37
$P(X = x)$	.594	.246	.072	.007	.081

See Figure 8.3 for the histogram. ∎

(© L.O.L. Inc./FPG
International Corporation.)

We will now use histograms to find the probability of events by measuring the areas under appropriate rectangles.

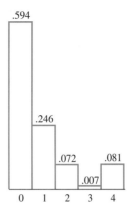

Figure 8.3

E X A M P L E   4   **Finding Probability Using Histograms**

Use the histogram in Figure 8.2 to find the probability that the sum of the top faces of two fair die when tossed will be at least 9 and less than 12.

Solution   We are seeking

$$P(X = 9) + P(X = 10) + P(X = 11).$$

The probability that $X = 9$, $P(X = 9)$, is the area of the rectangle above 9, $P(X = 10)$ is the area of the rectangle above 10, and so on. Thus, we are seeking the area of the shaded region in Figure 8.4. This is

$$\frac{4}{36} + \frac{3}{36} + \frac{2}{36} = \frac{1}{4}. \quad \blacksquare$$

A shorthand notation for the probability found in this example is $P(9 \leq X < 12)$.

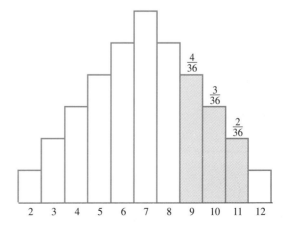

Figure 8.4

E X A M P L E 5 | **Finding Probability Using Histograms**

Using Figure 8.1a, find the (empirical) probability that a fund had at most 2 years of above-average returns.

Solution

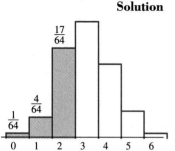

Figure 8.5

We are seeking $P(X \le 2)$. This is the area of the shaded region under the first 3 rectangles shown in Figure 8.5. This is

$$P(X \le 2) = \frac{1}{64} + \frac{4}{64} + \frac{17}{64} = \frac{11}{32}. \quad \blacksquare$$

SELF-HELP EXERCISE SET 8.1

1. Use the histogram in Figure 8.3 to find the probability that Ted Williams obtained an extra-base hit (double, triple, or home run) during 1941.

2. Find the probability distribution of the binomial distribution $b(k; 6, .75)$ and draw a histogram.

EXERCISE SET 8.1

In Exercises 1 through 4, determine the possible values of the given random variable and indicate whether the random variable is finite discrete, infinite discrete, or continuous.

1. The number of times a coin must be flipped before two heads appear in succession

2. The number of times two heads appear in succession when a coin is flipped 1,000,000 times

3. The hours spent studying math each week

4. The number of rotten eggs in a dozen

5. The grade distribution in a certain math class is given in the following table.

Event	A	B	C	D	F
Random Variable x	0	1	2	3	4
Frequency	6	10	19	11	4
$P(X = x)$					

Fill in the last row. Draw a histogram.

6. In 1961 Roger Maris set the record for most home runs in a season. Complete the following table.

Event	*No Hit*	*Single*	*Double*	*Triple*	*Home Run*
Random Variable x	0	1	2	3	4
Frequency	431	78	16	4	61
$P(X = x)$					

7. The student ratings of a particular mathematics professor are given in the following table. Complete the last line and draw a histogram.

Event	1	2	3	4	5	6	7	8	9	10
Frequency	5	2	3	0	4	10	10	11	3	2
$P(X = x)$										

8. A basketball player has a probability of .8 of sinking a foul shot. Assume that sinking a shot is independent of what occurred before. Let X be the random variable given by the number of foul shots sunk by this player in 5 successive shots. Find the probability distribution. Draw a histogram.

9. A pair of fair dice is tossed. Let X be the random variable given by the absolute value of the difference of the numbers on the top faces of the dice. Find the probability distribution. Draw a histogram.

10. On a true–false test with 5 questions, let X denote the random variable given by the total number of questions correctly answered by guessing. Find the probability distribution. Draw a histogram.

11. In the previous problem let X denote the random variable given by twice the total number of correct answers minus the number of incorrect answers. Find the probability distribution. Draw a histogram.

12. Let X denote the random variable given by the number of girls in a family of 4 children. Find the probability distribution if a girl is as likely as a boy. Draw a histogram.

13. Two balls are selected at random from an urn that contains 3 white and 7 red balls. Let the random variable X denote the number of white balls drawn. Find the probability distribution. Draw a histogram.

14. In the previous problem let the random variable X denote the number of white balls times the number of red balls drawn. Find the probability distribution.

15. The probability distribution of the random variable X is given in the following table. What must z be?

Random Variable x	0	1	2	3	4
$P(X = x)$	.2	.1	.1	z	.3

16. The probability distribution of the random variable X is given in the following table.

Random Variable x	-2	-1	0	1	2	3
$P(X = x)$	.20	.15	.05	.35	.15	.10

Find (a) $P(X = 0)$, (b) $P(X \leq 0)$, (c) $P(-1 < X \leq 4)$, (d) $P(X \geq 1)$. Draw a histogram and identify each of the probabilities as an area.

17. The probability distribution of the random variable X is given in the following table.

Random Variable x	4	5	6	7	8	9
$P(X = x)$	.15	.26	.14	.22	.18	.05

Find (a) $P(X = 4)$, (b) $P(X \leq 2)$, (c) $P(5 \leq X \leq 6)$, (d) $P(X \geq 8)$. Draw a histogram and identify each of the probabilities as an area.

18. Let Y be the random variable given by X^2 where X is given in Exercise 16. Find the probability distribution of Y by completing the following table.

Random Variable y	0	1	4	9
$P(X^2 = y)$				

19. Let Y be the random variable given by $X + 2$ where X is given in Exercise 16. Find the probability distribution of Y.

Applications

In Exercises 20 through 23 use a histogram to represent the requested data.

Height	*Percent Distribution of Population by Height*			
	Males 25–34 yrs	Males 35–44 yrs	Females 25–34 yrs	Females 35–44 yrs
<5′2″	.42	.87	19.11	18.24
5′2″–5′4″	1.13	2.81	28.32	31.66
5′4″–5′6″	7.96	8.79	27.49	28.17
5′6″–5′8″	17.18	14.01	18.56	15.36
5′8″–5′10″	28.66	30.61	6.21	5.76
5′10″–6′	26.23	24.14	.31	.81
>6′	18.42	18.77	.00	.00

20. Height Distribution. Males ages 25–34

21. Height Distribution. Males ages 35–44

22. Height Distribution. Females ages 25–34

23. Height Distribution. Females ages 35–44

24. Sales. A car dealership tracks the number of cars sold each week. During the past year they sold from zero to six cars per week with the frequencies indicated in the following table. Find the probability distribution.

Number Sold in One Week	0	1	2	3	4	5	6
Frequency	4	11	8	10	10	5	2

25. Quality Control. Suppose 2 light bulbs are chosen at random from an assembly line that has 5% defective light bulbs. Let X denote the random variable that gives the number of defective light bulbs chosen. Find the probability distribution of X. Draw a histogram.

Solutions to Self-Help Exercise Set 8.1

1. The appropriate area in the accompanying histogram has been shaded. The area and thus the probability that Ted Williams obtained an extra-base hit in 1941 is then

$$P(X > 1) = .072 + .007 + .081 = .160.$$

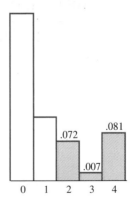

2. From Section 7.7,

$$b(k:6, .75) = C(6, k)\left(\frac{3}{4}\right)^k\left(\frac{1}{4}\right)^{6-k} = C(6, k)3^k\left(\frac{1}{4}\right)^6.$$

Let $h = (.25)^6 \approx .000244$. Then for $k = 0, 1, \ldots, 6$, $b(k:6, .75)$ are successively $1 \cdot 3^0 h$, $6 \cdot 3^1 h$, $15 \cdot 3^2 h$, $20 \cdot 3^3 h$, $15 \cdot 3^4 h$, $6 \cdot 3^5 h$, $1 \cdot 3^6 h$, or h, $18h$, $135h$, $540h$, $1215h$, $1458h$, $729h$. The histogram is shown in the figure. Notice how the probability distribution is skewed to the right with $p > .5$.

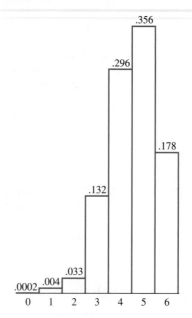

8.2 MEASURES OF CENTRAL TENDENCY

▶ *Expected Value*

▶ *Applications*

▶ *Mean of the Binomial Distribution*

▶ *Median and Mode*

APPLICATION

Expected Profit of an Insurance Policy

An insurance company sells a 65-year-old man a 1-year life insurance policy. The policy, which costs $100, pays $5000 in the event the man dies during the next year. Suppose there is a 1% chance that the man will die in the next year. If the insurance company were to sell a large number of the same insurance policies under the same conditions, what should they expect for an average profit per such policy? See Example 5 for the answer.

Expected Value

In this section we introduce several statistics that will be used to measure the "center" of the data.

In the last section we studied probability distributions, that is, how probability was distributed over the various outcomes. We also saw the need for comparing two probability distributions. The *expected value* will now be discussed. We will see that this number is a generalization of the average and conveys certain very useful information about the data. The expected value will be seen to be one way of locating the "center" of the data. In another typical application, we will use the expected value to compare two investments in order to determine which will be more profitable.

We begin with a notion of *average* which is familiar to most. If you have three test grades of 95, 85, and 93, the average is the sum of all the numbers divided by the total number or

$$\frac{95 + 85 + 93}{3} = \frac{273}{3} = 91.$$

In general we have the following definition of *average* or *mean*.

Average or Mean

The **average** or **mean** of the n numbers $x_1, x_2, \ldots, x_n$, denoted by $\bar{x}$ or the Greek letter μ, is given by

$$\mu = \frac{x_1 + x_2 + \cdots + x_n}{n}.$$

E X A M P L E   1   **Calculating the Mean**

You are thinking of buying a small manufacturer of quality yachts and have obtained the following data from the owners that give how often various numbers of yacht sales occur per quarter.

Number Sold in One Quarter	1	2	3	4	5
Frequency of Occurrence	10	15	16	7	2

Find the average or mean number sold per quarter.

Solution   We first need to find the total number of quarters. This is

$$10 + 15 + 16 + 7 + 2 = 50.$$

Now we need to find the total sold and divide by 50. To find the total sold, notice that on 10 occasions exactly 1 was sold, on 15 occasions 2 were sold, and so on. Thus we must sum up the number one 10 times plus sum up the number two 15 times plus . . . and then divide by 50, that is, we have

$$\frac{\overbrace{1 + \cdots + 1}^{10 \text{ times}} + \overbrace{2 + \cdots + 2}^{15 \text{ times}} + \overbrace{3 + \cdots + 3}^{16 \text{ times}} + \overbrace{4 + \cdots + 4}^{7 \text{ times}} + \overbrace{5 + 5}^{2 \text{ times}}}{50}$$

or

$$\mu = \frac{1 \cdot 10 + 2 \cdot 15 + 3 \cdot 16 + 4 \cdot 7 + 5 \cdot 2}{50} = 2.52. \quad \blacksquare$$

We can gain considerable insight into this mean if we write the above fraction in a different form. We can write

$$\mu = \frac{1 \cdot 10 + 2 \cdot 15 + 3 \cdot 16 + 4 \cdot 7 + 5 \cdot 2}{50}$$

$$= 1 \frac{10}{50} + 2 \frac{15}{50} + 3 \frac{16}{50} + 4 \frac{7}{50} + 5 \frac{2}{50}$$

$$= 1 \cdot p_1 + 2 \cdot p_2 + 3 \cdot p_3 + 4 \cdot p_4 + 5 \cdot p_5$$

where p_1, p_2, p_3, p_4, p_5, are, respectively, the (empirical) probabilities of the 5 possible outcomes 1, 2, 3, 4, 5. If we now let $x_1 = 1, x_2 = 2, x_3 = 3, x_4 = 4, x_5 = 5$, this can be written as

$$x_1 p_1 + x_2 p_2 + x_3 p_3 + x_4 p_4 + x_5 p_5.$$

This is the form we are seeking. Notice that each term is the product of an outcome with the probability of that outcome.

In general suppose there are the n outcomes $x_1, x_2, \ldots, x_n$, with respective probabilities $p_1, p_2, \ldots, p_n$. Now suppose the experiment has been repeated a large number of times N so that x_1 is observed to occur a frequency of f_1 times, x_2 is observed f_2 times, and so on. (We must have $f_1 + f_2 + \cdots + f_n = N$.) Then the mean is

$$\text{mean} = \frac{\overbrace{x_1 + \cdots + x_1}^{f_1 \text{ times}} + \overbrace{x_2 + \cdots + x_2}^{f_2 \text{ times}} + \cdots + \overbrace{x_n + \cdots + x_n}^{f_n \text{ times}}}{N}$$

$$= \frac{x_1 f_1 + x_2 f_2 + \cdots + x_n f_n}{N}$$

$$= x_1 \frac{f_1}{N} + x_2 \frac{f_2}{N} + \cdots + x_n \frac{f_n}{N}$$

But as N gets larger we expect the relative frequency $\dfrac{f_1}{N}$ to approach p_1. Similarly for the other relative frequencies. Thus the right-hand side of the last displayed line approaches

$$x_1 p_1 + x_2 p_2 + \cdots + x_n p_n.$$

We call this the **expected value**. In the language of random variables we then have the following.

Expected Value or Mean

Let X denote the random variable that takes on the values $x_1, x_2, \ldots, x_n$, and let the associated probabilities be $p_1, p_2, \ldots, p_n$, then the **expected value** or **mean** of the random variable X, denoted by $E(X)$ or by μ, is

$$\mu = E(X) = x_1 p_1 + x_2 p_2 + \cdots + x_n p_n.$$

Often we do not know the probabilities, but an experiment has been performed N times. Then the probabilities $p_1, p_2, \ldots, p_n$, are the relative frequencies (empirical probabilities). This was illustrated in Example 1.

Notice that the expected value is an average in some sense. As we saw above, if the random phenomenon is repeated a large number of times, the average of the observed outcomes will approach the expected value.

$E(X)$ is what we "expect" over the long term. But we must realize that $E(X)$ need not be an actual outcome. In Example 1, the expected value of sales per quarter was 2.52. We cannot sell 2.52 yachts. Nonetheless, it is extremely useful to think as though we could. (Actually, if the yacht manufacturer were incorporated with 100 shares outstanding and you owned one of these shares, it would make a great deal of sense to say you sold .0252 yachts.)

In the case that all the n outcomes are equally likely, the probabilities are just $\dfrac{1}{n}$ and then

$$E(X) = x_1 \frac{1}{n} + x_2 \frac{1}{n} + \cdots + x_n \frac{1}{n}$$

$$= \frac{x_1 + x_2 + \cdots + x_n}{n},$$

which is the average.

Applications

E X A M P L E   2   **Calculating the Expected Value**

Redo Example 1 using the formula for $E(X)$.

Solution   Using the relative frequencies calculated earlier as the (empirical) probability yields Table 8.3.

Table 8.3

Number Sold in One Quarter	1	2	3	4	5
Probability of Occurrence	$\frac{10}{50} = .20$	$\frac{15}{50} = .30$	$\frac{16}{50} = .32$	$\frac{7}{50} = .14$	$\frac{2}{50} = .04$

Then

$$E(X) = 1(.20) + 2(.30) + 3(.32) + 4(.14) + 5(.04) = 2.52. \quad \blacksquare$$

Suppose in considering buying the yacht manufacturer in Example 1, you wish to take into account a principal competitor who also manufactures yachts. Knowledge of the competitor's profits would be useful in determining how competitive the two manufacturers are and would heavily influence whether you will actually make the purchase. It would be convenient to know the details of the competitor's operations, but this is privileged information you are not likely to obtain. You then hire some analysts who have considerable knowledge of the yacht business and ask them to prepare a table like the one in Table 8.3 that you can use as a comparison. Naturally they are unable to give you exact figures, but they give you the data in Table 8.4.

Table 8.4

Number Sold in One Quarter	0	1	2	3	4	5
Probability of Occurrence	.10	.20	.30	.25	.10	.05

Notice that they have assessed from their collective knowledge a *probability* that the various outcomes will occur.

E X A M P L E 3   **Calculating the Expected Value**

Solution   Suppose in Example 1 you are told that the average profit per yacht is $20,000 and for the competitor is $21,000. Can you decide which company will have the higher profits per quarter?

For Example 1, the average quarterly sales are 2.52 and the average profit per yacht is $20,000, thus the average quarterly profit is

$$(2.52)(\$20,000) = \$50,400.$$

Looking now at the competitor, the "expected" quarterly sales is

$$E(X) = 0(.10) + 1(.20) + 2(.30) + 3(.25) + 4(.10) + 5(.05) = 2.2.$$

(William J. Kennedy/The Image Bank.)

Thus you expect the competitor to sell 2.2 yachts per quarter. Their quarterly profit is then expected to be

$$(2.2)(\$21,000) = \$46,200.$$

The profits of the competitor are less, so you feel more comfortable about the purchase. ■

There are a number of versions of roulette. In the next example, we will find the expected loss of one type of bet for one version. In this version the wheel has the numbers from 0 to 36 located at 37 equally spaced slots. Half of the numbers from 1 to 36 are red and the other half are black. You place a bet. The wheel spins, and a ball falls randomly into one of the 37 slots.

E X A M P L E 4   **Calculating the Expected Value**

In this bet select any of the 37 numbers. Making a $1 bet results in winning back $36 if your number occurs and, of course, you lose your $1 if your number does not occur. Find the expected return.

Solution   A tree diagram is shown in Figure 8.6. There is a probability of $\frac{1}{37}$ of winning and of $\frac{36}{37}$ of losing. Let the random variable X be $35 if you win and $-\$1$ if you lose. Then the expected return is

$$E(X) = 35\,\frac{1}{37} - 1\,\frac{36}{37} \approx -.027.$$

On this bet you expect to lose about $.027 or 2.7% per bet in the long run. ■

$\frac{1}{37}$ $W: \$35$

$\frac{36}{37}$

$L: -\$1$

Figure 8.6

E X A M P L E 5   **Calculating the Expected Value**

An insurance company sells a 65-year-old man a 1-year life insurance policy. The policy, which costs $100, pays $5000 in the event the man dies during the next year. If there is a 1% chance that the man will die in the next year, find the probability distribution for this financial transaction for the insurance company and find their expected return. What meaning does this expected value have to the insurance company?

Solution   If the man lives for the next year, $100 is made. If the man dies within the next year, $5000 − $100 = $4900 is lost. The random variable and the associated probability is described in the following table.

Event	*Lives for One Year*	*Dies within the Next Year*
Random Variable x	100	− 4900
$P(X = x)$	.99	.01

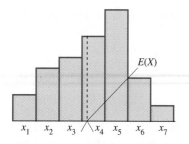

Figure 8.7

Thus

$$E(X) = 100(.99) - 4900(.01) = 50$$

dollars. If the insurance company were to sell a large number of the same insurance policies under the same conditions, they would expect a $50 return per such policy. ∎

There is a physical interpretation of expected value that is sometimes useful. Draw a histogram of the probability distribution as indicated in Figure 8.7. Turn the rectangle above x_1, the area of which is the probability p_1 of x_1 occurring, into a thin rectangle with uniform thickness with *weight* equal to p_1. Do the same for each of the points x_i. Then the point $E(X)$ is the point at which the histogram will be balanced.

Mean of the Binomial Distribution

Figure 8.1b of the last section gives the histogram for the binomial distribution $b(k : 6, .50)$. Notice the perfect symmetry about the point $x = 3$. By the physical interpretation of expected value, we should have $E(X) = 3$. This is indeed the case.

The histogram for the binomial distribution $b(k : 6, .75)$ is shown in the Solution to Self-Help Exercise 2 in the last section. One expects from the physical interpretation of $E(X)$ that $E(X)$ should be greater than 3; it is. Indeed the following is true.

Expected Value of Binomial Distribution

The expected value of the binomial distribution

$$p_k = b(k : n, p) = C(n, k)p^k q^{n-k} \qquad p + q = 1$$

is

$$E(X) = np.$$

E X A M P L E   6   **Finding the Expected Value of a Binomial Distribution**

Find the expected value of the binomial distribution $b(k: 6, .75)$ mentioned above. Is it greater than 3?

Solution   From (1) we obtain

$$E(X) = np = 6(.75) = 4.5,$$

which is greater than 3. ∎

E X A M P L E   7   **Finding the Expected Value of a Binomial Distribution**

A manufacturer of light bulbs produces 1% of them with defects. In a case of 100 bulbs, what will be the average number of defective bulbs?

Solution   If $p = .01$, then choosing a defective light bulb is a binomial distribution where ''success'' is picking a defective light bulb. Using (1) yields

$$E(X) = np = 100(.01) = 1. \quad \blacksquare$$

Median and Mode

Suppose 10 families live on Millionaire Avenue. One family has an annual income of $10 million, while the other 9 have annual incomes of $20,000 each. The mean annual income is over $1 million. Does this number convey what we would like the ''average'' to convey? Hardly. This simple example illustrates what happens to the mean when one or a few extreme values exist in a set of numbers.

Two other measures of the center of a set of numbers are the **median** and the **mode**. Roughly speaking, the median of a collection of numbers is the middle number when arranged in increasing (or decreasing) order, while the mode is the number that occurs most often. A more precise definition, covering all possible cases, is given in the following.

Median and Mode

   1. The **median** of a set of numbers is the middle number when the numbers are arranged in order of size and there are an odd number in the set. In the case that the number in the set is even, the median is the mean of the two middle numbers.
   2. The **mode** of a set of numbers is the number that occurs more frequently than the others. If the frequency of occurrence of two numbers is the same and also greater than the frequency of occurrence of all the other numbers, then we say the set is **bimodal** and has two modes. If no one or two numbers occurs more frequently than the others, we say that the set has no mode.

E X A M P L E   8   **Finding the Mean, Median, and the Mode**

Find the mean, median, and the mode of the set of data $\{1, 1, 1, 2, 3, 6, 6, 6\}$.

Solution   The mean is $(1 + 1 + 1 + 2 + 3 + 6 + 6 + 6)/8 = 3.25$. Since there is an even number in the collection, the median is the mean of the middle two numbers 2 and 3. This is 2.5. The set is bimodal with modes 1 and 6. $\blacksquare$

The following example illustrates that the mean is very sensitive to extreme values whereas the median and mode are not.

E X A M P L E   9   **Finding the Mean, Median, and the Mode**

Find the mean, median, and the mode of the set of data {1, 1, 1, 2, 3, 6, 6, 6, 8974} and compare your answers to the answers in the previous example.

Solution   The mean is $(1 + 1 + 1 + 2 + 3 + 6 + 6 + 6 + 8974)/9 = 1000$, which is considerably different from the mean in the previous example. Since there is an odd number in the collection, the median is the middle number 3. This has changed only slightly. The set is still bimodal with modes 1 and 6. ∎

SELF-HELP EXERCISE SET 8.2

1. Find the expected value of the random variable having the probability distribution given in the following table.

Random Variable x	-2	-1	0	1	3	5
$P(X = x)$	.20	.15	.30	.05	.20	.10

2. In this version of roulette you bet on a color, say red. If red occurs, you win back twice your bet. If black occurs, you lose your stake. If 0 occurs, the wheel is spun again until 0 does not occur. If red occurs, then you win your original stake back. If black comes up, you lose your original stake. Find the expected return.

EXERCISE SET 8.2

1. If the random variable X denotes the real valued outcomes given in the following table with the given frequency of occurrence of each of the outcomes, fill in the last line of the table and find $E(X)$.

Outcome	0	2	4	6	8	10
Frequency	50	5	10	5	20	10
$P(X = x)$			.4			

2. If the random variable X denotes the real valued outcomes given in the following table with the given frequency of occurrence of each of the outcomes, fill in the last line of the table and find $E(X)$.

Outcome	-2	-1	0	1	2	10
Frequency	10	11	9	9	5	6
$P(X = x)$						

3. Find the expected value of the random variable having the probability distribution given in the following table.

Random Variable x	− 30	− 10	0	5	10	20
P(X = x)	.15	.25	.05	.10	.25	.20

4. Find the expected value of the random variable having the probability distribution given in the following table.

Random Variable x	− 4	− 1	0	1	2	3
P(X = x)	.10	.10	.40	.10	.10	.20

5. Babe Ruth established the all-time highest "slugging average" for a single season in 1920. With the random variable X assigned as indicated in the following table with the associated probability, find E(X) rounded to 3 decimal places. This is the "slugging average."

Outcome	No Hit	Single	Double	Triple	Home Run
Random Variable x	0	1	2	3	4
P(X = x)	.6245	.1594	.0786	.0197	.1179

6. A student takes a single course during summer school and decides that the probability of obtaining each grade is as shown in the following table. Find the expected grade-point-average for this summer; that is, find the expected value of the random variable X.

Outcome	A	B	C	D	F
Random Variable x	4	3	2	1	0
P(X = x)	.6	.2	.1	.05	.05

7. The annual returns for an investment over the last 3 years are 50%, 0%, − 50%. Find the average annual return. Find the actual return over the 3-year period.

8. A 2-year investment breaks even at the end of two years but yielded a 50% loss in the first year. Find the average annual rate of return for this investment.

9. For an advertising promotion a company is giving away $10,000. All you have to do is submit a letter with your address on the entry form. Your entry will then be randomly selected with your chance of winning the same as anyone else's. It turns out that 100,000 people enter this contest. If your only cost is a 29 cent stamp, what is your expected return?

10. In the daily numbers game, a person picks any of the 1000 numbers from 0 to 999. A $1 wager on one number will return $600 if that number hits; otherwise nothing is returned. If the probability of any one number is the same as any other, find the expected return of this wager.

11. One version of roulette has a wheel with 38 numbers (1 through 36 plus 0 and 00) in 38 equally spaced slots. Half of the numbers from 1 to 36 are red, and the other half are black. The numbers 0 and 00 are green. Consider a $1 bet on red. If red comes up, you are returned $2, otherwise you lose the $1 wagered. Find the expected return for this bet.

12. In the version of roulette in the previous exercise, consider a $1 bet on a number from 1 to 36. If your number hits, you have $36 returned to you; otherwise you lose your $1 bet. Find the expected return for this bet.

13. In the version of roulette in the previous exercise, where the 0 and 00 are adjacent, consider a $1 bet that can be made on two adjacent numbers from 1 to 36. If one of the numbers comes up, the player receives $18; otherwise nothing is received. What is the expected value of this bet?

14. A fair die is tossed. If X denotes the random variable giving the number of the top face of the die, find $E(X)$.

15. A lottery has a grand prize of $500,000, 2 runner-up prizes of $100,000 each, and 100 consolation prizes of $1000 each. If 1 million tickets are sold for $1 each and the probability of any ticket winning is the same as any other, find the expected return on a $1 ticket.

16. A lottery has a grand prize of $50,000, 5 runner-up prizes of $5,000 each, 10 third-place prizes of $1000, and 100 consolation prizes of $10 each. If 100,000 tickets are sold for $1 each and the probability of any ticket winning is the same as any other, find the expected return on a $1 ticket.

17. Two fair dice are tossed. If you roll a total of 7, you win $6; otherwise you lose $1. What is the expected return of this game?

18. Two coins are taken at random (without replacement) from a bag containing 5 nickels, 4 dimes, and 1 quarter. Let X denote the random variable given by the total value of the two coins. Find $E(X)$.

19. Two people are asked to find the average of 200 numbers. They decide that one will find the average of the first 100 and the other will find the average of the second 100. They will then take the average of these two averages. Is this the average of the 200 numbers? Why or why not?

20. A person is asked to find the average of the 100 numbers 1234(1), 1234(2), ..., 1234(100). They decide to factor out the number 1234, then find the average of the integers from 1 to 100, and then multiply this last average by 1234. Are they obtaining the correct average? Why or why not?

21. A person is asked to find the average of the 100 integers from 1234 to 1333. They decide to "move each data point 1234 units to the left, find the average of the resulting data, then move the average over 1234." In other words, they find the average of all the integers from 0 to 99, and then add 1234 to this average. Is this the correct average? Why or why not?

22. Let X be a discrete random variable and c a constant. Define Y to be the random variable given by $Y = cX$.
    a. Show $E(Y) = cE(X)$, that is, show $E(cX) = cE(X)$.
    b. Given the random variable X with the probability distribution shown in the following table, find $E(X)$.

Random Variable x	12	24	36
$P(X = x)$	.1	.6	.3

    c. Now using the fact that the probability distribution for $\frac{1}{12}X$ must be the same as that for X, find $E(\frac{1}{12}X)$. Verify that $E(X) = 12E(\frac{1}{12}X)$.

23. Let X be a discrete random variable and c a constant. Define Y to be the random variable given by $Y = X - c$.
    a. Show $E(Y) = E(X) - c$, that is, show $E(X - c) = E(X) - c$, or $E(X) = E(X - c) + c$
    b. Given the random variable X with the probability distribution shown in the following table, find $E(X)$.

Random Variable x	12	13	14
$P(X = x)$	.2	.5	.3

    c. Now find the probability distribution for $X - 13$ and then find $E(X - 13)$. Verify that $E(X) = E(X - 13) + 13$.

24. Let X and Y be two discrete random variables associated with the outcomes of the same experiment. Define the random variable Z by $Z = X + Y$. Show $E(Z) = E(X) + E(Y)$, that is, show $E(X + Y) = E(X) + E(Y)$.

25. If the mean for a binomial distribution $b(k; 10, p)$ is .9, find p.

Applications

26. **Insurance.** An insurance company sells a $10,000, 5-year term life insurance policy to an individual for $700. Find the expected return for the company if the probability that the individual will live for the next 5 years is .95.

27. **Sales.** The number of sales per week and the associated probabilities of a car salesman are given in the following table. Find the expected number of sales per week.

Number of Sales per Week	0	1	2	3	4
Probability	.50	.30	.10	.07	.03

28. **Employee Attendance.**   A bank has 4 tellers. The following table gives the probabilities that a given number will be at work on any given day.

Number at Work	4	3	2	1	0
Probability	.80	.10	.07	.02	.01

Find the expected number that show up for work on a given day.

29. **Investment Returns.**   The following tables give all the possible returns and the associated probabilities of two investments, A and B. Find the expected value of each investment and compare the two.

Outcome of A	$1000	$2000
Probability	.2	.8

Outcome of B	−$1000	$0	$9000
Probability	.8	.1	.1

30. **Investment Advisor.**   An investment advisor informs you that his average annual return for the last 3 years is 100%. Furthermore, he says his annual return during each of the first 2 years of this 3-year period was 200%. Find how the people did who followed his advice by finding the return for the third year of this 3-year period.

31. **Quality Control.**   Electrical switches are manufactured with the probability of .05 that any one is defective. If 50 are chosen at random, what is the expected number of defective switches in this batch?

32. **Medicine.**   For a certain heart operation, the probability of survival is .95. If 10 of these operations are performed every week, what is the expected number of deaths due to this operation?

33. **Comparing Investments.**   Two car dealerships are up for sale. The following two tables give the number of cars sold per day together with the associated probabilities.

	First				*Second*				
Number Sold in One Day	0	1	2	3	0	1	2	3	4
Probability of Occurrence	.50	.30	.15	.05	.60	.20	.05	.05	.10

The average profit per car at the first dealership is $400 and at the second is $300. Which dealership will yield the highest daily profit?

34. **Comparing Investments.** Two motels are up for sale. The following two tables give the number of rooms rented per day together with the associated probabilities.

Number Rented in One Day	5	6	7	8	9	10
Probability of Occurrence	.10	.30	.40	.10	.05	.05

Number Rented in One Day	3	4	5	6	7
Probability of Occurrence	.05	.05	.10	.20	.60

The average profit per room rented at the first motel is $20 and at the second is $21. Which motel will yield the highest daily profit?

35. **Cases Argued Before Supreme Court.** Given below is the number of cases argued before the U.S. Supreme Court for 5 selected years. Find the mean of these 5 numbers.

Year	1970	1975	1980	1985	1990
Cases argued	151	179	154	171	125

36. **Voter Turnout.** The following table gives the voter turnout as a percentage of the voting-age population in the 5 presidential elections from 1972 to 1988. Find the mean percentage turnout for these five presidential elections.

Year	1972	1976	1980	1984	1988
% of VAP voting	55	54	53	53	50

37. **Number of Federal Employees.** The following table gives the number of federal employees as a percentage of the population. Find the mean of these percentages.

Year	1960	1965	1970	1975	1980	1985	1990
Percentage	1.32	1.28	1.44	1.32	1.23	1.24	1.20

38. **Oil Pollution.** The number of oil polluting incidents in and around U.S. waters for recent years is given in the following table. Find the average number per year of such incidents.

Year	1986	1987	1988	1989	1990
Incidents	6539	6352	6791	8225	7114

39. **Oil Pollution.**   The number of millions of gallons of oil polluted in and around U.S. waters for recent years is given in the following table. Find the average amount per year.

Year	1986	1987	1988	1989	1990
Millions of gallons	4.6	3.9	6.6	13.6	4.3

40. **Health Care Unit Location.**   A state government is trying to locate a new health care unit to serve the needs of the city of Adams and the four smaller cities indicated in the figure. The table gives the population and coordinates of the centers of the five cities. For computational purposes we assume that *all* residents of a city are located precisely at the center of each city and that there are no residents between cities. The health care unit should be located at the population "center" of these five cities. The probability that a randomly selected individual in this community is in North Adams is just the population of North Adams divided by the total population of this community of five cities. Similarly for the other four cities. If X and Y are the random variables that take on the values of x_i and y_i, respectively, then the population "center" and therefore the location of the health unit should be located at the coordinate $(E(X), E(Y))$. Find this point.

City	Population	Population Coordinates	
		x_i	y_i
North Adams	11,000	3	7
West Adams	12,000	1	4
Adams	42,000	4	3
East Adams	17,000	6	5
South Adams	18,000	5	1

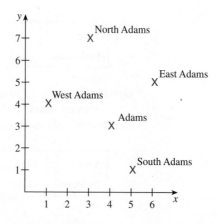

Graphing Calculator Exercises

Solve Exercises 41 through 42 by using the statistical routines on your graphing calculator.

41. Public Expenditure on Education. Following is a table of the public expenditure on education as a percentage of GNP in the United States. Find the mean and also find the mean by using the results of Exercise 23 with $c = 6.7$.

Year	1980	1981	1982	1983	1984	1985	1986	1987	1988	1989	1990
% of GNP	6.6	6.7	6.5	6.7	6.7	6.6	6.7	6.9	6.9	6.9	6.9

42. Lives Lost in Tornadoes. The following table gives the number of lives lost from tornadoes in the United States. Find the mean number of lives lost for these 10 years.

Year	1981	1982	1983	1984	1985	1986	1987	1988	1989	1990
Deaths	24	64	34	122	94	15	59	32	50	53

Solutions to Self-Help Exercise Set 8.2

1. $E(X) = -2(.20) - 1(.15) + 0(.30) + 1(.05) + 3(.20) + 5(.10) = 0.6$.

2. A tree diagram is shown in the figure. On a \$1 bet there is a probability on the first spin of $\frac{18}{37}$ of red occurring and of winning \$1 (\$2 less \$1 bet). There is a probability of $\frac{18}{37}$ of black occurring and of losing \$1. Then there is a probability of $\frac{1}{2}\frac{1}{37}$ of 0 and then red, which results in breaking even and a probability of $\frac{1}{2}\frac{1}{37}$ of 0 and then black with a subsequent loss of \$1. The expected value is

$$1\frac{18}{37} - 1\frac{18}{37} + 0\frac{1}{2}\frac{18}{37} - 1\frac{1}{2}\frac{18}{37} \approx -.0135.$$

On this bet you expect to lose about 1.35% per bet in the long run.

Clearly then, in general, this bet will permit one to play twice as long as the bet in Example 4 with the same stake. Many people will play the first bet because of the excitement of occasionally obtaining a substantially larger winning amount. But notice that the casino makes the customer pay for this excitement by taking a bigger cut of the money bet.

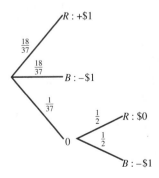

8.3   MEASURES OF DISPERSION

▶ *Variance and Standard Deviation*

▶ *An Alternate Form of the Variance*

▶ *Variance as a Measure of Risk*

▶ *Chebyshev's Inequality*

▶ *Variance of the Binomial Distribution*

APPLICATION

Quality Control

A manufacturer of precision ball bearings has just started up two assembly lines. Industry standards require these bearings to be 1 cm in diameter with at most a 1% error. In other words, the diameter of a ball bearing must lie in the interval [.99, 1.01] or be rejected. The first 4 ball bearings coming off each assembly line are very carefully measured in the lab. Initial word comes back that the average diameter of the 4 off the first assembly line have an average right at 1 cm, which looks very good, but that the average diameter of the 4 off the second assembly line is somewhat in error at 1.005 cm. At this point would you accept the products from the first assembly line? The second? Why or why not? For the answers see below.

Variance and Standard Deviation

In the last section we saw that the mean or the expected value is one measure of the "center" of the data. In this section, we will use the variance and the standard deviation to measure the extent of the dispersion of the data from the mean. We will then use Chebyshev's inequality to give crude estimates of any probability distribution, knowing only the mean and variance.

Further word comes from the lab of the manufacturer of ball bearings mentioned above indicating that none of the first 4 ball bearings are acceptable, having diameters of 1.020, .980, 1.030, and .970. Apparently the first assembly line produces an average right on target, but the actual diameters are so widely dispersed that the first assembly line is apparently useless for production of acceptable ball bearings.

Word also arrives that all the ball bearings from the second assembly line meet the standards, having diameters 1.006, 1.004, 1.007, 1.003. The average 1.005 is a little off being exactly 1, but the actual diameters are very close to this average and are dispersed only a very small amount.

The above example indicates how the *dispersion* of the data can play as important a role as the average. In this section we will define numbers called the **variance** and the **standard deviation** that measure the dispersion of data about the expected value. If the data is clustered close to the expected value, then the variance and the standard deviation will be small. If the data is widely dispersed away from the expected value, then the variance and standard deviation will be large.

(Barry L. Runk/Grant Heilman Photography, Inc.)

Suppose an experiment has the n outcomes $x_1, x_2, \ldots, x_n$, which we assume for the present are all equally likely and μ is the mean. Then it is tempting to measure the dispersion by taking the average of all the differences

$$x_1 - \mu, x_2 - \mu, \ldots, x_n - \mu.$$

However, the average of these differences must be zero. (Why?) To avoid this we take the average of the *square* of the above differences and call this the **variance**. Thus

$$\text{Var} = \frac{(x_1 - \mu)^2 + (x_2 - \mu)^2 + \cdots + (x_n - \mu)^2}{n}.$$

E X A M P L E   1   **Finding the Variance**

Find the variance of each of the sets of 4 ball bearings that were the first to come off the assembly line mentioned above.

Solutions   (a)   Here the data are 1.020, .980, 1.030, .970. The mean is $\mu = 1.000$. Thus

$$\text{Var} = \frac{(x_1 - \mu)^2 + (x_2 - \mu)^2 + (x_3 - \mu)^2 + (x_4 - \mu)^2}{4}$$

$$= \frac{(1.020 - 1)^2 + (.980 - 1)^2 + (1.030 - 1)^2 + (.970 - 1)^2}{4}$$

$$= \frac{.0004 + .0004 + .0009 + .0009}{4}$$

$$= .00065.$$

(b)   Here the data are 1.006, 1.004, 1.007, 1.003. The mean is $\mu = 1.005$. Thus

$$\text{Var} = \frac{(x_1 - \mu)^2 + (x_2 - \mu)^2 + (x_3 - \mu)^2 + (x_4 - \mu)^2}{4}$$

$$= \frac{(1.006 - 1.005)^2 + (1.004 - 1.005)^2 + (1.007 - 1.005)^2 + (1.003 - 1.005)^2}{4}$$

$$= \frac{.000\ 001 + .000\ 001 + .000\ 004 + .000\ 004}{4}$$

$$= .000\ 0025. \quad \blacksquare$$

As one can see, the variance in the second case is substantially smaller than in the first case.

There is one unsettling aspect of the variance we just found. The units are *squared*. Thus in the first case above, the variance is .00065 and is in cm^2 and thus is difficult to relate back to the original problem, which is in cm. To remedy this we define the **standard deviation** to be the square root of the variance and denote it by σ. Thus $\sigma = \sqrt{\text{Var}}$.

E X A M P L E 2   **Finding the Standard Deviation**

Find the standard deviation of the two sets of data in the previous example.

Solutions   (a) $\sigma = \sqrt{.00065} \approx .025$

Now this number .025 cm can be compared to the acceptable error of the ball bearings, which was .01 cm, and thus viewed as large.

(b) $\sigma = \sqrt{.000\ 0025} \approx .0016$

Now this number .0016 cm can be compared to the acceptable error of the ball bearings, which was .01 cm, and thus viewed as small. ∎

In general the outcomes of the experiment are not equally likely. We now consider this general situation.

In general suppose there are the n outcomes $x_1, x_2, \ldots, x_n$, with respective probabilities $p_1, p_2, \ldots, p_n$. Now suppose the experiment has been repeated a large number of times N so that x_1 is observed to occur a frequency of f_1 times, x_2 is observed f_2 times, and so on. (We must have $f_1 + f_2 + \cdots + f_n = N$.) Then if we set $\mu = E(X)$, the variance is

$$\text{Var} = \frac{\overbrace{(x_1 - \mu)^2 + \cdots + (x_1 - \mu)^2}^{f_1 \text{ times}} + \cdots + \overbrace{(x_n - \mu)^2 + \cdots + (x_n - \mu)^2}^{f_n \text{ times}}}{N}$$

$$= \frac{(x_1 - \mu)^2 f_1 + (x_2 - \mu)^2 f_2 + \cdots + (x_n - \mu)^2 f_n}{N}$$

$$= (x_1 - \mu)^2 \frac{f_1}{N} + (x_2 - \mu)^2 \frac{f_2}{N} + \cdots + (x_n - \mu)^2 \frac{f_n}{N}.$$

But as N gets larger, we expect the relative frequency $\dfrac{f_1}{N}$ to approach p_1, $\dfrac{f_2}{N}$ to approach p_2, and so forth. Thus the right-hand side of the last displayed line approaches

$$(x_1 - \mu)^2 p_1 + (x_2 - \mu)^2 p_2 + \cdots + (x_n - \mu)^2 p_n.$$

We call this term the **variance**. In the language of random variables, we then have the following.

Variance and Standard Deviation

Let X denote the random variable that takes on the values $x_1, x_2, \ldots, x_n$, and let the associated probabilities be $p_1, p_2, \ldots, p_n$. Then if $\mu = E(X)$, the **variance** of the random variable X, denoted by $\text{Var}(X)$, is

$$\text{Var}(X) = (x_1 - \mu)^2 p_1 + (x_2 - \mu)^2 p_2 + \cdots + (x_n - \mu)^2 p_n.$$

The **standard deviation**, denoted by $\sigma(X)$ is

$$\sigma(X) = \sqrt{\text{Var}(X)}.$$

E X A M P L E 3 **Comparing Variances of Two Probability Distributions**

Figure 8.8 shows the histograms of 2 probability distributions both with means equal to 1. Find the variance of each and compare the two.

Solutions The variance of the first probability distribution shown in Figure 8.8 is

$$(0 - 1)^2(0.10) + (1 - 1)^2(.80) + (2 - 1)^2(0.10) = 0.20.$$

The variance of the second probability distribution shown in Figure 8.8 is

$$(0 - 1)^2 \frac{1}{3} + (1 - 1)^2 \frac{1}{3} + (2 - 1)^2 \frac{1}{3} = \frac{2}{3}.$$

The variance of the first probability distribution is much smaller than the variance of the second. This just reflects the fact, seen from Figure 8.8, that the first probability distribution is less dispersed from the mean than the second one. ■

Often the probabilities are not known, but an experiment has been performed N times. Then the probabilities $p_1, p_2, \ldots, p_n$, are the relative frequencies (empirical probabilities). This is illustrated in the next problem.

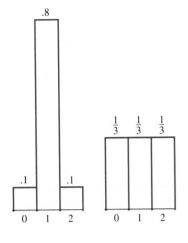

Figure 8.8

E X A M P L E 4 **Finding the Standard Deviation**

You have completed 40 courses, each with the same number of credits, with the frequency of each grade given in the following table.

Outcome	A	B	C	D	F
Random Variable X	4	3	2	1	0
Frequency	25	11	3	1	0

Find the expected value of the random variable X; that is, find the grade point average, and then find Var(X).

Solution We first notice that the total number is given by

$$f_1 + f_2 + f_3 + f_4 + f_5 = 25 + 11 + 3 + 1 + 0 = 40.$$

Dividing each of the frequencies by 40 gives the relative frequencies or the empirical probability. Doing this gives the following table.

Outcome	A	B	C	D	F
Random Variable x	4	3	2	1	0
Frequency	25	11	3	1	0
Relative Frequency	$\frac{25}{40}$	$\frac{11}{40}$	$\frac{3}{40}$	$\frac{1}{40}$	0

The expected value is then

$$\mu = 4\frac{25}{40} + 3\frac{11}{40} + 2\frac{3}{40} + 1\frac{1}{40} + 0\frac{0}{40} = 3.5.$$

Then

$$Var(X) = (4 - 3.5)^2\frac{25}{40} + (3 - 3.5)^2\frac{11}{40} + (2 - 3.5)^2\frac{3}{40}$$
$$+ (1 - 3.5)^2\frac{1}{40} + 0 = .55. \blacksquare$$

An Alternate Form of the Variance

The computation of

$$Var(X) = (x_1 - \mu)^2p_1 + (x_2 - \mu)^2p_2 + \cdots + (x_n - \mu)^2p_n$$

can be simplified by using an alternative form. To see how, write the terms on the right-hand side as

$$(x_1 - \mu)^2p_1 = x_1^2p_1 - 2x_1\mu p_1 + \mu^2p_1$$

$$\cdots$$

$$(x_n - \mu)^2p_n = x_n^2p_n - 2x_n\mu p_n + \mu^2p_n.$$

Summing up the left-hand sides of these equations gives $Var(x)$. Doing this and rearranging the sum of the right-hand sides gives

$$Var(X) = (x_1^2p_1 + \cdots + x_n^2p_n) - 2\mu(x_1p_1 + \cdots + x_np_n)$$
$$+ \mu^2(p_1 + \cdots + p_n).$$

But recall that $\mu = x_1p_1 + \cdots + x_np_n$

while $p_1 + \cdots + p_n = 1.$

Using these last two facts then yields

$$Var(X) = (x_1^2p_1 + \cdots + x_n^2p_n) - 2\mu\mu + \mu^2 =$$
$$(x_1^2p_1 + \cdots + x_n^2p_n) - \mu^2.$$

This will save a substantial amount of work. We have the following.

Alternate Form of Variance

An alternate form of the variance is given by

$$Var(X) = (x_1^2p_1 + x_2^2p_2 + \cdots + x_n^2p_n) - \mu^2.$$

EXAMPLE 5 **Using the Alternate Form of the Variance**

Use this last form to find $\text{Var}(X)$ in the last example.

Solution   Form the following table.

x_1	p_i	$x_i p_i$	$x_i^2 p_i$
4	$\frac{25}{40}$	2.500	10.000
3	$\frac{11}{40}$	0.825	2.475
2	$\frac{3}{40}$	0.150	0.300
1	$\frac{1}{40}$	0.025	0.025
0	$\frac{0}{40}$	0.000	0.000
		Sum =	Sum =
		3.5	12.8

Then

$$\text{Var}(X) = (x_1^2 p_1 + x_2^2 p_2 + \cdots + x_n^2 p_n) - \mu^2 = 12.8 - (3.5)^2 = 0.55. \quad \blacksquare$$

Variance as a Measure of Risk

It is a basic tenet in the modern theory of finance that the amount that the value of an asset varies over time is a measure of the risk of the asset. The standard measure of this variability is the standard deviation.

   If everyone has a great deal of confidence what the future value of an investment is, then the value of the investment should vary little. For example, everyone has essentially complete confidence what the future performance of money in the bank will be. Thus the value of this asset (the principal) does not vary in time. This reflects the fact that this investment is (essentially) risk-free. Everyone has very high confidence in the future performance of a high quality bond. However, there is some variability in price as investors change somewhat from week to week their assessment of the economy, inflation, and other factors that may affect the price of the bond. This reflects the fact that a bond does have some risk.

   Moving on to stocks, the future earnings and dividend performance of a very large and regulated company such as AT&T is estimated by many analysts with reasonable confidence and so the price of this security does not vary significantly. However, the future earnings of a small bio-tech firm is likely to be very unpredictable. Will it be able to come up with new products? Will it be able to fend off competitors? Will some key personnel leave? Thus the weekly assessment of such a firm is likely to change substantially more than that of AT&T. As a consequence, the weekly price changes of this company will vary significantly more, reflecting the fact that an investment in the small company is more risky.

Therefore the risk of a security is defined as the standard deviation of the daily (or perhaps weekly) prices of the security.

Table 8.5 gives the weekly price changes (during 5 consecutive weeks in 1991) in percent of 4 assets: money in the bank, a high quality bond (IBM), AT&T, and Scan Optics.

Table 8.5

Principal in Bank	IBM $7\frac{7}{8}04$ Bond	AT&T Stock	Scan Optics Stock
0.0	0.0	0.3	4.6
0.0	-0.2	0.0	4.7
0.0	-0.1	3.1	0.0
0.0	0.0	-2.0	8.7
0.0	-0.5	2.1	-4.2

According to Exercises 17 and 18 at the end of this section, the standard deviation of the four assets over this period of time was 0, 0.19, 1.77, 4.44, respectively. This reflects the fact that these assets are increasingly risky in the order listed.

Chebyshev's Inequality

Knowing only the mean and variance of a probability law is not sufficient in general to determine the probability law. However, crude estimates of the probability law can be made, which suffice for many purposes, from a knowledge of just the mean and variance. One such estimate is Chebyshev's inequality.

Chebyshev's Inequality

Let X be a random variable with expected value μ and standard deviation σ. Then the probability that the random variable associated with a random outcome lies between $\mu - h\sigma$ and $\mu + h\sigma$ is at least $1 - 1/h^2$, that is,

$$P(\mu - h\sigma \leq X \leq \mu + h\sigma) \geq 1 - \frac{1}{h^2}.$$

A proof is not given here.

For Chebyshev's inequality to give new information, clearly we must have $h \geq 1$.

E X A M P L E 6   **Using Chebyshev's Inequality**

A probability distribution of a random variable X has mean $\mu = 14$ and standard deviation $\sigma = 4$. Use Chebyshev's inequality to give a lower bound for the probability that X is within 12 units of μ.

Solution   Setting $12 = h\sigma = h \cdot 4$, yields $h = 3$. Then $\mu - h\sigma = 2$ and $\mu + h\sigma = 26$ and

$$P(2 \leq X \leq 26) \geq 1 - \frac{1}{3^2} \approx .89.$$

Thus there is at least an 89% chance that the random variable lies on the interval $[2, 26]$. ∎

E X A M P L E 7   **Using Chebyshev's Inequality**

When cereal is packaged, the actual amount of cereal varies from box to box. Suppose a sample indicates that the average in a 16-ounce box is 16.0 ounces with a standard deviation of 0.5. Using Chebyshev's inequality, find the minimum percent of boxes that will have 15 to 17 ounces of cereal.

Solution   Let X be the random variable equal to the number of ounces in a box of cereal. We are given $\mu = 16$ and $\sigma = 0.5$. We are seeking an estimate of $P(15 \leq X \leq 17)$ or in terms of $\mu = 16$, $P(\mu - 1 \leq X \leq \mu + 1)$. In the terminology of Chebyshev's inequality, we then must have $1 = h\sigma = h(0.5)$. This implies $h = 2$. Thus

$$P(15 \leq X \leq 17) \geq 1 - \frac{1}{h^2} = 1 - \frac{1}{4} = .75.$$

Thus at least 75% of these boxes will have between 15 and 17 ounces of cereal. ∎

Variance of the Binomial Distribution

We give without proof the following additional useful fact about binomial distributions.

Variance of the Binomial Distribution

The variance of the binomial distribution with n trials and the probability of "success" equal to p and of "failure" equal to q is

$$\text{Var}(X) = npq.$$

SELF-HELP EXERCISE SET 8.3

1. The following table gives two probability distributions associated with the same outcomes.

Outcome	1	2	3	4	5
(a) Probability	.1	.2	.4	.2	.1
(b) Probability	.4	.1	0	.1	.4

Draw a histogram of each probability distribution. By inspecting these histograms (making no calculations), determine the means and determine which one has the largest variance. Now calculate the means, variances, and standard deviations.

2. Use Chebyshev's inequality to estimate the minimum probability that the outcome is between 2 and 4 for a probability distribution with $\mu = 3$ and $\sigma = 0.25$.

EXERCISE SET 8.3

In Exercises 1 through 2 you are given a table with two probability distributions associated with the same outcomes. For each exercise draw a histogram of each probability distribution. Using only these histograms and making no calculations, determine the means and determine which one has the largest variance. Now calculate the means, variances, and standard deviations.

1.

Outcome	*1*	*2*	*3*	*4*	*5*
(a) Probability	.2	.2	.2	.2	.2
(b) Probability	.1	.2	.4	.2	.1

2.

Outcome	*5*	*6*	*7*	*8*	*9*
(a) Probability	.1	.1	.6	.1	.1
(b) Probability	.3	.1	.2	.1	.3

3. In the June 1991 Atlantic City Golf Classic, the top 3 finishers had the following scores for the three rounds of the tournament.

Jane Geddes	71	68	69
Cindy Schreger	70	70	69
Amy Alcott	69	68	72

Find the mean, variance, and standard deviation for the three rounds for each of the players. Which had the lowest average? Which was the most consistent?

4. In the June 1991 McDonald's Golf Championship, the top 5 finishers had the following scores for the four rounds of the tournament.

Beth Daniel	67	71	67	68
Pat Bradley	69	67	70	71
Sally Little	67	69	67	74
M. McGann	70	66	72	70
A. Okamoto	70	65	73	70

Find the mean, variance, and standard deviation for the four rounds for each of the players. Who had the lowest average? Who was the most consistent? Least consistent?

5. Find the mean, variance, and standard deviation of the random variable with the probability distribution given in the following table.

Random Variable x	-3	0	1
$P(X = x)$	.2	.5	.3

6. Find the mean, variance, and standard deviation of the random variable with the probability distribution given in the following table.

Random Variable x	-2	1	3
$P(X = x)$	.1	.3	.6

7. Find the mean, variance, and standard deviation of the random variable with the probability distribution given in the following table.

Random Variable x	-2	0	1	2	4
$P(X = x)$	.1	.3	.1	.2	.3

8. Find the mean, variance, and standard deviation of the random variable with the probability distribution given in the following table.

Random Variable x	-2	-1	1	2	5
$P(X = x)$	.2	.1	.25	.25	.2

9. By inspection of the following two histograms (and making no calculations), determine the means and determine which one has the largest variance.

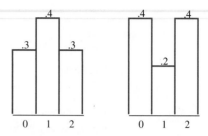

10. By inspection of the following two histograms (and making no calculations), determine the means and determine which one has the largest variance.

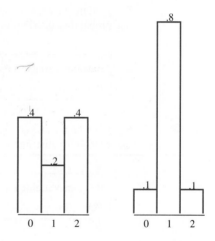

11. Let X be a discrete random variable and c a constant. Define Y to be the random variable given by $Y = cX$.

    a. Show $\text{Var}(Y) = c^2\text{Var}(X)$, that is, show $\text{Var}(cX) = c^2\text{Var}(X)$. *Hint:* Use the fact given in Exercise 22 of the last section that $E(cX) = cE(X)$, or $\mu(cX) = c\mu(X)$, and the second form of the variance.

    b. Given the random variable X with the probability distribution shown in the following table, find $\text{Var}(X)$.

Random Variable x	3	6	9
$P(X = x)$	.1	.6	.3

    c. Now using the fact that the probability distribution for $\frac{1}{3}X$ must be the same as for X, find $\text{Var}(\frac{1}{3}X)$. Verify that $\text{Var}(X) = 9\,\text{Var}(\frac{1}{3}X)$.

12. Let X be a discrete random variable and c a constant. Define Y to be the random variable given by $Y = X - c$.

    a. Show Var(Y) = Var(X), that is, show Var$(X - c)$ = Var(X). *Hint:* Use the fact given in Exercise 23 of the last section that $E(X) - c = E(X - c)$, or $\mu(X - c) = \mu(X) - c$, and the first form of the variance.

    b. Given the random variable X with the probability distribution shown in the following table, find Var(X).

Random Variable x	12	13	14
$P(X = x)$	.2	.5	.3

    c. Now find the probability distribution for $X - 13$ and then find Var$(X - 13)$. Verify that Var(X) = Var$(X - 13)$.

13. A probability distribution has a mean of 30 and a standard deviation of 2. Use Chebyshev's inequality to find the minimum probability that an outcome is between 20 and 40.

14. A probability distribution has a mean of 90 and a standard deviation of 3. Use Chebyshev's inequality to find the minimum probability that an outcome is between 60 and 120.

15. A probability distribution has a mean of 60 and a standard deviation of 1. Use Chebyshev's inequality to find the number c so that the probability that the numerical outcome is between $60 - c$ and $60 + c$ is at least .99.

16. A probability distribution has a mean of μ and a standard deviation of σ. Use Chebyshev's inequality to find a lower bound for the percentage of outcomes between $\mu - 5\sigma$ and $\mu + 5\sigma$.

Applications

17. Finance. Find the mean, variance, and standard deviation of the weekly percentage change in the value of the principal in the bank account and for the bond in Table 8.5 of the text. Which is varying the most? How do you correlate your answer with the risk of each investment?

18. Finance. Find the mean, variance, and standard deviation of the weekly percentage change in the values of the 2 stocks given in Table 8.5 of the text. Which is varying the most? How do you correlate your answer with the risk of each investment?

19. Finance. During a 5-week period in 1991, the stock of a utility company and the stock of a small company showed the following weekly percentage changes.

Company	*Weekly Price Change in %*				
United Illuminating	-1.1	0.4	1.5	-2.5	-2.2
U. S. Surgical	5.4	-1.2	5.9	1.0	-3.2

Find the variance of the weekly price changes of each. Relate the two variances found to the riskiness of the two stocks.

20. **Finance.**   During a 5-week period in 1991, the stock of a large insurance company and the stock of a small hi-tech company showed the following weekly percentage change.

Company	Weekly Price Change in %				
Aetna Life	−0.3	−2.8	1.2	1.8	0.0
Data Switch	2.9	−5.9	6.3	−11.8	−3.2

Find the variance of the weekly price changes of each. Relate the two variances found to the riskiness of the two stocks.

21. **Sales.**   The number of sales per week and the associated probabilities of 2 car salesmen, A and B, are given in the following table. Find the mean, variance, and standard deviation for the sales per week of each. Which one will sell the most cars? Which one is the most consistent?

Number of Sales per Week	0	1	2	3	4
Probability A	.50	.30	.10	.07	.03
Probability B	.40	.20	.20	.10	.10

22. **Employee Absences.**   The following table gives the probabilities that two employees, A and B, will have the given number of absences from work per month. Find the mean, variance, and standard deviation for each. Which attendance is the best? Which attendance is the most consistent?

Number of Absences per Month	0	1	2	3	4
Probability A	.90	.04	.03	.02	.01
Probability B	.85	.05	.05	.05	.00

23. **Venereal Disease.**   The following table gives the number of thousands of reported cases of gonorrhea and syphilis in the United States. Find the average, variance, and standard deviation of each. Which is varying the most?

Year	1985	1986	1987	1988	1989	1990
Gonorrhea	911	901	781	720	733	690
Syphilis	68	68	87	103	111	134

24. **Mortality Rates from Legal Abortions and Childbirth.**   The following table gives the mortality rates in numbers per 100,000 from legal abortions and child-

birth in the United States. Find the average, variance, and standard deviation of each. Which is varying the most?

Year	1981	1982	1983	1984	1985
Abortion	0.4	0.8	0.7	0.7	0.4
Childbirth	7.2	7.9	8.0	7.9	7.8

25. **Temperatures.** The following table gives the average monthly temperatures for two cities for four selected months. Find the average, variance, and standard deviation of each. Which is varying the most?

Month	Jan	Apr	Jul	Oct
San Diego	65	68	76	75
Chicago	29	59	83	64

26. **Health.** The following table gives the number of reported cases of plague and polio in the United States. Find the average, variance, and standard deviation of each. Which is varying the most?

Year	1985	1986	1987	1988	1989	1990
Plague	17	10	12	15	4	2
Polio, acute	7	8	6	9	5	7

27. **Drug Use.** The following table gives the percentage of current users of marihuana and cocaine in the 18 to 25-year-old age group in the United States. Find the average, variance, and standard deviation of each. Which is varying the most?

Year	1974	1979	1982	1985	1988	1991
Marihuana	25	35	27	22	16	13
Cocaine	3	9	7	8	5	2

28. **Drug Use.** The following table gives the percentage of current users of alcohol and cigarettes in the 18 to 25-year-old age group in the United States. Find the average, variance, and standard deviation of each. Which is varying the most?

Year	1974	1979	1982	1985	1988	1991
Alcohol	69	76	71	71	65	64
Cigarettes	49	43	40	37	35	32

29. **Scholastic Aptitude Test Scores.**   The following table gives the verbal SAT scores for recent years. Find the average, variance, and standard deviation of each. Which is varying the most?

Year	1987	1988	1989	1990	1991
Males	435	435	434	429	426
Females	425	422	421	419	418

30. **Scholastic Aptitude Test Scores.**   The following table gives the math SAT scores for recent years. Find the average, variance, and standard deviation of each. Which is varying the most?

Year	1987	1988	1989	1990	1991
Males	500	498	500	499	497
Females	453	455	454	455	453

31. **Air Pollutant Concentration.**   The following table gives the concentration of two pollutants, in appropriate units, for recent years. Find the average, variance, and standard deviation of each. Which is varying the most?

Year	1986	1987	1988	1989	1990
Carbon Monoxide	7.1	6.7	6.4	6.3	5.9
Ozone	.12	.13	.14	.12	.11

32. **Air Pollutant Concentration.**   The following table gives the concentration of two pollutants, in appropriate units, for recent years. Find the average, variance, and standard deviation of each. Which is varying the most?

Year	1986	1987	1988	1989	1990
Sulfur Dioxide	9	9	9	8	8
Nitrogen Dioxide	24	24	24	23	22

33. **Death Rates by Causes.**   The death rate per 100,000 in the United States for viral hepatitis and meningitis for selected years is given in the following table. Find the average, variance, and standard deviation of each. Which is varying the most?

Year	1970	1980	1985	1990
Viral Hepatitis	.5	.4	.4	.7
Meningitis	.8	.6	.5	.5

34. **Death Rates by Causes.** The death rate per 100,000 in the United States for accidents from automobiles and for all other accidents for selected years is given in the following table. Find the average, variance, and standard deviation of each. Which is varying the most?

Year	1970	1980	1985	1990
Automobiles	27	24	19	19
All Others	30	23	20	18

35. **Sales.** The expected amount of sales for each person on a sales force of a company is $100,000 per month with a standard deviation of $20,000. Find the minimum probability using Chebyshev's inequality that a salesman will sell between $70,000 and $130,000 in a particular month.

36. **Quality Control.** The expected number of defective parts produced on an assembly line per shift is 40 with a standard deviation of 10. Find the minimum probability using Chebyshev's inequality that the number of defective parts on a particular shift will be between 20 and 60.

37. **Useful Life of a Product.** The expected useful life of a certain brand of VCR is 5 years with standard deviation equal to 1 year. Find the minimum probability using Chebyshev's inequality that the useful life of one of these VCRs will be between 3 and 7 years.

38. **Weights.** If the mean weight of a delivered ton of top soil is 2000 pounds with a standard deviation 200 pounds, find the minimum probability using Chebyshev's inequality that the weight of a delivered "ton" is between 1700 and 2300 pounds.

Graphing Calculator Exercises

In Exercises 39 through 40 solve by using the statistical routines on your graphing calculator.

39. **Birth and Death Rates.** The following table gives the birth and death rates in the United States per 1000 total population for each of the 10 years from 1980 to 1989. Find the mean, variance, and standard deviation of each. Which is varying the most? Also use the results of Exercise 23 of the last section and Exercise 12 above with $c = 15.5$ for the birthrate and $c = 8.7$ for the death rate.

Year	1980	1981	1982	1983	1984	1985	1986	1987	1988	1989
Birth rate	15.9	15.8	15.9	15.5	15.5	15.8	15.6	15.7	15.9	16.2
Death rate	8.8	8.6	8.5	8.6	8.6	8.7	8.7	8.7	8.8	8.7

40. **Marriage and Divorce Rates.** The following table gives the marriage and divorce rates in the United States per 1000 total population for each of the 10 years from 1980 to 1989. Find the mean, variance, and standard deviation of

each. Which is varying the most? Use the results of Exercise 23 of the last section and Exercise 12 above with $c = 10.6$ for the marriage rate and $c = 5.0$ for the divorce rate.

Year	1980	1981	1982	1983	1984	1985	1986	1987	1988	1989
Marriage rate	10.6	10.6	10.6	10.5	10.5	10.1	10.0	9.9	9.7	9.7
Divorce rate	5.2	5.3	5.0	4.9	5.0	5.0	4.8	4.8	4.8	4.7

Solutions to Self-Help Exercise Set 8.3

1. The following figure shows the histograms.

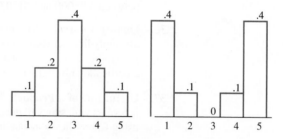

From the histograms one sees that the probability is evenly distributed about 3 in each case. Thus the mean should be 3. In the first case the probability is clustered about the mean, whereas in the second case the probability is more dispersed away from the mean. Thus the variance in the second case should be larger.

To calculate the mean and variance, make the following table.

		Probability (a)			Probability (b)	
x_i	p_i	$x_i p_i$	$x_i^2 p_i$	p_i	$x_i p_i$	$x_i^2 p_i$
1	.1	0.1	0.1	.4	0.4	0.4
2	.2	0.4	0.8	.1	0.2	0.4
3	.4	1.2	3.6	.0	0.0	0.0
4	.2	0.8	3.2	.1	0.4	1.6
5	.1	0.5	2.5	.4	2.0	10.0
		Sum = 3.0	Sum = 10.2		Sum = 3.0	Sum = 12.4

a. Thus from the table $\mu = 3$ and using the second form for variance gives

$$\text{Var} = 10.2 - 3^2 = 1.2$$

$$\sigma = \sqrt{1.2} \approx 1.095.$$

b. From the table $\mu = 3$ and using the second form for variance gives

$$\text{Var} = 12.4 - 3^2 = 3.4$$

$$\sigma = \sqrt{3.4} \approx 1.84.$$

As we see, the means are the same, and the variance of the second is larger.

2. If X is the random variable equal to the outcome, then we are seeking an estimate of $P(2 \le X \le 4)$. Since $\mu = 3$, this can be written as $P(\mu - 1 \le X \le \mu + 1)$. This implies that $1 = h\sigma = h(0.25)$. Thus $h = 4$ and, from Chebyshev's inequality,

$$P(\mu - h\sigma \le X \le \mu + h\sigma) \ge 1 - \frac{1}{h^2} = 1 - \frac{1}{(4)^2} = .9375.$$

Thus there is a probability of at least .9375 that the outcome will lie between 2 and 4.

8.4 THE NORMAL DISTRIBUTION

▸*Introduction*

▸*Standard Normal Distribution*

▸*Normal Distribution*

APPLICATION

Determination of Warranty Periods

A manufacturer of washing machines has collected data indicating that the time before one of its washers needs its first repair is normally distributed with mean 2 years and standard deviation 0.608 years. What length of time should the manufacturer set for the warranty period so that at least 95% of the machines will get through the warranty period without need of a repair? See Example 5 for the answer.

Introduction

In this section we will study the normal distribution, arguably the most important probability distribution. One reason for its importance is that many natural phenomena obey a normal distribution or at least have a probability distribution that closely approximates a normal distribution. Examples of such random variables are: the weights of house cats, the heights of women in Spain, the waiting time in a line at a bank, the diameter of a ball bearing coming off an assembly line, the pounds of fertilizer in a 100-pound bag.

Another reason for the importance of the normal distribution, is the "central limit theorem of probability theory." This theorem can be stated approximately

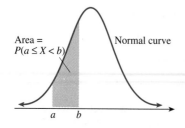

Figure 8.9

as follows. Suppose X is a random variable with an unknown probability distribution. Given a random sample of n observations with n large, the means

$$\bar{x} = \frac{x_1 + x_2 + \cdots + x_n}{n}.$$

have a distribution that is approximately normal. An important consequence of this theorem is that, under various conditions, the normal probability distribution closely approximates many other probability distributions. For example, we shall see in the next section how to approximate a binomial distribution with a normal distribution.

A normally distributed random variable has the property that it can take any value in the interval $(-\infty, +\infty)$ and thus is a continuous random variable. We are familiar with viewing probabilities associated with a discrete random variable as areas in a histogram. In a similar fashion we can view probabilities associated with a continuous random variable that is normally distributed as areas under a **normal curve**. Figure 8.9 shows a typical normal curve. The probability $P(a \le X \le b)$ that the random variable X is between a and b is the area under the normal curve on the interval $[a, b]$. The area under the entire curve is 1. This is merely due to the fact that the probability that the random variable X takes some value is 1.

Standard Normal Distribution

We now consider the *standard normal distribution*. As we shall see later in this section, to find the probability for a normally distributed random variable we find the probability of a certain standard normally distributed random variable. We thus need to work first with the standard normal distribution. We state the following definition.

Standard Normal Distribution

The random variable X has a standard normal distribution on the interval $(-\infty, +\infty)$ if the probability $P(a \le X \le b)$ that X is between a and b is the area under the standard normal curve given by

$$y = \frac{1}{\sqrt{2\pi}} e^{-0.5z^2}$$

on the interval $[a, b]$, where $\pi \approx 3.14159$ and $e \approx 2.71828$.

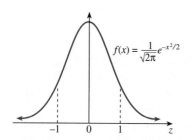

Figure 8.10   Standard normal
distribution

A graph of this curve is shown in Figure 8.10.
The curve has a number of important characteristics.

Characteristics of the Standard Normal Curve

The graph of the standard normal curve shown in Figure 8.10 has the following characteristics.

1. It is bell-shaped.
2. It is symmetric about $z = 0$.
3. It lies above the z-axis.
4. It is "asymptotic" to both the positive and negative z-axis.
5. The curve "bends downward" on the interval $(-1, 1)$ and "bends upward" outside this interval.
6. The area under the entire curve is exactly 1.

From the symmetry, it appears that the mean is zero. In fact, it can be shown that for the standard normal distribution, $\mu = 0$. It can also be shown that the standard deviation $\sigma = 1$.

Since the normal distribution is so extraordinarily important, a table has been constructed that gives the areas under the standard normal curve. Given any value b, the table gives the area under the curve to the left of b, shown as the shaded region in Figure 8.11. We designate this area as $\mathscr{A}(b)$. Table B.1 in Appendix B can be used to find this number.

C O N V E N T I O N. The random variable associated with the standard normal distribution is designated by Z.

If we are asked for $P(Z \le b)$, this is just $\mathscr{A}(b)$. Thus

$$P(Z \le b) = \mathscr{A}(b),$$

where we can find $\mathscr{A}(b)$ in Table B.1 in Appendix B.

The probability $P(b \le Z)$ is the area under the curve to the right of b, which is also the unshaded area under the curve in Figure 8.11. This area is also the entire area under the curve (which is 1) less the area under the curve to the left of b. Thus

$$P(b \le Z) = 1 - \mathscr{A}(b).$$

Finally, $P(a \le Z \le b)$ is the area under the curve from a to b shown in Figure 8.12. This can be obtained as the area to the left of b, $\mathscr{A}(b)$, less the area to the left of a, $\mathscr{A}(a)$. Thus

$$P(a \le Z \le b) = \mathscr{A}(b) - \mathscr{A}(a),$$

where both the numbers $\mathscr{A}(b)$ and $\mathscr{A}(a)$ are obtained in Table B.1 in the Appendix.

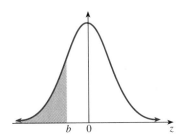

Figure 8.11

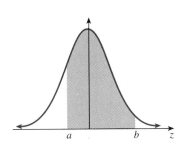

Figure 8.12

E X A M P L E   1   **Finding Probabilities for the Standard Normal Distribution**

Let Z be a random variable with a standard normal distribution. Use Table B.1 in Appendix B for the standard normal distribution to find (a) $P(Z \le 1.23)$ (b) $P(Z \ge 1.23)$ (c) $P(-1 \le Z \le 1)$ (d) $P(-2 \le Z \le 2)$.

Solutions   (a)   Here

$$P(Z \le 1.23) = \mathscr{A}(1.23),$$

which according to Table B.1 in Appendix B is .8907.

(b)   Here

$$P(Z \ge 1.23) = 1 - \mathscr{A}(1.23) = 1 - .8907 = .1093.$$

(c)   Here

$$P(-1 \le Z \le 1) = \mathscr{A}(1) - \mathscr{A}(-1),$$

which according to Table B.1 in Appendix B is

$$.8413 - .1587 = .6826.$$

(d)   Here

$$P(-2 \le Z \le 2) = \mathscr{A}(2) - \mathscr{A}(-2),$$

which according to Table B.1 in Appendix B is

$$.9772 - .0228 = .9544. \quad \blacksquare$$

It is instructive to notice that the answer to (c) is approximately two thirds and to (d) is approximately .95. That is, if the random variable is normally distributed with $\mu = 0$ and $\sigma = 1$, the probability that the random variable is within one standard deviation of the mean is about two thirds, and the probability that the random variable is within two standard deviations of the mean is about .95.

E X A M P L E   2   **Finding a Symmetric Interval About the Mean with Given Probability**

Let Z be a random variable with a standard normal distribution. Using Table B.1 for the standard normal distribution found in Appendix B, find the interval $[-c, c]$ so that the probability that the random variable is in this interval is .99.

Solution   Given the symmetry of the normal distribution, we need only find c such that the probability that the random variable is less than c is .995, that is, $P(Z \le c) = .995$. The area of the "tail" in Figure 8.13 is then .005. Symmetry then implies that the area of the "tail" to the left of $-c$ is also .005. Thus since the area under the entire curve is 1, the area under the curve from $-c$ to $+c$ is then $1 - .005 - .005 = .99$.

If we look up the value of c in Table B.1 for which $\mathscr{A}(c) = .995$, we obtain 2.575 after extrapolation. Thus $c = 2.575$, that is, $P(-2.575 \le Z \le 2.575) = .99$.

This also says that there is a 99% chance that the random variable Z is within about 2.5 standard deviations of the mean. $\blacksquare$

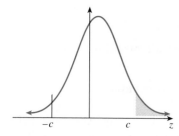

Figure 8.13

This gives a good indication of just how fast the standard normal curve is approaching the x-axis. The area under the curve on the interval $[-2.575, 2.575]$ is .99, so that the remaining area under all the rest of the curve is just .01.

Normal Distribution

We now consider the general normal distribution.

The Normal Probability Distribution

The random variable X has a **normal probability distribution with mean μ and standard deviation σ** on $(-\infty, +\infty)$ if the probability $P(a \leq X \leq b)$ that X is between a and b is the area under the normal curve given by

$$y = \frac{1}{\sigma\sqrt{2\pi}} e^{-0.5[(x-\mu)/\sigma]^2}$$

on the interval $[a, b]$.

Figure 8.14 indicates a typical such curve. The curve has a number of important characteristics.

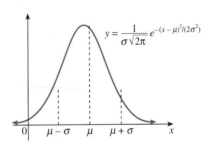

Figure 8.14   Normal distribution

Characteristics of the Normal Curve

The graph of the normal curve shown in Figure 8.14 has the following characteristics.

1. It is bell shaped.
2. It is symmetric about $x = \mu$.
3. It lies above the x-axis.
4. It is "asymptotic" to both the positive and negative x-axis.
5. The curve "bends downward" on the interval $(\mu - \sigma, \mu + \sigma)$ and "bends upward" outside this interval.
6. The area under the entire curve is exactly one.

Figure 8.15 indicates 3 such curves with the same standard deviation $\sigma = 1$ but different means. The curves with $\mu = -2$ and 2 are identical to the one with $\mu = 0$, except that they have been shifted over by an amount equal to μ.

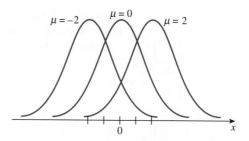

Figure 8.15

Figure 8.16 indicates 3 normal curves all with $\mu = 0$ and different standard deviations. Notice that the larger σ is the more spread out the curve is.

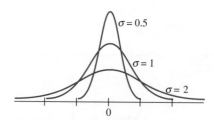

Figure 8.16

One might think that finding the area under a normal curve with mean μ and standard deviation σ would require a different table for different values of μ and σ. Fortunately this is not the case. We can readily find $P(a \le X \le b)$ in terms of certain probabilities of the standard normal distribution.

It turns out that if X is normally distributed with mean μ and standard deviation σ, and

$$Z = \frac{X - \mu}{\sigma},$$

then Z is normally distributed with mean $\mu = 0$ and $\sigma = 1$. The situation is summarized as follows.

> **Finding the Probability of a Normal Distribution**
>
> Let X be a random variable with mean μ and standard deviation σ. Then
>
> $$P(a \leq X \leq b) = P\left(\frac{a - \mu}{\sigma} \leq Z \leq \frac{b - \mu}{\sigma}\right)$$
>
> $$= \mathscr{A}\left(\frac{b - \mu}{\sigma}\right) - \mathscr{A}\left(\frac{a - \mu}{\sigma}\right).$$

E X A M P L E   3   **Finding the Probability for a Normal Distribution**

Suppose that the number of pounds in 100-pound bags of fertilizer is normally distributed with a mean of 100 and a standard deviation of 4. What is the probability that the bag will contain less than 95 pounds?

Solution   We are being asked for $P(X \leq 95)$ for a random variable that is normally distributed with mean $\mu = 100$ and standard deviation $\sigma = 4$. We form the Z variable as

$$Z = \frac{95 - \mu}{\sigma} = \frac{95 - 100}{4} = -1.25.$$

Then

$$P(X \leq 95) = P(Z \leq -1.25) = .1056$$

from Table B.1 in Appendix B. Thus we expect about 10.56% of the bags to have less than 95 pounds of fertilizer in them.   ■

E X A M P L E   4   **Quality Control Using the Normal Distribution**

The diameters of ball bearings produced at a plant are normally distributed with a mean of 1 cm and standard deviation of .004 cm. A ball bearing is acceptable if the diameter lies in the interval [.99, 1.01]. What percentage of ball bearings will be rejected?

Solution   We will first find the probability that one will be accepted. If X is the random variable that denotes the diameters of the ball bearings, then this probability is $P(.99 \leq X \leq 1.01)$. Then

$$P(.99 \leq X \leq 1.01) = P\left(\frac{a - \mu}{\sigma} \leq Z \leq \frac{b - \mu}{\sigma}\right)$$

$$= \mathscr{A}\left(\frac{b - \mu}{\sigma}\right) - \mathscr{A}\left(\frac{a - \mu}{\sigma}\right)$$

$$= \mathscr{A}\left(\frac{1.01 - 1.00}{.004}\right) - \mathscr{A}\left(\frac{.99 - 1.00}{.004}\right)$$

$$= \mathscr{A}(2.5) - \mathscr{A}(-2.5)$$

$$= .9938 - .0062 = .9876. \text{ From Table B.1}$$

To find the probability that one will be rejected is $1 - .9876 = .0124$. So about 1.24% of these ball bearings will be rejected.   ∎

E X A M P L E   5   **Determining a Warranty Period**

A manufacturer of washing machines has collected data indicating that the time before one of its washers needs its first repair is normally distributed with mean 2 years and standard deviation 0.608 years. What length of time should the manufacturer set for the warranty period so that at least 95% of the machines will get through the warranty period without need of a repair?

Solution   If X denotes the life of a washer and L denotes the length of the warranty period, we are asking that the probability that $X \le L$ should be .95; that is, $P(X \le L) = .95$. Since X is normally distributed with mean $\mu = 2$ and standard deviation $\sigma = 0.608$, using the Z statistic, this is the same as

$$P\left(Z \le \frac{L - \mu}{\sigma}\right).$$

We wish this probability to be .95. Using Table B.1 in Appendix B, we see that this will occur if

$$\frac{L - \mu}{\sigma} = 1.645.$$

Solving for L, we obtain

$$L = 1.645\sigma + \mu,$$

or

$$L = 1.645(.608) + 2 \approx 3.$$

Thus the manufacturer should set the warranty period at 3 years, since 95% of the washers will go through a 3-year period without any need for repairs.   ∎

SELF-HELP EXERCISE SET 8.4

1. Let X be a normally distributed random variable with mean $\mu = 4$ and standard deviation $\sigma = 2$. Find (a) $P(X \le 3)$. (b) $P(X \ge 3)$. (c) $P(1 \le X \le 3)$.

2. Let Z be a random variable with a standard normal distribution. Find the value Z_0 such that $P(Z \ge Z_0) = .10$.

3. Suppose the daily production X, in tons, of a steel manufacturer is normally distributed with $\mu = 100$ tons and $\sigma = 10$ tons. Management decides to offer the employees a bonus whenever the daily production level is in the top 10% of the daily production distribution. What is the level of production that will produce the bonus?

EXERCISE SET 8.4

In Exercises 1 through 12 find the probabilities given that Z is a random variable with a standard normal distribution.

1. $P(Z \leq 0.7)$

2. $P(Z \leq 0.9)$

3. $P(Z \leq -0.5)$

4. $P(Z \leq -1.2)$

5. $P(Z \geq 1.2)$

6. $P(Z \geq 0.7)$

7. $P(Z \geq -0.6)$

8. $P(Z \geq -1.22)$

9. $P(1.0 \leq Z \leq 2)$

10. $P(0 \leq Z \leq 1.25)$

11. $P(-1.5 \leq Z \leq -1.0)$

12. $P(-0.5 \leq Z \leq 0.75)$

In Exercises 13 through 20 find the probabilities assuming that X is a random variable with a normal distribution with the indicated mean and standard deviation.

13. $P(X \leq 26)$, $\mu = 10$, $\sigma = 8$

14. $P(X \leq 150)$, $\mu = 100$, $\sigma = 50$

15. $P(X \geq 0.01)$, $\mu = 0.006$, $\sigma = 0.002$

16. $P(X \geq -10)$, $\mu = 20$, $\sigma = 30$

17. $P(100 \leq X \leq 200)$, $\mu = 150$, $\sigma = 100$

18. $P(5 \leq X \leq 15)$, $\mu = 5$, $\sigma = 10$

19. $P(0.01 \leq X \leq 0.02)$, $\mu = 0.005$, $\sigma = 0.01$

20. $P(-2 \leq X \leq 8)$, $\mu = 4$, $\sigma = 10$

Applications

21. **Sales.**   Suppose the sales per day in thousands of dollars of a certain company is a normal random variable X with mean $\mu = 20$ and $\sigma = 4$. Find the probability that sales on a certain day are
     a.   between \$16,000 and \$24,000,   **b.**   over \$30,000.

22. **Temperature.**   Suppose the thermostat in a room is set at 70 degrees and that the actual temperature in the room is given by a normal random variable with $\mu = 70$ and $\sigma = 0.5$. What is the probability that the temperature in the room is
     a.   less than 69 degrees?   **b.**   between 69 degrees and 71 degrees?

23. **Temperature.**   Suppose if the thermostat in a room is set at t degrees, then the actual temperature in the room is given by a normal random variable with $\mu = t$ and $\sigma = 0.5$. What is the lowest setting of the thermostat that will maintain a temperature of at least 70 degrees with a probability of .99?

24. **Packaging.**   A certain package of wrapping tape comes in 10-foot lengths. However, the actual length in feet of any tape in any package is a normal random

variable with $\mu = 10$ and $\sigma = 0.25$. Find the probability that the length of tape in a package is

a. at least 9.5 feet, **b.** at most 9.5 feet, **c.** between 9.5 and 10.5 feet.

25. **Construction.** The finished inside diameter of a piston ring is normally distributed with a mean of 3.50 inches and $\sigma = 0.005$ inches. What is the probability of obtaining a piston ring with a diameter exceeding 3.51 inches?

26. **Quality Control.** The inside diameter of a certain cast-iron pipe is normally distributed with $\mu = 4.02$ inches and $\sigma = 0.04$ inches. The pipe will be rejected unless the actual diameter is within 0.05 inches of 4.00. What percentage of pipes will be rejected?

27. **IQ Scores.** Suppose the IQ scores of a new group of employees is normally distributed with $\mu = 115$ and $\sigma = 10$. Find the percentage that have IQs
a. above 125, **b.** below 100, **c.** between 110 and 125.

28. **Life of a Product.** Suppose the life of a pair of stockings worn once a week is normally distributed with $\mu = 12$ months and $\sigma = 3$ months. What should the manufacturer set as the warranty period in order to have at most 5% wear out during this period?

29. **Management.** Refer to Self-Help Exercise 3. Suppose management gives a bonus when daily production reaches the 95% level. What should the production level be then?

30. **Management.** Redo the previous exercise changing 95% to 99%.

31. **Grades.** Suppose the scores on a mathematics exam are assumed to be normally distributed. The instructor calculates the mean to be 75 and the standard deviation to be 15. If the instructor wishes to have 10% of the class obtain an A, 20% a B, 40% a C, 20% a D, and 10% a F, find the numerical grade corresponding to each letter grade.

Solutions to Self-Help Exercise Set 8.4

1. If X is a normally distributed random variable with mean $\mu = 4$ and standard deviation $\sigma = 2$, then if Z is the random variable for the standard normal distribution,

a. $P(X \le 3) = P\left(Z \le \dfrac{3 - \mu}{\sigma}\right) = P\left(Z \le \dfrac{3 - 4}{2}\right) = \mathscr{A}(-0.5) = .3085,$

b. $P(X \ge 3) = 1 - P(X \le 3) = 1 - .3085 = .6915,$

c. $P(1 \le X \le 3) = \mathscr{A}\left(\dfrac{3 - 4}{2}\right) - \mathscr{A}\left(\dfrac{1 - 4}{2}\right) = .3085 - .0668 = .2417.$

2. Let Z be a random variable with a standard normal distribution. The value Z_0, such that $P(Z \ge Z_0) = .10$, is given by the value Z_0 such that $\mathscr{A}(Z_0) = .90$. Looking through the Table B.1 in Appendix B for this Z_0, we find that $Z_0 = 1.28$.

3. With the daily production, X, in tons of a steel manufacturer normally distributed with $\mu = 100$ tons and $\sigma = 10$ tons, the Z statistic is given by

$$Z = \frac{X - \mu}{\sigma}.$$

We know from the previous example that $\mathcal{A}(1.28) = .90$. Thus we have

$$X = \mu + Z\sigma = 100 + 1.28(10) = 112.8.$$

Thus when the daily production is 112.8 tons, 90% of the time the daily production is less than this number.

8.5 **NORMAL APPROXIMATION TO THE BINOMIAL DISTRIBUTION**

- ► *Normal Approximation to the Binomial Distribution*
- ► *Law of Large Numbers*

APPLICATION
Lot Acceptance Sampling

An electronics manufacturer randomly selects 200 stamped circuits from the production for one day where a 5% rate of defectives is acceptable. Determine approximately the probability that 18 or more defectives are discovered in this sample. Is this alarming? The answer is given in Example 4.

Normal Approximation to the Binomial Distribution

Finding probabilities that arise using the binomial distribution can be extremely tedious. In this section we will see how, under certain conditions, the normal distribution can be used as an effective approximation to the binomial distribution.

According to the National Cancer Institute, the 5-year survival rate for cancer of the cervix diagnosed in the 1980–85 period was approximately 66%. Suppose at a particular hospital, 15 patients were diagnosed for cancer of the cervix. What is the probability that at least 10 of these patients will survive for 5 years using the above probability of survival?

If success for a patient diagnosed as having cervical cancer is survival for 5 years, then the situation above can be described as a repeated Bernoulli trial with $n = 15$ and $p = .66$, $q = .34$. The answer can readily be written down in terms of the binomial distribution as

$$C(15, 10)(.66)^{10}(.34)^5 + \mathbf{C}(15, 11)(.66)^{11}(.34)^4$$
$$+ \cdots + \mathbf{C}(15, 15)(.66)^{15}(.34)^0.$$

Unfortunately, this will be a lengthy calculation. Figure 8.17 shows a histogram of the probability of the number of successes for 15 repeated Bernoulli trials with the probability of success equal to .66. As we have noted in a previous section, the answer to the question posed above is the area of the shaded rectangles shown in the histogram of Figure 8.17. In this section we will see that this area is very close to a certain area under a certain normal curve, which will be easy to find using tables. We will now see how to do this.

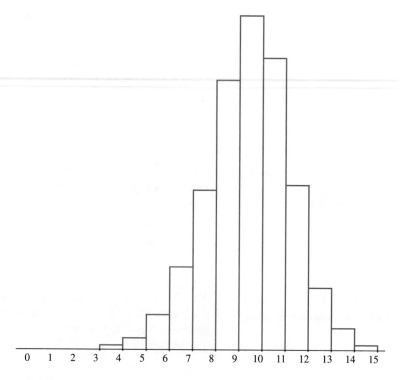

Figure 8.17

Figure 8.18, Figure 8.19, and Figure 8.20 show histograms of the binomial distributions for $p = .66$ and $n = 3$, 5, and 10, respectively. Notice how as n increases, the histograms are more bell shaped and appear more similar to a normal curve. This is indeed the case. Figure 8.21 shows the binomial distribution

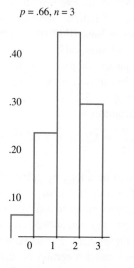

Figure 8.18

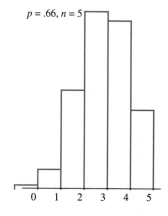

Figure 8.19

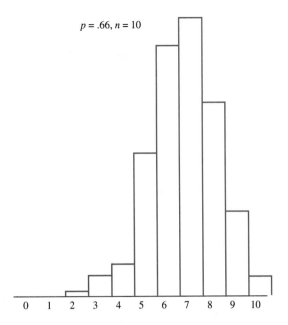

Figure 8.20

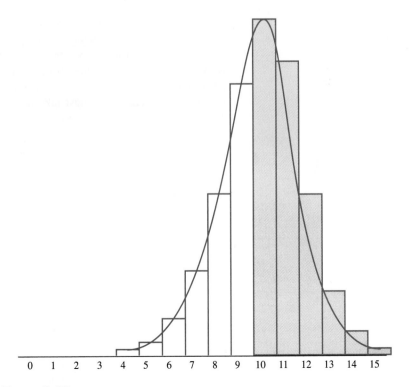

Figure 8.21

of Figure 8.17 overlaid with the normal distribution with the same mean $\mu = np = 15(.66) = 9.9$ and the same standard deviation $\sigma = \sqrt{npq} = \sqrt{15(.66)(.34)} \approx 1.835$ as the binomial distribution shown. Notice how good an approximation the normal curve is to the histogram for the binomial distribution. This is true under a wide set of conditions, as the following theorem indicates.

Normal Approximation to the Binomial Distribution

In a sequence of n repeated (independent) Bernoulli trials with probability of success equal to p and probability of failure equal to q, let the random variable X denote the number of successes. Then the histogram for the probability distribution of X is approximated by the normal curve with mean $\mu = np$ and standard deviation $\sigma = \sqrt{npq}$. This is a particularly close approximation when $np \geq 5$ and $nq \geq 5$.

E X A M P L E 1 **Using the Normal Approximation**

Using the normal approximation, find the probability that at least 10 of the patients mentioned above will survive for 5 years.

Solution First notice that $np = 9.9 \geq 5$ and $nq = 5.1 \geq 5$. Thus the normal approximation should be excellent.

To find the area of the shaded rectangles, we use the normal approximation with the normal curve with mean $\mu = np = 15(.66) = 9.9$ and standard deviation $\sigma = \sqrt{npq} = \sqrt{15(.66)(.34)} \approx 1.835$. Refer to Figure 8.21. Notice that since we want to include all of the histogram to the right of 10 and including 10, we must include the entire rectangle corresponding to 10. Thus to use the approximating normal curve we must take the area to the right of $10 - .50 = 9.5$. That is, using the Z substitution,

$$P(X \geq 9.5) = 1 - \mathscr{A}\left(\frac{9.5 - \mu}{\sigma}\right)$$

$$= 1 - \mathscr{A}\left(\frac{9.5 - 9.9}{1.835}\right)$$

$$= 1 - \mathscr{A}(-.22)$$

$$= 1 - .4129 = .5871. \text{ From Table B.1}$$

This compares with the actual value of .5968. ∎

E X A M P L E 2 **Using the Normal Approximation**

A coin has been flipped 900 times and heads has been observed 496 times. Is there some justification in claiming that this is an unfair coin?

Solution

First notice that $np = nq = 450 \geq 5$ so the normal approximation can be used.

If we assume the coin is fair, then we have 900 repeated Bernoulli trials with $p = .5 = q$. The mean is $\mu = np = 450$. We will calculate the probability that the number of heads of a fair coin could deviate by more than 45 from the mean. This is one less the probability that it could deviate by as much as 46 from the mean. If X is the random variable that denotes the number of heads, then this latter probability is denoted by $P(405 \leq X \leq 495)$ and is

$$C(900, 405)(.5)^{405}(.5)^{495} + C(900, 406)(.5)^{406}(.5)^{494}$$
$$+ \cdots + C(900, 495)(.5)^{495}(.5)^{405}.$$

The need for approximating this by the area under a normal curve is apparent. We use the normal curve with $\mu = 450$ and $\sigma = \sqrt{npq} = \sqrt{900(.5)(.5)} = 15$ and let the random variable Y denote the continuous random value associated with this. From Figure 8.22 we see that we must include the rectangles corresponding to 405 and 495. Then $P(405 \leq X \leq 495)$ is approximately the area under this normal curve from $405 - .50 = 404.5$ to $495 + .50 = 495.5$ and is denoted by $P(404.5 \leq Y \leq 495.5)$. Then

$$P(404.5 \leq Y \leq 495.5) = \mathcal{A}\left(\frac{495.5 - 450}{15}\right) - \mathcal{A}\left(\frac{404.5 - 450}{15}\right)$$

$$= \mathcal{A}(3.03) - \mathcal{A}(-3.03)$$

$$\approx .9988 - .0012 = .9976. \text{ From Table B.1}$$

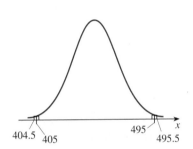

Figure 8.22

One less this is .0024 and is the probability that the number of heads can deviate by more than 45 from the mean when a fair coin is flipped 900 times. Since this is so low, we can conclude that the coin is not a fair coin. ∎

We saw in the last example that a fair coin has a very small probability of deviating very far from the mean. We now examine how likely it is for the number of heads to be very close to the mean.

E X A M P L E   3   **Using the Normal Approximation**

For the experiment in the previous example, find the probability that the number of heads varies at most by 1 from the mean.

Solution

With X denoting the number of heads in 900 flips of the fair coin, the mean is 450, and we are seeking the probability $P(449 \leq X \leq 451)$, which is the area under the three rectangles shown in Figure 8.23. If Y is the same as in the previous example, we see from Figure 8.23 that we approximate this by $P(448.5 \leq Y \leq 451.5)$, the area under the approximating normal curve from 448.5 to 451.5. We then obtain

$$P(448.5 \leq Y \leq 451.5) = \mathcal{A}\left(\frac{451.5 - 450}{15}\right) - \mathcal{A}\left(\frac{448.5 - 450}{15}\right)$$

$$= \mathcal{A}(0.10) - \mathcal{A}(-0.10)$$

$$\approx .5398 - .4602 = .0796. \text{ From Table B.1}$$

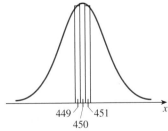

Figure 8.23

So there is about an 8% chance of the number of heads being within 1 of the mean. Thus we see that just as there is a small probability of the number of heads deviating far from the mean, so too is the probability that the number of heads will be very close to the mean. ∎

Many electronic items can be mass produced cheaply because the solid-state circuitry used can be stamped by a machine. Controlling the quality of the items is a serious problem and is often handled by using **lot acceptance sampling**. In this method a sample is taken from a large lot of items and inspected. The entire lot will be accepted or rejected on the basis of the number of defective items in the sample.

E X A M P L E 4   Lot Acceptance Sampling

An electronics manufacturer randomly selects 200 stamped circuits from the production for one day where a 5% rate of defectives is acceptable. Determine approximately the probability that 18 or more defectives are discovered in this sample. Should this discovery be cause for concern?

Solution   First, suppose that the defective rate is 5%. Then the random variable that gives the number of defective circuits is binomially distributed with $n = 200, p = .05$, and $q = .95$. Since $np = 10 \geq 5$ and $nq = 190 \geq 5$, we expect the normal approximation with mean $\mu = np = 10$ and standard deviation $\sigma = \sqrt{npq} = \sqrt{200(.05)(.95)} \approx 3.1$ to be a good approximation for the binomial distribution.

To find the approximating area corresponding to $X \geq 18$, refer to Figure 8.24. Since we must include the rectangle that corresponds to 18, we take the area under the approximating curve to start at $18 - .50 = 17.5$. Then

$$P(X \geq 17.5) = 1 - \mathscr{A}\left(\frac{17.5 - 10}{3.1}\right)$$

$$\approx 1 - .9922 = .0078. \text{ From Table B.1}$$

Thus the approximate probability that at least 18 defective circuits are in the sample of 200 is .0078 *if in fact the true defective rate is 5%*. Since the probability of .0078 is extremely small, the manufacturer should be concerned that the defective rate of the sample is, in fact, much higher than 5%. ∎

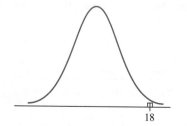

Figure 8.24

The Law of Large Numbers

Assume an experiment with n repeated independent Bernoulli trials with the probability of success p equal to the probability of failure q. Then $p = q = \frac{1}{2}$. Let X denote the random variable given by the number of successes. We seek an estimate of the probability that X differs from the mean $\mu = np = \frac{1}{2}n$ by at most h standard deviations.

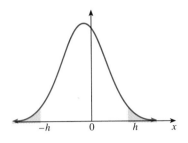

Figure 8.25

Since the standard deviation is $\sigma = \sqrt{npq} = \frac{1}{2}\sqrt{n}$, we are seeking a good estimate of

$$P\left(\left|X - \frac{1}{2}n\right| \leq \frac{h}{2}\sqrt{n}\right) = P(|X - \mu| \leq h\sigma) = P\left(-h \leq \frac{X - \mu}{\sigma} \leq h\right).$$

If Y is the random variable given by the normal distribution with mean $\mu = \frac{1}{2}n$ and standard deviation $\sigma = \frac{1}{2}\sqrt{n}$, then by the normal approximation, this probability is approximated by $P\left(-h \leq \frac{Y - \mu}{\sigma} \leq h\right)$. But using the Z statistic for the standard normal distribution, this is $P(-h \leq Z \leq h)$. From Figure 8.25 we note that the area under the entire standard normal curve is 1 and that by symmetry the area under each "tail" is the same. Thus

$$P(-h \leq Z \leq h) = 1 - 2P(Z \leq -h).$$

Thus we have established the important estimate that

$$P\left(\left|X - \frac{1}{2}n\right| \leq \frac{h}{2}\sqrt{n}\right) \approx 1 - 2P(Z \leq -h) = 1 - 2\mathscr{A}(-h). \quad (1)$$

Now if we are given a specific value of h, we can look up $P(Z \leq -h) = \mathscr{A}(-h)$ in Table B.1 in Appendix B. For example, if $h = 2$, then looking in this table, we have

$$P(-2 \leq Z \leq 2) = 1 - 2P(Z \leq -2) = 1 - 2(.0228) = .9544.$$

Thus we can say that there is about a 95% chance that the number of successes will be within $2\sigma = \sqrt{n}$ of $\mu = \frac{1}{2}n$.

E X A M P L E   5   **The Law of Large Numbers**

In the experiment of flipping a fair coin let H be the event that a head occurs. Estimate how large n must be so that there is a probability of .95 that the relative frequency $\dfrac{f(H)}{n}$ is within .01 of $\frac{1}{2}$. Within .001.

Solution   If we let the random variable $X = f(H)$, then we can use the estimate given by equation (1). We already noticed above that if $h = 2$, then $1 - 2\mathscr{A}(-h) \approx .95$. Thus with $h = 2$ equation (1) becomes

$$P\left(\left|X - \frac{1}{2}n\right| \leq \sqrt{n}\right) \approx .95.$$

Setting $X = f(H)$ and dividing by n, this can be rewritten as

$$P\left(\left|\frac{f(H)}{n} - \frac{1}{2}\right| \leq \frac{1}{\sqrt{n}}\right) \approx .95. \quad (2)$$

If we want the relative frequency $\dfrac{f(H)}{n}$ to be within ϵ of $\frac{1}{2}$, we want

$$\left| \frac{f(H)}{n} - \frac{1}{2} \right| \le \epsilon.$$

Looking at equation (2), we then want

$$\epsilon = \frac{1}{\sqrt{n}}$$

or

$$n = \frac{1}{\epsilon^2}. \tag{3}$$

Now using equation (3) if $\epsilon = .01$, then $n = 10{,}000$. If $\epsilon = .001$, then $n = 1{,}000{,}000$. This means, for example, that if $n = 10{,}000$, the relative frequency $\dfrac{f(H)}{n}$ will be within the interval $(.49, .51)$ with a probability of .95. More precisely, this means that 95% of the time that a fair coin is flipped 10,000 times, the relative frequency will be within the interval $(.49, .51)$. The second answer means that 95% of the time that a fair coin is flipped 1,000,000 times, the relative frequency will be within the interval $(.499, .501)$.

SELF-HELP EXERCISE SET 8.5

1. About 3% of the flashlights produced by a manufacturer are defective. Using the normal curve approximation, find the probability that in a shipment of 1000 to a distributor, at least 35 will be defective.

2. Assume an experiment with n repeated independent Bernoulli trials with the probability of success equal to p and the probability of failure equal to $q = 1 - p$. Let X denote the random variable given by the number of successes. Find the value of h so that the probability is about .99 that X differs from the mean $\mu = np$ by at most $h\sigma = h\sqrt{npq}$.

EXERCISE SET 8.5

1. A fair coin has been tossed 1600 times. Use the normal approximation to find the probability that (a) the number of heads is at least 825, (b) the number of heads is between 775 and 825.

2. A fair die has been tossed 180 times. Use the normal approximation to find the probability that (a) the number 1 appears at most 35 times, (b) the number 3 appears between 25 and 35 times.

3. A baseball player has a batting average of .300. What is the probability that he will have a batting average of at least .400 over the next 100 official times at bat?

4. A basketball player hits 80% of her free-throws. What is the probability that she will hit 90% of her next 50 free-throws?

Applications

5. **Bladder Cancer.** According to the National Cancer Institute, the 5-year survival rate for cancer of the bladder diagnosed in 1980–85 was 78%. Among 1000 such diagnosed cases, find the probability that (a) 750 or more survived 5 years, (b) between 750 and 800 survived 5 years.

6. **Prostate Cancer.** Repeat the previous exercise for cancer of the prostate with a 5-year survival rate of 76%.

7. **Heart Disease.** Aproximately 43% of deaths in this country are attributed to heart disease. Among 1000 randomly chosen deaths, what is the probability that heart disease was the cause in at least 400 instances? Between 400 and 450?

8. **Accidental Deaths.** Approximately 4.5% of deaths in this country are attributed to accidents. Among 1000 randomly chosen deaths, what is the probability that an accident was the cause in at most 50 instances? Between 30 and 60?

9. **Use of Marihuana.** Recently, 37% of college students reported using marihuana during the previous year. What is the probability that, among 300 randomly chosen college students, (a) at least 120 used marihuana during the previous year? (b) between 100 and 120?

10. **Use of Cocaine.** Recently, 8% of high school seniors reported using cocaine during the last year. What is the probability that, among 300 randomly chosen high school students, (a) at least 30 used cocaine during the previous year? (b) between 20 and 40?

11. **Immigration.** In recent years about 41% of immigrants admitted to this country were Asian. Among 100 randomly selected immigrants to this country, what is the probability that (a) at least 30 are Asian? (b) between 30 and 50 are Asian?

12. **Arrests.** Approximately 82% of those arrested each year in this country are males. Given 100 randomly chosen arrests, what is the probability that at least 10 were female? Between 10 and 20 were female?

13. **Heart Disease.** In Exercise 7 find the probability that the number of deaths due to heart disease differs from the mean by at most 2.5 standard deviations.

14. **Accidental Deaths.** In Exercise 8 find the probability that the number of deaths due to accidents differs from the mean by at most 3 standard deviations.

15. **Quality Control.** About 3% of a certain product of a manufacturer has been found to be defective. A new system of quality control has been introduced with 18 defective products found in the first 1000 after the new system went into effect. Is the new system effective? *Hint:* Find the probability of at most 18 defectives under the old system.

16. **Safety.** In a certain plant about 5% of the workers were suffering some accident each year. New safety procedures were instituted that resulted in 30 accidents among 1000 workers during the next year. Are the safety procedures working?

17. **Advertising.** A company received a 15% response to its mail advertisements. After changing to a new advertisement, it noticed a 20% response among the next 1000 mailings. Is the new advertisement effective?

18. **Drug Effectiveness.** For a certain disease, 20% of individuals with the disease will recover without any treatment. A new experimental drug is administered to 40 individuals with the disease and 16 recover. Is this drug promising? *Hint:* Find the probability that at least 16 patients would recover without the drug.

19. **The Law of Large Numbers.** In the experiment of flipping a fair coin let H be the event that a head occurs. Estimate how large n must be so that there is a probability of .50 that the relative frequency $\dfrac{f(H)}{n}$ is within .01 of $\frac{1}{2}$. Within .001?

20. **The Law of Large Numbers.** In the experiment of flipping a fair coin, let H be the event that a head occurs. Estimate how large n must be so that there is a probability of .50 that the relative frequency $\dfrac{f(H)}{n}$ is within .0001 of $\frac{1}{2}$. Within .00001?

21. **The Law of Large Numbers.** In the experiment of flipping a fair coin, let H be the event that a head occurs. Estimate how large n must be so that there is a probability of .90 that the relative frequency $\dfrac{f(H)}{n}$ is within .01 of $\frac{1}{2}$. Within .001?

22. **The Law of Large Numbers.** In the experiment of flipping an unfair coin, let H be the event that a head occurs and $p = p(H) = .60$. Estimate how large n must be so that there is a probability of .95 that the relative frequency $\dfrac{f(H)}{n}$ is within .01 of $p = .60$. Within .001?

Solutions to Self-Help Exercise Set 8.5

1. The situation is a repeated Bernoulli trial with $n = 1000$ and $p = .03$. If X denotes the number of defective flashlights, then we are seeking $P(X \geq 35)$. If Y denotes the random variable associated with the normal distribution with mean $\mu = np = 1000(.03) = 30$ and standard deviation $\sigma = \sqrt{npq} = \sqrt{1000(.03)(.97)} = 5.3944$, then

$$P(X \geq 35) \approx P(Y \geq 34.5)$$

$$= P\left(Z \geq \frac{34.5 - 30}{5.3944}\right)$$

$$= P(Z \geq .83)$$

$$= 1 - \mathcal{A}(.83)$$

$$= 1 - .7967 = .2033.$$

Thus there is about a 20% chance that at least 35 flashlights will be defective in a shipment of 1000.

2. Equation (1) says that

$$P(|X - np| \leq h\sqrt{npq}) \approx 1 - 2P(Z \leq -h) = 1 - 2\mathcal{A}(-h).$$

Thus we are seeking the value of h for which

$$.99 = 1 - 2\mathcal{A}(-h).$$

This can be written as

$$\mathcal{A}(-h) = .005.$$

Looking for an h in Table B.1 in Appendix B for which this is true, we find that $h = 2.575$. Thus we can say that X differs from the mean $\mu = np$ by at most $2.575\sigma = 2.575\sqrt{npq}$ with probability of .99.

8.6 THE POISSON DISTRIBUTION

▸ *The Poisson Distribution*

▸ *Comparing the Poisson with the Binomial*

A P P L I C A T I O N

Plant Ecology

Plant ecologists sometimes use the Poisson distribution to determine if a plant species is randomly distributed on the ground or if this species has a tendency to clump. To see how this is done, refer to Example 2 that contains the results of an actual study.

The Poisson Distribution

The Poisson probability arises frequently in the fields of operations research and management science and in the life and social sciences. The Poisson probability is useful in counting the number of occurrences in a given unit of measurement such as time, distance, or weight. This applies to the number of airplanes landing at an airport per hour, the number of telephone lines being used per minute, the number of items removed from inventory per week, the number of fungus spores per leaf, the number of red blood cells per cubic centimeter, the number of fish caught per hour, the number of vehicles arriving at a toll booth per minute during rush hour, and the number of accidents at a plant per year.

In order to understand the circumstances under which the Poisson probability arises, it is useful to compare these circumstances to those under which the binomial probability arises. The binomial probability law applies to situations in which we know the number of times an event occurs and the number of times it does not occur. However, there are random events that do not occur as outcomes of repeated experiments but rather occur as isolated events in time or space. We might observe the number of vehicles arriving at a toll booth, but it is not meaningful to ask for the number not arriving. We might count the number of fungus spores on a leaf, but it is not practical to wish to count the number not on the leaf. Thus a characteristic feature of the situations in which the Poisson distribution arises is that only the occurrence of an event can be counted; the nonoccurrence cannot. Thus the total number of events cannot be measured, and as a result, the binomial distribution is not applicable.

The following is more precise.

When the Poisson Probability Applies

The Poisson probability law will apply if there is a number λ such that within a small fractional unit of measurement, such as time or space,

  1. Probability of one count $\approx \lambda \cdot$ (size of the small unit).
  2. Probability of two or more counts per some small size of the unit is ≈ 0.
  3. The number of occurrences of an event in any one interval of time or space is independent of the number in any other disjoint interval of time or space.

The first condition basically says that if 3 vehicles arrive at a toll booth per minute, then on average 6 will arrive per 2 minutes. The second condition basically says that 2 cars cannot arrive at the same toll booth at the same time. The third condition says that the number arriving in one particular minute does not depend on the number of arrivals before then.

Sometimes the failure of the Poisson distribution to compare favorably with the actual data can be used as evidence that the events in space or time are *not* independent. We will indicate such an example given by an ecologist studying the random dispersion of certain plants.

We now give the Poisson probability distribution function.

Poisson Distribution with Mean λ

If $P_\lambda(X = x) = p_\lambda(x)$ is the probability of x number of occurrences per unit of measure, then the Poisson distribution with mean λ is given by

$$P_\lambda(X = x) = \frac{\lambda^x}{x!} e^{-\lambda}, \qquad e \approx 2.7182818.$$

The mean is $\mu = \lambda$ and the standard deviation is $\sigma = \sqrt{\lambda}$.

E X A M P L E   1   **Using the Poisson Distribution**

Suppose during a sale at a large department store, 120 customers entered the store per hour. Assuming that the number that enter per hour is described by a Poisson distribution, find the probability that in the next minute (a) nobody will enter, (b) exactly 2 will enter, (c) at most 2 will enter, (d) at least 3 will enter.

Solutions The average number of customers entering the store per minute is $\frac{120}{60} = 2$. Thus we have a Poisson distribution with $\lambda = 2$. Thus the probability that x customers will enter in the next minute is given by

$$p_2(x) = \frac{2^x}{x!} e^{-2}.$$

(a) Then for $x = 0$,

$$p_2(0) = \frac{2^0}{0!} e^{-2} = e^{-2} \approx .135.$$

(b) Then for $x = 2$,

$$p_2(2) = \frac{2^2}{2 \, !} e^{-2} = 2e^{-2} \approx .271.$$

(c) This is $p_2(0) + p_2(1) + p_2(2)$. We already have $p_2(0)$ and $p_2(2)$. Then

$$p_2(1) = \frac{2^1}{1!} e^{-2} = 2e^{-2} \approx .271.$$

Thus

$$P_2(0 \le X \le 2) = p_2(0) + p_2(1) + p_2(2)$$
$$= .135 + .271 + .271 = .677.$$

(d) For any event E and any probability function p, $p(E^c) = 1 - p(E)$. Thus

$$P_2(X \ge 3) = 1 - P_2(0 \le X \le 2) = 1 - .677 = .323. \quad \blacksquare$$

E X A M P L E   2   **Using Poisson Distributions in Ecology**

Plant ecologists sometimes use the Poisson distribution to determine if a plant species is randomly distributed on the ground. In an actual study 100 random quadrats, that is, small square areas, are selected and researchers count the number of plants of each particular species studied found in each quadrat. If the numbers in each quadrat follow a Poisson distribution, then the species is assumed to be dispersed randomly. Otherwise, the species is assumed to have a tendency for clumping. The following table gives the results of this study of two plants: Senecio jacobaea and Geranium robertianum. What conclusions can be drawn?

	Number per Quadrat						
	0	1	2	3	4	5	6
Frequency for Senecio jacobaea	46	36	15	3	0	0	0
Probability for Senecio jacobaea	.46	.36	.15	.03	0	0	0
Frequency for Geranium robertianum	36	9	5	10	14	11	15
Probability for Geranium robertianum	.36	.09	.05	.10	.14	.11	.15

Solution   The mean number of Senecio jacobaea per quadrat is

$$\frac{0 \cdot 46 + 1 \cdot 36 + 2 \cdot 15 + 3 \cdot 3}{100} = 0.75$$

Thus in the Poisson distribution we take $\lambda = 0.75$ and

$$p_{0.75}(x) = \frac{(.75)^x}{x!} e^{-.75}.$$

Thus

$$p_{0.75}(0) = \frac{(.75)^0}{0!} e^{-0.75} \approx .4724$$

$$p_{0.75}(1) = \frac{(.75)^1}{1!} e^{-0.75} \approx .3543$$

$$p_{0.75}(2) = \frac{(.75)^2}{2!} e^{-0.75} \approx .1329$$

$$p_{0.75}(3) = \frac{(.75)^3}{3!} e^{-0.75} \approx .0332$$

$$p_{0.75}(4) = \frac{(.75)^4}{4!} e^{-0.75} \approx .0062.$$

Since this is in close agreement with the data, we conclude that the species Senecio jacobaea is distributed randomly on the ground.

The distribution of the other plant, Geranium robertianum, will be considered in Self-Help Exercise 1. ■

It is interesting to notice from the above probabilities that

$$P_{0.75}(0 \le X \le 4) = .4724 + .3543 + .1329 + .0332 + .0062 = .9990.$$

Thus virtually all the probability is distributed over the first 5 outcomes. A histogram is drawn in Figure 8.26.

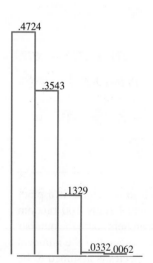

.4724

.3543

.1329

.0332 .0062

Figure 8.26

Comparing the Poisson with the Binomial

If $p = \lambda/n$ is small, then it can be shown that

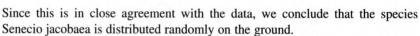

$$C(n, x)p^x q^{n-x} \approx \frac{\lambda^x}{x!} e^{-\lambda}.$$

This implies that the binomial and Poisson distributions will be in good agreement if p is small.

Approximating the Binomial Probability by the Poisson

As a rule of thumb, if $p \leq .10$, the Poisson distribution $\frac{\lambda^x}{x!} e^{-\lambda}$ will be approximately equal to the binomial distribution $C(n, x)p^x q^{n-x}$.

The following table compares the binomial distribution with the Poisson approximation for $p = .01$ and $n = 50$. We have $\lambda = pn = 0.5$ and are comparing

$$C(50, x)(.01)^x(.99)^{50-x} \qquad \text{with} \qquad \frac{(.5)^x}{x!} e^{-0.5}.$$

Notice the close agreement.

x	*Binomial*	*Poisson*
0	.6050	.6065
1	.3056	.3033
2	.0756	.0758
3	.0122	.0126
4	.0015	.0016

It is interesting to sum up the 5 Poisson probabilities in the table. One obtains .9998, so that virtually all the probability comes from these first 5 outcomes.

E X A M P L E   3   **Approximating the Binomial Probability by the Poisson**

A plant that manufactures light bulbs finds that 1% are defective. Find the probability that in a box of 10, none will be defective. Use both the binomial probability and the approximation by the Poisson probability.

Solution   Using the binomial probability, we have

$$C(10, 0)(.01)^0(.99)^{10} \approx .9044.$$

For the Poisson approximation, $\lambda = pn = (.01)(10) = .10$. Then

$$P_{0.10}(0) = \frac{(.1)^0}{0!} e^{-.1} \approx .9048.$$

This is in close agreement.   ∎

SELF-HELP EXERCISE SET 8.6

1. For the species Geranium robertianum found in Example 2, determine λ and the corresponding Poisson distribution. Draw a histogram. Compare the Poisson distribution to the data to see the degree of agreement and draw a conclusion with regard to the random distribution of this plant.

2. In order to save money, a certain manufacturing plant has neglected maintenance on all 20 of its lathes so that the probability that one will fail in the next month is .03. Find the probability that more than one will fail in the next month. Find the answer using the binomial distribution and compare this to the Poisson approximation.

EXERCISE SET 8.6

In Exercises 1 through 12 find the requested Poisson probability.

1. $p_3(0)$ 2. $p_4(0)$

3. $p_4(1)$ 4. $p_5(1)$

5. $p_{0.7}(2)$ 6. $p_{1.3}(3)$

7. $P_3(X \le 2)$ 8. $P_4(X \le 3)$

9. $P_{2.3}(X \ge 1)$ 10. $P_3(X \le 2)$

11. $P_{1.3}(1 \le X \le 3)$ 12. $P_5(2 \le X \le 4)$

In Exercises 13 through 18 find the binomial probabilities and their Poisson approximation.

13. $b(2 : 100, .01)$ 14. $b(3 : 100, .01)$

15. $b(3 : 100, .02)$ 16. $b(1 : 200, .03)$

17. $b(2 : 1000, .0003)$ 18. $b(1 : 2000, .005)$

19. Create a table that compares the values of $b(x; 200, .01)$ with its Poisson approximation for $x = 0, 1, 2, 3$.

20. Create a table that compares the values of $b(x; 2000, .001)$ with its Poisson approximation for $x = 0, 1, 2, 3$.

Applications

In all of the following problems, use the Poisson distribution, unless otherwise indicated.

21. **Publishing.** A publisher of textbooks has 1 misprint per 50 pages. In 100 pages, what is the probability of (a) no misprints? (b) at most 2? (c) more than 2?

22. **Emergency Room.** Suppose on average there are 16 emergency patients on the 8:00 A.M. to 4:00 P.M. shift of a certain hospital. What is the probability that the number of emergency patients arriving during any one hour of this shift is (a) zero? (b) at most 1? (c) more than 1?

23. **Quality Control.** Suppose there are, on average, 3 garbage trucks per work week of 5 days in a large city that must go to the garage for repairs. What is the probability that on any working day (a) no garbage truck will go to the garage? (b) at most 3?

24. **Quality Control.** A utility company's records show that trees fall on their power lines and require a repair crew, on average, 3 times a week. What is the

probability of such an occurrence (a) not happening in a given day? (b) happening exactly once in a given day?

(Dennis Drenner.)

25. **Traffic.** The average number of cars passing through a certain toll booth is 18 per hour. Find the probability that during a given 5-minute period (a) no cars go through the toll booth, (b) more than 1 goes through.

26. **Quality Control.** Suppose on average there are 2 flaws in every 3 bolts of a certain cloth. Find the probability that there is exactly 1 flaw in 2 bolts of cloth.

27. **Quality Control.** Suppose there are, on average, 4 raisins in a raisin cookie from a certain bakery. What is the probability that there will be at most 2 raisins in a raisin cookie?

28. **Services.** On average, there were 90 hamburgers sold during the lunch hour of a certain fast-food restaurant. What is the probability that during a certain minute of lunch hour there will be (a) no hamburgers sold? (b) more than 2 sold?

29. **Accidents.** Accidents occur at the rate of 2 per week along a certain highway. What is the probability that there will be 2 or fewer accidents during (a) one week? (b) 2 weeks?

30. **Quality Control.** A manufacturer of wire notes that on average there are 4 flaws per 10 rods of wire. Each day one rod of wire is inspected for flaws. If more than one flaw is found in this rod, the plant shuts down for repairs. Suppose nothing at all is wrong with the machinery on a certain day. What is the probability that the plant will be closed for repairs after an inspection on this day?

31. **Ecology.** An ecologist suspects that a certain type of tree has seedlings that are dispersed randomly in a certain area about the tree. An average of 6 seedlings per square yard are found near one of these trees. What is the probability that a randomly selected 1-square-yard sample region reveals no seedlings?

32. **Typing Errors.** A typist produces on average 2 errors per page. If more than 2 errors per page are produced, then the page must be retyped. What is the probability that a certain page will not need to be retyped?

33. **Quality Control.** A certain type of wood has on average 2.5 knots per 10 cubic feet of wood. What is the probability that in 10 cubic feet of this wood at most 1 knot will be found?

34. **Accidents.** Records of deaths due to vehicle accidents in 10 army corps over a period of 20 years are as follows.

No. deaths/corps/yr x	0	1	2	3	4	>4
No. corps/yr with x deaths	109	65	22	3	1	0

Can you conclude that the distribution of deaths is due to chance alone and not due to factors such as where the corps was located or who was the commander?

35. **Accidents.** The number of accidents per month at the plants of a certain company are given as follows

Number of accidents	0	1	2	3	4	5	>5
Number of days	18	27	26	16	9	4	0

Can you conclude that the number of accidents occurs randomly?

36. **Insurance Claims.** The probability that an insurance company must pay a medical benefit for a policy is .0001. If the company has 10,000 such policy holders, what is the probability that the insurance company will have to pay at least one medical benefit in this group? Use the Poisson approximation to the binomial distribution.

37. **Deaths.** The probability that people of a certain age will die within the next year is .004. In a city with 1000 such people, what is the probability that 10 of them will die within the next year? Use the Poisson approximation to the binomial distribution.

38. **Medicine.** The probability that an individual inoculated with a serum contracts a certain disease is .30. Use the Poisson approximation to find the probability that at most 2 of 20 inoculated individuals contract the disease.

Solutions to Self-Help Exercise Set 8.6

1. The mean number of Geranium robertianum per quadrat is

$$\frac{0 \cdot 36 + 1 \cdot 9 + 2 \cdot 5 + 3 \cdot 10 + 4 \cdot 14 + 5 \cdot 11 + 6 \cdot 15}{100} = 2.5.$$

Thus in the Poisson distribution we take $\lambda = 2.5$ and

$$p_{2.5}(x) = \frac{(2.5)^x}{x!} = e^{-2.5}.$$

Thus

$$p_{2.5}(0) = \frac{(2.5)^0}{0!} e^{-2.5} \approx .082$$

$$p_{2.5}(1) = \frac{(2.5)^1}{1!} e^{-2.5} \approx .205$$

$$p_{2.5}(2) = \frac{(2.5)^2}{2!} e^{-2.5} \approx .257$$

$$p_{2.5}(3) = \frac{(2.5)^3}{3!} e^{-2.5} \approx .214$$

$$p_{2.5}(4) = \frac{(2.5)^4}{4!} e^{-2.5} \approx .134$$

$$p_{2.5}(5) = \frac{(2.5)^5}{5!} e^{-2.5} \approx .067$$

$$p_{2.5}(6) = \frac{(2.5)^6}{6!} e^{-2.5} \approx .028.$$

Since this does not approximate the data, we conclude that the species Geranium robertianum has a strong tendency for clumping.

2. We first find the probability that at most 1 will fail. This is

$$\binom{20}{0}(.03)^0(.97)^{20} + \binom{20}{1}(.03)^1(.97)^{19} \approx .544 + .336 = .880.$$

The probability that more than 1 will fail is then $1 - .880 = .120$.

For the Poisson approximation, use $\lambda = pn = (.03)20 = 0.6$. Then

$$p_{0.6}(x) = \frac{(.6)^x}{x!} e^{-0.6},$$

and

$$p_{0.6}(0) + p_{0.6}(1) = \frac{(.6)^0}{0!} e^{-0.6} + \frac{(.6)^1}{1!} e^{-0.6} \approx .549 + .329 = .878.$$

Then $1 - .878 = .122$. The agreement is very close.

SUMMARY OUTLINE OF CHAPTER 8

- A **random variable** is a rule that assigns precisely one real number to each outcome of an experiment.

- A random variable is **finite discrete** if it assumes only a finite number of values. A random variable is **infinite discrete** if it takes on an infinite number of values that can be listed in a sequence, so that, there is a first one, a second one, a third one, and so on. A random variable is said to be **continuous** if it can take any of the infinite number of values in some interval of real numbers.

- Suppose the random variable X can take the values $x_1, \ldots, x_n$. The **probability distribution** of the random variable X is a listing of all the probabilities associated with all possible values of the random variable, that is, $p_1, \ldots, p_n$, where $p_1 = p(x_1), \ldots, p_n = p(x_n)$.

- Given a sequence of n Bernoulli trials with the probability of success p and the probability of failure q, the **binomial distribution** is given by
$$P(X = k) = b(k : n, p) = C(n, k)p^k q^{n-k}.$$

- The **average** or **mean** of the n numbers $x_1, x_2, \ldots, x_n$, denoted by $\bar{x}$ or the Greek letter μ, is given by
$$\mu = \frac{x_1 + x_2 + \cdots + x_n}{n}.$$

- Let X denote the random variable that takes on the values $x_1, x_2, \ldots, x_n$, and let the associated probabilities be $p_1, p_2, \ldots, p_n$, then the **expected value** or **mean** of the random variable X, denoted by $E(X)$ or by μ, is
$$E(X) = x_1 p_1 + x_2 p_2 + \cdots + x_n p_n.$$

- The expected value of the binomial distribution
$$p_k = b(k : n, p) = C(n, k)p^k q^{n-k} \qquad p + q = 1$$
is $E(X) = np$.

- The **median** of a set of numbers is the middle number when the numbers are arranged in order of size and there are an odd number in the set. If the number in the set is even, the median is the mean of the two middle numbers.

- The **mode** of a set of numbers is the number that occurs more frequently than the others. If the frequency of occurrence of two numbers is the same and also greater than the frequency of occurrence of all the other numbers, then we say the set is **bimodal** and has two modes. If no one or two numbers occur more frequently than the others, we say that the set has no mode.

- Let X denote the random variable that takes on the values $x_1, x_2, \ldots, x_n$, and let the associated probabilities be $p_1, p_2, \ldots, p_n$. Then if $\mu = E(X)$, the **variance** of the random variable X, denoted by $\mathrm{Var}(X)$, is

$$\mathrm{Var}(X) = (x_1 - \mu)^2 p_1 + (x_2 - \mu)^2 p_2 + \cdots + (x_n - \mu)^2 p_n.$$

The **standard deviation**, denoted by $\sigma(X)$ is

$$\sigma(X) = \sqrt{\mathrm{Var}(X)}.$$

- An alternate form of the variance is given by

$$\mathrm{Var}(X) = (x_1^2 p_1 + x_2^2 p_2 + \cdots + x_n^2 p_n) - \mu^2.$$

- **Chebyshev's Inequality.** Let X be a random variable with expected value μ and standard deviation σ. Then the probability that the random variable associated with a random outcome lies between $\mu - h\sigma$ and $\mu + h\sigma$ is at least $1 - 1/h^2$, that is,

$$P(\mu - h\sigma \leq X \leq \mu + h\sigma) \geq 1 - \frac{1}{h^2}.$$

- The variance of the binomial distribution with n trials and the probability of "success" equal to p and of "failure" equal to q is $\mathrm{Var}(X) = npq$.

- The random variable X has a **standard normal distribution** on the interval $(-\infty, +\infty)$ if the probability $P(a \leq X \leq b)$ that X is between a and b is the area under the normal curve given by

$$y = \frac{1}{\sqrt{2\pi}} e^{-0.5x^2}$$

on the interval $[a, b]$, where $\pi \approx 3.14159$ and $e \approx 2.71828$.

- The random variable X has a **normal probability distribution with mean μ and standard deviation σ** on $(-\infty, +\infty)$ if the probability $P(a \leq X \leq b)$ that X is between a and b is the area under the normal curve given by

$$y = \frac{1}{\sigma\sqrt{2\pi}} e^{-0.5[(x-\mu)/\sigma]^2}$$

on the interval $[a, b]$.

- Let X be a normal random variable with mean μ and standard deviation σ. Then

$$P(a \leq X \leq b) = P\left(\frac{a - \mu}{\sigma} \leq Z \leq \frac{b - \mu}{\sigma}\right) = \mathcal{A}\left(\frac{b - \mu}{\sigma}\right) - \mathcal{A}\left(\frac{a - \mu}{\sigma}\right)$$

where $\mathcal{A}(b)$ is the area under the standard normal curve to the left of b, which can be found in Table B.1 in Appendix B.

- In a sequence of n repeated (independent) Bernoulli trials with probability of success equal to p and probability of failure equal to q, let the random variable X denote the number of successes. Then the histogram for the probability distribution of X is approximated by the normal curve with mean $\mu = np$ and standard deviation $\sigma = \sqrt{npq}$. This is a particularly close approximation when $np \geq 5$ and $nq \geq 5$.

- If $P_\lambda(X = x) = p_\lambda(x)$ is the probability of x number of occurrences per unit of measure, then the Poisson distribution with mean λ is given by

$$P_\lambda(X = x) = \frac{\lambda^x}{x!} e^{-\lambda}, \quad e \approx 2.7182818.$$

The mean is $\mu = \lambda$ and the standard deviation is $\sigma = \sqrt{\lambda}$.

- As a rule of thumb, if $p \leq .10$, the Poisson distribution $\dfrac{\lambda^x}{x!} e^{-\lambda}$ will be approximately equal to the binomial distribution $C(n, x) p^x q^{n-x}$.

Chapter 8 Review Exercises

1. A baseball player has a batting average of .300. Let X be the number of hits in the next 5 official times at bat. Find the probability distribution.

2. The probability distribution of the random variable X is given in the following table.

Random Variable x	0	1	2	3	4	5
$P(X = x)$	.20	.10	.05	.15	.18	.32

Find **(a)** $P(X = 0)$, **(b)** $P(X \leq 2)$, **(c)** $P(0 < X \leq 3)$, **(d)** $P(X \geq 2)$. Identify each of the probabilities as an area.

3. Find the expected value of the random variable given in the previous exercise.

4. **Lottery.** A lottery has a grand prize of $1,000,000, a second prize of $100,000, and 10 consolation prizes of $2000 each. If 1 million tickets are sold and the probability of any ticket winning is the same as any other, find the expected return on a $1 ticket.

5. **Life Insurance.** An insurance company sells a $10,000, 5-year term life insurance policy to an individual for $800. Find the expected return for the company if the probability that the individual will live for the next 5 years is .96.

6. The pitcher Cy Young holds the all-time record for the most wins in a lifetime with 511, a record that is unlikely to be matched. The following table gives his win and loss record for the 5-year period beginning with 1900.

Year	*1900*	*1901*	*1902*	*1903*	*1904*
No. of Wins	20	33	32	28	27
No. of Losses	18	10	10	9	16

Find the mean, variance, and standard deviation for his wins and also for his losses for this 5-year period. Which is varying the most?

7. **AIDS.** The following table gives the number of deaths due to acquired immunodeficiency syndrome (AIDS) for recent years in the U.S. by two age groups. Find the average, variance, and standard deviation of each. Which is varying the most?

Year	*1987*	*1988*	*1989*	*1990*	*1991*
13–29	2864	3531	4598	4745	3574
30–39	6535	8091	11308	11927	8830

8. Find the variance and the standard deviation of the random variable given in Exercise 2 above.

9. A probability distribution has a mean of 20 and a standard deviation of 2. Use Chebyshev's inequality to find the minimum probability that an outcome is between 10 and 30.

10. Let Z be a random variable with a standard normal distribution. Find (a) $P(Z \le .87)$, (b) $P(Z \ge .87)$, (c) $P(-.50 \le Z \le .87)$.

11. Let Z be a random variable with a standard normal distribution. Find the interval $[-c, c]$ so that the probability that the random variable is in this interval is .98.

12. **Manufacturing.** Suppose that in a production run of 100 twelve-ounce sodas, the mean number of ounces in these cans was 12 and the standard deviation was 0.25 ounces. What is the probability that a can will have less than 11.5 ounces if the number of ounces in the cans is normally distributed?

13. The heights of professional soccer players are normally distributed with a mean of 71 inches and a standard deviation of 2 inches. What percentage of players are within 1 inch of the mean?

14. A coin has been flipped 900 times and heads has been observed 485 times. What is the probability of obtaining at least 485 heads if the coin were fair?

15. **Quality Control.** About 2% of a certain product is found to be defective. What is the probability that in a shipment of 1000 of these products at least 25 will be defective?

16. **Crime.** If the probability is .001 that any one passenger on a subway in a large city will be robbed, what is the probability that 10 of the next 5000 passengers will be robbed? Use the Poisson approximation to the binomial distribution.

17. **Pollution.** The amount of plant emissions of a certain pollutant is limited to 2 parts per million by the EPA. Suppose a particular plant emits 1 part per million on an average day. What is the probability on a given day that this plant will exceed the EPA emissions standard?

Which restaurant will you go to this week? (Dennis Drenner.)

9

Markov Processes

In this chapter we consider a sequence of finite experiments, all with the same possible outcomes, in which the outcomes and associated probabilities of each experiment depend only on the outcome of the immediately preceding one. Such a sequence of experiments is called a Markov process. The basic results are presented together with a number of applications including scheduling and charge accounts.

9.1   MARKOV PROCESSES

- *Two-State Markov Processes*
- *Three-State Markov Processes*
- *m-State Markov Processes*

APPLICATION

**Percentage of Sons
that Go to College**

A survey of a certain ethnic group indicated that among the sons of college-educated fathers, 85% went to college, while among the sons of fathers that did not attend college, 35% went to college. What percentage of the third generation went to college if initially 30% of the male population was college educated? For the answer see Example 3.

Two-State Markov Processes

In an earlier chapter, we considered finite stochastic processes. Recall that a finite stochastic process is a finite sequence of experiments in which the outcomes and associated probabilities of each experiment depend on the outcomes of all the preceding experiments. Recall that probability trees are effective in describing such processes.

We now wish to consider finite stochastic processes in which the outcomes and associated probabilities depend only on the preceding experiment. Such a process is called a **Markov process**. The outcome of any experiment is called the **state** of the experiment.

Markov Processes

A finite stochastic process in which the outcomes and associated probabilities depend only on the preceding experiment is called a **Markov process**.

We first consider Markov processes with two states.

E X A M P L E   1   **A Markov Process with Two States**

A survey of a certain ethnic group indicated that 85% of the sons of fathers who attended college also attended college, and 65% of the sons of fathers who did not attend college also did not attend college. Show that this is a Markov process and use a probability tree to describe the transition from one state to another.

Solution   This is a two-state Markov process. Each stage represents a generation. There are 2 states: "attended college," "did not attend college." The probabilities of each depend only on those of the previous stage.

Since 85% of the sons of college-educated fathers go to college and 15% do not, the probability that a randomly selected son of a college-educated father in this group goes to college is .85 and the probability that he does not is .15. This is shown in the tree diagram in Figure 9.1a. Notice that these probabilities must sum to one.

Since 35% of the sons of non–college-educated fathers go to college and 65% do not, the probability that a randomly selected son of a non–college-educated father in this group goes to college is .35 and the probability that he

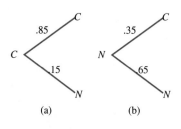

(a)   (b)

Figure 9.1

does not is .65. This is shown in the tree diagram in Figure 9.1b. Notice that these probabilities must sum to one. ■

E X A M P L E 2

A Markov Process with Two States

Use the results of Example 1 and a tree diagram to find the probability that a son in this group is college educated and the probability that a son is not college educated if 30% of the fathers were college educated.

Solution   The initial probability that a father from this group is college educated is given as .30 and thus of not being college educated is .70. We then draw the tree diagram shown in Figure 9.2. The probability that a son will be college educated is then seen to be

$$(.30)(.85) + (.70)(.35) = .50,$$

while the probability that a son will not be college educated is then seen to be

$$(.30)(.15) + (.70)(.65) = .50. \quad ■$$

As will be seen shortly, it is very useful to describe a Markov process by a **transition matrix**. The transition matrix gives the probabilities of going from one state to another. For example, if we create a matrix

$$T = \begin{bmatrix} p_{11} & p_{12} \\ p_{21} & p_{22} \end{bmatrix}$$

with the 4 probabilities in Figure 9.1 taken in the order that they appear there, we have

Current state

State 1   State 2

$$\begin{bmatrix} .85 & .35 \\ .15 & .65 \end{bmatrix} \begin{matrix} \text{State 1} \\ \text{State 2} \end{matrix} \quad \text{Next state}$$

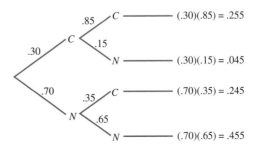

Figure 9.2

where State 1 is college educated and State 2 is not college educated. Thus the probability of going from State 1 to State 1, that is from college-educated father to college-educated son, is $p_{11} = .85$. As a conditional probability this is

$$p_{11} = p(\text{next in state } 1 \mid \text{currently in state } 1).$$

The probability of going from State 1 to State 2, or from college-educated father to non-college-educated son, is $p_{21} = .15$. As a conditional probability this is

$$p_{21} = p(\text{next in state } 2 \mid \text{currently in state } 1).$$

Similarly

$$.35 = p_{12} = p(\text{next in state } 1 \mid \text{currently in state } 2),$$

$$.65 = p_{22} = p(\text{next in state } 2 \mid \text{currently in state } 2).$$

For notational convenience, the fathers are described as generation 0; the current or initial generation. The sons are then described as the first generation.

Now let us see why the transition matrix is so useful. Suppose the percent of the initial generation (the fathers) that was college educated was $100x_0$ and the percent that was not was $100y_0$. If x_1 is the fraction of the first generation (the sons) that is college educated, then, according to Figure 9.3,

$$x_1 = .85x_0 + .35y_0.$$

If y_1 is the fraction of the first generation that are not college educated, then according to Figure 9.3,

$$y_1 = .15x_0 + .65y_0.$$

These last 2 equations can be written in matrix terms as

$$\begin{bmatrix} x_1 \\ y_1 \end{bmatrix} = \begin{bmatrix} .85 & .35 \\ .15 & .65 \end{bmatrix} \begin{bmatrix} x_0 \\ y_0 \end{bmatrix} = \begin{bmatrix} .85x_0 + .35y_0 \\ .15x_0 + .65y_0 \end{bmatrix}$$

If we let

$$X_0 = \begin{bmatrix} x_0 \\ y_0 \end{bmatrix} \quad \text{and} \quad X_1 = \begin{bmatrix} x_1 \\ y_1 \end{bmatrix},$$

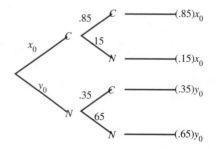

Figure 9.3

then the last matrix equation can be written as

$$X_1 = TX_0. \tag{1}$$

Now suppose we wish the distribution

$$X_2 = \begin{bmatrix} x_2 \\ y_2 \end{bmatrix}$$

of college- and non-college-educated males in the second generation. We can view the distribution in the first generation as the "initial" distribution and use formula (1) and obtain

$$\begin{aligned} X_2 &= TX_1 \\ &= T(TX_0) \\ &= T^2 X_0 \end{aligned}$$

Similarly, if X_3 is the distribution in the third generation, then X_2 can be viewed as the "initial" distribution in the second generation and we can use formula (1) again. This gives

$$\begin{aligned} X_3 &= TX_2 \\ &= T(T^2 X_0) \\ &= T^3 X_0 \end{aligned}$$

We now see that if X_n is the distribution in the nth generation, then

$$X_n = T^n X_0. \tag{2}$$

We also saw above that

$$X_n = TX_{n-1}, \tag{3}$$

where X_{n-1} can be viewed as the "initial" distribution in generation $(n - 1)$. We shall need equation (3) in the next section.

E X A M P L E   3   **A Markov Process with Two States**

Determine the population distribution in the second and third generations if the initial distribution was the same as in Example 2.

Solution   Since

$$T^2 = \begin{bmatrix} .85 & .35 \\ .15 & .65 \end{bmatrix} \begin{bmatrix} .85 & .35 \\ .15 & .65 \end{bmatrix} = \begin{bmatrix} .775 & .525 \\ .225 & .475 \end{bmatrix},$$

$$X_2 = T^2 X_0 = \begin{bmatrix} .775 & .525 \\ .225 & .475 \end{bmatrix} \begin{bmatrix} .30 \\ .70 \end{bmatrix} = \begin{bmatrix} .60 \\ .40 \end{bmatrix}.$$

Thus, in the second generation 60% of the males are college educated. Also since

$$T^3 = \begin{bmatrix} .775 & .525 \\ .225 & .475 \end{bmatrix} \begin{bmatrix} .85 & .35 \\ .15 & .65 \end{bmatrix} = \begin{bmatrix} .7375 & .6125 \\ .2625 & .3875 \end{bmatrix},$$

$$X_3 = T^3 X_0 = \begin{bmatrix} .7375 & .6125 \\ .2625 & .3875 \end{bmatrix} \begin{bmatrix} .30 \\ .70 \end{bmatrix} = \begin{bmatrix} .65 \\ .35 \end{bmatrix}.$$

Thus, in the third generation, 65% of the males are college educated and 35% are not. ■

Three-State Markov Processes

We now look at Markov processes with 3 states.

E X A M P L E 4   **A Markov Process with Three States**

(Dennis Drenner.)

There are 3 restaurants in town: Abby's, Barney's, and Cathy's. The 3 restaurants start a vigorous advertising campaign for Sunday business. A survey indicates that among those who eat their Sunday meal at Abby's, 50% return next Sunday, whereas 30% go to Barney's and 20% go to Cathy's. Among those who eat their Sunday meal at Barney's, 60% return next Sunday, whereas 30% go to Abby's and 10% go to Cathy's. Finally, among those who eat their Sunday meal at Cathy's, 80% return next Sunday, whereas 10% go to Abby's and 10% go to Barney's. Show that this is a Markov process. Draw a tree diagram describing the movements from one state to the other and give a transition matrix for this Markov process.

Solution   This is a 3-state Markov process. Each stage is a Sunday. The 3 states are "ate at Abby's," "ate at Barney's," "ate at Cathy's." The probabilities depend only on those of the previous stage.

The tree diagram is shown in Figure 9.4. Notice that the probabilities of each of the 3 branches must sum to one.

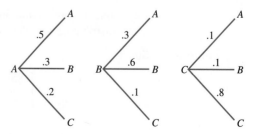

Figure 9.4

The transition matrix is read from Figure 9.4 to be

$$T = \begin{bmatrix} .50 & .30 & .10 \\ .30 & .60 & .10 \\ .20 & .10 & .80 \end{bmatrix}. \quad \blacksquare$$

Notice that each of the columns in the transition matrix sums to 1.

E X A M P L E   5   A Markov Process with Three States

Suppose initially Abby's has $100x_0\%$ of the Sunday business, Barney's had $100y_0\%$, and Cathy's had $100z_0\%$. (a) Draw a tree diagram indicating the distribution of business for the next Sunday. (b) Find the equations that give the distribution for next Sunday in terms of x_0, y_0, and z_0. (c) Verify that these equations can be written in the form of formula (1).

Solution
(a) Figure 9.5 gives the tree diagram.
(b) Let x_1, y_1, and z_1 be the distribution of business for the next Sunday for the respective 3 restaurants. Then, from Figure 9.5,

$$x_1 = .50x_0 + .30y_0 + .10z_0$$
$$y_1 = .30x_0 + .60y_0 + .10z_0$$
$$z_1 = .20x_0 + .10y_0 + .80z_0.$$

(c) If

$$X_0 = \begin{bmatrix} x_0 \\ y_0 \\ z_0 \end{bmatrix} \quad \text{and} \quad X_1 = \begin{bmatrix} x_1 \\ y_1 \\ z_1 \end{bmatrix}$$

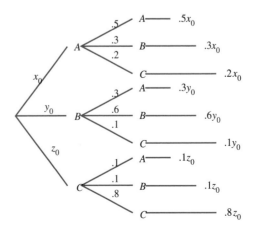

Figure 9.5

then

$$X_1 = \begin{bmatrix} x_1 \\ y_1 \\ z_1 \end{bmatrix} = \begin{bmatrix} .50x_0 + .30y_0 + .10z_0 \\ .30x_0 + .60y_0 + .10z_0 \\ .20x_0 + .10y_0 + .80z_0 \end{bmatrix}$$

$$= \begin{bmatrix} .50 & .30 & .10 \\ .30 & .60 & .10 \\ .20 & .10 & .80 \end{bmatrix} \begin{bmatrix} x_0 \\ y_0 \\ z_0 \end{bmatrix} = TX_0.$$

Since $X_1 = TX_0$, it then follows as before that

$$X_n = T^n X_0. \quad \blacksquare$$

E X A M P L E   6   **A Markov Process with Three States**

Assume initially that Abby's and Barney's each have half of the Sunday business. What is the distribution of business after 1 week? After 2 weeks?

Solution   After 1 week the distribution is given by

$$X_1 = TX_0 = \begin{bmatrix} .50 & .30 & .10 \\ .30 & .60 & .10 \\ .20 & .10 & .80 \end{bmatrix} \begin{bmatrix} .50 \\ .50 \\ 0 \end{bmatrix} = \begin{bmatrix} .40 \\ .45 \\ .15 \end{bmatrix}$$

To find the distribution for the second Sunday, first notice that

$$T^2 = \begin{bmatrix} .50 & .30 & .10 \\ .30 & .60 & .10 \\ .20 & .10 & .80 \end{bmatrix} \begin{bmatrix} .50 & .30 & .10 \\ .30 & .60 & .10 \\ .20 & .10 & .80 \end{bmatrix} = \begin{bmatrix} .36 & .34 & .16 \\ .35 & .46 & .17 \\ .29 & .20 & .67 \end{bmatrix}.$$

Then

$$X_2 = T^2 X_0 = \begin{bmatrix} .36 & .34 & .16 \\ .35 & .46 & .17 \\ .29 & .20 & .67 \end{bmatrix} \begin{bmatrix} .50 \\ .50 \\ 0 \end{bmatrix} = \begin{bmatrix} .350 \\ .405 \\ .245 \end{bmatrix}. \quad \blacksquare$$

E X A M P L E   7   **A Markov Process with Three States**

What are the probabilities that an individual will go to each of the 3 restaurants two Sundays after going to Barney's?

Solution   In this case the initial distribution is

$$X_0 = \begin{bmatrix} 0 \\ 1 \\ 0 \end{bmatrix}$$

Then

$$X_2 = T^2 X_0 = \begin{bmatrix} .36 & .34 & .16 \\ .35 & .46 & .17 \\ .29 & .20 & .67 \end{bmatrix} \begin{bmatrix} 0 \\ 1 \\ 0 \end{bmatrix} = \begin{bmatrix} .34 \\ .46 \\ .20 \end{bmatrix}$$

Thus there is a 34%, 46%, 20% chance he will go to respectively Abby's, Barney's, and Cathy's in two Sundays.

Another way of viewing this is to realize that T^2 represents the transition matrix from the initial stage to the stage after next. Thus

Current state

	State 1	State 2	State 3	
	.36	.34	.16	State 1
	.35	.46	.17	State 2 State after next
	.29	.20	.67	State 3

Then, for example, the entry .34 in the first row and second column represents the probability of going initially from state 2 (Barney's) to state 1 (Abby's) for the stage after next. ∎

m-State Markov Processes

The above analysis works for a Markov process with any number of (finite) states. This is summarized in the following.

---

m-State Markov Process

If a Markov process has m states and the probabilities of moving from one state to another are given by the tree diagrams in Figure 9.6, then the **transition matrix** is given by

$$T = \begin{bmatrix} p_{11} & p_{12} & \cdots & p_{1m} \\ p_{21} & p_{22} & \cdots & p_{2m} \\ & & \cdots & \\ p_{m1} & p_{m2} & \cdots & p_{mm} \end{bmatrix},$$

where the element in the ith row and jth column, p_{ij}, is the conditional probability

$$p_{ij} = p(\text{next in state } i \mid \text{currently in state } j).$$

If the column matrix X_n of dimension m

$$X_n = \begin{bmatrix} x_1 \\ x_2 \\ \ldots \\ x_m \end{bmatrix}$$

gives the state after the nth experiment, then

$$X_n = T^n X_0,$$

where the column matrix X_0 of dimension m is the initial state.

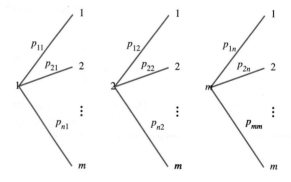

Figure 9.6

We also have the following definition.

Stochastic Matrix

We say that a matrix T is a **stochastic matrix** if

1. The matrix T is square.
2. All elements in the matrix T are nonnegative.
3. The sum of all the elements in any column is 1.

SELF-HELP EXERCISE SET 9.1

1. Consider migration between the North and the South of the United States. Suppose every year 6% of Northerners migrate south and 2% of Southerners migrate

north. For simplicity, assume the populations of the two regions change only due to migration between the regions and that the population is constant. Show that this is a Markov process. Use a probability tree to describe the transition from one state to another and find the transition matrix.

2. Determine the population distribution after 1 year and after 2 years if initially 70% of the population was in the North.

EXERCISE SET 9.1

In Exercises 1 through 12 determine whether or not the given matrix is stochastic.

1. $\begin{bmatrix} .1 & .4 \\ .2 & .4 \\ .7 & .2 \end{bmatrix}$

2. $\begin{bmatrix} .5 & .4 & .3 \\ .5 & .6 & .7 \end{bmatrix}$

3. $\begin{bmatrix} .5 & .2 & .7 \\ .8 & .5 & -.5 \\ -.3 & .3 & .8 \end{bmatrix}$

4. $\begin{bmatrix} 2 & .5 \\ -1 & .5 \end{bmatrix}$

5. $\begin{bmatrix} .87 & .7 \\ .2 & .2 \end{bmatrix}$

6. $\begin{bmatrix} .9 & .5 \\ .1 & .6 \end{bmatrix}$

7. $\begin{bmatrix} .6 & .4 \\ .3 & .7 \end{bmatrix}$

8. $\begin{bmatrix} .4 & .6 \\ .1 & .9 \end{bmatrix}$

9. $\begin{bmatrix} .4 & .7 \\ .5 & .3 \end{bmatrix}$

10. $\begin{bmatrix} .5 & .1 \\ .5 & .9 \end{bmatrix}$

11. $\begin{bmatrix} 0 & 1 \\ 1 & 0 \end{bmatrix}$

12. $\begin{bmatrix} .5 & 0 \\ .5 & 1 \end{bmatrix}$

In Exercises 13 through 18 is given a transition matrix for a Markov process and an initial distribution. Find X_1, X_2, and X_4.

13. $T = \begin{bmatrix} .5 & .4 \\ .5 & .6 \end{bmatrix}$, $X_0 = \begin{bmatrix} 1 \\ 0 \end{bmatrix}$

14. $T = \begin{bmatrix} .5 & 1 \\ .5 & 0 \end{bmatrix}$, $X_0 = \begin{bmatrix} 0 \\ 1 \end{bmatrix}$

15. $T = \begin{bmatrix} .3 & .2 \\ .7 & .8 \end{bmatrix}$, $X_0 = \begin{bmatrix} .5 \\ .5 \end{bmatrix}$

16. $T = \begin{bmatrix} .1 & .4 \\ .9 & .6 \end{bmatrix}$, $X_0 = \begin{bmatrix} .5 \\ .5 \end{bmatrix}$

17. $T = \begin{bmatrix} .5 & 0 & .5 \\ .5 & .5 & 0 \\ 0 & .5 & .5 \end{bmatrix}$, $X_0 = \begin{bmatrix} 1 \\ 0 \\ 0 \end{bmatrix}$

18. $T = \begin{bmatrix} .4 & .3 & 0 \\ .6 & 0 & .2 \\ 0 & .7 & .8 \end{bmatrix}$, $X_0 = \begin{bmatrix} 0 \\ 1 \\ 0 \end{bmatrix}$

19. A town has only two electricians. Of those who call the first electrician, 70% will call again next time. Of those who call the second electrician, 50% will call again next time. Write the transition matrix. If a person calls the first electrician, what is the probability he or she will call the first electrician again the call after next?

20. It is known that a certain businessman always wears a white shirt or a blue shirt to work and that 60% of the time when wearing a white shirt he changes the color the next business day, and 70% of the time when wearing a blue shirt he changes the color the next business day. Write the transition matrix. He was

observed wearing a white shirt on Monday. What is the probability that he will be wearing a white shirt again 4 business days later?

21. Every Mother's Day, a certain individual sends her mother either roses or carnations. If she sends roses one year, 30% of the time she will send roses again the next year. If she sends carnations one year, 80% of the time she will send roses the next year. Write the transition matrix. If she gave roses this year, what is the probability that she will give roses again 2 years from now?

22. A certain person always buys either vanilla or chocolate ice cream. If he buys vanilla, 30% of the time he will buy vanilla again the next time. If he buys chocolate, 10% of the time he will buy chocolate again the next time. Write the transition matrix. If vanilla was bought last time, what is the probability of buying vanilla again 4 times from now?

23. Suppose a person moves into the town mentioned in Exercise 19 and that she is just as likely to call one electrician as the other. What is the probability that she will call the first electrician on her third call?

24. On the first day back to work after vacation, the businessman mentioned in Exercise 20 is just as likely to wear a white shirt as a blue one. What is the probability that he will be wearing a white shirt 4 business days later?

25. The individual in Exercise 21 forgot to send flowers to her Mother last year. The probability that she will send roses this year is .90. What is the probability that she will send roses 2 Mother's Days after this year?

26. The person mentioned in Exercise 22 was given a free strawberry ice cream the last time he had ice cream and now has the same chance of buying vanilla as chocolate the next time he buys ice cream. What is the probability he will buy chocolate 4 times from now?

Applications

27. Marketing. A person always buys either a Ford, GM, or Chrysler product. If she buys a Ford, next time she will buy a Ford again 60% of the time and will be just as likely to buy a GM as a Chrysler. If she buys a GM, next time she will buy a GM again 80% of the time and will be just as likely to buy a Ford as a Chrysler. If she buys a Chrysler, next time she will buy a Chrysler again 70% of the time and will be twice as likely to buy a GM as a Ford. Find the transition matrix. The car that she now owns is a Nissan that she won in a contest. For her next car, she is equally likely to buy a Ford, GM, or Chrysler. What is the probability that she will buy a GM 3 purchases from now?

28. Politics. In a certain state, when a Democrat is governor 30% of the time a Democrat will be governor next time, 10% of the time an Independent will be, and 60% of the time a Republican will be. When an Independent is governor, 20% of the time an Independent will be governor next time, 50% of the time a Democrat, and 30% of the time a Republican. When a Republican is governor, 40% of the time a Republican will be governor next time, 50% of the time a Democrat, and 10% of the time an Independent. Write the transition matrix. For the next election, the experts are giving the Democrat the same chance as the Republican of being elected governor, with the Independent given a 20%

(Robert Houser/Comstock, Inc.)

chance. What is the probability that an Independent will be elected 2 elections from now?

29. **Biology.** A house cat tries to catch a mouse every night in one of 3 neighboring fields. If some night she visits the first field, then she has an 80% chance of returning to the first field the next night and an equal chance of going to one of the other 2 fields. If she visits the second field, she has a 90% chance of returning to the second field the next night and a 10% chance of going to the first. If she visits the third field, she has a 70% chance of returning to the third field the next night and a 30% chance of going to the second. If initially she has the same chance of visiting one field as any other, what is the chance she will be in each of the fields two nights later?

30. **Marketing.** A survey indicates that people in a certain area take their summer vacations either at the beach, at the lake, or in the mountains. The survey finds that among people who have gone to the beach, 20% go to the beach next summer, 30% go to the lake, and 50% go to the mountains. The survey finds also that among people who have gone to the lake, 20% go to the beach next summer, 20% go to the lake, and 60% go to the mountains. Finally the survey finds that among people who have gone to the mountains, 50% go to the beach next summer and 50% go to the lake. Last year the beach suffered an oil spill and nobody went there. Half went to the lake and half went to the mountains. What is the probability that an individual will go to each of the 3 places for vacation two summers from now? Four summers from now?

Solutions to Self-Help Exercise Set 9.1

1. This is a 2-state Markov process. The two states are: "live in the North," "live in the South." The probabilities of each state depend only on those of the previous stage. Each stage is one year.

   Since 94% of Northerners stay in the North and 6% go to the South, the probability that a randomly selected Northerner will stay in the North is .94 and will go to the South is .06. This is shown in the accompanying tree diagram.

   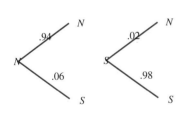

   Since 2% of Southerners go to the North and 98% stay in the South, the probability that a randomly selected Southerner will head to the North is .02 and will stay in the South is .98. This is shown in the accompanying tree diagram.

   The transition matrix is then seen from the tree diagrams to be

   $$\begin{bmatrix} .94 & .02 \\ .06 & .98 \end{bmatrix}.$$

2. The distribution after the first generation is

   $$X_1 = TX_0 = \begin{bmatrix} .94 & .02 \\ .06 & .98 \end{bmatrix}\begin{bmatrix} .70 \\ .30 \end{bmatrix} = \begin{bmatrix} .664 \\ .336 \end{bmatrix}$$

   Since

   $$T^2 = \begin{bmatrix} .94 & .02 \\ .06 & .98 \end{bmatrix}\begin{bmatrix} .94 & .02 \\ .06 & .98 \end{bmatrix} = \begin{bmatrix} .8848 & .0384 \\ .1152 & .9616 \end{bmatrix}$$

   the distribution after the second generation is

$$X_2 = T^2 X_0 = \begin{bmatrix} .8848 & .0384 \\ .1152 & .9616 \end{bmatrix} \begin{bmatrix} .70 \\ .30 \end{bmatrix} = \begin{bmatrix} .6309 \\ .3691 \end{bmatrix}$$

After the second generation about 63% of the population is in the North and about 37% in the South.

9.2   REGULAR MARKOV PROCESSES

▸ *The Long Run*

▸ *Finding the Steady-State Distribution*

APPLICATION

What Happens in the Long-Term

In the last section we studied a Markov process involving three restaurants: Abby's, Barney's, and Cathy's. What happens in the long-term? For the answer see Example 3.

The Long Run

In this section we will consider a certain type of Markov process that has the remarkable property that no matter what the initial distribution, the distribution at the nth stage will tend toward the very same limiting distribution as n becomes large.

Consider again the first Markov process mentioned in the last section. Recall that initially 30% of the fathers of this group attended college. The subsequent distributions over the next 3 generations were calculated to be

$$X_1 = \begin{bmatrix} .50 \\ .50 \end{bmatrix}, \quad X_2 = \begin{bmatrix} .60 \\ .40 \end{bmatrix}, \quad X_3 = \begin{bmatrix} .65 \\ .35 \end{bmatrix}$$

Let us continue the calculations further and obtain

$$X_4 = \begin{bmatrix} .675 \\ .325 \end{bmatrix}, \quad X_5 = \begin{bmatrix} .6875 \\ .3125 \end{bmatrix}, \quad X_6 = \begin{bmatrix} .69375 \\ .30625 \end{bmatrix}, \quad X_7 = \begin{bmatrix} .6969 \\ .3031 \end{bmatrix}$$

$$X_8 = \begin{bmatrix} .6984 \\ .3016 \end{bmatrix}, \quad X_9 = \begin{bmatrix} .6992 \\ .3008 \end{bmatrix}, \quad X_{10} = \begin{bmatrix} .6996 \\ .3004 \end{bmatrix}, \quad X_{11} = \begin{bmatrix} .6998 \\ .3002 \end{bmatrix}$$

We see that X_n is approaching

$$\overline{X} = \begin{bmatrix} .7 \\ .3 \end{bmatrix}$$

That is, in the long run, 70% of the males will attend college and 30% of the males will not attend college. The limiting column matrix $\overline{X}$ is referred to as the **steady-state distribution** for the Markov process.

In order to understand what is happening, we need to look at what is happening to the powers of the transition matrix T, that is, to T^n. Recall from the last section that

$$X_n = T^n X_0 \tag{1}$$

Since X_0 is fixed and X_n approaches the specific matrix $\overline{X}$, one suspects from equation (1) that T^n must approach some particular matrix. Calculating the powers of T, we obtain the following

$$T = \begin{bmatrix} .85 & .35 \\ .15 & .65 \end{bmatrix}, \quad T^2 = \begin{bmatrix} .775 & .525 \\ .225 & .475 \end{bmatrix}, \quad T^3 = \begin{bmatrix} .7375 & .6125 \\ .2625 & .3875 \end{bmatrix}$$

$$T^4 = \begin{bmatrix} .71875 & .65625 \\ .28125 & .34375 \end{bmatrix}, \quad T^5 = \begin{bmatrix} .7094 & .6781 \\ .2906 & .3219 \end{bmatrix}, \quad T^6 = \begin{bmatrix} .7047 & .6891 \\ .2953 & .3109 \end{bmatrix}$$

$$T^7 = \begin{bmatrix} .7023 & .6945 \\ .2977 & .3055 \end{bmatrix}, \quad T^8 = \begin{bmatrix} .7012 & .6973 \\ .2988 & .3027 \end{bmatrix}, \quad T^9 = \begin{bmatrix} .7006 & .6986 \\ .2994 & .3014 \end{bmatrix}$$

$$T^{10} = \begin{bmatrix} .7003 & .6993 \\ .2997 & .3007 \end{bmatrix}, \quad T^{11} = \begin{bmatrix} .7001 & .6997 \\ .2999 & .3003 \end{bmatrix}.$$

Thus the matrix T^n is heading for the limiting matrix

$$L = \begin{bmatrix} .7 & .7 \\ .3 & .3 \end{bmatrix}.$$

The matrix L is referred to as the **steady-state matrix** for the Markov process. Notice that all the columns of L are just $\overline{X}$.

Since $X_n = T^n X_0$ and $T^n \to L$, this shows why X_n is also heading for some limiting column matrix. In fact as n becomes large without bound,

$$T^n X_0 \to \begin{bmatrix} .7 & .7 \\ .3 & .3 \end{bmatrix} \begin{bmatrix} x_0 \\ y_0 \end{bmatrix} = \begin{bmatrix} 0.7 \, (x_0 + y_0) \\ 0.3 \, (x_0 + y_0) \end{bmatrix} = \begin{bmatrix} 0.7 \\ 0.3 \end{bmatrix}$$

since $x_0 + y_0$ must equal one. We have thus discovered that *no matter what the initial probability distribution, the limiting distribution is always the same.* Thus no matter how the system starts initially, the system will head for the limiting distribution $\overline{X}$. Markov processes that have this property are called **regular** and the associated transition matrix T is called a **regular stochastic matrix**.

Since not all Markov processes are regular, it would be convenient if there were a way of easily recognizing when a Markov process was regular. We first give the following definition.

Regular Stochastic Matrix

A stochastic matrix is said to be **regular** if some power has all positive entries.

The following is then true.

> **Theorem 1**
>
> A Markov process is regular if the associated transition matrix is regular.

E X A M P L E   1   **Identifying Regular Stochastic Matrices**

Determine which of the following stochastic matrices are regular.

$$A = \begin{bmatrix} .2 & 1 \\ .8 & 0 \end{bmatrix} \qquad B = \begin{bmatrix} 0 & 1 \\ 1 & 0 \end{bmatrix}$$

Solution   The matrix A does not have all entries positive, but

$$A^2 = \begin{bmatrix} .2 & 1 \\ .8 & 0 \end{bmatrix}\begin{bmatrix} .2 & 1 \\ .8 & 0 \end{bmatrix} = \begin{bmatrix} .84 & .20 \\ .16 & .80 \end{bmatrix}$$

does. Thus A is regular.

The matrix B does not have all entries positive. Furthermore,

$$B^2 = \begin{bmatrix} 0 & 1 \\ 1 & 0 \end{bmatrix}\begin{bmatrix} 0 & 1 \\ 1 & 0 \end{bmatrix} = \begin{bmatrix} 1 & 0 \\ 0 & 1 \end{bmatrix} = I$$

also does not have all entries positive. But $B^3 = B$ and $B^4 = B^2 = I$ and $B^5 = B$, and so on. Thus no power of B will have all entries positive, and B is not regular. ■

Finding the Steady-State Solution

Suppose now we are dealing with a regular Markov process and thus that the associated transition matrix T is a regular stochastic matrix. No matter what the initial state X_0 we know that there is a limiting state, $\overline{X}$. There is an easier way to find $\overline{X}$ than by calculating many powers of T.

Recall that

$$X_n = TX_{n-1} \tag{2}$$

Since as n becomes large without bound, $X_n \rightarrow \overline{X}$, then also $X_{n-1} \rightarrow \overline{X}$. Thus the left-hand side of equation (2) approaches $\overline{X}$, while the right-hand side approaches $T\overline{X}$. We conclude that

$$\overline{X} = T\overline{X} \tag{3}$$

Equation (3) together with the fact that the sum of the entries in the column matrix $\overline{X}$ must be one will then determine $\overline{X}$.

Finding the Steady-State Distribution

If T is a regular stochastic matrix, one can find the steady-state distribution $\overline{X}$, by solving the equations

$$\overline{X} = T\overline{X}$$

together with the fact that the sum of the entries in $\overline{X}$ must be one.

Let us now apply this procedure to the example considered earlier.

E X A M P L E 2   **Finding the Steady-State Distribution**

Find the steady-state distribution $\overline{X}$ for the Markov process with the regular stochastic transition matrix

$$\begin{bmatrix} .85 & .35 \\ .15 & .65 \end{bmatrix}.$$

Solution   Let

$$\overline{X} = \begin{bmatrix} x \\ y \end{bmatrix},$$

then we wish to solve

$$T\overline{X} = \overline{X}$$
$$x + y = 1$$

or

$$x + y = 1$$
$$.85x + .35y = x$$
$$.15x + .65y = y$$

This can be written as

$$x + y = 1$$
$$-.15x + .35y = 0$$
$$.15x - .35y = 0$$

Since the last equation is just the negative of the second equation this is just

$$x + y = 1$$
$$-.15x + .35y = 0$$

Now use the Gauss–Jordan method. Replace the second equation with .15 times the first plus the second and obtain

$$x + y = 1$$
$$.50y = .15.$$

Divide the last equation by .60 and obtain

$$x + y = 1$$
$$y = .3.$$

Now replace the first equation with the first minus the second and obtain

$$x = .7$$
$$y = .3.$$

This agrees with what we found earlier. ∎

The following is a summary of what has been done.

Regular Markov Processes

Let T be a regular stochastic matrix associated with a regular Markov process. Then

1. For any initial distribution X_0, $X_n = T^n X_0$ approaches the same steady-state distribution $\overline{X}$, as n becomes large without bound.
2. As n becomes large without bound, T^n approaches the steady-state matrix L.
3. Every column of L is just $\overline{X}$.
4. The steady-state distribution can be found by solving

    $$\overline{X} = T\overline{X}$$

    and using the fact that the entries in $\overline{X}$ sum to one.
5. A stochastic matrix T will be regular if some power of T has all positive entries.

E X A M P L E 3   **Determining the Steady-State**

Find the steady-state distribution $\overline{X}$ and the steady-state matrix L for the problem in Example 5 in the last section concerning the 3 restaurants in town where the transition matrix was

$$T = \begin{bmatrix} .50 & .30 & .10 \\ .30 & .60 & .10 \\ .20 & .10 & .80 \end{bmatrix}$$

Solution If

$$\overline{X} = \begin{bmatrix} x \\ y \\ z \end{bmatrix},$$

then we must solve

$$x + y + z = 1$$
$$T\overline{X} = \overline{X}$$

or

$$x + y + z = 1$$
$$.50x + .30y + .10z = x$$
$$.30x + .60y + .10z = y$$
$$.20x + .10y + .80z = z$$

The following summarizes the Gauss–Jordan method.

$$\begin{bmatrix} 1 & 1 & 1 & | & 1 \\ -.5 & .3 & .1 & | & 0 \\ .3 & -.4 & .1 & | & 0 \\ .2 & .1 & -.2 & | & 0 \end{bmatrix} \rightarrow \begin{bmatrix} 1 & 1 & 1 & | & 1 \\ 0 & .8 & .6 & | & .5 \\ 0 & -.7 & -.2 & | & -.3 \\ 0 & -.1 & -.4 & | & -.2 \end{bmatrix} \rightarrow$$

$$\begin{bmatrix} 1 & 1 & 1 & | & 1 \\ 0 & 1 & \frac{3}{4} & | & \frac{5}{8} \\ 0 & 0 & \frac{13}{40} & | & \frac{11}{80} \\ 0 & 0 & -\frac{13}{40} & | & -\frac{11}{80} \end{bmatrix} \rightarrow \begin{bmatrix} 1 & 1 & 1 & | & 1 \\ 0 & 1 & \frac{3}{4} & | & \frac{5}{8} \\ 0 & 0 & 1 & | & \frac{11}{26} \\ 0 & 0 & 0 & | & 0 \end{bmatrix} \rightarrow$$

$$\begin{bmatrix} 1 & 1 & 0 & | & \frac{15}{26} \\ 0 & 1 & 0 & | & \frac{8}{26} \\ 0 & 0 & 1 & | & \frac{11}{26} \\ 0 & 0 & 0 & | & 0 \end{bmatrix} \rightarrow \begin{bmatrix} 1 & 0 & 0 & | & \frac{7}{26} \\ 0 & 1 & 0 & | & \frac{8}{26} \\ 0 & 0 & 1 & | & \frac{11}{26} \\ 0 & 0 & 0 & | & 0 \end{bmatrix}$$

Thus after a large number of Sundays, approximately $\frac{7}{26}$ will eat at Abby's, approximately $\frac{8}{26}$ will eat at Barney's, and approximately $\frac{11}{26}$ at Cathy's. The limiting matrix L is

$$L = \begin{bmatrix} \frac{7}{26} & \frac{7}{26} & \frac{7}{26} \\ \frac{8}{26} & \frac{8}{26} & \frac{8}{26} \\ \frac{11}{26} & \frac{11}{26} & \frac{11}{26} \end{bmatrix}. \quad \blacksquare$$

SELF-HELP EXERCISE SET 9.2

1. Determine if the stochastic matrix

$$T = \begin{bmatrix} 0 & .50 \\ 1 & .50 \end{bmatrix}$$

   is regular.

2. Find the steady-state solution for the Markov process found in Self-Help Exercise Set 9.1.

EXERCISE SET 9.2

In Exercises 1 through 10, determine if the given stochastic matrix is regular.

1. $\begin{bmatrix} 0 & .4 \\ 1 & .6 \end{bmatrix}$

2. $\begin{bmatrix} .7 & 1 \\ .3 & 0 \end{bmatrix}$

3. $\begin{bmatrix} 1 & .5 \\ 0 & .5 \end{bmatrix}$

4. $\begin{bmatrix} .25 & 0 \\ .75 & 1 \end{bmatrix}$

5. $\begin{bmatrix} .5 & .2 & 0 \\ .5 & .5 & .4 \\ 0 & .3 & .6 \end{bmatrix}$

6. $\begin{bmatrix} 0 & .5 & .1 \\ .4 & .5 & .4 \\ .6 & 0 & .5 \end{bmatrix}$

7. $\begin{bmatrix} 0 & .5 & 0 \\ 1 & 0 & 1 \\ 0 & .5 & 0 \end{bmatrix}$

8. $\begin{bmatrix} 0 & .3 & 0 \\ 1 & .4 & 1 \\ 0 & .3 & 0 \end{bmatrix}$

9. $\begin{bmatrix} 0 & 0 & .25 \\ 1 & 0 & 0 \\ 0 & 1 & .75 \end{bmatrix}$

10. $\begin{bmatrix} 0 & .1 & 1 \\ 1 & .8 & 0 \\ 0 & .1 & 0 \end{bmatrix}$

Exercises 11 through 20 give regular stochastic matrices. Find the steady-state distribution for each.

11. $\begin{bmatrix} .8 & .5 \\ .2 & .5 \end{bmatrix}$

12. $\begin{bmatrix} .5 & .4 \\ .5 & .6 \end{bmatrix}$

13. $\begin{bmatrix} .7 & .2 \\ .3 & .8 \end{bmatrix}$

14. $\begin{bmatrix} .6 & .7 \\ .4 & .3 \end{bmatrix}$ **15.** $\begin{bmatrix} .7 & .4 \\ .3 & .6 \end{bmatrix}$ **16.** $\begin{bmatrix} .3 & 1 \\ .7 & 0 \end{bmatrix}$

17. $\begin{bmatrix} .6 & .3 & 0 \\ .4 & .4 & .6 \\ 0 & .3 & .4 \end{bmatrix}$ **18.** $\begin{bmatrix} .4 & .5 & .2 \\ .3 & .4 & .6 \\ .3 & .1 & .2 \end{bmatrix}$ **19.** $\begin{bmatrix} 0 & 0 & 1 \\ 1 & .4 & 0 \\ 0 & .6 & 0 \end{bmatrix}$

20. $\begin{bmatrix} .1 & .4 & .1 \\ .1 & .4 & .2 \\ .8 & .2 & .7 \end{bmatrix}$

21. Find the steady-state distribution for Exercise 19 of the last section.

22. Find the steady-state distribution for Exercise 20 of the last section.

23. Find the steady-state distribution for Exercise 21 of the last section.

24. Find the steady-state distribution for Exercise 22 of the last section.

Applications

25. Marketing. Find the steady-state distribution for Exercise 27 of the last section.

26. Politics. Find the steady-state distribution for Exercise 28 of the last section.

27. Biology. Find the steady-state distribution for Exercise 29 of the last section.

28. Marketing. Find the steady-state distribution for Exercise 30 of the last section.

Graphing Calculator Exercises

29. Find the steady-state distribution for the stochastic matrix A given in Exercise 17 by finding on your calculator A^2, A^4, A^8, A^{16}, ..., until your answers are repeated.

30. Repeat the previous exercise for the stochastic matrix given in Exercise 18.

Solutions to Self-Help Exercise Set 9.2

1. The matrix T does not have all positive entries, but

$$T^2 = \begin{bmatrix} 0 & .50 \\ 1 & .50 \end{bmatrix}\begin{bmatrix} 0 & .50 \\ 1 & .50 \end{bmatrix} = \begin{bmatrix} .50 & .25 \\ .50 & .75 \end{bmatrix}$$

has all entries positive. Thus T is regular.

2. Since

$$T = \begin{bmatrix} .94 & .02 \\ .06 & .98 \end{bmatrix},$$

the steady-state solution will be the solution to the equations

$$x + y = 1$$
$$.94x + .02y = x$$
$$.06x + .98y = y$$

or

$$x + y = 1$$
$$-.06x + .02y = 0$$
$$.06x - .02y = 0$$

The last equation is just the negative of the second; therefore, it can be eliminated. Using the Gauss–Jordan method, replace the second equation with the second equation plus .06 times the first and obtain

$$x + y = 1$$
$$.08y = .06$$

or

$$x + y = 1$$
$$y = .75$$

Subtracting gives

$$x = .25$$
$$y = .75$$

Thus, in the long run, 25% of the population will be in the North and 75% will be in the South.

9.3   ABSORBING MARKOV PROCESSES

► *Absorbing Markov Processes*

► *Long-Term Behavior*

► *Sketch of Proof of Theorem 2 (Optional)*

► *Additional Applications (Optional)*

APPLICATION
Long-Term Behavior

Suppose in the restaurant example of the last section that Cathy's responds to the competition by making the food, service, ambiance, and price so extraordinary that once anyone eats at Cathy's, they always eat at Cathy's. Assume furthermore that eating at Abby's results in eating next at Cathy's, Abby's, or Barney's, with probability .20, .70, and .10, respectively. Eating at Barney's results in eating at one of the 3 restaurants next time with probability .10, .10, .80, respectively. What happens in the long-term? For the answer see Example 3.

Absorbing Markov Processes

We now consider Markov processes with at least one state that has the property that when entered it is impossible to leave. Such a state is called **absorbing**. As we shall shortly see, Markov processes with at least one absorbing state are very different from the regular Markov processes studied in the last section. Some applications are given including an important application to charge accounts and production lines in the Exercises.

We will study Markov processes with at least one **absorbing state**, that is, a state that is impossible to leave.

We begin with the following definition.

Absorbing Markov Process

A Markov process is called **absorbing** if the following two conditions are satisfied.

1. There is at least one absorbing state.
2. It is possible to move from any nonabsorbing state to one of the absorbing states in a finite number of stages.

The transition matrix for an absorbing Markov process is said to be an **absorbing stochastic matrix**.

We begin with the following example of an absorbing Markov process.

EXAMPLE 1 **An Absorbing Markov Process**

Consider a particle in Figure 9.7 that can be at 0, 1, 2, or 3. If at 1 or 2, the particle moves left one unit with probability .50 and right one unit with probability .50. If the particle is at 0 or 3, it can never leave. Assume each stage takes one second. Draw a tree diagram that describes the situation. Indicate why this is an absorbing Markov process. Write the transition matrix.

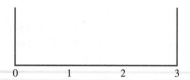

Figure 9.7

Solution   Figure 9.8 is a tree diagram that describes the situation. Since the particle can never leave states 0 and 3, these are absorbing states. From Figure 9.8 one can see that the transition matrix is

$$
\begin{array}{c}
\textbf{State} \\
\begin{array}{cccc}
\textbf{0} & \textbf{1} & \textbf{2} & \textbf{3}
\end{array} \\
\begin{bmatrix}
1 & .50 & 0 & 0 \\
0 & 0 & .50 & 0 \\
0 & .50 & 0 & 0 \\
0 & 0 & .50 & 1
\end{bmatrix}
\begin{array}{c}
\textbf{0} \\
\textbf{1} \\
\textbf{2} \\
\textbf{3}
\end{array}
\end{array}
\quad \textbf{State}
$$

One can also see that, starting at any of the 2 nonabsorbing states, the system can go to at least one of the absorbing states in a finite number of stages. For example, starting at 1, one can go $1 \to 0$. (One can also go $1 \to 2 \to 3$). This is then an absorbing Markov process. ■

This is an example of a *random walk*. Random walk not only plays an important role in physics but also in many other areas, including finance. Random walk is sometimes used to describe the behavior of movements of individual stocks or other financial securities. In these applications the probability that the particle (or price) moves one unit up or down need not be equal.

Example 1 can also be considered as ''gambler's ruin.'' In this interpretation, two gamblers are playing each other with the states representing the number of dollars that the first gambler has. When the number is zero, the first gambler is

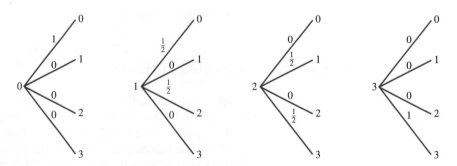

Figure 9.8

broke and the game cannot continue. When the number is 3, the second gambler is broke and again the game cannot continue.

E X A M P L E   2   **Transition Matrix**

Determine if the stochastic matrix

$$\begin{bmatrix} .10 & 0 & .30 \\ .40 & 1 & .20 \\ .50 & 0 & .50 \end{bmatrix}$$

is a transition matrix for an absorbing Markov process.

Solution   The first state is clearly not absorbing, since the first column indicates that it is possible to leave state 1. In the same way, the third column indicates that the third state is not absorbing. The second column, however, indicates that the second state is absorbing, since the probability of going from state 2 to state 2 is one. Thus, once in state 2, always in state 2. Finally, notice that the system can go from either of the nonabsorbing states 1 and 3 to the absorbing state 2 in just one stage. Thus this Markov process is absorbing. ∎

The above analysis indicates that the ith state will be absorbing if the ith column has all zeros except for the ith, which is 1.

We continue with another example of an absorbing Markov process.

E X A M P L E   3   **An Absorbing Markov Process**

Suppose in the restaurant example of the last section that Cathy's responds to the competition by making the food, service, ambiance, and price so extraordinary that once anyone eats at Cathy's, they always eat at Cathy's. Assume furthermore that eating at Abby's results in eating next at Cathy's, Abby's, or Barney's, with probability .20, .70, and .10, respectively. Eating at Barney's results in eating at one of the 3 restaurants next time with probability .10, .10, .80, respectively. Write the transition matrix and show why this is an absorbing Markov process.

Solution   Since Cathy's is absorbing, we place Cathy's first. The transition matrix is then

$$\begin{matrix} & \text{Cathy's} & \text{Abby's} & \text{Barney's} & \\ & \begin{bmatrix} 1 & .20 & .10 \\ 0 & .70 & .10 \\ 0 & .10 & .80 \end{bmatrix} & & & \begin{matrix} \textbf{Cathy's} \\ \textbf{Abby's} \\ \textbf{Barney's} \end{matrix} \end{matrix}$$

The first column indicates that the first state is an absorbing state. The other

2 columns indicate that the other two states are not absorbing and one can go from any of the 2 nonabsorbing states to the absorbing state in just one stage. Thus the Markov process is absorbing.   ■

Long-Term Behavior

Consider now the long-term behavior of the Markov process given above in Example 1. If a particle starts in state 1 or 2, then the only way for the particle to avoid being absorbed is to bounce back and forth between 1 and 2. The probability that this can occur will naturally decrease steadily toward zero in time. Thus in the long-term we expect that **the particle will be absorbed** by either 0 or 3. This is typical of absorbing Markov processes. In fact we have the following theorem.

Theorem 1

In an absorbing Markov process, the probability of going from any state to some absorbing state is one.

Another important observation is that if the particle is initially at 1, the probability that the particle will be absorbed by 0 is certainly larger than the probability it will be absorbed by 3. On the other hand, if the particle is initially at 2, then the probability it will be absorbed by 3 is greater than the probability it will be absorbed by 0. Thus **the long-term behavior may be different for different initial states**. This is in sharp contrast with regular Markov processes.

In order to see analytically exactly what is happening in the long-term, reorganize the states into two groups: the first is the absorbing states, the second is the nonabsorbing states. Doing this we have

$$
\begin{array}{cc}
\textbf{Absorbing} & \textbf{Nonabsorbing}
\end{array}
$$

$$
\begin{array}{cccc}
\textbf{State 0} & \textbf{State 3} & \textbf{State 1} & \textbf{State 2}
\end{array}
$$

$$
\begin{bmatrix}
1 & 0 & .50 & 0 \\
0 & 1 & 0 & .50 \\
0 & 0 & 0 & .50 \\
0 & 0 & .50 & 0
\end{bmatrix}
\begin{array}{l}
\textbf{State 0} \\
\textbf{State 3} \\
\textbf{State 1} \\
\textbf{State 2}
\end{array}
$$

Now to see the long-term behavior we need to find T^n for large values of n. Before doing just that, let us anticipate at least part of the answer. Consider the element in the third row and third column of T^n. This is the probability that after n stages the particle will have moved from state 1 to state 1. This can only happen if the particle bounces back and forth between 1 and 2. We have already indicated that the probability that this occurs should be steadily decreasing toward zero as

n becomes larger. Similarly, the element in the fourth row and third column of T^n represents the probability that the particle has moved from state 1 to state 2. Again, this can happen only if the particle bounces back and forth between 1 and 2. Thus we expect the element in the third row and fourth column of T^n to steadily approach zero. The same for the third and fourth elements in the fourth column.

Also the first column of T^n should always be

$$\begin{bmatrix} 1 \\ 0 \\ 0 \\ 0 \end{bmatrix}$$

since the probability that the particle moves from state 0 to states 0, 3, 1, and 2 must always be 1, 0, 0, 0, respectively, since 0 is an absorbing state. In the same way the second column of the matrices T^n must be

$$\begin{bmatrix} 0 \\ 1 \\ 0 \\ 0 \end{bmatrix}$$

Now actually calculating some of the T^n gives

$$T^2 = \begin{bmatrix} 1 & 0 & .50 & .25 \\ 0 & 1 & .25 & .50 \\ 0 & 0 & .25 & 0 \\ 0 & 0 & 0 & .25 \end{bmatrix}, \quad T^4 = \begin{bmatrix} 1 & 0 & .625 & .3125 \\ 0 & 1 & .3125 & .625 \\ 0 & 0 & .0625 & 0 \\ 0 & 0 & 0 & .0625 \end{bmatrix}$$

$$T^8 = \begin{bmatrix} 1 & 0 & .6641 & .3320 \\ 0 & 1 & .3320 & .6641 \\ 0 & 0 & .0039 & 0 \\ 0 & 0 & 0 & .0039 \end{bmatrix}, \quad T^{16} = \begin{bmatrix} 1 & 0 & .6667 & .3333 \\ 0 & 1 & .3333 & .6667 \\ 0 & 0 & .00002 & 0 \\ 0 & 0 & 0 & .00002 \end{bmatrix}$$

It then appears that this has the limiting value

$$L = \begin{bmatrix} 1 & 0 & \frac{2}{3} & \frac{1}{3} \\ 0 & 1 & \frac{1}{3} & \frac{2}{3} \\ 0 & 0 & 0 & 0 \\ 0 & 0 & 0 & 0 \end{bmatrix}$$

The number $\frac{2}{3}$ in the first row and third column can be interpreted to mean that the probability (in the long-term) that a particle will go from state 1 to state 0 is $\frac{2}{3}$. The number $\frac{1}{3}$ in the first row and fourth column can be interpreted to mean that the probability (in the long-term) that a particle will go from state 2 to state 0 is $\frac{1}{3}$. To take another example, the zero in the fourth row and third column can be interpreted to mean that the probability (in the long-term) of the particle going from state 1 to state 2 is zero.

Later in this section we give a sketch of the proof of how to obtain the other elements in the limiting matrix. The main result is given in the following general theorem.

Theorem 2

Let T be the transition matrix of an absorbing Markov process with a absorbing states and b nonabsorbing states. Reorder the states so that the first a are absorbing and the remaining b are nonabsorbing. Partition the matrix as follows

$$\begin{array}{cc} \textbf{Absorbing} & \textbf{Nonabsorbing} \end{array}$$

$$\left[\begin{array}{c|c} I_{a \times a} & A_{a \times b} \\ \hline O_{b \times a} & B_{b \times b} \end{array}\right] \begin{array}{l} \textbf{Absorbing} \\ \textbf{Nonabsorbing} \end{array}$$

Then,

1. As n becomes large without bound, the matrix T^n, which is the transition matrix from the initial stage to the nth stage, heads for the limiting matrix

$$L = \left[\begin{array}{c|c} I_{a \times a} & A(I - B)^{-1} \\ \hline O_{b \times a} & O_{b \times b} \end{array}\right]$$

where the identity matrix, I, in the expression $(I - B)^{-1}$ is the same dimension as B, that is, $b \times b$.
2. The entry in the ith row and jth column of the matrix $A(I - B)^{-1}$ gives the probability that the system will end up in the ith absorbing state when initially in the jth nonabsorbing state.
3. All column sums of $A(I - B)^{-1}$ are 1, and thus everything is expected to be absorbed.

E X A M P L E   4   **Determining the Long-Term Behavior**

Use the above formulas to find the long-term behavior in Example 3.

Solution   There is 1 absorbing state and 2 nonabsorbing states. Thus $a = 1$ and $b = 2$. We then write the transition matrix in partitioned form as

$$\left[\begin{array}{c|c} I_{1\times1} & A_{1\times2} \\ \hline O_{2\times1} & B_{2\times2} \end{array}\right] = \left[\begin{array}{c|cc} 1 & .20 & .10 \\ \hline 0 & .70 & .10 \\ 0 & .10 & .80 \end{array}\right]$$

Thus

$$A = [.20 \quad .10], \quad B = \begin{bmatrix} .70 & .10 \\ .10 & .80 \end{bmatrix}$$

Then

$$I - B = \begin{bmatrix} .30 & -.10 \\ -.10 & .20 \end{bmatrix}$$

and

$$(I - B)^{-1} = \begin{bmatrix} 4 & 2 \\ 2 & 6 \end{bmatrix}$$

Thus

$$A(I - B)^{-1} = [.20 \quad .10]\begin{bmatrix} 4 & 2 \\ 2 & 6 \end{bmatrix} = [1 \quad 1]$$

Then the limiting matrix is

$$L = \begin{bmatrix} 1 & 1 & 1 \\ 0 & 0 & 0 \\ 0 & 0 & 0 \end{bmatrix}. \quad \blacksquare$$

Notice that this limiting matrix indicates that no matter which state is the initial state, the limiting state will be state 1. That is, eventually everyone will be eating at Abby's and nobody will eat at the other two restaurants.

Sketch of Proof of Theorem 2*

We now give a sketch of the proof of Theorem 2 in the case that the matrices A and B are both of order 2×2. The general case follows in exactly the same fashion. An indication has already been given for why we end up with three of the blocks in the limiting matrix. We will now look at the upper-right block that contains the term $A(I - B)^{-1}$.

First write the matrix T in the block form

---

*Optional

$$\left[\begin{array}{c|c} I_{2\times2} & A_{2\times2} \\ \hline O_{2\times2} & B_{2\times2} \end{array}\right].$$

Then one can multiply T times T algebraically by treating the blocks just as if they were numbers and the matrix as if it were a 2×2 matrix and obtain

$$T^2 = \left[\begin{array}{c|c} I & A \\ \hline O & B \end{array}\right]\left[\begin{array}{c|c} I & A \\ \hline O & B \end{array}\right]$$

$$= \left[\begin{array}{c|c} II + AO & IA + AB \\ \hline OI + BO & OA + BB \end{array}\right]$$

$$= \left[\begin{array}{c|c} I & A(I + B) \\ \hline O & B^2 \end{array}\right]$$

Repeating this will give

$$T^3 = \left[\begin{array}{c|c} I & A(I + B + B^2) \\ \hline O & B^3 \end{array}\right]$$

and in general will give

$$T^n = \left[\begin{array}{c|c} I & A(I + B + \cdots B^{n-1}) \\ \hline O & B^n \end{array}\right]$$

The summation formula in Section 5.3 indicates that if $|x| < 1$, then

$$1 + x + x^2 + x^3 + \cdots = (1 - x)^{-1}$$

It is surprising that under appropriate conditions a similar formula holds for matrices. That is, we then expect

$$I + B + B^2 + B^3 + \cdots = (I - B)^{-1}.$$

This turns out to be the case. Thus the matrix in the upper right block is

$$A(I - B)^{-1}$$

Thus the limiting matrix L can be written as

$$\left[\begin{array}{c|c} I & A(I - B)^{-1} \\ \hline O & O \end{array}\right]$$

Additional Applications*

The matrix $F = (I - B)^{-1}$ is called the **fundamental matrix**. The elements of this matrix, together with the sums of the elements of its columns, convey interesting information. We summarize these facts in the following.

---

*Optional

> **Theorem 2 (Continued)**
>
> Given the same hypothesis and notation used in Theorem 2, then,
>
> 4. The entry f_{ij} of the fundamental matrix $F = (I - B)^{-1}$ gives the expected number of times that the system will be in the ith nonabsorbing state if it is initially in the jth nonabsorbing state.
> 5. The sum of the entries in the jth column of F is the expected number of stages before absorption if the system was initially in the jth nonabsorbing state.

E X A M P L E 5 **Finding the Expected Number of Times the System will be in a Nonabsorbing State**

For the absorbing Markov process in Example 3, find the expected number of times the system will be in the ith nonabsorbing state given it is initially in the jth nonabsorbing state for $i, j = 1, 2$.

Solution According to item (4) above, the answer to the question is given by f_{ij}, which is the entry in the ith row and jth column of the fundamental matrix $F = (I - B)^{-1}$. Letting Cathy's be state 1, Abby's state 2, and Barney's state 3, then Abby's is the first nonabsorbing state and Barney's is the second nonabsorbing state, and we have

$$\begin{array}{cc} \textbf{State 2} & \textbf{State 3} \end{array}$$

$$F = (I - B)^{-1} = \begin{bmatrix} 4 & 2 \\ 2 & 6 \end{bmatrix} \begin{array}{l} \textbf{State 2} \\ \textbf{State 3.} \end{array}$$

Since $f_{11} = 4$, the system is expected to spend 4 stages in state 2 if initially in state 2. That is, an individual who initially eats a Sunday meal at Abby's will eat an average of 4 Sunday meals at Abby's before always eating at Cathy's.

Since $f_{12} = 2$, the system is expected to spend 2 stages in state 2 if initially in state 3. That is, an individual who initially eats a Sunday meal at Barney's will eat an average of 2 Sunday meals at Abby's before always eating at Cathy's.

Since $f_{21} = 2$, the system is expected to spend 2 stages in state 3 if initially in state 2. That is, an individual who initially eats a Sunday meal at Abby's will eat an average of 2 Sunday meals at Barney's before always eating at Cathy's.

Since $f_{22} = 6$, the system is expected to spend 6 stages in state 3 if initially in state 3. That is, an individual who initially eats a Sunday meal at Barney's will eat an average of 6 Sunday meals at Barney's before always eating at Cathy's. ∎

E X A M P L E 6 **Expected Number of Times in a State**

For the absorbing Markov process in Example 3, what is the expected number of Sundays a person will eat at Abby's or Barney's before eating forevermore at Cathy's if that person ate initially at Abby's? Initially at Barney's?

Solution   According to item (5) above, the expected number of stages before absorption if the system was initially in state 2 (Abby's) is the sum of the entries in the first column of the fundamental matrix F. This is $4 + 2 = 6$. Thus a person who initially eats at Abby's, will eat an average of 6 meals at Abby's or Barney's before eating at Cathy's.

According to item (3) above, the expected number of stages before absorption if the system was initially in state 3 (Barney's) is the sum of the entries in the second column of the fundamental matrix F. This is $2 + 6 = 8$. Thus, a person who initially eats at Barney's will eat an average of 8 meals at Abby's or Barney's before eating at Cathy's.   ∎

SELF-HELP EXERCISE SET 9.3

1. Determine if the stochastic matrix

$$\begin{bmatrix} 1 & 0 & 0 \\ 0 & 0 & 1 \\ 0 & 1 & 0 \end{bmatrix}$$

is a transition matrix for an absorbing Markov process.

2. Find the fundamental matrix for the absorbing Markov process of Example 1 using the formula $F = (I - B)^{-1}$.

3. For the absorbing Markov process in Example 1, find the expected number of times the system will be in the state 1 given it is initially in state 2.

4. For the absorbing Markov process in Example 1, what is the expected number of stages before absorption if the system was initially in state 2?

EXERCISE SET 9.3

In Exercises 1 through 4 determine if the given matrix is an absorbing stochastic matrix.

1. $\begin{bmatrix} 0 & 0 & .10 \\ .30 & 1 & .20 \\ .70 & 0 & .70 \end{bmatrix}$ 2. $\begin{bmatrix} 1 & .20 & 0 \\ 0 & .20 & 0 \\ 0 & .60 & 1 \end{bmatrix}$

3. $\begin{bmatrix} 0 & 1 & 0 \\ 1 & 0 & 0 \\ 0 & 0 & 1 \end{bmatrix}$ 4. $\begin{bmatrix} 0 & 0 & 1 \\ 0 & 1 & 0 \\ 1 & 0 & 0 \end{bmatrix}$

Exercises 5 through 12 give a transition matrix for an absorbing Markov process. Find the limiting matrix.

5. $\begin{bmatrix} 1 & 0 & .10 \\ 0 & 1 & .20 \\ 0 & 0 & .70 \end{bmatrix}$
     6. $\begin{bmatrix} 1 & 0 & .30 \\ 0 & 1 & .20 \\ 0 & 0 & .50 \end{bmatrix}$
     7. $\begin{bmatrix} 1 & 0 & 0 & .10 \\ 0 & 1 & 0 & .20 \\ 0 & 0 & 1 & .30 \\ 0 & 0 & 0 & .40 \end{bmatrix}$

8. $\begin{bmatrix} 1 & 0 & 0 & .20 \\ 0 & 1 & 0 & .30 \\ 0 & 0 & 1 & .40 \\ 0 & 0 & 0 & .10 \end{bmatrix}$
     9. $\begin{bmatrix} 1 & .20 & .10 \\ 0 & .50 & .50 \\ 0 & .30 & .40 \end{bmatrix}$
     10. $\begin{bmatrix} 1 & .10 & .30 \\ 0 & .20 & .20 \\ 0 & .70 & .50 \end{bmatrix}$

11. $\begin{bmatrix} 1 & 0 & .10 & .20 \\ 0 & 1 & .30 & .50 \\ 0 & 0 & .20 & .20 \\ 0 & 0 & .40 & .10 \end{bmatrix}$
     12. $\begin{bmatrix} 1 & 0 & .20 & .20 \\ 0 & 1 & .20 & .30 \\ 0 & 0 & .30 & .10 \\ 0 & 0 & .30 & .40 \end{bmatrix}$

13. Find the expected number of times that the system in Exercise 9 will be in the ith nonabsorbing state given that it is initially in the jth nonabsorbing state for $i, j = 1, 2$.

14. Find the expected number of times that the system in Exercise 10 will be in the ith nonabsorbing state given that it is initially in the jth nonabsorbing state for $i, j, = 1, 2$.

15. Find the expected number of times that the system in Exercise 11 will be in the ith nonabsorbing state given that it is initially in the jth nonabsorbing state for $i, j, = 1, 2$.

16. Find the expected number of times that the system in Exercise 12 will be in the ith nonabsorbing state given that it is initially in the jth nonabsorbing state for $i, j = 1, 2$.

17. Find the expected number of times until the system in Exercise 9 is absorbed if initially in the jth nonabsorbing state for $j = 1, 2$.

18. Find the expected number of times until the system in Exercise 10 is absorbed if initially in the jth nonabsorbing state for $j = 1, 2$.

19. Find the expected number of times until the system in Exercise 11 is absorbed if initially in the jth nonabsorbing state for $j = 1, 2$.

20. Find the expected number of times until the system in Exercise 12 is absorbed if initially in the jth nonabsorbing state for $j = 1, 2$.

Applications

21. **Estes Learning Model.**   Assume an individual is in state 1 if he has learned a particular task and in state 2 if he has not. Once the individual learns a task, he never forgets it. If a person has not learned the task, there is a probability of .25 that he will learn it during the next time period. Write the transition matrix for this Markov process. Show that this Markov process is absorbing. Determine the long-term behavior.

22. **Estes Learning Model.** Repeat the previous problem if for the individual who has not learned the task, the probability that he will learn the task in the next time period is p, with $0 < p < 1$.

23. **Gambler's Ruin.** A person plays a game in which the probability of winning \$1 is .40 and the probability of losing \$1 is .60. If he goes broke or reaches \$3, he quits. Find the long-term behavior if he starts with \$1 or \$2.

24. **Gambler's Ruin.** A person plays a game in which the probability of winning \$1 is .50 and the probability of losing \$1 is .50. If she goes broke or reaches \$4, she quits. Find the long-term behavior if she starts with \$1, \$2, or \$3. *Hint:*

$$\begin{bmatrix} 1 & -\frac{1}{2} & 0 \\ -\frac{1}{2} & 1 & -\frac{1}{2} \\ 0 & -\frac{1}{2} & 1 \end{bmatrix}^{-1} = \begin{bmatrix} \frac{3}{2} & 1 & \frac{1}{2} \\ 1 & 2 & 1 \\ \frac{1}{2} & 1 & \frac{3}{2} \end{bmatrix}$$

(Courtesy of TropWorld Casino and Entertainment Resort.)

25. **Production Line.** A certain manufacturing process consists of 2 manufacturing states and a completion state together with a fourth state in which the item is scrapped if improperly manufactured. At the end of each of the 2 manufacturing states each item is inspected. At each inspection there is a probability of $\frac{2}{3}$ that the item will be passed on to the next state (the first manufacturing state to the second or the second manufacturing state to the completion state), probability $\frac{1}{6}$ that it will be sent back to the same state for reworking, and probability $\frac{1}{6}$ that the item will be scrapped. Naturally, an item that is complete stays complete, and an item that is scrapped stays scrapped. Write a transition matrix for this Markov process. Show that this is an absorbing Markov process with 2 absorbing states and 2 nonabsorbing states. Determine the long-term behavior.

26. **Production Line.** Repeat the previous problem with the following changes. At each inspection there is a probability of .70 that the item will be passed on to the next stage, probability .20 that it will be sent back to the same stage for reworking, and probability .10 that the item will be scrapped.

27. **Charge Accounts.** A charge account is classified at the end of each month as "paid," "overdue less than 30 days," "overdue less than 60 days," or "overdue 60 days or more." In the latter case, the debt is judged "bad" and the charge account is canceled. Each month the debt can age by at most one month, keep the same or lesser age by means of a payment, or be paid off. If paid off, the account remains paid off. Over a period of time, the company has obtained figures that give the following transition matrix

$$\begin{array}{cccc} \textbf{Paid} & \textbf{<30} & \textbf{<60} & \textbf{Bad} \end{array}$$
$$\begin{bmatrix} 1 & \frac{1}{3} & \frac{1}{6} & 0 \\ 0 & \frac{1}{3} & \frac{1}{3} & 0 \\ 0 & \frac{1}{3} & \frac{1}{3} & 0 \\ 0 & 0 & \frac{1}{6} & 1 \end{bmatrix} \begin{array}{l} \textbf{Paid} \\ \textbf{<30} \\ \textbf{<60} \\ \textbf{Bad} \end{array}$$

By looking at the long-term behavior, determine the probability that the debt will be paid off and the probability that the debt will default in each of the cases: <30 and <60.

28. Charge Accounts. The charge card company in the previous exercise includes a threat in the "less than 60" category debt that going to 60 days or more will result in "serious consequences" for the customer. As a result the transition matrix changes to

$$\begin{array}{cccc} \textbf{Paid} & \textbf{<30} & \textbf{<60} & \textbf{Bad} \end{array}$$
$$\begin{bmatrix} 1 & \frac{1}{3} & \frac{1}{6} & 0 \\ 0 & \frac{1}{3} & \frac{1}{3} & 0 \\ 0 & \frac{1}{3} & \frac{5}{12} & 0 \\ 0 & 0 & \frac{1}{12} & 1 \end{bmatrix} \begin{array}{l} \textbf{Paid} \\ \textbf{<30} \\ \textbf{<60} \\ \textbf{Bad} \end{array}$$

Now answer the same questions as in the previous exercise. Carefully compare how the probability of default has now gone down dramatically with this small change.

29. Production Line. Refer to Exercise 25. Find the average number of times each item passes through the first and second manufacturing processes, respectively.

30. Production Line. Refer to Exercise 26. Find the average number of times each item passes through the first and second manufacturing processes, respectively.

31. Charge Accounts. Refer to Exercise 27. For each of the 2 categories of debt, find the expected number of months until paid off or defaulting.

32. Charge Accounts. Refer to Exercise 28. For each of the 2 categories of debt, find the expected number of months until paid off or defaulting.

33. A mouse is placed in room 1 in the maze shown. The mouse will move to an adjoining room at each stage, and the probability it will pass through any door in a given room is the same for all doors in that room. If the mouse enters the room with the trap, it will be trapped. What is the expected number of visits to room 1 before the mouse is trapped? To room 2?

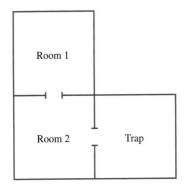

Graphing Calculator Exercises

34. Using the stochastic matrix A in Example 3 of the text, determine the long-term behavior by calculating A^2, A^4, A^8, A^{16}, ..., until your answers repeat themselves.

35. Repeat the previous exercise using the 4×4 stochastic matrix discussed in the text after Theorem 1.

36. Repeat Exercise 34 using the stochastic matrix found in Exercise 9.

37. Repeat Exercise 34 using the stochastic matrix found in Exercise 10. Determine the fraction that each decimal in each entry represents.

38. Charge Accounts. A charge account is classified at the end of each month as "paid," "overdue less than 30 days," "overdue less than 60 days," "overdue less than 90 days," or "overdue 90 days or more." In the latter case, the debt is judged "bad" and the charge account is canceled. Each month the debt can age by at most one month, keep the same or lesser age by means of a payment, or be paid off. If paid off, the account remains paid off. Over a period of time, the company has obtained figures that give the following transition matrix.

$$
\begin{array}{ccccc}
\textbf{Paid} & \textbf{<30} & \textbf{<60} & \textbf{<90} & \textbf{Bad} \\
\end{array}
$$

$$
\begin{bmatrix}
1 & .6 & .1 & 0 & 0 \\
0 & .3 & .2 & .1 & 0 \\
0 & .1 & .4 & .4 & 0 \\
0 & 0 & .3 & .2 & 0 \\
0 & 0 & 0 & .3 & 1
\end{bmatrix}
\begin{array}{l}
\textbf{Paid} \\
\textbf{<30} \\
\textbf{<60} \\
\textbf{<90} \\
\textbf{Bad}
\end{array}
$$

By looking at the long-term behavior, determine the probability that the debt will be paid off and the probability that the debt will default in each of the cases: <30, <60, and <90.

39. **Charge Accounts.** The charge card company in the previous exercise includes a threat in the "less than 60" category debt that going to 60 days or more will result in "serious consequences" for the customer. As a result the transition matrix changes to

$$
\begin{array}{ccccc}
\textbf{Paid} & \textbf{<30} & \textbf{<60} & \textbf{<90} & \textbf{Bad} \\
\end{array}
$$

$$
\begin{bmatrix}
1 & .7 & .25 & 0 & 0 \\
0 & .2 & .25 & .1 & 0 \\
0 & .1 & .25 & .4 & 0 \\
0 & 0 & .25 & .2 & 0 \\
0 & 0 & 0 & .3 & 1
\end{bmatrix}
\begin{array}{l}
\textbf{Paid} \\
\textbf{<30} \\
\textbf{<60} \\
\textbf{<90} \\
\textbf{Bad}
\end{array}
$$

Now answer the same questions as in the previous exercise. Carefully compare how the probability of default has now gone down dramatically with this small change.

Solutions to Self-Help Exercise Set 9.3

1. The first state is absorbing. Thus this Markov process has at least one absorbing state. But notice from the transition matrix that if the system is in state 2, it must go to state 3, while if the system is in state 3, it must go to state 2. Thus, if the system starts in either of the nonabsorbing states, state 2 or state 3, the system cannot enter the absorbing state 1. Thus the system fails to satisfy condition (2) above and is not an absorbing Markov process

2. We have

$$
I - B = \begin{bmatrix} 1 & 0 \\ 0 & 1 \end{bmatrix} - \begin{bmatrix} 0 & \frac{1}{2} \\ \frac{1}{2} & 0 \end{bmatrix} = \begin{bmatrix} 1 & -\frac{1}{2} \\ -\frac{1}{2} & 1 \end{bmatrix}
$$

and

$$(I - B)^{-1} = \begin{bmatrix} \frac{4}{3} & \frac{2}{3} \\ \frac{2}{3} & \frac{4}{3} \end{bmatrix}$$

3. According to item (4) of Theorem 2, the answer to the question is given by f_{12}, which is the entry in the first row and second column of the fundamental matrix $F = (I - B)^{-1}$. Since $f_{12} = \frac{2}{3}$ from the previous exercise, the system will spend an average of $\frac{2}{3}$ of a stage in state 1 if initially in state 2.

4. According to item (3) of Theorem 2, the expected number of stages before absorption if the system was initially in state 2 is the sum of the entries in the second column of the fundamental matrix F. This is

$$\frac{2}{3} + \frac{4}{3} = 2.$$

Thus if initially in state 2, the system is expected to spend an average of 2 stages in the two nonabsorbing states before absorption.

SUMMARY OUTLINE OF CHAPTER 9

- A finite stochastic process in which the outcomes and associated probabilities depend only on the preceding experiment is called a **Markov process**.
- The outcome of any experiment is called the **state** of the experiment.
- A **transition matrix** for a Markov process gives the probabilities of going from one state to another.
- If the column matrix X_n gives the state after the nth experiment, then $X_n = T^n X_0$, where T is the transition matrix and X_0 is the initial state.
- A Markov process is said to be **regular** if no matter what the initial state, the limiting distribution is always the same. This limiting distribution is called the **steady-state distribution**.
- A stochastic matrix is said to be **regular** if some power has all positive entries.
- A Markov process is regular if the associated transition matrix is regular.
- If T is a regular stochastic matrix, one can find the steady-state distribution $\overline{X}$, by solving the equations $\overline{X} = T\overline{X}$ together with the fact that the sum of the entries in $\overline{X}$ must be one.
- **Regular Markov Processes.** Let T be a regular stochastic matrix associated with a Markov process. Then
  1. For any initial distribution X_0, $X_n = T^n X_0$ approaches the same steady-state distribution $\overline{X}$, as n becomes large without bound.
  2. As n becomes large without bound, T^n approaches the steady-state matrix L.
  3. Every column of L is just $\overline{X}$.

4. The steady-state distribution can be found solving

$$\overline{X} = T\overline{X}$$

and using the fact that the entries in $\overline{X}$ sum to one.

5. A stochastic matrix T will be regular if some power of T has all positive entries.

- A Markov process is called **absorbing** if there is at least one absorbing state and it is possible to move from any nonabsorbing state to one of the absorbing states in a finite number of stages.

- The transition matrix for an absorbing Markov process is said to be an **absorbing stochastic matrix**.

- In an absorbing Markov process, the probability of going from any state to some absorbing state is one.

- Let T be the transition matrix of an absorbing Markov process with a absorbing states and b nonabsorbing states. Reorder the states so that the first a are absorbing and the remaining b are nonabsorbing. Partition the matrix as follows

Absorbing   Nonabsorbing

$$\left[\begin{array}{c|c} I_{a\times a} & A_{a\times b} \\ \hline O_{b\times a} & B_{b\times b} \end{array}\right] \begin{array}{l} \textbf{Absorbing} \\ \textbf{Nonabsorbing} \end{array}$$

Then,

1. As n becomes large without bound, the matrix T^n, which is the transition matrix from the initial stage to the nth stage, heads for the limiting matrix

$$L = \left[\begin{array}{c|c} I_{a\times a} & A(I - B)^{-1} \\ \hline O_{b\times a} & O_{b\times b} \end{array}\right]$$

where the identity matrix, I, in the expression $(I - B)^{-1}$ is the same dimension as B, that is, $b \times b$.

2. The entry in the ith row and jth column of the matrix $A(I - B)^{-1}$ gives the probability that the system will end up in the ith absorbing state when initially in the jth nonabsorbing state.

3. All column sums of $A(I - B)^{-1}$ are 1, and thus, everything is expected to be absorbed.

4. The entry f_{ij} of the fundamental matrix $F = (I - B)^{-1}$ gives the expected number of times that the system will be in the ith nonabsorbing state if it is initially in the jth nonabsorbing state.

5. The sum of the entries in the jth column of F is the expected number of stages before absorption if the system was initially in the jth state.

Chapter 9 Review Exercises

In Exercises 1 through 6 determine if the given matrix is stochastic, regular stochastic, absorbing stochastic, or none of these.

1. $\begin{bmatrix} .10 & .90 \\ .50 & .50 \end{bmatrix}$

2. $\begin{bmatrix} .20 & .40 \\ .80 & .60 \end{bmatrix}$

3. $\begin{bmatrix} 0 & .50 & .30 \\ .40 & 0 & .70 \\ .60 & .50 & 0 \end{bmatrix}$

4. $\begin{bmatrix} 1 & 0 & 0 \\ 0 & .40 & .50 \\ 0 & .60 & .50 \end{bmatrix}$

5. $\begin{bmatrix} 0 & .5 & 0 \\ 1 & .5 & 0 \\ 0 & 0 & 1 \end{bmatrix}$

6. $\begin{bmatrix} 1 & .10 & 0 \\ 0 & .30 & .50 \\ 0 & .60 & .50 \end{bmatrix}$

Exercises 7 through 8 give a regular stochastic matrix, T. Find T^2 and T^4 and interpret what these matrices represent.

7. $\begin{bmatrix} .10 & .60 \\ .90 & .40 \end{bmatrix}$

8. $\begin{bmatrix} .30 & .20 & .10 \\ .30 & .20 & .40 \\ .40 & .60 & .50 \end{bmatrix}$

9. Find the steady-state distribution for the regular stochastic matrix found in Exercise 7.

10. Find the steady-state distribution for the regular stochasitc matrix found in Exercise 8.

11. Find the limiting matrix for the absorbing stochastic matrix

$$\begin{bmatrix} 1 & 0 & .20 \\ 0 & 1 & .30 \\ 0 & 0 & .50 \end{bmatrix}.$$

12. Find the limiting matrix for the absorbing stochastic matrix

$$\begin{bmatrix} 1 & .10 & .20 \\ 0 & .50 & .50 \\ 0 & .40 & .30 \end{bmatrix}.$$

13. **Car Rental.** A car rental company has two locations in a certain city. A car rented at the first location will be returned to this location 70% of the time and 30% of the time to the second location. A car rented at the second location will be returned to this location 80% of the time and 20% of the time to the first location. If initially the cars are evenly distributed between the two locations, what will the distribution of cars eventually be?

14. **Social Mobility.** Suppose that the son of a father in the upper class has a probability of .50 of being in the upper class, .40 of being in the middle class, and .10 of being in the lower class. For a son of a father in the middle class, suppose the probabilities are .20, .70, and .10, respectively, while for a son of a father in the lower class, the probabilities are .10, .30, and .60, respectively. Determine what happens in the long-term.

15. **Personnel.** A firm has 2 types of mechanics: mechanic and master mechanic. Each year a mechanic has a 20% chance of leaving the firm and a 20% chance

of being promoted to a master mechanic. A master mechanic has a 10% chance of leaving. If a mechanic or master mechanic leave, they never return. A master mechanic is never demoted to a mechanic. Consider the 4 states: leave as master mechanic, leave as mechanic, mechanic, master mechanic. Determine the long-term behavior.

16. **Personnel.**   What is the expected number of years a person will be a master mechanic, given that they were initially a mechanic?

17. **Personnel.**   What is the expected number of years for a mechanic to stay with the firm?

Game theory can help a firm develop the best competitive strategy. (Gregory Heisler/The Image Bank.)

Game Theory

"The theory of games rests on the notion that there is a close analogy between parlor games of skill, on the one hand, and conflict situations in economics, political, and military life, on the other. In any of these situations there are a number of participants with incompatible objectives, and the extent to which each participant attains his objectives depends upon what all the participants do. The problem faced by each participant is to lay his plans so as to take account of the actions of his opponents, each of whom, of course, is laying his own plans so as to take account of the first participant's actions. Thus each participant must surmise what each of his opponents will expect him to do and how these opponents will react to these expectations.'' Dorfman, Samuelson, Solow in *Linear Programming and Economic Analysis*.

10.1   STRICTLY DETERMINED GAMES

▶ *Strictly Determined Games*

▶ *A Nonstrictly Determined Game*

John von Neuman,
1903–1957

(© The Bettmann Archive)

Born in Hungary, John von Neuman began his teaching career in Germany. In 1933 he became a permanent member of the Institute for Advanced Study at Princeton. Von Neuman was one of the most creative and versatile mathematicians of the 20th century. He made deep contributions to logic and set theory and in 1928 developed the central theorem of game theory. The impact of game theory on economics was delayed until the publication of *The Theory of Games and Economic Behavior* by von Neumann and Morgenstern in 1944. Von Neuman was also deeply involved in quantum theory and in 1955 was appointed to the Atomic Energy Commission. He was most responsible for initiating the first fully electronic calculator and for the concept of a stored digital computer.

Strictly Determined Games

We use the term **game** to refer to situations of conflict between players. The players can be individuals, coalitions, companies or even the weather. We first consider the simplest games: strictly determined games. In these games the strategies of both players and the outcome are all precisely determined.

To obtain an idea of what is involved in game theory, consider the following very simplified decision problem.

E X A M P L E   1   Analysis of a Strictly Determined Game

(Dennis Drenner.)

Two car dealers, R and C, in a certain town need to clear out their inventory of new cars in August to make room for the new models that will arrive in September. They decide to have either a $300 or $500 over-the-cost sale. From past price wars they know that if both set prices at $500 above cost, dealer R will obtain 70% of the total sales. If both set prices at $300 above cost, dealer R will obtain 50% of sales. Also if dealer R sets prices $300 above cost while dealer C sets prices $500 above cost, dealer R will obtain 75% of sales. Finally if dealer R sets prices $500 above cost while dealer C sets price $300 above cost, dealer R will obtain 35% of sales. What is the best strategy for both dealers to clear out their inventory if we make the important assumption that *both dealers are rational and intelligent?*

Solution　　　We summarize the situation as follows:

Dealer C

$500　　$300

$$\text{Dealer R} \quad \begin{array}{c} \$500 \\ \$300 \end{array} \begin{bmatrix} 70\% & 35\% \\ 75\% & 50\% \end{bmatrix} \quad (1)$$

Consider now the strategy for dealer R. Suppose dealer R sets his prices at $500 above cost (pick row 1 in matrix (1)). As soon as dealer C got wind of this, dealer C would set his prices at $300 above cost, resulting in dealer R obtaining only 35% of sales. This would be a disaster for dealer R. On the other hand, if dealer R sets his prices $300 above cost (pick row 2), then he knows that dealer C will respond in kind, and dealer R will obtain 50% of the sales. The best strategy for dealer R is to set his prices at $300 above cost (pick row 2) and obtain 50% of the sales.

　　Now consider the point of view of dealer C. If he sets his prices at $500 above cost (picks column 1), then dealer R will set prices at $300 above cost, resulting in dealer R obtaining 75% of sales. This is a disaster for dealer C. If dealer C sets his prices at $300 above cost (picks column 2 in matrix (1)), then dealer R will be expected to set his prices at $300 above cost, resulting in dealer R obtaining 50% of the sales. Clearly then, the best strategy for dealer C is to set prices at $300 above cost (pick column 2).

　　The final result is that both dealers will set their prices at $300 above cost and each will obtain 50% of the sales.　■

　　The above game is an example of a **two-person zero-sum game**.

Two-Person Zero-Sum Game

A game is called a **two-person zero-sum game** if the following are satisfied.

1.　There are two players (called the row player and the column player).
2.　The row player must choose one of m strategies and simultaneously, the column player must choose one of n strategies.
3.　If the row player chooses her ith strategy and the column player chooses his jth strategy, the row player receives a reward of a_{ij} and the column player loses an amount a_{ij}.

　　The game is called zero-sum since the row player's gain is the column player's loss, and vice versa. Thus the 2 players are in total conflict, and no cooperation between the two is possible.

　　The following indicates the general situation.

Column's strategy

$$
\begin{array}{c}
\\
\text{Row's strategy}
\end{array}
\begin{array}{cccc}
\text{Column 1} & \text{Column 2} & & \text{Column } n \\
\end{array}
$$

$$
\begin{array}{c}
\text{Row 1} \\
\text{Row 2} \\
\vdots \\
\text{Row } m
\end{array}
\begin{bmatrix}
a_{11} & a_{12} & \cdots & a_{1n} \\
a_{21} & a_{22} & \cdots & a_{2n} \\
\vdots & \vdots & \cdots & \vdots \\
a_{m1} & a_{m2} & \cdots & a_{mn}
\end{bmatrix} \quad (2)
$$

Payoff Matrix

The $m \times n$ matrix $A = (a_{ij})$ in (2) is called the **payoff matrix**.

In Example 1 each of the elements a_{ij} in the payoff matrix represented a percent. We now consider a game in which each of the elements a_{ij} in the payoff matrix represents a dollar amount.

E X A M P L E 2   **Determining the Best Strategies**

Consider the game with the following payoff matrix

$$
\begin{bmatrix}
1 & -2 & -5 & 1 \\
6 & 1 & 2 & 3 \\
0 & -3 & 3 & -6
\end{bmatrix}
$$

where the element in the ith row and jth column represents the amount in dollars that the row player wins from the column player. (A negative win means the row player loses this amount.) Thus if the row player chooses row 1 and the column player chooses column 1, then the row player wins \$1 from the column player. If the row player chooses row 1 and the column player chooses column 2, then the row player loses \$2 to the column player. Find the best strategies for each player.

Solution   We first look at the row player's strategy. If the row player chooses the first row, that is, row strategy 1, then she can count on the column player picking column 3, which contains the minimum of the numbers in row 1. This results in a loss of \$5 to the row player. We place a small "m" after the -5 in the first row. We refer to the -5 as the minimum of row 1. See matrix (3). If the row player chooses the second row, that is, row strategy 2, then she can count on the column player picking the column that contains the smallest number in the second row. This is column 2. This results in a gain of \$1 to the row player. We place a small "m" after the 1 in the second row. If the row player chooses the third row, then this row minimum is -6 and is in column 4. Thus the column player will pick column 4. This results in a loss of \$6 to the row player. We place a small "m" after the -6 in the third row. The best strategy for the row player is then to pick the row containing the *maximum* of the row minimums. The row minimums are the numbers with a small "m" after them. Since the maximum of $\{-5, 1, -6\}$ is 1 and is in row 2, the best strategy for the row player is to pick row 2. This results in a certain gain of at least \$1.

$$\begin{bmatrix} +1 & -2 & -5m & +1 \\ +6 & +1m & +2 & +3 \\ 0 & -3 & +3 & -6m \end{bmatrix} \qquad (3)$$

If the column player chooses column 1, then the row player will pick row 2 since 6 is the column maximum for the first column. This results in a gain of \$6 for the row player. This is indicated in matrix (4) by a capital "M" after the 6. If the column player chooses column 2, then the row player will pick row 2, since the 1 in column 2 is that column's maximum. This is indicated in matrix (2) by a capital "M" after the 1. Selecting column 3 results in the row player selecting row 3 since 3 is the column maximum for that column. Selecting column 4 results in the row player selecting row 2, since 3 is the fourth column's maximum. The best strategy for the column player is to then pick the column that contains the minimum of the column maximums. This is column 2. This results in a gain of \$1 for the row player.

$$\begin{bmatrix} +1 & -2 & -5 & +1 \\ +6M & +1M & +2 & +3M \\ 0 & -3 & +3M & -6 \end{bmatrix} \qquad (4)$$

Thus the only rational outcome of this game is for the row player to pick the second row and the column player to pick the second column, resulting in row winning \$1. ∎

The value \$1 in the previous game is called the **payoff value**. If the payoff value is zero, the game is called **fair**.

Let us now see what happens when we overlay matrix (3) with matrix (4). This is shown in matrix (5).

$$\begin{bmatrix} +1 & -2 & -5m & +1 \\ +6M & +1mM & +2 & +3M \\ 0 & -3 & +3M & -6m \end{bmatrix} \qquad (5)$$

Notice the payoff value 1 in the second row and second column has both a small "m" and a capital "M" next to it. This indicates that this element is both the row minimum and the column maximum for the row and column that it is in. Such an element is called a **saddle point**. One can think of riding a horse in the direction →. Directly in front of you and behind you, the saddle rises. To each side, where your legs are, the saddle drops. In the same way, as we move horizontally away (left or right) from the saddle point 1 in the previous example, the values increase, whereas, as we move vertically (up or down) away from 1, the values decrease. This is in analogy with the saddle. As may be suspected from the previous example, saddle points play an important role in game theory.

Let the payoff matrix of a game have a_{ij} in the ith row and jth column.

Saddle Point

If there is an entry, a_{hk}, that is the minimum in its row (row h) and also the maximum in its column (column k), we call the entry a_{hk} a **saddle point**.

Strictly Determined Game

A game with a saddle point is said to be **strictly determined**.

If a game has a saddle point a_{hk}, then the number a_{hk} is called the **value** of the game. If there is more than one saddle point, they will both yield the same value for the game. (This is not proven here.) In a game with a saddle, the optimal strategy for the row player is to choose row R_h (the row containing a saddle). The optimal strategy for the column player is to choose column C_k (the column containing a saddle).

Locating a Saddle Point (The m & M method)

1. For each row, find the minimum value and place a small "m" next to it.
2. For each column, find the maximum value and place a capital "M" next to it.
3. Any entry with both a small "m" and a capital "M" next to it is a saddle point.

A saddle point can also be thought of as an **equilibrium point**. The word equilibrium is used because the row player's payoff will decrease if the row player unilaterally changes to another row other than the optimal row. Thus the row player will move back to the optimal row. If the column player changes the column from the optimal column, the result is an increased gain for the row player. Thus the column player would return to the optimal choice. The saddle point is then in "equilibrium" since if either player moves away from it, he or she will quickly return.

E X A M P L E   3   **Determining Optimal Strategies and the Value of a Game**

Determine if the game with the following payoff matrix is strictly determined. If it is, find the value of the game and the optimal strategies of each of the players.

$$A = \begin{bmatrix} -3 & 0 & -2 \\ 1 & 2 & 0 \\ 3 & 1 & -1 \end{bmatrix}$$

Solution   We place a small "m" next to the -3 in the first row of A, the 0 in the second row, and the -1 in the third row. Then place a capital "M" next to the 3 in the first column, the 2 in the second column, and the 0 in the third column. Notice that there is one entry with both a small m and a capital M next to it. This is the saddle point, which indicates that the game is strictly determined with value 0. This game is a fair game. The saddle point is in the second row and third column. Thus the optimal strategy for the row player is to select row 2 and the optimal strategy for the column player is to select column 3.

$$\begin{bmatrix} -3m & 0 & -2 \\ +1 & +2M & 0mM \\ +3M & +1 & -1m \end{bmatrix} \quad ■$$

The next example has more than 2 saddle points.

EXAMPLE 4   Finding Saddle Points

Find all saddle points for the game with the following payoff matrix

$$\begin{bmatrix} 0 & 1 & 1 & 0 \\ 1 & 1 & 2 & 1 \\ 1 & 2 & 3 & 1 \end{bmatrix}.$$

Solution   Placing the m's and the M's yields

$$\begin{bmatrix} 0m & +1 & +1 & 0m \\ +1mM & +1m & +2 & +1mM \\ +1mM & +2M & +3M & +1mM \end{bmatrix}.$$

There are 4 saddle points. The value of this game is 1.   ■

A Nonstrictly Determined Game

Consider the game of Odds and Evens.

EXAMPLE 5   A Game that is not Strictly Determined

Two players, called Odd and Even, simultaneously put out either 1 or 2 fingers. If the sum of all the fingers from both players is odd, then Odd wins \$1 from Even. If the sum of all the fingers from both players is even, then Odd loses \$1 to Even. Let Odd be the row player and Even the column player. Find the payoff matrix and determine if there is a saddle point.

Solution   The result of the various outcomes are summarized in the following.

$$
\begin{array}{cc}
 & \textbf{Even} \\
 & \begin{array}{cc} \textbf{1 finger} & \textbf{2 fingers} \end{array} \\
\textbf{Odd}\begin{array}{c} \textbf{1 finger} \\ \textbf{2 fingers} \end{array} & \begin{bmatrix} -1 & +1 \\ +1 & -1 \end{bmatrix}
\end{array}
$$

Now adding the m's and M's gives

$$
\begin{bmatrix} -1m & +1M \\ +1M & -1m \end{bmatrix}.
$$

Since no entry has both a m and a M next to it, there are no saddle points and the game is not strictly determined. ■

We will study games that are not strictly determined in the next section.

SELF-HELP EXERCISE SET 10.1

Determine if any of the following matrices are payoff matrices for strictly determined games. If one is, then find the value of the game and the optimal strategies for each player.

1. $\begin{bmatrix} -2 & 2 \\ 1 & -1 \\ -1 & 3 \end{bmatrix}$

2. $\begin{bmatrix} -3 & -1 & 0 \\ -2 & 0 & 2 \\ -4 & 0 & -3 \end{bmatrix}$

EXERCISE SET 10.1

In Exercises 1 through 12 determine if any of the matrices are payoff matrices for strictly determined games. If one is, then find the value of the game and the optimal strategies for each player.

1. $\begin{bmatrix} 1 & 0 \\ 2 & 3 \end{bmatrix}$

2. $\begin{bmatrix} -1 & 0 \\ -2 & -3 \end{bmatrix}$

3. $\begin{bmatrix} -3 & -2 & -4 \\ 0 & -2 & -1 \end{bmatrix}$

4. $\begin{bmatrix} 3 & 2 & 4 \\ 0 & 1 & 1 \end{bmatrix}$

5. $\begin{bmatrix} 1 & 2 \\ 3 & 4 \\ 5 & 6 \end{bmatrix}$

6. $\begin{bmatrix} -1 & -2 \\ -3 & -4 \\ -5 & -6 \end{bmatrix}$

7. $\begin{bmatrix} -3 & -2 & -4 \\ 0 & -1 & -1 \end{bmatrix}$

8. $\begin{bmatrix} 1 & 1 & 3 \\ 1 & 1 & 2 \end{bmatrix}$

9. $\begin{bmatrix} 1 & 0 \\ 0 & 1 \end{bmatrix}$

10. $\begin{bmatrix} -1 & 0 \\ 0 & -2 \end{bmatrix}$

11. $\begin{bmatrix} 2 & -1 & -5 & -2 \\ 4 & -1 & 8 & 7 \\ 3 & 0 & 1 & 4 \end{bmatrix}$

12. $\begin{bmatrix} 4 & -4 & -2 & 0 \\ 0 & 0 & -1 & 0 \\ 9 & 3 & -3 & -2 \end{bmatrix}$

13. Show that the game with the payoff matrix

$$\begin{bmatrix} 0 & a \\ 1 & 2 \end{bmatrix}$$

is strictly determined no matter what a is.

14. Show that the game with the payoff matrix

$$\begin{bmatrix} 2 & 4 \\ 1 & b \end{bmatrix}$$

is strictly determined no matter what b is.

In Exercises 15 through 18, write the payoff matrix for the given game and determine if the game is strictly determined. If it is, find the optimal strategies for each player.

15. Each of two players R and C has, in their respective pockets, 3 coins: a penny, a nickel, and a dime. Each one selects a coin, and simultaneously each lays a coin on the table. If the coins are the same, no payment is made. If the coins are different, the one who played the coin with the smallest denomination wins both coins.

16. In this version of three-finger Morra, each player simultaneously shows either 1, 2, or 3 fingers. Player C agrees to pay player R an amount of dollars equal to the number of fingers shown by player R less the number shown by player C.

17. A person secretly places a penny, nickel, or dime in his fist. If you guess the correct coin, you win the coin. If you guess incorrectly, you give him the difference between your guess and the coin held.

18. In this version of four-finger Morra, each player simultaneously shows either 1, 2, 3, or 4 fingers. If the sum of fingers shown is even, you win an amount in dollars equal to the sum. If the sum of fingers is odd, you lose an amount equal to the sum.

Solutions to Self-Help Exercise Set 10.1

1. For this matrix we have

$$\begin{bmatrix} -2m & 2 \\ 1M & -1m \\ -1m & 3M \end{bmatrix}.$$

Since there is no entry with both a m and a M next to it, there is no saddle. Thus the game is not strictly determined.

2. For this matrix we have

$$\begin{bmatrix} -3m & -1 & 0 \\ -2mM & 0M & 2M \\ -4m & 0M & -3 \end{bmatrix}.$$

The entry -2 has both a m and a M next to it, indicating that this entry is a saddle point. The game is strictly determined with value -2. The optimal strategy is for the row player to choose row 2 and the column player to choose column 1.

10.2   MIXED STRATEGY GAMES

▶ *Mixed Strategies*

▶ *Expectation*

▶ *A Geometric Method*

▶ *Domination*

Mixed Strategies

Games that are not strictly determined are considered in this section. We will find the value and the optimal strategies for nonstrictly determined two-person, zero-sum games.

Consider again the game of Odds and Evens. This game is simple enough that a strategy can easily be developed and the value of the game determined.

Two players, called Odd and Even, simultaneously put out either 1 or 2 fingers. If the sum of all the fingers from both players is odd, then Odd wins \$1 from Even. If the sum of all the fingers from both players is even, then Odd loses \$1 to Even. Let Odd be the row player and Even the column player. Find the payoff matrix and determine if there is a saddle point.

The results of the various outcomes are summarized in the following.

Even

1 finger   2 fingers

$$\begin{array}{cc} & \begin{array}{cc} \text{1 finger} & \text{2 fingers} \end{array} \\ \textbf{Odd} \begin{array}{c} \text{1 finger} \\ \text{2 fingers} \end{array} & \left[\begin{array}{cc} -1 & +1 \\ +1 & -1 \end{array} \right] \end{array}$$

We saw in the last section that this was not a strictly determined game. Consider how such a game might be played. If the row player always plays 1 finger, then the column player can counter by always playing 1 finger and win all the time. If the row player always plays 2 fingers, then the column player can counter with always playing 2 fingers, and win all the time.

Consider now the possibility of the row player selecting either 1 finger or 2 fingers in some **random** fashion. For example, in playing the game a number of times, the row player might select 1 finger one-third of the time and 2 fingers two-thirds of the time with the choice being made in a random fashion. Once the column player realizes that this particular strategy has been selected, one that favors 2 fingers, then the column player will counter by always putting out 2 fingers. Such a strategy for the row player will then result in losses. If the row player makes a random selection by favoring 1 finger, then the column player can counter by always putting out 1 finger, again resulting in losses for the row player.

Thus for this simple game, it is clear that if the row player is to avoid losses, she must not favor either of the two strategies. In other words, she will need to select 1 finger or 2 fingers equally often and in a random manner. This should lead to her breaking even. Similar reasoning implies a similar strategy for the column player. Thus we suspect that this game has a value of zero. Each player can assure that they are selecting each of the two strategies randomly and with equal probability by determining their selections by, say, flipping a fair coin. This way, the column player, for example, cannot possibly guess how many fingers will be put out by the row player since the row player will not know herself until the coin is flipped.

A **mixed strategy**

$$p = \begin{bmatrix} p_1 \\ p_2 \end{bmatrix},$$

or written in the more convenient transposed form, $p^T = [p_1 \quad p_2]$, is a strategy in which the first and second strategy is selected randomly with probability p_1 and p_2, respectively. Thus a strategy of $p^T = [\frac{1}{3} \quad \frac{2}{3}]$, mentioned earlier, means selecting randomly the first strategy with probability $\frac{1}{3}$ and the second with probability $\frac{2}{3}$.

We will always use the letter p to denote the strategy of the row player and the letter q for the column player. Thus a strategy of $q^T = [\frac{1}{4} \quad \frac{3}{4}]$ means that the column player will randomly select his first strategy with probability .25 and the second with probability .75.

Expectation

In the discussion of the Odds and Evens game, we decided that if the row player used a strategy that favored selecting 2 fingers, then the column player could counter with a strategy that favored selecting 2 fingers and cause the row player to lose money. Let us in fact now consider the expected gain or loss for the row player if the row player used the strategy $p^T = [\frac{1}{3} \quad \frac{2}{3}]$ in the game of Odds and Evens and the column player used the strategy $q^T = [\frac{1}{4} \quad \frac{3}{4}]$.

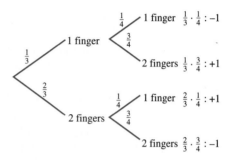

Figure 10.1

Figure 10.1 is a probability tree showing the probability of each of the 4 possible outcomes using these two strategies of the row and column player. The expected value for these two selections of strategies is then

$$E = \frac{1}{3} \cdot \frac{1}{4}(-1) + \frac{1}{3} \cdot \frac{3}{4}(+1) + \frac{2}{3} \cdot \frac{1}{4}(+1) + \frac{2}{3} \cdot \frac{3}{4}(-1) = -\frac{1}{6}.$$

As we strongly suspected, this favoring of 2 fingers by both players will result in losses for the row player.

To generalize, consider a game with a payoff matrix

$$A = \begin{bmatrix} a_{11} & a_{12} \\ a_{21} & a_{22} \end{bmatrix}$$

and assume the row player selects the mixed strategy $p^T = [p_1 \quad p_2]$ and the column player selects the mixed strategy $q^T = [q_1 \quad q_2]$. Then since the probabilities of the 4 possible outcomes are given by the tree diagram in Figure 10.2, the expectation is

$$E(p, q) = p_1 q_1 a_{11} + p_1 q_2 a_{12} + p_2 q_1 a_{21} + p_2 q_2 a_{22}.$$

This can be written in the more convenient form

$$E(p, q) = p^T A q,$$

as the reader can easily verify.

Similar calculations work for a general $m \times n$ payoff matrix A. We then have

Expected Value

Suppose a game has a payoff matrix

$$A = \begin{bmatrix} a_{11} & a_{12} & \cdots & a_{1n} \\ a_{21} & a_{22} & \cdots & a_{2n} \\ & & \cdots & \\ a_{m1} & a_{m2} & \cdots & a_{mn} \end{bmatrix}$$

and row plays the strategy $p^T = [p_1 \quad p_2 \quad \cdots \quad p_m]$, and column selects $q^T = [q_1 \quad q_2 \quad \cdots \quad q_n]$, then the expected value for these two strategies is

$$E(p, q) = p^T A q = [p_1 \quad p_2 \quad \cdots \quad p_m] \begin{bmatrix} a_{11} & a_{12} & \cdots & a_{1n} \\ a_{21} & a_{22} & \cdots & a_{2n} \\ & & \cdots & \\ a_{m1} & a_{m2} & \cdots & a_{mn} \end{bmatrix} \begin{bmatrix} q_1 \\ q_2 \\ \vdots \\ q_n \end{bmatrix}.$$

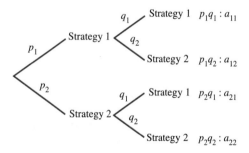

Figure 10.2

We say that the strategy $[p_1 \quad p_2 \quad \cdots \quad p_m]$ is a **pure strategy** if any one of the p_i equals 1. (In such a case all the other p_i's must be zero.) Thus in the Odds and Evens game the strategy $[0 \quad 1]$ means always selecting 2 fingers. Notice that pure strategies were the only kind considered in the last section and that a pure strategy is just a special case of a mixed strategy.

E X A M P L E   1   **Finding Expected Values**

Consider the game with the payoff matrix

$$A = \begin{bmatrix} -1 & 1 \\ 2 & -4 \end{bmatrix}.$$

Find the expected value if (a) $p^T = [.2 \quad .8]$, $q^T = [1 \quad 0]$. (b) $p^T = [.7 \quad .3]$, $q^T = [1 \quad 0]$.

Solutions   (a)   In this case

$$E(p, q) = p^T A q = [.2 \quad .8] \begin{bmatrix} -1 & 1 \\ 2 & -4 \end{bmatrix} \begin{bmatrix} 1 \\ 0 \end{bmatrix} = 1.4.$$

(b)   In this case

$$E(p, q) = p^T A q = [.7 \quad .3] \begin{bmatrix} -1 & 1 \\ 2 & -4 \end{bmatrix} \begin{bmatrix} 1 \\ 0 \end{bmatrix} = -0.1. \quad \blacksquare$$

We will see in the next section, using our knowledge of linear programming, that for any game with a $m \times n$ payoff matrix A, there exists an optimal row-strategy $\bar{p}$ and a value v_r, such that

$$E(\bar{p}, q) = \bar{p}^T A q \geq v_r$$

for any q. Thus $\bar{p}$ is optimal in the sense that by selecting the mixed strategy $\bar{p}$, row can assure herself a profit of at least v_r, if $v_r \geq 0$ (or a loss of at most v_r if $v_r \leq 0$), no matter what strategy column selects.

Also we shall see in the next section that there exists an optimal column strategy $\bar{q}$ and a value v_c, such that

$$E(p, \bar{q}) = p^T A \bar{q} \leq v_c$$

for any p. Thus $\bar{q}$ is optimal in the sense that by selecting the mixed strategy $\bar{q}$, the column player can assure himself that the row player will obtain a profit of at most v_c, if $v_c \geq 0$ (or a loss of at least v_c if $v_c \leq 0$), no matter what strategy the row player selects.

The remarkable fact is that $v_r = v_c$, as will be seen in the next section. This common value, denoted by v, is called the **value** of the game. This is all summarized in the following.

Fundamental Theorem of Game Theory

Given any game with a $m \times n$ payoff matrix A, there exist strategies, $\bar{p}$ and $\bar{q}$, for row and column respectively, such that

1. $E(\bar{p}, q) = \bar{p}^T A q \geq v$

for all q and

2. $E(p, \bar{q}) = p^T A \bar{q} \leq v$

for all p.

Since $E(\bar{p}, q) = \bar{p}^T A q \geq v$ for any q, then

$$E(\bar{p}, \bar{q}) = \bar{p}^T A \bar{q} \geq v. \tag{1}$$

Also since $E(p, \overline{q}) = p^T A \overline{q} \leq v$ for any p, then

$$E(\overline{p}, \overline{q}) = \overline{p}^T A \overline{q} \leq v. \tag{2}$$

Relationships (1) and (2) then imply that

$$E(\overline{p}, \overline{q}) = \overline{p}^T A \overline{q} = v. \tag{3}$$

Thus if both players use their optimal strategies, the expected outcome is v. Furthermore, from parts 1 and 2 of the fundamental theorem, this is the best each player can hope to do. This is why we call v the value of the game.

A Geometric Method

We shall now give a geometric method of finding optimal strategies and the value of the game. This technique works readily if the payoff matrix has at most 2 rows, or at most 2 columns, or if it can be reduced to such a case. The more general situation will be solved in the next section using linear programming. The advantage of considering the geometric method given here is that it gives considerable insight into an important result that will be needed in the next section. (This result says that adding the value k to every entry in the payoff matrix A results in a new game that has the *same* optimal strategies as the original game, with the value of the new game $v + k$ where v is the value of the original game.)

E X A M P L E 2 **Finding Optimal Strategies**

Find the optimal strategies and the value of the game given in Example 1.

Solution We will find Row's optimal strategy and reserve finding Column's for a Self-Help Exercise. Since $p_1 + p_2 = 1$, $p_2 = 1 - p_1$. Thus any mixed strategy $[p_1 \quad p_2]$ can be written as $[p_1 \quad 1 - p_1]$. Suppose then that Row chooses a particular mixed strategy $[p_1 \quad 1 - p_1]$. Let us find the expected value that will result for each of the possible 2 selections that Column can make.

If Column selects strategy 1, that is, $q^T = [1 \quad 0]$, then

$$E(p, q) = [p_1 \quad 1 - p_1] \begin{bmatrix} -1 & 1 \\ 2 & -4 \end{bmatrix} \begin{bmatrix} 1 \\ 0 \end{bmatrix} = 2 - 3p_1.$$

If Column selects strategy 2, that is, $q^T = [0 \quad 1]$, then

$$E(p, q) = [p_1 \quad 1 - p_1] \begin{bmatrix} -1 & 1 \\ 2 & -4 \end{bmatrix} \begin{bmatrix} 0 \\ 1 \end{bmatrix} = -4 + 5p_1.$$

Both of these are straight lines, as shown in Figure 10.3.

Column will naturally always select the strategy that yields the smallest expected return for Row. This is shown as the two line segments in color in Figure 10.3. Row will naturally maximize her expected return and choose the point p_1 that yields the highest value on the line segments marked in color.

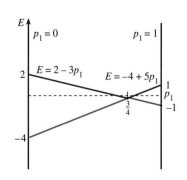

Figure 10.3

This can be determined as the intersection of the two lines. Setting

$$2 - 3p_1 = -4 + 5p_1$$

gives $p_1 = \frac{3}{4}$. Thus the optimal strategy for Row is $p^T = [.75 \quad .25]$. The value can be computed using either one of the lines. Using the first gives

$$v = 2 - 3(.75) = -.25. \quad \blacksquare$$

E X A M P L E   3   **Using the Geometric Technique to Find Optimal Strategies**

Use the geometric technique to find the value of the Odds and Evens game and the optimal strategies for Column.

Solution   Recall that the payoff matrix is given by

$$\begin{bmatrix} -1 & +1 \\ +1 & -1 \end{bmatrix}.$$

Since $q_1 + q_2 = 1$, $q_2 = 1 - q_1$. Thus any mixed strategy $[q_1 \quad q_2]$ can be written as $[q_1 \quad 1 - q_1]$. Suppose then that Column chooses a particular mixed strategy $[q_1 \quad 1 - q_1]$. Let us find the expected value that will result for each of the possible 2 selections that Row can make.

If Row selects strategy 1, that is, $p^T = [1 \quad 0]$, then

$$E(p, q) = [1 \quad 0] \begin{bmatrix} -1 & +1 \\ +1 & -1 \end{bmatrix} \begin{bmatrix} q_1 \\ 1 - q_1 \end{bmatrix} = 1 - 2q_1.$$

If Row selects strategy 2, that is, $p^T = [0 \quad 1]$, then

$$E(p, q) = [0 \quad 1] \begin{bmatrix} -1 & +1 \\ +1 & -1 \end{bmatrix} \begin{bmatrix} q_1 \\ 1 - q_1 \end{bmatrix} = -1 + 2q_1.$$

These lines are shown in Figure 10.4.

Since Row will select the strategy that maximizes her return, she will select the strategy shown as the colored line segments. Column will then minimize Row's expected return by choosing the point q_1 that gives the lowest value on the colored line segments in Figure 10.4. The value of q_1 can be found as the intersection of the two lines. Thus setting

$$1 - 2q_1 = -1 + 2q_1$$

yields $q_1 = .50$ and a value of

$$v = 1 - 2q_1 = 1 - 2(.50) = 0.$$

Thus the optimal strategy for Column is $[.50 \quad .50]$, and the value of the game is 0. $\blacksquare$

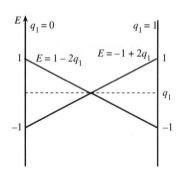

Figure 10.4

We will establish the fundamental theorem of game theory in the next section, using linear programming. The proof requires that all entries in the payoff matrix be positive. Consider now a game with a payoff matrix with some nonpositive elements. One can always select a number k so that adding this number to every entry in the payoff matrix will result in a new payoff matrix that has all entries positive. The following theorem indicates that the game with the new payoff matrix has the same optimal strategies and a value that differs by k.

Translated Games

Consider a game with the $m \times n$ payoff matrix A and value v. Let E be the $m \times n$ matrix with every entry equal to 1 and let k be any constant. Then the game with payoff matrix $A + kE$ has value $v + k$ and the same optimal strategies as the game with payoff matrix A.

The proof of this theorem is clear from the geometric methods used in this section. Adding k to every entry in the matrix A will merely move all the line segments in the previous discussion up by k units if k is positive and down by k units if k is negative. Thus it is apparent that the optimal strategies will remain the same and the value will be changed by k.

Domination

There are certain rows (columns) for some payoff matrices that the row (column) player should never choose. These rows and column can then be deleted from consideration and the payoff matrix then reduced to a lower order.

Consider the game with the following payoff matrix.

$$\begin{bmatrix} 2 & -1 & 3 \\ 1 & -2 & 1 \\ -1 & 1 & 2 \end{bmatrix}$$

Let us compare the first and second rows. No matter what choice Column makes, Row will do better by selecting the first row over the second row. We say from Row's point of view that the first row **dominates** the second row. The second row can be eliminated obtaining

$$\begin{bmatrix} 2 & -1 & 3 \\ -1 & 1 & 2 \end{bmatrix}.$$

Now from Column's point of view, the first or second column is always a better choice than the third column. Thus the third column can be eliminated, leaving the game

$$\begin{bmatrix} 2 & -1 \\ -1 & 1 \end{bmatrix}.$$

We have the following definition.

Domination

Row i is said to **dominate** row j if every element in row i is greater than or equal to the corresponding element in row j. Also column i is said to dominate column j if every element in column i is less than or equal to the corresponding element in column j.

E X A M P L E   4   **Finding the Value of a Game and the Optimal Strategy**

Find the value and the optimal strategy for Row for the game with the payoff matrix

$$B = \begin{bmatrix} -2 & -1 & 0 \\ -1 & 0 & 1 \\ 2 & 3 & -4 \end{bmatrix}.$$

Solution   The second row dominates the first row since every element in the second row is greater than or equal to the corresponding element in the first row. Thus the first row can be eliminated to obtain

$$\begin{bmatrix} -1 & 0 & 1 \\ 2 & 3 & -4 \end{bmatrix}.$$

Now the first column dominates the second column since every element in the first column is smaller or equal to every corresponding element in the second column. Thus the second column can be eliminated to obtain the payoff matrix

$$A = \begin{bmatrix} -1 & 1 \\ 2 & -4 \end{bmatrix}.$$

We recognize this payoff matrix as the same one in as in Example 2. Since the value of the game with payoff matrix A is $-.25$, the value of the game with payoff matrix B is also $-.25$. Recall that the optimal strategy for Row in the game with payoff matrix A is $\overline{p}^T = [.75 \quad .25]$. Thus, since the first row is never chosen, the optimal strategy for Row in the game with payoff matrix B is $\overline{p}^T = [0 \quad .75 \quad .25]$. ■

SELF-HELP EXERCISE SET 10.2

1. For the game of Odds and Evens, determine the optimal strategy for Row using the geometric methods presented in this section.

2. For the game in Example 2, determine the optimal strategy for Column using the geometric methods presented in this section.

3. Given the game with the payoff matrix

$$\begin{bmatrix} 0 & -1 & 1 & 1 \\ 1 & 0 & -1 & -2 \\ -1 & 0 & -2 & -2 \end{bmatrix},$$

reduce it to the smallest order possible by eliminating any row or column that is dominated.

EXERCISE SET 10.2

In Exercises 1 through 4 find the expected return for the given mixed strategies, using the game with the indicated payoff matrix

$$\begin{bmatrix} 1 & -1 \\ -2 & 3 \end{bmatrix}.$$

1. $p^T = [.50 \quad .50]$, $q^T = [.25 \quad .75]$

2. $p^T = [.50 \quad .50]$, $q^T = [.50 \quad .50]$

3. $p^T = [.30 \quad .70]$, $q^T = [.80 \quad .20]$

4. $p^T = [.60 \quad .40]$, $q^T = [.30 \quad .70]$

In Exercises 5 through 8 find the expected return for the given mixed strategies, using the game with the indicated payoff matrix

$$\begin{bmatrix} 2 & -1 & 1 \\ -1 & 3 & 4 \\ 2 & -1 & 3 \end{bmatrix}$$

5. $p^T = [.50 \quad .50 \quad 0]$, $q^T = [.20 \quad 0 \quad .80]$

6. $p^T = [.10 \quad .20 \quad .70]$, $q^T = [0 \quad .50 \quad .50]$

7. $p^T = [.30 \quad .30 \quad .40]$, $q^T = [.40 \quad .20 \quad .40]$

8. $p^T = [.30 \quad .60 \quad .10]$, $q^T = [0 \quad .70 \quad .30]$

Exercises 9 through 18 are payoff matrices. Find the value and the optimal strategies using the geometric methods of this section. Use dominance to eliminate any rows or columns.

9. $\begin{bmatrix} 1 & -2 \\ -1 & 2 \end{bmatrix}$ 10. $\begin{bmatrix} 0 & -1 \\ -2 & 1 \end{bmatrix}$

11. $\begin{bmatrix} 0 & 2 \\ 1 & -1 \end{bmatrix}$ 12. $\begin{bmatrix} -1 & 0 \\ 1 & -2 \end{bmatrix}$

13. $\begin{bmatrix} 1 & 1 & 0 \\ -1 & 0 & 2 \end{bmatrix}$ 14. $\begin{bmatrix} 1 & 1 & -1 \\ -1 & -2 & 1 \end{bmatrix}$

15. $\begin{bmatrix} -1 & 0 \\ 2 & -1 \\ 2 & -1 \end{bmatrix}$ **16.** $\begin{bmatrix} -2 & 0 \\ 1 & 0 \\ 1 & -1 \end{bmatrix}$

17. $\begin{bmatrix} 2 & 0 & 0 \\ 2 & 1 & 0 \\ 0 & -1 & 2 \end{bmatrix}$ **18.** $\begin{bmatrix} 0 & -1 & 1 \\ 4 & 3 & -2 \\ 3 & 3 & 0 \end{bmatrix}$

The geometric theory given in the text can be used to find the optimal strategy for Row for any game with a payoff matrix with two rows. Find the optimal strategy for Row in Exercises 19 through 20.

19. $\begin{bmatrix} 3 & 0 & -1 \\ -5 & -4 & 2 \end{bmatrix}$ **20.** $\begin{bmatrix} 3 & 2 & -1 \\ -2 & 0 & 1 \end{bmatrix}$

The geometric theory given in the text can be used to find the optimal strategy for Column in any game with a payoff matrix with two columns. Find the optimal strategy for Column in Exercises 21 through 22.

21. $\begin{bmatrix} 4 & -2 \\ 1 & 2 \\ 3 & 1 \end{bmatrix}$ **22.** $\begin{bmatrix} 3 & -5 \\ -2 & 0 \\ -1 & -3 \end{bmatrix}$

Applications

(Dennis Drenner.)

23. **Setting Prices.**   Every August and January two carpet retailers in town run their sales. They must decide whether to have a 20% or 30% sale. From many previous years of experience with these sales, they know that if both set prices at 20% off the first retailer will get 80% of the business, and if both set prices at 30% off the first retailer will get 70% of the business. If the first retailer takes 20% off and the second 30% off, then the first retailer gets 30% of the business, while if the first takes 30% off and the second 20% off, the first retailer gets 60% of the business. What should their strategies be and what is the value of this game?

24. **Price Wars.**   Every day two movie theaters in town must decide on the price to set on their movies for the next day so the advertisement can go into the morning paper for the next day. They know from long experience that if they both set their prices at $5 or both at $6, then the first theater will get 70% of the business. If the first theater sets prices at $5 and the second at $6, then the first theater gets 60% of the business, while if the first theater charges $6 and the second $5, then the first theater gets 40% of the business. What should their strategies be and what is the value of this game?

(Dennis Drenner.)

25. **Advertising.**   Two competing electronics stores must decide each week whether to advertise in one and only one of the three media: TV, radio, or newspaper. From past experience they know that the payoff matrix in terms of percent of the market gained or lost to the other is

	TV	Radio	Paper
TV	1	1	0
Radio	−1	0	−1
Paper	2	−1	2

Find the value of this game and the optimal strategies.

26. **Marketing.** Early in each year two local growers compete for the freshly cut gladiola market, committing themselves entirely to one and only one color of gladiola. Past experience indicates the following payoff matrix given in terms of the percentage of the market the first grower gains or loses to the second.

	Pink	Yellow	Red
Pink	20	10	−20
Orange	−10	−20	−10
Red	30	20	−20

Find the value of this game and the optimal strategies.

Solutions to Self-Help Exercise Set 10.2

1. As usual any mixed strategy $[p_1 \quad p_2]$ can be written as $[p_1 \quad 1 - p_1]$. Suppose then that Row chooses a particular mixed strategy $[p_1 \quad 1 - p_1]$. Let us find the expected value that will result for each of the possible 2 selections that Column can make.

    If Column selects strategy 1, that is, $q^T = [1 \quad 0]$, then

    $$E(p, q) = [p_1 \quad 1 - p_1] \begin{bmatrix} -1 & +1 \\ +1 & -1 \end{bmatrix} \begin{bmatrix} 1 \\ 0 \end{bmatrix} = 1 - 2p_1.$$

    If Column selects strategy 2, that is, $q^T = [0 \quad 1]$, then

    $$E(p, q) = [p_1 \quad 1 - p_1] \begin{bmatrix} -1 & +1 \\ +1 & -1 \end{bmatrix} \begin{bmatrix} 0 \\ 1 \end{bmatrix} = -1 + 2p_1.$$

    Both of these are straight lines and are shown in the figure.
    Column will always select the strategy that yields the smallest expected return for Row. This is shown in color in the figure. Row will naturally maximize her expected return and choose the point p_1 that yields the highest value on the line segments marked in color. This can be determined as the intersection of the two lines. Setting

    $$1 - 2p_1 = -1 + 2p_1$$

    gives $p_1 = .50$. Thus the optimal strategy for Row is $p = [.50 \quad .50]$.

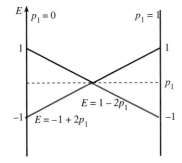

2. As usual, the mixed strategy $[q_1 \quad q_2]$ can be written as $[q_1 \quad 1 - q_1]$. Suppose then that Column chooses a particular mixed strategy $[q_1 \quad 1 - q_1]$. Let us find

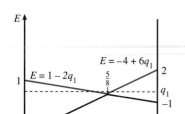

the expected value that will result for each of the possible 2 selections that Row can make.

If Row selects strategy 1, that is, $p^T = [1 \quad 0]$, then

$$E(p, q) = [1 \quad 0]\begin{bmatrix} -1 & +1 \\ +2 & -4 \end{bmatrix}\begin{bmatrix} q_1 \\ 1 - q_1 \end{bmatrix} = 1 - 2q_1.$$

If Row selects strategy 2, that is, $p^T = [0 \quad 1]$, then

$$E(p, q) = [0 \quad 1]\begin{bmatrix} -1 & +1 \\ +2 & -4 \end{bmatrix}\begin{bmatrix} q_1 \\ 1 - q_1 \end{bmatrix} = -4 + 6q_1.$$

These lines are shown in the figure.

Since Row will select the strategy that maximizes her return, she will select the strategy shown as the colored line segments. Column will then minimize Row's expected return by choosing the point q_1 that gives the lowest value on the colored line segments in the figure. The value of q_1 can be found as the intersection of the two lines. Thus setting

$$1 - 2q_1 = -4 + 6q_1$$

yields $q_1 = \frac{5}{8}$.

Thus the optimal strategy for Column is $[\frac{5}{8} \quad \frac{3}{8}]$.

3. The second row dominates the third row. Removing the third row gives

$$\begin{bmatrix} 0 & -1 & 1 & 1 \\ 1 & 0 & -1 & -2 \end{bmatrix}.$$

Now notice that the second column dominates the first column. Thus the first column can be eliminated. Also notice that the fourth column dominates the third column. Thus the third column can be eliminated. This leaves

$$\begin{bmatrix} -1 & 1 \\ 0 & -2 \end{bmatrix}.$$

10.3   LINEAR PROGRAMMING AND GAME THEORY

▶ *Matrices with Positive Entries*

▶ *Linear Programming and Game Theory*

In the last section we saw how to find the optimal strategies and the value of nonstrictly determined games using mixed strategies. A geometric method was given to find the optimal strategy for Row in a game with a payoff matrix with at most 2 rows, or the optimal strategy for Column for any game with a payoff matrix with at most 2 columns. We need to use linear programming, however, to solve games with payoff matrices of larger order. The theory of linear programming is also used to establish the basic theorems of game theory.

Matrices with Positive Entries

To solve a game using the techniques of linear programming presented here, we will see that the payoff matrix must have all entries that are positive. This turns out not to be a restriction. To see why, consider the game given in the last section with payoff matrix

$$A = \begin{bmatrix} -1 & 1 \\ 2 & -4 \end{bmatrix}.$$

We know from the last section that adding, for example 5, to every entry in the payoff matrix gives the payoff matrix

$$B = \begin{bmatrix} 4 & 6 \\ 7 & 1 \end{bmatrix}$$

which will have the very same optimal strategies as A and value equal to the value of A plus 5.

In the same way we can always add some positive constant k to all the entries of any payoff matrix A with value v_A that have some nonnegative entries and obtain a new payoff matrix B with the same optimal strategies and value $v_B = v_A + k$. If we then solve the game with payoff matrix B, the very same optimal strategies for this game will also be optimal for the original game and the value of the original game will be the value of the game with payoff matrix B *minus* k. This is, $v_A = v_B - k$.

Linear Programming and Game Theory

We now consider a game with payoff matrix

$$A = \begin{bmatrix} a_{11} & a_{12} \\ a_{21} & a_{22} \end{bmatrix}$$

for which all the entries are positive. We will show how to use linear programming to solve this game.

Consider first the point of view of Row. Row is interested in finding a strategy $\bar{p}$ and the largest possible number v_R such that

$$E(\bar{p}, q) = \bar{p}^T A q \geq v_R \tag{1}$$

for any choice of mixed strategy q.

Since (1) must hold for all mixed strategies q, it certainly must hold for the two particular strategies $q_1^T = [1 \quad 0]$ and $q_2^T = [0 \quad 1]$. Replacing q with q_1 and q_2 successively in (1) then yields

$$[p_1 \quad p_2] \begin{bmatrix} a_{11} & a_{12} \\ a_{21} & a_{22} \end{bmatrix} \begin{bmatrix} 1 \\ 0 \end{bmatrix} \geq v_R, \qquad [p_1 \quad p_2] \begin{bmatrix} a_{11} & a_{12} \\ a_{21} & a_{22} \end{bmatrix} \begin{bmatrix} 0 \\ 1 \end{bmatrix} \geq v_R$$

or

$$a_{11}p_1 + a_{21}p_2 \geq v_R$$

$$a_{12}p_1 + a_{22}p_2 \geq v_R.$$

Although it is tempting to use the substitution $p_2 = 1 - p_1$, we take a more useful tactic and divide each inequality by v_R. Since all entries in the matrix A are positive, $v_R > 0$. We then have

$$a_{11}\frac{p_1}{v_R} + a_{21}\frac{p_2}{v_R} \geq 1 \tag{2}$$

$$a_{12}\frac{p_1}{v_R} + a_{22}\frac{p_2}{v_R} \geq 1.$$

It is more convenient to let

$$y_1 = \frac{p_1}{v_R}, \quad \text{and} \quad y_2 = \frac{p_2}{v_R}.$$

Notice that $y_1 \geq 0$ and $y_2 \geq 0$. Inequality (2) then becomes

$$a_{11}y_1 + a_{21}y_2 \geq 1$$

$$a_{12}y_1 + a_{22}y_2 \geq 1 \tag{3}$$

$$y_1, y_2 \geq 0.$$

Row then wishes to maximize v_R subject to the constraints given in inequality (3). But for a linear programming problem, the function to be maximized or minimized must be a linear function of y_1 and y_2. To see how to obtain this function, notice that

$$y_1 + y_2 = \frac{p_1}{v_R} + \frac{p_2}{v_R} = \frac{p_1 + (1 - p_1)}{v_R} = \frac{1}{v_R}.$$

We will then maximize v_R by minimizing $1/v_R = y_1 + y_2$. We then have the following linear programming problem:

$$\text{Minimize} \quad w = \frac{1}{v_R} = y_1 + y_2$$

$$\text{Subject to} \quad a_{11}y_1 + a_{21}y_2 \geq 1 \tag{4}$$

$$a_{12}y_1 + a_{22}y_2 \geq 1$$

$$y_1, y_2 \geq 0.$$

Now consider the point of view of the column player. He wishes to find a strategy $\overline{q}$ and the smallest number v_C such that

$$E(p, \overline{q}) = p^T A \overline{q} \leq v_C \tag{5}$$

for all strategies p.

Since inequality (5) must hold for all p, it must hold for the two particular values of p given by $p_1^T = [1 \quad 0]$ and $p_2^T = [0 \quad 1]$. Substituting these two values for p in inequality (5) then yields

$$a_{11}q_1 + a_{12}q_2 \leq v_C \tag{6}$$
$$a_{21}q_1 + a_{22}q_2 \leq v_C.$$

In a similar fashion as before, we divide both inequalities in equation (6) by v_C and set

$$x_1 = \frac{q_1}{v_C} \quad \text{and} \quad x_2 = \frac{q_2}{v_C}.$$

This gives

$$a_{11}x_1 + a_{12}x_2 \leq 1$$
$$a_{21}x_1 + a_{22}x_2 \leq 1 \tag{7}$$
$$x_1, x_2 \geq 0.$$

Now notice that

$$x_1 + x_2 = \frac{q_1}{v_C} + \frac{q_2}{v_C} = \frac{q_1 + (1 - q_1)}{v_C} = \frac{1}{v_C}.$$

We then minimize v_C by maximizing $1/v_C$. We then have the following linear programming problem

$$\text{Maximize} \quad z = \frac{1}{v_C} = x_1 + x_2$$

$$\text{Subject to} \quad a_{11}x_1 + a_{12}x_2 \leq 1 \tag{8}$$
$$a_{21}x_1 + a_{22}x_2 \leq 1$$
$$x_1, x_2 \geq 0.$$

We notice that this last linear programming problem is a standard linear programming problem and that the linear programming problems in (4) and (8) are the duals of each other. From our knowledge of the duals, we know that the maximum v_C found in (8) must equal the minimum v_R found in (4). This common value, denoted by v, is then the value of the game.

The general case of a $m \times n$ payoff matrix A will follow in a similar manner. We then have the following.

Optimal Strategy for Column

Let a game have the $m \times n$ payoff matrix

$$A = \begin{bmatrix} a_{11} & a_{12} & \cdots & a_{1n} \\ a_{21} & a_{22} & \cdots & a_{2n} \\ & & \cdots & \\ a_{m1} & a_{m2} & \cdots & a_{mn} \end{bmatrix}.$$

Let $x_1, x_2, \ldots, x_n$, be the solution to the standard linear programming problem

Maximize $\qquad\qquad z = x_1 + x_2 + \cdots + x_n$

Subject to $\qquad a_{11}x_1 + a_{12}x_2 + \cdots + a_{1n}x_n \leq 1$

$\qquad\qquad\qquad a_{21}x_1 + a_{22}x_2 + \cdots + a_{2n}x_n \leq 1$

$$\vdots$$

$\qquad\qquad a_{m1}x_1 + a_{m2}x_2 + \cdots + a_{mn}x_n \leq 1$

$$x_1, x_2, \ldots, x_n \geq 0.$$

Then the value v of the game is the reciprocal of this maximum, and the optimal strategy for Column is

$$\overline{q} = \begin{bmatrix} q_1 \\ q_2 \\ \vdots \\ q_n \end{bmatrix} = v \begin{bmatrix} x_1 \\ x_2 \\ \vdots \\ x_n \end{bmatrix} = \begin{bmatrix} vx_1 \\ vx_2 \\ \vdots \\ vx_n \end{bmatrix}.$$

Optimal Strategy for Row

Let a game have the $m \times n$ payoff matrix

$$A = \begin{bmatrix} a_{11} & a_{12} & \cdots & a_{1n} \\ a_{21} & a_{22} & \cdots & a_{2n} \\ & & \cdots & \\ a_{m1} & a_{m2} & \cdots & a_{mn} \end{bmatrix}.$$

Let $y_1, y_2, \ldots, y_m$, be the solution to the linear programming problem

Minimize $\qquad\qquad w = y_1 + y_2 + \cdots + y_m$

Subject to $\qquad a_{11}y_1 + a_{21}y_2 + \cdots + a_{m1}y_m \geq 1$

$\qquad\qquad\qquad a_{12}y_1 + a_{22}y_2 + \cdots + a_{m2}y_m \geq 1$

$$\vdots$$

$\qquad\qquad a_{1n}y_1 + a_{2n}y_2 + \cdots + a_{mn}y_m \geq 1$

$$y_1, y_2, \ldots, y_m \geq 0.$$

Then the value v of the game is the reciprocal of this minimum, and the optimal strategy for Row is

$$\overline{p} = \begin{bmatrix} p_1 \\ p_2 \\ \vdots \\ p_m \end{bmatrix} = v \begin{bmatrix} y_1 \\ y_2 \\ \vdots \\ y_m \end{bmatrix} = \begin{bmatrix} vy_1 \\ vy_2 \\ \vdots \\ vy_m \end{bmatrix}.$$

The minimum value of w is the same as the maximum value of z found in solving the linear programming problem for Column. In fact the two linear programming problems are the duals of each other. If Row and Column play their optimal strategies $\overline{p}$ and $\overline{q}$, respectively, then the expected return for the game is

$$E(\overline{p}, \overline{q}) = v.$$

For any p, $E(p, \overline{q}) \leq v$. While for any q, $E(\overline{p}, q) \geq v$.

E X A M P L E   1   **Using Linear Programming in Game Theory**

Use linear programming to find the value of the game and the optimal strategy for Row in the game with payoff matrix

$$A = \begin{bmatrix} -1 & 1 \\ 2 & -4 \end{bmatrix}.$$

Solution   Since some entries are nonnegative, the first step is to add 5 to every entry in the matrix to obtain the payoff matrix

$$B = \begin{bmatrix} 4 & 6 \\ 7 & 1 \end{bmatrix}.$$

Row then must solve the following linear programming problem:

$$\text{Minimize} \qquad w = \frac{1}{v_R} = y_1 + y_2$$

$$\text{Subject to} \qquad 4y_1 + 7y_2 \geq 1$$
$$6y_1 + y_2 \geq 1$$
$$y_1, y_2 \geq 0.$$

Since the number of constraints and the number of variables is only 2, we will solve by using the geometric method of linear programming rather than the simplex method.

Figure 10.5 shows the feasible region and the 3 corners $(0, 1)$, $(\frac{3}{19}, \frac{1}{19})$, $(\frac{1}{4}, 0)$. The function $w = y_1 + y_2$ takes the values of 1, $\frac{4}{19}$, and $\frac{1}{4}$, respectively, at these corners. Thus the minimum of w is $\frac{4}{19}$ at the point $(\frac{3}{19}, \frac{1}{19})$. The value of the game with payoff matrix B is the reciprocal of this minimum, that is $v = \frac{19}{4}$.

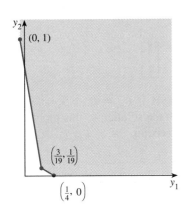

Figure 10.5

The optimal strategy for Row for the game with the payoff matrix B or A is then

$$\overline{p} = v \begin{bmatrix} y_1 \\ y_2 \end{bmatrix} = \frac{19}{4} \begin{bmatrix} \frac{3}{19} \\ \frac{1}{19} \end{bmatrix} = \begin{bmatrix} \frac{3}{4} \\ \frac{1}{4} \end{bmatrix}.$$

The value of the original game with payoff matrix A is

$$v_A = v - 5 = -\frac{1}{4}.$$

This all agrees with the solution to this problem found in the last section, using the method of that section. ∎

E X A M P L E   2   Using Linear Programming in Game Theory

Use linear programming to find the value of the game and the optimal strategy for Column in the game in the previous example.

Solution   Using the matrix B in Example 1, Column must solve the following linear programming problem:

$$\begin{aligned} \text{Maximize} \quad & z = x_1 + x_2 \\ \text{Subject to} \quad & 4x_1 + 6x_2 \le 1 \\ & 7x_1 + x_2 \le 1 \\ & x_1,\, x_2 \ge 0. \end{aligned}$$

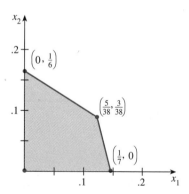

x_2
.2
$\left(0, \frac{1}{6}\right)$
.1
$\left(\frac{5}{38}, \frac{3}{38}\right)$
$\left(\frac{1}{7}, 0\right)$
.1   .2 x_1

Figure 10.6

The feasible region is shown in Figure 10.6. The 4 corner points are $(0, 0)$, $(0, \frac{1}{6})$, $(\frac{5}{38}, \frac{3}{38})$, $(\frac{1}{7}, 0)$. The value of z at these 4 corners is, respectively, 0, $\frac{1}{6}$, $\frac{4}{19}$, and $\frac{1}{7}$. The largest of these is $\frac{4}{19}$, occurring at the corner point $(\frac{5}{38}, \frac{3}{38})$. The value of the game with payoff matrix B is then the reciprocal of $\frac{4}{19}$, or $\frac{19}{4}$. Thus the optimal strategy for Column is

$$\overline{q} = v \begin{bmatrix} x_1 \\ x_2 \end{bmatrix} = \frac{19}{4} \begin{bmatrix} \frac{5}{38} \\ \frac{3}{38} \end{bmatrix} = \begin{bmatrix} \frac{5}{8} \\ \frac{3}{8} \end{bmatrix}.$$

The value of the original game with payoff matrix A is $v_A = v - 5 = -\frac{1}{4}$, which agrees with Example 1. This all agrees with the solution to this problem found in the last section, using the method of that section. ∎

Using the simplex method has one considerable advantage. One need solve only one of the linear programming problems. Since the other one is the dual, the solution can be read off as the numbers in the last row under the slack variables of the final tableau.

E X A M P L E   3 **Using the Simplex Method in Game Theory**

Use the simplex method to find the value of the game and the optimal strategy for Column in the game with the payoff matrix in the previous example. Use the theory for the dual problem to find the optimal stragegy for Row by using the information in the final tableau.

Solution The initial tableau for the game with the payoff matrix B is

x_1	x_2	s_1	s_2	z		
4	6	1	0	0	1	$\frac{1}{4}$
⑦	1	0	1	0	1	$\frac{1}{7}$
-1	-1	0	0	1	0	

Since there is a tie, we arbitrarily pick the first column as the pivot column. The fractions then indicate that the pivot row is the second row. Pivoting on the element in the second row and first column gives

x_1	x_2	s_1	s_2	z		
0	$\frac{38}{7}$	1	$-\frac{4}{7}$	0	$\frac{3}{7}$	$\frac{3/7}{38/7} = \frac{3}{38}$
1	$\frac{1}{7}$	0	$\frac{1}{7}$	0	$\frac{1}{7}$	1
0	$-\frac{6}{7}$	0	$\frac{1}{7}$	1	$\frac{1}{7}$	

The pivot column is the second, and the fractions to the right of the tableau indicate that the pivot row is the first. Pivoting about the first row and second column gives the final tableau.

x_1	x_2	s_1	s_2	z	
0	1	$\frac{7}{38}$	$-\frac{2}{19}$	0	$\frac{3}{38}$
1	0	$-\frac{1}{38}$	$\frac{3}{19}$	0	$\frac{5}{38}$
0	0	$\frac{3}{19}$	$\frac{1}{19}$	1	$\frac{4}{19}$

The solution is $(\frac{5}{38}, \frac{3}{38})$ with maximum $\frac{4}{19}$. This all agrees with the answer in the previous exercise. The remaining part of the problem then proceeds as in the previous problem.

Now reading off the numbers in the last row under the slack variables gives $(\frac{3}{19}, \frac{1}{19})$. This is the solution found in Exercise 1. ∎

SELF-HELP EXERCISE SET 10.3

1. Consider the game with payoff matrix

$$A = \begin{bmatrix} 5 & -1 & 2 \\ -2 & 5 & 1 \end{bmatrix}.$$

Find the value of the game and the optimal strategy for Row using linear programming (without using the simplex method).

2. For the game with the payoff matrix in the previous exercise, find the value of the game and the optimal strategy for Column using the simplex method. Using the final tableau, find the solution to the dual problem and check this with the solution to the linear programming problem solved in the previous exercise.

EXERCISE SET 10.3

In Exercises 1 through 8 find the value and the optimal strategies for both Row and Column for the game with the indicated payoff matrix. Use geometric linear programming.

1. $\begin{bmatrix} 1 & -2 \\ -3 & 2 \end{bmatrix}$

2. $\begin{bmatrix} 0 & -1 \\ -2 & 2 \end{bmatrix}$

3. $\begin{bmatrix} 0 & 3 \\ 1 & -1 \end{bmatrix}$

4. $\begin{bmatrix} -1 & 0 \\ 1 & -3 \end{bmatrix}$

5. $\begin{bmatrix} 1 & -1 \\ -1 & 0 \end{bmatrix}$

6. $\begin{bmatrix} 1 & 2 \\ 3 & -2 \end{bmatrix}$

7. $\begin{bmatrix} -1 & 0 \\ 2 & -1 \end{bmatrix}$

8. $\begin{bmatrix} -2 & 4 \\ 1 & 0 \end{bmatrix}$

In Exercises 9 through 10 find the value and the optimal strategy for Row for the game with the indicated payoff matrix. Use geometric linear programming.

9. $\begin{bmatrix} 3 & 0 & -1 \\ -5 & -4 & 2 \end{bmatrix}$

10. $\begin{bmatrix} 3 & 2 & -1 \\ -2 & 0 & 1 \end{bmatrix}$

In Exercises 11 through 12 find the value and the optimal strategy for Column for the game with the indicated payoff matrix. Use geometric linear programming.

11. $\begin{bmatrix} 4 & -2 \\ 1 & 2 \\ 3 & 1 \end{bmatrix}$

12. $\begin{bmatrix} 3 & -5 \\ -2 & 0 \\ -1 & -3 \end{bmatrix}$

In Exercises 13 through 16 use the simplex method and dual theory in the indicated problems to find the optimal strategies for both row and column.

13. Exercise 1 above. **14.** Exercise 2 above.

15. Exercise 3 above. **16.** Exercise 4 above.

17. Find the optimal strategy for Column in Exercise 9 by using the simplex method. Use the final tableau and dual theory to find the optimal strategy for Row. Compare with Exercise 9.

18. Find the optimal strategy for Column by using the simplex method in Exercise 10. Use the final tableau and dual theory to find the optimal strategy for Row. Compare with Exercise 10.

19. Find the optimal strategy for Row in Exercise 11 by using the simplex method in Section 4.4. Use the final tableau and dual theory to find the optimal strategy for Column. Compare with Exercise 11.

20. Find the optimal strategy for Row by using the simplex method in Exercise 12. Use the final tableau and dual theory to find the optimal strategy for Column. Compare with Exercise 12.

21. Solve the Odds and Evens game given in the last section by linear programming.

22. In the stone, paper, scissors game, two players each simultaneously say one of the words: stone, paper, scissors. The game is a draw if both players say the same word. A player will win $1 from the other player according to the following rules: scissors defeats (cuts) paper, paper defeats (covers) stone, stone defeats (breaks) scissors. Determine the payoff matrix and find the value and optimal strategies for both players.

Applications

Solve Exercises 23 through 26 using linear programming. Note that these are Exercises 23 through 26 of the last section.

23. **Setting Prices.** Every August and January two carpet retailers in town run their sales. They must decide whether to have a 20% or 30% sale. From many previous years of experience with these sales, they know that if they both set prices at 20% off, the first retailer will get 80% of the business and if they both set prices at 30% off, the first retailer will get 70% of the business. If the first retailer takes 20% off and the second 30% off, then the first retailer gets 30% of the business, while if the first takes 30% off and the second 20% off, the first retailer gets 60% of the business. What should their strategies be, and what is the value of this game?

24. **Price Wars.** Every day two movie theaters in town must decide on the price to set on their movies for the next day so the advertisement can go into the morning paper for the next day. They know from long experience that if both set their prices at $5 or both at $6, then the first theater will get 70% of the business. If the first theater sets prices at $5 and the second at $6, then the first theater gets 60% of the business, while if the first theater charges $6 and the second $5, then the first theater gets 40% of the business. What should their strategies be and what is the value of this game?

25. **Advertising.**   Two competing electronics stores must decide each week whether to advertise in one and only one of the three media: TV, radio, or newspaper. From past experience they know that the payoff matrix in terms of percent of the market gained or lost to the other is

$$
\begin{array}{c}
\\
\textbf{TV} \\
\textbf{Radio} \\
\textbf{Paper}
\end{array}
\begin{array}{ccc}
\textbf{TV} & \textbf{Radio} & \textbf{Paper} \\
\left[\begin{array}{ccc}
1 & 1 & 0 \\
-1 & 0 & -1 \\
2 & -1 & 2
\end{array}\right].
\end{array}
$$

Find the value of this game and the optimal strategies.

26. **Marketing.**   Early in each year two local growers compete for the freshly cut gladiola market, committing themselves entirely to one and only one color of gladiola. Past experience indicates the following payoff matrix given in terms of the percentage of the market the first grower gains or loses to the second.

$$
\begin{array}{c}
\\
\textbf{Pink} \\
\textbf{Orange} \\
\textbf{Red}
\end{array}
\begin{array}{ccc}
\textbf{Pink} & \textbf{Yellow} & \textbf{Red} \\
\left[\begin{array}{ccc}
20 & 10 & -20 \\
-10 & -20 & -10 \\
30 & 20 & -20
\end{array}\right].
\end{array}
$$

Find the value of this game and the optimal strategies.

Solutions to Self-Help Exercise Set 10.3

1. First add 3 to each entry in the payoff matrix A and obtain the new payoff matrix

$$
B = \begin{bmatrix} 8 & 2 & 5 \\ 1 & 8 & 4 \end{bmatrix}.
$$

Row must then solve the following linear programming problem:

$$
\begin{array}{ll}
\text{Minimize} & w = y_1 + y_2 \\
\text{Subject to} & 8y_1 + y_2 \geq 1 \\
& 2y_1 + 8y_2 \geq 1 \\
& 5y_1 + 4y_2 \geq 1 \\
& y_1, y_2 \geq 0.
\end{array}
$$

The feasible region is shown in the figure. The 4 corners are $(0, 1)$, $(\frac{1}{9}, \frac{1}{9})$, $(\frac{1}{8}, \frac{3}{32})$, $(\frac{1}{2}, 0)$. The values of $w = y_1 + y_2$ at these 4 corners are, respectively, 1, $\frac{2}{9} = .22222$, $\frac{7}{32} = .21875$, $\frac{1}{2}$. The minimum is then $\frac{7}{32}$ at the point $(\frac{1}{8}, \frac{3}{32})$. The value of the game with payoff matrix B is the reciprocal of the minimum value of w or $v_B = \frac{32}{7}$. The optimal strategy for Row is then

$$
\bar{p} = \frac{32}{7} \begin{bmatrix} \frac{1}{8} \\ \frac{3}{32} \end{bmatrix} = \begin{bmatrix} \frac{4}{7} \\ \frac{3}{7} \end{bmatrix}.
$$

Since we added 3 to each entry in A, the value of A is

$$v_A = v_B - 3 = \frac{32}{7} - 3 = \frac{11}{7}.$$

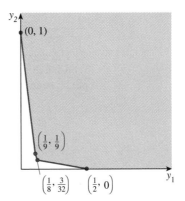

2. Referring to the payoff matrix B in the previous exercise, Column must first solve the linear programming problem:

$$\text{Maximize} \qquad z = x_1 + x_2 + x_3$$
$$\text{Subject to} \qquad 8x_1 + 2x_2 + 5x_3 \le 1$$
$$x_1 + 8x_2 + 4x_3 \le 1$$
$$x_1, x_2 \ge 0.$$

The initial tableau is

x_1	x_2	x_3	s_1	s_2	z	
8	2	⑤	1	0	0	1	$\frac{1}{5}$
1	8	4	0	1	0	1	$\frac{1}{4}$
-1	-1	-1	0	0	1	0	

Arbitrarily pick the pivot column to be the third. Then the pivot row must be the first. Pivoting yields

x_1	x_2	x_3	s_1	s_2	z	
$\frac{8}{5}$	$\frac{2}{5}$	1	$\frac{1}{5}$	0	0	$\frac{1}{5}$	$\frac{1}{2}$
$-\frac{27}{5}$	$\frac{32}{5}$	0	$-\frac{4}{5}$	1	0	$\frac{1}{5}$	$\frac{1}{32}$
$\frac{3}{5}$	$-\frac{3}{5}$	0	$\frac{1}{5}$	0	1	$\frac{1}{5}$	

Pivoting about the second row and second column yields the final tableau.

x_1	x_2	x_3	s_1	s_2	z	
$\frac{31}{16}$	0	1	$\frac{1}{4}$	$-\frac{1}{16}$	0	$\frac{3}{16}$
$-\frac{27}{32}$	1	0	$-\frac{1}{8}$	$\frac{5}{32}$	0	$\frac{1}{32}$
$\frac{3}{32}$	0	0	$\frac{1}{8}$	$\frac{3}{32}$	1	$\frac{7}{32}$

The maximum of $\frac{7}{32}$ occurs at the point $(0, \frac{1}{32}, \frac{3}{16})$. Then v_B is the reciprocal of this maximum or $\frac{32}{7}$. This agrees with that found in the previous exercise. The optimal strategy for Column is then

$$\bar{q} = \frac{32}{7} \begin{bmatrix} 0 \\ \frac{1}{32} \\ \frac{3}{16} \end{bmatrix} = \begin{bmatrix} 0 \\ \frac{1}{7} \\ \frac{6}{7} \end{bmatrix}.$$

The solution to the dual problem can be found under the slack variables on the last line in the final tableau. This is $(\frac{1}{8}, \frac{3}{32})$. This agrees with the solution of the linear programming problem in the previous exercise, since this linear programming problem is the dual.

SUMMARY OUTLINE OF CHAPTER 10

- **Two-Person Zero-Sum Game.** A game is called a **two-person zero-sum game** if the following are satisfied.
    1. There are two players (called the row player and the column player).
    2. The row player must choose one of m strategies, and, simultaneously, the column player must choose one of n strategies.
    3. If the row player chooses her ith strategy and the column player chooses his jth strategy, the row player receives a reward of a_{ij} and the column player loses an amount a_{ij}.

- The matrix $A = (a_{ij})$ given in the prior item is called the **payoff matrix**.

- If there is an entry, a_{hk}, that is the minimum in its row (row h) and also the maximum in its column (column k), we call the entry a_{hk} a **saddle point**.

- A game with a saddle point is said to be **strictly determined**.

- **Locating a Saddle Point (The m & M method)**
    1. For each row, find the minimum value and place a small ''m'' next to it.
    2. For each column, find the maximum value and place a capital ''M'' next to it.
    3. Any entry with both a small ''m'' and a capital ''M'' next to it is a saddle point.

- A **mixed strategy**

$$p = \begin{bmatrix} p_1 \\ p_2 \end{bmatrix}$$

is a strategy in which the first and second strategies are selected randomly with probability p_1 and p_2, respectively.

- Suppose a game has a payoff matrix A and Row plays the strategy $p^T = [p_1 \quad p_2 \quad \cdots \quad p_m]$ and Column selects $q^T = [q_1 \quad q_2 \quad \cdots \quad q_n]$, then the **expected value** for these two strategies is $E(p, q) = p^T Aq$.

- **Fundamental Theorem of Game Theory.** Given any game with a $m \times n$ payoff matrix A, there exist strategies $\bar{p}$ and $\bar{q}$ for row and column, respectively, such that

$$E(\bar{p}, q) = \bar{p}^T Aq \geq v \text{ for all } q \quad \text{and} \quad E(p, \bar{q}) = p^T A\bar{q} \leq v \text{ for all } p.$$

- **Translated Games.** Consider a game with the $m \times n$ payoff matrix A and value v. Let E be the $m \times n$ matrix with every entry equal to 1 and let k be any constant. Then the game with payoff matrix $A + kE$ has value $v + k$ and the same optimal strategies as the game with payoff matrix A.

- Row i is said to **dominate** row j if every element in row i is greater than or equal to the corresponding element in row j. Also column i is said to dominate column j if every element in column i is less than or equal to the corresponding element in column j.

- **Optimal Strategies for Column and Row.** Let a game have the $m \times n$ payoff matrix $A = (a_{ij})$ and let $x_1, x_2, \ldots, x_n$, be the solution to the standard linear programming problem

$$\text{Maximize} \quad\quad\quad z = x_1 + x_2 + \cdots + x_n$$

$$\text{Subject to} \quad\quad a_{11}x_1 + a_{12}x_2 + \cdots + a_{1n}x_n \leq 1$$
$$a_{21}x_1 + a_{22}x_2 + \cdots + a_{2n}x_n \leq 1$$
$$\vdots$$
$$a_{m1}x_1 + a_{m2}x_2 + \cdots + a_{mn}x_n \leq 1$$
$$x_1, x_2, \ldots, x_n \geq 0$$

Then the value v of the game is also reciprocal of this maximum.
Let $y_1, y_2, \ldots, y_m$, be the solution to the dual of the above linear programming problem. Then the value v of the game is also the reciprocal of the minimum to the dual problem. The optimal strategy for Column is $\bar{q} = vX$ while the optimal strategy for Row is $\bar{p} = vY$.

Chapter 10 Review Exercises

In Exercises 1 through 8 find the value of the game and the optimal strategies for both players for the games with the given payoff matrix.

1.
$$\begin{bmatrix} 3 & 2 & 0 \\ 2 & -2 & -1 \end{bmatrix}$$

2.
$$\begin{bmatrix} 4 & -1 & 2 & 1 \\ 3 & -2 & 1 & 2 \end{bmatrix}$$

3.
$$\begin{bmatrix} 2 & 1 & 0 \\ 2 & 2 & 1 \\ 1 & 3 & 3 \end{bmatrix}$$

4.
$$\begin{bmatrix} 3 & 2 & 1 \\ 1 & 1 & 0 \\ 2 & 1 & 2 \end{bmatrix}$$

5. $\begin{bmatrix} 2 & -1 \\ -1 & 1 \end{bmatrix}$

6. $\begin{bmatrix} 0 & 2 \\ 3 & -3 \end{bmatrix}$

7. $\begin{bmatrix} -1 & 0 & 1 \\ 2 & -1 & 0 \end{bmatrix}$

8. $\begin{bmatrix} -2 & 3 & 5 \\ 1 & 0 & 1 \end{bmatrix}$

9. Determine the value of the game and the optimal strategy for Row for the game with the payoff matrix given by

$$\begin{bmatrix} 1 & 0 & -1 \\ -3 & -1 & 2 \end{bmatrix}.$$

10. Determine the value of the game and the optimal strategy for Column for the game with the payoff matrix given by

$$\begin{bmatrix} 6 & 1 \\ 5 & 4 \\ 0 & 5 \end{bmatrix}.$$

11. Determine the value of the game and the optimal strategy for Column for the game with the payoff matrix in Exercise 9.

12. Determine the value of the game and the optimal strategy for Row for the game with the payoff matrix in Exercise 10.

13. Determine the value of the game and the optimal strategy for Row and for Column for the game with the payoff matrix

$$\begin{bmatrix} 1 & 0 & 0 \\ 0 & 1 & 0 \\ 0 & 0 & 1 \end{bmatrix}.$$

14. Show that every game with a payoff matrix

$$\begin{bmatrix} a & a \\ b & c \end{bmatrix}$$

must be strictly determined.

Biologists use dynamical systems to determine the number of fish to harvest without depleting the resource. (Agricultural Photography by Grant Heilman.)

Discrete Dynamical Systems and Mathematical Models

I n this chapter there are a number of formulas, including some basic formulas used every day in the banking and insurance industry. The data that can be calculated from these formulas are now given in tables and by hand-held calculators. Thus these formulas are much less important than they once were.

The important part of this chapter, then, is not the formulas. Rather, the most important part is to learn the skill of how to take a problem or situation and translate it into a mathematical problem that can be solved. This is called mathematical modeling. Specifically we will be interested in modeling using discrete dynamical systems.

In this chapter you will see that a "solution" to a discrete dynamical system is really a string of numbers and the graph a string of dots in the xy-plane. Another important skill is to recognize patterns made by these strings of numbers. Thus, for example, we will want to recognize when the string of numbers is increasing, decreasing, oscillating, or repeating (periodic).

11.1   LINEAR DYNAMICAL SYSTEMS AND MATHEMATICAL MODELS

▶ *Mathematical Modeling*

▶ *An Example from Finance*

▶ *Discrete Dynamical Systems and Mathematical Modeling*

▶ *Population Dynamics*

▶ *Radioactive Decay*

A P P L I C A T I O N
Population

What is an estimate of the human population of the earth in the year 2000? See Example 5 for the answer.

A P P L I C A T I O N
Safe Time
After Nuclear Accident

Due to an unfortunate nuclear accident, some radioactive iodine-131 has been released into the atmosphere and has been absorbed by the grass in some fields, which then has been ingested by some cows. The milk from the cows has three times the maximum allowable amounts of iodine-131. How long until it is safe to consume the milk from these animals? (Such an accident occurred in Michigan.) See Example 7 for the answer.

A P P L I C A T I O N
Carbon Dating

An anthropologist discovers a bone fragment and is told from laboratory testing that exactly 20% of carbon-14 is missing. How old is the bone? See Example 8 for the answer.

Mathematical Modeling

In this section we will begin to develop the art of building mathematical models of phenomena that change over time. We will not only examine various areas of finance but also other phenomena, such as population growth and radioactive decay.

Building a mathematical model involves the following steps.

1. Make clear assumptions concerning the phenomena being studied.
2. Translate these assumptions into mathematical equations.
3. Use your knowledge of mathematics to solve these equations.
4. Translate your solution back to the real-life phenomena being studied.
5. Determine if the mathematical solution makes sense. If the solution does not make sense, recheck the mathematical calculations, and if these are correct, then reconsider the original assumptions.

An Example from Finance

Let us first consider an example from finance.

E X A M P L E   1   **Determining a Savings Account Balance Based on the Balance of the Previous Quarter**

Suppose $1000 is deposited into a savings account earning an annual rate of 8% interest, compounded quarterly. (a) Determine an equation that will give the amount of money in the account at the end of any quarter if given the amount at the end of the previous quarter. (b) Use your equation to find the amount at the end of the first year.

Solutions

(a) Since there are 4 quarters in a year, the interest per quarter is obtained by dividing the annual interest 8% by 4, giving 2% per quarter. Let us designate the initial amount in the account as F_0. Thus $F_0 = 1000$. Let us also designate the amount in the account at the end of the first quarter as F_1, at the end of the second quarter as F_2, and so forth.

Since the amount F_1 at the end of the first quarter is the original amount F_0 plus interest, we have

$$F_1 = F_0 + \text{interest} = F_0 + 0.02F_0 = (1.02)F_0.$$

Since the amount F_2 at the end of the second quarter is the amount F_1 at the beginning of the second quarter, plus interest earned during the second quarter, we have

$$F_2 = F_1 + \text{second quarter interest} = F_1 + 0.02F_1 = (1.02)F_1.$$

And in a similar fashion we have

$$F_3 = 1.02F_2.$$

We then see that if F_n is the amount in the account at the end of n quarters, then F_{n+1} is the amount in the account at the end of the next quarter and

$$F_{n+1} = 1.02F_n. \tag{1}$$

This last equation is the equation we are seeking.

(b) Using this equation we then have

$$F_1 = 1.02F_0 = 1.02(1000) = 1020$$

$$F_2 = 1.02F_1 = 1.02(1020) = 1040.40$$

$$F_3 = 1.02F_2 = 1.02(1040.40) = 1061.21$$

$$F_4 = 1.02F_3 = 1.02(1061.21) = 1082.43.$$

The amount at the end of the year is the amount at the end of the fourth quarter, which is $1082.43. ■

Often we wish to know the balance in a savings account based only on the initial amount. The following example illustrates how this can be done without calculating the balances for all the intermediate periods.

E X A M P L E   2   **Determining a Savings Account Balance Based on the Initial Amount**

(a) For the savings account in Example 1, find a formula for the balance at the end of any quarter based only on the initial amount. (b) Use this formula to find the balance at the end of the tenth year.

Solutions   (a) We can make the following calculations:

$$F_1 = 1.02F_0$$

$$F_2 = 1.02F_1 = 1.02(1.02F_0) = (1.02)^2F_0$$

$$F_3 = 1.02F_2 = 1.02(1.02)^2F_0 = (1.02)^3F_0.$$

From this pattern we can see that the amount F_n at the end of n quarters is given by

$$F_n = (1.02)^nF_0. \tag{2}$$

The graph is shown in Figure 11.1.

(b) The amount at the end of 10 years is the amount at the end of $4 \times 10 = 40$ quarters. We then use equation (2) to find F_{40}. Recalling that the initial amount was $F_0 = 1000$, and using a calculator, we obtain

$$F_{40} = (1.02)^{40}F_0$$

$$= (1.02)^{40}1000$$

$$= 2208.04. \quad \blacksquare$$

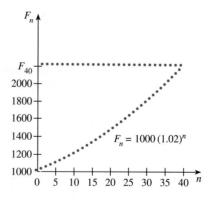

Figure 11.1

Discrete Dynamical Systems and Mathematical Modeling

Our work in Examples 1 and 2 above represents a thorough analysis of a particular problem in finance. Our next step is to generalize. There are two important reasons for this. The first obvious reason is that we need formulas for the balances in savings accounts no matter what the initial amounts, interest rates, and periods of compounding. The second reason is that the mathematical models we create will be used to solve problems in other areas such as population growth and radioactive decay simply by giving a different interpretation to the unknown variables.

Suppose now that a savings account earns interest at an annual rate of r where r is a decimal. Suppose furthermore that the rate is compounded m times a year. Thus there are m time periods per year. The interest rate per time period is then $i = r/m$. For example, if the annual rate is $r = .09$ and compounding is monthly, then $m = 12$ and the rate per month is $i = .09/12 = .0075$.

E X A M P L E   3   **Finding the Balance in Savings Accounts**

Suppose an initial amount of $F_0 = P$ dollars is deposited into a savings account earning an annual rate of r compounded m times per year, where r is a decimal. Thus there are m time periods per year. (a) Determine an equation that will give the amount of money in the account at the end of any time period if given the amount at the end of the previous time period. (b) Find a formula for the balance at the end of any time period based only on the initial amount.

Solutions   Let F_n be the amount at the end of n time periods.

(a)   The interest per time period is $i = r/m$ and

$$F_1 = F_0 + \text{first time period interest} = F_0 + iF_0 = (1 + i)F_0$$

$$F_2 = F_1 + \text{second time period interest} = F_1 + iF_1 = (1 + i)F_1$$

$$F_3 = F_2 + \text{third time period interest} = F_2 + iF_2 = (1 + i)F_2.$$

From this pattern we see that

$$F_{n+1} = (1 + i)F_n. \tag{3}$$

(b)   We can make the following calculations:

$$F_1 = (1 + i)F_0$$

$$F_2 = (1 + i)F_1 = (1 + i)[(1 + i)F_0] = (1 + i)^2 F_0$$

$$F_3 = (1 + i)F_2 = (1 + i)[(1 + i)^2 F_0] = (1 + i)^3 F_0.$$

From this pattern we can see that the amount F_n at the end of n time periods is given by

$$F_n = (1 + i)^n F_0 \quad \blacksquare$$

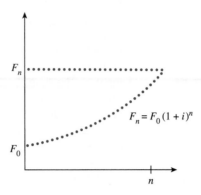

Figure 11.2

Since $F_0 = P$, we can also write this as

$$F_n = (1 + i)^n P. \tag{4}$$

See Figure 11.2.

Compound Interest

Suppose a savings account has an initial amount of P and earns interest at the annual rate of r, expressed as a decimal, and interest is compounded m times a year. Then the amount F_n in the account after n time periods is

$$F_n = (1 + i)^n P$$

where $i = r/m$ is the interest per time period.

E X A M P L E   4   **Finding the Balance in Savings Accounts**

Suppose an initial amount of $100 is deposited into a savings account earning an annual rate of 9% compounded monthly. Find the amount in this account at the end of the fifth year.

Solution   The monthly interest rate is $i = .09/12 = .0075$. In five years there are $n = 5 \times 12 = 60$ months. We then have

$$F_{60} = (1.0075)^{60}(\$100) = \$156.57. \quad \blacksquare$$

Equations (1) and (3) are examples of **discrete dynamical systems** or difference equations.

> **Linear Discrete Dynamical System**
>
> A **linear discrete dynamical system** or **linear difference equation** is an equation of the form
>
> $$y_{n+1} = ay_n + b. \tag{5}$$
>
> The term y_0 is called the **initial value**. If $b = 0$, then equation (5) is called **homogeneous**. If $b \neq 0$, then equation (5) is called **nonhomogeneous**.

In equation (1) $a = 1.02$ and $b = 0$, while in equation (3), $a = 1 + i$ and $b = 0$. Since $b = 0$ in equations (1) and (3), these two equations are homogeneous. We will consider nonhomogeneous dynamical systems in the next section.

If $b = 0$, equation (5) becomes

$$y_{n+1} = ay_n.$$

This is the same as equation (3) with $a = 1 + i$. Thus from equation (4) we must have

$$y_n = a^n y_0.$$

> **Solution to a Linear Discrete Dynamical System**
>
> A **solution** to (5) is an equation that expresses y_n in terms of the constants n, the initial value y_0, and the constants a and b. A solution of the homogeneous equation
>
> $$y_{n+1} = ay_n \tag{6}$$
>
> is
>
> $$y_n = a^n y_0 \tag{7}$$
>
> where y_0 is the initial value.

Population Dynamics

In modeling we often encounter the term *proportional.* We first give the definition of this term.

> **Proportional**
>
> Two quantities P and Q are said to be **proportional** (to one another) if one quantity is a positive constant times the other, that is, there exists a positive constant k such that
>
> $$P = kQ.$$

> Two quantities P and Q are said to be **negatively proportional** (to one another) if one quantity is a negative constant times the other, that is, there exists a positive constant k such that
>
> $$P = -kQ.$$

We now turn to an application in population dynamics. Let P_n be the population at the end of n time periods. The quantity $P_{n+1} - P_n$ is then the increase in the population during the $(n + 1)$st time period. Biologists argue that under a wide range of circumstances this quantity is proportional to the population at the beginning of this time period. This then implies that there exists a positive constant k, called the **growth constant**, such that

$$P_{n+1} - P_n = kP_n.$$

This can be written as

$$P_{n+1} = P_n + kP_n,$$

or

$$P_{n+1} = (1 + k)P_n. \tag{8}$$

Notice that this is precisely the same as equation (6) with $a = 1 + k$. Thus from (7) we can see that the solution of (8) is given by

$$P_n = (1 + k)^n P_0,$$

where P_0 is the initial population.

Now notice that since $P_{n+1} - P_n = kP_n$, we can write

$$\frac{P_{n+1} - P_n}{P_n} = k.$$

The term $\dfrac{P_{n+1} - P_n}{P_n}$ is the increase in population during the $(n + 1)$st time period, divided by the population at the beginning of this time period. Since this term is always k, the population increases by $100k\%$ each time period.

E X A M P L E   5   **Future Population of the Earth**

According to the U.S. Bureau of the Census, the human population of the earth was 5.333 billion in 1990 and growing at a rate of 1.7% per year. Assuming that the human population of the earth satisfies equation (8), determine the population of the earth in the year 2000.

Solution   Let us take the year 1990 as the initial time and n the number of years from 1990. Then if P_n is the human population of the earth in billions n years from 1990,

$$P_n = (1 + .017)^n P_0 = (1.017)^n 5.333.$$

The year 2000 is given by $t = 10$. Thus, using a calculator, we obtain

$$P_{10} = (1.017)^{10} 5.333 = 6.312.$$

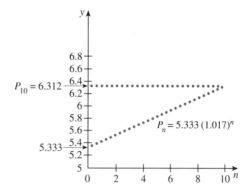

Figure 11.3

This gives the population of the earth in the year 2000 to be about 6.312 billion. See Figure 11.3. ∎

E X A M P L E 6   **Time to Double the Population of the Earth**

Assuming that the human population of the earth satisfies equation (8), determine the year in which the population will double that of 1990.

Solution   We have from Example 4 that

$$P_n = (1.017)^n P_0 = (1.017)^n 5.333,$$

where $n = 0$ corresponds to the year 1990. We wish to find N so that $10.666 = P_N$. Thus

$$10.666 = P_N = (1.017)^N 5.333$$

$$2 = (1.017)^N$$

$$\log 2 = \log(1.017)^N = N \log 1.017$$

$$N = \frac{\log 2}{\log 1.017} = \frac{.30103}{.007321} \approx 41$$

Thus the population of the earth will double that of 1990 in the year 2031 if the population satisfies equation (8) throughout this period. See Figure 11.4. ∎

Radioactive Decay

Some elements exhibit radioactive decay. For these elements, on occasion, one nucleus will spontaneously divide into two or more nuclei. Thus, over a period of time, the amount of the substance will decrease or decay. The rate at which the decay takes place depends on the substance.

Let A_n be the amount of a radioactive substance at the end of n time periods. The quantity $A_{n+1} - A_n$ is the decrease in the amount of the substance during

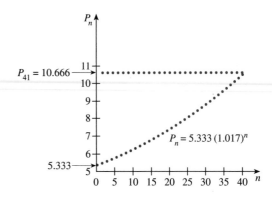

Figure 11.4

the $(n + 1)$st time period. Physicists argue that this quantity is *negatively* proportional to the amount A_n at the beginning of this time period. This means that there is a positive constant k, called the **decay constant**, such that

$$A_{n+1} - A_n = -kA_n.$$

This can be written as

$$A_{n+1} = (1 - k)A_n. \tag{9}$$

Since this is a homogeneous discrete dynamical system in the form of (6) where $a = 1 - k$, the solution is

$$A_n = (1 - k)^n A_0 \tag{10}$$

where A_0 is the initial amount of the radioactive substance. See Figure 11.5.

A common way of measuring how fast a radioactive substance is decaying is to use the **half-life**. This is the time that must elapse for half of the substance to decay.

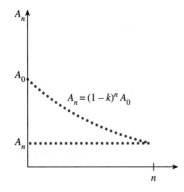

Figure 11.5

E X A M P L E   7   **Safe Time after Nuclear Accident**

Due to an unfortunate nuclear accident, some radioactive iodine-131 has been released into the atmosphere and in some fields has been absorbed by the grass that then has been ingested by some cows. The milk from the cows has three times the maximum allowable amounts of iodine-131. How long until it is safe to consume the milk from these animals? The half-life of iodine-131 is eight days.

Solution   Let A_0 be the amount of iodine-131 in the grass immediately after the accident. Then the amount after n days is $A_n = (1 - k)^n A_0$. We are given that the half-life of iodine-131 is 8 days. Thus the amount of iodine-131 after 8 days is one-half the initial amount or $A_8 = \frac{1}{2}A_0$. We need to find $1 - k$. We have

$$\frac{1}{2} A_0 = A_8 = (1 - k)^8 A_0$$

$$\frac{1}{2} = (1 - k)^8$$

$$1 - k = \sqrt[8]{\frac{1}{2}} \approx .917$$

Let T be the time when the iodine-131 returns to a safe level. A safe level in terms of A_0 is $\frac{1}{3}A_0$, that is, one-third the initial unsafe amount. Then

$$\frac{1}{3} A_0 = A_T = (1 - k)^T A_0 = (.917)^T A_0$$

$$\frac{1}{3} = (.917)^T$$

$$\log \frac{1}{3} = \log(.917)^T = T \log .917$$

$$T = \frac{\log \frac{1}{3}}{\log .917} \approx 12.7$$

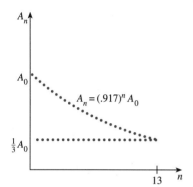

$A_n = (.917)^n A_0$

Figure 11.6

Thus about 13 days need to elapse before the milk is again safe. Refer to Figure 11.6. ■

Radioactive decay of carbon-14 was first used by the Nobel Prize–winner, Willard Libby, about 1950 to date archaeological finds. It is known that the ratio of radioactive carbon-14 to nonradioactive carbon-12 exists in all animals in a fixed constant percent while the animal is alive. When the animal dies the carbon-14 decays, while the amount of nonradioactive carbon-12 remains constant. From this and the known half-life of carbon-14, one can then estimate the date when the animal died.

E X A M P L E 8   **Carbon Dating**

Solution

An anthropologist discovers a bone fragment and is told from laboratory testing that exactly 20% of carbon-14 is missing. The half-life of carbon-14 is known to be 5570 years. How old is the bone?

We first need to find $1 - k$ where k is the decay constant for carbon-14. We are given that the half-life of carbon-14 is 5570 years. Thus $A_{5570} = \frac{1}{2}A_0$, where A_0 is the initial amount. Thus

$$\frac{1}{2} A_0 = A_{5570} = (1 - k)^{5570} A_0$$

$$\frac{1}{2} = (1 - k)^{5570}$$

$$1 - k = \sqrt[5570]{\frac{1}{2}} \approx .9998756$$

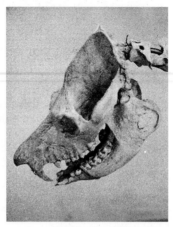

(© Russ Kinne/Comstock, Inc.)

Thus the amount of carbon-14 in the bone n years after its death is

$$A_n = (.9998756)^n A_0,$$

where A_0 is the amount of carbon-14 present in the animal at the time of death. The laboratory reports that $A_n = .80A_0$. Thus

$$.80A_0 = (.9998756)^n A_0$$

$$.80 = (.9998756)^n$$

$$\log .80 = \log(.9998756)^n = n \log .9998756$$

$$n = \frac{\log .80}{\log .9998756} \approx 1794.$$

Thus the bone is about 1794 years old. Refer to Figure 11.7. ∎

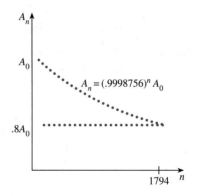

Figure 11.7

SELF-HELP EXERCISE SET 11.1

1. A thousand dollars is deposited into an account growing at an annual rate of 6% per year compounded monthly. How long before the account has $1500?

2. Plutonium-239 is a product of nuclear reactors with a half-life of about 24,000 years. What percentage of a given sample will remain after 10,000 years?

EXERCISE SET 11.1

1. Suppose $100 is deposited into an account earning interest at an annual rate of 6%. Find the amount in the account at the end of the first year if the compounding is (a) semiannual, (b) quarterly, (c) monthly, (d) weekly, (e) daily.

2. Suppose $1000 is deposited into an account earning interest at an annual rate of 8%. Find the amount in the account at the end of the first year if the compounding is (a) semiannual, (b) quarterly, (c) monthly, (d) weekly, (e) daily.

In Exercises 3 through 8, suppose $1000 is deposited into an account that earns interest at the given annual rate with the given compounding. Find the amount in the account after 10 years.

3. 8% compounded annually

4. 7% compounded semiannually

5. 6% compounded quarterly

6. 5% compounded monthly

7. 9% compounded weekly

8. 6.5% compounded daily

9. How long does it take an account growing at an annual rate of 7% compounded annually to double?

10. How long does it take an account growing at an annual rate of 8% compounded annually to double?

11. What annual rate r when compounded annually is necessary to double an account in 6 years?

12. What annual rate r when compounded annually is necessary to triple an account in 5 years?

13. An account earns an annual rate of r compounded annually. What must r be in order for this account to increase from $1000 to $1400 in 5 years?

14. An account earns an annual rate of r compounded annually. What must r be in order for this account to increase from $1000 to $1600 in 7 years?

Applications

15. **Population.** Suppose a population satisfies $P_n = (1 + k)^n P_0$ with $k > 0$. If N is the time it takes for the population to double in size, show that

$$N = \frac{\log 2}{\log(1 + k)}.$$

16. **Population.**   Suppose a population satisfies $P_n = (1 + k)^n P_0$ with $k > 0$. If N is the time it takes for the population to double in size, show that $1 + k = \sqrt[N]{2}$.

17. **Population.**   According to the U.S. Bureau of Statistics, Nigeria is growing at a rate that will result in the population doubling in 23 years. Assuming that this population is growing according to equation (8) in the text, what is the growth constant?

18. **Population.**   According to the U.S. Bureau of Statistics, Japan is growing at a rate that will result in the population doubling in 150 years. Assuming that this population is growing according to equation (8) of the text, what is the growth constant?

19. **Population.**   According to the U.S. Bureau of Statistics, Sub-Saharan Africa has a growth constant of about 0.03. At this rate and assuming the population satisfies equation (8) of the text, how long will it take for this population to double?

20. **Population.**   According to the U.S. Bureau of Statistics, the developed world has a growth constant of about 0.005. At this rate and assuming the population satisfies equation (8) of the text, how long will it take for this population to double?

21. **Population.**   According to the U.S. Bureau of Statistics, the world's population increased from 2 billion to 3 billion from 1925 to 1959. Assuming the population satisfied equation (8) of the text, what was the growth constant for this period?

22. **Population.**   According to the U.S. Bureau of Statistics, the world's population increased from 4 billion to 5 billion from 1974 to 1986. Assuming the population satisfied equation (8) of the text, what was the growth constant for this period?

23. **Population.**   If the world's population during 1960 to 1986 had continued at the same rate as found in Exercise 21 and satisfies equation (8) of the text, what would have been the population in 1986?

24. **Population.**   If the world's population during 1987 to 2000 continues at the same rate as found in Exercise 22 and satisfies equation (8) of the text, what would have been the population in 2000?

25. **Radioactive Half-Life.**   The decay constant for cobalt-60 is $k = 0.13$ when time is measured in years. Find the half-life of cobalt-60.

26. **Radioactivity.**   Radioactive potassium has a half-life of 1.3 billion years. Find its decay constant.

27. **Radioactivity.**   The decay constant for cobalt-60 is $k = 0.13$ when time is measured in years. Find the time it takes for this substance to decay to one-third of its original amount.

28. **Radioactive Half-Life.**   Show that if N is the half-life of a radioactive substance and k is the decay constant, then

$$N = \frac{\log \frac{1}{2}}{\log(1 - k)}.$$

29. Radioactive Half-Life.   Show that if N is the half-life of a radioactive substance and k is the decay constant, then

$$1 - k = \sqrt[N]{\frac{1}{2}}.$$

30. Age of Bone Fragment.   An anthropologist discovers a bone fragment and learns from laboratory testing that exactly 10% of carbon-14 is missing. How old is the bone?

31. Age of Bone Fragment.   An anthropologist discovers a bone fragment and learns from laboratory testing that exactly 90% of carbon-14 is missing. How old is the bone?

32. Overdose.   Mathium-105 is a radioactive substance with a half-life of only ten days. If an individual receives four times the maximum safe dose of Mathium 105, how much time must elapse before a safe level is reached?

33. Radioactive Contamination.   In Example 7 of the text, suppose the milk from the cows had twenty times the maximum allowable amount of iodine-131. How long until it is safe to consume the milk from these animals?

34. Medical Tracers.   Radioactive tracers are used for medical diagnosis. For example iodine-131 with a half-life of 8 days is used in the diagnosis of the thyroid gland. Suppose 20 units of iodine-131 are shipped and take two days to arrive. How much of the substance arrives?

35. Medical Tracers.   In the previous problem suppose 20 units are needed in two days. How much should be ordered?

36. Mice Population.   A population of mice was noted to quadruple in 3 years. What is the growth constant assuming the population satisfies equation (8) of the text?

37. Population.   According to the U.S. Bureau of Statistics, the population of West Virginia has been decreasing by about 1% per year in recent years. Let P_n be the population of West Virginia where n is the number of years from 1990. Find a discrete dynamical system that allows you to find P_{n+1} if given P_n. Assuming this trend continues, find the population in the year 2010, without finding the population for the intermediate years. The population in 1990 was 1.8 million.

38. Inflation.   Inflation has resulted in the dollar losing its value by about 4% a year. Let V_n be the value of the dollar in n years. Find a discrete dynamical system that allows you to find V_{n+1} if given V_n. Assuming this trend continues, find the value of the dollar in 10 years without finding its value for the intermediate years.

39. Drug Absorption.   Suppose 200 milligrams of glucose is initially injected intravenously into a patient and that each minute the body of the patient absorbs 1% of the amount of the glucose present in the blood at the beginning of that minute. Let A_n be the amount of glucose in the blood of the patient after n minutes. Find a discrete dynamical system that allows you to find A_{n+1} if given A_n. Find the amount of glucose in the blood one hour later without finding the amount for the intermediate time.

40. **Light Intensity.**   Let I_n be the intensity of light n meters below the surface of the ocean. Suppose in this particular water the difference $I_{n+1} - I_n$ in the intensity of light at $n + 1$ meters from the intensity at n meters is negatively proportional to I_n with the constant of proportionality $k = 1.4$. Find a discrete dynamical system that allows you to find I_{n+1} if given I_n. At what depth will the intensity be one fourth that of at the surface?

Graphing Calculator Exercises

Solve the following by using the zoom feature of your graphing calculator to find the intersection of two appropriate curves.

41. Find the time it takes for an account growing at an annual rate of 9% compounded annually to double in value.

42. Find the time it takes for an account growing at an annual rate of 12% compounded annually to increase by 50% in value.

43. Find the half-life of a radioactive substance with decay constant $k = .85$, where time is measured in years.

44. Find the time for a radioactive substance with decay constant $k = .85$ (time is measured in years) to decay to 25% of the original amount.

45. According to the U.S. Bureau of the Census, the population of Nigeria was 113 million in 1990 and growing at about 3% a year, while the population of Japan was 123 million in 1990 and growing at about 0.5% a year. Find when the populations of these two countries will be the same assuming the two growth rates continue at the given values.

46. According to the U.S. Bureau of the Census, the population of Indonesia was 181 million in 1990 and growing at about 1.9% a year while the population of the U.S. was 249 million in 1990 and growing at about 0.9% a year. Find when the populations of these two countries will be the same, assuming the two growth rates continue at the given values.

Solutions to Self-Help Exercise Set 11.1

1. The account earns a monthly rate of $i = \dfrac{r}{m} = \dfrac{.06}{12} = .005$. The amount in the account after n months is $A_n = 1000(1.005)^n$. Let N be the number of months needed for the amount in the account to become \$1500. Then $A_N = 1500$ and

$$1500 = A_N = 1000(1.005)^N$$

$$1.5 = (1.005)^N$$

$$\log 1.5 = \log(1.005)^N = N \log 1.005$$

$$N = \frac{\log 1.5}{\log 1.005} \approx 81.3$$

or about 81 months.

2. We first need to find $1 - k$ where k is the decay constant for plutonium-239. We have

$$\frac{1}{2} A_0 = (1 - k)^{24,000} A_0$$

$$\frac{1}{2} = (1 - k)^{24,000}$$

$$1 - k = \sqrt[24,000]{\frac{1}{2}} \approx .9999711$$

Thus $A_n = (.9999711)^n A_0$ and the percentage of the given sample after 10,000 years is

$$100 \frac{A_{10,000}}{A_0} = 100(.9999711)^{10,000} \approx 75.$$

That is, 75% of this material will be left after 10,000 years. (This is why this material poses such a long-term concern.)

11.2   NONHOMOGENEOUS LINEAR DYNAMICAL SYSTEMS

- ► *Some Problems in Personal Finance*
- ► *Solutions in the Nonhomogeneous Case*
- ► *Additional Problems in Personal Finance*
- ► *Learning Theory*

APPLICATION
**Regular Deposits Into
a Savings Account**

An individual is saving money for a down payment on a house to be purchased in 5 years. She can deposit $250 at the end of each month into an account that pays interest at an annual rate of 6% compounded monthly. How much is in this account after 5 years? See Example 4 for the answer.

APPLICATION
**Calculating Payments
on a Loan**

You wish to borrow $10,000 from a bank to purchase a car. The bank charges interest at an annual rate of 12%, and there are to be 48 equal monthly payments with the first to begin in one month. What must the payments be so that the loan will be paid off after 48 months? See Example 6 for the answer.

Some Problems in Personal Finance

If interest is paid only on the initial amount (and not on any of the interest that has been earned), then the interest is called **simple**. For example if $100 earns simple interest of 5% a year, then the interest *every* year is always $i = .05 \times 100 = 5$ dollars. Thus the amount in the account at the end of the first year is $100 + \$5 = \$105$, the second is $\$105 + \$5 = \$110$, the third is $\$110 + \$5 = \$115$, and so forth.

E X A M P L E   1   **Simple Interest**

Suppose an amount of $F_0 = P$ is initially deposited into an account that earns simple interest at an annual rate of i expressed as a decimal. Find a discrete dynamical system that models this situation.

Solution   Interest every year is always iF_0 dollars. Let F_n be the amount at the end of n years. Since the amount next year is just the amount this year plus the interest iF_0, we have

$$F_{n+1} = F_n + iF_0. \tag{1}$$

This can also be written as

$$F_{n+1} = F_n + iP.$$

Recall that a discrete linear dynamical system is an equation of the form

$$y_{n+1} = ay_n + b. \tag{2}$$

Equation (1) is of this form with $a = 1$ and $b = iF_0$.   ∎

Since $b \neq 0$, equation (1) is called nonhomogeneous. We will find the solutions of general discrete nonhomogeneous linear dynamical systems later in this section.

Now let us consider regular deposits into a savings account. But first we give some common terminology.

An **annuity** is a sequence of equal payments made at equal time periods. An **ordinary annuity** is one in which the payments are made at the *end* of the time periods and the periods of compounding are the same time periods. The **term** of an annuity is the time from the beginning of the first period to the end of the last period. The total amount in the account, including interest, at the end of the term of an annuity is called the **future value of the annuity**. Examples of annuities are regular deposits into a saving account, monthly home mortgage payments, and monthly insurance payments.

E X A M P L E   2   **Regular Deposits into a Savings Account**

Suppose a savings account earns interest at an annual rate of r expressed as a decimal with interest compounded m times a year. Thus we have m time periods per year. Suppose furthermore we make a deposit of d dollars at the end of each period. Use a discrete dynamical system to model this situation.

Solution   Let F_n be the amount in the account at the end of n periods. Notice that this is the amount immediately after the deposit has been made and that the initial amount $F_0 = 0$. Now the amount F_{n+1} in the account after $(n + 1)$ periods is the amount F_n at the end of the previous period plus the interest plus the deposit. The interest during the $(n + 1)$st period is iF_n where $i = \dfrac{r}{m}$. Thus

$$F_{n+1} = F_n + iF_n + d,$$

or

$$F_{n+1} = (1 + i)F_n + d \tag{3}$$

with $F_0 = 0$. Notice that this equation is in the form of the linear dynamical system (2) with $a = 1 + i$ and $b = d$. ∎

Solution in the Nonhomogeneous Case

We will now find the solution of the nonhomogeneous dynamical system (2).

Solution of Nonhomogeneous Linear Dynamical Systems.

The solution of the linear dynamical system

$$y_{n+1} = ay_n + b$$

is given by

$$y_n = \frac{b}{1 - a} + \left(y_0 - \frac{b}{1 - a}\right)a^n \tag{4}$$

if $a \neq 1$ and by

$$y_n = y_0 + bn \tag{5}$$

if $a = 1$.

To see why these are the solutions notice that

$$y_1 = ay_0 + b$$
$$y_2 = ay_1 + b = a(ay_0 + b) + b = a^2y_0 + (b + ba)$$
$$y_3 = ay_2 + b = a[a^2y_0 + (b + ba)] + b$$
$$= a^3y_0 + (b + ba + ba^2)$$

From this pattern we see that

$$y_n = a^ny_0 + (b + ba + ba^2 + \cdots + ba^{n-1}) \tag{6}$$

We now must consider two cases, $a \neq 1$ and $a = 1$. If $a \neq 1$, recall from Section 5.3 that

$$b + ba + ba^2 + \cdots + ba^{n-1} = b \frac{1 - a^n}{1 - a}$$

Using this in equation (6) then yields

$$y_n = a^n y_0 + b \frac{1 - a^n}{1 - a}$$

$$= a^n y_0 + b \frac{1}{1 - a} - b \frac{a^n}{1 - a}$$

$$= \frac{b}{1 - a} + \left(y_0 - \frac{b}{1 - a} \right) a^n$$

This is (4).

Now if $a = 1$, equation (6) becomes

$$y_n = 1^n y_0 + \overbrace{(b + b + b + \cdots + b)}^{n \text{ terms}}$$

$$= y_0 + bn.$$

This is (5).

Additional Problems in Personal Finance

E X A M P L E   3   **Simple Interest**

Suppose an amount of $1000 is initially deposited into an account that earns simple interest at an annual rate of 6%. (a) Find the amount in this account after 10 years. (b) Draw a graph.

Solutions   (a)   If F_n is the amount in the account after n years, then we saw in Example 1 that equation (1) must be satisfied. Since $F_0 = 1000$ and $iF_0 = (.06)1000 = 60$, this equation is

$$F_{n+1} = F_n + 60, \qquad F_0 = 1000.$$

The solution from formula (5) is then

$$F_n = 1000 + 60n.$$

Substituting $n = 10$ then gives

$$F_{10} = 1000 + 60(10) = 1600.$$

(b)   A graph of $F_n = 1000 + 60n$ is shown in Figure 11.8.   ∎

In general we can see from (5) that the following is true by noting that equation (1) is the dynamical system (2) with $a = 1$ and $b = iF_0$.

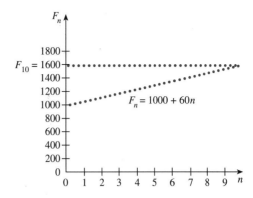

Figure 11.8

> **Simple Interest**
>
> If an amount of $F_0 = P$ dollars is deposited into an account earning simple interest at an annual rate of i expressed as a decimal, then the amount F_n in the account at the end of n years is given by
>
> $$F_n = P + iPn.$$

E X A M P L E 4 **Regular Deposits into a Savings Account**

An individual is saving money for a downpayment on a house to be purchased in 5 years. She can deposit \$250 at the end of each month into an account that pays interest at an annual rate of 6% compounded monthly. How much is in this account after 5 years?

Solution Let F_n be the amount in the account after n months. The interest rate per month is $i = .06/12 = .005$. Then since F_n satisfies equation (3) with $1 + i = 1.005$ and $d = 250$, we have

$$F_{n+1} = 1.005F_n + 250.$$

This is in the form of equation (2) with $a = 1.005$ and $b = 250$. We also have $n = 5 \times 12 = 60$. Since $F_0 = 0$ and

$$\frac{b}{1-a} = \frac{250}{-.005} = -50{,}000,$$

equation (4) indicates that

$$F_n = \frac{b}{1-a} + \left(F_0 - \frac{b}{1-a}\right)a^n$$

$$= -50{,}000 + 50{,}000(1.005)^n$$

$$F_{60} = -50{,}000 + 50{,}000(1.005)^{60} = 17{,}442.51.$$

Thus she will save \$17,442.51 in 5 years. ∎

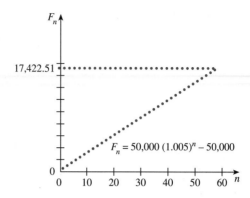

Figure 11.9

The graph of $F_n = 50{,}000(1.005)^n - 50{,}000$ is given in Figure 11.9. The graph shows the amount in the account steadily increasing.

Often an individual or corporation knows that at some future date, a certain amount F of money will be needed. The question is how much money should be put away periodically so that the deposits plus interest will be F at the needed point in time. Such an account is referred to as a **sinking fund**. In such a case F, n and i are known and one wishes to calculate the periodic payment or deposit d. The following is such an example.

E X A M P L E   5

Determining a Deposit to Meet a Future Obligation

A corporation wishes to set up a sinking fund in order to replace a grinding machine in 6 years at a cost of $100,000. How much per quarter should be deposited into an account with an annual interest rate of 8% compounded quarterly to meet this future obligation?

Solution

Let F_n be the amount in the account after n quarters. The interest rate per quarter is $i = .08/4 = .02$. In $n = 4 \times 6 = 24$ quarters, the corporation needs $F_{24} = 100{,}000$. We have $a = 1 + i = 1.02$, and the deposit d is unknown. From (4) we have

$$F_n = \frac{b}{1-a} + \left(F_0 - \frac{b}{1-a} \right) a^n$$

$$100{,}000 = F_{24} = \frac{d}{-0.02} - \left(0 + \frac{d}{-.02} \right)(1.02)^{24}$$

$$= d[50(1.02)^{24} - 50] \approx d(30.42)$$

$$d = 3287.11.$$

So the company should deposit $3287.11 each quarter for the next 6 years in order to have $100,000 in this account at that time. ∎

We now consider loans and payment schedules. Suppose you borrow F_0 dollars and agree to make a periodic payment (for example, every month). Then initially you owe F_0 dollars. Let F_n be the amount that you owe after n periods of time and suppose the interest charged per time period is i. Then the balance F_{n+1} of the loan after $n + 1$ periods is the balance after n periods plus the interest iF_n charged during this period less the payment, which we denote by PMT. This gives the following discrete dynamical system

$$F_{n+1} = F_n + iF_n - PMT$$

or

$$F_{n+1} = (1 + i)F_n - PMT \tag{7}$$

Comparing this with equation (2), we have $a = 1 + i$ and $b = -PMT$. Thus the solution from equation (4) is

$$F_n = \frac{PMT}{i} + \left(F_0 - \frac{PMT}{i} \right)(1 + i)^n \tag{8}$$

The process of paying off a debt is called **amortization** (or killing) the debt.

EXAMPLE 6   Calculating Payments on a Loan

You wish to borrow $10,000 from a bank to purchase a car. The bank charges interest at an annual rate of 12%, and there are to be 48 equal monthly payments with the first to begin in one month. Determine the payments so that the loan will be amortized after 48 months.

Solution   The monthly interest rate is $.12/12 = .01$ and $F_0 = 10,000$. The balance of the loan after 48 months must be zero, that is, we must have $F_{48} = 0$. Thus from equation (8) we have

$$0 = F_{48} = \frac{PMT}{.01} + \left(10,000 - \frac{PMT}{.01} \right)(1.01)^{48}$$

$$\frac{PMT}{.01}\left((1.01)^{48} - 1 \right) = 10,000(1.01)^{48}$$

$$\frac{PMT}{.01}\left(1 - \frac{1}{(1.01)^{48}} \right) = 10,000$$

$$PMT = \frac{(.01)10,000}{1 - \frac{1}{(1.01)^{48}}} = 263.34.$$

Thus a monthly payment of $263.34 will pay off the $10,000 loan in 48 months. ∎

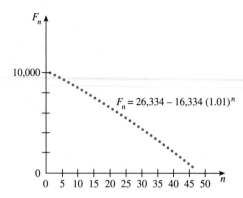

Figure 11.10

With a monthly payment of \$263.34, equation (8) becomes

$$F_n = 26{,}334 - 16{,}334(1.01)^n.$$

A graph is shown in Figure 11.10. Notice how the amount owed F_n steadily decreases.

E X A M P L E   7   **Determining the Amount You Can Afford to Borrow**

Suppose you can afford monthly payments of \$800 on a mortgage for a house. The bank charges annual interest of 9% compounded monthly, and the loan is for 30 years. How large a mortgage can you handle?

The interest per month is $.09/12 = .0075$, and $PMT = 800$. Since the mortgage is to be paid off in 30 years or 360 months, we must have $F_{360} = 0$. We then have from equation (8) that

$$0 = F_{360} = \frac{800}{.0075} + \left(F_0 - \frac{800}{.0075}\right)(1.0075)^{360}$$

$$F_0(1.0075)^{360} = \frac{800}{.0075}(1.0075)^{360} - \frac{800}{.0075}$$

$$F_0 = \frac{800}{.0075}\left(1 - \frac{1}{(1.0075)^{360}}\right) = 99{,}425.49.$$

Thus the mortgage can be as much as \$99,425.49.   ■

Learning Theory

Nonhomogeneous dynamical systems arise in other areas beside personal finance. We now give an example from the theory of learning.

E X A M P L E   8

Learning

It has been observed that a well trained employee on a certain assembly line can handle 20 items per hour and that the increase in the number of items that a new employee can handle per hour is proportional to the difference between 20 and the number they handled in the previous hour.

   (a) Find a dynamical system that models this situation.

   (b) Suppose a typical new employee can handle 4 items per hour with no training and 12 items per hour after 10 hours of training. How many items per hour can the new employee handle after 30 hours of experience?

   (c) Graph and determine what is happening as n becomes larger and larger.

Solutions

   (a) Let N_0 be the number of items handled by the new employee in the first hour and N_n the number handled during the nth hour. The increase in the number of items handled in the $(n + 1)$st hour is $N_{n+1} - N_n$. This quantity is given to be proportional to $20 - N_n$. Thus there must exist a positive constant k so that

$$N_{n+1} - N_n = k(20 - N_n).$$

This can be written as

$$N_{n+1} = N_n - kN_n + 20k$$

or finally as the nonhomogeneous dynamical system

$$N_{n+1} = (1 - k)N_n + 20k.$$

   (b) Referring to equation (4),

$$\frac{b}{1 - a} = \frac{20k}{k} = 20,$$

thus the solution is

$$N_n = 20 + (N_0 - 20)(1 - k)^n. \tag{9}$$

We need to find $1 - k$. We are given that $N_0 = 4$ and $N_{10} = 12$. Putting this into equation (9) gives

$$12 = N_{10} = 20 + (4 - 20)(1 - k)^{10}$$

$$16(1 - k)^{10} = 8$$

$$1 - k = \sqrt[10]{0.5} \approx .933.$$

Thus

$$N_n = 20 - 16(.933)^n$$

$$N_{30} = 20 - 16(.933)^{30} \approx 18.$$

Thus after 30 hours of experience on the assembly line, the new employee can handle 18 items per hour.

   (c) The graph of $N_n = 20 - 16(.933)^n$ is shown in Figure 11.11. Notice that N_n is increasing toward the limiting value of 20. ∎

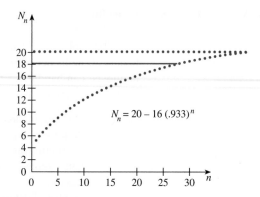

Figure 11.11

SELF-HELP EXERCISE SET 11.2

1. You hold the winning ticket to a $1 million lotto prize, which entitles you to $50,000 now and $50,000 at the end of each of the next 19 years. Someone offers to buy this "$1 million" ticket from you for just $600,000. Would you accept if current interest rates were at 7% per year? Answer the question by determining if you can receive more or less than $50,000 now and at the end of each of the next 19 years by investing the money.

2. What would be a fair price for the winning ticket in the previous exercise?

EXERCISE SET 11.2

1. An account with an initial amount of $1000 pays simple interest at an annual rate of 8%. How much is in the account after 15 years?

2. An account with an initial amount of $3000 pays simple interest at an annual rate of 6%. How much is in the account after 20 years?

Applications

3. **Retirement.** An individual earns an extra $2000 each year and places this money at the end of each year into an Individual Retirement Account (IRA) in which both the original earnings and the interest in the account are not subject to taxation. If the account has an annual interest rate of 7% compounded annually, how much is in the account at the end of 40 years?

4. **Retirement.** Repeat the previous problem if the annual interest rate is 9% per year.

5. **Retirement.** Suppose the personal income tax (federal plus state) is $33\frac{1}{3}\%$. An individual earns an extra $2000 each year and pays income taxes on these earn-

ings and at the end of each year places the remaining funds $2000(\frac{2}{3})$ into a regular savings account in which the interest in the account is subject to the personal income tax mentioned before. If the account earns an annual interest rate of 7% compounded annually and the individual pays income taxes owed on the interest out of these funds, how much is in the account at the end of 40 years? Compare your answer to the exercise before last.

6. **Retirement.** Repeat the previous problem if the annual interest rate is 9% per year. Compare your answer to the exercise before last.

7. **Education Fund.** New parents wish to save for their newborn's education and wish to have $40,000 at the end of 18 years. How much should they place at the end of each year into a savings account that earns an annual rate of 8% compounded annually?

8. **House Down Payment.** A couple will need $20,000 at the end of 6 years for a down payment on a house. How much should they place at the end of each month into a savings account earning an annual rate of 7% compounding monthly to meet this goal?

9. **Equipment.** A corporation creates a fund in order to have $1 million to replace some machinery in 10 years. How much should be placed into this account at the end of each quarter if the annual interest rate is 8% compounded quarterly?

10. **Pension.** Find the amount needed to deposit into an account today that will yield a pension payment of $25,000 at the end of each of the next 20 years if the account pays interest at an annual rate of 7% compounded annually.

11. **Loans.** Find the annual payment needed to pay off a $5000 loan in 6 payments if the annual interest is 12% compounded annually.

12. **Mortgage.** Find the monthly payment needed to pay off a $100,000 mortgage over 30 years if the annual interest is 8% compounded monthly. Then find the total interest paid.

13. **Car Loan.** Find the monthly payment needed to pay off a car loan for $8000 in four years if the annual interest is 12%.

14. **Credit Card Payment.** Find the monthly payment needed to pay off a credit card debt of $1000 in three years if the annual interest is 18%.

15. **Comparing Prices.** Two oil wells are for sale. The first will yield payments of $10,000 at the end of each of the next 10 years, while the second will yield payments of $6500 at the end of each of the next 20 years. Interest rates are assumed to hold steady at 8% per year over the next 20 years. Which well is worth the most? Why?

16. **Comparing Prices.** Redo the previous problem if the second oil well pays $7000 per year for 20 years.

17. **Comparing Retirement Options.** You are being offered a ''half a million dollar'' retirement package to be given in $50,000 payments at the end of each of the next 10 years. You are also given the option of accepting a $350,000 lump sum payment now. Interest rates are at 9% per year. Which looks better to you? Why?

18. **Population.**   In 1990 the population of the U.S. was 249 million and growing (without considering immigration) at about 0.6% a year. Legal immigration amounted to about 0.5 million per year. Let P_n be the population of the U.S. n years after 1990. Find a discrete dynamical system that incorporates these assumptions and allows you to find P_{n+1} if given P_n. Assuming this trend continues, give the population in the year 2020 without finding the population of the intermediate years.

19. **Drug Absorption.**   Suppose 10 milligrams of glucose is injected intravenously each minute into a patient and that each minute the body of the patient absorbs 1% of the amount of the glucose present in the blood at the beginning of that minute. Let A_n be the amount of glucose in the blood of the patient after n minutes. Find a discrete dynamical system that allows you to find A_{n+1} if given A_n. Find the amount of glucose in the blood one hour later without finding the amount for the intermediate time.

20. **Pollutants.**   Suppose 30 tons of pollutants are dumped into a lake each year and that each year the lake is able to absorb 20% of the pollutants that were there at the beginning of the year. Let A_n be the amount of tons of pollutants in the lake after n years. Find a discrete dynamical system that allows you to find A_{n+1} if given A_n. Without finding the amount for the intermediate time, find the amount of pollutants in the lake after 10 years if initially the lake was free of pollution.

21. **Spread of Rumors.**   In a corporation, rumor spreads of large impending layoffs. The increase in the number of employees who have heard the rumor on any day is proportional to the number who have not heard the rumor the previous day. Let P_n be the number who have heard the rumor after n days. (a) Write a discrete dynamical system that gives P_{n+1} in terms of P_n. (b) Suppose there are 5000 employees and that 10 are aware of the rumor initially. If 200 hear of the rumor by the end of the first day, how many hear of the rumor after 10 days?

22. **Heating**   Suppose a jug of cold milk at 40°F is placed on a table in a room kept at 70°F. According to Newton's law of heating, the increase in the temperature of the milk in any minute is proportional to the difference of 70° and the temperature of the milk at the end of the previous minute. Let T_n be the temperature of the milk after n minutes. (a) Write a discrete dynamical system that gives T_{n+1} in terms of T_n. (b) If the temperature of the milk is 55°F after the first hour on the table, find the temperature after the second hour without finding the temperature for the intermediate times.

23. **Cooling.**   Let T_n be the temperature of a cup of coffee n minutes after being poured. According to Newton's law of cooling, the difference in temperature $T_{n+1} - T_n$ is negatively proportional to $T_n - C$, where C is the temperature of the room. Find a discrete dynamical system that gives T_{n+1} if T_n is known. Suppose the temperature of the coffee when poured was 200°F and one minute later was 190°F. If the room is at 72°F, find the time it will take for the coffee to reach 160°F.

24. **Determination of the Time of Death.**   Let T_n be the temperature of the body of a murdered man n minutes after his death. Suppose the corpse has lain in a room kept at a constant temperature of 68°F. According to Newton's law of cooling, the difference $T_{n+1} - T_n$ in temperature during the $(n + 1)$st minute is negatively proportional to $T_n - 68$. When the police discover the corpse at

3:00 P.M., the body temperature was measured to be 85°F, and two hours later it was 75°F. Assuming at the time of death the body temperature was a normal 98.6°F, find the time of death.

Solutions to Self-Help Exercise Set 11.2

1. Suppose you sold the ticket for \$600,000, kept \$50,000, invested the remaining 550,000 at 7% and withdrew from this account an amount equal to W at the end of each of the next 19 years. If F_n is the amount in the account after n years, then the amount F_{n+1} next year will be F_n plus the interest $.07F_n$ less the withdrawal W or $F_{n+1} = F_n + .07F_n - W$ which can be also written as

$$F_{n+1} = (1.07)F_n - W.$$

The solution to this is

$$F_n = \frac{W}{.07} + \left(550,000 - \frac{W}{.07} \right)(1.07)^n.$$

Since we want $F_{19} = 0$, we then have

$$0 = F_{19} = \frac{W}{.07} + \left(550,000 - \frac{W}{.07} \right)(1.07)^{19}$$

$$0 = \frac{W}{.07(1.07)^{19}} + 550,000 - \frac{W}{.07}$$

$$\frac{W}{.07}\left(1 - \frac{1}{(1.07)^{19}} \right) = 550,000$$

$$W = \frac{(.07)550,000}{1 - \frac{1}{(1.07)^{19}}} = 53,214.16.$$

This gives \$53,214.16 per year for the next 19 years, which is more than the \$50,000 you obtain from the lotto prize. This indicates that the "\$1 million ticket" is actually worth less than \$600,000.

2. We take \$50,000 now and \$50,000 at the end of each of the next 19 years. Using the same notation as in the previous example, we have

$$F_{n+1} = (1.07)F_n - 50,000,$$

with the initial amount F_0 unknown. The solution to the dynamical system is

$$F_n = \frac{50,000}{.07} + \left(F_0 - \frac{50,000}{.07} \right)(1.07)^n.$$

Since $F_{19} = 0$, we have

$$0 = F_{19} = \frac{50,000}{.07} + \left(F_0 - \frac{50,000}{.07} \right)(1.07)^{19}$$

$$F_0(1.07)^{19} = \frac{50,000}{.07}\left((1.07)^{19} - 1 \right)$$

$$F_0 = \frac{50,000}{.07}\left(1 - \frac{1}{(1.07)^{19}} \right) = 516,779.76$$

Adding the $50,000 first payment to this gives a fair value of $566,779.76 to the winning ticket. This value is almost half the $1 million the ticket is often assumed to be worth. This value is dependent on the obtainable interest rate.

11.3   QUALITATIVE BEHAVIOR

▶ *Introduction*

▶ *The Case $0 < a < 1$*

▶ *The Case $a > 1$*

▶ *More Applications*

▶ *Cobwebs*

APPLICATION

**Temperature of a Body
in the Long-Term**

Suppose a cup of hot tea, a cup of cold tea, and a cup of tea at room temperature are set on a table in a room kept at constant temperature. What discrete dynamical system will have solutions that determines the temperature of each of these cups of tea? See the discussion under the heading ''The Case $0 < a < 1$.''

APPLICATION

**The Effect of Harvesting
the Minke Whale**

Suppose the International Whaling Commission lifts its moratorium on commercial whaling instituted in 1986 and permits the harvesting of 36,000 minke whales a year. What will happen to the long-term population of these whales? For the answer see the discussion after the heading ''The Case $a > 1$.''

*(© Jen and Des Bartlett/Photo
Researchers, Inc.)*

Introduction

We again consider the discrete dynamical system

$$y_{n+1} = ay_n + b, \tag{1}$$

where $a \neq 1$. In this section we are less interested in the numerical values of the solution y_n but rather more interested in the general type of behavior that the solution exhibits. For example, we will want to know if the values of y_n are approaching some limiting value or if they are becoming unbounded. Applications will be given.

The qualitative behavior of the solutions to (1) will depend dramatically on the values of the constant a in that equation. In this section we consider the two cases when $a > 0$.

The Case $0 < a < 1$

Consider first a familiar situation. Suppose you place a cup of hot tea, a cup of iced or cold tea, and a cup of tea made from tap water that is exactly at room temperature, all on a table in a room at constant temperature. Then we all know that the hot tea steadily cools, that is, the temperature of the hot tea steadily decreases and approaches room temperature. The cold tea steadily warms up, that is, the temperature of the cold tea steadily increases and also approaches room temperature. The tea that was initially at room temperature stays at room temperature, that is, the temperature of this tea remains constant at the temperature of the room.

Let us now create a mathematical model that describes this situation and uses a discrete dynamical system. First, Newton's law of heating and cooling states that the quantity given by the temperature at the end of any minute less the temperature at the beginning of this minute is proportional to the quantity given by the room temperature less the temperature at the beginning of the minute. Thus if we let T_n be the temperature of the tea after n minutes, then the quantity $T_{n+1} - T_n$ is proportional to $C - T_n$, where C is the constant room temperature. This means there is a positive constant k such that $T_{n+1} - T_n = k(C - T_n)$. Suppose $k = 0.02$ and the room temperature is $C = 70°F$. Then $T_{n+1} - T_n = 0.02(70 - T_n)$, or

$$T_{n+1} = 0.98T_n + 1.4. \tag{2}$$

This is the dynamical system (1) with $a = 0.98$ and $b = 1.4$.

Recall that the solution to the dynamical system (1) is

$$y_n = \frac{b}{1-a} + \left(y_0 - \frac{b}{1-a}\right)a^n. \tag{3}$$

Since in equation (2)

$$\frac{b}{1-a} = \frac{1.4}{1-(1-0.98)} = 70,$$

the solution to (2) is given by

$$T_n = 70 + (T_0 - 70)(0.98)^n \tag{4}$$

where T_0 is the initial temperature.

Let us now consider the three cases: hot tea, cold tea, and tea at room temperature. First consider the case when the temperature of the tea is initially greater than room temperature, that is, $T_0 > 70$. Then the term $T_0 - 70$ in equation (4) is *positive*. Notice also that $(0.98)^n = (0.98)^{n-1}(0.98)$. This means that each of the terms $(0.98)^n$ is 0.98 times the previous. Therefore these values are decreasing. Since $T_0 - 70$ is positive, the term $(T_0 - 70)(0.98)^n$ is decreasing, and then it follows from equation (4) that T_n also must decrease. Table 11.1 indicates what is happening for selected values of n in the case that the initial temperature of the tea is $T_0 = 200$. The values of the temperature are given in the third column.

Finally, for $n = 512$, $(0.98)^{512} = .0000322$ and $T_{512} = 70.004$. We make three very important observations from the first three columns of Table 11.1.

Table 11.1

n	$(0.98)^n$	Hot Tea $T_n = 70 + 130(0.98)^n$	Cold Tea $T_n = 70 - 30(0.98)^n$
0	1.000	200.0	40.0
1	.980	197.4	40.6
2	.960	194.9	41.2
3	.941	192.4	41.8
4	.922	189.9	42.3
8	.851	180.6	44.5
16	.724	164.1	48.3
32	.524	138.1	54.3
64	.274	105.7	61.8
128	.075	79.8	67.7
256	.006	70.7	69.8

First, the terms $(0.98)^n$ are approaching 0 as n becomes larger. Second, the temperature T_n of the hot tea becomes successively smaller for larger values of n. Finally, the temperature T_n of the hot tea approaches the room temperature of 70°. See Figure 11.12. Since the number T_n, indicated in Figure 11.12, approaches the constant line $T = 70$, we say that the graph of $T = T_n$ is **asymptotic** to the line $y = 70$.

Consider now the case when the temperature of the tea is initially smaller than room temperature, that is, $T_0 < 70$. Then the term $T_0 - 70$ in equation (4) is *negative*. Since $(0.98)^n$ is decreasing, this implies that the term $(T_0 - 70)(0.98)^n$ becomes less negative for larger values of n, that is, this term is increasing. Thus from equation (4) T_n must be increasing.

The fourth column in Table 11.1 indicates what is happening for selected values of n in the case that the initial temperature of the tea is $T_0 = 40$. Finally for $n = 512$, $T_{512} = 69.999$. We make two very important observations from the fourth column of Table 11.1. First, the temperature T_n of the cold tea becomes successively larger for larger values of n. Second, the temperature T_n of the cold tea approaches the room temperature of 70°. See Figure 11.12. Again the graph of $T = T_n$ is asymptotic to the line $T = 70$.

The third case in which the tea is initially at room temperature follows from the following general result.

Constant Solution

The solution to the dynamical system (1) is the constant $y_n = \dfrac{b}{1 - a}$ if the initial value is $y_0 = \dfrac{b}{1 - a}$.

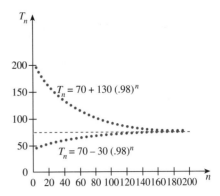

Figure 11.12

This can easily be seen by noticing that if $y_0 = \dfrac{b}{1-a}$ in the solution (3), then the coefficient of a^n in that equation is 0. Thus the result follows.

For our problem with the tea recall that $\dfrac{b}{1-a} = 70$. Thus our mathematical model gives the constant solution $T_n = 70$ in the case that the initial temperature is also 70. Notice how the solutions to our mathematical model all agree with our knowledge of the physical setup.

We notice from Figure 11.12 that the graphs of solutions are asymptotic to the line $T = 70$ whether the solutions are initially above this line or below this line. In such a case we say that the line $T = 70$ is an **attractor**.

If the numbers y_n are either always increasing or always decreasing, then we say that y_n is **monotonic**.

Increasing, Decreasing, Monotonic

1. If $y_n < y_{n+1}$ for every n, we say that y_n is **increasing**.
2. If $y_n > y_{n+1}$ for every n, we say that y_n is **decreasing**.
3. If y_n is either always increasing or always decreasing, then we say that y_n is **monotonic**.

By using an analysis similar to that which we used in studying the temperature of the tea, we can obtain the following general qualitative result.

Qualitative Behavior when $0 < a < 1$

In the case $0 < a < 1$, the solutions given by (3) of the dynamical system (1) exhibits the following 3 types of behavior. (See also Figure 11.13.)

1. If $y_0 > \dfrac{b}{1-a}$, then the solution y_n is decreasing and approaches $\dfrac{b}{1-a}$.

2. If $y_0 < \dfrac{b}{1 - a}$, then the solution y_n is increasing and approaches $\dfrac{b}{1 - a}$.

3. If $y_0 = \dfrac{b}{1 - a}$, then the solution y_n is the constant $y_n = \dfrac{b}{1 - a}$.

We can more briefly describe this by saying that the solutions are monotonic with the line $y = \dfrac{b}{1 - a}$ an attractor.

The Case $a > 1$

For a general second case we consider the case that the term a in equation (1) satisfies $a > 1$. Consider the specific problem of harvesting some renewable resource such as fish. This would include halibut, tuna, salmon, and so forth. Suppose then P_n is the population of a species of fish in thousands after n years. As we have seen earlier in this chapter, biologists argue that the difference in the population $P_{n+1} - P_n$ in any year is proportional to the population P_n at the beginning of the year. Thus there exists a positive constant k such that $P_{n+1} - P_n = kP_n$.

   To be more specific, suppose we consider the smallest of all the whales, the minke whale[1], and assume that the growth constant is $k = 0.04$. Then the last equation can be written as $P_{n+1} = 1.04P_n$. Now suppose 36,000 of these whales are harvested every year[2], then since population is being measured in thousands

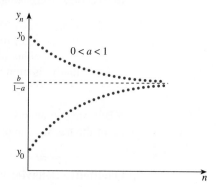

Figure 11.13

---

[1]Whales are not fish but aquatic mammals.

[2]The International Whaling Commission in 1986 banned the harvesting of any species of whale but recently has seriously considered allowing a limited resumption of commercial whaling.

we have the discrete dynamical system

$$P_{n+1} = 1.04P_n - 36. \tag{5}$$

Comparing this with (1) we see that $a = 1.04$ which is larger than 1. Since $\dfrac{b}{1-a} = \dfrac{-36}{-0.04} = 900$, the solution from equation (3) is

$$P_n = 900 + (P_0 - 900)(1.04)^n. \tag{6}$$

Now let us see what this mathematical model will predict will happen for some different estimates of the total minke whale population. Keep in mind that environmental factors such as disease and food supply can change quickly and result in dramatic changes in the population. Estimating the total minke whale population[3] at 1,100,000 and setting $P_0 = 1100$ in equation (6) then yields the values given in the third column of Table 11.2.

A number of observations can be made from the first three columns of Table 2. Looking at the first two columns we see that as n becomes large without bound $(1.04)^n$ becomes large without bound. From the third column we see that P_n is increasing and as n becomes large without bound P_n becomes large without bound. See Figure 11.14.

Estimating the total minke whale population at 700,000 and setting $P_0 = 700$ in equation (6) then yields the values given in the fourth column of Table

Table 11.2

n	$(1.04)^n$	$P_0 = 1100$	$P_0 = 700$
0	1.000	1100	700
1	1.040	1108	692
2	1.082	1116	684
3	1.125	1125	675
4	1.170	1134	666
8	1.369	1174	626
16	1.873	1275	525
32	3.508	1602	198
38	4.439	1788	12
39	4.616	1823	-23
64	12.306	3361	-1561
128	151.449	31,189	-29,390
256	22936.907	4,588,281	-4,586,481

---

[3]The International Whaling Commission estimated the population in 1992 to be 874,000.

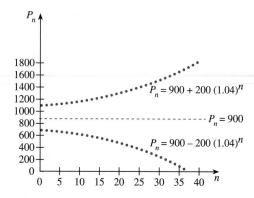

Figure 11.14

11.2. Notice from the fourth column that P_n is decreasing and as n becomes large without bound P_n becomes (negatively) large without bound. See Figure 11.14.

If the minke population is actually 900,000, then $P_0 = 900$ in equation (6), and this leads to the constant solution $P_n = 900$ shown in Figure 11.14.

In Figure 11.14 we notice that the solution above and below the constant solution $P_0 = 900$ move *away* from this line. In such a case we say that the line $P = P_0$ is **repelling**.

We will interpret what these numbers mean in terms of the minke whale population shortly. But first we note that it can be shown in a similar fashion that the same behavior as shown in Figure 11.14 holds for any discrete dynamical system of the form (1) when $a > 1$.

Qualitative Behavior When $a > 1$

In the case $a > 1$, the solutions given by (3) of the dynamical system (1) exhibit the following 3 types of behavior. (See also Figure 11.15.)

1. If $y_0 > \dfrac{b}{1 - a}$, then the solution y_n is increasing and becomes large without bound as n becomes larger without bound.

2. If $y_0 < \dfrac{b}{1 - a}$, then the solution y_n is decreasing and becomes (negatively) large without bound as n becomes large without bound.

3. If $y_0 = \dfrac{b}{1 - a}$, then the solution y_n is the constant $y_n = \dfrac{b}{1 - a}$.

We can more briefly describe this by saying that the solutions are monotonic with the line $y = \dfrac{b}{1 - a}$ being a repeller.

We now interpret the meaning of the numbers in Table 11.2 for the future population of the minke whale under the assumed harvesting policy. First notice

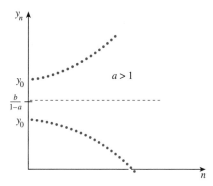

Figure 11.15

from the fourth column of Table 11.2 that P_n turns *negative* for $n = 39$. This means that the minke whale population will become extinct in 39 years. Since such a policy would lead to the rapid extinction of the minke whale we would hope that the international community would not permit such a catastrophic occurrence.[4] The most obvious solution would be to lower the level of harvesting.

In the third column we see the minke whale population increasing in such a manner that in 256 years there will be about 4.5 *billion* of these mammals in the oceans of the world. Any whaling scientist knows that such numbers are completely absurd. The oceans can not possibly provide the necessary food to sustain such a population. We therefore need to question our mathematical model given by (5). The mathematical model given by equation (5) does not include any term that takes into consideration the obvious fact that there are only a *finite* amount of available resources for the whales. Such a more realistic term will be added later in this chapter leading to a substantially improved model for long-term predictions. Unfortunately this mathematical model will be a *non*linear dynamical system and more difficult to solve and analyze. Nonetheless, the model we are using in this chapter is adequate for shorter time periods and has the advantage of being easier to solve and analyze.

More Applications

E X A M P L E 1   **How Much Can You Borrow?**

Suppose that the annual rate on a mortgage is 9% compounded monthly and you are able to make monthly payments of $600. How large of a mortgage can you afford?

Solution   Let B_n be the balance of the loan after n months with B_0 the initial amount and

---

[4]The International Whaling Commission in 1992 approved a formula for setting catch limits on minke whales but put off sanctioning commercial catches at least until the next year. The catch formula would be set to ensure that the population does not go into decline.

also the amount of the loan. Interest per month is $.09/12 = .0075$. The balance B_{n+1} the next month is equal to the balance B_n plus the interest $0.0075B_n$ on the balance, less the payment of 600. This gives

$$B_{n+1} = B_n + .0075B_n - 600,$$

or

$$B_{n+1} = (1.0075)B_n - 600 \tag{7}$$

This is the dynamical system (1) with $a = 1.0075 > 1$. Thus the graph of the solutions will be similar to that shown in Figure 11.15. Since $b = -600$,

$$\frac{b}{1-a} = \frac{-600}{1-1.0075} = 80,000.$$

Thus we see from Figure 11.16 that a loan of less than $B_0 = 80,000$ will result in eventually paying off the loan. But if $B_0 \geq 80,000$, the mortgage cannot be paid off in any amount of time.   ∎

Lewis Fry Richardson[5] wrote a series of papers that developed mathematical models of an arms race between two countries. Suppose there are two countries, A and B, who distrust each other sufficiently that they both must spend money on military arms for their defense. Let A_n and B_n be, respectively, the amount (in some appropriate currency) spent on arms by the corresponding countries n years from now. We shall determine an equation that might reasonably give the increase $A_{n+1} - A_n$ in expenditures of country A. We have

$$A_{n+1} - A_n = -eA_n + dB_n + f,$$

where d, e, and f are all constants. The term d is a "distrust" coefficient. The larger this number the more distrust country A has of B, resulting in an increase in arms given by dB_n, where B_n is the arms expenditure on country B. Now

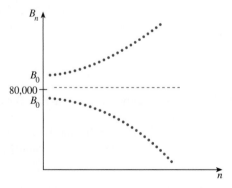

Figure 11.16

[5]James R. Newman, *The World of Mathematics* (New York: Simon and Schuster, 1956), pp. 1240–1253.

country A has some constraints on the amount it can spend on arms. This is primarily determined by its economy. The term $-eA_n$ measures the restraint due to the economy. Richardson used actual data to try to estimate these numbers.

We then make a similar analysis for country B. Although Richardson considered a more general situation, let us make a simplifying assumption that the coefficients of distrust and restraints from the economy of country B are the same as for country A. Thus we have

$$B_{n+1} = -eB_n + dA_n + g.$$

Adding the two equations gives

$$[A_{n+1} + B_{n+1}] - [A_n + B_n] = (d - e)[A_n + B_n] + c,$$

where $c = f + g$. Now consider the total expenditures $T_n = A_n + B_n$ of the two countries. Then from the last equation we have

$$T_{n+1} = (1 + d - e)T_n + c. \tag{8}$$

This is equation (1) with $a = 1 + d - e$.

EXAMPLE 2   When the Distrust is Small

Determine what will happen to total expenditures in the above arms race if $0 < 1 + d - e < 1$.

Solution   If $0 < 1 + d - e < 1$, then the solutions of (8) are monotonic with the constant solution $T_n = \dfrac{b}{1 - a} = \dfrac{c}{e - d}$ an attractor. The total expenditures will go monotonically to the constant $\dfrac{c}{e - d}$. See Figure 11.17. Notice that this will be the case if the distrust coefficient d is less than the coefficient e of restraint from the economy. Thus if the distrust is small, relative to the restraints in the economy, this model predicts a *stable* arms race. ∎

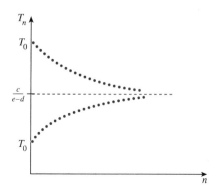

Figure 11.17

Cobwebs

We now consider a graphical technique that will give additional insight into discrete dynamical systems.

Consider the dynamical system

$$y_{n+1} = 0.50y_n + 2,$$

with $y_0 = 1$. Since $y_n = 4 + (y_0 - 4)(0.50)^n$, it follows that

$$y_1 = 4 - 3(0.50)^1 = 2.5$$

$$y_2 = 4 - 3(0.50)^2 = 3.25$$

$$y_3 = 4 - 3(0.50)^3 = 3.625.$$

In Figure 11.18 we have graphed the two linear equations $y = 0.50x + 2$ and $y = x$. We first notice that the intersection of these two lines will give the constant solution to the dynamical system. To see why this is the case notice that a *constant* solution must have $y_{n+1} = y_n$ for all n. This implies that $y_n = y_{n+1} = 0.50y_n + 2$. The solution of this last equation is $y_n = 4$ and must also be the solution to $x = y = 0.50x + 2$.

Now for a cobweb interpretation of how the subsequent values of y_n are obtained. In Figure 11.19 a vertical line is drawn from $x = 1$ until this line strikes the line $y = 0.50x + 2$. The y-coordinate of this point is $y_1 = 2.5$. Now move horizontally until you strike the line $y = x$. The x-coordinate is the same as the y-coordinate, which is 2.5. Now begin the cycle over again. Notice from Figure 11.19 that this indicates that the values y_n are approaching the value 4. In the same way Figure 11.19 indicates that if $y_0 > 4$, then the solution decreases and approaches the value 4.

Now consider the dynamical system

$$y_{n+1} = 2y_n - 2,$$

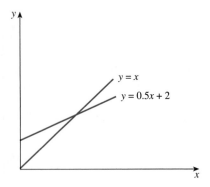

Figure 11.18

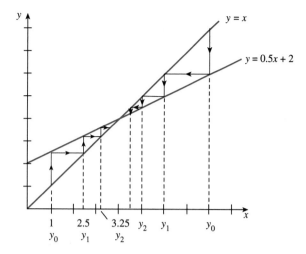

Figure 11.19

with $y_0 = 2.5$. Since $y_n = 2 + (y_0 - 2)2^n$, it follows that $y_1 = 3$, $y_2 = 4$, $y_3 = 6$.

Figure 11.20 shows the cobweb interpretation. Here we see that since the slope of the line $y = 2x - 2$ is $a = 2 > 1$, the values of y_n must move away from 2.

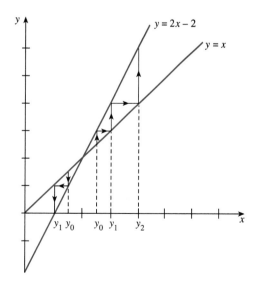

Figure 11.20

SELF-HELP EXERCISE SET 11.3

1. Graph the solutions of $y_{n+1} = .80y_n + 10$ for various values of y_0. Determine if the constant solution is an attractor or repeller.

2. Determine what will happen to total expenditures in the above arms race if distrust is large, that is, $1 + d - e > 1$.

EXERCISE SET 11.3

In the following discrete dynamical systems (a) find the constant solution, (b) decide if the constant solution is an attractor or repeller, (c) graph the solutions.

1. $y_{n+1} = 2y_n - 10$ 2. $y_{n+1} = 1.05y_n + 5$

3. $y_{n+1} = 0.90y_n + 20$ 4. $y_{n+1} = 0.50y_n - 10$

5. $y_{n+1} = 0.999y_n + 100$ 6. $y_{n+1} = 1.001y_n + 1$

Use a cobweb diagram to determine graphically what is happening to the values of y_n in the following dynamical systems.

7. $y_{n+1} = 0.75y_n + 2$ 8. $y_{n+1} = 0.50y_n + 3$

9. $y_{n+1} = 1.5y_n - 2$ 10. $y_{n+1} = 2y_n - 4$

11. $y_{n+1} = .90y_n + 3$ 12. $y_{n+1} = 1.1y_n - 3$

Applications

13. **Borrowing Limit.**   Suppose the annual interest rate on a mortgage is 10% compounded monthly. If you can make a monthly payment of $500, what can you borrow?

14. **Borrowing Limit.**   A corporation can take a loan at an annual rate of 8% compounded quarterly. If it can afford to make a quarterly payment of $1 million, how much can the corporation borrow?

15. **Perpetual Income.**   A person wishes to deposit money into an account that pays annual interest of 5% compounded annually and make a withdrawal of $4000 at the end of every year. How much must be deposited so that the money never runs out?

16. **Perpetual Income.**   A wealthy individual wishes to make a large gift to a certain university to create a fund from which the university may withdraw $1 million at the end of each year in perpetuity. If interest rates are assumed to always be at an annual rate of 6% compounded annually, what must be the size of the gift?

17. **Population.**   A certain country has the percentage of births less deaths in any year equal to 4% of the population at the beginning of the year. In addition 2 million of its population leave each year and 1 million immigrants arrive each year. What initial population level will keep this population the same at the

end of each year? What initial population levels will lead to the population eventually becoming zero? To becoming unbounded? Draw a graph of these situations.

18. **Medicine.**   Suppose each minute the body of an individual removes 1% of the glucose in its blood at the beginning of the minute and that each minute 10 milligrams of glucose is being given intravenously. If this continues, determine what will happen for various initial values of glucose in the blood. Draw a graph that illustrates what is happening.

19. **Biology.**   Suppose each year a certain lake is able to absorb 10% of the amount of a certain pollutant at the beginning of the year and that each year 20 tons of this pollutant is dumped into the lake. If this continues, determine what will happen for various initial values of the pollutant in the lake. Draw a graph that illustrates what is happening. What always happens in the long-term?

20. **Radioactive Disposal.**   A certain waste disposal site accepts low-level radio-active waste with a decay constant of $k = 0.10$, where time is measured in years. If 5 tons of this material is deposited at the site each year, what should be the total amount of this radioactive waste at the site in the long-term?

21. **Harvesting Equation.**   If k and H are positive constants, show that the solution of $P_{n+1} = (1 + k)P_n - H$, is

$$P_n = \frac{H}{k} + \left(P_0 - \frac{H}{k} \right)(1 + k)^n.$$

22. **Harvesting Equation.**   If k and I are positive constants, show that the solution of $A_{n+1} = (1 - k)A_n - I$, is

$$A_n = -\frac{I}{k} + \left(A_0 + \frac{I}{k} \right)(1 - k)^n.$$

23. **Antarctic Blue Whale.**   Whaling scientists estimate that the population of Ant-arctic blue whales before commercial whaling began was about 150,000 and that these mammals grew at about a rate of 3% a year. Using these estimates and the result of Exercise 21, determine the number of these whales that could be harvested without sending the population into a decline.

24. **Antarctic Fin Whale.**   It has been estimated that the optimal population of the Antarctic fin whale (a smaller whale than the blue whale) for commercial whal-ing is about 200,000 and that under ideal circumstances the population of these whales grows about 8% a year. Using these figures and the result of Exercise 21, determine the number of these whales that could be harvested without send-ing the population into a decline.

25. **The Peruvian Anchovy Fishery.**   The Peruvian anchovy fishery became the largest fishery in the world during the 1960's with an annual catch of about 10 million tons. In 1973 an unusual incursion of warm tropical waters (called El Niño) occurred, dropping the population precipitously. Since harvesting did not abate, the population of the fishery declined catastrophically. Mathematically, the following models what happened. Assume the anchovy biomass is 35 million tons, has a growth constant of $k = .40$, and that 10 million tons are harvested annually. Using the result of Exercise 21, graph a solution and indicate what

happens. Now suppose following this El Niño the biomass drops to 15 million tons and the same annual harvest continues. Now graph the solution and indicate what happens.

Solutions to Self-Help Exercise Set 11.3

1. Since $a = .80$ is between 0 and 1, the constant solution $y_n = \dfrac{b}{1 - a} = \dfrac{10}{.20} = 50$ is an attractor. If $y_0 > 50$, then solutions decrease and approach 50. If $y_0 < 50$, then solutions increase and approach 50.

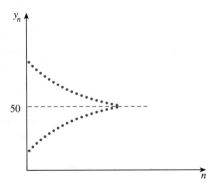

2. If $a = 1 + d - e > 1$, then the solutions of (8) are monotonic, with the constant solution $\dfrac{c}{e - d}$ a repeller. See the figure. If the initial total expenditure T_0 is less than $\dfrac{c}{e - d}$, then from the figure the total expenditure T_n will eventually go to zero. If however the initial total expenditure is greater than this number, then the total expenditure becomes unbounded. Since this is impossible, the only alternative is war or negotiations.

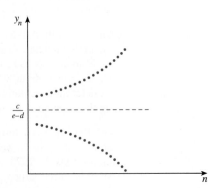

11.4   QUALITATIVE BEHAVIOR (CONTINUED)

- ▸ *Introduction*
- ▸ *Supply and Demand*
- ▸ *The Case* $-1 < a < 0$
- ▸ *The Case* $a < -1$
- ▸ *The Case* $a = -1$

Introduction

We continue to study the qualitative behavior of solutions to the discrete dynamical system

$$y_{n+1} = ay_n + b. \tag{1}$$

Recall that the solution is

$$y_n = \frac{b}{1-a} + \left(y_0 - \frac{b}{1-a}\right) a^n. \tag{2}$$

Supply and Demand

We now consider the cases for which $a < 0$. One place that this situation occurs is in linear supply and demand curves. The demand for a commodity such as soybeans depends on the price. If the price increases, naturally consumers will buy less. Thus the demand curve must slope downward. We now assume a linear demand equation given by

$$d = 10 - Ap,$$

where p is the price per bushel of soybeans, A is as yet an unspecified positive constant, and d is the amount of soybeans measured in some appropriate units. See Figure 11.21.

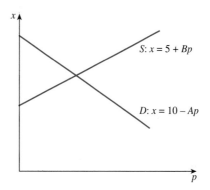

Figure 11.21

In the same way, the amount s of soybeans supplied by farmers will also depend on the price. If the price increases, naturally this will induce the farmers to produce more soybeans. Thus the supply curve should slope upwards. We now assume a linear supply equation

$$s = 5 + Bp,$$

where B is a positive constant that we do not specify at this time. See Figure 11.21.

In Chapter 1 we noted that in equilibrium demand must equal supply, which corresponds to the point where the two curves in Figure 11.21 intersect. This rarely happens. What actually happens is the farmer bases his production on the price p_n of soybeans that prevails at planting time. The supply equation will then determine the supply of soybeans. Due to the time lag in the growing process, the resulting supply of soybeans is not available until the next period. At that time the price will be determined by the demand equation. The price will adjust so that all the soybeans are sold. (We assume no government interference.) The new price p_{n+1} is noted by the farmer at planting time, and a new cycle begins.

The supply s_{n+1} is determined by the equation

$$s_{n+1} = 5 + Bp_n,$$

while the demand d_{n+1} is determined by the equation

$$d_{n+1} = 10 - Ap_{n+1}.$$

Now we impose the condition that the price (at the time the soybeans are brought to market) is adjusted so that the demand equals the supply, that is, so that all the soybeans are sold. Imposing the condition that supply must equal demand gives the equation

$$5 + Bp_n = s_{n+1} = d_{n+1} = 10 - Ap_{n+1}.$$

This gives the discrete dynamical system

$$p_{n+1} = \frac{5}{A} - \frac{B}{A}p_n. \tag{3}$$

Since from (1) $a = -B/A$ and $b = 5/A$,

$$\frac{b}{1 - a} = \frac{5/A}{1 + B/A} = \frac{5}{A + B},$$

and the solution of (3) is given by

$$p_n = \frac{5}{A + B} + \left(p_0 - \frac{5}{A + B}\right)\left(-\frac{B}{A}\right)^n. \tag{4}$$

Recall that A and B are both positive constants.

The Case $-1 < a < 0$

There are three cases, each of which lead to different behavior. Consider the first case for which $-1 < a < 0$ or since $a = -B/A$, $B/A < 1$. To be more specific, consider the case that $A = \frac{2}{3}$ and $B = \frac{1}{3}$. Then equation (3) becomes

$$p_{n+1} = 7.5 - 0.50p_n \tag{5}$$

with solution

$$p_n = 5 + (p_0 - 5)(-0.50)^n. \tag{6}$$

If the initial price is $p_0 = 7$, then Table 11.3 gives the values of the prices in the subsequent times.

From the second column notice how the values of $a^n = (-0.50)^n$ are changing sign. Also notice that the values of $a^n = (-0.50)^n$ are approaching 0 as n becomes large. Notice from the third column how the price oscillates about 5, first larger than 5, then smaller than 5, and so on. Finally notice from the third or fourth column that the price p_n approaches 5 as n becomes large. See Figure 11.22.

Table 11.3

n	$(-0.50)^n$	$p_n = 5 + 2(-0.50)^n$	*Rounded*
0	1.0000	7.000	7.00
1	-0.5000	4.000	4.00
2	0.2500	5.500	5.50
3	-0.1250	4.750	4.75
4	0.0625	5.125	5.13
5	-0.0313	4.938	4.94
6	0.0156	5.031	5.03
7	-0.0078	4.984	4.98
8	0.0039	5.008	5.01
9	-0.0020	4.996	5.00
10	0.0010	5.002	5.00

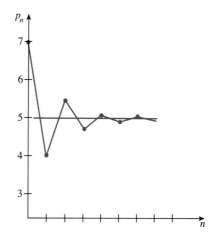

Figure 11.22

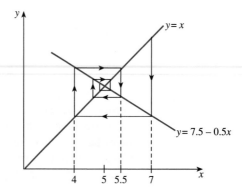

Figure 11.23

Notice that $p_n = 5$ is the constant solution. In a case such as shown in Figure 11.22, for which the solution is alternately above and below the line $p = 5$, we say that the solution **oscillates**. Since the values of p_n approach 5, we still call the line $p = 5$ an attractor.

Figure 11.23 shows the cobweb interpretation.

A similar behavior occurs for the general equation (1) in the case that $-1 < a < 0$.

Qualitative Behavior When $-1 < a < 0$

In the case $-1 < a < 0$, the solutions given by (2) of the dynamical system (1) oscillate alternately above and below the limiting value of $\dfrac{b}{1 - a}$. We can describe this by saying that the solutions are oscillatory with the line $y = \dfrac{b}{1 - a}$ being an attractor. Refer to Figure 11.24.

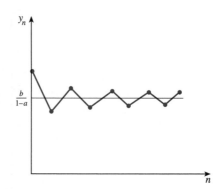

Figure 11.24

The Case $a < -1$

Now consider the case for which $B/A > 1$. Since $a = -B/A$, this is the case for which $a < -1$. To be specific let $A = \frac{1}{3}$, $B = \frac{2}{3}$, and $p_0 = 5.5$. Then equation (3) becomes

$$p_{n+1} = 15 - 2p_n \tag{7}$$

with solution

$$p_n = 5 + 0.50(-2)^n. \tag{8}$$

Table 11.4 indicates what is happening.

Notice that the term $(-2)^n$ is alternately positive and negative and that $|(-2)^n|$ is becoming large without bound. The prices p_n then oscillate about $p = 5$ with $|p_n|$ becoming large without bound. See Figure 11.25.

Table 11.4

n	$(-2)^n$	$p_n = 5 + 0.50(-2)^n$
0	1	5.5
1	-2	4
2	4	7
3	-8	1
4	16	13
5	-32	-11
6	64	37
7	-128	-59
8	256	133

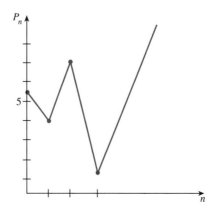

Figure 11.25

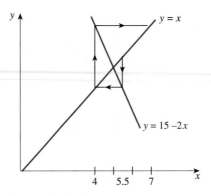

Figure 11.26

Figure 11.26 shows the cobweb interpretation.
The same behavior is true for any equation (1) with $a < -1$.

The Case $a < -1$

If in equation (1) $a < -1$, then the following is true.

1. The solutions oscillate.
2. The solutions are unbounded (except for the constant solution).
3. The constant solution is a repeller.

Refer to Figure 11.27.

The Case $a = -1$

Consider finally the case for which $a = -1$. For example let $A = B = 2.5$ in equation (3). This gives

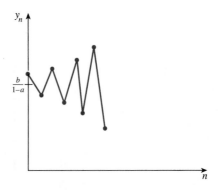

Figure 11.27

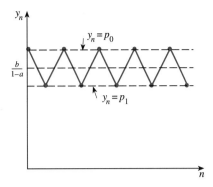

Figure 11.28

$$p_{n+1} = -p_n + 2 \qquad (9)$$

with solution

$$p_n = 1 + (p_0 - 1)(-1)^n \qquad (10)$$

Notice that $p_1 = 1 + (p_0 - 1)(-1) = 2 - p_0$ and that

$$p_2 = 1 + (p_0 - 1)(-1)^2 = 1 + p_0 - 1 = p_0.$$

Thus the price returns to its original price after every two periods. This is called a **2-cycle**. As usual there is the constant solution

$$p_n = \frac{b}{1-a} = \frac{2}{1 - (-1)} = 1.$$

The 2-cycles then oscillate about the constant solution. See Figure 11.28.

Figure 11.29 gives the cobweb interpretation.

This is easily seen to hold in general for (1). We then have the following.

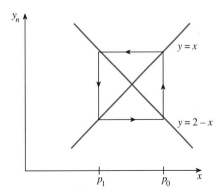

Figure 11.29

> **The Case $a = -1$**
>
> If $a = -1$ in (1), then all solutions are (bounded) 2-cycles and oscillate about the constant solution $y_n = \dfrac{b}{1-a}$.

Recalling the results of the two cases in the last section and the three cases in this section, we have the following.

> **Summary of the Qualitative Behavior**
>
> The solutions of the discrete dynamical system $y_{n+1} = ay_n + b$ have the following properties, which are also indicated in Figure 11.30.
>
> 1.   If $|a| < 1$, then the constant solution is an attractor.
> 2.   If $|a| > 1$, then the constant solution is a repeller.
> 3.   If $a > 0$, then the solutions are monotonic.
> 4.   If $a < 0$, then the solutions are oscillatory.
> 5.   If $a = -1$, then the solutions are (bounded) 2-cycles and oscillate.

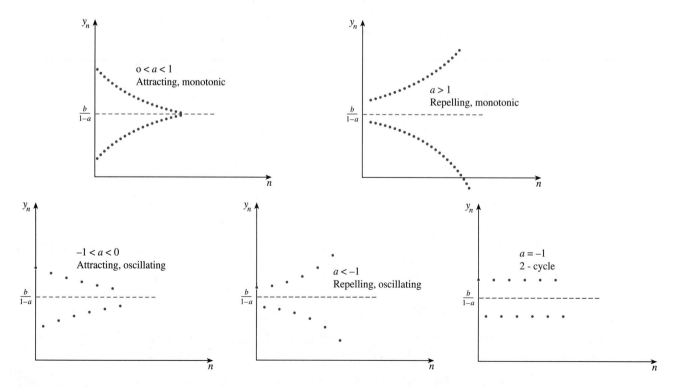

Figure 11.30

SELF-HELP EXERCISE SET 11.4

Graph the solutions of $y_{n+1} = (c - 1)y_n + 30$ if

1. $c = 0.50$ **2.** $c = 0$ **3.** $c = -1$

EXERCISE SET 11.4

Exercises 1 through 10 give discrete dynamical systems. (a) Find the constant solution. (b) Decide if the constant solution is an attractor or repeller or neither. (c) Decide if the nonconstant solutions are monotonic, oscillatory, or 2-cycles. (d) Graph the solutions.

1. $y_{n+1} = 2y_n - 10$ **2.** $y_{n+1} = 1.05y_n + 5$

3. $y_{n+1} = 0.90y_n + 20$ **4.** $y_{n+1} = 0.50y_n - 10$

5. $y_{n+1} = -0.99y_n + 100$ **6.** $y_{n+1} = -.80y_n + 1$

7. $y_{n+1} = -1.001y_n + 100$ **8.** $y_{n+1} = 1.01y_n + 1$

9. $y_{n+1} = -y_n + 100$ **10.** $y_{n+1} = -y_n + 1$

In Exercises 11 through 20 determine the behavior of the dynamical systems graphically by using cobwebs.

11. $y_{n+1} = 3y_n - 6$ **12.** $y_{n+1} = 2y_n - 4$

13. $y_{n+1} = 0.25y_n + 2$ **14.** $y_{n+1} = 0.50y_n + 1$

15. $y_{n+1} = -2y_n + 10$ **16.** $y_{n+1} = -1.5y_n + 6$

17. $y_{n+1} = -0.50y_n + 4$ **18.** $y_{n+1} = -0.75y_n + 8$

19. $y_{n+1} = -y_n + 3$ **20.** $y_{n+1} = -y_n + 1$

Applications

21. Supply and Demand. Set $A = .2$ and $B = .8$ in equation (3) of the text and obtain

$$p_{n+1} = -4p_n + 25.$$

Graph the solutions.

22. Supply and Demand. Set $A = .8$ and $B = .2$ in equation (3) of the text and obtain

$$p_{n+1} = -0.25p_n + 6.25.$$

Graph the solutions.

23. Supply and Demand. Set $A = 2$ and $B = 2$ in equation (3) of the text and obtain

$$p_{n+1} = -p_n + 2.5.$$

Graph the solutions.

24. **Arms Race.**   In the last section we developed a model of an arms race between two countries given by

$$T_{n+1} = (1 + d - e)T_n + c,$$

where T_n is the total expenditures on the military and where d, e, and c are constants, d is positive and measures the mutual distrust of the two countries, and e is a positive constant that measures the restraints of the economy on the arms race. Suppose that distrust is such that $d = e - 2$. Graph the solution and interpret your answer.

25. **Arms Race.**   In the previous exercise suppose distrust is small so that $-1 < 1 + d - e < 0$. Graph the solution and interpret your answer.

26. **Arms Race.**   Richardson studied the arms race prior to World War I between the two alliances consisting of France and Russia against Germany and Austria-Hungary. Using available data he estimated that the total expenditures in millions of pounds sterling T_n of the two alliances satisfied

$$T_{n+1} = \frac{5}{3} T_n - \frac{38}{3}.$$

He estimated that $T_0 = 199$. What should happen?

Solutions to Self-Help Exercise Set 11.4

1. In this case we have $y_{n+1} = -0.50y_n + 30$. The constant solution is $y_n = \dfrac{b}{1 - a} = 20$. Since $a = -0.50 < 0$, the nonconstant solutions oscillate. Since $|a| = |-0.50| = 0.50 < 1$, the constant solution is an attractor.

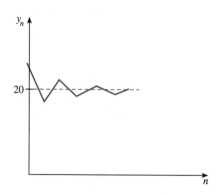

2. In this case we have $y_{n+1} = -y_n + 30$. The constant solution is $y_n = \dfrac{b}{1 - a} = 15$. Since $a = -1$, the nonconstant solutions oscillate and are 2-cycles.

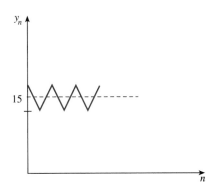

3. In this case we have $y_{n+1} = -2y_n + 30$. The constant solution is $y_n = \dfrac{b}{1-a} = 10$. Since $a = -2 < 0$, the nonconstant solutions oscillate. Since $|a| = |-2| = 2 > 1$, the constant solution is a repeller.

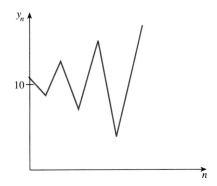

11.5 NONLINEAR DYNAMICAL SYSTEMS

▶ *The Logistic Equation*

▶ *Stability*

▶ *Management of Renewable Natural Resources*

APPLICATION

Population

Estimate the future population of the world. See Example 2 for an estimate.

The Logistic Equation

Earlier in this chapter we noted that if n is the number of years from 1990 and if the human population P_n of the earth in billions continues to grow at 1.7% a year, then P_n satisfies the following dynamical system

$$P_{n+1} = (1 + k)P_n \tag{1}$$

where $k = 0.017$. Since the population of the world in 1990 was 5.333 billion, $P_0 = 5.333$ and the solution of (1) is given by

$$P_n = 5.333(1.017)^n. \tag{2}$$

But this implies that the population will be unbounded! In fact, using this model, the following table gives the population in trillions.

| Date | 2390 | 2490 | 2590 | 2690 | 2790 |
n	400	500	600	700	800
P_n	4.5	24	132	710	3834

At this rate the population in the year 2790 will be 3834 trillion. But the surface of the earth has fewer than 2000 trillion square feet with 80% of this water! Clearly we have come to an absurd conclusion.

Biologists have noted that populations tend to grow exponentially when the population is small, but then level off in time due to the finite amount of available resources. If the population is larger than the available resources, then the population has been observed to decrease to some limiting value. See Figure 11.31 for a typical graph of a population in the long-term. The Dutch mathematical-biologist P. Verhulst in 1838 replaced the growth model (1) with a more realistic model that incorporates this finite resource factor.

It is helpful to consider the annual change in the population given by $P_{n+1} - P_n$. Subtracting P_n from both sides of equation (1) then gives

$$P_{n+1} - P_n = kP_n. \tag{3}$$

If the constant k in equation (3) is *positive*, we call k the **natural growth rate**. We will only consider populations with positive natural growth rates. Notice that equation (3) says that the increase in the population will become larger as the population becomes larger. This is reasonable when the population is small. But as the population becomes large enough to be affected by the limited available

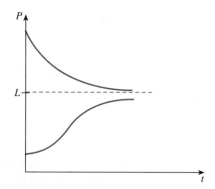

Figure 11.31

resources, this increase should slow. Thus the equation should be modified to reflect this fact.

Verhulst suggested subtracting a positive term from the right-hand side of equation (3) that would become larger if the population became larger. The simplest such term is cP_n^2 where c is a positive constant. Doing this gives

$$P_{n+1} - P_n = kP_n - cP_n^2. \tag{4}$$

The term c is assumed very small compared to k. In fact for the population of the earth in billions $c \approx .002$. Thus if the population is less than a billion or two, the term cP_n^2 in equation (4) is much smaller than the term kP_n. In this case the right-hand side of equation (4) is approximately kP_n, which just gives equation (1) or (3) again. However, as the population of P_n increases, the term cP_n^2 becomes larger and substantially slows the change in population.

It is helpful to factor the right-hand side of equation (4), obtaining

$$P_{n+1} - P_n = kP_n \left(1 - \frac{c}{k} P_n \right).$$

It is furthermore useful to let $L = k/c$ and finally obtain

$$P_{n+1} - P_n = k \left(1 - \frac{P_n}{L} \right) P_n. \tag{5}$$

This equation is called the **logistic equation**. From this equation we can see that if $P_n < L$, then the right-hand side is *positive* and the population increases. Furthermore as the population increases, the term in front of P_n in equation (5) becomes smaller, that is, the growth rate slows. Thus it is instructive to view the term

$$r_n = k \left(1 - \frac{P_n}{L} \right)$$

as a changing growth rate. As P_n increases toward L, this growth rate r_n slows to zero. Should P_n ever exceed L, then this growth rate becomes *negative* and the population decreases. This all coincides with the observations made in Figure 11.31.

Also notice that there are *two* constant solutions of equation (5). The first is just $P_n = 0$. The second is $P_n = L$. The constant solutions are also referred to as **equilibrium points** or **fixed points**. It is very important to determine if these constant solutions are attractors or repellers.

When we studied *linear* discrete dynamical systems in previous sections of this chapter, we were able to find an explicit formula that gave a solution. This enabled us to calculate any value y_n without calculating all the previous values. One very unsettling aspect of the *nonlinear* discrete dynamical system (5) is that we cannot find a simple algebraic expression for y_n in terms of n. In order to find some value P_n for the nonlinear dynamical system (5), we will unfortunately need to calculate all the previous values. This makes the determination of the qualitative behavior substantially more difficult. In fact, at this time, a great deal of knowledge is lacking concerning the behavior of the dynamical system (5), and research continues to fill in the gaps.

For computational purposes, it is much more convenient to place equation (5) in a different form. We do this now.

$$P_{n+1} - P_n = kP_n - \frac{k}{L} P_n^2$$

$$
\begin{aligned}
P_{n+1} &= (1 + k)P_n - \frac{k}{L} P_n^2 \\
&= P_n \left((1 + k) - \frac{k}{L} P_n \right) \\
&= (1 + k)P_n \left(1 - \frac{k}{L(k + 1)} P_n \right) \\
&= aP_n(1 - bP_n),
\end{aligned}
$$

where $a = k + 1$ and $b = \dfrac{k}{L(k + 1)}$. In the next example we will show why this form is so convenient for computational purposes.

E X A M P L E 1   **Solutions of a Logistic Equation**

Determine the solutions P_n to the logistic equation (5) when $k = 0.25$ and $L = 10$ in the two cases for which $P_0 = 2$ and $P_0 = 15$. If P_n represents a population in thousands after n years, interpret the results.

Solution   Since $a = k + 1 = .25 + 1 = 1.25$ and

$$b = \frac{k}{L(k + 1)} = \frac{.25}{10(1.25)} = .02,$$

the logistic equation (5) can be written as

$$P_{n+1} = 1.25P_n(1 - .02P_n).$$

Consider the case $P_0 = 2$. To easily obtain the answers shown in Table 11.5 (even with a cheap calculator) perform the following steps.

- Enter $P_0 = 2$ into your calculator.
- Multiply this by .02.
- Change the sign.
- Add 1.
- Multiply by $P_0 = 2$.
- Multiply by 1.25.
- The answer is $P_1 = 2.4$.
- Write this down.
- Multiply this by .02.
- Change the sign.
- Add 1.
- Multiply by $P_1 = 2.4$.

Table 11.5

n	P_n	$P_n - P_{n-1}$	P_n	n	P_n	$P_n - P_{n-1}$	P_n
0	2.000		15.000	8	6.374	.610	10.301
1	2.400	.400	13.125	9	6.952	.578	10.224
2	2.856	.456	12.100	10	7.482	.530	10.166
3	3.366	.510	11.464	11	7.953	.471	10.124
4	3.924	.558	11.045	12	8.360	.407	10.093
5	4.520	.596	10.756	15	9.213	.228	10.039
6	5.140	.619	10.553	20	9.797	.065	10.009
7	5.764	.624	10.407	25	9.951	.016	10.002

- Multiply by 1.25.
- The answer is $P_2 = 2.856$.
- Write this down.

Continue in this manner to obtain the second column of Table 11.5. This is also graphed in Figure 11.32. Notice P_n increases steadily and approaches 10. Thus the population steadily increases and approaches 10,000.

The third column of Table 11.5 shows the yearly change in P_n. Notice how this change is increasing at first, then is slowing, and after 25 years is nearing zero. During the first 6 or 7 years, the term in equation (5) with P_n^2 is small, and so the population is increasing exponentially. During this period the population is "exploding." However, after this, the finite resource term begins playing an increasing role, and the increase in the population slows.

You can also obtain the solution in the case $P_0 = 15$ by the above procedure, starting this time by entering into your calculator $P_0 = 15$. The results are shown in the fourth column of Table 11.5. Notice that P_n is initially larger than L and steadily decreases and approaches L. Thus if the population is initially larger than

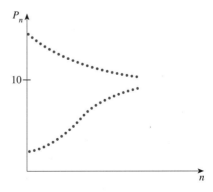

Figure 11.32

the available resources can support, the population will *decrease* toward a limiting value that the resources can support. ∎

The term L in the logistic equation (5) is called the **carrying capacity**. This is the population that the available resources can support. If the population is at L, then the population stays at L. If the population is less than L, then the population increases in the next period. If the population is greater than L, then the population decreases in the next period.

From the above example and analysis and from observation of the growth of populations, we then expect a typical population to initially explode with a fast exponential growth and then slow to some limiting value in time. Let us estimate what this limiting value might be for the population of the earth. Using historical records to estimate the past population of the earth, biologists estimate that the correct value of the growth constant is $k = .03$. We are now prepared to do the following example.

E X A M P L E 2   **Estimating the Future Population of the World**

We noted above that biologists estimate that the correct value of the growth constant in the logistic equation (4) to be $k = 0.03$. The Census Bureau notes that in 1990 the human population of the earth was 5.333 billion and growing at 1.7% per year. Estimate the constant c in the logistic equation and obtain an estimate for the limiting value of the human population of the earth.

Solution   Divide each side of the logistic equation (4) by P_n and obtain

$$\frac{P_{n+1} - P_n}{P_n} = k - cP_n.$$

The left-hand side of this equation is the change in the population during a year divided by the population at the beginning of the year. This is the fractional change in the annual population and, as noted above, the Census Bureau estimated this to be 0.017 in 1990, at which time the total population was 5.333 billion. Setting the left-hand side equal to 0.017 and recalling that $k = .03$, we obtain

$$0.017 = \frac{P_{n+1} - P_n}{P_n} = k - cP_n = 0.03 - c(5.333).$$

This yields

$$c = \frac{.03 - .017}{5.333} = .00244.$$

The limiting value of the population of the earth per this model is then

$$L = \frac{k}{c} = \frac{0.03}{.00244} = 12.3,$$

or about 12 billion. ∎

Stability

We now consider the question of determining the **stability** of the constant solutions, that is, whether the constant solutions are attractors or repellers. We now state the following fundamental result, a proof of which is not given.

Stability of the Constant Solutions

The constant solution $P_n = L$ of the logistic equation (5) is an attractor if $0 < k < 2$ and a repeller if $k > 2$ or if $k < 0$.

    The constant solution $P_n = 0$ is a repeller if $k > 0$ or if $k < -2$ and an attractor if $-2 < k < 0$.

In this section we only consider cases for which $0 < k < 2$. Such moderate growth constants include virtually all cases other than some insect populations and certain small mammals. We will consider these latter cases and their environmental importance in the next section. Then in this section we only consider cases for which $0 < k < 2$. Thus the constant solution $P_n = L$ is an attractor, and the other constant solution $P_n = 0$ is a repeller, and the graph looks like the one shown in Figure 11.31.

    One can also use a carefully drawn cobweb diagram to determine stability.

E X A M P L E   3   **Cobweb Diagram**

Use a cobweb diagram to show that $L = 10$ in the previous example is an attractor.

Solution   Figure 11.33 gives the diagram. ■

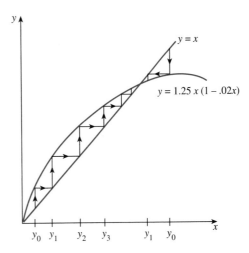

Figure 11.33

Management of Renewable Natural Resources

We now give an important application to fisheries, although the ideas presented here can also be applied to other renewable natural resources. To avoid confusion we emphasize that we are interested in the maximization of productive resources rather than the preservation of natural environments.

A fishery can consist of all or part of an ocean, a lake, a river, a natural or artificial pond, or even an area of water that has been fenced in. To fix a particular situation in our minds, we assume that a corporation has the sole rights to the fish in a fairly small lake. Of course the corporation wishes to harvest the maximum number of fish with the minimum effort and cost. We assume that the corporation does no active management of the fish other than simply harvesting.

Let P_n represent the total number of fish of some species at the beginning of the nth year and assume that P_n satisfies the logistic equation (5).

For convenience we always harvest at the end of the year. We are interested in the *increase* in population since it is this increase that we wish to harvest. We would like to predict how much of the resource we can harvest and still allow the resource to replenish itself. The largest possible increase is called the **maximum sustainable yield**. We want to know what the number of fish should be to obtain the maximum sustainable yield. Thus we wish to find the value of P_n that makes the right-hand side of equation (5) the largest.

In order to determine this, replace P_n with x in equation (5) and let y denote the increase in the population $P_{n+1} - P_n$. Thus

$$y = k \left(1 - \frac{x}{L} \right) x.$$

Clearly $y = 0$ when $x = 0$ or $x = L$. Notice also that this is just a quadratic equation since

$$y = kx - \frac{k}{L} x^2.$$

The graph of this quadratic is shown in Figure 11.34 where we see that the

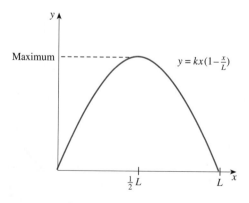

Figure 11.34

maximum value of y occurs when $x = \frac{1}{2}L$. Thus we see that the maximum annual increase in the population will occur if the number at the beginning of the year is one-half the carrying capacity L.

Maximum Sustainable Yield

If the population P_n satisfies the logistic equation (5), where L is the carrying capacity, then the maximum sustainable yield (MSY) occurs when P_n is $\frac{1}{2}L$.

E X A M P L E   4   **Maximum Sustainable Yield**

Whaling scientists have estimated that the Antarctic blue whale has a natural growth rate of $k = .03$ and a carrying capacity of 150,000. What is the population that gives the maximum sustainable yield (MSY), and what is MSY?

Solution   Since the carrying capacity is $L = 150,000$, the population that gives the maximum sustainable yield is $\frac{1}{2}L = 75,000$.

Since $k = .03$ and $L = 150,000$, the logistic equation is

$$P_{n+1} - P_n = .03 \left(1 - \frac{P_n}{150,000} \right) P_n.$$

Evaluating the right-hand side of the last expression when $P_n = 75,000$ gives

$$\text{MSY} = .03 \left(1 - \frac{75,000}{150,000} \right) 75,000 = 1125.$$

The maximum sustainable yield is then 1125 whales.   ∎

In the 1960s the Antarctic blue whale was hunted almost to extinction. Under the protection of the International Whaling Commission, the population increased to about 10,000 in ten years.

As Table 11.6 indicates, the fishing industry plays an important role in the feeding of the world's population, with about 200 billion pounds of fish harvested each year. To sustain such harvests the international community will need to manage this important renewable resource wisely, especially in light of the fact that human greed and ignorance has almost always led to overexploitation.

The Coastal Shark Fishery

Rising demand for shark fins in Asia and for shark meat in the United States has caused a sharp rise in the killing of sharks over the last decade, with the result that large coastal sharks are being killed in numbers that outstrip their capacity to reproduce. As a result the National Marine Fisheries Service of the Federal Government has imposed controls on shark fishing within 200 miles of the Atlantic and Gulf coasts by issuing regulations that will place a total quota of 2436 metric tons, dressed weight, on the catch of large coastal sharks by commercial fishermen in 1993 and 2570 metric tons in 1994. Twenty-two shark species are in this management category, including the tiger, lemon, hammerhead, bull, and great white. The total catch of the large coastal sharks in the United States waters has substantially exceeded 4000 metric tons in each of the last five years, according to the fisheries service. The service has calculated that the maximum sustainable yield of these sharks is 3800 metric tons a year and that the 1993 and 1994 quotas will enable that level to be achieved in 1996.

Table 11.6

Fish Catches by Region, 1988
(in metric tons, live weight catches)

Region	Total Catch
Africa	5,310,300
Asia	43,601,200
Europe	12,877,400
North America	9,567,900
Oceania	887,300
South America	14,412,200
(Former) USSR	11,332,100
World Total	**97,985,400**

SELF-HELP EXERCISE SET 11.5

1. We decided that the logistic equation

$$P_{n+1} = 1.03P_n - .00244P_n^2,$$

modeled the population of the earth in billions, where the population P_0 in 1990 was 5.333 billion. Use this equation to determine the population in billions to five decimal places for each year up to the year 2002. Determine the increase in population each year. When does the yearly increase in the population stop increasing? Determine the percentage increase in population for each year. What is happening to this growth rate?

2. Determine the stability of the two constant solutions of Example 1 in the text.

EXERCISE SET 11.5

Exercises 1 through 6 give logistic equations. First determine the limiting value of P_n. Then take initial populations of one-fifth of this limiting value and also 50% in excess of the limiting value and create a table like Table 11.5 and graph the solutions.

1. $P_{n+1} = 1.25P_n(1 - .04P_n)$

2. $P_{n+1} = 1.5P_n(1 - .04P_n)$

3. $P_{n+1} = \frac{4}{3}P_n(1 - .025P_n)$

4. $P_{n+1} = 2P_n(1 - .05P_n)$

5. $P_{n+1} = \frac{10}{9}P_n(1 - .02P_n)$

6. $P_{n+1} = 2.9P_n(1 - .05P_n)$

7. Find the constant solutions for the dynamical systems given in Exercises 1, 3, and 5 and determine the stability of each.

8. Find the constant solutions for the dynamical systems given in Exercises 2, 4, and 6 and determine the stability of each.

Applications

9. **Limiting Population of the United States.** Estimate the limiting population of the United States if the growth constant $k = .031$ in the logistic equation. The population in 1990 was 250 million and growing at 1% a year.

10. **Limiting Population of Nigeria.** Estimate the limiting population of Nigeria if the growth constant $k = .03$ in the logistic equation. The population in 1990 was 113 million and growing at 3.0% a year.

11. **Limiting Population of India.** Estimate the limiting population of India if the growth constant $k = .03$ in the logistic equation. The population in 1990 was 853 million and growing at 2.0% a year.

12. **Limiting Population of Japan.** Estimate the limiting population of Japan if the growth constant $k = .03$ in the logistic equation. The population in 1990 was 123 million and growing at 0.5% a year.

13. **Population Model of United States.** Using the figures given in Exercise 9, write a logistic equation that models the population of the United States. Find the population and the annual growth rate for each of the next 10 years.

14. **Population Model of Nigeria.** Using the figures given in Exercise 10, write a logistic equation that models the population of Nigeria. Find the population and the annual growth rate for each of the next 10 years.

15. **Population Model of India.** Using the figures given in Exercise 11, write a logistic equation that models the population of India. Find the population and the annual growth rate for each of the next 10 years.

16. **Population Model of Japan.** Using the figures given in Exercise 12, write a logistic equation that models the population of Japan. Find the population and the annual growth rate for each of the next 10 years.

17. **Harvesting.** Suppose the population in millions of a fish species satisfies the logistic equation

$$P_{n+1} = 1.8P_n - 0.8P_n^2 - H,$$

where the constant H represents an annual harvest of fish in millions. If there are initially one million fish, calculate the terms P_n and determine what happens with this harvesting policy when (a) $H = .1$ (b) $H = .3$.

18. **Harvesting.** Repeat the previous exercise for (a) $H = .15$ (b) $H = .22$.

19. **Profits.** The profits from the introduction of a new product often explode at

first and then level off and thus can sometimes be modeled by a logistic equation. Suppose the annual profits P_n in millions of dollars n years after the introduction of a certain new product satisfies

$$P_{n+1} = 1.25P_n - .05P_n^2,$$

where $P_0 = 1$. Determine P_n over the next number of years, graph $P = P_n$, and determine what is happening.

20. **Spread of Technological Innovation.**   When a new technology is introduced, its growth often explodes at first and then levels off when adopted by most firms. Plot the following data on the sales of minicomputers in the U.S. and see if the graph looks like the graph of a logistic curve.

Year	1965	1970	1975	1980	1985	1990
Units in millions	0.6	6.1	27.0	95.9	153.8	157.3

21. **Spread of Technological Innovations.**   Suppose an innovation has been introduced into a fixed number of firms and that the percentage of firms that adopt this innovation after n years is given by P_n. Economists[1] have used the logistic equation to model the spread of this innovation throughout the industry. Suppose initially 10% of the firms are using the innovation and that P_n satisfies the logistic equation

$$P_{n+1} = 1.20P_n - 0.20P_n^2.$$

Find P_n over a number of years. Graph $P = P_n$. What seems to be happening?

22. **Spread of AIDS.**   The number of individuals obtaining an infectious disease typically explodes and then levels off and thus often can be modeled by a logistic equation. The following table gives the reported cases of AIDS in the United States in recent years. Graph this data and determine if the resulting curve looks like a logistic equation.

Year	1983	1984	1985	1986	1987	1988	1989	1990	1991
Cases	2117	4445	8249	13,166	21,070	31,001	33,722	41,595	43,672

23. **Spread of an Infectious Disease.**   The number of individuals obtaining an infectious disease typically explodes and then levels off and thus often can be modeled by a logistic equation. Suppose in an isolated community of 1000

---

[1] E. Mansfield, ''Technical Change and the Rate of Imitation,'' *Econometrica*, vol. 29, no. 4 (Oct. 1961), pp. 741–766.

people one person has an infectious disease. Let P_n be the number of infected people after n days and suppose P_n satisfies

$$P_{n+1} = 2P_n - 0.001P_n^2.$$

Find P_n for the next several weeks and graph. What is happening?

24. **Forest Management.** The following table gives the net stumpage value of a stand of British Columbia Douglas fir trees. Graph and notice how the graph looks like a logistic curve.

Age	30	40	50	60	70	80	90	100	110	120
Value	0	43	143	303	497	650	805	913	1000	1075

Source: P. Pearse, "The Optimal Forest Rotation," *Forestry Chronicle*, vol. 43 (1967), pp. 178–195.

25. **Antarctic Fin Whale.** Whaling scientists estimate the carrying capacity for the Antarctic fin whale to be 200,000 and the growth rate to be $k = .08$. Assuming the population satisfies the logistic equation (5), determine the population that gives the maximum sustainable yield (MSY) and find MSY.

26. **Pacific Halibut.** The carrying capacity of the Pacific halibut in "area 2" in the classification of the Pacific Halibut Commission is estimated to be about 80 million kilograms. Assuming the population satisfies the logistic equation (5), determine the kilograms that give the maximum sustainable yield (MSY) and find MSY if the growth constant $k = .71$.

27. **Eastern Pacific Yellowfin Tuna.** The carrying capacity of the Eastern Pacific yellowfin tuna is estimated to be about 130 million kilograms. Assuming the population satisfies the logistic equation (5), determine the kilograms that give the maximum sustainable yield (MSY) and find MSY if the growth constant $k = 2.61$.

28. **Maximum Sustainable Yield.** If a population satisfies the logistic equation (5) with growth constant k and carrying capacity L, show that the maximum sustainable yield (MSY) is

$$\text{MSY} = \frac{1}{4} kL.$$

Solutions to Self-Help Exercise Set 11.5

1. Table 11.7 summarizes the calculations.

    Notice that each year the annual growth rate as a percentage decreases. Also notice that the yearly increase in population itself increases until the year 2001 when the yearly increase in population begins to decrease.

2. For this equation $k = .25$. Thus $0 < k < 2$ and therefore the constant solution $P_n = 0$ is a repeller and the constant solution $P_n = 10$ is an attractor.

Table 11.7

n	P_n	$P_n - P_{n-1}$	*Percent Increase*	n	P_n	$P_n - P_{n-1}$	*Percent Increase*
0	5.33300			7	5.97292	.09204	1.57
1	5.42359	.09059	1.70	8	6.06507	.09214	1.54
2	5.51453	.09093	1.68	9	6.15725	.09220	1.52
3	5.60576	.09124	1.65	10	6.24947	.09221	1.50
4	5.69726	.09150	1.63	11	6.34165	.09219	1.48
5	5.78898	.09172	1.61	12	6.43378	.09212	1.45
6	5.88088	.09190	1.59				

11.6   NONLINEAR DYNAMICAL SYSTEMS WITH CYCLES

- ▶ *Introduction*
- ▶ *Large Growth Rates*
- ▶ *Chaos*
- ▶ *Finding* k-*Cycles Algebraically*

R. M. May, biologist   " *The world would be a better place if every young student were given a pocket calculator and encouraged to play with the logistic equation.* "

Introduction

We continue to study the logistic equation

$$P_{n+1} = P_n + k \left(1 - \frac{P_n}{L} \right) P_n \tag{1}$$

In the last section we studied this equation when the growth constant was not too "large." We will see in this section that the behavior of the logistic equation (1) changes dramatically when the growth constant k becomes larger than 2. We will give some indication of what implications this has in dealing with complex practical problems.

Recall that the logistic equation (1) has two constant solutions, $P_n = 0$ and $P_n = L$. We can choose the size of the units that measure population to be in thousands, millions, or anything we wish. For computational purposes it is helpful to choose the units so that $L = \dfrac{k}{k + 1}$ in equation (1). Equation (1) then becomes

$$P_{n+1} = (k + 1)P_n - (k + 1)P_n^2.$$

Setting $r = k + 1$ then gives

$$P_{n+1} = rP_n(1 - P_n) \tag{2}$$

This has constant solutions $P_n = 0$ and $P_n = \dfrac{r-1}{r}$.

We now consider the stability of the constant solutions $P_n = 0$ and $P_n = \dfrac{r-1}{r}$. Using the stability result of the last section, we then have the following.

Stability of the Constant Solutions

For the equation (2) the constant solution $P_n = 0$ is a repeller if $|r| > 1$ and an attractor if $|r| < 1$.

If $1 < r < 3$, then the constant solution $P_n = \dfrac{r-1}{r}$ is an attractor.

If $r < 1$ or $r > 3$, then the constant solution $P_n = \dfrac{r-1}{r}$ is a repeller.

When $r > 1$, $k = r - 1 > 0$. This means the population is growing. Thus it is readily understood why the constant solution $P_n = 0$ should be a repeller. We are all familiar with human populations and animal populations growing at rates for which the growth constant $k = r - 1 \le 2$. For example, $k = 2$ ($r = 3$) means the population triples every time period. There are insect populations that do have growth constants that exceed $k = 2$ ($r = 3$), and many of these insects play critical roles in the environment. We now wish to see what happens in this case. Biologists have long noticed that the populations of certain insects seem to rise and fall in cycles that are often regular.

Large Growth Rates

When the biologist R. M. May first studied the logistic equation (2), he was very surprised to find that the solutions behaved very differently when r was greater than 3. He carried out a program of intense numerical exploration into the behavior of the solutions to this seemingly simple nonlinear equation. He was surprised at the complexity of the behavior of the solutions and naturally began to think what implications this had in ecology.

Recall that when $r > 3$, the constant solution $P_n = L = \dfrac{r-1}{r}$ of equation (2) is a repeller. The following indicates what happens in one such case.

E X A M P L E   1   **Behavior When $r = 3.3$**

Determine the behavior of the logistic equation (2) when $r = 3.3$.

Solution   With $r = 3.3$ the equation (2) becomes

$$P_{n+1} = 3.3P_n(1 - P_n)$$

We take the initial value $P_0 = 0.4$. Table 11.8, Figure 11.35 and also the cobweb diagram in Figure 11.36 indicate what is happening. Notice how P_n oscillates about the line $P = \dfrac{r - 1}{r} \approx 0.697$ and seems to approach a 2-cycle. We see in fact that P_n approaches 0.824 (to three decimal places) for odd values of n and approaches 0.479 for even values of n. Thus, after about 10 periods, P_n moves back and forth between approximately 0.824 and 0.479.   ■

Table 11.8

n	0	1	2	3	4	5	6	7	8	9	10	11	12	13
P_n	.400	.792	.544	.819	.490	.825	.477	.823	.480	.824	.479	.824	.479	.824

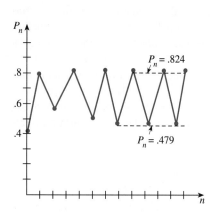

Figure 11.35

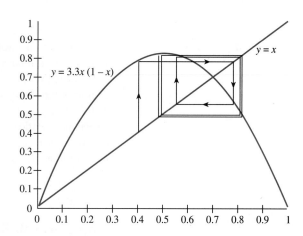

Figure 11.36

The following result is known and indicates when a 2-cycle is an attractor and when it is a repeller. A solution of a dynamical system is a 2-cycle if the solution alternately equals two numbers p_1 and p_2. Thus the values of the 2-cycle are

$$p_1, \quad p_2, \quad p_1, \quad p_2, \quad p_1, \quad p_2, \quad \cdots$$

Stability of a 2-Cycle

Suppose the dynamical system

$$y_{n+1} = ay_n - cy_n^2$$

has a 2-cycle, that is, a solution that alternately equals the two numbers p_1 and p_2. Then this 2-cycle is

1. An attractor if $|(a - 2cp_1)(a - 2cp_2)| < 1$,
2. A repeller if $|(a - 2cp_1)(a - 2cp_2)| > 1$.

A corresponding result exists for any k-cycle.

E X A M P L E 2 **Determining the Stability of a 2-Cycle**

Determine the stability of the 2-cycle found in Example 1.

Solution Let $a = 3.3$ and $c = 3.3$, then since

$$|(3.3 - 6.6(0.479))(3.3 - 6.6(.824))| \approx 0.296 < 1,$$

this 2-cycle is an attractor. ∎

It can be shown (but we do not do so here) that if $r > 3$ but less than $1 + \sqrt{6} \approx 3.449$, then the dynamical system (2) has a 2-cycle that is an attractor. For values of r slightly larger than $1 + \sqrt{6} \approx 3.449$, the solutions approach an attracting 4-cycle, that is, a solution that repeats itself every 4 time periods. The following example illustrates this.

E X A M P L E 3 **Behavior When $r = 3.5$**

Determine the behavior of the logistic equation (2) when $r = 3.5$.

Solution With $r = 3.5$ the equation (2) becomes

$$P_{n+1} = 3.5P_n(1 - P_n)$$

We take the initial value $P_0 = 0.4$. Table 11.9, the graph in Figure 11.37, and also the cobweb diagram in Figure 11.38 indicate what is happening. Notice that P_n approaches a 4-cycle .875, .383, .827, .501 (to 3 decimal places). (If Table 11.9 were continued, it would merely show the four numbers .875, .383, .827, .501 repeating themselves.) ∎

Table 11.9

n	0	1	2	3	4	5	6	7	8	9	10	11	12	13	14
P_n	.400	.840	.470	.872	.391	.833	.486	.874	.385	.828	.498	.875	.383	.827	.501

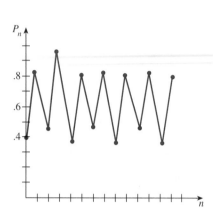

Figure 11.37

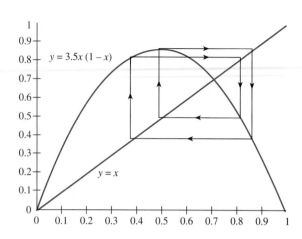

Figure 11.38

It can be shown that when r is between the two numbers 3.449 and 3.544 (rounded to three decimal places), there always exists a 4-cycle and the 4-cycle is attracting.

We also call a 2-cycle a **periodic solution with period 2** and a 4-cycle a **periodic solution of period 4** and so forth. Thus when the parameter r passes through the number 3.449 the solutions are said to **bifurcate** into period-doubling solutions.

The story does not end here. There are numbers $c_1, c_2, c_3, \ldots$, all less than the number 3.57 (to two decimal places) for which

$$c_1 < c_2 < c_3 < \cdots$$

that have the following properties. These numbers approach 3.57 and when $c_n < r < c_{n+1}$, the dynamical system (2) has an attracting 2^n-cycle. In other words, no matter how large a value of n you pick, you can find a number r close and to the left of 3.57 so that the dynamical system has an attracting 2^n-cycle. We have already noted that $c_1 = 3$, $c_2 = 3.449$, and $c_3 = 3.544$.

This model may explain the cyclic fluctuations in populations with large growth rates such as for certain insects and small mammals[1]. Similar behavior has been observed in mechanics, astronomy, meteorology, and chemistry.

If we then imagine an insect population satisfying the logistic equation and the growth rate k and thus $r = k + 1$ increases, then as r passes through 3, the insect population would refuse to settle down, but oscillate between two values in alternate time periods. Now as r passes through the value 3.449, the population settles down into eventually repeating every four years. As r passes through the value 3.544, the cycle would double again and again be stable. Now as r increases, the period-doubling bifurcations appear ever faster.

(© Runk/Schoenberger/Grant Heilman Photography, Inc.)

[1]R. M. May, ''Biological Populations Obeying Difference Equations: Stable Points, Stable Cycles, and Chaos,'' *Journal of Theoretical Biology*, vol. 51 (June, 1975), pp. 511–524.

Chaos

We now consider what happens in equation (2) when $r > 3.57$. Take the case that $r = 3.7$. this gives

$$P_{n+1} = 3.7P_n(1 - P_n) \qquad (3)$$

Although the solutions do not seem to have any pattern, it is true that all solutions eventually fall into a certain group of several intervals. Also it is known that there is a repelling 2-cycle since $r = 3.7 > 3.45$. It is also known that there is a repelling 4-cycle since $r = 3.7 > 3.54$. In fact, for every value of n there is a repelling 2^n-cycle, and these cycles tend to fill up the several intervals. As a consequence, two solutions may begin very close to each other but in a modest period of time be far apart and exhibit a very different type of behavior. This is referred to as **chaos**.

Figure 11.39 shows two solutions, one starting at 0.2 and the other at 0.2001. In a relatively short period of time they begin to exhibit very different behavior. In the study of weather, for example (where the equations are much more complex than studied here), this means that a very slight change in conditions at some time can result in a very large change in the weather later, making prediction very difficult.

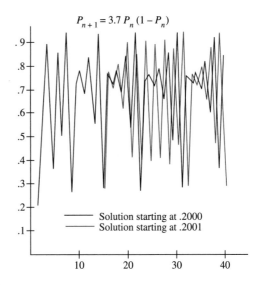

Figure 11.39

Finding k-Cycles Algebraically[2]

Finding k-cycles simply by calculating large numbers of values of P_n and examining the data is not always entirely satisfactory or practical. The following is a useful algebraic way of finding k-cycles.

---

[2]This subsection requires the use of a graphics calculator or a computer.

Algebraic Method of Finding 2-Cycles

Let the numbers p, p_1, and p_2 satisfy

$$p_1 = p(a - cp), \qquad p_2 = p_1(a - cp_1), \qquad p_2 = p.$$

Then p is either a constant solution or part of a 2-cycle for the dynamical system

$$P_{n+1} = P_n(a - cP_n).$$

With p, p_1, and p_2 defined above, we have

$$p = p_2 = p_1(a - cp_1) = p(a - cp)[a - cp(a - cp)].$$

This is a fourth-order polynomial. We then are seeking the solutions of this equation. Notice, however, that the constants $p = 0$ and $p = \dfrac{a - 1}{c}$ are always solutions.

There is a corresponding result for any k-cycle. For example, to find 3-cycles, look for solutions of

$$p_1 = p(a - cp), \qquad p_2 = p_1(a - cp_1), \qquad p_3 = p_2(a - cp_2), \qquad p_3 = p.$$

E X A M P L E 4   **Using a Graphics Calculator to Find k-Cycles**

Find a 3-cycle for the dynamical system

$$P_{n+1} = 3.84P_n(1 - P_n)$$

using the above result and a graphics calculator.

Solution   First notice that $a = c = 3.84$. Set $r = 3.84$. Since

$$p_1 = rp(1 - p), \qquad p_2 = rp_1(1 - p_1), \qquad p_3 = rp_2(1 - p_2), \qquad p_3 = p,$$

we have

$$
\begin{aligned}
p &= p_3 \\
&= rp_2(1 - p_2) \\
&= rrp_1(1 - p_1)(1 - rp_1(1 - p_1)) \\
&= r^2rp(1 - p)(1 - rp(1 - p))(1 - rrp(1 - p)(1 - rp(1 - p))) \\
&= r^3p(1 - p)[1 - rp(1 - p)][r^3p^2(1 - p)^2 - r^2p(1 - p) + 1]
\end{aligned}
$$

where $r = 3.84$. Enter this function and also $y = x$ and graph on the graphics calculator. After several zooms obtain Figure 11.40 on the screen.

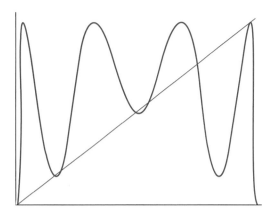

Figure 11.40

Figure 11.41

One clearly sees that the eighth-order polynomial intersects the line $y = x$ at $x = 0$ and $x = p = \dfrac{3.84 - 1}{3.84} \approx .7396$. By using the zoom feature, one can locate the other six places that the graphs cross. For example, using the zoom feature on the smallest of these, we can obtain Figure 11.41. This shows that there are two crossings close together. Further use of the zoom will indicate that these values are .1494 and .1694. Doing the same thing at the other points yields .4880, .5404, .9537, 9594.

Setting $P_0 = .1494$ and using a calculator we find the next three successive values, to four decimal places, to be .4880, .9594, .1494. Thus, the 3-cycle is .1494, .4880, .9594. Continuing to calculate P_n on our calculator indicates that this 3-cycle is stable. We can check this by using a stability result for 3-cycles. Since

$$|3.84(1 - 2[.1494])3.84(1 - 2[.4880])3.84(1 - 2[.9594])| = .88 < 1,$$

we conclude that this 3-cycle is definitely stable.

Setting $P_0 = .1694$ and using a calculator we find the next three successive values, to four decimal places, to be .5403, .9538, .1693. Continuing to calculate P_n on our calculator indicates that this 3-cycle is not stable. This can be verified by using the stability result for 3-cycles, since

$$|3.84(1 - 2[.1693])3.84(1 - 2[.5403])3.84(1 - 2[.9538])| = 2.74 > 1. \quad \blacksquare$$

SELF-HELP EXERCISE SET 11.6

1. Find a 2-cycle of the dynamical system $P_{n+1} = 3.58P_n(1 - P_n)$.

2. Determine the stability of the 2-cycle you found in the previous exercise.

EXERCISE SET 11.6

According to the text the equation (2) has a bifurcation point at $r = 3$, in which for $r < 3$, there are no 2-cycles, but for $3 < r < 3.449$ there are. Explore this in Exercise 1 through 8 by taking r to be the indicated values. Start at .20.

1. 2.9   **2.** 2.95   **3.** 2.98

4. 2.99   **5.** 3.03   **6.** 3.1

7. 3.01   **8.** 3.02

According to the text the equation (2) has a bifurcation point at $r = 3.449$, in which for $3 < r < 3.449$ there are no 4-cycles, but for $3.449 < r < 3.544$ there are. Explore this in Exercises 9 through 16 by taking r to be the indicated values. Start at .20.

9. 3.4   **10.** 3.445   **11.** 3.44

12. 3.447   **13.** 3.52   **14.** 3.451

15. 3.46   **16.** 3.455

17. Determine the two solutions to equation (2) for $r = 3.7$ with $P_0 = .3$ and $P_0 = .301$. See if these solutions that are initially close, stay close.

18. Determine the two solutions to equation (2) for $r = 3.65$ with $P_0 = .3$ and $P_0 = .301$. See if these solutions that are initially close, stay close.

19. In Table 11.8 we found the 2-cycle to three decimal places. Continue the table and determine the 2-cycle to four decimal places.

20. In Table 11.9 we found the 4-cycle to three decimal places. Continue the table and determine the 4-cycle to four decimal places.

21. Find the 2-cycle (to three decimal places) of equation (2) with $r = 3.1$. Show from using the 2-cycle stability theorem that this two cycle is an attractor.

22. Find the 2-cycle (to three decimal places) of equation (2) with $r = 3.3$. Show from using the 2-cycle stability theorem that this two cycle is an attractor.

23. Find the 2-cycle (to three decimal places) of equation (2) with $r = 3.35$. Show from using the 2-cycle stability theorem that this two cycle is an attractor.

Graphing Calculator Exercises

24. Explore the behavior of

$$P_{n+1} = 3.5P_n(1 - P_n).$$

further. Find a 2-cycle and 4-cycle using the algebraic method and a graphics calculator. Find the two constant solutions and show that they are repelling.

According to the discussion in the text if $r > 3.57$, then the dynamical system given by (2) in the text has a 2^k-cycle for any k. Explore this for the dynamical system

$$P_{n+1} = 3.83P_n(1 - P_n)$$

in Exercises 25–26.

25. Using the graphics calculator and the algebraic method, find to three decimal places (a) a 2-cycle, and (b) two 3-cycles.

26. Using the graphics calculator and the algebraic method, find a 4-cycle.

27. Consider the dynamical system $P_{n+1} = 3.6\,P_n(1 - P_n)$. Using the graphics calculator and the algebraic method, show that no 3-cycle exists.

28. Repeat Exercises 25 and 26 for the dynamical system $P_{n+1} = 3.86\,P_n(1 - P_n)$.

Solutions to Self-Help Exercise Set 11.6

1. To do this algebraically, we must find the solutions to

$$p = r^2 p(1 - p)[rp^2 - rp + 1],$$

where $r = 3.58$. Two solutions are the constant solutions $p = 0$ and $p = \dfrac{2.58}{3.58} \approx .7207$. To find the other two, graph on the graphics calculator the two curves $y = 3.58^2 x(1 - x)(3.58^2 x^2 - 3.58x + 1)$ and $y = x$. Using the zoom feature, we can find the other two solutions to be .4120 and .8673 to four decimal places.

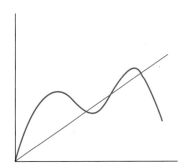

2. To determine the stability, notice that

$$|3.58[1 - 2(.4120)]3.58[1 - 2(.8673)]| = |-1.66| = 1.66.$$

Since this is larger than 1, the 2-cycle is a repeller.

SUMMARY OUTLINE OF CHAPTER 11

- A **linear discrete dynamical system** or **linear difference equation** is an equation of the form $y_{n+1} = ay_n + b$. The term y_0 is called the **initial value**. If $b = 0$, then this equation is called **homogeneous**. If $b \neq 0$, the equation is called **nonhomogeneous**.

- A **solution** to a linear discrete dynamical system is an equation that expresses y_n in terms of the constants n, the initial value y_0, and the constants a and b. A solution of the homogeneous equation $y_{n+1} = ay_n$ is $y_n = a^n y_0$, where y_0 is the initial value.

- The solution of the linear dynamical system $y_{n+1} = ay_n + b$ is given by

$$y_n = \frac{b}{1-a} + \left(y_0 - \frac{b}{1-a}\right)a^n$$

  if $a \neq 1$ and by $y_n = y_0 + bn$ if $a = 1$.

- If $y_n < y_{n+1}$ for every n, we say that y_n is **increasing**.

- If $y_n > y_{n+1}$ for every n, we say that y_n is **decreasing**.

- If y_n is either always increasing or always decreasing, then we say that y_n is **monotonic**.

- A constant solution is said to be an **attractor** if solutions near to this constant solution approach the constant solution for large values of n.

- A constant solution is said to be a **repeller** if solutions near to this constant solution move away from the constant solution.

- A solution y_n is said to **oscillate** if this solution always moves alternately above and below a horizontal line.

- A solution y_n is said to be a **two-cycle** if the solution alternately takes two different values forever.

- **Summary of the Qualitative Behavior.**   The solutions of the discrete dynamical system $y_{n+1} = ay_n + b$ have the following properties.

  1. If $|a| < 1$, then the constant solution is an attractor.
  2. If $|a| > 1$, then the constant solution is a repeller.
  3. If $a > 0$, then the solutions are monotonic.
  4. If $a < 0$, then the solutions are oscillatory.
  5. If $a = -1$, then the solutions are (bounded) 2-cycles and oscillate.

- An equation of the form $P_{n+1} = k\left(1 - \dfrac{P_n}{L}\right)P_n$ is called a **logistic equation**.

- **Stability of the Constant Solutions.**   The solutions $P_n = 0$ or $P_n = L$ are constant solutions for the above logistic equation. The constant solution $P_n = L$ is an attractor if $0 < k < 2$ and a repeller if $k > 2$ or if $k < 0$. The constant solution $P_n = 0$ is a repeller if $k > 0$ or if $k < -2$ and an attractor if $-2 < k < 0$.

- **Stability of the Constant Solutions.**   For the logistic equation in the form $P_{n+1} = rP_n(1 - P_n)$ the constant solution $P_n = 0$ is a repeller if $|r| > 1$ and an attractor if $|r| < 1$. If $1 < r < 3$, then the constant solution $P_n = \dfrac{r-1}{r}$ is an attractor. If $r < 1$ or $r > 3$, then the constant solution $P_n = \dfrac{r-1}{r}$ is a repeller.

- The largest possible increase per time period in a population that can be sustained for all time is called the **maximum sustainable yield**.

- If the population P_n satisfies the logistic equation, where L is the carrying capacity, then the maximum sustainable yield (MSY) occurs when P_n is $\frac{1}{2}L$.

- **Stability of a 2-Cycle.**   Suppose the logistic equation $y_{n+1} = ay_n - cy_n^2$ has a 2-cycle. This 2-cycle is an attractor if $|(a - 2cp_1)(a - 2cp_2)| < 1$, and a repeller if $|(a - 2cp_1)(a - 2cp_2)| > 1$.

- **Algebraic Method of Finding 2-Cycles.** Let the numbers p, p_1, and p_2 satisfy

  $$p_1 = p(a - cp), \qquad p_2 = p_1(a - cp_1), \qquad p_2 = p.$$

  Then p is either a constant solution or part of a 2-cycle for the logistic equation $P_{n+1} = P_n(a - cP_n)$.

Chapter 11 Review Exercises

1. Suppose $1000 is deposited into an account earning interest at an annual rate of 7%. Find the amount in the account at the end of the first year if the compounding is (a) semiannual, (b) quarterly, (c) monthly, (d) weekly, (e) daily.

2. Suppose $1000 is deposited into an account earning interest at an annual rate of 8% compounded monthly. Find the amount in the account at the end of 15 years.

3. How long does it take an account growing at an annual rate of 5% compounded annually to double?

4. What annual rate r when compounded annually is necessary to double an account in 12 years?

5. **Population.** Suppose a population satisfies $P_n = (1 + k)^n P_0$ with $k > 0$. If N is the time it takes for the population to triple in size, show that

$$N = \frac{\log 3}{\log(1 + k)}.$$

6. **Population.** Suppose a population satisfies $P_n = (1 + k)^n P_0$ with $k > 0$. If N is the time it takes for the population to triple in size, show that $1 + k = \sqrt[N]{3}$.

7. **Population.** According to the U.S. Bureau of Statistics, China is growing at a rate that will result in the population doubling in 44 years. Assuming that this population is growing according to the equation $P_{n+1} = (1 + k)P_n$, what is the growth constant k?

8. **Radioactivity.** A radioactive substance has a half-life of 15 days. Find its decay constant.

9. **Age of Bone Fragment.** An anthropologist discovers a bone fragment and learns from laboratory testing that exactly 25% of carbon-14 is missing. How old is the bone?

10. **Appreciation.** Suppose a piece of real estate now worth $100,000 increases in value by about 4% a year. Let V_n be the value in n years. Find a discrete dynamical system that allows you to find V_{n+1} if given V_n. Assuming this trend continues, find the value in 40 years without finding the value for the intermediate years.

11. **Barometric Pressure.** Let p_n be the barometric pressure (in inches of mercury) at an altitude of n miles above sea level. The difference $p_{n+1} - p_n$ in pressure at $n + 1$ miles and n miles is negatively proportional with the proportionality constant .20 to the pressure p_n. Find a discrete dynamical system that allows you to find p_{n+1} if given p_n. Find the pressure at 5 miles above sea level given that the pressure at sea level is 29.9.

12. An account with an initial amount of $1000 pays simple interest at an annual rate of 7%. How much is in the account after 20 years?

13. **Education Fund.**　New parents wish to save for their newborn's education and wish to have $50,000 at the end of 18 years. How much should they place at the end of each year into a savings account that earns an annual rate of 7% compounded annually?

14. **Mortgage.**　Find the monthly payment needed to pay off a $100,000 mortgage over 25 years if interest is 7.5% compounded monthly. Then find the total interest paid.

15. **Comparing Retirement Options.**　You are being offered a "million dollar" retirement package to be given in $50,000 payments at the end of each of the next 20 years. You are also given the option of accepting a $500,000 lump sum payment now. Interest rates are at 9% per year. Which looks better to you? Why?

16. **Population.**　The current population of a city is 200,000. Each year the number of births is 4%, and the number of deaths is 2% of the population at the beginning of the year. In addition, each year 5000 people move into the city and 9000 leave. Let P_n be the population after n years. Find a discrete dynamical system that incorporates the above assumptions and allows you to find P_{n+1} if given P_n. Assuming this trend continues, find the population 20 years later without finding the population of the intermediate years.

Exercises 17 through 21 give discrete dynamical systems. (a) Find the constant solution. (b) Decide if the constant solution is an attractor, repeller, or neither. (c) Decide if the nonconstant solutions are monotonic, oscillatory, or a 2-cycle. (d) Graph the solutions.

17. $y_{n+1} = 2.5y_n - 10$

18. $y_{n+1} = .80y_n + 5$

19. $y_{n+1} = -0.90y_n + 20$

20. $y_{n+1} = -1.1y_n + 10$

21. $y_{n+1} = -y_n + 100$

22. For each of the dynamical systems in the previous five exercises, determine the behavior of the solutions graphically by using cobwebs.

23. Suppose the annual interest rate on a mortgage is 7% compounded monthly. If you can make a monthly payment of $800, what can you borrow?

24. A person wishes to deposit money into an account that pays annual interest of 6% compounded annually and to make a withdrawal of $5000 at the end of every year. How much must be deposited so that the money never runs out?

25. **Population.**　Every month a mouse population increases by 10% of its population at the beginning of the month. Some predators move into the area and begin to remove 100 mice per month. Determine the mouse population at the time the predators move in that will lead to different long-term behavior.

26. **Real Estate.**　Without any upkeep or improvements a certain apartment building will depreciate in value by 5% a year. Suppose the owner makes improvements amounting to $10,000 at the end of every year. Find the initial values of

the building that will result in (a) the value of the property increasing, (b) staying constant, (c) decreasing. Draw a graph of the solutions.

27. **The Pacific Halibut Fishery.** The Pacific Halibut Commission estimates that in "area 2" the ideal biomass of halibut is about 80 million killograms and that the growth rate (under ideal conditions) is $k = .71$. Using these estimates and assuming a linear model, determine the number of kilograms that could be harvested without sending the biomass into a decline.

Exercises 28 through 29 give logistic equations. First determine the limiting value of P_n. Then take initial populations of one-fifth of this limiting value and also 50% in excess of the limiting value and create a table that includes P_n and, in the first instance, the increase in P_n each year.

28. $P_{n+1} = 1.5P_n (1 - \frac{10}{3}P_n)$ 29. $P_{n+1} = 1.25P_n (1 - .025\, P_n)$

30. Determine the stability of the two solutions in the previous two exercises.

31. Estimate the limiting population of China if the growth constant $k = .03$ in the logistic equation. The population in 1990 was 1.135 billion and growing at 1.6% a year.

32. Using the figures given in the previous exercise, write a logistic equation that models the population of China. Find the population and the annual growth rate for each of the next 10 years.

33. **Profits.** Suppose the annual profit P_n in millions of dollars n years after the introduction of a certain new product satisfies

$$P_{n+1} = 1.5P_n - .04P_n^2,$$

where $P_0 = 1$. Determine P_n over the next number of years, graph, and determine what is happening.

34. **Population.** An isolated lake was stocked with 50 lake trout. One year later it is estimated that there are 100 of these trout in this lake. It is also estimated that this lake can support 500 of these trout. Assuming that the trout population satisfies a logistic equation, find how many lake trout will be in this lake in 4 years.

35. **Coastal Sharks.** Biologists estimate the carrying capacity for coastal sharks to be 314,000 metric tons and the growth rate to be $k = .05$. Assuming the population satisfies the logistic equation, determine the population that gives the maximum sustainable yield (MSY) and find MSY.

36. Find the 2-cycle (to three decimal places) of the equation $y_{n+1} = ry_n(1 - y_n)$ with $r = 3.2$. Show from using the 2-cycle stability theorem that this two cycle is an attractor.

37. Find the 2-cycle (to three decimal places) of the equation $y_{n+1} = ry_n(1 - y_n)$ with $r = 3.25$. Show from using the 2-cycle stability theorem that this two-cycle is an attractor.

Consider the dynamical system

$$P_{n+1} = 3.85P_n(1 - P_n)$$

in Exercises 38 and 39.

38. Using the graphics calculator and the algebraic method, find a 2-cycle.

39. Using the graphics calculator and the algebraic method, find two 3-cycles.

40. Using the graphics calculator and the algebraic method, show that the dynamical system $P_{n+1} = 3.7 \, P_n(1 - P_n)$ has no 3-cycles.

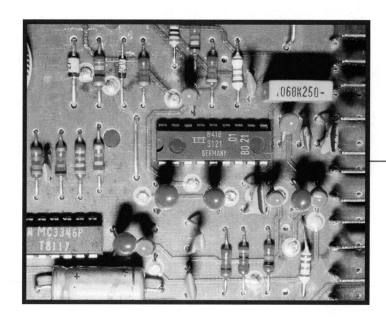

Logic is used in the design of circuit boards. (Bernard Roussel/Image Bank.)

Logic

12.1 INTRODUCTION TO LOGIC

► *Statements*

► *Conjunction, Disjunction, Negation*

**Aristotle
(384–322 BC)**

(Brown Brothers.)

A Very Short History of Logic

The Greek philosopher Aristotle (384–322 B.C.) is generally given credit for the first systematic study of logic. His work, however, used ordinary language. The second great period of logic came with Gottfried Leibnitz (1646–1716), who initiated the use of symbols to simplify complicated logical arguments. This treatment is referred to as **symbolic logic** or **mathematical logic**. In symbolic logic, symbols and prescribed rules are used very much as in ordinary algebra. This frees the subject from the ambiguities of ordinary

Gottfried Leibnitz
(1646–1716)

(The Bettmann Archive.)

language and permits the subject to proceed and develop in a methodical way. It was, however, Augustus De Morgan (1806–1871) and George Boole (1815–1864) who systematically developed symbolic logic. The "algebra" of logic that they developed removed logic from philosophy and attached it to mathematics.

Statements

Logic is the science of correct reasoning and of making valid conclusions. In logic conclusions must be inescapable. Every concept must be clearly defined. Thus dictionary definitions are often not sufficient since there can be no ambiguities or vagueness.

We restrict our study to declarative sentences that are unambiguous and that can be classified as true or false but not both. Such declarative sentences are called **statements** and form the basis of logic.

Statements

A **statement** is a declarative sentence that is either true or false but not both.

Thus commands, questions, exclamations, and ambiguous sentences cannot be statements.

E X A M P L E   1   **Determining if Sentences are Statements**

Decide which of the following sentences are statements and which are not.

(a)   Look at me.
(b)   Do you enjoy music?
(c)   What a beautiful sunset!
(d)   Two plus two equals four.
(e)   Two plus two equals five.
(f)   The author got out of bed after 6:00 A.M. today.
(g)   That was a great game.
(h) $x + 2 = 5$.

Solutions   The first three sentences are not statements since the first is a command, the second is a question, and the third is an exclamation. Sentences (d) and (e) are statements; (d) is a true statement while (e) is a false statement. Sentence (f) is a statement, but you do not know if it is true or not. Sentence (g) is not a statement since we are not told what "great" means. The sentence (h) is not a statement since it cannot be classified as true or false. For example, if $x = 3$ it is true. But if $x = 2$ it is false. ∎

Conjunction, Disjunction, Negation

A statement such as "I have money in my pocket" is called a **simple** statement since it expresses a single thought. But we need to also deal with **compound** statements such as "I have money in my pocket and my gas tank is full." We will let letters such as p, q, and r denote simple statements. To write compound statements, we need to introduce symbols for the **connectives**.

Connectives

A **connective** is a word or words, such as "and" or "if and only if," that is used to combine two or more simple statements into a compound statement.

We will consider the 5 connectives given in the following table. We will discuss the first 3 in this section and the last 2 in the next section.

Name	Connective	Symbol
Conjunction	and	$\wedge$
Disjunction	or	$\vee$
Negation	not	$\sim$
Conditional	if . . . then	$\rightarrow$
Biconditional	if and only if	$\leftrightarrow$

Logic does not concern itself with whether a simple statement is true or false. But if all the simple statements that make up a compound statement are known to be true or false, then the rules of logic will enable us to determine if the compound statement is true or false. We will do this in the next section.

We now carefully give the definitions of the three connectives "and," "or," and "not." Notice that the precise meanings of the three compound statements that involve these connectives are incomplete unless a clear statement is made as to when the compound statement is true and when it is false.

Conjunction

A **conjunction** is a statement of the form "p and q" and is written symbolically as

$$p \wedge q.$$

The conjunction $p \wedge q$ is true if both p and q are true, otherwise it is false.

E X A M P L E   1   **Writing Compound Statements in Symbolic Form**

Solution

Write the compound statement "I have money in my pocket and my gas tank is full" in symbolic form.

First let p be the statement "I have money in my pocket" and q be the statement "my gas tank is full." Since $\wedge$ represents the word "and," the compound statement can be written symbolically as $p \wedge q$.   ■

Disjunction

A **disjunction** is a statement of the form "p or q" and is written symbolically as

$$p \vee q.$$

The disjunction $p \vee q$ is false if both p and q are false and is true in all other cases.

R E M A R K .   The word "or" in this definition conveys the meaning "one or the other, or both." This is also called the **inclusive or**.

E X A M P L E   2   **Writing Compound Statements in Symbolic Form**

Solution

Write the compound statement "Janet is in the top 10% of her class or she lives on campus" in symbolic form.

First let p be the statement "Janet is in the top 10% of her class" and q the statement "She lives on campus." Since $\vee$ represents the word "or," the compound statement can be written as $p \vee q$.   ■

R E M A R K .   In everyday language the word "or" is not always used in the way indicated above. For example, if a car salesman tells you that for $10,000 you can have a new car with automatic transmission or a new car with air conditioning, he means "one or the other, but not both." This use of the word "or" is called **exclusive or**.

The last connective that we consider in this section is "not."

Negation

A **negation** is a statement of the form "not p" and is written symbolically as

$$\sim p.$$

The negation $\sim p$ is true if p is false and false if p is true.

For example, if p is the statement "Janet is smart," then $\sim p$ is the statement "Janet is not smart."

E X A M P L E 3   **Writing Compound Statements in Symbolic Form**

Let *p* and *q* be the following statements:

- p: Jose Canseco plays football for the Washington Redskins.
- q: The Dow Jones industrial average set a new record high last week.

Write the following statements in symbolic form.

(a) Jose Canseco does not play football for the Washington Redskins, and the Dow Jones industrial average set a new record high last week.

(b) Jose Canseco plays football for the Washington Redskins, or the Dow Jones industrial average did not set a new record high last week.

(c) Jose Canseco does not play football for the Washington Redskins, and the Dow Jones industrial average did not set a new record high last week.

(d) It is not true that Jose Canseco plays football for the Washington Redskins and that the Dow Jones industrial average set a new record high last week.

Solutions   (a) $(\sim p) \wedge q$, (b) $p \vee \sim q$, (c) $\sim p \wedge \sim q$, (d) $\sim(p \wedge q)$ ■

E X A M P L E 4   **Translating Symbolic Forms into Compound Statements**

Let *p* and *q* be the following statements:

- p: Philadelphia is the capital of New Jersey.
- q: General Electric lost money last year.

Write out the statements that correspond to each of the following.

(a) $p \vee q$, (b) $p \wedge q$, (c) $p \vee \sim q$, (d) $\sim p \wedge \sim q$

Solutions   (a) Philadelphia is the capital of New Jersey, or General Electric lost money last year.

(b) Philadelphia is the capital of New Jersey, and General Electric lost money last year.

(c) Philadelphia is the capital of New Jersey, or General Electric did not lose money last year.

(d) Philadelphia is not the capital of New Jersey, and General Electric did not lose money last year.   ■

SELF-HELP EXERCISE SET 12.1

1. Determine which of the following sentences are statements.
   (a) The Atlanta Braves won the World Series in 1992.
   (b) IBM makes oil tankers for Denmark.
   (c) Does IBM make oil tankers for Denmark?
   (d) Please pay attention.
   (e) I have a three-dollar bill in my purse, or I don't have a purse.

2. Let p be the statement ''George Washington was never president of the United States'' and q be the statement ''George Washington wore a wig.'' Write out the statements that correspond to the following.
    (a) $\sim p$, **(b)** $p \vee q$, **(c)** $\sim p \wedge q$, **(d)** $p \wedge \sim q$, **(e)** $\sim p \vee \sim q$

EXERCISE SET 12.1

In Exercises 1 through 14 decide which are statements.

1. Water freezes at 70°F.

2. It rained in St. Louis on May 4, 1992.

3. $5 > 10$.

4. This sentence is false.

5. The number 4 is not a prime.

6. How are you feeling?

7. I feel great!

8. $10 + 10 - 5 = 25$.

9. There is life on Mars.

10. Cleveland is the largest city in Ohio.

11. Who said Cleveland is the largest city in Ohio?

12. You don't say!

13. IBM lost money in 1947.

14. Groundhog Day is on February 12.

15. Let p and q denote the following statements:

    ▪ p: George Washington was the third president of the United States.
    ▪ q: Austin is the capital of Texas.

    Express the following compound statements in words:
    (a) $\sim p$, **(b)** $p \wedge q$, **(c)** $p \vee q$, **(d)** $\sim p \wedge q$, **(e)** $p \vee \sim q$, **(f)** $\sim (p \wedge q)$

16. Let p and q denote the following statements:

    ▪ p: Mount McKinley is the highest point in the United States.
    ▪ q: George Washington was a signer of the Declaration of Independence.

    Express the following compound statements in words.
    (a) $\sim q$, **(b)** $p \wedge q$, **(c)** $p \vee q$, **(d)** $p \wedge \sim q$, **(e)** $\sim p \wedge \sim q$, **(f)** $\sim (p \vee q)$

17. Let p and q denote the following statements:

- p: George Washington owned over 100,000 acres of property.
- q: The Exxon Valdez was a luxury liner.

(a) State the negation of these statements in words.
(b) State the disjunction for these statements in words.
(c) State the conjunction for these statements in words.

18. Let p and q denote the following statements:

- p: McDonald's Corporation operates large farms.
- q: Wendy's Corporation operates fast food restaurants.

(a) State the negation of these statements in words.
(b) State the disjunction for these statements in words.
(c) State the conjunction for these statements in words.

19. Let p and q denote the following statements:

- p: The Wall Street Journal has the highest daily circulation of any newspaper.
- q: *Advise and Consent* was written by Irving Stone.

Express the following statements symbolically.

(a) *Advise and Consent* was not written by Irving Stone.
(b) The Wall Street Journal has the highest daily circulation of any newspaper, and *Advise and Consent* was not written by Irving Stone.
(c) The Wall Street Journal has the highest daily circulation of any newspaper, or *Advise and Consent* was written by Irving Stone.
(d) The Wall Street Journal does not have the highest daily circulation of any newspaper, or *Advise and Consent* was not written by Irving Stone.

20. Let p and q denote the following statements:

- p: IBM makes computers.
- q: IBM makes trucks.

Express the following statement symbolically.

(a) IBM does not make trucks.
(b) IBM makes computers, or IBM makes trucks.
(c) IBM makes computers, or IBM does not make trucks.
(d) IBM does not make computers, and IBM does not make trucks.

Solutions to Self-Help Exercise Set 12.1

1. The sentences (a), (b), and (e) are statements, while (c) and (d) are not.

2. **(a)** George Washington was a President of the United States.
(b) George Washington was never President of the United States, or George Washington wore a wig.
(c) George Washington was a President of the United States, and George Washington wore a wig.

(d) George Washington was never President of the United States, and George Washington did not wear a wig.

(e) George Washington was a President of the United States, or George Washington did not wear a wig.

12.2.   TRUTH TABLES

► *Truth Tables*

► *Exclusive Disjunction*

► *Tautologies and Contradictions*

Truth Tables

The **truth value** of a statement is either true or false. Thus the statement ''Ronald H. Coarse won the Nobel Prize in Economics in 1991'' has truth value true since it is a true statement, whereas the statement ''Los Angeles is the capital of California'' has truth value false since it is a false statement.

Logic does not concern itself with the truth value of simple statements. But if we know the truth values of the simple statements that make up a compound statement, then logic can determine the truth value of the compound statement.

For example, to understand the very definition of $p \lor q$, one must know under what conditions the compound statement will be true. As defined in the last section $p \lor q$ is always true unless both p and q are false. A convenient way of summarizing this is by a truth table. This is done in Table 12.1.

The truth tables for the statements $p \land q$ and $\sim p$ are given in Table 12.2 and Table 12.3.

As Table 12.2 indicates, $p \land q$ is true only if both p and q are true.

Given a general compound statement, we wish to determine the truth value given any possible combination of truth values for the simple statements that are contained in the compound statement. We use a truth table for this purpose. The next examples illustrate how this is done.

Table 12.1

p	q	$p \lor q$
T	T	T
T	F	T
F	T	T
F	F	F

Table 12.2

p	q	$p \land q$
T	T	T
T	F	F
F	T	F
F	F	F

Table 12.3

p	$\sim p$
T	F
F	T

E X A M P L E 1   **Constructing a Truth Table**

Construct a truth table for the statement $p \vee \sim q$.

Solution   Place p and q at the head of the first two columns as indicated in Table 12.4 and list all possible truth values for p and q as indicated. It is strongly recommended that you always list the truth values in the first two columns in the same way. This will be particularly useful later when we will need to compare two truth tables. Now enter the truth values for $\sim q$ in the third column. Now using the first and third columns of the table, construct the fourth column using the definition of $\vee$ found in Table 12.1. ∎

Table 12.4

p	q	$\sim q$	$p \vee \sim q$
T	T	F	T
T	F	T	T
F	T	F	F
F	F	T	T

E X A M P L E 2   **Constructing a Truth Table**

Construct a truth table for the statement $\sim p \wedge (p \vee q)$.

Solution   Make the same first two columns as before. Now make a column for $\sim p$ and the corresponding truth values. Now make a fourth column for $p \vee q$. Now using the third and fourth columns and the definition of $\wedge$, fill in the fifth column (Table 12.5). Thus we see that $\sim p \wedge (p \vee q)$ is true only if p is false and q is true. ∎

We can construct a truth table for a compound statement with three simple statements.

Table 12.5

p	q	$\sim p$	$p \vee q$	$\sim p \wedge (p \vee q)$
T	T	F	T	F
T	F	F	T	F
F	T	T	T	T
F	F	T	F	F

E X A M P L E　3　　**Constructing a Truth Table**

Construct a truth table for the statement $(p \wedge q) \wedge [(r \vee \sim p) \wedge q]$.

Solution　Always use the same order of T's and F's that are indicated in the first three columns of Table 12.6. Fill in the rest of the columns in the order given.　■

Table 12.6

p	q	r	$p \wedge q$	$\sim p$	$r \vee \sim p$	$(r \vee \sim p) \wedge q$	$(p \wedge q) \wedge [(r \vee \sim p) \wedge q]$
T	T	T	T	F	T	T	T
T	T	F	T	F	F	F	F
T	F	T	F	F	T	F	F
T	F	F	F	F	F	F	F
F	T	T	F	T	T	T	F
F	T	F	F	T	T	T	F
F	F	T	F	T	T	F	F
F	F	F	F	T	T	F	F
1	2	3	4	5	6	7	8
			use columns 1 and 2	use column 1	use columns 3 and 5	use columns 2 and 6	use columns 4 and 7

Exclusive Disjunction

We now consider the exclusive or. Recall that this use of "or" means "one or the other, but not both." The truth table for the exclusive disjunction is given in Table 12.7 where we note that the symbol for the exclusive disjunction is $\underline{\vee}$.

Notice that $\underline{\vee}$ is true only if exactly one of the two statements is true.

Table 12.7

p	q	$p \underline{\vee} q$
T	T	F
T	F	T
F	T	T
F	F	F

R E M A R K .　Unless clearly specified otherwise, the word "or" will always be taken in the **in**clusive sense.

E X A M P L E   4   **Determining the Truth Value of a Statement**

Let p and q be the following statements:

    p: Aaron Copland was an American composer.

    q: Rudolf Serkin was a violinist.

Determine the truth value of each of the following statements.

    (a) $p \vee q$, (b) $\sim(p \vee q)$, (c) $p \veebar q$, (d) $\sim(p \veebar q)$,
    (e) $p \wedge \sim q$

Solution   First note that p is true and q is false (Serkin was a famous pianist). Both the disjunction in (a) and the exclusive disjunction in (c) are therefore true. Thus their negations in (b) and (d) are false. The statement in (e) is the conjunction of a true statement p with a true statement $\sim q$ and thus is true. ■

Table 12.8

p	$\sim p$	$p \wedge \sim p$
T	F	F
F	T	F

Tautology and Contradiction

The statement $p \wedge \sim p$ is *always* false according to the truth table in Table 12.8.
    In such a case, we say that the statement $p \wedge \sim p$ is a **contradiction**. If the statement is always true, we say that the statement is a **tautology**.

Contradiction and Tautology

We say that a statement is a **contradiction** if the truth value of the statement is always false no matter what the truth values of its simple component statements. We say that a statement is a **tautology** if the truth value of the statement is always true no matter what the truth values of its simple component statements.

E X A M P L E   5   **Determining if a Statement is a Tautology**

Determine if the statement $p \vee (\sim p \vee q)$ is a tautology.

Solution   The truth table in Table 12.9 indicates that the statement is true no matter what the truth values of p and q are. Thus this statement is a tautology. ■

Table 12.9

p	q	$\sim p$	$\sim p \vee q$	$p \vee (\sim p \vee q)$
T	T	F	T	T
T	F	F	F	T
F	T	T	T	T
F	F	T	T	T

SELF-HELP EXERCISE SET 12.2

1. Construct the truth table for the statement $(p \lor q) \lor (r \land \sim q)$.

2. Determine the truth values of each of the statements given in Self-Help Exercise 2 of the last section.

EXERCISE SET 12.2

In Exercises 1 through 20, construct a truth table for the given statement. Indicate if a statement is a tautology or a contradiction.

1. $p \land \sim q$
2. $p \underline{\lor} \sim q$
3. $\sim(\sim p)$
4. $\sim(p \land q)$
5. $(p \land \sim q) \lor q$
6. $(p \underline{\lor} q) \lor \sim q$
7. $\sim p \lor (p \land q)$
8. $\sim p \underline{\lor} (p \land q)$
9. $(p \lor q) \land (p \land q)$
10. $(p \land q) \lor (p \lor q)$
11. $(p \lor \sim q) \lor (\sim p \land q)$
12. $(p \land \sim q) \land (p \lor \sim q)$
13. $(p \lor q) \land r$
14. $p \lor (q \land r)$
15. $\sim[(p \land q) \land r]$
16. $\sim[p \land (q \land r)]$
17. $(p \lor q) \lor (q \land r)$
18. $(p \land q) \land (q \lor r)$
19. $(p \lor \sim q) \lor (\sim q \land r)$
20. $(\sim p \land q) \land (\sim q \underline{\lor} r)$

21. Let p and q be the statements:

    p: Roe v. Wade was a famous boxing match.
    q: Iraq invaded Kuwait in 1990.

    Determine the truth value of the following statements.
    (a) $\sim p$, **(b)** $p \land q$, **(c)** $p \lor q$, **(d)** $\sim p \land q$, **(e)** $p \lor \sim q$

22. Let p and q be the statements:

    p: The moon rises in the east.
    q: Procter & Gamble is a casino in Las Vegas.

    Determine the truth value of the following statements:
    (a) $\sim q$, **(b)** $p \land q$, **(c)** $p \underline{\lor} q$, **(d)** $p \lor \sim q$, **(e)** $p \lor \sim q$

23. Let p and q be the statements:

    p: The South Pole is the southernmost point on the Earth.
    q: The North Pole is a monument in Washington, D.C.

    Determine the truth value of the following statements.
    (a) $\sim q$, **(b)** $p \underline{\lor} \sim q$, **(c)** $\sim p \land q$, **(d)** $\sim(p \land q)$

24. Let p and q be the statements:

p: Stevie Wonder is a famous singer.
q: Simon & Garfunkel is a famous law firm.

Determine the truth value of the following statements:
(a) $p \wedge \sim q$, **(b)** $p \veebar q$, **(c)** $p \vee q$, **(d)** $\sim(p \vee q)$

Solutions to Self-Help Exercise Set 12.2

1.

p	q	r	$p \vee q$	$\sim q$	$r \wedge \sim q$	$(p \vee q) \vee (r \wedge \sim q)$
T	T	T	T	F	F	T
T	T	F	T	F	F	T
T	F	T	T	T	T	T
T	F	F	T	T	F	T
F	T	T	T	F	F	T
F	T	F	T	F	F	T
F	F	T	F	T	T	T
F	F	F	F	T	F	F

2. The statement p is false, while q is true. Thus $\sim p$, $p \vee q$, and $\sim p \wedge q$ are true. But $p \wedge \sim q$ is false, being the conjunction of false statements. The statement $\sim p \vee \sim q$ is true since $\sim p$ is true.

12.3 IMPLICATION AND EQUIVALENCE

▶ *Conditional*

▶ *Biconditional*

▶ *Logical Equivalence*

▶ *Proofs*

Conditional

Statements such as "If you do all your homework, then you will pass the course" and "If you work hard, then you will succeed" are familiar. They involve the **conditional** "if . . . , then." We will see that this connective is used as a basis of deductive reasoning.

Table 12.10

p	q	$p \rightarrow q$
T	T	T
T	F	F
F	T	T
F	F	T

Conditional Statement

A conditional statement is a compound statement of the form "if p, then q" and is written symbolically as

$$p \rightarrow q.$$

A conditional statement is false if p is true and q is false and is true in all other cases. See Table 12.10.

Hypothesis and Conclusion

In the conditional statement $p \rightarrow q$, the statement p is the hypothesis and the statement q is the conclusion.

Table 12.11

		Definition of $p \rightarrow q$		
p	q	1	2	3
T	T	T	T	T
T	F	F	F	F
F	T	F	T	F
F	F	T	F	F

The question naturally arises as to why the statement $p \rightarrow q$ is defined to be true whenever p is false. To see this, consider again the statement "If you do all your homework, then you will pass this course." This can be written symbolically as $p \rightarrow q$ if p is the statement "You do all your homework" and q is the statement "You pass this course." Now if, for example, you did almost all your homework, then surely you expect to pass. In such a case, p is false and q is true. You would certainly not want the statement $p \rightarrow q$ to have truth value false in such a situation. The alternative is then to give it a truth value true.

There is another reason for the definition of $p \rightarrow q$. Everyone certainly agrees with the first two lines of the truth table for $p \rightarrow q$. Table 12.11 gives the 3 possible alternative definitions of $p \rightarrow q$ by changing the last two lines of the truth table of $p \rightarrow q$.

Notice that the third definition given in the fifth column simply gives the definition of $p \wedge q$, while the second definition given in the fourth column is the same as q. We will see later in this section that the first definition given in the third column is the definition of the connective "if and only if." Thus if $p \rightarrow q$ were defined in any of these other ways, $\rightarrow$ would not represent a new connective.

There are a number of ways of stating the conditional $p \rightarrow q$ in English. Among them are

- p implies q
- p only if q
- q if p
- q whenever p
- q provided p
- p is sufficient for q
- q is necessary for p
- suppose p, then q

E X A M P L E 1

Recognizing $p \rightarrow q$ in English

Write each of the following as $p \rightarrow q$, identifying p and q.

    (a) Working hard is sufficient for passing this course.

    (b) To be wealthy, it is sufficient for your parents to be wealthy.

    (c) $x^2 = 4$, whenever $x = \pm 2$.

    (d) I will succeed provided I work hard.

Solutions

    (a) This is $p \rightarrow q$, if p is "I work hard" and q is "I pass this course."

    (b) This is $p \rightarrow q$, if p is "My parents are wealthy" and q is "I will be wealthy."

    (c) This is $p \rightarrow q$, if p is "$x = \pm 2$" and q is "$x^2 = 4$."

    (d) This is $p \rightarrow q$, if p is "I work hard" and q is "I will succeed." ∎

E X A M P L E 2

Determining the Truth Value of a Conditional Statement

Determine the truth value of each of the following statements.

    (a) If Alaska is a state, then Puerto Rico is also.

    (b) If Puerto Rico is a state, then Alaska is also.

    (c) $4 + 4 = 10$, whenever $2 + 2 = 6$.

    (d) Miami is in Georgia provided Atlanta is in Texas.

Solutions

(a) is false since the hypothesis is true while the conclusion is false. All the other statements are true since in each case the hypothesis is false. ∎

E X A M P L E 3

Constructing a Truth Table

Construct a truth table for $(p \rightarrow q) \wedge (q \rightarrow p)$.

Solution

See Table 12.12. ∎

Table 12.12

p	q	$p \rightarrow q$	$q \rightarrow p$	$(p \rightarrow q) \wedge (q \rightarrow p)$
T	T	T	T	T
T	F	F	T	F
F	T	T	F	F
F	F	T	T	T

Biconditional

We now come to the last of the 5 basic connectives, the biconditional $p \leftrightarrow q$. By this we mean $p \rightarrow q$ and simultaneously $q \rightarrow p$, that is, $(p \rightarrow q) \wedge (q \rightarrow p)$. We just considered this in Example 3, where we saw that $(p \rightarrow q) \wedge (q \rightarrow p)$ is true only when p and q are both true or both false. We then make the following definition.

Table 12.13

p	q	$p \leftrightarrow q$
T	T	T
T	F	F
F	T	F
F	F	T

Biconditional Statement

A biconditional statement is a compound statement of the form "p if and only if q" and is written symbolically as

$$p \leftrightarrow q.$$

The biconditional $p \leftrightarrow q$ is true only when p and q are both true or both false. See Table 12.13.

In mathematics a common alternative way of saying "if and only if" is to say "necessary and sufficient."

E X A M P L E   4   **Using Necessary and Sufficient**

Rewrite the statement "$x^2 = 4$ if and only if $x = \pm 2$" using *necessary and sufficient*.

Solution   "$x^2 = 4$ is necessary and sufficient for $x = \pm 2$." ∎

Logical Equivalence

If we compare the truth table for $p \leftrightarrow q$ found in Table 12.13 with the truth table for $(p \rightarrow q) \wedge (q \rightarrow p)$ found in Table 12.12, we see that they are identical. This should not be surprising since $p \leftrightarrow q$ means both $p \rightarrow q$ and $q \rightarrow p$. We thus see that the statements $p \leftrightarrow q$ and $(p \rightarrow q) \wedge (q \rightarrow p)$ are **logically equivalent**.

Logical Equivalence

Two statements p and q are logically equivalent, denoted by

$$p \Leftrightarrow q,$$

if they have identical truth tables.

We now will consider three logical variants of the conditional statements $p \rightarrow q$. The first is the **contrapositive**.

> **Contrapositive**
>
> The **contrapositive** of the statement "if p, then q" ($p \rightarrow q$) is of the form "if not q, then not p" and is written
>
> $$\sim q \rightarrow \sim p.$$

A truth table for the contrapositive is given in Table 12.14

By comparing Table 12.10 with Table 12.14, we can see that the conditional $p \rightarrow q$ is logically equivalent to the contrapositive $\sim q \rightarrow \sim p$.

Table 12.14

p	q	$\sim q$	$\sim p$	$\sim q \rightarrow \sim p$
T	T	F	F	T
T	F	T	F	F
F	T	F	T	T
F	F	T	T	T

E X A M P L E 5 **Using the Contrapositive**

Write the statement "If I do all my homework, then I will pass this course" using the logically equivalent contrapositive.

Solution Let p be the statement "I did all my homework" and q be the statement "I passed this course." Then the given compound statement can be written symbolically as the conditional $p \rightarrow q$. The contrapositive is $\sim q \rightarrow \sim p$ and is written as the statement "If I failed this course, then I did not do all my homework." ∎

We now consider the logical variant of the conditional $p \rightarrow q$ called the **converse**.

> **Converse**
>
> The **converse** of the conditional statement "if p, then q" ($p \rightarrow q$) is "if q, then p" and is written
>
> $$q \rightarrow p.$$

Table 12.15

p	q	$q \rightarrow p$
T	T	T
T	F	T
F	T	F
F	F	T

The truth table is given in Table 12.15

Comparing Table 12.15 with Table 12.10 we see that the converse $q \rightarrow p$ is not logically equivalent to the conditional $p \rightarrow q$.

E X A M P L E 6   **Using the Converse**

Write the converse to the conditional statement "If I do all my homework, then I pass this course."

Solution   The converse is "If I passed this course, then I did all my homework." ∎

The last example indicates why the conditional and its converse are not logically equivalent. It is reasonable to expect that doing all your homework will result in passing this course, but this course can easily be passed without doing all the homework.

The last logical variant of the conditional $p \to q$ that we consider is the **inverse**.

Inverse

The **inverse** of the conditional statement "if p, then q" ($p \to q$) is "if not p, then not q" and is written

$$\sim p \to \sim q.$$

The truth table for the inverse $\sim p \to \sim q$ is given in Table 12.16

Notice that Table 12.16 is identical to Table 12.15. Thus the converse is logically equivalent to the inverse.

Table 12.16

p	q	$\sim p$	$\sim q$	$\sim p \to \sim q$
T	T	F	F	T
T	F	F	T	T
F	T	T	F	F
F	F	T	T	T

Proofs

It is sometimes difficult to give a direct proof of a conditional statement but easy to give a proof of the contrapositive. Since the contrapositive is logically equivalent to the conditional, establishing the contrapositive will establish the conditional. This is indicated in the next example.

E X A M P L E 7   **Using the Contrapositive in a Proof**

Prove that if n^2 is an odd integer, then n is also.

Solution   Let p be the statement "n^2 is odd" and q be the statement "n is odd." Then we wish to show that $p \to q$. To do this we will show that $\sim q \to \sim p$. Assume then that n is not odd, that is, assume that n is even. Then we can write n as $n = 2k$ for some integer k and

$$n^2 = (2k)^2 = 2(2k^2).$$

Since this is twice an integer, n^2 is even and thus not odd. We have shown that if n is not odd, then n^2 is not odd. Since the contrapositive $\sim q \to \sim p$ is logically equivalent to the conditional $p \to q$, we have established the result.   ∎

SELF-HELP EXERCISE SET 12.3

1. Let the statements p and q be:

   p: We raise prices.
   q: Sales drop.

   Then the conditional statement $p \to q$ is "If we raise prices, then sales drop." Write the contrapositive, converse, and inverse statements.

2. Determine if the two statements $p \to q$ and $\sim p \lor q$ are logically equivalent.

EXERCISE SET 12.3

For Exercises 1 through 6 let p and q be the statements:

p: Interest rates drop.
q: The stock market goes up.

Write each of the following statements in English. Then write both in English and symbolically the contrapositive, inverse, and converse of the given statement.

1. $p \to q$ 2. $p \to \sim q$ 3. $\sim q \to \sim p$

4. $q \to p$ 5. $\sim p \to \sim q$ 6. $\sim p \to q$

For Exercises 7 through 14 let p and q be the statements:

p: The rate of inflation is increasing.
q: Interest rates are increasing.

Write each of the following statements in symbolic form using $\to$ or $\leftrightarrow$.

7. If the rate of inflation is increasing, then interest rates are increasing.

8. An increasing rate of inflation is necessary for interest rates to be increasing.

9. An increasing rate of inflation is sufficient for interest rates to be increasing.

10. If interest rates are increasing, then the rate of inflation is increasing.

11. Interest rates are increasing if and only if the rate of inflation is increasing.

12. Interest rates are increasing if the rate of inflation is increasing.

13. If interest rates are not increasing, then the rate of inflation is not increasing.

14. Interest rates are increasing only if the rate of inflation is increasing.

In Exercises 15 through 22 let p and q be the statements:

p: Elvis lives.
q: Barbra Streisand is a singer.

Determine the truth value of each of the following.

15. $p \rightarrow q$ **16.** $q \rightarrow p$ **17.** $\sim p \rightarrow \sim q$

18. $\sim q \rightarrow \sim p$ **19.** $(p \wedge q) \rightarrow p$ **20.** $(p \vee q) \rightarrow q$

21. $p \leftrightarrow q$ **22.** $(p \vee q) \rightarrow p$

Construct a truth table for each of the following statements.

23. $(p \rightarrow q) \wedge p$ **24.** $(p \rightarrow q) \vee p$

25. $(p \rightarrow q) \vee (q \rightarrow p)$ **26.** $\sim(p \rightarrow q)$

27. $[\sim(p \vee q)] \leftrightarrow \sim p \wedge \sim q$ **28.** $\sim(p \rightarrow \sim q)$

29. $(p \rightarrow q) \rightarrow r$ **30.** $p \rightarrow (q \rightarrow r)$

31. $[p \vee (q \wedge r)] \leftrightarrow [(p \vee q) \wedge (p \vee r)]$

32. $(p \vee q) \rightarrow r$ **33.** $(p \rightarrow q) \leftrightarrow (q \rightarrow p)$

34. $(p \rightarrow q) \wedge (\sim p \vee q)$ **35.** $(p \vee q) \vee r$

36. $p \vee (q \vee r)$

In Exercises 37 through 44 determine which are tautologies.

37. $(p \wedge q) \rightarrow p$ **38.** $p \rightarrow (p \vee q)$

39. $[p \wedge (p \rightarrow q)] \rightarrow q$ **40.** $[\sim p \wedge (p \vee q)] \rightarrow q$

41. $p \rightarrow [q \rightarrow (p \wedge q)]$ **42.** $[(p \rightarrow q) \wedge (q \rightarrow r)] \rightarrow (p \rightarrow r)$

43. $(p \rightarrow q) \rightarrow [(p \vee r) \rightarrow (q \vee r)]$ **44.** $(p \rightarrow q) \rightarrow [(p \wedge r) \rightarrow (q \wedge r)]$

45. Prove that if n^2 is even, then n is even.

46. Prove that if mn is odd, then both m and n are odd.

Solutions to Self-Help Exercise Set 12.3

1. The contrapositive is: "If sales did not drop, then we did not raise prices."
The converse is: "If sales drop, then we raise prices."
The inverse is: "If we did not raise prices, then sales did not drop."

2. From truth table, Table 12.17, we see that the two statements $p \rightarrow q$ and $\sim p \vee q$ have identical truth tables and therefore are logically equivalent.

Table 12.17

p	q	$p \rightarrow q$	$\sim p$	$\sim p \vee q$
T	T	T	F	T
T	F	F	F	F
F	T	T	T	T
F	F	T	T	T

12.4   LAWS OF LOGIC

► *Logical Equivalence*
► *Implication*

Logical Equivalence

We are all aware that $a(x + y) = ax + ay$ for all numbers a, x, and y. This represents a law of numbers. No matter what the numbers a, x, and y are, the two numbers $a(x + y)$ and $ax + ay$ are equal. In an analogous fashion we noticed in the last section that given any truth value for p and q, the truth values of the statements $p \leftrightarrow q$ and $(p \rightarrow q) \wedge (q \rightarrow p)$ are the same. Thus we said that the two statements $p \leftrightarrow q$ and $(p \rightarrow q) \wedge (q \rightarrow p)$ were logically equivalent and wrote $p \leftrightarrow q \Leftrightarrow (p \rightarrow q) \wedge (q \rightarrow p)$. The symbol $\Leftrightarrow$ in logic is thus analogous to the symbol $=$ in algebra.

We already defined two statements to be logically equivalent if their truth tables are identical. We now give an alternate definition.

Table 12.18

p	q	$p \leftrightarrow q$
T	T	T
T	F	F
F	T	F
F	F	T

Logical Equivalence

Two statements p and q are logically equivalent, written

$$p \Leftrightarrow q,$$

if the biconditional $p \leftrightarrow q$ is a tautology.

Table 12.19

p	q	$p \leftrightarrow q$
T	T	T
F	F	T

To see why these two definitions are the same, recall the truth table for the biconditional shown in Table 12.18.

The cases when the biconditional is true are listed in Table 12.19. Notice from Table 12.19 that the truth tables of p and q are the same.

Laws of Logic

1.	$\sim(\sim p) \Leftrightarrow p$	Double negation
2.	$(p \rightarrow q) \Leftrightarrow (\sim q \rightarrow \sim p)$	Contrapositive
3a.	$(p \rightarrow q) \Leftrightarrow (\sim p \vee q)$	Implication

3b. $(p \rightarrow q) \Leftrightarrow \sim(p \wedge \sim q)$ Implication

4. $(p \leftrightarrow q) \Leftrightarrow [(p \rightarrow q) \wedge (q \rightarrow p)]$ Equivalence

5a. $p \wedge p \Leftrightarrow p$ Idempotent law for conjunction

5b. $p \vee p \Leftrightarrow p$ Idempotent law for disjunction

6a. $(p \wedge q) \Leftrightarrow (q \wedge p)$ Commutative law for conjunction

6b. $(p \vee q) \Leftrightarrow (q \vee p)$ Commutative law for disjunction

6c. $(p \leftrightarrow q) \Leftrightarrow (q \leftrightarrow p)$ Commutative law for
                                                          biconditional

7a. $(p \wedge q) \wedge r \Leftrightarrow p \wedge (q \wedge r)$ Associative law for conjunction

7b. $(p \vee q) \vee r \Leftrightarrow p \vee (q \vee r)$ Associative law for disjunction

8a. $p \wedge (q \vee r) \Leftrightarrow (p \wedge q) \vee (p \wedge r)$ Distributive law for conjunction

8b. $p \vee (q \wedge r) \Leftrightarrow (p \vee q) \wedge (p \vee r)$ Distributive law for disjunction

9a. $\sim(p \vee q) \Leftrightarrow (\sim p \wedge \sim q)$ De Morgan's law

9b. $\sim(p \wedge q) \Leftrightarrow (\sim p \vee \sim q)$ De Morgan's law

10a. $p \vee (p \wedge q) \Leftrightarrow p$ Absorption law

10b. $p \wedge (p \vee q) \Leftrightarrow p$ Absorption law

In the following t is a tautology and c is a contradiction.

11a. $p \vee \sim p \Leftrightarrow t$ Inverse law

11b. $p \wedge \sim p \Leftrightarrow c$ Inverse law

12a. $p \vee t \Leftrightarrow t$ Identity law

12b. $p \wedge t \Leftrightarrow p$ Identity law

12c. $p \vee c \Leftrightarrow p$ Identity law

12d. $p \wedge c \Leftrightarrow c$ Identity law

Any one of the laws of logic can be verified by constructing truth tables. We illustrate this by verifying one of De Morgan's laws.

EXAMPLE 1   Establishing De Morgan's First Law

Show $\sim(p \vee q) \Leftrightarrow (\sim p \wedge \sim q)$.

Solution   Construct the truth table as shown in Table 12.20. Since $\sim(p \vee q) \leftrightarrow (\sim p \wedge \sim q)$ is a tautology, $\sim(p \vee q) \Leftrightarrow (\sim p \wedge \sim q)$. ∎

Table 12.20

p	q	$p \vee q$	$\sim(p \vee q)$	$\sim p$	$\sim q$	$\sim p \wedge \sim q$	$\sim(p \wedge q) \leftrightarrow (\sim p \wedge \sim q)$
T	T	T	F	F	F	F	T
T	F	T	F	F	T	F	T
F	T	T	F	T	F	F	T
F	F	F	T	T	T	T	T

E X A M P L E 2   **Using De Morgan's Laws**

Negate the statement: "The author's name is Mud, and Abraham Lincoln was President of the United States" using De Morgan's second law.

Solution   Let p and q be the statements:

p: The author's name is Mud.
q: Abraham Lincoln was a President of the United States.

Then the given statement is $p \wedge q$. The negation is $\sim(p \wedge q)$, which is logically equivalent to $\sim p \vee \sim q$ according to De Morgan's first law. This can be written as "The author's name is not Mud, or Abraham Lincoln was not President of the United States." ∎

Just as one can give a proof of a theorem based on the validity of other theorems, so also one can establish a law of logic based on the validity of other laws of logic. The following is an example.

E X A M P L E 3   **Establishing Additional Laws of Logic**

Show that $p \vee (q \vee r) \Leftrightarrow r \vee (q \vee p)$.

Solution
$$p \vee (q \vee r) \Leftrightarrow (p \vee q) \vee r \qquad \text{Law 7b}$$
$$\Leftrightarrow r \vee (p \vee q) \qquad \text{Law 6b}$$
$$\Leftrightarrow r \vee (q \vee p) \qquad \text{Law 6b} \quad ∎$$

In algebra a complex expression can sometimes be simplified. For example, the expression $-(x - a) - a$ is the same as $-x$. In a similar way in logic a complex statement can sometimes be simplified, that is, written as a logically equivalent simpler statement. The following is an example.

E X A M P L E 4   **Simplifying a Logical Statement**

Simplify $\sim(p \rightarrow q) \vee (p \wedge q)$.

Solution
$$\sim(p \rightarrow q) \vee (p \wedge q) \Leftrightarrow \sim(\sim p \vee q) \vee (p \wedge q) \qquad \text{Implication}$$
$$\Leftrightarrow [(\sim\sim p) \wedge \sim q] \vee (p \wedge q) \qquad \text{De Morgan's law}$$
$$\Leftrightarrow (p \wedge \sim q) \vee (p \wedge q) \qquad \text{Double negative}$$
$$\Leftrightarrow p \wedge (\sim q \vee q) \qquad \text{8a}$$
$$\Leftrightarrow p \wedge (q \vee \sim q) \qquad \text{6b}$$
$$\Leftrightarrow p \wedge t \qquad \text{11a}$$
$$\Leftrightarrow p \qquad \text{12b}$$

Thus the rather complicated expression $\sim(p \rightarrow q) \vee (p \wedge q)$ is logically equivalent to p. ∎

Implication

In the next section we will be making a logical argument. Certainly replacing a statement with its logical equivalence is very useful. However, what we really will need to do is replace a statement with one that is logically implied by it. We have the following definition.

Logical Implication

We say that the statement P logically implies the statement Q, denoted by

$$P \Rightarrow Q,$$

if Q is true whenever P is true.

We use capital letters P and Q to emphasize that P and Q are usually compound statements.

E X A M P L E 5   **Verifying a Logical Implication**

Show $\sim q \wedge (p \rightarrow q) \Rightarrow \sim p$.

Solution   The truth table for $[\sim q \wedge (p \rightarrow q)]$ (denoted also by P) and $\sim p$ (denoted also by Q) is given in Table 12.21. Notice from the table that Q is true whenever P is true. Thus $P \Rightarrow Q$ or $\sim q \wedge (p \rightarrow q) \Rightarrow \sim p$.   ∎

Table 12.21

				P	Q
p	q	$\sim q$	$p \rightarrow q$	$\sim q \wedge (p \rightarrow q)$	$\sim p$
T	T	F	T	F	F
T	F	T	F	F	F
F	T	F	T	F	T
F	F	T	T	T	T

Table 12.22

P	Q	$P \rightarrow Q$
T	T	T
T	F	F
F	T	T
F	F	T

The truth table for the conditional $P \rightarrow Q$ is reproduced in Table 12.22. Notice that $P \rightarrow Q$ will be a tautology if the second case never arises. But the second case never arises precisely when Q is true whenever P is. That is, precisely when $P \Rightarrow Q$. We have then established the following equivalent definition of logical equivalence.

> **Logical Implication**
>
> We say that the statement P logically implies the statement Q, denoted by
>
> $$P \Rightarrow Q,$$
>
> if $P \to Q$ is a tautology.

We suggest that one use this latter definition to verify logical implications.

E X A M P L E   6   **Verifying a Logical Implication**

Show $(p \to q) \wedge (q \to r) \Rightarrow p \to r$.

Solution   First construct the truth table for $[(p \to q) \wedge (q \to r)] \to (p \to r)$.
Table 12.23 indicates that $[(p \to q) \wedge (q \to r)] \to (p \to r)$ is a tautology.
Thus $(p \to q) \wedge (q \to r) \Rightarrow p \to r$. ∎

Table 12.23

p	q	r	$p \to q$	$q \to r$	$(p \to q) \wedge (q \to r)$	$p \to r$	$[(p \to q) \wedge (q \to r)] \to (p \to r)$
T	T	T	T	T	T	T	T
T	T	F	T	F	F	F	T
T	F	T	F	T	F	T	T
T	F	F	F	T	F	F	T
F	T	T	T	T	T	T	T
F	T	F	T	F	F	T	T
F	F	T	T	T	T	T	T
F	F	F	T	T	T	T	T

The following list gives some logical implications.

Logical Implications

1. $p \wedge (p \to q) \Rightarrow q$
2. $\sim p \wedge (p \vee q) \Rightarrow q$
3. $p \wedge q \Rightarrow p$
4. $p \Rightarrow p \vee q$
5. $(p \to q) \wedge (q \to r) \Rightarrow p \to r$
6. $(p \wedge q) \to r \Rightarrow p \to (q \to r)$

7. $p \to (q \to r) \Rightarrow (p \wedge q) \to r$
8. $p \to (q \wedge \sim q) \Rightarrow \sim p$
9. $(p \leftrightarrow q) \wedge (q \leftrightarrow r) \Rightarrow p \leftrightarrow r$
10. $(p \to q) \wedge (r \to s) \Rightarrow (p \wedge r) \to (q \wedge s)$

SELF-HELP EXERCISE SET 12.4

1. Prove that $\sim (p \wedge q) \Leftrightarrow \sim p \vee \sim q$.

2. Prove that $p \wedge (p \to q) \Rightarrow q$.

EXERCISE SET 12.4

Establish the laws of logic that are given in Exercises 1 through 14 by using truth tables. (The statement t is a tautology and the statement c is a contradiction.)

1. $p \wedge p \Rightarrow p$
2. $p \vee p \Rightarrow p$

3. $p \vee t \Rightarrow t$
4. $p \wedge t \Rightarrow p$

5. $p \wedge c \Rightarrow c$
6. $p \vee c \Rightarrow p$

7. $(p \wedge q) \wedge r \Rightarrow p \wedge (q \wedge r)$
8. $(p \vee q) \vee r \Rightarrow p \vee (q \vee r)$

9. $\sim p \wedge (p \vee q) \Rightarrow q$
10. $p \Rightarrow q \to (p \wedge q)$

11. $(p \leftrightarrow q) \wedge (q \leftrightarrow r) \Rightarrow p \leftrightarrow r$
12. $(p \wedge q) \to r \Rightarrow p \to (q \to r)$

13. $p \wedge (q \vee r) \Leftrightarrow (p \wedge q) \vee (p \wedge r)$

14. $p \vee (q \wedge r) \Leftrightarrow (p \vee q) \wedge (p \vee r)$

Establish the laws of logic given in Exercises 15 through 20.

15. $(p \vee q) \vee \sim q \Leftrightarrow t$
16. $p \vee (\sim p \wedge \sim q) \Leftrightarrow p \vee \sim q$

17. $p \wedge q \Leftrightarrow \sim (\sim p \vee \sim q)$
18. $p \vee q \Leftrightarrow \sim (\sim p \wedge \sim q)$

19. $(\sim p \wedge \sim q) \wedge \sim r \Leftrightarrow \sim [p \vee (q \vee r)]$

20. $\sim [p \vee (q \vee r)] \Leftrightarrow (\sim p \wedge \sim q) \wedge \sim r$

In Exercises 21 through 24 rewrite each using only the connectives $\sim$ and $\vee$.

21. $p \wedge q$
22. $\sim p \wedge \sim q$
23. $p \to q$
24. $p \leftrightarrow q$

In Exercises 25 through 28 rewrite using only the connectives $\sim$ and $\wedge$.

25. $p \vee q$
26. $\sim p \vee \sim q$
27. $p \to q$
28. $p \leftrightarrow q$

In Exercises 29 through 32 negate the statements using the De Morgan laws.

29. Jim likes Sue or Mary.

30. Jim likes Sue and Mary.

31. Sales are up and we all get raises.

32. Sales are up or we all get raises.

33. Show that $\sim(p \rightarrow q) \Leftrightarrow p \wedge \sim q$.

In Exercises 34 through 37 use the result in Exercise 33 to negate the given statement.

34. If I do all my homework, then I will pass this course.

35. If I work hard, then I will succeed.

36. If interest rates go down, then stock prices will go up.

37. If prices are raised, then sales will drop.

Solutions to Self-Help Exercise Set 12.4

1. The truth table for $[\sim(p \wedge q)] \leftrightarrow [\sim p \vee \sim q]$, Table 12.24, indicates that this biconditional is a tautology. Thus $\sim(p \wedge q) \Leftrightarrow \sim p \vee \sim q$.

Table 12.24

p	q	$p \wedge q$	$\sim(p \wedge q)$	$\sim p$	$\sim q$	$\sim p \vee \sim q$	$\sim(p \wedge q) \leftrightarrow (\sim p \vee \sim q)$
T	T	T	F	F	F	F	T
T	F	F	T	F	T	T	T
F	T	F	T	T	F	T	T
F	F	F	T	T	T	T	T

2. The truth table for $[p \wedge (p \rightarrow q)] \rightarrow q$, Table 12.25, indicates that this conditional is a tautology. Thus $p \wedge (p \rightarrow q) \Rightarrow q$.

Table 12.25

p	q	$p \rightarrow q$	$p \wedge (p \rightarrow q)$	$[p \wedge (p \rightarrow q)] \rightarrow q$
T	T	T	T	T
T	F	F	F	T
F	T	T	F	T
F	F	T	F	T

12.5   ARGUMENTS

► *Valid Arguments*

► *Using Implications and Equivalences in Arguments*

► *Indirect Proof*

Valid Arguments

Consider the following argument.

> If interest rates are falling, then stock prices are rising.
>
> Interest rates are falling.
> _____
>
> Therefore, stock prices are rising.

We refer to the first two statements as *hypothesis* and to the third statement as the *conclusion*. We do not wish to concern ourselves with whether any of these statements are true. Rather, we ask if the truth of the conclusion follows logically from the *assumed* truth of the hypothesis.

Let p be the statement "Interest rates are falling" and q the statement "Stock prices are rising." It is convenient to write the above argument in symbolic form as

$$p \rightarrow q$$
$$\underline{p}$$
$$\therefore q$$

where the symbol $\therefore$ means "therefore." Then we ask if q is true whenever both $p \rightarrow q$ and p are true. That is, we ask if q is true whenever $(p \rightarrow q) \wedge p$ is true. But this happens precisely when $[(p \rightarrow q) \wedge p] \rightarrow q$ is a tautology, since $[(p \rightarrow q) \wedge p] \rightarrow q$ is a tautology precisely when the case $(p \rightarrow q) \wedge p$ is true while q is false never occurs.

The truth table, Table 12.26, indicates that the statement $[(p \rightarrow q) \wedge p] \rightarrow q$ is a tautology. Recall from the last section that we denote this by $[(p \rightarrow q) \wedge p] \Rightarrow q$. In general we have the following.

Table 12.26

p	q	$p \rightarrow q$	$(p \rightarrow q) \wedge p$	$[(p \rightarrow q) \wedge p] \rightarrow q$
T	T	T	T	T
T	F	F	F	T
F	T	T	F	T
F	F	T	F	T

> **Argument, Hypothesis, Conclusion**
>
> An **argument** consists of a set of statements $h_1, h_2, \ldots, h_n$, called **hypotheses** and a statement q called the **conclusion**. An argument is **valid** if the conclusion is true whenever the hypotheses are all true, that is, whenever
>
> $$h_1 \wedge h_2 \wedge \cdots \wedge h_n \Rightarrow q.$$

It is very important to understand the distinction between the two words "valid" and "true." Consider the following argument.

> If George Washington was a man,
>   then he was never President of the United States.
>
> George Washington was a man.
> ___
> Therefore George Washington was never President of the United States.

Let p be "George Washington was a man" and q be "George Washington was never President of the United States." Then this argument can be written symbolically as

$$p \rightarrow q$$
$$\underline{p \qquad}.$$
$$\therefore q$$

As we saw above, this is a valid argument. The conclusion, however, we know in fact to be false. Perhaps this can be characterized as garbage-in, garbage-out. It is, of course, always possible that, despite garbage-in, the output is good. Consider the following argument.

> If George Washington was not a man,
>   then he was President of the United States.
>
> George Washington was not a man.
> ___
> Therefore George Washington was president of the United States.

If p is "George Washington was not a man" and q is "George Washington was President of the United States," then this argument can also be written symbolically as the previous two were, and thus is a valid argument. However, one of the hypotheses is false while the conclusion is true.

E X A M P L E   1   **Determining if an Argument is Valid**

Determine if the following argument is valid.

> If interest rates are falling, then stock prices are rising.
>
> Stock prices are rising.
> ___
> Therefore, interest rates are falling.

Solution   Let p be the statement "Interest rates are falling" and q the statement "Stock prices are rising." Then this argument can be written as

$$p \rightarrow q$$
$$\underline{q}.$$
$$\therefore p$$

Thus we wish to determine if $[(p \rightarrow q) \wedge q] \rightarrow p$ is a tautology. The truth table in Table 12.27 indicates that this is not a tautology and therefore the given argument is not valid.   ■

Table 12.27

p	q	$p \rightarrow q$	$(p \rightarrow q) \wedge q$	$[(p \rightarrow q) \wedge q] \rightarrow p$
T	T	T	T	T
T	F	F	F	T
F	T	T	T	F
F	F	T	F	T

E X A M P L E   2   **Determining if an Argument is Valid**

Determine if the following argument is valid.

Either Jim takes an economics course or Jim takes a math course.

Jim does not take an economics course.

Therefore Jim takes a math course.

Solution   Let p be the statement "Jim takes an economics course" and q be the statement "Jim takes a math course." Then this argument can be written as

$$p \vee q$$
$$\underline{\sim p}.$$
$$\therefore q$$

Thus we wish to determine if $[(p \vee q) \wedge \sim p] \rightarrow q$ is a tautology. Table 12.28 indicates that this statement is a tautology.   ■

Table 12.28

p	q	$p \vee q$	$\sim p$	$(p \vee q) \wedge \sim p$	$[(p \vee q) \wedge \sim p] \rightarrow q$
T	T	T	F	F	T
T	F	T	F	F	T
F	T	T	T	T	T
F	F	F	T	F	T

Using Implications and Equivalences in Arguments

Using truth tables to verify that an argument is valid is a mechanical process. Even if there are a large number of hypothesis, a computer should be able to handle the construction of the truth table and the verification of the appropriate tautology. But this is not enough. Mechanically constructing a truth table does not lead to insight into the argument, or to possible simplifications, or to generalizations. To obtain this greater insight and to get a better feel for logical arguments in mathematics in general, we need to construct proofs based on the logical inferences and equivalences that were established in the last section.

In the following problem, instead of creating a truth table, we establish a proof by using some of the logical implications and equivalences established in the last section. In the following list we present some of these again.

1. $p \Rightarrow p \vee q$ Addition
2. $p \wedge q \Rightarrow p$ Subtraction
3. $p \wedge (p \rightarrow q) \Rightarrow q$ Modus ponens
4. $(p \rightarrow q) \wedge {\sim}q \Rightarrow {\sim}p$ Modus tollens
5. $(p \vee q) \wedge {\sim}p \Rightarrow q$ Disjunctive syllogism
6. $(p \rightarrow q) \wedge (q \rightarrow r) \Rightarrow p \rightarrow r$ Hypothetical syllogism

E X A M P L E 3   **Constructing a Proof**

Suppose the following are true:

     Janet is healthy or she is not wealthy.

     If Janet is healthy, then she plays tennis.

     Janet does not play tennis.

Prove that Janet is not wealthy.

Solution   Let h, w, and p be the following statements:

     h: Janet is healthy.

     w: Janet is wealthy.

     p: Janet plays tennis.

Then we wish to establish the following.

     $h \vee {\sim}w$

     $h \rightarrow p$

     $\underline{{\sim}p}$

     $\therefore {\sim}w$

We can establish this without using a truth table as follows.

$$(h \vee {\sim}w) \wedge (h \rightarrow p) \wedge ({\sim}p) \Rightarrow (h \vee {\sim}w) \wedge ({\sim}h) \quad \text{Modus tollens}$$

$$\Rightarrow {\sim}w \quad \text{Disjunctive syllogism}$$

Thus ${\sim}w$ is true, and this means Janet is not wealthy. ∎

E X A M P L E   4   **Constructing a Proof**

Suppose the following are true:

Joe goes to the mountains, or he goes to the beach.

If Joe does not take his convertible, then he does not go to the beach.

Joe does not go to the mountains.

Prove that Joe takes his convertible.

Solution   Let m, b, and h be the following statements:

m: Joe goes to the mountains.

b: Joe goes to the beach.

h: Joe takes his convertible.

Then we wish to prove the following.

$m \lor b$

$\sim h \to \sim b$

$\underline{\sim m}$

$\therefore h$

This can be established without using truth table as follows.

$(m \lor b) \land (\sim h \to \sim b) \land (\sim m)$

$\Leftrightarrow (m \lor b) \land (\sim m) \land (\sim h \to \sim b)$	Commutative law
$\Rightarrow b \land (\sim h \to \sim b)$	Disjunctive syllogism
$\Leftrightarrow b \land (b \to h)$	Contrapositive
$\Rightarrow h$	Modus ponens

Thus h is true, and this means Joe takes his convertible.   ■

E X A M P L E   5   **Constructing a Proof**

Suppose the following are true.

If Joe does not play tennis, then Joe does not drive to the mountains.

If Joe drives to the beach, then Joe does not rent a car.

If it is sunny, then Joe rents a car.

If Joe does not drive to the mountains, then Joe drives to the beach.

It is sunny.

Prove that Joe plays tennis.

Solution　　　Let p, m, b, r, and s be the following statements:

> p: Joe plays tennis.
> m: Joe drives to the mountains.
> b: Joe drives to the beach.
> r: Joe rents a car.
> s: It is sunny.

The first five statements are then written symbolically as follows.

1. h_1: $\sim p \to \sim m$ 　　　Logically equivalent to $m \to p$
2. h_2: $b \to \sim r$ 　　　　　Logically equivalent to $r \to \sim b$
3. h_3: $s \to r$
4. h_4: $\sim m \to b$ 　　　　　Logically equivalent to $\sim b \to m$
5. h_5: s

The given logical equivalence for h_1 is just the contrapositive. The given logical equivalences for h_2 and h_4 use the contrapositive and the double negation. For example, the contrapositive for h_2 is $[\sim(\sim c)] \to \sim b$. Now, using the double negation, this becomes logically equivalent to $c \to \sim b$. Then

$$h_1 \wedge h_2 \wedge h_3 \wedge h_4 \wedge h_5$$

$$\Leftrightarrow (\sim p \to \sim m) \wedge (b \to \sim r) \wedge (s \to r) \wedge (\sim m \to b) \wedge s$$

$$\Leftrightarrow (m \to p) \wedge (r \to \sim b) \wedge (s \to r) \wedge (\sim b \to m) \wedge s$$

Now, using the commutivity laws, we can write

$$h_1 \wedge h_2 \wedge h_3 \wedge h_4 \wedge h_5$$

$$\Leftrightarrow h_5 \wedge h_3 \wedge h_2 \wedge h_4 \wedge h_1$$

$$\Leftrightarrow s \wedge (s \to r) \wedge (r \to \sim b) \wedge (\sim b \to m) \wedge (m \to p)$$

Now making repeated use of modus ponens yields

$$s \wedge (s \to r) \wedge (r \to \sim b) \wedge (\sim b \to m) \wedge (m \to p)$$

$$\Rightarrow r \wedge (r \to \sim b) \wedge (\sim b \to m) \wedge (m \to p)$$

$$\Rightarrow (\sim b) \wedge (\sim b \to m) \wedge (m \to p)$$

$$\Rightarrow m \wedge (m \to p)$$

$$\Rightarrow p.$$

Thus p is true, and this means that Joe plays tennis. ■

Indirect Proof

In order to establish

$$h_1 \wedge h_2 \wedge \cdots \wedge h_n \Rightarrow q,$$

it is sometimes easier to establish the contrapositive

$$\sim q \Rightarrow \sim(h_1 \wedge h_2 \wedge \cdots \wedge h_n).$$

Using the De Morgan law, this can be written as

$$\sim q \Rightarrow (\sim h_1) \vee (\sim h_2) \vee \cdots \vee (\sim h_n).$$

To show this, we must assume that $\sim q$ is true and then show that at least one of the statements $\sim h_1, \sim h_2, \ldots, \sim h_n$, is true. That is, we must assume that q is false and show that one of the statements $h_1, h_2, \ldots, h_n$, is false.

Indirect Proof

In order to give an indirect proof of

$$h_1 \wedge h_2 \wedge \cdots \wedge h_n \Rightarrow q,$$

we assume that the conclusion q is false and then prove that at least one of the hypotheses $h_1, h_2, \ldots, h_n$, is false.

E X A M P L E   6   **Indirect Proof**

Give an indirect proof for Example 5.

Solution   For convenience the hypothesis are listed again.

1. $h_1: \sim p \rightarrow \sim m$
2. $h_2: b \rightarrow \sim r$
3. $h_3: s \rightarrow r$
4. $h_4: \sim m \rightarrow b$
5. $h_5: s$

The conclusion is the statement p. For an indirect proof we assume that p is false, that is, that $\sim p$ is true. Naturally if $h_1, h_2, h_3,$ or h_4 are false, we are finished. We will complete the proof by assuming that $h_1, h_2, h_3,$ and h_4 are true, and then show that h_5 must be false. Then h_1 implies that $\sim m$ is true, and then h_4 implies that b is true, and then h_2 implies $\sim r$ is true. The contrapositive of h_3 is $\sim r \rightarrow \sim s$, and thus since $\sim r$ is true this indicates that $\sim s$ is true. But this states that the fifth hypothesis, h_5, is false. Thus one of the hypotheses has been shown to be false given that the conclusion is false. This then establishes the required results by an indirect proof.   ■

SELF-HELP EXERCISE SET 12.5

1. Establish the following argument using truth tables.

$p \wedge q$

$\dfrac{\sim p}{}$

$\therefore q$

2. Establish the following without using truth tables.

If it is raining, then I will not be at the beach. If I have money for gas for my car, then I will be at the beach. I have money for gas for my car. Therefore it is not raining.

3. Use an indirect proof to establish the argument in the previous exercise.

EXERCISE SET 12.5

In Exercises 1 through 10 use truth tables to establish whether or not the arguments are valid.

1. $p \lor q$
$\dfrac{\sim p}{}$
$\therefore q$

2. $p \to q$
$\dfrac{\sim q}{}$
$\therefore \sim p$

3. $p \to q$
$\dfrac{q \to r}{}$
$\therefore p \to r$

4. $p \to q$
$\dfrac{q \lor r}{}$
$\therefore p \lor r$

5. $p \leftrightarrow q$
$\dfrac{q}{}$
$\therefore p$

6. $p \to q$
$\dfrac{q \to p}{}$
$\therefore p \leftrightarrow q$

7. $p \to \sim q$
$r \to q$
$\dfrac{r}{}$
$\therefore \sim p$

8. $(\sim p \to \sim q)$
$r \to \sim p$
$\dfrac{q}{}$
$\therefore \sim r$

9. $p \to q$
$\dfrac{p \to r}{}$
$\therefore q \land r$

10. $p \to q$
$\dfrac{\sim q \to \sim r}{}$
$\therefore p \to \sim r$

In Exercises 11 through 14 without using truth tables, determine if the arguments are valid.

11. If I go out with my sister or my friend, then I do not drive my car. I am driving my car. Therefore I am not out with my sister.

12. If it is Saturday, then I sleep in. Today is Saturday or not Sunday. Today is Sunday. Therefore I sleep in.

13. You are happy if and only if you are healthy. You are healthy or you are smart. You are not happy. Therefore if you are not happy, then you are not smart.

14. If it is snowing, then I go skiing. If I go skiing, then I am happy. I am happy. Therefore it is snowing.

In Exercises 15 through 26 show that the argument is valid without using truth tables.

15. If I have an egg and orange juice for breakfast, then I do not have cereal. If I do not have orange juice, then I do not have grapefruit juice. I had an egg and grapefruit juice this morning for breakfast. Therefore I did not have cereal.

16. If I study hard, then I make good grades and am not depressed. I am not depressed. I did not make good grades. Therefore I did not study hard.

17.   If I clean my room, my mother is not mad. My mother is mad. Therefore I did not clean my room.

18.   If the price of lumber rises, then the price of houses rises. If the price of steel rises, then the price of cars rises. The price of lumber is rising or the price of steel is rising. Therefore the price of houses is rising or the price of cars is rising.

19.   Jane eats broccoli and spinach. If Jane eats broccoli, then she does not eat spinach. Therefore Jane eats spinach and does not eat broccoli.

20.   I eat an apple or an orange every day. If I do not eat lunch, then I do not eat an apple. I did not eat lunch today. Therefore I ate an orange.

21.   If I do not play baseball, then I play soccer. If I do not play soccer, then I play basketball. Therefore if I do not play soccer, then I play baseball or basketball.

22.   If I work hard or am smart, then I will not fail. If I do not work hard, then I fail and am poor. I failed. Therefore I am poor.

23.   If the rate of inflation increases, then the price of gold increases. If the price of gold increases, then the prices of bonds do not increase. The prices of bonds are increasing. Therefore the rate of inflation is not increasing.

24.   If the summer is hot and dry, then the crops will be poor. If the summer is dry, then there are no floods. Therefore, if there are floods and poor crops, then the summer is not hot.

25.   If the price of sugar rises, then the price of candy rises and I eat less candy. The price of candy is not rising. I am not eating less candy. Therefore the price of sugar is not rising.

26.   If I go out Friday night and Saturday night, then I sleep in on Sunday. I always go out on Saturday night. Therefore if I go out on Friday night, I will sleep in on Sunday.

27.   Use an indirect proof to establish the argument in Exercise 11.

28.   Use an indirect proof to establish the argument in Exercise 20.

29.   Use an indirect proof to establish the argument in Exercise 23.

30.   Use an indirect proof to establish the argument in Exercise 25.

Solutions to Self-Help Exercise Set 12.5

1.   The following truth table, Table 12.29, indicates that $[(p \land q) \land (\sim p)] \to q$ is a tautology. Thus the argument is valid.

Table 12.29

p	q	$p \land q$	$\sim p$	$(p \land q) \land (\sim p)$	$[(p \land q) \land (\sim p)] \to q$
T	T	T	F	F	T
T	F	F	F	F	T
F	T	F	T	F	T
F	F	F	T	F	T

2. Let p, q, and r be the following statements.

   r: It is raining.

   b: I am at the beach.

   g: I have money for gasoline for my car.

Then the argument can be written as follows.

   $r \rightarrow \sim b$

   $g \rightarrow b$

   $\underline{g}$

   $\therefore \sim r$

The proof can be given as follows.

$$(r \rightarrow \sim b) \wedge (g) \wedge (g \rightarrow b) \Rightarrow (r \rightarrow \sim b) \wedge (b) \qquad \text{Modus ponens}$$

$$\Rightarrow \sim r \qquad \text{Modus tollens}$$

The following is another proof.

$$(g) \wedge (g \rightarrow b) \wedge (r \rightarrow \sim b) \Rightarrow (b) \wedge (r \rightarrow \sim b) \qquad \text{Modus ponens}$$

$$\Rightarrow (b) \wedge (b \rightarrow \sim r) \qquad \text{Contrapositive}$$

$$\Rightarrow \sim r \qquad \text{Modus ponens}$$

3. We begin the indirect proof by assuming the conclusion $\sim r$ is false. This means that r is true. If $r \rightarrow \sim b$ is true, we then have that $\sim b$ is true. If $g \rightarrow b$ is true, we then have $\sim g$ is true. But this contradicts the third hypothesis g.

12.6 SWITCHING NETWORKS

We will now see how the principles of logic can be used in the design and analysis of switching networks. A **switching network** consists of an energy input, such as a battery, an output, such as a light bulb, and an arrangement of wires and switches connecting the input and output. A **switch** is a device that is either **closed** or **open**. If the switch is closed, current will flow through the wire. If the switch is open, current will not flow. Because a switch has two states, we can represent the switch by a proposition p that is true if the switch is closed and false if the switch is open.

Normally a house switch on a wall can move in a vertical direction. Electricians usually wire such a switch so that the up position turns the light on (the switch is closed) and the down position turns the light off (the switch is open). Figure 12.1 then indicates our schematic for a switch. Up indicates the switch is closed (the proposition p is true) and down indicates the switch is open (the proposition p is false).

Consider now a network of two switches arranged as shown in Figure 12.2. The switches p and q are said to be in **series**. For this network, current will flow from the input to the output if and only if both of the switches p and q are closed.

Input ———o ▪▪ ——— Output

Figure 12.1

Figure 12.2

(If one or more of the switches are open, then no current can flow.) If we think of p and q as propositions and T corresponds to a closed switch and F to an open switch, then we can write the following truth table, Table 12.30. We notice that this is the truth table for $p \wedge q$. Therefore two switches p and q connected in series corresponds to the conjunction $p \wedge q$.

Table 12.30

p	q	*Current Flow?*	*Truth Assignment*
T	T	yes	T
T	F	no	F
F	T	no	F
F	F	no	F

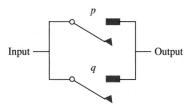

Figure 12.3

Consider now a network of two switches arranged as shown in Figure 12.3. The switches p and q are said to be in **parallel**. For this network, current will flow from the input to the output if, and only if, at least one of the switches p or q is closed. If we again think of p and q as propositions and T corresponds to a closed switch and F to an open switch, then we can write the following truth table, Table 12.31. We notice that this is the truth table for $p \vee q$. Therefore two switches p and q connected in parallel corresponds to the (inclusive) disjunction $p \vee q$.

We will use the standard convention that the input is on the left and the output is on the right and will drop the words ''input'' and ''output'' from the figures.

Table 12.31

p	q	*Current Flow?*	*Truth Assignment*
T	T	yes	T
T	F	yes	T
F	T	yes	T
F	F	no	F

E X A M P L E   1   **Using Truth Tables to Evaluate Networks**

Find a compound statement that represents the network shown in Figure 12.4. By constructing a truth table for this compound statement, determine the conditions under which current will flow.

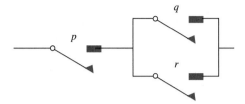

Figure 12.4

Solution   The compound statement is $p \wedge (q \vee r)$. Table 12.32 is the truth table. From the truth table we see that current will flow whenever p is closed and either q or r (or both) are closed. ■

Table 12.32

p	q	r	$q \vee r$	$p \wedge (q \vee r)$
T	T	T	T	T
T	T	F	T	T
T	F	T	T	T
T	F	F	F	F
F	T	T	T	F
F	T	F	T	F
F	F	T	T	F
F	F	F	F	F

Two different switches can be wired so that they open and close simultaneously. In such a case, we let the same letter designate each switch. Consider the network shown in Figure 12.5. By examining the network in Figure 12.5 we see that when the switch p is closed, current flows. When the switch p is open, current does not flow. Thus this network is equivalent to the one shown in Figure 12.1.

We can also use the laws of logic to show this.

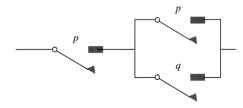

Figure 12.5

E X A M P L E   2   **Using Truth Tables to Evaluate Networks**

Find a compound statement that represents the network shown in Figure 12.5. By constructing a truth table for this compound statement, determine a simpler network that is equivalent.

Solution   The network in Figure 12.5 corresponds to the compound statement $p \wedge (p \vee q)$. Table 12.33 is a truth table. Since the compound statement $p \wedge (p \vee q)$ is equivalent to p, the network in Figure 12.5 is equivalent to the one in Figure 12.1. ∎

Table 12.33

p	q	$p \vee q$	$p \wedge (p \vee q)$
T	T	T	T
T	F	T	T
F	T	T	F
F	F	F	F

The previous two examples indicate how a network can be viewed as a compound statement. The last example indicates how the known laws of logic can sometimes be used to determine an equivalent and simpler network. This type of simplification is extremely important in the economical design of networks for practical problems.

Suppose we have a switch p. A second switch can be wired so that the second switch will be closed when p is open and open when p is closed. In such a case we designate the second switch by $\sim p$ since the statement $\sim p$ is true when p is false and $\sim p$ is false when p is true.

E X A M P L E   3   **Determining Equivalent Networks**

Find an equivalent and simpler network for each of the networks shown in Figure 12.6.

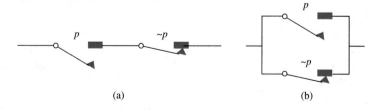

(a)   (b)

Figure 12.6

Solutions   (a) This network corresponds to the compound statement $p \wedge \sim p$, which is always false. An equivalent network is shown in Figure 12.7a. This is a network with a permanent break and current never flows.

(a)   (b)

Figure 12.7

(b) This network corresponds to the compound statement $p \vee \sim p$ which is always true. An equivalent network is shown in Figure 12.7b. This is a network in which current always flows. ■

E X A M P L E   4   **Simplifying a Network**

Simplify the network shown in Figure 12.8.

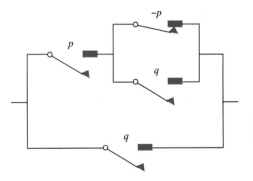

Figure 12.8

Solution   The corresponding compound statement is $[p \wedge (\sim p \vee q)] \vee q$. Using the rules of logic, we then have

$$[p \wedge (\sim p \vee q)] \vee q \Leftrightarrow [(p \wedge \sim p) \vee (p \wedge q)] \vee q \qquad \text{Distributive law}$$
$$\Leftrightarrow [c \vee (p \wedge q)] \vee q \qquad\qquad c \text{ a contradiction}$$
$$\Leftrightarrow (p \wedge q) \vee q \qquad\qquad\qquad \text{Identity law}$$
$$\Leftrightarrow q \qquad\qquad\qquad\qquad\qquad \text{Absorption law}$$

Thus this network can be replaced with the equivalent network shown in Figure 12.1. ■

E X A M P L E   5   **Simplifying a Network**

Simplify the network shown in Figure 12.9.

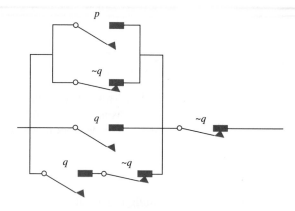

Figure 12.9

Solution   The corresponding compound statement is $(\sim q) \wedge [(p \vee \sim q) \vee q \vee (q \wedge \sim q)]$. Using the rules of logic we then have.

$(\sim q) \wedge [(p \vee \sim q) \vee q \vee (q \wedge \sim q)]$

$\Leftrightarrow (\sim q) \wedge [p \vee ((\sim q) \vee q) \vee (q \wedge \sim q)]$ Associative law

$\Leftrightarrow (\sim q) \wedge [(p \vee t) \vee (q \wedge \sim q)]$ Inverse law

$\Leftrightarrow (\sim q) \wedge [t \vee (q \wedge \sim q)]$ Identity law

$\Leftrightarrow (\sim q) \wedge t$ Identity law

$\Leftrightarrow \sim q$ Identity law

Hence the compound statement is equivalent to the statement $\sim q$. Thus this network can be replaced with the equivalent network shown in Figure 12.10.   ∎

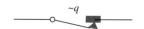

Figure 12.10

SELF-HELP EXERCISE SET 12.6

1.   Find a logic statement that corresponds to the following network.

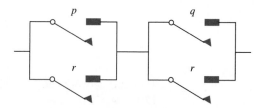

2. Simplify the statement in the previous exercise by using the laws of logic. Then draw a network that corresponds to the simplified statement.

EXERCISE SET 12.6

In Exercises 1 through 4, find a logic statement corresponding to the given network. Determine conditions under which current will flow.

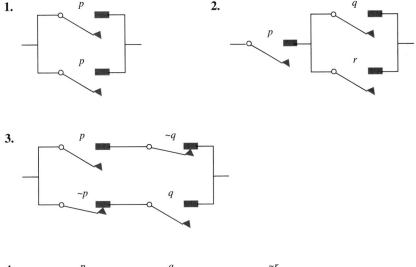

In Exercises 5 through 8, draw the network corresponding to the given logic statement.

5. $p \vee (q \wedge r)$ 6. $(p \wedge q) \wedge r$

7. $(p \vee \sim q) \vee (q \wedge r)$ 8. $[p \vee (\sim q \wedge r)] \wedge q$

In Exercises 9 through 16, find a logic statement corresponding to the given network. Then find a simpler but equivalent network.

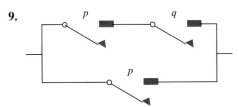

10.

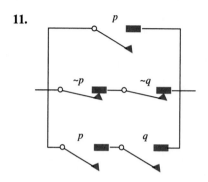

11.

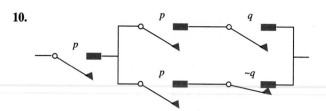

12.

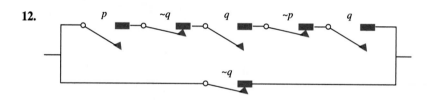

13.

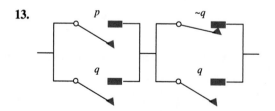

14.

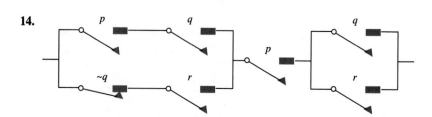

15.

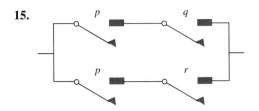

16.

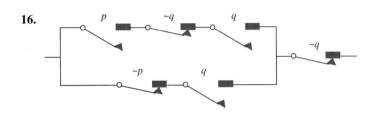

Solutions to Self-Help Exercise Set 12.6

1.   The network corresponds to the statement $(p \lor r) \land (q \lor r)$.

2.   Using the distributive law we have

$$(p \lor r) \land (q \lor r) \Leftrightarrow (p \land q) \lor r.$$

The following network corresponds to this simplified statement.

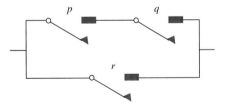

SUMMARY OUTLINE OF CHAPTER 11

- A **statement** is a declarative sentence that is either true or false but not both.

- A **connective** is a word or words, such as ''and'' or ''if and only if,'' that is used to combine two or more simple statements into a compound statement.

- A **conjunction** is a statement of the form ''p and q'' and is written symbolically as $p \land q$. The conjunction $p \land q$ is true if both p and q are true, otherwise it is false.

- A **disjunction** is a statement of the form "p or q" and is written symbolically as $p \lor q$. The disjunction $p \lor q$ is false if both p and q are false and is true in all other cases.

- A **negation** is a statement of the form "not p" and is written symbolically as $\sim p$. The negation $\sim p$ is true if p is false and false if p is true.

- **Truth tables of some important connectives**

p	q	$p \lor q$
T	T	T
T	F	T
F	T	T
F	F	F

p	q	$p \land q$
T	T	T
T	F	F
F	T	F
F	F	F

p	$\sim p$
T	F
F	T

p	q	$p \rightarrow q$
T	T	T
T	F	F
F	T	T
F	F	T

p	q	$p \leftrightarrow q$
T	T	T
T	F	F
F	T	F
F	F	T

- We say that a statement is a **contradiction** if the truth value of the statement is always false no matter what the truth values of its simple component statements.

- We say that a statement is a **tautology** if the truth value of the statement is always true no matter what the truth values of its simple component statements.

- A **conditional statement** is a compound statement of the form "if p, then q" and is written symbolically as $p \rightarrow q$. A conditional statement is false if p is true and q is false, and is true in all other cases.

- In the conditional statement $p \rightarrow q$, the statement p is the **hypothesis** and the statement q is the **conclusion**.

- A **biconditional statement** is a compound statement of the form "p if and only if q" and is written symbolically as $p \leftrightarrow q$. The biconditional $p \leftrightarrow q$ is true only when p and q are both true or both false.

- Two statements p and q are **logically equivalent**, denoted by $p \Leftrightarrow q$, if they have identical truth tables.

- The **contrapositive** of the statement "if p, then q" ($p \rightarrow q$) is of the form "if not q, then not p" and is written $\sim q \rightarrow \sim p$.

- The **converse** of the conditional statement "if p, then q" ($p \rightarrow q$) is "if q, then p" and is written $q \rightarrow p$.

- The **inverse** of the conditional statement "if p, then q" ($p \rightarrow q$) is "if not p, then not q" and is written $\sim p \rightarrow \sim q$.

- Two statements p and q are **logically equivalent**, written $p \Leftrightarrow q$, if the biconditional $p \leftrightarrow q$ is a tautology.

- The **laws of logic** are found on pages 707–708.

- We say that the statement P **logically implies** the statement Q, denoted by $P \Rightarrow Q$, if Q is true whenever P is true. Another way of stating this is that $P \rightarrow Q$ is a tautology.

- An **argument** consists of a set of statements $h_1, h_2, \ldots, h_n$, called **hypotheses** and a statement q called the **conclusion**. An argument is **valid** if the conclusion is true whenever the hypotheses are all true, that is, whenever $h_1 \land h_2 \land \cdots \land h_n \Rightarrow q$.

- In order to give an **indirect proof** of $h_1 \wedge h_2 \wedge \cdots \wedge h_n \Rightarrow q$, we assume that the conclusion q is false and then prove that at least one of the hypotheses h_1, $h_2, \ldots, h_n$, is false.

- A **switching network** consists of an energy input, such as a battery, an output, such as a light bulb, and an arrangement of wires and switches connecting the input and output.

- A **switch** is a device that is either **closed** or **open**.

- Switches can be in **series** (see Figure 12.2), or in **parallel** (see Figure 12.3).

Chapter 12 Review Exercises

In Exercises 1 through 4 decide which sentences are statements.

1. Water boils at 152°F.

2. George Washington never told a lie.

3. Is it true that $2 + 2 = 4$?

4. $x^2 = -1$ has two solutions.

5. Let p and q denote the following statements:

   p: Abraham Lincoln was the shortest President of the United States.
   q: Carson City is the capital of Nevada.

Express the following compound statements in words.

   (a) $\sim p$, **(b)** $p \wedge q$, **(c)** $p \vee q$, **(d)** $\sim p \wedge q$, **(e)** $p \vee \sim q$,
   (f) $\sim(p \wedge q)$

6. Let p and q denote the statements in the previous exercise.
   (a) State the negation of these statements in words.
   (b) State the disjunction for these statements in words.
   (c) State the conjunction for these statements in words.

7. Let p and q denote the following statements:

   p: GM makes trucks.
   q: GM makes toys.

Express the following statement symbolically.

   (a) GM does not make toys.
   (b) GM makes trucks, or GM makes toys.
   (c) GM makes trucks, or GM does not make toys.
   (d) GM does not make trucks, and GM does not make toys.

In Exercises 8 through 9 construct truth tables for the given statements.

8. $(\sim p \wedge q) \wedge (\sim p \vee q)$ **9.** $(p \vee q) \vee (q \wedge \sim r)$

10. Determine if the statement $[(p \vee \sim q) \vee q \vee (\sim p \vee q)] \vee \sim q$ is a tautology.

11. Determine if the statement $[(p \wedge \sim q) \wedge q] \vee (\sim p \wedge q)$ is a contradiction.

12. Let p and q be the statements:

p: China is in Asia.
q: Mexico is in South America.

Determine the truth value of the following statements.

(a) $\sim q$, (b) $p \wedge \sim q$, (c) $\sim p \wedge q$, (d) $\sim(p \wedge q)$

13. Let p and q be the statements:

p: Inflation is increasing.
q: The price of gold is rising.

Write each of the following statements in English. Then write both in English and symbolically the contrapositive, inverse, and converse of the given statement.

(a) $p \rightarrow q$, (b) $p \rightarrow \sim q$, (c) $\sim q \rightarrow \sim p$

In Exercises 14 through 15 construct a truth table for the given statement.

14. $\sim(\sim p \rightarrow q)$ **15.** $(p \wedge q) \rightarrow r$

In Exercises 16 through 19 establish the given law of logic.

16. $p \wedge (p \rightarrow q) \Rightarrow q$ **17.** $p \rightarrow (q \rightarrow r) \Rightarrow (p \wedge q) \rightarrow r$

18. $p \vee (p \wedge q) \Leftrightarrow p$ **19.** $(p \rightarrow q) \Leftrightarrow \sim(p \wedge \sim q)$

20. Negate the following using the De Morgan laws.
(a) Sue will major in math or physics.
(b) Sue enjoys math and physics.

21. Use truth tables to establish whether or not the following argument is valid.

$p \rightarrow \sim q$
$\sim r \rightarrow p$
q
_____
$\therefore r$

22. Show that the following argument is valid without using truth tables.
If it is sunny, then I do not study. If it is not sunny, then I am not at the beach. It is sunny and I am at the beach. Therefore I am not studying.

23. Find a logic statement for the following network. Then find a simpler but equivalent network.

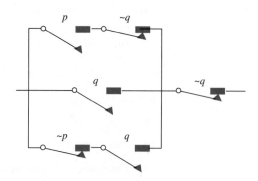

A

Special Topics

A.1 COMMON LOGARITHMS

▶ *Basic Properties of Common Logarithms*

▶ *Laws of Common Logarithms*

▶ *Solving Equations with Exponents and Logarithms*

Basic Properties of Common Logarithms

Given any number x, the solution y to the equation

$$10^y = x$$

is called the **common logarithm of x** and is denoted by $\log x$. Thus the common logarithm is an *exponent*. In fact $\log x$ is *the exponent to which 10 must be raised in order to obtain the number x*. We have the following definition.

> **Common Logarithm or Logarithm to the Base 10**
>
> If $x > 0$,
>
> $$y = \log x \quad \text{if, and only if,} \quad x = 10^y.$$

EXAMPLE 1   **Using the Definition of Common Logarithm**

Find (a) $\log 100$,   (b) $\log 0.001$,   (c) $\log 1$,   (d) $\log 10$.

Solutions

(a) $\log 100 = 2$ since $10^2 = 100$
(b) $\log 0.001 = -3$ since $10^{-3} = 0.001$
(c) $\log 1 = 0$ since $10^0 = 1$
(d) $\log 10 = 1$ since $10^1 = 10$ ∎

EXAMPLE 2   **Using the Definition of Common Logarithm**

Solve for x in (a) $\log x = 3$,   (b) $\log x = -3$,   (c) $\log 10^2 = x$.

Solutions

(a) $\log x = 3$ if, and only if, $x = 10^3 = 1000$
(b) $\log x = -3$ if, and only if, $x = 10^{-3} = 0.001$
(c) $\log 10^2 = x$ if, and only if, $10^x = 10^2$. This is true if, and only if, $x = 2$.   ∎

REMARK.   Notice that one can only have $x > 0$ in the definition of the common logarithm since $x = 10^y$ is always positive.

According to the definition of common logarithm, $\log x$ is that exponent to which 10 must be raised in order to obtain x. This can also be stated as the following important formula:

$$x = 10^{\log x}.$$

We now develop another similar important formula. According to the definition of common logarithm, for any $x > 0$, $y = \log 10^x$ if, and only if, $10^x = 10^y$. But this implies that $x = y = \log 10^x$, or $x = \log 10^x$. These two important formulas are now summarized.

> **Basic Properties of the Common Logarithm**
>
> 1. $10^{\log x} = x$ if $x > 0$.
> 2. $\log 10^x = x$ for all x.

E X A M P L E   3   **Using the Basic Properties of Common Logarithm**

Find (a) $\log 10^5$ (b) $10^{\log \pi}$

Solutions   (a)   Use Property 2 to obtain $\log 10^5 = 5$.

(b)   Use Property 1 to obtain $10^{\log \pi} = \pi$.   ∎

Laws of Common Logarithms

The common logarithm $\log x$ obeys the following laws.

Laws for the Common Logarithm

1. $\log xy = \log x + \log y$.

2. $\log \dfrac{x}{y} = \log x - \log y$.

3. $\log x^a = a \log x$.

To establish these laws, we need to use $x = 10^{\log x}$, $y = 10^{\log y}$, $xy = 10^{\log xy}$, $\dfrac{x}{y} = 10^{\log \frac{x}{y}}$. To establish Rule 1, notice that

$$10^{\log xy} = xy = 10^{\log x}10^{\log y} = 10^{\log x + \log y}.$$

Then equating exponents gives Rule 1.

One can establish Rule 2 in a similar manner. Notice that

$$10^{\log \frac{x}{y}} = \frac{x}{y} = \frac{10^{\log x}}{10^{\log y}} = 10^{\log x - \log y}.$$

Then equating exponents gives Rule 2.

To establish Rule 3, notice that

$$10^{\log x^a} = x^a = (10^{\log x})^a = 10^{a \log x}.$$

Equating exponents gives Rule 3.

John Napier (1550–1617) invented logarithms in the early 17th century so that the numerical calculations demanded by astronomy, navigation, engineering, and other areas could be done more quickly. This invention was enthusiastically adopted quickly throughout Europe. The great mathematician Laplace commented that the invention of logarithms ''by shortening the labors, doubled the life of the astronomer.''

Let us now see how logarithms can aid in the calculation of a complex expression.

E X A M P L E 4   **Using Logarithms to Simplify Calculations**

Express

$$\log \frac{x^2 \sqrt{y}}{z^3}$$

in terms of the common logarithms of x, y, and z.

Solution   Using the properties of logarithms

$$\log \frac{x^2 \sqrt{y}}{z^3} = \log \frac{x^2 y^{1/2}}{z^3}$$

$$= \log x^2 y^{1/2} - \log z^3$$

$$= \log x^2 + \log y^{1/2} - \log z^3$$

$$= 2 \log x + \frac{1}{2} \log y - 3 \log z. \quad \blacksquare$$

If the common logarithms of x, y, and z are known (from a table, for example), then this last expression v can be easily evaluated. One then looks in the table to find the number u for which $\log u = v$. The number u is then $\dfrac{x^2 \sqrt{y}}{z^3}$. This was how complex calculations were done in Napier's time and up until the advent of the modern computer. (Napier expended enormous amounts of time creating a table of common logarithms that was then circulated throughout the scientific community.)

Although tables of common logarithms are no longer used, the common logarithm is an important tool for solving equations involving exponents.

Solving Equations with Exponents and Logarithms

If one has an equation with an unknown exponent to the power 10, then at an appropriate point one must take the common logarithm, as in the following example.

E X A M P L E 5   **Solving an Equation with an Exponent**

Solve $15 = 5 \cdot 10^{4x-1}$.

Solution

$$15 = 5 \cdot 10^{4x-1}$$

$$3 = 10^{4x-1}$$

$$\log 3 = \log 10^{4x-1} = 4x - 1 \qquad \text{Property 2}$$

$$x = \frac{1}{4}(1 + \log 3) \quad \blacksquare$$

If one has an equation with a logarithm, then at an appropriate point one can use the definition of common logarithm, as in the following example.

E X A M P L E   6   **Solving an Equation with a Logarithm**

Solve $2 \log(2x + 5) + 6 = 0$.

Solution

$$2 \log(2x + 5) + 6 = 0$$
$$\log(2x + 5) = -3$$
$$2x + 5 = 10^{-3} \qquad \text{Definition of logarithm}$$
$$x = \frac{1}{2}(10^{-3} - 5) \quad \blacksquare$$

R E M A R K .   In Example 6, we had $\log(2x + 5) = -3$ at one point. At this point we could apply Property 1 and obtain

$$10^{-3} = 10^{\log(2x+5)} = 2x + 5.$$

SELF-HELP EXERCISE SET A.1

1. Solve for x in the equation $10^{x^3} = 5$.
2. Solve for x in the equation $1500 = 1000 \cdot 10^{5x}$.

EXERCISE SET A.1

In Exercises 1 through 6 solve for x.

1. $\log x = 6$	**2.** $\log x = -5$	**3.** $\log x = -4$
4. $\log x = 5$	**5.** $3 \log x = 9$	**6.** $2 \log x = 8$

In Exercises 7 through 16 simplify.

7. $\log 10^4$	**8.** $\log \sqrt{10}$	**9.** $\log \dfrac{1}{10}$
10. $\log \dfrac{1}{\sqrt{10}}$	**11.** $10^{\log 2\pi}$	**12.** $10^{\log \sqrt{2}}$
13. $10^{3 \log 2}$	**14.** $10^{0.5 \log 9}$	**15.** $(10^{\log 3})^4$
16. $\sqrt{10^{\log 2}}$		

In Exercises 17 through 22 write the given quantity in terms of log x, log y, and log z.

17. $\log x^2 \sqrt{y}\, z$

18. $\log \sqrt{xyz}$

19. $\log \dfrac{\sqrt{xy}}{z}$

20. $\log \dfrac{xy^2}{z^3}$

21. $\log \sqrt{\dfrac{xy}{z}}$

22. $\log \dfrac{x^2y^3}{\sqrt{z}}$

In Exercises 23 through 28 write the given quantity as one common logarithm.

23. $2 \log x + \log y$

24. $2 \log x - \log y$

25. $\frac{1}{2} \log x - \frac{1}{3} \log y$

26. $2 \log x - \frac{1}{2} \log y + \log z$

27. $3 \log x + \log y - \frac{1}{3} \log z$

28. $\sqrt{2} \log x - \log y$

In Exercises 29 through 40 solve the equation for x.

29. $10 \cdot 10^{5x} = 3$

30. $2 \cdot 10^{3x} = 5$

31. $6 = 2 \cdot 10^{-2x}$

32. $3 = 4 \cdot 10^{-0.5x}$

33. $2 \cdot 10^{3x-1} = 1$

34. $3 \cdot 10^{2-5x} = 4$

35. $10^{x^2} = 4$

36. $10^{\sqrt{x}} = 4$

37. $2 \log(x + 7) + 3 = 0$

38. $\log x^2 = 9$

39. $\log(\log 4x) = 0$

40. $\log 3x = \log 6$

Solutions to Self-Help Exercise Set A.1

1. Taking the common logarithm of each side gives

$$\log 5 = \log 10^{x^3} = x^3.$$

Thus $x = \sqrt[3]{\log 5}$.

2.
$$1500 = 1000 \cdot 10^{5x}$$
$$1.5 = 10^{5x}$$
$$\log 1.5 = \log 10^{5x} = 5x$$
$$x = \frac{1}{5} \log 1.5$$

A.2   ARITHMETIC AND GEOMETRIC SEQUENCES

► *Sequences*
► *Arithmetic Sequences*
► *Geometric Sequences*
► *Multiplier Effect*

Thomas Malthus,
1766–1834

Malthus was an English economist famous for his assertion that population, if unchecked, will tend to increase according to a geometric progression, whereas the means of subsistence will increase only according to an arithmetic progression. Thus he concluded that population will always expand to the limit of subsistence and will be held there by famine, war, and poor health. The two sequences below indicate how a geometric progression with the same first two terms as an arithmetic progression will grow substantially more rapidly.

1, 2, 3, 4, 5, 6, 7, 8, 9, . . .   Arithmetic progression
1, 2, 4, 8, 16, 32, 64, 128, 256, . . .   Geometric progression

A P P L I C A T I O N

Profits Increasing at a
Geometric Rate

A very small company had profits of \$200,000 last year, and profits are assumed to increase by 10% each year. What is the profit in the tenth year and the total profit over the next 10 years? (The answers are given in Examples 6 and 7.)

Sequences

If we were to list the positive even integers, we could write

$$2, 4, 6, 8, \ldots$$

where the three dots indicate that this list goes on forever. This is an example of a sequence.

If a company deposits \$5000 into an account that pays 10% simple interest per year, then the total amount in the account over subsequent years can be listed as

$$5000, 5050, 5100, 5150, 5200, \ldots$$

This is another example of a sequence. The definition of a sequence is given as follows.

Sequence

A **sequence** is a list of numbers

$$a_1, a_2, a_3, \ldots$$

in which there is a first one, a second one, a third one, and so on, so that given any positive integer n, there is an nth one.

The total amount in the account in the above example can be written as

$$a_n = 5000 + 50(n - 1),$$

since $a_1 = 5000 + 50(0) = 5000$, $a_2 = 5000 + 50(1) = 5050$, and so on. The values

$$a_1 = 5000, a_2 = 5050, a_3 = 5100, \ldots$$

are called the **terms** of the sequence, with a_1 the first term, a_2 the second term, and a_n the nth term.

EXAMPLE 1   **Finding the Terms of a Sequence**

Let the nth term of a sequence be given by $a_n = 4 + 5n$. Find the values of the first four terms and also the value of the 20th term.

Solution   We have

$$a_1 = 4 + 5(1) = 9$$

$$a_2 = 4 + 5(2) = 14$$

$$a_3 = 4 + 5(3) = 19$$

$$a_4 = 4 + 5(4) = 24$$

$$a_{20} = 4 + 5(20) = 104 \quad \blacksquare$$

Arithmetic Sequences

The sequence in Example 1 has the property that successive terms always differ by the same number. In fact

$$a_2 - a_1 = 14 - 9 = 5, \qquad a_3 - a_2 = 19 - 14 = 5,$$

$$a_4 - a_3 = 24 - 19 = 5.$$

In general

$$a_n - a_{n-1} = (5n + 4) - (5[n - 1] + 4) = 5n + 4 - 5n + 5 - 4 = 5.$$

The number 5 is called the **common difference**. A sequence in which successive terms differ by the same number is called an **arithmetic sequence**.

Arithmetic Sequence

A sequence a_n is an arithmetic sequence if successive terms have the same difference, d, that is,

$$a_n - a_{n-1} = d, \qquad n = 2, 3, \ldots$$

The number d is called the common difference.

For an arithmetic sequence a_n, we have

$$a_n = a_{n-1} + d, \quad n = 2, 3, \ldots$$

In other words each term (after the first) in an arithmetic sequence is obtained by adding the number d to the previous term. Thus

$$a_2 = a_1 + d$$

$$a_3 = a_2 + d = (a_1 + d) + d = a_1 + 2d$$

$$a_4 = a_3 + d = (a_2 + 2d) + d = a_1 + 3d$$

$$a_5 = a_4 + d = (a_4 + 3d) + d = a_1 + 4d$$

We see from this pattern that $a_n = a_1 + (n - 1)d$. We have the following.

---

nth Term of an Arithmetic Sequence

If a_n is the nth term of an arithmetic sequence, then

$$a_n = a_1 + (n - 1)d.$$

where d is the common difference.

E X A M P L E   2   **Finding the Terms of an Arithmetic Sequence**

Given the sequence

$$2, 5, 8, 11, \ldots$$

find the 16th and 31st terms.

Solution   First we notice that successive terms differ by 3. Thus this sequence is an arithmetic sequence with $d = 3$. Since $a_1 = 2$, we have

$$a_n = 2 + (n - 1)3$$

$$a_{16} = 2 + (16 - 1)3 = 47$$

$$a_{31} = 2 + (31 - 1)3 = 92 \quad \blacksquare$$

E X A M P L E   3   **Calculating a Term in an Arithmetic Sequence**

A firm will have profits of $3,000,000 this quarter with profits assumed to increase by $90,000 each quarter. What will be the profits in the 20th quarter?

Solution   Since profits increase by $90,000 each quarter, the quarterly profits form an arithmetic sequence with a common difference of $d = 90,000$ and a first term equal to 3,000,000. Thus if P_n represents the profits in the nth quarter, we have

$$P_n = 3,000,000 + (n - 1)90,000.$$

Then the profits in the 20th quarter are

$$P_{20} = 3,000,000 + (20 - 1)90,000 = 4,710,00. \quad \blacksquare$$

In the previous example, one might wish to know what were the total profits over a period of time. This would require finding the sum of terms of the arithmetic sequence. We now consider finding the sum of the first k terms of an arithmetic sequence.

Let a_n be the nth term of an arithmetic sequence with common difference d and let S_k be the sum of the first k terms of this sequence, that is,

$$S_k = a_1 + a_2 + a_3 + \cdots + a_{k-1} + a_k.$$

Since $a_n = (n - 1)d + a_1$, this becomes

$$S_k = [a_1] + [d + a_1] + [2d + a_1] + \cdots + [(k - 1 - 1)d + a_1] + [(k - 1)d + a_1]$$

$$= [a_1] + [d + a_1] + [2d + a_1] + \cdots + [(k - 1)d + a_1 - d] + [(k - 1)d + a_1]$$

$$= [a_1] + [d + a_1] + [2d + a_1] + \cdots + [a_k - d] + [a_k]$$

Now rewrite this sum in the reverse order to obtain

$$S_k = [a_k] + [a_k - d] + \cdots + [2d + a_1] + [d + a_1] + [a_1].$$

Now add the last two displayed lines and obtain

$$2S_k = [a_1 + a_k] + [d + a_1 + a_k - d] + \cdots + [a_k - d + d + a_1] + [a_k + a_1]$$

$$= [a_1 + a_k] + [a_1 + a_k] + \cdots + [a_1 + a_k] + [a_1 + a_k]$$

$$= k[a_1 + a_k]$$

Since $a_k = a_1 + (k - 1)d$, we can also write

$$S_k = \frac{k}{2}[a_1 + a_1 + (k - 1)d] = \frac{k}{2}[2a_1 + (k - 1)d]$$

We have thus shown the following.

Sum of k Terms of an Arithmetic Sequence

Let a_n be an arithmetic sequence. The sum of the first k terms, S_k is given by

$$S_k = \frac{k}{2}[a_1 + a_k] = \frac{k}{2}[2a_1 + (k - 1)d].$$

E X A M P L E   4   **Finding the Sum of an Arithmetic Series**

Find the total profits over the next 20 quarters for the company in Example 3.

Solution   From Example 3, we already know $P_{20} = 4{,}710{,}000$. Since $P_1 = 3{,}000{,}000$, the sum of the profits over the next 20 quarters is

$$S_{20} = \frac{20}{2}[P_1 + P_{20}] = \frac{20}{2}[3{,}000{,}000 + 4{,}710{,}000] = 77{,}100{,}000. \quad \blacksquare$$

Geometric Sequences

If \$100 is deposited into an account earning 6% a year compounding annually, the amount in the account at the beginning of each year is given by the sequence

$$100, \; 100(1.06), \; 100(1.06)^2, \; 100(1.06)^3, \; \ldots$$

This is an example of a *geometric sequence*.

Geometric Sequence

A sequence is a geometric sequence if the ratio of any term (after the first) to the preceding term is always the same. This ratio is called the common ratio.

Thus the above sequence has ratios

$$\frac{100(1.06)}{100} = 1.06, \; \frac{100(1.06)^2}{100(1.06)} = 1.06, \; \frac{100(1.06)^3}{100(1.06)^2} = 1.06, \ldots$$

all of which are the same and thus the sequence is a geometric sequence.

Suppose we have a geometric sequence with first term a_1 and common ratio r. Then

$$\frac{a_2}{a_1} = r \text{ implies } a_2 = a_1 r$$

$$\frac{a_3}{a_2} = r \text{ implies } a_3 = a_2 r = (a_1 r)r = a_1 r^2$$

$$\frac{a_4}{a_3} = r \text{ implies } a_4 = a_3 r = (a_1 r^2)r = a_1 r^3$$

$$\frac{a_5}{a_4} = r \text{ implies } a_5 = a_4 r = (a_1 r^3)r = a_1 r^4$$

And in general we can see that $a_n = a_1 r^{n-1}$. Thus we have the following.

---

nth Term of Geometric Sequence

If a_1 is the first term of a geometric sequence and r is the common ratio, then the nth term is given by

$$a_n = a_1 r^{n-1}.$$

Thus each term after the first is obtained by multiplying the previous term by r.

If an amount P is deposited into an account earning interest at a rate of i per period where i is expressed as a decimal, then the amount in the account at the beginning of each period is

$$P, P(1 + i), P(1 + i)^2, P(1 + i)^3, \ldots$$

This is a geometric sequence, the first term is $a_1 = P$, and the common ratio is $r = (1 + i)$. Thus the nth term is given by

$$a_n = a_1 r^{n-1} = P(1 + i)^{n-1},$$

and represents the amount in the account after the $(n - 1)$st period.

E X A M P L E   5   **Finding the nth Term of a Geometric Sequence**

Given the following geometric sequence

$$2, -6, 18, -54, \ldots,$$

find the nth term and the fifth term.

Solution   First we notice that the sequence is a geometric sequence since all the ratios

$$\frac{-6}{2} = -3, \frac{18}{-6} = -3, \frac{-54}{18} = -3, \ldots$$

are the same value -3. Since the first term is $a_1 = 2$,

$$a_n = a_1 r^{n-1} = 2(-3)^{n-1}.$$

Then

$$a_5 = 2(-3)^{5-1} = 162. \quad \blacksquare$$

E X A M P L E   6   **Finding the nth Term of a Geometric Sequence**

A very small company had profits of \$200,000 in the first year and profits are assumed to increase by 10% each year. Find a formula for the profit in the nth year. Find the profit in the tenth year.

Solution   The profit in the first year is \$200,000, and since the profit in any year is obtained by multiplying 1.10 times the profit in the previous year, the profit P_n in the nth year is a geometric sequence with first term $P_1 = 200,000$ and common ratio $r = 1.10$. Thus the profit in the nth year is given by

$$P_n = 200,000(1.10)^{n-1},$$

and in the tenth year is

$$P_{10} = 200,000(1.10)^9 \approx 471,589.54 \quad \blacksquare$$

As we shall see shortly, in order to solve many financial problems it will be necessary to find the sum of the first k terms in a geometric series. We do this now.

Let a_n be a geometric series with common ratio r. Let S_k be the sum of the first k terms of this geometric series. Then

$$S_k = a_1 + a_2 + a_3 + \cdots + a_{k-1} + a_k$$

$$= a_1 + a_1 r + a_1 r^2 + \cdots + a_1 r^{k-2} + a_1 r^{k-1}$$

If $r = 1$, then $S_k = ka_1$. If $r \neq 1$, then multiply S_k by r and obtain

$$rS_k = a_1 r + a_1 r^2 + a_1 r^3 + \cdots + a_1 r^{k-1} + a_1 r^k.$$

If we now substract S_k from rS_k, notice that all but two of the terms cancel and we have

$$rS_k - S_k = a_1 r^k - a_1$$

$$(r - 1)S_k = a_1(r^k - 1)$$

$$S_k = a_1 \frac{r^k - 1}{r - 1}$$

Thus we have the following.

Sum of First k Terms of a Geometric Sequence

If a_1 is the first term and r is the common ratio of a geometric sequence, then the sum of the first k terms of this geometric sequence is given by

$$S_k = a_1 \frac{r^k - 1}{r - 1} \qquad r \neq 1.$$

E X A M P L E   7   **Finding the Sum of k Terms of a Geometric Sequence**

Find the total profits earned by the company in the previous example over the next 10 years.

Solution   The profit, in thousands of dollars, of the company for the first year is 200 and is increasing by 10% each year. Since the yearly profits form a geometric sequence with first term $a_1 = 200$ and common ratio $r = 1.10$, the total profit over the next 10 years is given by

$$S_{10} = 200 \frac{(1.1)^{10} - 1}{1.1 - 1} = (200)(10)[(1.1)^{10} - 1] \approx 3187.4849$$

or $3,187,484.90.   ■

E X A M P L E   8   **Comparing Arithmetic and Geometric Sequences**

Solution   The company in the last example had profits of $200,000 in the first year and an increase in profits of $20,000 in the next year, with profits continuing to increase at a geometric rate. Suppose a second company had profits in the first year of $250,000 and that profits also increase by $20,000 the next year, but increased in subsequent years at an arithmetic rate. Find the profit of the second company in the tenth year and compare this to that of the first company.

Solution   The profit of the second company is given by an arithmetic sequence with first term equal to \$250,000 and common difference \$20,000. Thus the profit P_n^* in the nth year in thousands of dollars is given by $P_n^* = 20(n - 1) + 250$ and thus

$$P_{10}^* = 20(10 - 1) + 250 = 430,$$

or \$430,000. This compares with a profit of \$471,589.54 for the first company. ∎

Notice how initially the first company had profits substantially lower than the second company, with both companies having the same increase in profits the next year. Nonetheless, profits of the first company exceeded that of the second by the tenth year. Thus the figure below indicates how the profits of the company growing at a geometric rate easily overtake that of the second company growing only at an arithmetic rate. Notice that the graph of the profits of the second company with the arithmetic growth rate follows along a (straight) *line*.

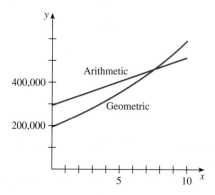

Self-Help Exercise 2 will further illustrate the rate at which a geometric sequence can increase over a number of time periods.

Multiplier Effect

One way the Federal Reserve exercises control over the banking system is by setting reserve requirements. Typically, the Federal Reserve may require banks to have on reserve at least 5% of any amount of money that is lent out. (The Federal Reserve changes this percent from time to time, depending on economic conditions.)

E X A M P L E   9   **Calculating the Effect of an Injection of Funds by the Federal Reserve**

Suppose the Federal Reserve injects \$1000 into the monetary system and this money is deposited in a bank. This bank loans out \$1000(.95) and still meets the reserve requirements. The money \$1000(.95) is then deposited in another bank.

This second bank then loans out 95% of this or $1000(.95)^2$. If this continues, how much money will enter the system after 5 turnovers?

Solution   The amount of money in the system after 5 such moves is

$$1000 + 1000(.95) + 1000(.95)^2 + 1000(.95)^3 + 1000(.95)^4.$$

But this is just the sum S_5 of the first 5 terms of a geometric sequence with first term 1000 and common ratio $r = 0.95$. This is

$$S_k = a_1 \frac{1 - r^k}{1 - r} = 1000 \frac{1 - (.95)^5}{.05} = 4521. \quad \blacksquare$$

If the number of turnovers is k, the amount of money entering the system would be

$$S_k = a_1 \frac{1 - (.95)^k}{.05}.$$

If k is large then $(.95)^k$ is small and thus S is approximately $\dfrac{1}{1 - r} = \dfrac{1}{1 - .95} = $ 20 times the original amount.

The fraction $\dfrac{1}{1 - r} = 20$ is called the "multiplier effect." This means that for each of the original dollars injected into the system, very nearly 20 dollars could be loaned out if the money went through many transactions as indicated above. In normal times one would only expect 5 or 6 turnovers in a year, so that reaching the upper limit of 20 times the original amount is not likely.

SELF-HELP EXERCISE SET A.2

1. An individual borrows $10,000 from a bank at 1% a month interest. Each month $500 is paid toward the principal plus interest on the balance. (a) Find the first three payments, and note that the payments form an arithmetic sequence. (b) Find the 20th payment. (c) Find the total amount paid to the bank.

2. In a classic fairy tale, the King asks the Wise Man what reasonable reward can he give for the great contributions the Wise Man has made to the Kingdom. The Wise Man responds that he only wishes one small copper coin (about $1) today, 2 tomorrow, 4 the day after, and so on, for the next month (30 days). The King grants what appears a modest request. Determine approximately (without using a calculator but using the approximation $2^{10} \approx 1000$) how much the Wise Man will receive.

EXERCISE SET A.2

In Exercises 1 through 6 a formula for the nth term of a sequence is given. Decide in each case if the sequence is an arithmetic sequence. If it is, find the sum of the first 20 terms.

1. $a_n = 3 + 2n$ 2. $a_n = 2 + 3n$

3. $a_n = n^2$ 4. $a_n = \sqrt{n}$

5. $a_n = 0.5 + 0.2n$ 6. $a_n = -100 - n/10$

In Exercises 7 through 12 you will find the first several terms of a sequence that continues in the implied way. Decide in each case if the sequence is an arithmetic sequence. If it is, find the 11th term and the sum of the first 200 terms.

7.   10, 14, 18, 22, ...   8.   30, 27, 24, 21, ...

9.   10, 13, 17, 22, ...   10.   11.1, 11.4, 11.7, 12.0, ...

11. $-10, -7, -4, -1, \ldots$ 12.   2, 4, 8, 16, ...

In each of the Exercises 13 through 18 you will find the nth term a_n of an arithmetic sequence and some additional information. Find the indicated quantity.

13.   If $a_1 = 10$, $d = 3$, then find a_{12}.

14.   If $a_1 = 100$, $d = -3$, then find a_{21}.

15.   If $a_1 = 10$, $a_{11} = 5$, then find d.

16.   If $a_{11} = -10$, $d = 5$, then find a_1.

17.   If $a_{13} = 20$, $a_3 = 5$, find a_1 and d.

18.   If $a_{24} = -10$, $a_4 = 10$, find a_1 and d.

In Exercises 19 through 24 you will find the first several terms of a sequence that continues in the implied way. Decide in each case if the sequence is geometric. If it is find the 11th term and the sum of the first 20 terms.

19.   3, 12, 48, 192, ...   20.   10, 1, 0.1, 0.01, ...

21. $3, -3, 3, -3, \ldots$ 22.   3, 1, 3, 1, ...

23.   27, 9, 3, 1, ...   24. $1, \sqrt{2}, 2, 2\sqrt{2}, \ldots$

Assume in Exercises 25 through 28 that the sequence is a geometric sequence a_n and find the indicated quantity.

25.   If $a_1 = 10$, $r = 3$, find a_4.   26.   If $a_1 = 1$, $r = (1.08)^{-1}$, find a_5.

27.   If $a_2 = 8$, $a_5 = 1$, find a_n.   28.   If $a_4 = 1$, $a_7 = 27$, find a_n.

Find the sum of the first five terms of each of the geometric sequences given in Exercises 29 through 32.

29. $a_1 = 3$, $r = 2$.   30. $a_1 = 3$, $r = -2$.

31. $a_1 = 2$, $r = \frac{2}{3}$.   32. $a_1 = 1$, $r = -1$.

33.   **Sum of Integers.**   For any positive integer k find

$$1 + 2 + 3 + \cdots + k.$$

(*Hint:* Consider the arithmetic sequence $a_n = n$).

Applications

34. **Loan Repayment.** An indvidual borrows $8000 from a bank at 1% interest per month. Each month $500 is paid toward the principal plus interest on the balance. (a) Find the first three payments, and note that the payments form an arithmetic sequence. (b) Find the 11th payment.

35. **Loan Repayment.** Find the total amount of interest paid to the bank in Exercise 34.

36. **Loan Repayment.** In the problem before last, what is the total amount paid to the bank over the life of this loan?

37. **Depreciation.** A piece of machinery depreciates by a fixed amount each year. Its value at the end of the third year is $45,000 and after the sixth year is $30,000. How much was the machinery originally worth?

38. **Depreciation.** A piece of machinery that originally cost $50,000 depreciates by a fixed amount each year. If the life of the machinery is 11 years and the scrap value is $2000, how much is it being depreciated each year?

39. **Salary Increases.** A person has a salary of $31,000 in the first year and receives the same raise (in dollars) every year. In the seventh year he makes $43,000. How much does he make in the 11th year?

40. **Salary Increases.** A person has a salary of $30,000 in the first year and receives a salary increase of $2000 each and every year. What is her total earnings over a forty-year period?

41. **Depreciation.** A computer is depreciated at 50% annually. If the computer originally cost $16,000 and has a scrap value of $1000, what is the life of this computer?

42. **Depreciation.** A computer is depreciated at 20% annually. Write a formula for its value V_n after n years if it originally cost $10,000.

43. **Depreciation.** Suppose a piece of machinery with an initial value of I is depreciated at an annual rate of i, where i is a decimal. Then the value V_n after n years is given by

$$V_n = I(1 - i)^n.$$

If the scrap value of the machinery is E, show that the life of the machinery is given by

$$N = \frac{\log(E/I)}{\log(1 - i)}.$$

44. **Depreciation.** A computer depreciates at 20% annually, has an initial value of $20,000, and a scrap value of $2000. Use the formula in the previous problem to find its useful life.

45. **Velocity of Money.** Suppose the Federal Reserve has set reserve requirements for banks at 5%. Suppose the Federal Reserve then injects $1 billion into the monetary system and the money goes through 6 transactions in one year of the type indicated in Example 9. What is the total effect in terms of dollars on the economy? Compare your answers to that found in Example 9. If the number of transactions per year is different, then the impact on the economy of the injected money has a different effect. This makes it difficult for the Federal Reserve to gauge its effect on the economy when it injects money into the monetary system.

46. **Velocity of Money.**   Suppose, in Example 9 of the text, the number of turnovers was (a) 100, (b) 200. How much money will enter the system? Compare your answers to the multiplier $\dfrac{1}{1 - r} = \dfrac{1}{1 - .95} = 20$.

47. **Multiplier Effect.**   A city is attempting to attract a convention to use its civic center. It believes that for every dollar spent by the people attending the convention and winding up in the hands of local businesses, these local businesses will spend 80% locally, and 80% of this locally spent money will in turn be spent locally, and so on. Find the effect that one dollar will have if turned over (a) three times, (b) six times, (c) nine times.

Solutions to Self-Help Exercise Set A.2

1. **a.**   If an individual borrows $10,000 from a bank at 1% a month interest, then the interest at the end of the first month is $10,000(.01) = $100. The first payment is this interest plus the principal of $500, which totals $600.

    The principal has been reduced by $500 and is now $10,000 − $500 = $9500. The interest on this for the next month is $9500(.01) = $95. The second payment is then $95 + $500 = $595.

    The principal has been reduced by another $500 and is now $9500 − $500 = $9000. The interest on this for the next month is $9000(.01) = $90. The third payment is then $90 + $500 = $590.

    b.   In summary, the three payments are $600, $595, $590. Since the difference between successive terms is the same, the sequence is arithmetic with common difference $d = -5$ and first term $a_1 = 600$. Thus

$$a_n = 600 + (n - 1)(-5),$$

    where a_n is the nth payment. Thus the 20th payment is

$$a_{20} = 600 + (20 - 1)(-5) = 505.$$

    c.   Twenty payments are needed to pay off the loan. If the sum of the first 20 payments is denoted by S_{20}, then

$$S_{20} = \frac{20}{2}\,[a_1 + a_{20}] = 10[600 + 505] = 11,050.$$

2.   The number of coins received on successive days is

$$1, 2, 4, 8, \ldots,$$

    which is a geometric series with first term 1 and common ratio $r = 2$. The sum of the first 30 terms is given by

$$S_{30} = a_1\,\frac{2^{30} - 1}{2 - 1} = 2^{30} - 1 \approx 2^{30}.$$

    Using $2^{10} = 1024 \approx 1000$, we have

$$S_{30} \approx 2^{30} = (2^{10})^3 \approx (1000)^3 = 1,000,000,000,$$

    or about 1 billion dollars.

A.3   CONTINUOUS COMPOUNDING

▸ *Continuous Compounding*
▸ *Present Value and Effective Yield*

A P P L I C A T I O N
Present Value
of a Future Balance

How much money must grandparents set aside at the birth of their grandchild if they wish to have $20,000 when the grandchild turns eighteen? They can earn 9% compounded continuously. The answer is given in Example 3.

Continuous Compounding

Let us first consider what happens to money in an account if the number of compoundings keeps increasing.

E X A M P L E   1

Compounding Over Smaller and Smaller Time Intervals

Suppose $1000 is deposited into an account that yields 6% annually. Find the future amount at the end of the first year if compounded (a) annually, (b) monthly, (c) weekly, (d) daily, (e) hourly, (f) by the minute.

Solutions   In all cases $P = \$1000$, $r = .06$, and $t = 1$. We then have

Table A.1

Compounded	Amount at End of 1 Year
(a) Yearly	$\$1000\,(1.06) = \$1060.00$
(b) Monthly	$\$1000\left(1 + \frac{.06}{12}\right)^{12} = \$1061.68$
(c) Weekly	$\$1000\left(1 + \frac{.06}{52}\right)^{52} = \$1061.80$
(d) Daily	$\$1000\left(1 + \frac{.06}{365}\right)^{365} = \$1061.83$
(e) Hourly	$\$1000\left(1 + \frac{.06}{8760}\right)^{8760} = \$1061.84$
(f) By the minute	$\$1000\left(1 + \frac{.06}{525600}\right)^{525600} = \$1061.84$

Thus as the number of compoundings per year increases, the amount of money in the account increases also and approaches a certain value. In the above case, this value was $1061.84.   ■

We then need to analyze in general what happens as the number of compoundings increases without bound. To do this set $k = m/r$ in the formula for the future value F. This gives $m = rk$ and $r/m = 1/k$. Then

$$F = P\left(1 + \frac{r}{m}\right)^{mt}$$

$$= P\left(1 + \frac{1}{k}\right)^{rkt}$$

$$= P\left[\left(1 + \frac{1}{k}\right)^{k}\right]^{rt}$$

Since $k = m/r$, as the number m of compoundings increases without bound, so will k. Thus we wish to discover what happens to the quantity

$$\left(1 + \frac{1}{k}\right)^{k}$$

as k becomes large without bound. The following table indicates that this quantity is approaching the number 2.71828 (to five decimal places).

Table A.2

k	10	100	1000	10,000	100,000	1,000,000
$\left(1 + \dfrac{1}{k}\right)^{k}$	2.593743	2.704814	2.716924	2.718146	2.718268	2.718281

This number is of such importance that a special letter is set aside for it. This letter is e. Using this in the above formula for F then yields

$$F = P\left[\left(1 + \frac{1}{k}\right)^{k}\right]^{rt}$$

$$\rightarrow P[e]^{rt} \qquad \text{as } k \text{ becomes large}$$

$$= Pe^{rt}$$

We refer to this as *continuous* compounding. The compounding is not done just every day, or every hour, or every minute, or every second, but *continuously*.

Continuous Compounding

If a principal of P earns interest at an annual rate of r, expressed as a decimal, and interest is compounded continuously, then the future amount F after t years is

$$F = Pe^{rt},$$

where the number e to five decimal places is given by

$$e \approx 2.71828.$$

E X A M P L E 2

Finding Amounts in a Continuously Compounded Account

Suppose $2000 is invested at an annual rate of 9% compounded continuously. How much is in the account after (a) 1 year, (b) 3 years?

Solutions

(a) Here $P = \$2000$, $r = .09$, and $n = 1$. Using the e^x key on a calculator we have

$$A = Pe^{rt} = \$2000e^{.09} = \$2000(1.09417) = \$2188.35$$

(b) Here $P = \$2000$, $r = .09$, and $n = 3$. Using the e^x key on a calculator we have

$$A = Pe^{rt} = \$2000e^{(.09)(3)} = \$2000(1.30996) = \$2619.93 \quad \blacksquare$$

Present Value and Effective Yield

If we wish to know how many (present) dollars P to set aside now in an account earning an annual interest rate of r, expressed as a decimal, so that we will have F dollars after t years, we simply solve the expression $F = Pe^{rt}$ for P and obtain

$$P = \frac{F}{e^{rt}}.$$

This is again called the **present value**.

To find the effective yield for such an account we let r_e be the effective annual yield. Then r_e must satisfy

$$Pe^r = P(1 + r_e).$$

Solving for r_e we obtain

$$r_e = e^r - 1.$$

We then have the following.

Present Value and Effective Yield when Compounding Continuously

Suppose an account earns an annual rate of r expressed as a decimal and is compounded continuously. Then the amount P, called the **present value**, needed presently in this account so that a future balance of F will be attained in t years is given by

$$P = \frac{F}{e^{rt}}.$$

The effective yield r_e for such an account is

$$r_e = e^r - 1.$$

E X A M P L E 3   **Calculating Present Value When Compounding Continuously**

How much money must grandparents set aside at the birth of their grandchild if they wish to have $20,000 when the grandchild turns eighteen? They can earn 9% compounded continuously.

Solution   Here $r = 0.09$, $t = 18$, and $A = \$20,000$. Using the e^x key on a calculator we have

$$P = Ae^{-rt} = \$20,000e^{-0.09(18)} = \$3957.97 \quad \blacksquare$$

E X A M P L E 4   **Finding the Effective Yield When Compounding Continuously**

Find the effective annual yield if 9% is compounded continuously.

Solution   Here $r = .09$. Then using the e^x key on a calculator we have

$$r_e = e^{0.09} - 1 = 0.09417$$

or as a percent, 9.417%. $\blacksquare$

SELF-HELP EXERCISE SET A.3

1. An account with $1000 earns interest at an annual rate of 8% compounded continuously. Find the amount in this account after 10 years.

2. Find the effective yield if the annual rate is 8% and the compounding is continuous.

3. How much money should be deposited in a bank account earning the annual interest rate of 8% compounded continuously in order that there be $10,000 in the account at the end of 10 years?

EXERCISE SET A.3

1. Suppose $1000 is deposited into an account that yields 5% annually. Find the future amount at the end of the first year if compounded (a) annually, (b) monthly, (c) weekly, (d) daily, (e) hourly, (f) by the minute, (g) continuously.

2. Suppose $1000 is deposited into an account that yields 5.5% annually. Find the future amount at the end of the first year if compounded (a) annually, (b) monthly, (c) weekly, (d) daily, (e) hourly, (f) by the minute, (g) continuously.

In Exercises 3 through 8 find how much is in the accounts after the given years where P is the initial principal, r is the annual rate given as a percent, and the compounding is continuous.

3. $P = \$1000$, $r = 6\%$, 1 year

4. $P = \$1000$, $r = 7\%$, 1 year

5. $P = \$1000, r = 5\%$, 10 years

6. $P = \$1000, r = 5.5\%$, 10 years

7. $P = \$1000, r = 6.5\%$, 40 years

8. $P = \$1000, r = 4.5\%$, 40 years

In Exercises 9 through 12 find the effective yield given the annual rate r with continuous compounding.

9. $r = 5.5\%$ **10.** $r = 4.3\%$ **11.** $r = 6.2\%$ **12.** $r = 5.8\%$

In Exercises 13 through 16 find the present value of the given amounts F for the indicated time with the indicated annual rate of return r and continuous compounding.

13. $F = \$10,000, r = 9\%, t = 20$ years

14. $F = \$10,000, r = 10\%, t = 20$ years

15. $F = \$100,000, r = 9\%, t = 40$ years

16. $F = \$100,000, r = 10\%, t = 40$ years

17. Your rich uncle has just given you a high school graduation present of "$1,000,000." The present is in the form of a 40-year bond with an annual interest rate of 9% compounded continuously. The bond says it will be worth $1,000,000 in 40 years. What is this "$1,000,000" gift worth at the present time?

18. Redo the previous exercise if the annual interest rate is 6%.

19. Your second rich uncle gives you a high school graduation present of "$2,000,000." The present is in the form of a 50-year bond with an annual interest rate of 9% compounded continuously. The bond says it will be worth $2,000,000 in 50 years. What is this "$2,000,000" gift worth at the present time? Compare your answer to the exercise before last.

20. Redo the previous exercise if the annual interest rate is 6%.

Applications

21. **Real Estate Appreciation.** The United States paid about 4 cents an acre for the Louisiana Purchase in 1803. Suppose the value of this property grew at an annual rate of 5% compounded continuously. What would an acre be worth in 1994? Does this seem realistic?

22. **Real Estate Appreciation.** Redo the previous problem using a rate of 6% instead of 5%. Compare your answer with the answer to the previous problem.

23. **Comparing Rates at Banks.** One bank advertises a nominal rate of 6.6% compounded quarterly. A second bank advertises a nominal rate of 6.5% compounded continuously. What are the effective yields? In which bank would you deposit your money?

24. **Comparing Rates at Banks.** One bank advertises a nominal rate of 8.1% compounded semiannually. A second bank advertises a nominal rate of 8%

compounded continuously. What are the effective yields? In which bank would you deposit your money?

25. **Saving for Machinery.**   How much money should a company deposit in an account with a nominal rate of 8% compounded continuously in order to have $100,000 for a certain piece of machinery in 5 years?

26. **Saving for Machinery.**   Repeat the previous exercise if the annual rate was 7%.

Solutions to Self-Help Exercise Set A.3

1. Here $P = \$1000$, $r = .08$, and $t = 10$. Therefore

$$F = Pe^{rt} = \$1000e^{(.08)(10)} = \$2225.54.$$

2. The effective yield is

$$r_{\text{eff}} = e^r - 1 = e^{.08} - 1 = .0833.$$

3. The present value is

$$P = \frac{F}{e^{rt}} = \frac{\$10,000}{e^{(.08)(10)}} = \$4493.29.$$

Tables

B.1 Area Under Standard Normal Curve to the Left of $z = \dfrac{x - \mu}{\sigma}$

B.2 Financial Tables

Table B.1

Area Under Standard Normal Curve to the Left of $z = \dfrac{x - \mu}{\sigma}$

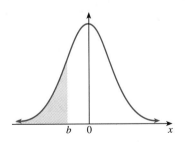

z	.00	.01	.02	.03	.04	.05	.06	.07	.08	.09
-3.4	.0003	.0003	.0003	.0003	.0003	.0003	.0003	.0003	.0003	.0002
-3.3	.0005	.0005	.0005	.0004	.0004	.0004	.0004	.0004	.0004	.0003
-3.2	.0007	.0007	.0006	.0006	.0006	.0006	.0006	.0005	.0005	.0005
-3.1	.0010	.0009	.0009	.0009	.0008	.0008	.0008	.0008	.0007	.0007
-3.0	.0013	.0013	.0013	.0012	.0012	.0011	.0011	.0011	.0010	.0010

Table B.1 (continued)

z	.00	.01	.02	.03	.04	.05	.06	.07	.08	.09
−2.9	.0019	.0018	.0017	.0017	.0016	.0016	.0015	.0015	.0014	.0014
−2.8	.0026	.0025	.0024	.0023	.0023	.0022	.0021	.0021	.0020	.0019
−2.7	.0035	.0034	.0033	.0032	.0031	.0030	.0029	.0028	.0027	.0026
−2.6	.0047	.0045	.0044	.0043	.0041	.0040	.0039	.0038	.0037	.0036
−2.5	.0062	.0060	.0059	.0057	.0055	.0054	.0052	.0051	.0049	.0048
−2.4	.0082	.0080	.0078	.0075	.0073	.0071	.0069	.0068	.0066	.0064
−2.3	.0107	.0104	.0102	.0099	.0096	.0094	.0091	.0089	.0087	.0084
−2.2	.0139	.0136	.0132	.0129	.0125	.0122	.0119	.0116	.0113	.0110
−2.1	.0179	.0174	.0170	.0166	.0162	.0158	.0154	.0150	.0146	.0143
−2.0	.0228	.0222	.0217	.0212	.0207	.0202	.0197	.0192	.0188	.0183
−1.9	.0287	.0281	.0274	.0268	.0262	.0256	.0250	.0244	.0239	.0233
−1.8	.0359	.0352	.0344	.0336	.0329	.0322	.0314	.0307	.0301	.0294
−1.7	.0446	.0436	.0427	.0418	.0409	.0401	.0392	.0384	.0375	.0367
−1.6	.0548	.0537	.0526	.0516	.0505	.0495	.0485	.0475	.0465	.0455
−1.5	.0668	.0655	.0643	.0630	.0618	.0606	.0594	.0582	.0571	.0559
−1.4	.0808	.0793	.0778	.0764	.0749	.0735	.0722	.0708	.0694	.0681
−1.3	.0968	.0951	.0934	.0918	.0901	.0885	.0869	.0853	.0838	.0823
−1.2	.1151	.1131	.1112	.1093	.1075	.1056	.1038	.1020	.1003	.0985
−1.1	.1357	.1335	.1314	.1292	.1271	.1251	.1230	.1210	.1190	.1170
−1.0	.1587	.1562	.1539	.1515	.1492	.1469	.1446	.1423	.1401	.1379
−0.9	.1841	.1814	.1788	.1762	.1736	.1711	.1685	.1660	.1635	.1611
−0.8	.2119	.2090	.2061	.2033	.2005	.1977	.1949	.1922	.1894	.1867
−0.7	.2420	.2389	.2358	.2327	.2296	.2266	.2236	.2206	.2177	.2148
−0.6	.2743	.2709	.2676	.2643	.2611	.2578	.2546	.2514	.2483	.2451
−0.5	.3085	.3050	.3015	.2981	.2946	.2912	.2877	.2843	.2810	.2776
−0.4	.3446	.3409	.3372	.3336	.3300	.3264	.3228	.3192	.3156	.3121
−0.3	.3821	.3783	.3745	.3707	.3669	.3632	.3594	.3557	.3520	.3483
−0.2	.4207	.4168	.4129	.4090	.4052	.4013	.3974	.3936	.3897	.3859
−0.1	.4602	.4562	.4522	.4483	.4443	.4404	.4364	.4325	.4286	.4247
−0.0	.5000	.4960	.4920	.4880	.4840	.4801	.4761	.4721	.4681	.4641
0.0	.5000	.5040	.5080	.5120	.5160	.5199	.5239	.5279	.5319	.5359
0.1	.5398	.5438	.5478	.5517	.5557	.5596	.5636	.5675	.5714	.5753
0.2	.5793	.5832	.5871	.5910	.5948	.5987	.6026	.6064	.6103	.6141
0.3	.6179	.6217	.6255	.6293	.6331	.6368	.6406	.6443	.6480	.6517
0.4	.6554	.6591	.6628	.6664	.6700	.6736	.6772	.6808	.6844	.6879
0.5	.6915	.6950	.6985	.7019	.7054	.7088	.7123	.7157	.7190	.7224
0.6	.7257	.7291	.7324	.7357	.7389	.7422	.7454	.7486	.7517	.7549
0.7	.7580	.7611	.7642	.7673	.7704	.7734	.7764	.7794	.7823	.7852
0.8	.7881	.7910	.7939	.7967	.7995	.8023	.8051	.8078	.8106	.8133
0.9	.8159	.8186	.8212	.8238	.8264	.8289	.8315	.8340	.8365	.8389
1.0	.8413	.8438	.8461	.8485	.8508	.8531	.8554	.8577	.8599	.8621
1.1	.8643	.8665	.8686	.8708	.8729	.8749	.8770	.8790	.8810	.8830
1.2	.8849	.8869	.8888	.8907	.8925	.8944	.8962	.8980	.8997	.9015
1.3	.9032	.9049	.9066	.9082	.9099	.9115	.9131	.9147	.9162	.9177
1.4	.9192	.9207	.9222	.9236	.9251	.9265	.9278	.9292	.9306	.9319

Table B.1 (*continued*)

z	.00	.01	.02	.03	.04	.05	.06	.07	.08	.09
1.5	.9332	.9345	.9357	.9370	.9382	.9394	.9406	.9418	.9429	.9441
1.6	.9452	.9463	.9474	.9484	.9495	.9505	.9515	.9525	.9535	.9545
1.7	.9554	.9564	.9573	.9582	.9591	.9599	.9608	.9616	.9625	.9633
1.8	.9641	.9649	.9656	.9664	.9671	.9678	.9686	.9693	.9699	.9706
1.9	.9713	.9719	.9726	.9732	.9738	.9744	.9750	.9756	.9761	.9767
2.0	.9772	.9778	.9783	.9788	.9793	.9798	.9803	.9808	.9812	.9817
2.1	.9821	.9826	.9830	.9834	.9838	.9842	.9846	.9850	.9854	.9857
2.2	.9861	.9864	.9868	.9871	.9875	.9878	.9881	.9884	.9887	.9890
2.3	.9893	.9896	.9898	.9901	.9904	.9906	.9909	.9911	.9913	.9916
2.4	.9918	.9920	.9922	.9925	.9927	.9929	.9931	.9932	.9934	.9936
2.5	.9938	.9940	.9941	.9943	.9945	.9946	.9948	.9949	.9951	.9952
2.6	.9953	.9955	.9956	.9957	.9959	.9960	.9961	.9962	.9963	.9964
2.7	.9965	.9966	.9967	.9968	.9969	.9970	.9971	.9972	.9973	.9974
2.8	.9974	.9975	.9976	.9977	.9977	.9978	.9979	.9979	.9980	.9981
2.9	.9981	.9982	.9982	.9983	.9984	.9984	.9985	.9985	.9986	.9986
3.0	.9987	.9987	.9987	.9988	.9988	.9989	.9989	.9989	.9990	.9990
3.1	.9990	.9991	.9991	.9991	.9992	.9992	.9992	.9992	.9993	.9993
3.2	.9993	.9993	.9994	.9994	.9994	.9994	.9994	.9995	.9995	.9995
3.3	.9995	.9995	.9995	.9996	.9996	.9996	.9996	.9996	.9996	.9997
3.4	.9997	.9997	.9997	.9997	.9997	.9997	.9997	.9997	.9997	.9998

Table B.2

i = **rate of interest per period** n = **number of periods** $i = \frac{1}{4}\%$

n	$(1 + i)^n$	$(1 + i)^{-n}$	$s_{\overline{n}\|i}$	$a_{\overline{n}\|i}$	$\dfrac{1}{s_{\overline{n}\|i}}$	$\dfrac{1}{a_{\overline{n}\|i}}$
1	1.0025 0000	0.9975 0623	1.0000 0000	0.9975 0623	1.0000 0000	1.0025 0000
2	1.0050 0625	0.9950 1869	2.0025 0000	1.9925 2492	0.4993 7578	0.5018 7578
3	1.0075 1877	0.9925 3734	3.0075 0625	2.9850 6227	0.3325 0139	0.3350 0139
4	1.0100 3756	0.9900 6219	4.0150 2502	3.9751 2446	0.2490 6445	0.2515 6445
5	1.0125 6266	0.9875 9321	5.0250 6258	4.9627 1766	0.1990 0250	0.2015 0250
6	1.0150 9406	0.9851 3038	6.0376 2523	5.9478 4804	0.1656 2803	0.1681 2803
7	1.0176 3180	0.9826 7370	7.0527 1930	6.9305 2174	0.1417 8928	0.1442 8928
8	1.0201 7588	0.9802 2314	8.0703 5110	7.9107 4487	0.1239 1035	0.1264 1035
9	1.0227 2632	0.9777 7869	9.0905 2697	8.8885 2357	0.1100 0462	0.1125 0462
10	1.0252 8313	0.9753 4034	10.1132 5329	9.8638 6391	0.0988 8015	0.1013 8015
11	1.0278 4634	0.9729 0807	11.1385 3642	10.8367 7198	0.0897 7840	0.0922 7840
12	1.0304 1596	0.9704 8187	12.1663 8277	11.8072 5384	0.0821 9370	0.0846 9370
13	1.0329 9200	0.9680 6171	13.1967 9872	12.7753 1555	0.0757 7595	0.0782 7595
14	1.0355 7448	0.9656 4759	14.2297 9072	13.7409 6314	0.0702 7510	0.0727 7510
15	1.0381 6341	0.9632 3949	15.2653 6520	14.7042 0264	0.0655 0777	0.0680 0777

Table B.2 *(continued)*

$i = \frac{1}{4}\%$

n	$(1 + i)^n$	$(1 + i)^{-n}$	$s_{\overline{n}\mid i}$	$a_{\overline{n}\mid i}$	$\dfrac{1}{s_{\overline{n}\mid i}}$	$\dfrac{1}{a_{\overline{n}\mid i}}$
16	1.0407 5882	0.9608 3740	16.3035 2861	15.6650 4004	0.0613 3642	0.0638 3642
17	1.0433 6072	0.9584 4130	17.3442 8743	16.6234 8133	0.0576 5587	0.0601 5587
18	1.0459 6912	0.9560 5117	18.3876 4815	17.5795 3250	0.0543 8433	0.0568 8433
19	1.0485 8404	0.9536 6700	19.4336 1727	18.5331 9950	0.0514 5722	0.0539 5722
20	1.0512 0550	0.9512 8878	20.4822 0131	19.4844 8828	0.0488 2288	0.0513 2288
21	1.0538 3352	0.9489 1649	21.5334 0682	20.4334 0477	0.0464 3947	0.0489 3947
22	1.0564 6810	0.9465 5011	22.5872 4033	21.3799 5488	0.0442 7278	0.0467 7278
23	1.0591 0927	0.9441 8964	23.6437 0843	22.3241 4452	0.0422 9455	0.0447 9455
24	1.0617 5704	0.9418 3505	24.7028 1770	23.2659 7957	0.0404 8121	0.0429 8121
25	1.0644 1144	0.9394 8634	25.7645 7475	24.2054 6591	0.0388 1298	0.0413 1298
26	1.0670 7247	0.9371 4348	26.8289 8619	25.1426 0939	0.0372 7312	0.0397 7312
27	1.0697 4015	0.9348 0646	27.8960 5865	26.0774 1585	0.0358 4736	0.0383 4736
28	1.0724 1450	0.9324 7527	28.9657 9880	27.0098 9112	0.0345 2347	0.0370 2347
29	1.0750 9553	0.9301 4990	30.0382 1330	27.9400 4102	0.0332 9093	0.0357 9093
30	1.0777 8327	0.9278 3032	31.1133 0883	28.8678 7134	0.0321 4059	0.0346 4059
31	1.0804 7773	0.9255 1653	32.1910 9210	29.7933 8787	0.0310 6449	0.0335 6449
32	1.0831 7892	0.9232 0851	33.2715 6983	30.7165 9638	0.0300 5569	0.0325 5569
33	1.0858 8687	0.9209 0624	34.3547 4876	31.6375 0262	0.0291 0806	0.0316 0806
34	1.0886 0159	0.9186 0972	35.4406 3563	32.5561 1234	0.0282 1620	0.0307 1620
35	1.0913 2309	0.9163 1892	36.5292 3722	33.4724 3126	0.0273 7533	0.0298 7533
36	1.0940 5140	0.9140 3384	37.6205 6031	34.3864 6510	0.0265 8121	0.0290 8121
37	1.0967 8653	0.9117 5445	38.7146 1171	35.2982 1955	0.0258 3004	0.0283 3004
38	1.0995 2850	0.9094 8075	39.8113 9824	36.2077 0030	0.0251 1843	0.0276 1843
39	1.1022 7732	0.9072 1272	40.9109 2673	37.1149 1302	0.0244 4335	0.0269 4335
40	1.1050 3301	0.9049 5034	42.0132 0405	38.0198 6336	0.0238 0204	0.0263 0204
41	1.1077 9559	0.9026 9361	43.1182 3706	38.9225 5697	0.0231 9204	0.0256 9204
42	1.1105 6508	0.9004 4250	44.2260 3265	39.8229 9947	0.0226 1112	0.0251 1112
43	1.1133 4149	0.8981 9701	45.3365 9774	40.7211 9648	0.0220 5724	0.0245 5724
44	1.1161 2485	0.8959 5712	46.4499 3923	41.6171 5359	0.0215 2855	0.0240 2855
45	1.1189 1516	0.8937 2281	47.5660 6408	42.5108 7640	0.0210 2339	0.0235 2339
46	1.1217 1245	0.8914 9407	48.6849 7924	43.4023 7047	0.0205 4022	0.0230 4022
47	1.1245 1673	0.8892 7090	49.8066 9169	44.2916 4137	0.0200 7762	0.0225 7762
48	1.1273 2802	0.8870 5326	50.9312 0842	45.1786 9463	0.0196 3433	0.0221 3433
49	1.1301 4634	0.8848 4116	52.0585 3644	46.0635 3580	0.0192 0915	0.0217 0915
50	1.1329 7171	0.8826 3457	53.1886 8278	46.9461 7037	0.0188 0099	0.0213 0099
51	1.1358 0414	0.8804 3349	54.3216 5449	47.8266 0386	0.0184 0886	0.0209 0886
52	1.1386 4365	0.8782 3790	55.4574 5862	48.7048 4176	0.0180 3184	0.0205 3184
53	1.1414 9026	0.8760 4778	56.5961 0227	49.5808 8953	0.0176 6906	0.0201 6906
54	1.1443 4398	0.8738 6312	57.7375 9252	50.4547 5265	0.0173 1974	0.0198 1974
55	1.1472 0484	0.8716 8391	58.8819 3650	51.3264 3656	0.0169 8314	0.0194 8314

Table B.2 (continued)

$i = \frac{1}{4}\%$

n	$(1 + i)^n$	$(1 + i)^{-n}$	$s_{\overline{n}|i}$	$a_{\overline{n}|i}$	$\dfrac{1}{s_{\overline{n}|i}}$	$\dfrac{1}{a_{\overline{n}|i}}$
56	1.1500 7285	0.8695 1013	60.0291 4135	52.1959 4669	0.0166 5858	0.0191 5858
57	1.1529 4804	0.8673 4178	61.1792 1420	53.0632 8847	0.0163 4542	0.0188 4542
58	1.1558 3041	0.8651 7883	62.3321 6223	53.9284 6730	0.0160 4308	0.0185 4308
59	1.1587 1998	0.8630 2128	63.4879 9264	54.7914 8858	0.0157 5101	0.0182 5101
60	1.1616 1678	0.8608 6911	64.6467 1262	55.6523 5769	0.0154 6869	0.0179 6869
61	1.1645 2082	0.8587 2230	65.8083 2940	56.5110 7999	0.0151 9564	0.0176 9564
62	1.1674 3213	0.8565 8085	66.9728 5023	57.3676 6083	0.0149 3142	0.0174 3142
63	1.1703 5071	0.8544 4474	68.1402 8235	58.2221 0557	0.0146 7561	0.0171 7561
64	1.1732 7658	0.8523 1395	69.3106 3306	59.0744 1952	0.0144 2780	0.0169 2780
65	1.1762 0977	0.8501 8848	70.4839 0964	59.9246 0800	0.0141 8764	0.0166 8764
66	1.1791 5030	0.8480 6831	71.6601 1942	60.7726 7631	0.0139 5476	0.0164 5476
67	1.1820 9817	0.8459 5343	72.8392 6971	61.6186 2974	0.0137 2886	0.0162 2886
68	1.1850 5342	0.8438 4382	74.0213 6789	62.4624 7355	0.0135 0961	0.0160 0961
69	1.1880 1605	0.8417 3947	75.2064 2131	63.3042 1302	0.0132 9674	0.0157 9674
70	1.1909 8609	0.8396 4037	76.3944 3736	64.1438 5339	0.0130 8996	0.0155 8996
71	1.1939 6356	0.8375 4650	77.5854 2345	64.9813 9989	0.0128 8902	0.0153 8902
72	1.1969 4847	0.8354 5786	78.7793 8701	65.8168 5774	0.0126 9368	0.0151 9368
73	1.1999 4084	0.8333 7442	79.9763 3548	66.6502 3216	0.0125 0370	0.0150 0370
74	1.2029 4069	0.8312 9618	81.1762 7632	67.4815 2834	0.0123 1887	0.0148 1887
75	1.2059 4804	0.8292 2312	82.3792 1701	68.3107 5146	0.0121 3898	0.0146 3898
76	1.2089 6191	0.8271 5523	83.5851 6505	69.1379 0670	0.0119 6385	0.0144 6385
77	1.2119 8532	0.8250 9250	84.7941 2797	69.9629 9920	0.0117 9327	0.0142 9327
78	1.2150 1528	0.8230 3491	86.0061 1329	70.7860 3411	0.0116 2708	0.0141 2708
79	1.2180 5282	0.8209 8246	87.2211 2857	71.6070 1657	0.0114 6511	0.0139 6511
80	1.2210 9795	0.8189 3512	88.4391 8139	72.4259 5169	0.0113 0721	0.0138 0721
81	1.2241 5070	0.8168 9289	89.6602 7934	73.2428 4458	0.0111 5321	0.0136 5321
82	1.2272 1108	0.8148 5575	90.8844 3004	74.0577 0033	0.0110 0298	0.0135 0298
83	1.2302 7910	0.8128 2369	92.1116 4112	74.8705 2402	0.0108 5639	0.0133 5639
84	1.2333 5480	0.8107 9670	93.3419 2022	75.6813 2072	0.0107 1330	0.0132 1330
85	1.2364 3819	0.8087 7476	94.5752 7502	76.4900 9548	0.0105 7359	0.0130 7359
86	1.2395 2928	0.8067 5787	95.8117 1321	77.2968 5335	0.0104 3714	0.0129 3714
87	1.2426 2811	0.8047 4600	97.0512 4249	78.1015 9935	0.0103 0384	0.0128 0384
88	1.2457 3468	0.8027 3915	98.2938 7060	78.9043 3850	0.0101 7357	0.0126 7357
89	1.2488 4901	0.8007 3731	99.5396 0527	79.7050 7581	0.0100 4625	0.0125 4625
90	1.2519 7114	0.7987 4046	100.7884 5429	80.5038 1627	0.0099 2177	0.0124 2177
91	1.2551 0106	0.7967 4859	102.0404 2542	81.3005 6486	0.0098 0004	0.0123 0004
92	1.2582 3882	0.7947 6168	103.2955 2649	82.0953 2654	0.0096 8096	0.0121 8096
93	1.2613 8441	0.7927 7973	104.5537 6530	82.8881 0628	0.0095 6446	0.0120 6446
94	1.2645 3787	0.7908 0273	105.8151 4972	83.6789 0900	0.0094 5044	0.0119 5044
95	1.2676 9922	0.7888 3065	107.0796 8759	84.4677 3966	0.0093 3884	0.0118 3884
96	1.2708 6847	0.7868 6349	108.3473 8681	85.2546 0315	0.0092 2957	0.0117 2957
97	1.2740 4564	0.7849 0124	109.6182 5528	86.0395 0439	0.0091 2257	0.0116 2257
98	1.2772 3075	0.7829 4388	110.8923 0091	86.8224 4827	0.0090 1776	0.0115 1776
99	1.2804 2383	0.7809 9140	112.1695 3167	87.6034 3967	0.0089 1508	0.0114 1508
100	1.2836 2489	0.7790 4379	113.4499 5550	88.3824 8346	0.0088 1446	0.0113 1446

Table B.2 *(continued)*

$i = \frac{1}{2}\%$

n	$(1 + i)^n$	$(1 + i)^{-n}$	$s_{\overline{n}|i}$	$a_{\overline{n}|i}$	$\dfrac{1}{s_{\overline{n}|i}}$	$\dfrac{1}{a_{\overline{n}|i}}$
1	1.0050 0000	0.9950 2488	1.0000 0000	0.9950 2488	1.0000 0000	1.0050 0000
2	1.0100 2500	0.9900 7450	2.0050 0000	1.9850 9938	0.4987 5312	0.5037 5312
3	1.0150 7513	0.9851 4876	3.0150 2500	2.9702 4814	0.3316 7221	0.3366 7221
4	1.0201 5050	0.9802 4752	4.0301 0013	3.9504 9566	0.2481 3279	0.2531 3279
5	1.0252 5125	0.9753 7067	5.0502 5063	4.9258 6633	0.1980 0997	0.2030 0997
6	1.0303 7751	0.9705 1808	6.0755 0188	5.8963 8441	0.1645 9546	0.1695 9546
7	1.0355 2940	0.9656 8963	7.1058 7939	6.8620 7404	0.1407 2854	0.1457 2854
8	1.0407 0704	0.9608 8520	8.1414 0879	7.8229 5924	0.1228 2886	0.1278 2886
9	1.0459 1058	0.9561 0468	9.1821 1583	8.7790 6392	0.1089 0736	0.1139 0736
10	1.0511 4013	0.9513 4794	10.2280 2641	9.7304 1186	0.0977 7057	0.1027 7057
11	1.0563 9583	0.9466 1487	11.2791 6654	10.6770 2673	0.0886 5903	0.0936 5903
12	1.0616 7781	0.9419 0534	12.3355 6237	11.6189 3207	0.0810 6643	0.0860 6643
13	1.0669 8620	0.9372 1924	13.3972 4018	12.5561 5131	0.0746 4224	0.0796 4224
14	1.0723 2113	0.9325 5646	14.4642 2639	13.4887 0777	0.0691 3609	0.0741 3609
15	1.0776 8274	0.9279 1688	15.5365 4752	14.4166 2465	0.0643 6436	0.0693 6436
16	1.0830 7115	0.9233 0037	16.6142 3026	15.3399 2502	0.0601 8937	0.0651 8937
17	1.0884 8651	0.9187 0684	17.6973 0141	16.2586 3186	0.0565 0579	0.0615 0579
18	1.0939 2894	0.9141 3616	18.7857 8791	17.1727 6802	0.0532 3173	0.0582 3173
19	1.0993 9858	0.9095 8822	19.8797 1685	18.0823 5624	0.0503 0253	0.0553 0253
20	1.1048 9558	0.9050 6290	20.9791 1544	18.9874 1915	0.0476 6645	0.0526 6645
21	1.1104 2006	0.9005 6010	22.0840 1101	19.8879 7925	0.0452 8163	0.0502 8163
22	1.1159 7216	0.8960 7971	23.1944 3107	20.7840 5896	0.0431 1380	0.0481 1380
23	1.1215 5202	0.8916 2160	24.3104 0322	21.6756 8055	0.0411 3465	0.0461 3465
24	1.1271 5978	0.8871 8567	25.4319 5524	22.5628 6622	0.0393 2061	0.0443 2061
25	1.1327 9558	0.8827 7181	26.5591 1502	23.4456 3803	0.0376 5186	0.0426 5186
26	1.1384 5955	0.8783 7991	27.6919 1059	24.3240 1794	0.0361 1163	0.0411 1163
27	1.1441 5185	0.8740 0986	28.8303 7015	25.1980 2780	0.0346 8565	0.0396 8565
28	1.1498 7261	0.8696 6155	29.9745 2200	26.0676 8936	0.0333 6167	0.0383 6167
29	1.1556 2197	0.8653 3488	31.1243 9461	26.9330 2423	0.0321 2914	0.0371 2914
30	1.1614 0008	0.8610 2973	32.2800 1658	27.7940 5397	0.0309 7892	0.0359 7892
31	1.1672 0708	0.8567 4600	33.4414 1666	28.6507 9997	0.0299 0304	0.0349 0304
32	1.1730 4312	0.8524 8358	34.6086 2375	29.5032 8355	0.0288 9453	0.0338 9453
33	1.1789 0833	0.8482 4237	35.7816 6686	30.3515 2592	0.0279 4727	0.0329 4727
34	1.1848 0288	0.8440 2226	36.9605 7520	31.1955 4818	0.0270 5586	0.0320 5586
35	1.1907 2689	0.8398 2314	38.1453 7807	32.0353 7132	0.0262 1550	0.0312 1550
36	1.1966 8052	0.8356 4492	39.3361 0496	32.8710 1624	0.0254 2194	0.0304 2194
37	1.2026 6393	0.8314 8748	40.5317 8549	33.7025 0372	0.0246 7139	0.0296 7139
38	1.2086 7725	0.8273 5073	41.7354 4942	34.5298 5445	0.0239 6045	0.0289 6045
39	1.2147 2063	0.8232 3455	42.9441 2666	35.3530 8900	0.0232 8607	0.0282 8607
40	1.2207 9424	0.8191 3886	44.1588 4730	36.1722 2786	0.0226 4552	0.0276 4552

Table B.2 *(continued)*
$i = \frac{1}{2}\%$

n	$(1 + i)^n$	$(1 + i)^{-n}$	$s_{\overline{n}|i}$	$a_{\overline{n}|i}$	$\dfrac{1}{s_{\overline{n}|i}}$	$\dfrac{1}{a_{\overline{n}|i}}$
41	1.2268 9821	0.8150 6354	45.3796 4153	36.9872 9141	0.0220 3631	0.0270 3631
42	1.2330 3270	0.8110 0850	46.6065 3974	37.7982 9991	0.0214 5622	0.0264 5622
43	1.2391 9786	0.8069 7363	47.8395 7244	38.6052 7354	0.0209 0320	0.0259 0320
44	1.2453 9385	0.8029 5884	49.0787 7030	39.4082 3238	0.0203 7541	0.0253 7541
45	1.2516 2082	0.7989 6402	50.3241 6415	40.2071 9640	0.0198 7117	0.0248 7117
46	1.2578 7892	0.7949 8907	51.5757 8497	41.0021 8547	0.0193 8894	0.0243 8894
47	1.2641 6832	0.7910 3390	52.8336 6390	41.7932 1937	0.0189 2733	0.0239 2733
48	1.2704 8916	0.7870 9841	54.0978 3222	42.5803 1778	0.0184 8503	0.0234 8503
49	1.2768 4161	0.7831 8250	55.3683 2138	43.3635 0028	0.0180 6087	0.0230 6087
50	1.2832 2581	0.7792 8607	56.6451 6299	44.1427 8635	0.0176 5376	0.0226 5376
51	1.2896 4194	0.7754 0902	57.9283 8880	44.9181 9537	0.0172 6269	0.0222 6269
52	1.2960 9015	0.7715 5127	59.2180 3075	45.6897 4664	0.0168 8675	0.0218 8675
53	1.3025 7060	0.7677 1270	60.5141 2090	46.4574 5934	0.0165 2507	0.0215 2507
54	1.3090 8346	0.7638 9324	61.8166 9150	47.2213 5258	0.0161 7686	0.0211 7686
55	1.3156 2887	0.7600 9277	63.1257 7496	47.9814 4535	0.0158 4139	0.0208 4139
56	1.3222 0702	0.7563 1122	64.4414 0384	48.7377 5657	0.0155 1797	0.0205 1797
57	1.3288 1805	0.7525 4847	65.7636 1086	49.4903 0505	0.0152 0598	0.0202 0598
58	1.3354 6214	0.7488 0445	67.0924 2891	50.2391 0950	0.0149 0481	0.0199 0481
59	1.3421 3946	0.7450 7906	68.4278 9105	50.9841 8855	0.0146 1392	0.0196 1392
60	1.3488 5015	0.7413 7220	69.7700 3051	51.7255 6075	0.0143 3280	0.0193 3280
61	1.3555 9440	0.7376 8378	71.1188 8066	52.4632 4453	0.0140 6096	0.0190 6096
62	1.3623 7238	0.7340 1371	72.4744 7507	53.1972 5824	0.0137 9796	0.0187 9796
63	1.3691 8424	0.7303 6190	73.8368 4744	53.9276 2014	0.0135 4337	0.0185 4337
64	1.3760 3016	0.7267 2826	75.2060 3168	54.6543 4839	0.0132 9681	0.0182 9681
65	1.3829 1031	0.7231 1269	76.5820 6184	55.3774 6109	0.0130 5789	0.0180 5789
66	1.3898 2486	0.7195 1512	77.9649 7215	56.0969 7621	0.0128 2627	0.0178 2627
67	1.3967 7399	0.7159 3544	79.3547 9701	56.8129 1165	0.0126 0163	0.0176 0163
68	1.4037 5785	0.7123 7357	80.7515 7099	57.5252 8522	0.0123 8366	0.0173 8366
69	1.4107 7664	0.7088 2943	82.1553 2885	58.2341 1465	0.0121 7206	0.0171 7206
70	1.4178 3053	0.7053 0291	83.5661 0549	58.9394 1756	0.0119 6657	0.0169 6657
71	1.4249 1968	0.7017 9394	84.9839 3602	59.6412 1151	0.0117 6693	0.0167 6693
72	1.4320 4428	0.6983 0243	86.4088 5570	60.3395 1394	0.0115 7289	0.0165 7289
73	1.4392 0450	0.6948 2829	87.8408 9998	61.0343 4222	0.0113 8422	0.0163 8422
74	1.4464 0052	0.6913 7143	89.2801 0448	61.7257 1366	0.0112 0070	0.0162 0070
75	1.4536 3252	0.6879 3177	90.7265 0500	62.4136 4543	0.0110 2214	0.0160 2214
76	1.4609 0069	0.6845 0923	92.1801 3752	63.0981 5466	0.0108 4832	0.0158 4832
77	1.4682 0519	0.6811 0371	93.6410 3821	63.7792 5836	0.0106 7908	0.0156 7908
78	1.4755 4622	0.6777 1513	95.1092 4340	64.4569 7350	0.0105 1423	0.0155 1423
79	1.4829 2395	0.6743 4342	96.5847 8962	65.1313 1691	0.0103 5360	0.0153 5360
80	1.4903 3857	0.6709 8847	98.0677 1357	65.8023 0538	0.0101 9704	0.0151 9704

Table B.2 *(continued)*

$i = \frac{1}{2}\%$

n	$(1 + i)^n$	$(1 + i)^{-n}$	$s_{\overline{n}|i}$	$a_{\overline{n}|i}$	$\dfrac{1}{s_{\overline{n}|i}}$	$\dfrac{1}{a_{\overline{n}|i}}$
81	1.4977 9026	0.6676 5022	99.5580 5214	66.4699 5561	0.0100 4439	0.0150 4439
82	1.5052 7921	0.6643 2858	101.0558 4240	67.1342 8419	0.0098 9552	0.0148 9552
83	1.5128 0561	0.6610 2346	102.5611 2161	67.7953 0765	0.0097 5028	0.0147 5028
84	1.5203 6964	0.6577 3479	104.0739 2722	68.4530 4244	0.0096 0855	0.0146 0855
85	1.5279 7148	0.6544 6248	105.5942 9685	69.1075 0491	0.0094 7021	0.0144 7021
86	1.5356 1134	0.6512 0644	107.1222 6834	69.7587 1135	0.0093 3513	0.0143 3513
87	1.5432 8940	0.6479 6661	108.6578 7968	70.4066 7796	0.0092 0320	0.0142 0320
88	1.5510 0585	0.6447 4290	110.2011 6908	71.0514 2086	0.0090 7431	0.0140 7431
89	1.5587 6087	0.6415 3522	111.7521 7492	71.6929 5608	0.0089 4837	0.0139 4837
90	1.5665 5468	0.6383 4350	113.3109 3580	72.3312 9958	0.0088 2527	0.0138 2527
91	1.5743 8745	0.6351 6766	114.8774 9048	72.9664 6725	0.0087 0493	0.0137 0493
92	1.5822 5939	0.6320 0763	116.4518 7793	73.5984 7487	0.0085 8724	0.0135 8724
93	1.5901 7069	0.6288 6331	118.0341 3732	74.2273 3818	0.0084 7213	0.0134 7213
94	1.5981 2154	0.6257 3464	119.6243 0800	74.8530 7282	0.0083 5950	0.0133 5950
95	1.6061 1215	0.6226 2153	121.2224 2954	75.4756 9434	0.0082 4930	0.0132 4930
96	1.6141 4271	0.6195 2391	122.8285 4169	76.0952 1825	0.0081 4143	0.0131 4143
97	1.6222 1342	0.6164 4170	124.4426 8440	76.7116 5995	0.0080 3583	0.0130 3583
98	1.6303 2449	0.6133 7483	126.0648 9782	77.3250 3478	0.0079 3242	0.0129 3242
99	1.6384 7611	0.6103 2321	127.6952 2231	77.9353 5799	0.0078 3115	0.0128 3115
100	1.6466 6849	0.6072 8678	129.3336 9842	78.5426 4477	0.0077 3194	0.0127 3194

$i = \frac{3}{4}\%$

n	$(1 + i)^n$	$(1 + i)^{-n}$	$s_{\overline{n}|i}$	$a_{\overline{n}|i}$	$\dfrac{1}{s_{\overline{n}|i}}$	$\dfrac{1}{a_{\overline{n}|i}}$
1	1.0075 0000	0.9925 5583	1.0000 0000	0.9925 5583	1.0000 0000	1.0075 0000
2	1.0150 5625	0.9851 6708	2.0075 0000	1.9777 2291	0.4981 3200	0.5056 3200
3	1.0226 6917	0.9778 3333	3.0225 5625	2.9555 5624	0.3308 4579	0.3383 4579
4	1.0303 3919	0.9705 5417	4.0452 2542	3.9261 1041	0.2472 0501	0.2547 0501
5	1.0380 6673	0.9633 2920	5.0755 6461	4.8894 3961	0.1970 2242	0.2045 2242
6	1.0458 5224	0.9561 5802	6.1136 3135	5.8455 9763	0.1635 6891	0.1710 6891
7	1.0536 9613	0.9490 4022	7.1594 8358	6.7946 3785	0.1396 7488	0.1471 7488
8	1.0615 9885	0.9419 7540	8.2131 7971	7.7366 1325	0.1217 5552	0.1292 5552
9	1.0695 6084	0.9349 6318	9.2747 7856	8.6715 7642	0.1078 1929	0.1153 1929
10	1.0775 8255	0.9270 0315	10.3443 3940	9.5995 7958	0.0966 7123	0.1041 7123

Table B.2 *(continued)*
$i = \frac{3}{4}\%$

n	$(1 + i)^n$	$(1 + i)^{-n}$	$s_{\overline{n}\rceil i}$	$a_{\overline{n}\rceil i}$	$\dfrac{1}{s_{\overline{n}\rceil i}}$	$\dfrac{1}{a_{\overline{n}\rceil i}}$
11	1.0856 6441	0.9210 9494	11.4219 2194	10.5206 7452	0.0875 5094	0.0950 5094
12	1.0938 0690	0.9142 3815	12.5075 8636	11.4349 1267	0.0799 5148	0.0874 5148
13	1.1020 1045	0.9074 3241	13.6013 9325	12.3423 4508	0.0735 2188	0.0810 2188
14	1.1102 7553	0.9006 7733	14.7034 0370	13.2430 2242	0.0680 1146	0.0755 1146
15	1.1186 0259	0.8939 7254	15.8136 7923	14.1369 9495	0.0632 3639	0.0707 3639
16	1.1269 9211	0.8873 1766	16.9322 8183	15.0243 1261	0.0590 5879	0.0665 5879
17	1.1354 4455	0.8807 1231	18.0592 7394	15.9050 2492	0.0553 7321	0.0628 7321
18	1.1439 6039	0.8741 5614	19.1947 1849	16.7791 8107	0.0520 9766	0.0595 9766
19	1.1525 4009	0.8676 4878	20.3386 7888	17.6468 2984	0.0491 6740	0.0566 6740
20	1.1611 8414	0.8611 8985	21.4912 1897	18.5080 1969	0.0465 3063	0.0540 3063
21	1.1698 9302	0.8547 7901	22.6524 0312	19.3627 9870	0.0441 4543	0.0516 4543
22	1.1786 6722	0.8484 1589	23.8222 9614	20.2112 1459	0.0419 7748	0.0494 7748
23	1.1875 0723	0.8421 0014	25.0009 6336	21.0533 1473	0.0399 9846	0.0474 9846
24	1.1964 1353	0.8358 3140	26.1884 7059	21.8891 4614	0.0381 8474	0.0456 8474
25	1.2053 8663	0.8296 0933	27.3848 8412	22.7187 5547	0.0365 1650	0.0440 1650
26	1.2144 2703	0.8234 3358	28.5902 7075	23.5421 8905	0.0349 7693	0.0424 7693
27	1.2235 3523	0.8173 0380	29.8046 9778	24.3594 9286	0.0335 5176	0.0410 5176
28	1.2327 1175	0.8112 1966	31.0282 3301	25.1707 1251	0.0322 2871	0.0397 2871
29	1.2419 5709	0.8051 8080	32.2609 4476	25.9758 9331	0.0309 9723	0.0384 9723
30	1.2512 7176	0.7991 8690	35.5029 0184	26.7750 8021	0.0298 4816	0.0373 4816
31	1.2606 5630	0.7932 3762	34.7541 7361	27.5683 1783	0.0287 7352	0.0362 7352
32	1.2701 1122	0.7873 3262	36.0148 2991	28.3556 5045	0.0277 6634	0.0352 6634
33	1.2796 3706	0.7814 7158	37.2849 4113	29.1371 2203	0.0268 2048	0.0343 2048
34	1.2892 3434	0.7756 5418	38.5645 7819	29.9127 7621	0.0259 3053	0.0334 3053
35	1.2989 0359	0.7698 8008	39.8538 1253	30.6826 5629	0.0250 9170	0.0325 9170
36	1.3086 4537	0.7641 4896	41.1527 1612	31.4468 0525	0.0242 9973	0.0317 9973
37	1.3184 6021	0.7584 6051	42.4613 6149	32.2052 6576	0.0235 5082	0.0310 5082
38	1.3283 4866	0.7528 1440	43.7798 2170	32.9580 8016	0.0228 4157	0.0303 4157
39	1.3383 1128	0.7472 1032	45.1081 7037	33.7052 9048	0.0221 6893	0.0296 6893
40	1.3483 4861	0.7416 4796	46.4464 8164	34.4469 3844	0.0215 3016	0.0290 3016
41	1.3584 6123	0.7361 2701	47.7948 3026	35.1830 6545	0.0209 2276	0.0284 2276
42	1.3686 4969	0.7306 4716	49.1532 9148	35.9137 1260	0.0203 4452	0.0278 4452
43	1.3789 1456	0.7252 0809	50.5219 4117	36.6389 2070	0.0197 9338	0.0272 9338
44	1.3892 5642	0.7198 0952	51.9008 5573	37.3587 3022	0.0192 6751	0.0267 6751
45	1.3996 7584	0.7144 5114	53.2901 1215	38.0731 8136	0.0187 6521	0.0262 6521
46	1.4101 7341	0.7091 3264	54.6897 8799	38.7823 1401	0.0182 8495	0.0257 8495
47	1.4297 4971	0.7038 5374	56.0999 6140	39.4861 6775	0.0178 2532	0.0253 2532
48	1.4314 0533	0.6986 1414	57.5207 1111	40.1847 8189	0.0173 8504	0.0248 8504
49	1.4421 4087	0.6934 1353	58.9521 1644	40.8781 9542	0.0169 6292	0.0244 6292
50	1.4529 5693	0.6882 5165	60.3942 5732	41.5664 4707	0.0165 5787	0.0240 5787

Table B.2 *(continued)*

$i = \frac{3}{4}\%$

n	$(1 + i)^n$	$(1 + i)^{-n}$	$s_{\overline{n}\rvert i}$	$a_{\overline{n}\rvert i}$	$\dfrac{1}{s_{\overline{n}\rvert i}}$	$\dfrac{1}{a_{\overline{n}\rvert i}}$
51	1.4638 5411	0.6831 2819	61.8472 1424	42.2495 7525	0.0161 6888	0.0236 6888
52	1.4748 3301	0.6780 4286	63.3110 6835	42.9276 1812	0.0157 9503	0.0232 9503
53	1.4858 9426	0.6729 9540	64.7859 0136	43.6006 1351	0.0154 3546	0.0229 3546
54	1.4970 3847	0.6679 8551	66.2717 9562	44.2685 9902	0.0150 8938	0.0225 8938
55	1.5082 6626	0.6630 1291	67.7688 3409	44.9316 1193	0.0147 5605	0.0222 5605
56	1.5195 7825	0.6580 7733	69.2771 0035	45.5896 8926	0.0144 3478	0.0219 3478
57	1.5309 7509	0.6531 7849	70.7966 7860	46.2428 6776	0.0141 2496	0.0216 2496
58	1.5424 5740	0.6483 1612	72.3276 5369	46.8911 8388	0.0138 2597	0.0213 2597
59	1.5540 2583	0.6434 8995	73.8701 1109	47.5346 7382	0.0135 3727	0.0210 3727
60	1.5656 8103	0.6386 9970	75.4241 3693	48.1733 7352	0.0132 5836	0.0207 5836
61	1.5774 2363	0.6339 4511	76.9898 1795	48.8073 1863	0.0129 8873	0.0204 8873
62	1.5892 5431	0.6292 2592	78.5672 4159	49.4365 4455	0.0127 2795	0.0202 2795
63	1.6011 7372	0.6245 4185	80.1564 9590	50.0610 8640	0.0124 7560	0.0199 7560
64	1.6131 8252	0.6198 9266	81.7576 6962	50.6809 7906	0.0122 3127	0.0197 3127
65	1.6252 8139	0.6152 7807	83.3708 5214	51.2962 5713	0.0119 9460	0.0194 9460
66	1.6374 7100	0.6106 9784	84.9961 3353	51.9069 5497	0.0117 6524	0.0192 6524
67	1.6497 5203	0.6061 5170	86.6336 0453	52.5131 0667	0.0115 4286	0.0190 4286
68	1.6621 2517	0.6016 3940	88.2833 5657	53.1147 4607	0.0113 2716	0.0188 2716
69	1.6745 9111	0.5971 6070	89.9454 8174	53.7119 0677	0.0111 1785	0.0186 1785
70	1.6871 5055	0.5927 1533	91.6200 7285	54.3046 2210	0.0109 1464	0.0184 1464
71	1.6998 0418	0.5883 0306	93.3072 2340	54.8929 2516	0.0107 1728	0.0182 1728
72	1.7125 5271	0.5839 2363	95.0070 2758	55.4768 4880	0.0105 2554	0.0180 2554
73	1.7253 9685	0.5795 7681	96.7195 8028	56.0564 2561	0.0103 3917	0.0178 3917
74	1.7383 3733	0.5752 6234	98.4449 7714	56.6316 8795	0.0101 5796	0.0176 5796
75	1.7513 7486	0.5709 7999	100.1833 1446	57.2026 6794	0.0099 8170	0.0174 8170
76	1.7645 1017	0.5667 2952	101.9346 8932	57.7693 9746	0.0098 1020	0.0173 1020
77	1.7777 4400	0.5625 1069	103.6991 9949	58.3319 0815	0.0096 4328	0.0171 4328
78	1.7910 7708	0.5583 2326	105.4769 4349	58.8902 3141	0.0094 8074	0.0169 8074
79	1.8045 1015	0.5541 6701	107.2680 2056	59.4443 9842	0.0093 2244	0.0168 2244
80	1.8180 4398	0.5500 4170	109.0725 3072	59.9944 4012	0.0091 6821	0.0166 6821
81	1.8316 7931	0.5459 4710	110.8905 7470	60.5403 8722	0.0090 1790	0.0165 1790
82	1.8454 1691	0.5418 8297	112.7222 5401	61.0822 7019	0.0088 7136	0.0163 7136
83	1.8592 5753	0.5378 4911	114.5676 7091	61.6201 1930	0.0087 2847	0.0162 2847
84	1.8732 0196	0.5338 4527	116.4269 2845	62.1539 6456	0.0085 8908	0.0260 8908
85	1.8872 5098	0.5298 7123	118.3001 3041	62.6838 3579	0.0084 5308	0.0159 5308
86	1.9014 0536	0.5259 2678	120.1873 8139	63.2097 6257	0.0083 2034	0.0158 2034
87	1.9156 6590	0.5220 1169	122.0887 8675	63.7317 7427	0.0081 9076	0.0156 9076
88	1.9300 3339	0.5181 2575	124.0044 5265	64.2499 0002	0.0080 6423	0.0155 6423
89	1.9445 0865	0.5142 6873	125.9344 8604	64.7641 6875	0.0079 4064	0.0154 4064
90	1.9590 9246	0.5104 4043	127.8789 9569	65.2746 0918	0.0078 1989	0.0153 1989

Table B.2 *(continued)*
$i = \frac{3}{4}\%$

n	$(1 + i)^n$	$(1 + i)^{-n}$	$s_{\overline{n}|i}$	$a_{\overline{n}|i}$	$\dfrac{1}{s_{\overline{n}|i}}$	$\dfrac{1}{a_{\overline{n}|i}}$
91	1.9737 8565	0.5066 4063	129.8380 8715	65.7812 4981	0.0077 0190	0.0152 0190
92	1.9885 8905	0.5028 6911	131.8118 7280	66.2841 1892	0.0075 8657	0.0150 8657
93	2.0035 0346	0.4991 2567	133.8004 6185	66.7832 4458	0.0074 7382	0.0149 7382
94	2.0185 2974	0.4954 1009	135.8039 6531	67.2786 5467	0.0073 6356	0.0148 6356
95	2.0336 6871	0.4917 2217	137.8224 9505	67.7703 7685	0.0072 5571	0.0147 5571
96	2.0489 2123	0.4880 6171	139.8561 6377	68.2584 3856	0.0071 5020	0.0146 5020
97	2.0642 8814	0.4844 2850	141.9050 8499	68.7428 6705	0.0070 4696	0.0145 4696
98	2.0797 7030	0.4808 2233	143.9693 7313	69.2236 8938	0.0069 4592	0.0144 4592
99	2.0953 6858	0.4772 4301	146.0491 4343	69.7009 3239	0.0068 4701	0.0143 4701
100	2.1110 8384	0.4736 9033	148.1445 1201	70.1746 2272	0.0067 5017	0.0142 5017

$i = 1\%$

n	$(1 + i)^n$	$(1 + i)^{-n}$	$s_{\overline{n}|i}$	$a_{\overline{n}|i}$	$\dfrac{1}{s_{\overline{n}|i}}$	$\dfrac{1}{a_{\overline{n}|i}}$
1	1.0100 0000	0.9900 9901	1.0000 0000	0.9900 9901	1.0000 0000	1.0100 0000
2	1.0201 0000	0.9802 9605	2.0100 0000	1.9703 9506	0.4975 1244	0.5075 1244
3	1.0303 0100	0.9705 9015	3.0301 0000	2.9409 8521	0.3300 2211	0.3400 2211
4	1.0406 0401	0.9609 8034	4.0604 0100	3.9019 6555	0.2462 8109	0.2562 8109
5	1.0510 1005	0.9514 6569	5.1010 0501	4.8534 3124	0.1960 3980	0.2060 3980
6	1.0615 2015	0.9420 4524	6.1520 1506	5.7954 7647	0.1625 4837	0.1725 4837
7	1.0721 3535	0.9327 1805	7.2135 3521	6.7281 9453	0.1386 2828	0.1486 2828
8	1.0828 5671	0.9234 8322	8.2856 7056	7.6516 7775	0.1206 9029	0.1306 9029
9	1.0936 8527	0.9143 3982	9.3685 2727	8.5660 1758	0.1067 4036	0.1167 4036
10	1.1046 2213	0.9052 8695	10.4622 1254	9.4713 0453	0.0955 8208	0.1055 8208
11	1.1156 6835	0.8963 2372	11.5668 3467	10.3676 2825	0.0864 5408	0.0964 5408
12	1.1268 2503	0.8874 4923	12.6825 0301	11.2550 7747	0.0788 4879	0.0888 4879
13	1.1380 9328	0.8786 6260	13.8093 2804	12.1337 4007	0.0724 1482	0.0824 1482
14	1.1494 7421	0.8699 6297	14.9474 2132	13.0037 0304	0.0669 0117	0.0769 0117
15	1.1609 6896	0.8613 4947	16.0968 9554	13.8650 5252	0.0621 2378	0.0721 2378
16	1.1725 7864	0.8528 2126	17.2578 6449	14.7178 7378	0.0579 4460	0.0679 4460
17	1.1843 0443	0.8443 7749	18.4304 4314	15.5622 5127	0.0542 5806	0.0642 5806
18	1.1961 4748	0.8360 1731	19.6147 4757	16.3982 6858	0.0509 8205	0.0609 8205
19	1.2081 0895	0.8277 3992	20.8108 9504	17.2260 0850	0.0480 5175	0.0580 5175
20	1.2201 9004	0.8195 4447	22.0190 0399	18.0455 5297	0.0454 1531	0.0554 1531

Table B.2 *(continued)*
$i = 1\%$

n	$(1 + i)^n$	$(1 + i)^{-n}$	$s_{\overline{n}|i}$	$a_{\overline{n}|i}$	$\dfrac{1}{s_{\overline{n}|i}}$	$\dfrac{1}{a_{\overline{n}|i}}$
21	1.2323 9194	0.8114 3017	23.2391 9403	18.8569 8313	0.0430 3075	0.0530 3075
22	1.2447 1586	0.8033 9621	24.4715 8598	19.6603 7934	0.0408 6372	0.0508 6372
23	1.2571 6302	0.7954 4179	25.7163 0183	20.4558 2113	0.0388 8584	0.0488 8584
24	1.2697 3465	0.7875 6613	26.9734 6485	21.2433 8726	0.0370 7347	0.0470 7347
25	1.2824 3200	0.7797 6844	28.2431 9950	22.0231 5570	0.0354 0675	0.0454 0675
26	1.2952 5631	0.7720 4796	29.5256 3150	22.7952 0366	0.0338 6888	0.0438 6888
27	1.3082 0888	0.7644 0392	30.8208 8781	23.5596 0759	0.0324 4553	0.0424 4553
28	1.3212 9097	0.7568 3557	32.1290 9669	24.3164 4316	0.0311 2444	0.0411 2444
29	1.3345 0388	0.7493 4215	33.4503 8766	25.0657 8530	0.0298 9502	0.0398 9502
30	1.3478 4892	0.7419 2292	34.7848 9153	25.8077 0822	0.0287 4811	0.0387 4811
31	1.3613 2740	0.7345 7715	36.1327 4045	26.5422 8537	0.0276 7573	0.0376 7573
32	1.3749 4068	0.7273 0411	37.4940 6785	27.2695 8947	0.0266 7089	0.0366 7089
33	1.3886 9009	0.7201 0307	38.8690 0853	27.9896 9255	0.0257 2744	0.0357 2744
34	1.4025 7699	0.7129 7334	40.2576 9862	28.7026 6589	0.0248 3997	0.0348 3997
35	1.4166 0276	0.7059 1420	41.6602 7560	29.4085 8009	0.0240 0368	0.0340 0368
36	1.4307 6878	0.6989 2495	43.0768 7836	30.1075 0504	0.0232 1431	0.0332 1431
37	1.4450 7647	0.6920 0490	44.5076 4714	30.7995 0994	0.0224 6805	0.0324 6805
38	1.4595 2724	0.6851 5337	45.9527 2361	31.4846 6330	0.0217 6150	0.0317 6150
39	1.4741 2251	0.6783 6967	47.4122 5085	32.1630 3298	0.0210 9160	0.0310 9160
40	1.4888 6373	0.6716 5314	48.8863 7336	32.8346 8611	0.0204 5560	0.0304 5560
41	1.5037 5237	0.6650 0311	50.3752 3709	33.4996 8922	0.0198 5102	0.0298 5102
42	1.5187 8989	0.6584 1892	51.8789 8946	34.1581 0814	0.0192 7563	0.0292 7563
43	1.5339 7779	0.6518 9992	53.3977 7936	34.8100 0806	0.0187 2737	0.0287 2737
44	1.5493 1757	0.6454 4546	54.9317 5715	35.4554 5352	0.0182 0441	0.0282 0441
45	1.5648 1075	0.6390 5492	56.4810 7472	36.0945 0844	0.0177 0505	0.0277 0505
46	1.5804 5885	0.6327 2764	58.0458 8547	36.7272 3608	0.0172 2775	0.0272 2775
47	1.5962 6344	0.6264 6301	59.6263 4432	37.3536 9909	0.0167 7111	0.0267 7111
48	1.6122 2608	0.6202 6041	61.2226 0777	37.9739 5949	0.0163 3384	0.0263 3384
49	1.6283 4834	0.6141 1921	62.8348 3385	38.5880 7871	0.0159 1474	0.0259 1474
50	1.6446 3182	0.6080 3882	64.4631 8218	39.1961 1753	0.0155 1273	0.0255 1273
51	1.6610 7814	0.6020 1864	66.1078 1401	39.7981 3617	0.0151 2680	0.0251 2680
52	1.6776 8892	0.5960 5806	67.7688 9215	40.3941 9423	0.0147 5603	0.0247 5603
53	1.6944 6581	0.5901 5649	69.4465 8107	40.9843 5072	0.0143 9956	0.0243 9956
54	1.7114 1047	0.5843 1336	71.1410 4688	41.5686 6408	0.0140 5658	0.0240 5658
55	1.7285 2457	0.5785 2808	72.8524 5735	42.1471 9216	0.0137 2637	0.0237 2637
56	1.7458 0982	0.5728 0008	74.5809 8192	42.7199 9224	0.0134 0824	0.0234 0824
57	1.7632 6792	0.5671 2879	76.3267 9174	43.2871 2102	0.0131 0156	0.0231 0156
58	1.7809 0060	0.5615 1365	78.0900 5966	43.8486 3468	0.0128 0573	0.0228 0573
59	1.7987 0960	0.5559 5411	79.8709 6025	44.4045 8879	0.0125 2020	0.0225 2020
60	1.8166 9670	0.5504 4962	81.6696 6986	44.9550 3841	0.0122 4445	0.0222 4445

Table B.2 *(continued)*
$i = 1\%$

n	$(1 + i)^n$	$(1 + i)^{-n}$	$s_{\overline{n}|i}$	$a_{\overline{n}|i}$	$\dfrac{1}{s_{\overline{n}|i}}$	$\dfrac{1}{a_{\overline{n}|i}}$
61	1.8348 6367	0.5449 9962	83.4863 6655	45.5000 3803	0.0119 7800	0.0219 7800
62	1.8532 1230	0.5396 0358	85.3212 3022	46.0396 4161	0.0117 2041	0.0217 2041
63	1.8717 4443	0.5342 6097	87.1744 4252	46.5739 0258	0.0114 7125	0.0214 7125
64	1.8904 6187	0.5289 7126	89.0461 8695	47.1028 7385	0.0112 3013	0.0212 3013
65	1.9093 6649	0.5237 3392	90.9366 4882	47.6266 0777	0.0109 9667	0.0209 9667
66	1.9284 6015	0.5185 4844	92.8460 1531	48.1451 5621	0.0107 7052	0.0207 7052
67	1.9477 4475	0.5134 1429	94.7744 7546	48.6585 7050	0.0105 5136	0.0205 5136
68	1.9672 2220	0.5083 3099	96.7222 2021	49.1669 0149	0.0103 3889	0.0203 3889
69	1.9868 9442	0.5032 9801	98.6894 4242	49.6701 9949	0.0101 3280	0.0201 3280
70	2.0067 6337	0.4983 1486	100.6763 3684	50.1685 1435	0.0099 3282	0.0199 3282
71	2.0268 3100	0.4933 8105	102.6831 0021	50.6618 9539	0.0097 3870	0.0197 3870
72	2.0470 9931	0.4884 9609	104.7099 3121	51.1503 9148	0.0095 5019	0.0195 5019
73	2.0675 7031	0.4836 5949	106.7570 3052	51.6340 5097	0.0093 6706	0.0193 6706
74	2.0882 4601	0.4788 7078	108.8246 0083	52.1129 2175	0.0091 8910	0.0191 8910
75	2.1091 2847	0.4741 2949	110.9128 4684	52.5870 5124	0.0090 1609	0.0190 1609
76	2.1302 1975	0.4694 3514	113.0219 7530	53.0564 8638	0.0088 4784	0.0188 4784
77	2.1515 2195	0.4647 8726	115.1521 9506	53.5212 7364	0.0086 8416	0.0186 8416
78	2.1730 3717	0.4601 8541	117.3037 1701	53.9814 5905	0.0085 2488	0.0185 2488
79	2.1947 6754	0.4556 2912	119.4767 5418	54.4370 8817	0.0083 6983	0.0183 6983
80	2.2167 1522	0.4511 1794	121.6715 2172	54.8882 0611	0.0082 1885	0.0182 1885
81	2.2388 8237	0.4466 5142	123.8882 3694	55.3348 5753	0.0080 7179	0.0180 7179
82	2.2612 7119	0.4422 2913	126.1271 1931	55.7770 8666	0.0079 2851	0.0179 2851
83	2.2838 8390	0.4378 5063	128.3883 9050	56.2149 3729	0.0077 8887	0.0177 8887
84	2.3067 2274	0.4335 1547	130.6722 7440	56.6484 5276	0.0076 5273	0.0176 5273
85	2.3297 8997	0.4292 2324	132.9789 9715	57.0776 7600	0.0075 1998	0.0175 1998
86	2.3530 8787	0.4249 7350	135.3087 8712	57.5026 4951	0.0073 9050	0.0173 9050
87	2.3766 1875	0.4207 6585	137.6618 7499	57.9234 1535	0.0072 6418	0.0172 6418
88	2.4003 8494	0.4165 9985	140.0384 9374	58.3400 1520	0.0071 4089	0.0171 4089
89	2.4243 8879	0.4124 7510	142.4388 7868	58.7524 9030	0.0070 2056	0.0170 2056
90	2.4486 3267	0.4083 9119	144.8632 6746	59.1608 8148	0.0069 0306	0.0169 0306
91	2.4731 1900	0.4043 4771	147.3119 0014	59.5652 2919	0.0067 8832	0.0167 8832
92	2.4978 5019	0.4003 4427	149.7850 1914	59.9655 7346	0.0066 7624	0.0166 7624
93	2.5228 2869	0.3963 8046	152.2828 6933	60.3619 5392	0.0065 6673	0.0165 6673
94	2.5480 5698	0.3924 5590	154.8056 9803	60.7544 0982	0.0064 5971	0.0164 5971
95	2.5735 3755	0.3885 7020	157.3537 5501	61.1429 8002	0.0063 5511	0.0163 5511
96	2.5992 7293	0.3847 2297	159.9272 9256	61.5277 0299	0.0062 5284	0.0162 5284
97	2.6252 6565	0.3809 1383	162.5265 6548	61.9086 1682	0.0061 5284	0.0161 5284
98	2.6515 1831	0.3771 4241	165.1518 3114	62.2857 5923	0.0060 5503	0.0160 5503
99	2.6780 3349	0.3734 0832	167.8033 4945	62.6591 6755	0.0059 5936	0.0159 5936
100	2.7048 1383	0.3697 1121	170.4813 8294	63.0288 7877	0.0058 6574	0.0158 6574

Table B.2 *(continued)*

$i = 1\frac{1}{2}\%$

n	$(1 + i)^n$	$(1 + i)^{-n}$	$s_{\overline{n}|i}$	$a_{\overline{n}|i}$	$\dfrac{1}{s_{\overline{n}|i}}$	$\dfrac{1}{a_{\overline{n}|i}}$
1	1.0150 0000	0.9852 2167	1.0000 0000	0.9852 2167	1.0000 0000	1.0150 0000
2	1.0302 2500	0.9706 6175	2.0150 0000	1.9558 8342	0.4962 7792	0.5112 7792
3	1.0456 7838	0.9563 1699	3.0452 2500	2.9122 0042	0.3283 8296	0.3433 8296
4	1.0613 6355	0.9421 8423	4.0909 0338	3.8543 8465	0.2444 4479	0.2594 4479
5	1.0772 8400	0.9282 6033	5.1522 6693	4.7826 4497	0.1940 8932	0.2090 8932
6	1.0934 4326	0.9145 4219	6.2295 5093	5.6971 8717	0.1605 2521	0.1755 2521
7	1.1098 4491	0.9010 2679	7.3229 9419	6.5982 1396	0.1356 5616	0.1515 5616
8	1.1264 9259	0.8877 1112	8.4328 3911	7.4959 2508	0.1185 8402	1.1335 8402
9	1.1433 8998	0.8745 9224	9.5593 3169	8.3605 1732	0.1046 0982	0.1196 0982
10	1.0605 4083	0.8616 6723	10.7027 2167	9.2221 8455	0.0934 3518	0.1084 3418
11	1.1779 4894	0.8489 3323	11.8632 6249	10.0711 1779	0.0842 9384	0.0992 9384
12	1.1956 1817	0.8363 8742	13.0412 1143	10.9075 0521	0.0766 7999	0.0916 7999
13	1.2135 5244	0.8240 2702	14.2368 2960	11.7315 3222	0.0702 4036	0.0852 4036
14	1.2317 5573	0.8118 4928	15.4503 8205	12.5433 8150	0.0647 2332	0.0797 2332
15	1.2502 3207	0.7998 5150	16.6821 3778	13.3432 3301	0.0599 4436	0.0749 4436
16	1.2689 8555	0.7880 3104	17.9323 6984	14.1312 6405	0.0557 6508	0.0707 6508
17	1.2880 2033	0.7763 8526	19.2013 5539	14.9076 4931	0.0520 7966	0.0670 7966
18	1.3073 4064	0.7649 1159	20.4893 7572	15.6725 6089	0.0488 0578	0.0638 0578
19	1.3269 5075	0.7536 0747	21.7967 1636	16.4261 6837	0.0458 7847	0.0608 7847
20	1.3468 5501	0.7424 7042	23.1236 6710	17.4686 3879	0.0432 4574	0.0582 4574
21	1.3670 5783	0.7314 9795	24.4705 2211	17.9001 3673	0.0408 6550	0.0558 6550
22	1.3875 6370	0.7206 8763	25.8375 7994	18.6208 2437	0.0387 0332	0.0537 0332
23	1.4083 7715	0.7100 3708	27.2251 4364	19.3308 6145	0.0367 3075	0.0517 3075
24	1.4295 0281	0.6995 4392	28.6335 2080	20.0304 0537	0.0349 2410	0.0499 2410
25	1.4509 4535	0.6892 0583	30.0630 2361	20.7196 1120	0.0332 6345	0.0482 6345
26	1.4727 0953	0.6790 2052	31.5139 6896	21.3986 3172	0.0317 3196	0.0467 3196
27	1.4948 0018	0.6689 8574	32.9866 7850	22.0676 1746	0.0303 1527	0.0453 1527
28	1.5172 2218	0.6590 9925	34.4814 7867	22.7267 1671	0.0290 0108	0.0440 0108
29	1.5399 8051	0.6493 5887	35.9987 0085	23.3760 7558	0.0277 7878	0.0427 7878
30	1.5630 8022	0.6397 6243	37.5386 8137	24.0158 3801	0.0266 3919	0.0416 3919
31	1.5865 2642	0.6303 0781	39.1017 6159	24.6461 4582	0.0255 7430	0.0405 7430
32	1.6103 2432	0.6209 9292	40.6882 8801	25.2671 3874	0.0245 7710	0.0395 7710
33	1.6344 7918	0.6118 1568	42.2986 1233	25.8789 5442	0.0236 4144	0.0386 4144
34	1.6589 9637	0.6027 7407	43.9330 9152	26.4817 2849	0.0227 6189	0.0377 6189
35	1.6838 8132	0.5938 6608	45.5920 8789	27.0755 9458	0.0219 3363	0.0369 3363
36	1.7091 3954	0.5850 8974	47.2759 6921	27.6606 8431	0.0211 5240	0.0361 5240
37	1.7347 7663	0.5764 4309	48.9851 0874	28.2371 2740	0.0204 1437	0.0354 1437
38	1.7607 9828	0.5679 2423	50.7198 8538	28.8050 5163	0.0197 1613	0.0347 1613
39	1.7872 1025	0.5595 3126	52.4806 8366	29.3645 8288	0.0190 5463	0.0340 5463
40	1.8140 1841	0.5512 6232	54.2678 9391	29.9158 4520	0.0184 2710	0.0334 2710

Table B.2 (*continued*)

$i = 1\frac{1}{2}\%$

n	$(1 + i)^n$	$(1 + i)^{-n}$	$s_{\overline{n}|i}$	$a_{\overline{n}|i}$	$\dfrac{1}{s_{\overline{n}|i}}$	$\dfrac{1}{a_{\overline{n}|i}}$
41	1.8412 2868	0.5431 1559	56.0819 1232	30.4589 6079	0.0178 3106	0.0328 3106
42	1.8688 4712	0.5350 8925	57.9231 4100	30.9940 5004	0.0172 6426	0.0322 6426
43	1.8968 7982	0.5271 8153	59.7919 8812	31.5212 3157	0.0167 2465	0.0317 2465
44	1.9253 3302	0.5193 9067	61.6888 6794	32.0406 2223	0.0162 1038	0.0312 1038
45	1.9542 1301	0.5117 1494	63.6142 0096	32.5523 3718	0.0157 1976	0.0307 1976
46	1.9835 2621	0.5041 5265	65.5684 1398	33.0564 8983	0.0152 5125	0.0302 5125
47	2.0132 7910	0.4967 0212	67.5519 4018	33.5531 9195	0.0148 0342	0.0298 0342
48	2.0434 7829	0.4893 6170	69.5652 1929	34.0425 5365	0.0143 7500	0.0293 7500
49	2.0741 3046	0.4821 2975	71.6086 9758	34.5246 8339	0.0139 6478	0.0289 6478
50	2.1052 4242	0.4750 0468	73.6828 2804	34.9996 8807	0.0135 7168	0.0285 7168
51	2.1368 2106	0.4679 8491	75.7880 7046	35.4676 7298	0.0131 9469	0.0281 9469
52	2.1688 7337	0.4610 6887	77.9248 9152	35.9287 4185	0.0128 3287	0.0278 3287
53	2.2014 0647	0.4542 5505	80.0937 6489	36.3829 9690	0.0124 8537	0.0274 8537
54	2.2344 2757	0.4475 4192	82.2951 7136	36.8305 3882	0.0121 5138	0.0271 5138
55	2.2679 4398	0.4409 2800	84.5295 9893	37.2714 6681	0.0118 3018	0.0268 3018
56	2.3019 6314	0.4344 1182	86.7975 4292	37.7058 7863	0.0115 2106	0.0265 2106
57	2.3364 9259	0.4279 9194	89.0995 0606	38.1338 7058	0.0112 2341	0.0262 2341
58	2.3715 3998	0.4216 6694	91.4359 9865	38.5555 3751	0.0109 3661	0.0259 3661
59	2.4071 1308	0.4154 3541	93.8075 3863	38.9709 7292	0.0106 6012	0.0256 6012
60	2.4432 1978	0.4092 9597	96.2146 5171	39.3802 6889	0.0103 9343	0.0253 9343
61	2.4798 6807	0.4032 4726	98.6578 7149	39.7835 1614	0.0101 3604	0.0251 3604
62	2.5170 6609	0.3972 8794	101.1377 3956	40.1808 0408	0.0098 8751	0.0248 8751
63	2.5548 2208	0.3914 1669	103.6548 0565	40.5722 2077	0.0096 4741	0.0246 4741
64	2.5931 4442	0.3856 3221	106.2096 2774	40.9578 5298	0.0094 1534	0.0244 1534
65	2.6320 4158	0.3799 3321	108.8027 7215	41.3377 8618	0.0091 9094	0.0241 9094
66	2.6715 2221	0.3743 1843	111.4348 1374	41.7121 0461	0.0089 7386	0.0239 7386
67	2.7115 9504	0.3687 8663	114.1063 3594	42.0808 9125	0.0087 6376	0.0237 6376
68	2.7522 6896	0.3633 3658	116.8179 3098	42.4442 2783	0.0085 6033	0.0235 6033
69	2.7935 5300	0.3579 6708	119.5701 9995	42.8021 9490	0.0083 6329	0.0233 6329
70	2.8354 5629	0.3526 7692	122.3637 5295	43.1548 7183	0.0081 7235	0.0231 7235
71	2.8779 8814	0.3474 6495	125.1992 0924	43.5023 3678	0.0079 8727	0.0229 8727
72	2.9211 5796	0.3423 3000	128.0771 9738	43.8446 6677	0.0078 0779	0.0228 0779
73	2.9649 7533	0.3372 7093	130.9983 5534	44.1819 3771	0.0076 3368	0.0226 3368
74	3.0094 4996	0.3322 8663	133.9633 3067	44.5142 2434	0.0074 6473	0.0224 6473
75	3.0545 9171	0.3273 7599	136.9727 8063	44.8416 0034	0.0073 0072	0.0223 0072
76	3.1004 1059	0.3225 3793	140.0273 7234	45.1641 3826	0.0071 4146	0.0221 4146
77	3.1469 1674	0.3177 7136	143.1277 8292	45.4819 0962	0.0069 8676	0.0219 8676
78	3.1941 2050	0.3130 7523	146.2746 9967	45.7949 8485	0.0068 3645	0.0218 3645
79	3.2420 3230	0.3084 4850	149.4688 2016	46.1034 3335	0.0066 9036	0.0216 9036
80	3.2906 6279	0.3038 9015	152.7108 5247	46.4073 2349	0.0065 4832	0.0215 4832

Table B.2 *(continued)*

$i = 1\frac{1}{2}\%$

n	$(1 + i)^n$	$(1 + i)^{-n}$	$s_{\overline{n}|i}$	$a_{\overline{n}|i}$	$\dfrac{1}{s_{\overline{n}|i}}$	$\dfrac{1}{a_{\overline{n}|i}}$
81	3.3400 2273	0.2993 9916	156.0015 1525	46.7067 2265	0.0064 1019	0.0214 1019
82	3.3901 2307	0.2949 7454	159.3415 3798	47.0016 9720	0.0062 7583	0.0212 7583
83	3.4409 7492	0.2906 1531	162.7316 6105	47.2923 1251	0.0061 4509	0.0211 4509
84	3.4925 8954	0.2863 2050	166.1726 3597	47.5786 1784	0.0060 1784	0.0210 1784
85	3.5449 7838	0.2820 8917	169.6652 2551	47.8607 2218	0.0058 9396	0.0208 9396
86	3.5981 5306	0.2779 2036	173.2101 0389	48.1386 4254	0.0057 7333	0.0207 7333
87	3.6521 2535	0.2738 1316	176.8083 5695	48.4124 5571	0.0056 5584	0.0206 5584
88	3.7069 0723	0.2697 6666	180.4604 8230	48.6822 2237	0.0055 4138	0.0205 4138
89	3.7625 1084	0.2657 7996	184.1673 8954	48.9480 0234	0.0054 2984	0.0204 2984
90	3.8189 4851	0.2618 5218	187.9299 0038	49.2098 5452	0.0053 2113	0.0203 2113
91	3.8762 3273	0.2579 8245	191.7488 4889	49.4678 3696	0.0052 1516	0.0202 1516
92	3.9343 7622	0.2541 6990	195.6250 8162	49.7220 0686	0.0051 1182	0.0201 1182
93	3.9933 9187	0.2504 1369	199.5594 5784	49.9724 2055	0.0050 1104	0.0200 1104
94	4.0532 9275	0.2467 1300	203.5528 4971	50.2191 3355	0.0049 1273	0.0199 1273
95	4.1140 9214	0.2430 6699	207.6061 4246	50.4622 0054	0.0048 1681	0.0198 1681
96	4.1758 0352	0.2394 7487	211.7202 3459	50.7016 7541	0.0047 2321	0.0197 2321
97	4.2384 4057	0.2359 3583	215.8960 3811	50.9376 1124	0.0046 3186	0.0196 3186
98	4.3020 1718	0.2324 4909	220.1344 7868	51.1700 6034	0.0045 4268	0.0195 4268
99	4.3665 4744	0.2290 1389	224.4364 9586	51.3990 7422	0.0044 5560	0.0194 5560
100	4.4320 4565	0.2256 2944	228.8030 4330	51.6247 0367	0.0043 7057	0.0193 7057

$i = 2\%$

n	$(1 + i)^n$	$(1 + i)^{-n}$	$s_{\overline{n}|i}$	$a_{\overline{n}|i}$	$\dfrac{1}{s_{\overline{n}|i}}$	$\dfrac{1}{a_{\overline{n}|i}}$
1	1.0200 0000	0.9803 9216	1.0000 0000	0.9803 9216	1.0000 0000	1.0200 0000
2	1.0404 0000	0.9611 6878	2.0200 0000	1.9415 6094	0.4950 4950	0.5150 4950
3	1.0612 0800	0.9423 2233	3.0604 0000	2.8838 8327	0.3267 5467	0.3467 5467
4	1.0824 3216	0.9238 4543	4.1216 0800	3.8077 2870	0.2426 2375	0.2626 2375
5	1.1040 8080	0.9057 3081	5.2040 4016	4.7134 5951	0.1921 5839	0.2121 5839
6	1.1261 6242	0.8879 7138	6.3081 2096	5.6014 3089	0.1585 2581	0.1785 2581
7	1.1486 8567	0.8705 6018	7.4342 8338	6.4719 9107	0.1345 1196	0.1545 1196
8	1.1716 5938	0.8534 9037	8.5829 6905	7.3254 8144	0.1165 0980	0.1365 0980
9	1.1950 9257	0.8367 5527	9.7546 2843	8.1622 3671	0.1025 1544	0.1225 1544
10	1.2189 9442	0.8203 4830	10.9497 2100	8.9825 8501	0.0913 2653	0.1113 2653
11	1.2433 7431	0.8042 6304	12.1687 1542	9.7868 4805	0.0821 7794	0.1021 7794
12	1.2682 4179	0.7884 9318	13.4120 8973	10.5753 4122	0.0745 5960	0.0945 5960
13	1.2936 0663	0.7730 3253	14.6803 3152	11.3483 7375	0.0681 1835	0.0881 1835
14	1.3194 7876	0.7578 7502	15.9739 3815	12.1062 4877	0.0626 0197	0.0826 0197
15	1.3458 6834	0.7430 1473	17.2934 1692	12.8492 6350	0.0578 2547	0.0778 2547

Table B.2 *(continued)*

$i = 2\%$

n	$(1 + i)^n$	$(1 + i)^{-n}$	$s_{\overline{n}|i}$	$a_{\overline{n}|i}$	$\dfrac{1}{s_{\overline{n}|i}}$	$\dfrac{1}{a_{\overline{n}|i}}$
16	1.3727 8571	0.7284 4581	18.6392 8525	13.5777 0931	0.0536 5013	0.0736 5013
17	1.4002 4142	0.7141 6256	20.0120 7096	14.2918 7188	0.0499 6984	0.0699 6984
18	1.4282 4625	0.7001 5937	21.4123 1238	14.9920 3125	0.0467 0210	0.0667 0210
19	1.4568 1117	0.6864 3076	22.8405 5863	15.6784 6201	0.0437 8177	0.0637 8177
20	1.4859 4740	0.6729 7133	24.2973 6980	16.3514 3334	0.0411 5672	0.0611 5672
21	1.5156 6634	0.6597 7582	25.7833 1719	17.0112 0916	0.0387 8477	0.0587 8477
22	1.5459 7967	0.6468 3904	27.2989 8354	17.6580 4820	0.0366 3140	0.0566 3140
23	1.5768 9926	0.6341 5592	28.8449 6321	18.2922 0412	0.0346 6810	0.0546 6810
24	1.6084 3725	0.6217 2149	30.4218 6247	18.9139 2560	0.0328 7110	0.0528 7110
25	1.6406 0599	0.6095 3087	32.0302 9972	19.5234 5647	0.0312 2044	0.0512 2044
26	1.6734 1811	0.5975 7928	33.6709 0572	20.1210 3576	0.0296 9923	0.0496 9923
27	1.7068 8648	0.5858 6204	35.3443 2383	20.7068 9780	0.0282 9309	0.0482 9309
28	1.7410 2421	0.5743 7455	37.0512 1031	21.2812 7236	0.0269 8967	0.0469 8967
29	1.7758 4469	0.5631 1231	38.7922 3451	21.8443 8466	0.0257 7836	0.0457 7836
30	1.8113 6158	0.5520 7089	40.5680 7921	22.3964 5555	0.0246 4992	0.0446 4992
31	1.8475 8882	0.5412 4597	42.3794 4079	22.9377 0152	0.0235 9635	0.0435 9635
32	1.8845 4059	0.5306 3330	44.2270 2961	23.4683 3482	0.0226 1061	0.0426 1061
33	1.9222 3140	0.5202 2873	46.1115 7020	23.9885 6355	0.0216 8653	0.0416 8653
34	1.9606 7603	0.5100 2817	48.0338 0160	24.4985 9172	0.0208 1867	0.0408 1867
35	1.9998 8955	0.5000 2761	49.9944 7763	24.9986 1933	0.0200 0221	0.0400 0221
36	2.0398 8734	0.4902 2315	51.9943 6719	25.4888 4248	0.0192 3285	0.0392 3285
37	2.0806 8509	0.4806 1093	54.0342 5453	25.9694 5341	0.0185 0678	0.0385 0678
38	2.1222 9879	0.4711 8719	56.1149 3962	26.4406 4060	0.0178 2057	0.0378 2057
39	2.1647 4477	0.4619 4822	58.2372 3841	26.9025 8883	0.0171 7114	0.0371 7114
40	2.2080 3966	0.4528 9042	60.4019 8318	27.3554 7924	0.0165 5575	0.0365 5575
41	2.2522 0046	0.4440 1021	62.6100 2284	27.7994 8945	0.0159 7188	0.0359 7188
42	2.2972 4447	0.4353 0413	64.8622 2330	28.2347 9358	0.0154 1729	0.0354 1729
43	2.3431 8936	0.4267 6875	67.1594 6777	28.6615 6233	0.0148 8993	0.0348 8993
44	2.3900 5314	0.4184 0074	69.5026 5712	29.0799 6307	0.0143 8794	0.0343 8794
45	2.4378 5421	0.4101 9680	71.8927 1027	29.4901 5987	0.0139 0962	0.0339 0962
46	2.4866 1129	0.4021 5373	74.3305 6447	29.8923 1360	0.0134 5342	0.0334 5342
47	2.5363 4352	0.3942 6836	76.8171 7576	30.2865 8196	0.0130 1792	0.0330 1792
48	2.5870 7039	0.3865 3761	79.3535 1927	30.6731 1957	0.0126 0184	0.0326 0184
49	2.6388 1179	0.3789 5844	81.9405 8966	31.0520 7801	0.0122 0396	0.0322 0396
50	2.6915 8803	0.3715 2788	84.5794 0145	31.4236 0589	0.0118 2321	0.0318 2321
51	2.7454 1979	0.3642 4302	87.2709 8948	31.7878 4892	0.0114 5856	0.0314 5856
52	2.8003 2819	0.3571 0100	90.0164 0927	32.1449 4992	0.0111 0909	0.0311 0909
53	2.8563 3475	0.3500 9902	92.8167 3746	32.4950 4894	0.0107 7392	0.0307 7392
54	2.9134 6144	0.3432 3433	95.6730 7221	32.8382 8327	0.0104 5226	0.0304 5226
55	2.9717 3067	0.3365 0425	98.5865 3365	33.1747 8752	0.0101 4337	0.0301 4337

Table B.2 *(continued)*

$i = 2\%$

n	$(1 + i)^n$	$(1 + i)^{-n}$	$s_{\overline{n}\rvert i}$	$a_{\overline{n}\rvert i}$	$\dfrac{1}{s_{\overline{n}\rvert i}}$	$\dfrac{1}{a_{\overline{n}\rvert i}}$
56	3.0311 6529	0.3299 0613	101.5582 6432	33.5046 9365	0.0098 4656	0.0298 4656
57	3.0917 8859	0.3234 3738	104.5894 2961	33.8281 3103	0.0095 6120	0.0295 6120
58	3.1536 2436	0.3170 9547	107.6812 1820	34.1452 2650	0.0092 8667	0.0292 8667
59	3.2166 9685	0.3108 7791	110.8348 4257	34.4561 0441	0.0090 2243	0.0290 2243
60	3.2810 3079	0.3047 8227	114.0515 3942	34.7608 8668	0.0087 6797	0.0287 6797
61	3.3466 5140	0.2988 0614	117.3325 7021	35.0596 9282	0.0085 2278	0.0285 2278
62	3.4135 8443	0.2929 4720	120.6792 2161	35.3526 4002	0.0082 8643	0.0282 8643
63	3.4818 5612	0.2872 0314	124.0928 0604	35.6398 4316	0.0080 5848	0.0280 5848
64	3.5514 9324	0.2815 7170	127.5746 6216	35.9214 1486	0.0078 3855	0.0278 3855
65	3.6225 2311	0.2760 5069	131.1261 5541	36.1974 6555	0.0076 2624	0.0276 2624
66	3.6949 7357	0.2706 3793	134.7486 7852	36.4681 0348	0.0074 2122	0.0274 2122
67	3.7688 7304	0.2653 3130	138.4436 5209	36.7334 3478	0.0072 2316	0.0272 2316
68	3.8442 5050	0.2601 2873	142.2125 2513	36.9935 6351	0.0070 3173	0.0270 3173
69	3.9211 3551	0.2550 2817	146.0567 7563	37.2485 9168	0.0068 4665	0.0268 4665
70	3.9995 5822	0.2500 2761	149.9779 1114	37.4986 1929	0.0066 6765	0.0266 6765
71	4.0795 4939	0.2451 2511	153.9774 6937	37.7437 4441	0.0064 9446	0.0264 9446
72	4.1611 4038	0.2403 1874	158.0570 1875	37.9840 6314	0.0063 2683	0.0263 2683
73	4.2443 6318	0.2356 0661	162.2181 5913	38.2196 6975	0.0061 6454	0.0261 6454
74	4.3292 5045	0.2309 8687	166.4625 2231	38.4506 5662	0.0060 0736	0.0260 0736
75	4.4158 3546	0.2264 5771	170.7917 7276	38.6771 1433	0.0058 5508	0.0258 5508
76	4.5041 5216	0.2220 1737	175.2076 0821	38.8991 3170	0.0057 0751	0.0257 0751
77	4.5942 3521	0.2176 6408	179.7117 6038	39.1167 9578	0.0055 6447	0.0255 6447
78	4.6861 1991	0.2133 9616	184.3059 9558	39.3301 9194	0.0054 2576	0.0254 2576
79	4.7798 4231	0.2092 1192	188.9921 1549	39.5394 0386	0.0052 9123	0.0252 9123
80	4.8754 3916	0.2051 0973	193.7719 5780	39.7445 1359	0.0051 6071	0.0251 6071
81	4.9729 4794	0.2010 8797	198.6473 9696	39.9456 0156	0.0050 3405	0.0250 3405
82	5.0724 0690	0.1971 4507	203.6203 4490	40.1427 4663	0.0049 1110	0.0249 1110
83	5.1738 5504	0.1932 7948	208.6927 5180	40.3360 2611	0.0047 9173	0.0247 9173
84	5.2773 3214	0.1894 8968	213.8666 0683	40.5255 1579	0.0046 7581	0.0246 7581
85	5.3828 7878	0.1857 7420	219.1439 3897	40.7112 8999	0.0045 6321	0.0245 6321
86	5.4905 3636	0.1821 3157	224.5268 1775	40.8934 2156	0.0044 5381	0.0244 5381
87	5.6003 4708	0.1785 6036	230.0173 5411	41.0719 8192	0.0043 4750	0.0243 4750
88	5.7123 5402	0.1750 5918	235.6177 0119	41.2470 4110	0.0042 4416	0.0242 4416
89	5.8266 0110	0.1716 2665	241.3300 5521	41.4186 6774	0.0041 4370	0.0241 4370
90	5.9431 3313	0.1682 6142	247.1566 5632	41.5869 2916	0.0040 4602	0.0240 4602
91	6.0619 9579	0.1649 6217	253.0997 8944	41.7518 9133	0.0039 5101	0.0239 5101
92	6.1832 3570	0.1617 2762	259.1617 8523	41.9136 1895	0.0038 5859	0.0238 5859
93	6.3069 0042	0.1585 5649	265.3450 2093	42.0721 7545	0.0037 6868	0.0237 6868
94	6.4330 3843	0.1554 4754	271.6519 2135	42.2276 2299	0.0036 8118	0.0236 8118
95	6.5616 9920	0.1523 9955	278.0849 5978	42.3800 2254	0.0035 9602	0.0235 9602

Table B.2 (*continued*)

$i = 2\%$

n	$(1+i)^n$	$(1+i)^{-n}$	$s_{\overline{n}|i}$	$a_{\overline{n}|i}$	$\dfrac{1}{s_{\overline{n}|i}}$	$\dfrac{1}{a_{\overline{n}|i}}$
96	6.6929 3318	0.1494 1132	284.6466 5898	42.5294 3386	0.0035 1313	0.0235 1313
97	6.8267 9184	0.1464 8169	291.3395 9216	42.6759 1555	0.0034 3242	0.0234 3242
98	6.9633 2768	0.1436 0950	298.1663 8400	42.8195 2505	0.0033 5383	0.0233 5383
99	7.1025 9423	0.1407 9363	305.1297 1168	42.9603 1867	0.0032 7729	0.0232 7729
100	7.2446 4612	0.1380 3297	312.2323 0591	43.0983 5164	0.0032 0274	0.0232 0274

$i = 2\frac{1}{2}\%$

n	$(1+i)^n$	$(1+i)^{-n}$	$s_{\overline{n}|i}$	$a_{\overline{n}|i}$	$\dfrac{1}{s_{\overline{n}|i}}$	$\dfrac{1}{a_{\overline{n}|i}}$
1	1.0250 0000	0.9756 0976	1.0000 0000	0.9756 0976	1.0000 0000	1.0250 0000
2	1.0506 2500	0.9518 1440	2.0250 0000	1.9274 2415	0.4938 2716	0.5188 2716
3	1.0768 9063	0.9285 9941	3.0756 2500	2.8560 2356	0.3251 3717	0.3501 3717
4	1.1038 1289	0.9059 5064	4.1525 1563	3.7619 7421	0.2408 1788	0.2658 1788
5	1.1314 0821	0.8838 5429	5.2563 2852	4.6458 2850	0.1902 4686	0.2152 4686
6	1.1596 9342	0.8622 9687	6.3877 3673	5.5081 2536	0.1565 4997	0.1815 4997
7	1.1886 8575	0.8412 6524	7.5474 3015	6.3493 9060	0.1324 9543	0.1574 9543
8	1.2184 0290	0.8207 4657	8.7361 1590	7.1701 3717	0.1144 6735	0.1394 6735
9	1.2488 6297	0.8007 2836	9.9545 1880	7.9708 6553	0.1004 5689	0.1254 5689
10	1.2800 8454	0.7811 9840	11.2033 8177	8.7520 6393	0.0892 5876	0.1142 5876
11	1.3120 8666	0.7621 4478	12.4834 6631	9.5142 0871	0.0801 0596	0.1051 0596
12	1.3448 8882	0.7435 5589	13.7955 5297	10.2577 6460	0.0724 8713	0.0974 8713
13	1.3785 1104	0.7254 2038	15.1404 4179	10.9831 8497	0.0660 4827	0.0910 4827
14	1.4129 7382	0.7077 2720	16.5189 5284	11.6909 1217	0.0605 3652	0.0855 3652
15	1.4482 9817	0.6904 6556	17.9319 2666	12.3813 7773	0.0557 6646	0.0807 6646
16	1.4845 0562	0.6736 2493	19.3802 2483	13.0550 0266	0.0515 9899	0.0765 9899
17	1.5216 1826	0.6571 9506	20.8647 3045	13.7121 9772	0.0479 2777	0.0729 2777
18	1.5596 5872	0.6411 6591	22.3863 4871	14.3533 6363	0.0446 7008	0.0696 7008
19	1.5986 5019	0.6255 2772	23.9460 0743	14.9788 9134	0.0417 6062	0.0667 6062
20	1.6386 1644	0.6102 7094	25.5446 5761	15.5891 6229	0.0391 4713	0.0641 4713
21	1.6795 8185	0.5953 8629	27.1832 7405	16.1845 4857	0.0367 8733	0.0617 8733
22	1.7215 7140	0.5808 6467	28.8628 5590	16.7654 1324	0.0346 4661	0.0596 4661
23	1.7646 1068	0.5666 9724	30.5844 2730	17.3321 1048	0.0326 9638	0.0576 9638
24	1.8087 2595	0.5528 7535	32.3490 3798	17.8849 8583	0.0309 1282	0.0559 1282
25	1.8539 4410	0.5393 9059	34.1577 6393	18.4243 7642	0.0292 7592	0.0542 7592
26	1.9002 9270	0.5262 3472	36.0117 0803	18.9506 1114	0.0277 6875	0.0527 6875
27	1.9478 0002	0.5133 9973	37.9120 0073	19.4640 1087	0.0263 7687	0.0513 7687
28	1.9964 9502	0.5008 7778	39.8598 0075	19.9648 8866	0.0250 8793	0.0500 8793
29	2.0464 0739	0.4886 6125	41.8562 9577	20.4535 4991	0.0238 9127	0.0488 9127
30	2.0975 6758	0.4767 4269	43.9027 0316	20.9302 9259	0.0227 7764	0.0477 7764

Table B.2 *(continued)*

$i = 2\frac{1}{2}\%$

n	$(1 + i)^n$	$(1 + i)^{-n}$	$s_{\overline{n}|i}$	$a_{\overline{n}|i}$	$\dfrac{1}{s_{\overline{n}|i}}$	$\dfrac{1}{a_{\overline{n}|i}}$
31	2.1500 0677	0.4651 1481	46.0002 7074	21.3954 0741	0.0217 3900	0.0467 3900
32	2.2037 5694	0.4537 7055	48.1502 7751	21.8491 7796	0.0207 6831	0.0457 6831
33	2.2588 5086	0.4427 0298	50.3540 3445	22.2918 8094	0.0198 5938	0.0448 5938
34	2.3153 2213	0.4319 0534	52.6128 8531	22.7237 8628	0.0190 0675	0.0440 0675
35	2.3732 0519	0.4213 7107	54.9282 0744	23.1451 5734	0.0182 0558	0.0432 0558
36	2.4325 3532	0.4110 9372	57.3014 1263	23.5562 5107	0.0174 5158	0.0424 5158
37	2.4933 4870	0.4010 6705	59.7339 4794	23.9573 1812	0.0167 4090	0.0417 4090
38	2.5556 8242	0.3912 8492	62.2272 9664	24.3486 0304	0.0160 7012	0.0410 7012
39	2.6195 7448	0.3817 4139	64.7829 7906	24.7303 4443	0.0154 3615	0.0404 3615
40	2.6850 6384	0.3724 3062	67.4025 5354	25.1027 7505	0.0148 3623	0.0398 3623
41	2.7521 9043	0.3633 4695	70.0876 1737	25.4661 2200	0.0142 6786	0.0392 6786
42	2.8209 9520	0.3544 8483	72.8398 0781	25.8206 0683	0.0137 2876	0.0387 2876
43	2.8915 2008	0.3458 3886	75.6608 0300	26.1664 4569	0.0132 1688	0.0382 1688
44	2.9638 0808	0.3374 0376	78.5523 2308	26.5038 4945	0.0127 3037	0.0377 3037
45	3.0379 0328	0.3291 7440	81.5161 3116	26.8330 2386	0.0122 6751	0.0372 6751
46	3.1138 5086	0.3211 4576	84.5540 3443	27.1541 6962	0.0118 2676	0.0368 2676
47	3.1916 9713	0.3133 1294	87.6678 8530	27.4674 8255	0.0114 0669	0.0364 0669
48	3.2714 8956	0.3056 7116	90.8595 8243	27.7731 5371	0.0110 0599	0.0360 0599
49	3.3532 7680	0.2982 1576	94.1310 7199	28.0713 6947	0.0106 2348	0.0356 2348
50	3.4371 0872	0.2909 4221	97.4843 4879	28.3623 1168	0.0102 5806	0.0352 5806
51	3.5230 3664	0.2838 4606	100.9214 5751	28.6461 5774	0.0099 0870	0.0349 0870
52	3.6111 1235	0.2769 2298	104.4444 9395	28.9230 8072	0.0095 7446	0.0345 7446
53	3.7013 9016	0.2701 6876	108.0556 0629	29.1932 4948	0.0092 5449	0.0342 5449
54	3.7939 2491	0.2635 7928	111.7569 9645	29.4568 2876	0.0089 4799	0.0339 4799
55	3.8887 7303	0.2571 5052	115.5509 2136	29.7139 7928	0.0086 5419	0.0336 5419
56	3.9859 9236	0.2508 7855	119.4396 9440	29.9648 5784	0.0083 7243	0.0333 7243
57	4.0856 4217	0.2447 5956	123.4256 8676	30.2096 1740	0.0081 0204	0.0331 0204
58	4.1877 8322	0.2387 8982	127.5113 2893	30.4484 0722	0.0078 4244	0.0328 4244
59	4.2924 7780	0.2329 6568	131.6991 1215	30.6813 7290	0.0075 9307	0.0325 9307
60	4.3997 8975	0.2272 8359	135.9915 8995	30.9086 5649	0.0073 5340	0.0323 5340
61	4.5097 8449	0.2217 4009	140.3913 7970	31.1303 9657	0.0071 2294	0.0321 2294
62	4.6225 2910	0.2163 3179	144.9011 6419	31.3467 2836	0.0069 0126	0.0319 0126
63	4.7380 9233	0.2110 5541	149.5236 9330	31.5577 8377	0.0066 8790	0.0316 8790
64	4.8565 4464	0.2059 0771	154.2617 8563	31.7636 9148	0.0064 8249	0.0314 8249
65	4.9779 5826	0.2008 8557	159.1183 3027	31.9645 7705	0.0062 8463	0.0312 8463
66	5.1024 0721	0.1959 8593	164.0962 8853	33.1605 6298	0.0060 9398	0.0310 9398
67	5.2299 6739	0.1912 0578	169.1986 9574	32.3517 6876	0.0059 1021	0.0309 1021
68	5.3607 1658	0.1865 4223	174.4286 6314	32.5383 1099	0.0057 3300	0.0307 3300
69	5.4947 3449	0.1819 9241	179.7893 7971	32.7203 0340	0.0055 6206	0.0305 6206
70	5.6321 0286	0.1775 5358	185.2841 1421	32.8978 5698	0.0053 9712	0.0303 9712

Table B.2 *(continued)*

$i = 2\frac{1}{2}\%$

n	$(1 + i)^n$	$(1 + i)^{-n}$	$s_{\overline{n}|i}$	$a_{\overline{n}|i}$	$\dfrac{1}{s_{\overline{n}|i}}$	$\dfrac{1}{a_{\overline{n}|i}}$
71	5.7729 0543	0.1732 2300	190.9162 1706	33.0710 7998	0.0052 3790	0.0302 3790
72	5.9172 2806	0.1689 9805	196.6891 2249	33.2400 7803	0.0050 8417	0.0300 8417
73	6.0651 5876	0.1648 7615	202.6063 5055	33.4049 5417	0.0049 3568	0.0299 3568
74	6.2167 8773	0.1608 5478	208.6715 0931	33.5658 0895	0.0047 9222	0.0297 9222
75	6.3722 0743	0.1569 3149	214.8882 9705	33.7227 4044	0.0046 5358	0.0296 5358
76	6.5315 1261	0.1531 0389	221.2605 0447	33.8758 4433	0.0045 1956	0.0295 1956
77	6.6948 0043	0.1493 6965	227.7920 1709	34.0252 1398	0.0043 8997	0.0293 8997
78	6.8621 7044	0.1457 2649	234.4868 1751	34.1709 4047	0.0042 6463	0.0292 6463
79	7.0337 2470	0.1421 7218	241.3489 8795	34.3131 1265	0.0041 4338	0.0291 4338
80	7.2095 6782	0.1387 0457	248.3827 1265	34.4518 1722	0.0040 2605	0.0290 2605
81	7.3898 0701	0.1353 2153	255.5922 8047	34.5871 3875	0.0039 1248	0.0289 1248
82	7.5745 5219	0.1320 2101	262.9820 8748	34.7191 5976	0.0038 0254	0.0288 0254
83	7.7639 1599	0.1288 0098	270.5566 3966	34.8479 6074	0.0036 9608	0.0286 9608
84	7.9580 1389	0.1256 5949	278.3205 5566	34.9736 2023	0.0035 9298	0.0285 9298
85	8.1569 6424	0.1225 9463	286.2785 6955	35.0962 1486	0.0034 9310	0.0284 9310
86	8.3608 8834	0.1196 0452	294.4355 3379	35.2158 1938	0.0033 9633	0.0283 9633
87	8.5699 1055	0.1166 8733	302.7964 2213	35.3325 0671	0.0033 0255	0.0283 0255
88	8.7841 5832	0.1138 4130	311.3663 3268	35.4463 4801	0.0032 1165	0.0282 1165
89	9.0037 6228	0.1110 6468	320.1504 9100	35.5574 1269	0.0031 2353	0.0281 2353
90	9.2288 5633	0.1083 5579	329.1542 5328	35.6657 6848	0.0030 3809	0.0280 3809
91	9.4595 7774	0.1057 1296	338.3831 0961	35.7714 8144	0.0029 5523	0.0279 5523
92	9.6960 6718	0.1031 3460	347.8426 8735	35.8746 1604	0.0028 7486	0.0278 7486
93	9.9384 6886	0.1006 1912	357.5387 5453	35.9752 3516	0.0027 9690	0.0277 9690
94	10.1869 3058	0.0981 6500	367.4772 2339	36.0734 0016	0.0027 2126	0.0277 2126
95	10.4416 0385	0.0957 7073	377.6641 5398	36.1691 7089	0.0026 4786	0.0276 4786
96	10.7026 4395	0.0934 3486	388.1057 5783	36.2626 0574	0.0025 7662	0.0275 7662
97	10.9702 1004	0.0911 5596	398.8084 0177	36.3537 6170	0.0025 0747	0.0276 0747
98	11.2444 6530	0.0889 3264	409.7786 1182	36.4426 9434	0.0024 4034	0.0274 4034
99	11.5255 7693	0.0867 6355	421.0230 7711	36.5294 5790	0.0023 7517	0.0273 7517
100	11.8137 1635	0.0846 4737	432.5486 5404	36.6141 0526	0.0023 1188	0.0273 1188

$i = 3\%$

n	$(1 + i)^n$	$(1 + i)^{-n}$	$s_{\overline{n}|i}$	$a_{\overline{n}|i}$	$\dfrac{1}{s_{\overline{n}|i}}$	$\dfrac{1}{a_{\overline{n}|i}}$
1	1.0300 0000	0.9708 7379	1.0000 0000	0.9708 7379	1.0000 0000	1.0300 0000
2	1.0609 0000	0.9425 9591	2.0300 0000	1.9134 6970	0.4926 1084	0.5226 1084
3	1.0927 2700	0.9151 4166	3.0909 0000	2.8286 1135	0.3235 3036	0.3535 3036
4	1.1255 0881	0.8884 8705	4.1836 2700	3.7170 9840	0.2390 2705	0.2690 2705
5	1.1592 7407	0.8626 0878	5.3091 3581	4.5797 0719	0.1883 5457	0.2183 5457

Table B.2 *(continued)*

$i = 3\%$

n	$(1 + i)^n$	$(1 + i)^{-n}$	$s_{\overline{n}\rvert i}$	$a_{\overline{n}\rvert i}$	$\dfrac{1}{s_{\overline{n}\rvert i}}$	$\dfrac{1}{a_{\overline{n}\rvert i}}$
6	1.1940 5230	0.8374 8426	6.4684 0988	5.4171 9144	0.1545 9750	0.1845 9750
7	1.2298 7837	0.8130 9151	7.6624 6218	6.2302 8296	0.1305 0635	0.1605 0635
8	1.2667 7008	0.7894 0923	8.8923 3605	7.0196 9219	0.1124 5639	0.1424 5639
9	1.3047 7318	0.7664 1673	10.1591 0613	7.7861 0892	0.0984 3386	0.1284 3386
10	1.3439 1638	0.7440 9391	11.4638 7931	8.5302 0284	0.0872 3051	0.1172 3051
11	1.3842 3387	0.7224 2128	12.8077 9569	9.2526 2411	0.0780 7745	0.1080 7745
12	1.4257 6089	0.7013 7988	14.1920 2956	9.9540 0399	0.0704 6209	0.1004 6209
13	1.4685 3371	0.6809 5134	15.6177 9045	10.6349 5533	0.0640 2954	0.0940 2954
14	1.5125 8972	0.6611 1781	17.0863 2416	11.2960 7314	0.0585 2634	0.0885 2634
15	1.5579 6742	0.6418 6195	18.5989 1389	11.9379 3509	0.0537 6658	0.0837 6658
16	1.6047 0644	0.6231 6694	20.1568 8130	12.5611 0203	0.0496 1085	0.0796 1085
17	1.6528 4763	0.6050 1645	21.7615 8774	13.1661 1847	0.0459 5253	0.0759 5253
18	1.7024 3306	0.5873 9461	23.4144 3537	13.7535 1308	0.0427 0870	0.0727 0870
19	1.7535 0605	0.5702 8603	25.1168 6844	14.3237 9911	0.0398 1388	0.0698 1388
20	1.8061 1123	0.5536 7575	26.8703 7449	14.8774 7486	0.0372 1571	0.0672 1571
21	1.8602 9456	0.5375 4928	28.6764 8572	15.4150 2414	0.0348 7178	0.0648 7178
22	1.9161 0341	0.5218 9520	30.5367 8030	15.9369 1664	0.0327 4739	0.0627 4739
23	1.9735 8651	0.5066 9175	32.4528 8370	16.4436 0839	0.0308 1390	0.0608 1390
24	2.0327 9411	0.4919 3374	34.4264 7022	16.9355 4212	0.0290 4742	0.0590 4742
25	2.0937 7793	0.4776 0557	36.4592 6432	17.4131 4769	0.0274 2787	0.0574 2787
26	2.1565 9127	0.4636 9473	38.5530 4225	17.8678 4242	0.0259 3829	0.0559 3829
27	2.2212 8901	0.4501 8906	40.7096 3352	18.3270 3147	0.0245 6421	0.0545 6421
28	2.2879 2768	0.4370 7675	42.9309 2252	18.7641 0823	0.0232 9323	0.0532 9323
29	2.3565 6551	0.4243 4636	45.2188 5020	19.1884 5459	0.0221 1467	0.0524 1467
30	2.4272 6247	0.4119 8676	47.5754 1571	19.6004 4135	0.0210 1926	0.0510 1926
31	2.5000 8035	0.3999 8715	50.0026 7818	20.0004 2849	0.0199 9893	0.0499 9893
32	2.5750 8276	0.3883 3703	52.5027 5852	20.3887 6553	0.0190 4662	0.0490 4662
33	2.6523 3524	0.3770 2625	55.0778 4128	20.7657 9178	0.0181 5612	0.0481 5612
34	2.7319 0530	0.3660 4490	57.7301 7652	21.1318 3668	0.0173 2196	0.0473 2196
35	2.8138 6245	0.3553 8340	60.4620 8181	21.4872 2007	0.0165 3929	0.0465 3929
36	2.8982 7833	0.3450 3243	63.2759 4427	21.8322 5250	0.0158 0379	0.0458 0379
37	2.9852 2668	0.3349 8294	66.1742 2259	22.1672 3544	0.0151 1162	0.0451 1162
38	3.0747 8348	0.3252 2615	69.1594 4927	22.4924 6159	0.0144 5934	0.0444 5934
39	3.1670 2698	0.3157 5355	72.2342 3275	22.8082 1513	0.0138 4385	0.0428 4385
40	3.2620 3779	0.3065 5684	75.4012 5973	23.1147 7197	0.0132 6238	0.0432 6238
41	3.3598 9893	0.2976 2800	78.6632 9753	23.4123 9997	0.0127 1241	0.0427 1241
42	3.4606 9589	0.2889 5922	82.0231 9645	23.7013 5920	0.0121 9167	0.0421 9167
43	3.5645 1677	0.2805 4294	85.4838 9234	23.9819 0213	0.0116 9811	0.0416 9811
44	3.6714 5227	0.2723 7178	89.0484 0911	24.2542 7392	0.0112 2985	0.0412 2985
45	3.7815 9584	0.2644 3862	92.7198 6139	24.5187 1254	0.0107 8518	0.0407 8518

Table B.2 *(continued)*

$i = 3\%$

n	$(1 + i)^n$	$(1 + i)^{-n}$	$s_{\overline{n}|i}$	$a_{\overline{n}|i}$	$\dfrac{1}{s_{\overline{n}|i}}$	$\dfrac{1}{a_{\overline{n}|i}}$
46	3.8950 4372	0.2567 3653	96.5014 5723	24.7754 4907	0.0103 6254	0.0403 6254
47	4.0118 9503	0.2492 5876	100.3965 0095	25.0247 0783	0.0099 6051	0.0399 6051
48	4.1322 5188	0.2419 9880	104.4083 9598	25.2667 0664	0.0095 7777	0.0395 7777
49	4.2562 1944	0.2349 5029	108.5406 4785	25.5016 5693	0.0092 1314	0.0392 1314
50	4.3839 0602	0.2281 0708	112.7968 6729	25.7297 6401	0.0088 6549	0.0388 6549
51	4.5154 2320	0.2214 6318	117.1807 7331	25.9512 2719	0.0085 3382	0.0385 3382
52	4.6508 8590	0.2150 1280	121.6961 9651	26.1662 3999	0.0082 1718	0.0382 1718
53	4.7904 1247	0.2087 5029	126.3470 8240	26.3749 9028	0.0079 1471	0.0349 1471
54	4.9341 2485	0.2026 7019	131.1374 9488	26.5776 6047	0.0076 2558	0.0376 2558
55	5.0821 4859	0.1967 6717	136.0716 1972	26.7744 2764	0.0073 4907	0.0373 4907
56	5.2346 1305	0.1910 3609	141.1537 6831	26.9654 6373	0.0070 8447	0.0370 8447
57	5.3916 5144	0.1854 8193	146.3883 8136	27.1509 3566	0.0068 3114	0.0368 3114
58	5.5534 0098	0.1800 6984	151.7800 3280	27.3310 0549	0.0065 8848	0.0365 8848
59	5.7200 0301	0.1748 2508	157.3334 3379	27.5058 3058	0.0063 5593	0.0363 5593
60	5.8916 0310	0.1697 3309	163.0534 3680	27.6755 6367	0.0061 3296	0.0361 3296
61	6.0683 5120	0.1647 8941	168.9450 3991	27.8403 5307	0.0059 1908	0.0359 1908
62	6.2504 0173	0.1599 8972	175.0133 9110	28.0003 4279	0.0057 1385	0.0357 1385
63	6.4379 1379	0.1553 2982	181.2637 9284	28.1556 7261	0.0055 1682	0.0355 1682
64	6.6310 5120	0.1508 0565	187.7017 0662	28.3064 7826	0.0053 2760	0.0353 2760
65	6.8299 8273	0.1464 1325	194.3327 5782	28.4528 9152	0.0051 4581	0.0351 4581
66	7.0348 8222	0.1421 4879	201.1627 4055	28.5950 4031	0.0049 7110	0.0349 7110
67	7.2459 2868	0.1380 0853	208.1976 2277	28.7330 4884	0.0048 0313	0.0348 0313
68	7.4633 0654	0.1339 8887	215.4435 5145	28.8670 3771	0.0046 4159	0.0346 4159
69	7.6872 0574	0.1300 8628	222.9068 5800	28.9971 2399	0.0044 8618	0.0344 8618
70	7.9178 2191	0.1262 9736	230.5940 6374	29.1234 2135	0.0043 3663	0.0343 3663
71	8.1553 5657	0.1226 1880	238.5118 8565	29.2460 4015	0.0041 9266	0.0341 9266
72	8.4000 1727	0.1190 4737	246.6672 4222	29.3650 8752	0.0040 5404	0.0340 5404
73	8.6520 1778	0.1155 7998	255.0672 5949	29.4806 6750	0.0039 2053	0.0339 2053
74	8.9115 7832	0.1122 1357	263.7192 7727	29.5928 8107	0.0037 9191	0.0337 9191
75	9.1789 2567	0.1089 4521	272.6308 5559	29.7018 2628	0.0036 6796	0.0336 6796
76	9.4542 9344	0.1057 7205	281.8097 8126	29.8075 9833	0.0035 4849	0.0335 4849
77	9.7379 2224	0.1026 9131	291.2640 7469	29.9102 8964	0.0034 3331	0.0334 3331
78	10.0300 5991	0.0997 0030	301.0019 9693	30.0099 8994	0.0033 2224	0.0333 2224
79	10.3309 6171	0.0967 9641	311.0320 5684	30.1067 8635	0.0032 1510	0.0335 1510
80	10.6408 9056	0.0939 7710	321.3630 1855	30.2007 6345	0.0031 1175	0.0331 1175
81	10.9601 1727	0.0912 3990	332.0039 0910	30.2920 0335	0.0030 1201	0.0330 1201
82	11.2889 2079	0.0885 8243	342.9640 2638	30.3805 8577	0.0029 1576	0.0329 1576
83	11.6275 8842	0.0860 0236	354.2529 4717	30.4665 8813	0.0028 2284	0.0328 2284
84	11.9764 1607	0.0834 9743	365.8805 3558	30.5500 8556	0.0027 3313	0.0327 3313
85	12.3357 0855	0.0810 6547	377.8569 5165	30.6311 5103	0.0026 4650	0.0326 4650

Table B.2 *(continued)*

$i = 3\%$

n	$(1 + i)^n$	$(1 + i)^{-n}$	$s_{\overline{n}|i}$	$a_{\overline{n}|i}$	$\dfrac{1}{s_{\overline{n}|i}}$	$\dfrac{1}{a_{\overline{n}|i}}$
86	12.7057 7981	0.0787 0434	390.1926 6020	30.7098 5537	0.0025 6284	0.0325 6284
87	13.0869 5320	0.0764 1198	402.8984 4001	30.7862 6735	0.0024 8202	0.0324 8202
88	13.4795 6180	0.0741 8639	415.9853 9321	30.8604 5374	0.0024 0393	0.0324 0393
89	13.8839 4865	0.0720 2562	429.4649 5500	30.9324 7936	0.0023 2848	0.0323 2848
90	14.3004 6711	0.0699 2779	443.3489 0365	31.0024 0714	0.0022 5556	0.0322 5556
91	14.7294 8112	0.0678 9105	457.6493 7076	31.0702 9820	0.0021 8508	0.0321 8508
92	15.1713 6556	0.0659 1364	472.3788 5189	31.1362 1184	0.0021 1694	0.0321 1694
93	15.6265 0652	0.0639 9383	487.5502 1744	31.2002 0567	0.0020 5107	0.0320 5107
94	16.0953 0172	0.0621 2993	503.1767 2397	31.2623 3560	0.0019 8737	0.0319 8737
95	16.5781 6077	0.0603 2032	519.2720 2568	31.3226 5592	0.0019 2577	0.0319 2577
96	17.0755 0559	0.0585 6342	535.8501 8645	31.3812 1934	0.0018 6619	0.0318 6619
97	17.5877 7076	0.0568 5769	552.9256 9205	31.4380 7703	0.0018 0856	0.0318 0856
98	18.1154 0388	0.0552 0164	570.5134 6281	31.4932 7867	0.0017 5281	0.0317 5281
99	18.6588 6600	0.0535 9383	588.6288 6669	31.5468 7250	0.0016 9886	0.0316 9886
100	19.2186 3198	0.0520 3284	607.2877 3269	31.5989 0534	0.0016 4667	0.0316 4667

$i = 3\frac{1}{2}\%$

n	$(1 + i)^n$	$(1 + i)^{-n}$	$s_{\overline{n}|i}$	$a_{\overline{n}|i}$	$\dfrac{1}{s_{\overline{n}|i}}$	$\dfrac{1}{a_{\overline{n}|i}}$
1	1.0350 0000	0.9661 8357	1.0000 0000	0.9661 8357	1.0000 0000	1.0350 0000
2	1.0712 2500	0.9335 1070	2.0350 0000	1.8996 9428	0.4914 0049	0.5264 0049
3	1.1087 1788	0.9019 4271	3.1062 2500	2.8016 3698	0.3219 3418	0.3569 3418
4	1.1475 2300	0.8714 4223	4.2149 4288	3.6730 7921	0.2372 5114	0.2722 5114
5	1.1876 8631	0.8419 7317	5.3624 6588	4.5150 5238	0.1864 8137	0.2214 8137
6	1.2292 5533	0.8135 0064	6.5501 5218	5.3285 5302	0.1526 6821	0.1876 6821
7	1.2722 7926	0.7859 9096	7.7794 0751	6.1145 4398	0.1285 4449	0.1635 4449
8	1.3168 0904	0.7594 1156	9.0516 8677	6.8739 5554	0.1104 7665	0.1454 7665
9	1.3628 9735	0.7337 3097	10.3684 9581	7.6076 8651	0.0964 4601	0.1314 4601
10	1.4105 9876	0.7089 1881	11.7313 9316	8.3166 0532	0.0852 4137	0.1202 4137
11	1.4599 6972	0.6849 4571	13.1419 9192	9.0015 5104	0.0760 9197	0.1110 9197
12	1.5110 6866	0.6617 8330	14.6019 6164	9.6633 3433	0.0684 8395	0.1034 8395
13	1.5639 5606	0.6394 0415	16.1130 3030	10.3027 3849	0.0620 6157	0.0970 6157
14	1.6186 9452	0.6177 8179	17.6769 8636	10.9205 2028	0.0565 7073	0.0915 7073
15	1.6753 4883	0.5968 9062	19.2956 8088	11.5174 1090	0.0518 2507	0.0868 2507
16	1.7339 8604	0.5767 0591	20.9710 2971	12.0941 1681	0.0476 8483	0.0826 8483
17	1.7946 7555	0.5572 0378	22.7050 1575	12.6513 2059	0.0440 4313	0.0790 4313
18	1.8574 8920	0.5383 6114	24.4996 9130	13.1896 8173	0.0408 1684	0.0758 1684
19	1.9225 0132	0.5201 5569	26.3571 8050	13.7098 3742	0.0379 4033	0.0729 4033
20	1.9897 8886	0.5025 6588	28.2796 8181	14.2124 0330	0.0353 6108	0.0703 6108

Table B.2 *(continued)*
$i = 3\frac{1}{2}\%$

n	$(1 + i)^n$	$(1 + i)^{-n}$	$s_{\overline{n}|i}$	$a_{\overline{n}|i}$	$\dfrac{1}{s_{\overline{n}|i}}$	$\dfrac{1}{a_{\overline{n}|i}}$
21	2.0594 3147	0.4855 7090	30.2694 7068	14.6979 7420	0.0330 3659	0.0680 3659
22	2.1315 1158	0.4691 5063	32.3289 0215	15.1671 2484	0.0309 3207	0.0659 3207
23	2.2061 1448	0.4532 8563	34.4604 1373	15.6204 1047	0.0290 1880	0.0640 1880
24	2.2833 2849	0.4379 5713	36.6665 2821	16.0583 6760	0.0272 7283	0.0622 7283
25	2.3632 4498	0.4231 4699	38.9498 5669	16.4815 1459	0.0256 7404	0.0606 7404
26	2.4459 5856	0.4088 3767	41.3131 0168	16.8903 5226	0.0242 0540	0.0592 0540
27	2.5315 6711	0.3950 1224	43.7590 6024	17.2853 6451	0.0228 5241	0.0578 5241
28	2.6201 7196	0.3816 5434	46.2906 2734	17.6670 1885	0.0216 0265	0.0566 0265
29	2.7118 7798	0.3687 4815	48.9107 9930	18.0357 6700	0.0204 4538	0.0554 4538
30	2.8067 9370	0.3562 7841	51.6226 7728	18.3920 4541	0.0193 7133	0.0543 7133
31	2.9050 3148	0.3442 3035	54.4294 7098	18.7362 7576	0.0183 7240	0.0533 7240
32	3.0067 0759	0.3325 8971	57.3345 0247	19.0688 6547	0.0174 4150	0.0524 4150
33	3.1119 4235	0.3213 4271	60.3412 1005	19.3902 0818	0.0165 7242	0.0515 7242
34	3.2208 6033	0.3104 7605	63.4531 5240	19.7006 8423	0.0157 5966	0.0507 5966
35	3.3335 9045	0.2999 7586	66.6740 1274	20.0006 6110	0.0149 9835	0.0499 9835
36	3.4502 6611	0.2898 3272	70.0076 0318	20.2904 9381	0.0142 8416	0.0492 8416
37	3.5710 2543	0.2800 3161	73.4578 6930	20.5705 2542	0.0136 1325	0.0486 1325
38	3.6960 1132	0.2705 6194	77.0288 9472	20.8410 8736	0.0129 8214	0.0479 8214
39	3.8253 7171	0.2614 1250	80.7249 0604	21.1024 9987	0.0123 8775	0.0473 8775
40	3.9592 5972	0.2525 7247	84.5502 7775	21.3550 7234	0.0118 2728	0.0468 2728
41	4.0978 3381	0.2440 3137	88.5095 3747	21.5991 0371	0.0112 9822	0.0462 9822
42	4.2412 5799	0.2357 7910	92.6073 7128	21.8348 8281	0.0107 9828	0.0457 9828
43	4.3897 0202	0.2278 0590	96.8486 2928	22.0626 8870	0.0103 2539	0.0453 2539
44	4.5433 4160	0.2201 0231	101.2383 3130	22.2827 9102	0.0098 7768	0.0448 7768
45	4.7023 5855	0.2126 5924	105.7816 7290	22.4954 5026	0.0094 5343	0.0444 5343
46	4.8669 4110	0.2054 6787	110.4840 3145	22.7009 1813	0.0090 5108	0.0440 5108
47	5.0372 8404	0.1985 1968	115.3509 7255	22.8994 3780	0.0086 6919	0.0436 6919
48	5.2135 8898	0.1918 0645	120.3882 5659	23.0912 4425	0.0083 0646	0.0433 0646
49	5.3960 6459	0.1853 2024	125.6018 4557	23.2765 6450	0.0079 6167	0.0429 6167
50	5.5849 2686	0.1790 5337	130.9979 1016	23.4556 1787	0.0076 3371	0.0426 3371
51	5.7803 9930	0.1729 9843	136.5828 3702	23.6286 1630	0.0073 2156	0.0423 2156
52	5.9827 1327	0.1671 4824	142.3632 3631	23.7957 6454	0.0070 2429	0.0420 2429
53	6.1921 0824	0.1614 9589	148.3459 4958	23.9572 6043	0.0067 4100	0.0417 4100
54	6.4088 3202	0.1560 3467	154.5380 5782	24.1132 9510	0.0064 7090	0.0414 7090
55	6.6331 4114	0.1507 5814	160.9468 8984	24.2640 5323	0.0062 1323	0.0412 1323
56	6.8653 0108	0.1456 6004	167.5800 3099	24.4097 1327	0.0059 6730	0.0409 6730
57	7.1055 8662	0.1407 3433	174.4453 3207	24.5504 4760	0.0057 3245	0.0407 3245
58	7.3542 8215	0.1359 7520	181.5509 1869	24.6864 2281	0.0055 0810	0.0405 0810
59	7.6116 8203	0.1313 7701	188.9052 0085	24.8177 9981	0.0052 9366	0.0402 9366
60	7.8780 9090	0.1269 3431	196.5168 8288	29.9447 3412	0.0050 8862	0.0400 8862

Table B.2 *(continued)*
$i = 3\frac{1}{2}\%$

n	$(1 + i)^n$	$(1 + i)^{-n}$	$s_{\overline{n}|i}$	$a_{\overline{n}|i}$	$\dfrac{1}{s_{\overline{n}|i}}$	$\dfrac{1}{a_{\overline{n}|i}}$
61	8.1538 2408	0.1226 4184	204.3949 7378	25.0673 7596	0.0048 9249	0.0398 9249
62	8.4392 0793	0.1184 9453	212.5487 9786	25.1858 7049	0.0047 0480	0.0397 0480
63	8.7345 8020	0.1144 8747	220.9880 0579	25.3003 5796	0.0045 2513	0.0395 2513
64	9.0402 9051	0.1106 1591	229.7225 8599	25.4109 7388	0.0043 5308	0.0393 5308
65	9.3567 0068	0.1068 7528	238.7628 7650	25.5178 4916	0.0041 8826	0.0391 8826
66	9.6841 8520	0.1032 6114	248.1195 7718	25.6211 1030	0.0040 3031	0.0390 3031
67	10.0231 3168	0.0997 6922	257.8037 6238	25.7208 7951	0.0038 7892	0.0388 7892
68	10.3739 4129	0.0963 9538	267.8268 9406	25.8172 7489	0.0037 3375	0.0387 3375
69	10.7370 2924	0.0931 3563	278.2008 3535	25.9104 1052	0.0035 9453	0.0385 9453
70	11.1128 2526	0.0899 8612	288.9378 6459	26.0003 9664	0.0034 6095	0.0384 6095
71	11.5017 7414	0.0869 4311	300.0506 8985	26.0873 3975	0.0033 3277	0.0383 3277
72	11.9043 3624	0.0840 0300	311.5524 6400	26.1713 4275	0.0032 0973	0.0382 0973
73	12.3209 8801	0.0811 6232	323.4568 0024	26.2525 0508	0.0030 9160	0.0380 9160
74	12.7522 2259	0.0784 1770	335.7777 8824	26.3309 2278	0.0029 7816	0.0379 7816
75	13.1985 5038	0.0757 6590	348.5300 1083	26.4066 8868	0.0028 6919	0.0378 6919
76	13.6604 9964	0.0732 0376	361.7285 6121	26.4798 9244	0.0027 6450	0.0377 6450
77	14.1386 1713	0.0707 2827	375.3890 6085	26.5506 2072	0.0026 6390	0.0376 6390
78	14.6334 6873	0.0683 3650	389.5276 7798	26.6189 5721	0.0025 6721	0.0375 6721
79	15.1456 4013	0.0660 2560	404.1611 4671	26.6849 8281	0.0024 7426	0.0374 7426
80	15.6757 3754	0.0637 9285	419.3067 8685	26.7487 7567	0.0023 8489	0.0373 8489
81	16.2243 8835	0.0616 3561	434.9825 2439	26.8104 1127	0.0022 9894	0.0372 9894
82	16.7922 4195	0.0595 5131	451.2069 1274	26.8699 6258	0.0022 1628	0.0372 1628
83	17.3799 7041	0.0575 3750	467.9991 5469	26.9275 0008	0.0021 3676	0.0371 3676
84	17.9882 6938	0.0555 9178	485.3791 2510	26.9830 9186	0.0020 6025	0.0370 6025
85	18.6178 5881	0.0537 1187	503.3673 9448	27.0368 0373	0.0019 8662	0.0369 8662
86	19.2694 8387	0.0518 9553	521.9852 5329	27.0886 9926	0.0019 1576	0.0369 1576
87	19.9439 1580	0.0501 4060	541.2547 3715	27.1388 3986	0.0018 4756	0.0368 4756
88	20.6419 5285	0.0484 4503	561.1986 5295	27.1872 8489	0.0017 8190	0.0367 8190
89	21.3644 2120	0.0468 0679	581.8406 0581	27.2340 9168	0.0017 1868	0.0367 1868
90	22.1121 7595	0.0452 2395	603.2050 2701	27.2793 1564	0.0016 5781	0.0366 5781
91	22.8861 0210	0.0436 9464	625.3172 0295	27.3230 1028	0.0015 9919	0.0365 9919
92	23.6871 1568	0.0422 1704	648.2033 0506	27.3652 2732	0.0015 4273	0.0365 4273
93	24.5161 6473	0.0407 8941	671.8904 2073	27.4060 1673	0.0014 8834	0.0364 8834
94	25.3742 3049	0.0349 1006	696.4065 8546	27.4454 2680	00.014 3594	0.0364 3594
95	26.2623 2856	0.0380 7735	721.7808 1595	27.4835 0415	0.0013 8546	0.0363 8546
96	27.1815 1006	0.0367 8971	748.0431 4451	27.5202 9387	0.0013 3692	0.0363 3682
97	28.1328 6291	0.0355 4562	775.2246 5457	27.5558 3948	0.0012 8995	0.0362 8995
98	29.1175 1311	0.0343 4359	803.3575 1748	27.5901 8308	0.0012 4478	0.0362 4478
99	30.1366 2607	0.0331 8221	832.4750 3059	27.6233 6529	0.0012 0124	0.0362 0124
100	31.1914 0798	0.0320 6011	862.6116 5666	27.6554 2540	0.0011 5927	0.0361 5927

Table B.2 *(continued)*
$i = 4\%$

n	$(1 + i)^n$	$(1 + i)^{-n}$	$s_{n\rceil i}$	$a_{n\rceil i}$	$\dfrac{1}{s_{n\rceil i}}$	$\dfrac{1}{a_{n\rceil i}}$
1	1.0400 0000	0.9615 3846	1.0000 0000	0.9615 3846	1.0000 0000	1.0400 0000
2	1.0816 0000	0.9245 5621	2.0400 0000	1.8860 9467	0.4901 9608	0.5301 9608
3	1.1248 6400	0.8889 9636	3.1216 0000	2.7750 9103	0.3203 4854	0.3603 4854
4	1.1698 5856	0.8548 0419	4.2464 6400	3.6298 9522	0.2354 9005	0.2754 9005
5	1.2166 5290	0.8219 2711	5.4163 2256	4.4518 2233	0.1846 2711	0.2246 2711
6	1.2653 1902	0.7903 1453	6.6329 7546	5.2421 3686	0.1507 6190	0.1907 6190
7	1.3159 3178	0.7599 1781	7.8982 9448	6.0020 5467	0.1266 0961	0.1666 0961
8	1.3685 6905	0.7306 9021	9.2142 2626	6.7327 4487	0.1085 2783	0.1485 2783
9	1.4233 1181	0.7025 8674	10.5827 9531	7.4353 3161	0.0944 9299	0.1344 9299
10	1.4802 4428	0.6755 6417	12.0061 0712	8.1108 9578	0.0832 9094	0.1232 9094
11	1.5394 5406	0.6495 8093	13.4863 5141	8.7604 7671	0.0741 4904	0.1141 4904
12	1.6010 3222	0.6245 9705	15.0258 0546	9.3850 7376	0.0665 5217	0.1065 5217
13	1.6650 7351	0.6005 7409	16.6268 3768	9.9856 4785	0.0601 4373	0.1001 4373
14	1.7316 7645	0.5774 7508	18.2919 1119	10.5631 2293	0.0546 6897	0.0946 6897
15	1.8009 4351	0.5552 6450	20.0235 8764	11.1183 8743	0.0499 4110	0.0899 4110
16	1.8729 8125	0.5339 0818	21.8245 3114	11.6522 9561	0.0458 2000	0.0858 2000
17	1.9479 0050	0.5133 7325	23.6975 1239	12.1656 6885	0.0421 9852	0.0821 9852
18	2.0258 1652	0.4936 2812	25.6454 1288	12.6592 9697	0.0389 9333	0.0789 9333
19	2.1068 4918	0.4746 4242	27.6712 2940	13.1339 3940	0.0361 3862	0.0761 3862
20	2.1911 2314	0.4563 8695	29.7780 7858	13.5903 2634	0.0335 8175	0.0735 8175
21	2.2787 6807	0.4388 3360	31.9692 0172	14.0291 5995	0.0312 8011	0.0712 8011
22	2.3699 1879	0.4219 5539	34.2479 6979	14.4511 1533	0.0291 9881	0.0691 9881
23	2.4647 1554	0.4057 2633	36.6178 8858	14.8568 4167	0.0273 0906	0.0673 0906
24	2.5633 0416	0.3901 2147	39.0826 0412	15.2469 6314	0.0255 8683	0.0655 8683
25	2.6658 3633	0.3751 1680	41.6459 0829	15.6220 7994	0.0240 1196	0.0640 1196
26	2.7724 6978	0.3606 8923	44.3117 4462	15.9827 6918	0.0225 6738	0.0625 6738
27	2.8833 6858	0.3468 1657	47.0842 1440	16.3295 8575	0.0212 3854	0.0612 3854
28	2.9987 0332	0.3334 7747	49.9675 8298	16.6630 6322	0.0200 1298	0.0600 1298
29	3.1186 5145	0.3206 5141	52.9662 8630	16.9837 1463	0.0188 7993	0.0588 7993
30	3.2433 9751	0.3083 1867	56.0849 3775	17.2920 3330	0.0178 3010	0.0578 3010
31	3.3731 3341	0.2964 6026	59.3283 3526	17.5884 9356	0.0168 5535	0.0568 5535
32	3.5080 5875	0.2850 5794	62.7014 6867	17.8735 5150	0.0159 4859	0.0559 4859
33	3.6483 8110	0.2740 9417	66.2095 2742	18.1476 4567	0.0151 0357	0.0551 0357
34	3.7943 1634	0.2635 5209	69.8579 0851	18.4111 9776	0.0143 1477	0.0543 1477
35	3.9460 8899	0.2534 1547	73.6522 2486	18.6646 1323	0.0135 7732	0.0535 7732
36	4.1039 3255	0.2436 6872	77.5983 1385	18.9082 8195	0.0128 8688	0.0528 8688
37	4.2680 8986	0.2342 9685	81.7022 4640	19.1425 7880	0.0122 3957	0.0522 3957
38	4.4388 1345	0.2252 8543	85.9703 3626	19.3678 6423	0.0116 3192	0.0516 3192
39	4.6163 6599	0.2166 2061	90.4091 4971	19.5844 8484	0.0110 6083	0.0510 6083
40	4.8010 2063	0.2082 8904	95.0255 1570	19.7927 7388	0.0105 2349	0.0505 2349

Table B.2 *(continued)*
$i = 4\%$

n	$(1 + i)^n$	$(1 + i)^{-n}$	$s_{\overline{n}|i}$	$a_{\overline{n}|i}$	$\dfrac{1}{s_{\overline{n}|i}}$	$\dfrac{1}{a_{\overline{n}|i}}$
41	4.9930 6145	0.2002 7793	99.8265 3633	19.9930 5181	0.0100 1738	0.0500 1738
42	5.1927 8391	0.1925 7493	104.8195 9778	20.1856 2674	0.0095 4020	0.0495 4020
43	5.4004 9527	0.1851 6820	110.0123 8169	20.3707 9494	0.0090 8989	0.0490 8989
44	5.6165 1508	0.1780 4635	115.4128 7696	20.5488 4129	0.0086 6454	0.0486 6454
45	5.8411 7568	0.1711 9841	121.0293 9204	20.7200 3970	0.0082 6246	0.0482 6246
46	6.0748 2271	0.1646 1386	126.8705 6772	20.8846 5356	0.0078 8205	0.0478 8205
47	6.3178 1562	0.1582 8256	132.9453 9043	21.0429 3612	0.0075 2189	0.0475 2189
48	6.5705 2824	0.1521 9476	139.2632 0604	21.1951 3088	0.0071 8065	0.0471 8065
49	6.8333 4937	0.1463 4112	145.8337 3429	21.3414 7200	0.0068 5712	0.0468 5712
50	7.1066 8335	0.1407 1262	152.6670 8366	21.4821 8462	0.0065 5020	0.0465 5020

$i = 4\frac{1}{2}\%$

n	$(1 + i)^n$	$(1 + i)^{-n}$	$s_{\overline{n}|i}$	$a_{\overline{n}|i}$	$\dfrac{1}{s_{\overline{n}|i}}$	$\dfrac{1}{a_{\overline{n}|i}}$
1	1.0450 0000	0.9569 3780	1.0000 0000	0.9569 3780	1.0000 0000	1.0450 0000
2	1.0920 2500	0.9157 2995	2.0450 0000	1.8726 6775	0.4889 9756	0.5339 9756
3	1.1411 6613	0.8762 9660	3.1370 2500	2.7489 6435	0.3187 7336	0.3637 7336
4	1.1925 1860	0.8385 6134	4.2781 9113	3.5875 2570	0.2337 4365	0.2787 4365
5	1.2461 8194	0.8024 5105	5.4707 0973	4.3899 7674	0.1827 9164	0.2277 9164
6	1.3022 6012	0.7678 9574	6.7168 9166	5.1578 7248	0.1488 7839	0.1938 7839
7	1.3608 6183	0.7348 2846	8.0191 5179	5.8927 0094	0.1247 0147	0.1697 0147
8	1.4221 0061	0.7031 8513	9.3800 1362	6.5958 8607	0.1066 0965	0.1516 0965
9	1.4860 9514	0.6729 0443	10.8021 1423	7.2687 9050	0.0925 7447	0.1375 7447
10	1.5529 6942	0.6439 2768	12.2882 0937	7.9127 1818	0.0813 7882	0.1263 7882
11	1.6228 5305	0.6161 9874	13.8411 7879	8.5289 1692	0.0722 4818	0.1172 4818
12	1.6958 8143	0.5896 6386	15.4640 3184	9.1185 8078	0.0646 6619	0.1096 6619
13	1.7721 9610	0.5642 7164	17.1599 1327	9.6828 5242	0.0582 7535	0.1032 7535
14	1.8519 4492	0.5399 7286	18.9321 0937	10.2228 2528	0.0528 2032	0.0978 2032
15	1.9352 8244	0.5167 2044	20.7840 5429	10.7395 4573	0.0481 1381	0.0931 1381
16	2.0223 7015	0.4944 6932	22.7193 3673	11.2340 1505	0.0440 1537	0.0890 1537
17	2.1133 7681	0.4731 7639	24.7417 0689	11.7071 9143	0.0404 1758	0.0854 1758
18	2.2084 7877	0.4528 0037	26.8550 8370	12.1599 9180	0.0372 3690	0.0822 3690
19	2.3078 6031	0.4333 0179	29.0635 6246	12.5932 9359	0.0344 0734	0.0794 0734
20	2.4117 1402	0.4146 4286	31.3714 2277	13.0079 3645	0.0318 7614	0.0768 7614
21	2.5202 4116	0.3967 8743	33.7831 3680	13.4047 2388	0.0296 0057	0.0746 0057
22	2.6336 5201	0.3797 0089	36.3033 7795	13.7844 2476	0.0275 4565	0.0725 4565
23	2.7521 6635	0.3633 5013	38.9370 2996	14.1477 7489	0.0256 8249	0.0706 8249
24	2.8760 1383	0.3477 0347	41.6891 9631	14.4954 7837	0.0239 8703	0.0689 8703
25	3.0054 3446	0.3327 3060	44.5652 1015	14.8282 0896	0.0224 3903	0.0674 3903

Table B.2 *(continued)*
$i = 4\frac{1}{2}\%$

n	$(1 + i)^n$	$(1 + i)^{-n}$	$s_{\overline{n}|i}$	$a_{\overline{n}|i}$	$\dfrac{1}{s_{\overline{n}|i}}$	$\dfrac{1}{a_{\overline{n}|i}}$
26	3.1406 7901	0.3184 0248	47.5706 4460	15.1466 1145	0.0210 2137	0.0660 2137
27	3.2820 0956	0.3046 9137	50.7113 2361	15.4513 0282	0.0197 1946	0.0647 1946
28	3.4296 9999	0.2915 7069	53.9933 3317	15.7428 7351	0.0185 2081	0.0635 2081
29	3.5840 3649	0.2790 1502	57.4230 3316	16.0218 8853	0.0174 1461	0.0624 1461
30	3.7453 1813	0.2670 0002	61.0070 6966	16.2888 8854	0.0163 9154	0.0613 9154
31	3.9138 5745	0.2555 0241	64.7523 8779	16.5443 9095	0.0154 4345	0.0604 4345
32	4.0899 8104	0.2444 9991	68.6662 4524	16.7888 9086	0.0145 6320	0.0595 6320
33	4.2740 3018	0.2339 7121	75.7562 2628	17.0228 6207	0.0137 4453	0.0587 4453
34	4.4663 6154	0.2238 9589	77.0302 5646	17.2467 5796	0.0129 8191	0.0579 8191
35	4.6673 4781	0.2142 5444	81.4966 1800	17.4610 1240	0.0122 7045	0.0572 7045
36	4.8773 7846	0.2050 2817	86.1639 6581	17.6660 4058	0.0116 0578	0.0566 0578
37	5.0968 6049	0.1961 9921	91.0413 4427	17.8622 3979	0.0109 8402	0.0559 8402
38	5.3262 1921	0.1877 5044	96.1382 0476	18.0499 9023	0.0104 0169	0.0554 0169
39	5.5658 9908	0.1796 6549	101.4644 2398	18.2296 5572	0.0098 5567	0.0548 5567
40	5.8163 6454	0.1719 2870	107.0303 2306	18.4015 8442	0.0093 4315	0.0543 4315
41	6.0781 0094	0.1645 2507	112.8466 8760	18.5661 0949	0.0088 6158	0.0538 6158
42	6.3516 1548	0.1574 4026	118.9247 8854	18.7235 4975	0.0084 0868	0.0534 0868
43	6.6374 3818	0.1506 6054	125.2764 0402	18.8742 1029	0.0079 8235	0.0529 8235
44	6.9361 2290	0.1441 7276	131.9138 4220	19.0183 8305	0.0075 8071	0.0525 8071
45	7.2482 4843	0.1379 6437	138.8499 6510	19.1563 4742	0.0072 0202	0.0522 0202
46	7.5744 1961	0.1320 2332	146.0982 1353	19.2883 7074	0.0068 4471	0.0518 4471
47	7.9152 6849	0.1263 3810	153.6726 3314	19.4147 0884	0.0065 0734	0.0515 0734
48	8.2714 5557	0.1208 9771	161.5879 0163	19.5356 0654	0.0061 8858	0.0511 8858
49	8.6436 7107	0.1156 9158	169.8593 5720	19.6512 9813	0.0058 8722	0.0508 8722
50	9.0326 3627	0.1107 0965	178.5030 2828	19.7620 0778	0.0056 0215	0.0506 0215

$i = 5\%$

n	$(1 + i)^n$	$(1 + i)^{-n}$	$s_{\overline{n}|i}$	$a_{\overline{n}|i}$	$\dfrac{1}{s_{\overline{n}|i}}$	$\dfrac{1}{a_{\overline{n}|i}}$
1	1.0500 0000	0.9523 8095	1.0000 0000	0.9523 8095	1.0000 0000	1.0500 0000
2	1.1025 0000	0.9070 2948	2.0500 0000	1.8594 1043	0.4878 0488	0.5378 0488
3	1.1576 2500	0.8638 3760	3.1525 0000	2.7232 4803	0.3172 0856	0.3672 0856
4	1.2155 0625	0.8227 0247	4.3101 2500	3.5459 5050	0.2320 1183	0.2820 1183
5	1.2762 8156	0.7835 2617	5.5256 3125	4.3294 7667	0.1809 7480	0.2309 7480
6	1.3400 9564	0.7462 1540	6.8019 1281	5.0756 9207	0.1470 1747	0.1970 1747
7	1.4071 0042	0.7106 8133	8.1420 0845	5.7863 7340	0.1228 1982	0.1728 1982
8	1.4774 5544	0.6768 3936	9.5491 0888	6.4632 1276	0.1047 2181	0.1547 2181
9	1.5513 2822	0.6446 0892	11.0265 6432	7.1078 2168	0.0906 9008	0.1406 9008
10	1.6288 9463	0.6139 1325	12.5778 9254	7.7217 3493	0.0795 0457	0.1295 0457

Table B.2 *(continued)*
$i = 5\%$

n	$(1 + i)^n$	$(1 + i)^{-n}$	$s_{\overline{n}\rvert i}$	$a_{\overline{n}\rvert i}$	$\dfrac{1}{s_{\overline{n}\rvert i}}$	$\dfrac{1}{a_{\overline{n}\rvert i}}$
11	1.7103 3936	0.5846 7929	14.2067 8716	8.3064 1422	0.0703 8889	0.1203 8889
12	1.7958 5633	0.5568 3742	15.9171 2652	8.8632 5164	0.0628 2541	0.1128 2541
13	1.8856 4914	0.5303 2135	17.7129 8285	9.3935 7299	0.0564 5577	0.1064 5577
14	1.9799 3160	0.5050 6795	19.5986 3199	9.8986 4094	0.0510 2397	0.1010 2397
15	2.0789 2818	0.4810 1710	21.5785 6359	10.3796 5804	0.0463 4229	0.0963 4229
16	2.1828 7459	0.4581 1152	23.6574 9177	10.8377 6956	0.0422 6991	0.0922 6991
17	2.2920 1832	0.4362 9669	25.8403 6636	11.2740 6625	0.0386 9914	0.0886 9914
18	2.4066 1923	0.4155 2065	28.1323 8467	11.6895 8690	0.0355 4622	0.0855 4622
19	2.5269 5020	0.3957 3396	30.5390 0391	12.0853 2086	0.0327 4501	0.0827 4501
20	2.6532 9771	0.3768 8948	33.0659 5410	12.4622 1034	0.0302 4259	0.0802 4259
21	2.7859 6259	0.3589 4236	35.7192 5181	12.8211 5271	0.0279 9611	0.0779 9611
22	2.9252 6072	0.3418 4987	38.5052 1440	13.1630 0258	0.0259 7051	0.0759 7051
23	3.0715 2376	0.3255 7131	41.4304 7512	13.4885 7388	0.0241 3682	0.0741 3682
24	3.2250 9994	0.3100 6791	44.5019 9887	13.7986 4179	0.0224 7090	0.0724 7090
25	3.3863 5494	0.2953 0277	47.7270 9882	14.0939 4457	0.0209 5246	0.0709 5246
26	3.5556 7269	0.2812 4073	51.1134 5376	14.3751 8530	0.0195 6432	0.0695 6432
27	3.7334 5632	0.2678 4832	54.6691 2645	14.6430 3362	0.0182 9186	0.0682 9186
28	3.9201 2914	0.2550 9364	58.4025 8277	14.8981 2726	0.0171 2253	0.0671 2253
29	4.1161 3560	0.2429 4632	62.3227 1191	15.1410 7358	0.0160 4551	0.0660 4551
30	4.3219 4238	0.2313 7745	66.4388 4750	15.3724 5103	0.0150 5144	0.0650 5144
31	4.5380 3949	0.2203 5947	70.7607 8988	15.5928 1050	0.0141 3212	0.0641 3212
32	4.7649 4147	0.2098 6617	75.2988 2937	15.8026 7667	0.0132 8042	0.0632 8042
33	5.0031 8854	0.1998 7254	80.0637 7084	16.0025 4921	0.0124 9004	0.0624 9004
34	5.2533 4797	0.1903 5480	85.0669 5938	16.1929 0401	0.0117 5545	0.0617 5545
35	5.5160 1537	0.1812 9029	90.3203 0735	16.3741 9429	0.0110 7171	0.0610 7171
36	5.7918 1614	0.1726 5741	95.8363 2272	16.5468 5171	0.0104 3446	0.0604 3446
37	6.0814 0694	0.1644 3563	101.6281 3886	16.7112 8734	0.0098 3979	0.0598 3979
38	6.3854 7729	0.1566 0536	107.7095 4580	16.8678 9271	0.0092 8423	0.0592 8423
39	6.7047 5115	0.1491 4797	114.0950 2309	17.0170 4067	0.0087 6462	0.0587 6462
40	7.0399 8871	0.1420 4568	120.7997 7424	17.1590 8635	0.0082 7816	0.0582 7816
41	7.3919 8815	0.1352 8160	127.8397 6295	17.2943 6796	0.0078 2229	0.0578 2229
42	7.7615 8756	0.1288 3962	135.2317 5110	17.4232 0758	0.0073 9471	0.0573 9471
43	8.1496 6693	0.1227 0440	142.9933 3866	17.5459 1198	0.0069 9333	0.0569 9333
44	8.5571 5028	0.1168 6133	151.1430 0559	17.6627 7331	0.0066 1625	0.0566 1625
45	8.9850 0779	0.1112 9651	159.7001 5587	17.7740 6982	0.0062 6173	0.0562 6173
46	9.4342 5818	0.1059 9668	168.6851 6366	17.8800 6650	0.0059 2820	0.0559 2820
47	9.9059 7109	0.1009 4921	178.1194 2185	17.9810 1571	0.0056 1421	0.0556 1421
48	10.4012 6965	0.0961 4211	188.0253 9294	18.0771 5782	0.0053 1843	0.0553 1843
49	10.9213 3313	0.0915 6391	198.4266 6259	18.1687 2173	0.0050 3965	0.0550 3965
50	11.4673 9979	0.0872 0373	209.3479 9572	18.2559 2546	0.0047 7674	0.0547 7674

Table B.2 *(continued)*
$i = 6\%$

n	$(1 + i)^n$	$(1 + i)^{-n}$	$s_{\overline{n}|i}$	$a_{\overline{n}|i}$	$\dfrac{1}{s_{\overline{n}|i}}$	$\dfrac{1}{a_{\overline{n}|i}}$
1	1.0600 0000	0.9433 9623	1.0000 0000	0.9433 9623	1.0000 0000	1.0600 0000
2	1.1236 0000	0.8899 9644	2.0600 0000	1.8333 9267	0.4854 3689	0.5454 3689
3	1.1910 1600	0.8396 1928	3.1836 0000	2.6730 1195	0.3141 0981	0.3741 0981
4	1.2624 7696	0.7920 9366	4.3746 1600	0.4651 0561	0.2285 9149	0.2885 9149
5	1.3382 2558	0.7472 5817	5.6370 9296	4.2123 6379	0.1773 9640	0.2373 9640
6	1.4185 1911	0.7049 6054	6.9753 1854	4.9173 2433	0.1433 6463	0.2033 6263
7	1.5036 3026	0.6650 5711	8.3938 3765	5.5823 8144	0.1191 3502	0.1791 3502
8	1.5938 4807	0.6274 1237	9.8974 6791	6.2097 9381	0.1010 3594	0.1610 3594
9	1.6894 7896	0.5918 9846	11.4913 1598	6.8016 9227	0.0870 2224	0.1470 2224
10	1.7908 4770	0.5583 9478	13.1807 9494	7.3600 8705	0.0758 6796	0.1358 6796
11	1.8982 9856	0.5267 8753	14.9716 4264	7.8868 7458	0.0667 9294	0.1267 9294
12	2.0121 9647	0.4969 6936	16.8699 4120	8.3838 4394	0.0592 7703	0.1192 7703
13	2.1329 2826	0.4688 3902	18.8821 3767	8.8526 8296	0.0529 6011	0.1129 6011
14	2.2609 0396	0.4423 0096	21.0150 6593	9.2949 8393	0.0475 8491	0.1075 8491
15	2.3965 5819	0.4172 6506	23.2759 6988	9.7122 4899	0.0429 6276	0.1029 6276
16	2.5403 5168	0.3936 4628	25.6725 2808	10.1058 9527	0.0389 5214	0.0989 5214
17	2.6927 7279	0.3713 6442	28.2128 7976	10.4772 5969	0.0354 4480	0.0954 4480
18	2.8543 3915	0.3503 4379	30.9056 5255	10.8276 0348	0.0323 5654	0.0923 5654
19	3.0255 9950	0.3305 1301	33.7599 9170	11.1581 1649	0.0296 2086	0.0896 2086
20	3.2071 3547	0.3118 0473	36.7855 9120	11.4699 2122	0.0271 8456	0.0871 8456
21	3.3995 6360	0.2941 5540	39.9927 2668	11.7640 7662	0.0250 0455	0.0850 0455
22	3.6035 3742	0.2775 0510	43.3922 9028	12.0415 8172	0.0230 4557	0.0830 4557
23	3.8197 4966	0.2617 9726	46.9958 2769	12.3033 7898	0.0212 7848	0.0812 7848
24	4.0489 3464	0.2469 7855	50.8155 7735	12.5503 5753	0.0196 7900	0.0796 7900
25	4.2918 7072	0.2329 9863	54.8645 1200	12.7833 5616	0.0182 2672	0.0782 2672
26	4.5493 8296	0.2198 1003	59.1563 8272	13.0031 6619	0.0169 0435	0.0769 0435
27	4.8223 4594	0.2073 6795	63.7057 6568	13.2105 3414	0.0156 9717	0.0756 9717
28	5.1116 8670	0.1956 3014	68.5281 1162	13.4061 6428	0.0145 9255	0.0745 9255
29	5.4183 8790	0.1845 5674	73.6397 9832	13.5907 2102	0.0135 7961	0.0735 7961
30	5.7434 9117	0.1741 1013	79.0581 8622	13.7648 3115	0.0126 4891	0.0726 7891
31	6.0881 0064	0.1642 5484	84.8016 7739	13.9290 8599	0.0117 9222	0.0717 9222
32	6.4533 8668	0.1549 5740	90.8897 7803	14.0840 4339	0.0110 0234	0.0710 0234
33	6.8405 8988	0.1461 8622	97.3431 6471	14.2302 2961	0.0102 7293	0.0702 7293
34	7.2510 2528	0.1379 1153	104.1837 5460	14.3681 4114	0.0095 9843	0.0695 9843
35	7.6860 8679	0.1301 0522	111.4347 7987	14.4982 4636	0.0089 7386	0.0689 7386
36	8.1472 5200	0.1227 4077	119.1208 6666	14.6209 8713	0.0083 9483	0.0683 9483
37	8.6360 8712	0.1157 9318	127.2681 1866	14.7367 8031	0.0078 5743	0.0678 5743
38	9.1542 5235	0.1092 3885	135.9042 0578	14.8460 1916	0.0073 5812	0.0673 5812
39	9.7035 0749	0.1030 5552	145.0584 5813	14.9490 7468	0.0068 9377	0.0668 9377
40	10.2857 1794	0.0972 2219	154.7619 6562	15.0462 9687	0.0064 6154	0.0664 6154

Table B.2 *(continued)*
$i = 6\%$

n	$(1 + i)^n$	$(1 + i)^{-n}$	$s_{\overline{n}|i}$	$a_{\overline{n}|i}$	$\dfrac{1}{s_{\overline{n}|i}}$	$\dfrac{1}{a_{\overline{n}|i}}$
41	10.9028 6101	0.0917 1905	165.0476 8356	15.1380 1592	0.0060 5886	0.0660 5886
42	11.5570 3267	0.0865 2740	175.9505 4457	15.2245 4332	0.0056 8342	0.0656 8342
43	12.2504 5463	0.0816 2962	187.5075 7724	15.3061 7294	0.0053 3312	0.0653 3312
44	12.9854 8191	0.0770 0908	199.7580 3188	15.3831 8202	0.0050 0606	0.0650 0606
45	13.7646 1083	0.0726 5007	212.7435 1379	15.4558 3209	0.0047 0050	0.0647 0050
46	14.5904 8748	0.0685 3781	226.5081 2462	15.5243 6990	0.0044 1485	0.0644 1485
47	15.4659 1673	0.0646 5831	241.0986 1210	15.5890 2821	0.0041 4768	0.0641 4768
48	16.3938 7173	0.0609 9840	256.5645 2882	15.6500 2661	0.0038 9765	0.0638 9765
49	17.3775 0403	0.0575 4566	272.9584 0055	15.7075 7227	0.0036 6356	0.0636 6356
50	18.4201 5427	0.0542 8836	290.3359 0458	15.7618 6064	0.0034 4429	0.0634 4429

$i = 7\%$

n	$(1 + i)^n$	$(1 + i)^{-n}$	$s_{\overline{n}|i}$	$a_{\overline{n}|i}$	$\dfrac{1}{s_{\overline{n}|i}}$	$\dfrac{1}{a_{\overline{n}|i}}$
1	1.0700 0000	0.9345 7944	1.0000 0000	0.9345 7944	1.0000 0000	1.0700 0000
2	1.1449 0000	0.8734 3873	2.0700 0000	1.8080 1817	0.4830 9179	0.5530 9179
3	1.2250 4300	0.8162 9788	3.2149 0000	2.6243 1604	0.3110 5167	0.3810 5167
4	1.3107 9601	0.7628 9521	4.4399 4300	3.3872 1126	0.2252 2812	0.2952 2812
5	1.4025 5173	0.7129 8618	5.7507 3901	4.1001 9744	0.1738 9069	0.2438 9069
6	1.5007 3035	0.6663 4222	7.1532 9074	4.7665 3966	0.1397 9580	0.2097 9580
7	1.6057 8148	0.6227 4974	8.6540 2109	5.3892 8940	0.1155 5322	0.1855 5322
8	1.7181 8618	0.5820 0910	10.2598 0257	5.9712 9851	0.0974 6776	0.1674 6776
9	1.8384 5921	0.5439 3374	11.9779 8875	6.5152 3225	0.0834 8647	0.1534 8647
10	1.9671 5136	0.5083 4929	13.8164 4796	7.0235 8154	0.0723 7750	0.1423 7750
11	2.1048 5195	0.4750 9280	15.7835 9932	7.4986 7434	0.0633 5690	0.1333 5690
12	2.2521 9159	0.4440 1196	17.8884 5127	7.9426 8630	0.0559 0199	0.1259 0199
13	2.4098 4500	0.4149 6445	20.1406 4286	8.3576 5074	0.0496 5085	0.1196 5085
14	2.5785 3415	0.3878 1724	22.5504 8786	8.7454 6799	0.0443 4494	0.1143 4494
15	2.7590 3154	0.3624 4602	25.1290 2201	9.1079 1401	0.0397 9462	0.1097 9462
16	2.9521 6375	0.3387 3460	27.8880 5355	9.4466 4860	0.0358 5765	0.1058 5765
17	3.1588 1521	0.3165 7439	30.8402 1730	9.7632 2299	0.0324 2519	0.1024 2519
18	3.3799 3228	0.2958 6392	33.9990 3251	10.0590 8691	0.0294 1260	0.0994 1260
19	3.6165 2754	0.2765 0833	37.3789 6479	10.3355 9524	0.0267 5301	0.0967 5301
20	3.8696 8446	0.2584 1900	40.9954 9232	10.5940 1425	0.0243 9293	0.0943 9293
21	4.1405 6237	0.2415 1309	44.8651 7678	10.8355 2733	0.0222 8900	0.0922 8900
22	4.4304 0174	0.2257 1317	49.0057 3916	11.0612 4050	0.0204 0577	0.0904 0577
23	4.7405 2986	0.2109 4688	53.4361 4090	11.2721 8738	0.0187 1393	0.0887 1393
24	5.0723 6695	0.1971 4662	58.1766 7076	11.4693 3400	0.0171 8902	0.0871 8902
25	5.4274 3264	0.1842 4918	63.2490 3772	11.6535 8318	0.0158 1052	0.0858 1052

Table B.2 *(continued)*
$i = 7\%$

n	$(1 + i)^n$	$(1 + i)^{-n}$	$s_{\overline{n}\rvert i}$	$a_{\overline{n}\rvert i}$	$\dfrac{1}{s_{\overline{n}\rvert i}}$	$\dfrac{1}{a_{\overline{n}\rvert i}}$
26	5.8073 5292	0.1721 9549	68.6764 7036	11.8257 7867	0.0145 6103	0.0845 6103
27	6.2138 6763	0.1609 3037	74.4838 2328	11.9867 0904	0.0134 2573	0.0834 2573
28	6.6488 3836	0.1504 0221	80.6976 9091	12.1371 1125	0.0123 9193	0.0823 9193
29	7.1142 5705	0.1405 6282	87.3465 2927	12.2776 7407	0.0114 4865	0.0814 4865
30	7.6122 5504	0.1313 6712	94.4607 8632	12.4090 4118	0.0105 8640	0.0805 8640
31	8.1451 1290	0.1227 7301	102.0730 4137	12.5318 1419	0.0097 9691	0.0797 9691
32	8.7152 7080	0.1147 4113	110.2181 5426	12.6465 5532	0.0090 7292	0.0790 7292
33	9.3253 3975	0.1072 3470	118.9334 2506	12.7537 9002	0.0084 0807	0.0784 0807
34	9.9781 1354	0.1002 1934	128.2587 6481	12.8540 0936	0.0077 9674	0.0777 9674
35	10.6765 8148	0.0936 6294	138.2368 7835	12.9476 7230	0.0072 3396	0.0772 3396
36	11.4239 4219	0.0875 3546	148.9134 5984	13.0352 0776	0.0067 1531	0.0767 1531
37	12.2236 1814	0.0818 0884	160.3374 0202	13.1170 1660	0.0062 3685	0.0762 3685
38	13.0792 7141	0.0764 5686	172.5610 2017	13.1934 7345	0.0057 9505	0.0757 9505
39	13.9948 2041	0.0714 5501	185.6402 9158	13.2649 2846	0.0053 8676	0.0753 8676
40	14.9744 5784	0.0667 8038	199.6351 1199	13.3317 0884	0.0050 0914	0.0750 0914
41	16.0226 6989	0.0624 1157	214.6095 6983	13.3941 2041	0.0046 5962	0.0746 5962
42	17.1442 5678	0.0583 2857	230.6322 3972	13.4524 4898	0.0043 3591	0.0743 3591
43	18.3443 5475	0.0545 1268	247.7764 9650	13.5069 6167	0.0040 3590	0.0740 3590
44	19.6284 5959	0.0509 4643	266.1208 5125	13.5579 0810	0.0037 5769	0.0737 5769
45	21.0024 5176	0.0476 1349	285.7493 1084	13.6055 2159	0.0034 9957	0.0734 9957
46	22.4726 2338	0.0444 9859	306.7517 6260	13.6500 2018	0.0032 5996	0.0732 5996
47	24.0457 0702	0.0415 8747	329.2243 8598	13.6916 0764	0.0030 3744	0.0730 3744
48	25.7289 0651	0.0388 6679	353.2700 9300	13.7304 7443	0.0028 3070	0.0728 3070
49	27.5299 2997	0.0363 2413	378.9989 9951	13.7667 9853	0.0026 3853	0.0726 3853
50	29.4570 2506	0.0339 4776	406.5289 2947	13.8007 4629	0.0024 5985	0.0724 5985

$i = 8\%$

n	$(1 + i)^n$	$(1 + i)^{-n}$	$s_{\overline{n}\rvert i}$	$a_{\overline{n}\rvert i}$	$\dfrac{1}{s_{\overline{n}\rvert i}}$	$\dfrac{1}{a_{\overline{n}\rvert i}}$
1	1.0800 0000	0.9259 2593	1.0000 0000	0.9259 2593	1.0000 0000	1.0800 0000
2	1.1664 0000	0.8573 3882	2.0800 0000	1.7832 6475	0.4807 6923	0.5607 6923
3	1.2597 1200	0.7938 3224	3.2464 0000	2.5770 9699	0.3080 3351	0.3880 3351
4	1.3604 8896	0.7350 2985	4.5061 1200	3.3121 2684	0.2219 2080	0.3019 2080
5	1.4693 2808	0.6805 8320	5.8666 0096	3.9927 1004	0.1704 5645	0.2504 5645
6	1.5868 7432	0.6301 6963	7.3359 2904	4.6228 7966	0.1363 1539	0.2163 1539
7	1.7138 2427	0.5834 9040	8.9228 0336	5.2063 7006	0.1120 7240	0.1920 7240
8	1.8509 3021	0.5402 6888	10.6366 2763	5.7466 3894	0.0940 1476	0.1740 1476
9	1.9990 0463	0.5002 4897	12.4875 5784	6.2468 8791	0.0800 7971	0.1600 7971
10	2.1589 2500	0.4631 9349	14.4865 6247	6.7100 8140	0.0690 2949	0.1490 2949

Table B.2 *(continued)*
$i = 8\%$

n	$(1 + i)^n$	$(1 + i)^{-n}$	$s_{\overline{n}|i}$	$a_{\overline{n}|i}$	$\dfrac{1}{s_{\overline{n}|i}}$	$\dfrac{1}{a_{\overline{n}|i}}$
11	2.3316 3900	0.4288 8286	16.6454 8746	7.1389 6426	0.0600 7634	0.1400 7634
12	2.5181 7012	0.3971 1376	18.9771 2646	7.5360 7802	0.0526 9502	0.1326 9502
13	2.7196 2373	0.3676 9792	21.4952 9658	7.9037 7594	0.0465 2181	0.1265 2181
14	2.9371 9362	0.3404 6104	24.2149 2030	8.2442 3698	0.0412 9685	0.1212 9685
15	3.1721 6911	0.3152 4170	27.1521 1393	8.5594 7869	0.0368 2954	0.1168 2954
16	3.4259 4264	0.2918 9047	30.3242 8304	8.8513 6916	0.0329 7687	0.1129 7687
17	3.7000 1805	0.2702 6895	33.7502 2569	9.1216 3811	0.0296 2943	0.1096 2943
18	3.9960 1950	0.2502 4903	37.4502 4374	9.3718 8714	0.0267 0210	0.1067 0210
19	4.3157 0106	0.2317 1206	41.4462 6324	9.6035 9920	0.0241 2763	0.1041 2763
20	4.6609 5714	0.2145 4821	45.7619 6430	9.8181 4741	0.0218 5221	0.1018 5221
21	5.0338 3372	0.1986 5575	50.4229 2144	10.0168 0316	0.0198 3225	0.0998 3225
22	5.4365 4041	0.1839 4051	55.4567 5516	10.2007 4366	0.0180 3207	0.0980 3207
23	5.8714 6365	0.1703 1528	60.8932 9557	10.3710 5895	0.0164 2217	0.0964 2217
24	6.3411 8074	0.1576 9934	66.7647 5922	10.5287 5828	0.0149 7796	0.0949 7796
25	6.8484 7520	0.1460 1790	73.1059 3995	10.6747 7619	0.0136 7878	0.0936 7878
26	7.3963 5321	0.1352 0176	79.9544 1515	10.8099 7795	0.0125 0713	0.0925 0713
27	7.9880 6147	0.1251 8682	87.3507 6836	10.9351 6477	0.0114 4810	0.0914 4810
28	8.6271 0639	0.1159 1372	95.3388 2983	11.0510 7849	0.0104 8891	0.0904 8891
29	9.3172 7490	0.1073 2752	103.9659 3622	11.1584 0601	0.0096 1854	0.0896 1854
30	10.0626 5689	0.0993 7733	113.2832 1111	11.2577 8334	0.0088 2743	0.0888 2743
31	10.8676 6944	0.0920 1605	123.3458 6800	11.3497 9939	0.0081 0728	0.0881 0728
32	11.7370 8300	0.0852 0005	134.2135 3744	11.4349 9944	0.0074 5081	0.0874 5081
33	12.6760 4964	0.0788 8893	145.9506 2044	11.5138 8837	0.0068 5163	0.0868 5163
34	13.6901 3361	0.0730 4531	158.6266 7007	11.5869 3367	0.0063 0411	0.0863 0411
35	14.7853 4429	0.0676 3454	172.3168 0368	11.6545 6822	0.0058 0326	0.0858 0326
36	15.9681 7184	0.0626 2458	187.1021 4797	11.7171 9279	0.0053 4467	0.0853 4467
37	17.2456 2558	0.0579 8572	203.0703 1981	11.7751 7851	0.0049 2440	0.0849 2440
38	18.6252 7563	0.0536 9048	220.3159 4540	11.8288 6899	0.0045 3894	0.0845 3894
39	20.1152 9768	0.0497 1341	238.9412 2103	11.8785 8240	0.0041 8513	0.0841 8513
40	21.7245 2150	0.0460 3093	259.0565 1871	11.9246 1333	0.0038 6016	0.0838 6016
41	23.4624 8322	0.0426 2123	280.7810 4021	11.9672 3457	0.0035 6149	0.0835 6149
42	25.3394 8187	0.0394 6411	304.2435 2342	12.0066 9867	0.0032 8684	0.0832 8684
43	27.3666 4042	0.0365 4084	329.5830 0530	12.0432 3951	0.0030 3414	0.0830 3414
44	29.5559 7166	0.0338 3411	356.9496 4572	12.0770 7362	0.0028 0152	0.0828 0152
45	31.9204 4939	0.0313 2788	386.5056 1738	12.1084 0150	0.0025 8728	0.0825 8728
46	34.4740 8534	0.0290 0730	418.4260 6677	12.1374 0880	0.0023 8991	0.0823 8991
47	37.2320 1217	0.0268 5861	452.9001 5211	12.1642 6741	0.0022 0799	0.0822 0799
48	40.2105 7314	0.0248 6908	490.1321 6428	12.1891 3649	0.0020 4027	0.0820 4027
49	46.4274 1899	0.0230 2693	530.3427 3742	12.2121 6341	0.0018 8557	0.0818 8557
50	43.9016 1251	0.0213 2123	573.7701 5642	12.2334 8464	0.0017 4286	0.0817 4286

The Graphing Calculator

C.1   THE GRAPHING CALCULATOR

► *Introduction*
► *Graphing*
► *Matrices and the Gauss–Jordan Method*
► *Matrix Inverse*
► *Statistical Calculations*

Introduction

Inexpensive hand-held graphing calculators that are available today have a number of powerful and useful features. First of all, these graphing calculators can graph functions on a small screen. They also have a zoom feature that permits the user to magnify a particular part of the graph to essentially any desired amount. One considerable advantage of this zoom feature is the ability to find, to a considerable degree of accuracy, the zeros of a function or the solution of two linear or nonlinear equations.

Another important feature of these calculators is their ability to manipulate matrices of orders as large as 6×6. The calculator can not only add, subtract, and multiply matrices of these orders, but can also find the inverses. These calculators also have the capability to perform elementary row operations on a matrix, permitting the user to apply the Gauss–Jordan method to systems of equations.

These calculators also can perform a variety of statistical calculations, such as finding the mean, standard deviation, linear and nonlinear regression, least square equations, and correlation coefficients.

Finally, these calculators have a programming feature that gives the user the flexibility to create individual programs much like those of a small personal computer. A number of such programs are given in the second section of this appendix.

The most popular graphing calculators are the Texas Instruments TI-81 graphing calculator (or the more powerful TI-85 pocket computer), the Casio power graphic fx-7700G, the Sharp EL-5200, and the Hewlett-Packard HP-28S and HP-48. In this appendix we will demonstrate the features of the TI-81. Since all of the calculators have similar features, you should be able to translate the instructions to the other calculators by carefully reading the appropriate manual.

Graphing

We will demonstrate the graphing capabilities of the TI-81 by showing how to graph each of the equations in the system

$$3x + 9y = 45$$

$$2x + y = 10$$

that we considered at the beginning of Section 2.1. In order to graph these equations on the TI-81, we first need to solve each of the linear equations for y and obtain

$$y = -\frac{1}{3}x + 5$$

$$y = -2x + 10$$

After turning on the TI-81, press the $Y =$ key and the first line will read

$$:Y_1 =$$

with a blinking cursor set right after the $=$ sign. To obtain the term $-x/3 + 5$, depress the following keys

$$\boxed{(-)} \ \boxed{X|T} \ \boxed{\div} \ \boxed{3} \ \boxed{+} \ \boxed{5} \ \boxed{\text{ENTER}}.$$

The $\boxed{X|T}$ key is the same as x. Now move the cursor to the point after the equal sign of the expression

$$:Y =$$

on the second line. To input the expression $-2x + 10$ depress the keys

$$\boxed{(-)} \ \boxed{2} \ \boxed{X|T} \ \boxed{+} \ \boxed{10} \ \boxed{\text{ENTER}}.$$

Now press the $\boxed{\text{ZOOM}}$ key and then $\boxed{6}$. A graph of each of the two equations will appear on the screen.

In order to find the solution of the system, we need to find the coordinates of the point of intersection of the two lines. To do this depress the $\boxed{\text{ZOOM}}$ key again and then depress $\boxed{2}$. The graph will reappear on the screen with a flashing cursor. By using the arrow buttons near the top right of the calculator, move the cursor to the point of intersection and again depress the $\boxed{\text{ENTER}}$ key. The graph reappears, but this time in magnified form. The flashing cursor should be near the point of intersection. Place the cursor at the point of intersection. The numbers at the bottom of the screen are the coordinates of the cursor and represent an approximation of the solution to the system. Depress the $\boxed{\text{ENTER}}$ key again and repeat this process until you have obtained the coordinates to a sufficient accuracy. After only three or four repetitions, the coordinates should already be approximately

$$x = 3.001 \qquad y = 4.001.$$

After only several more repetitions, you should obtain an answer something like

$$x = 3.0000064 \qquad y = 4.0000019.$$

After several more repetitions, you should finally obtain

$$x = 3 \qquad y = 4.$$

Notice that the graph was drawn on a screen with the x-axis and y-axis going from -10 to $+10$. Often the range of the x- or y-axis needs to be adjusted. To change the range depress the $\boxed{\text{RANGE}}$ key and set the range as desired.

Matrices and the Gauss–Jordan Method

We will now see how to use the TI-81 to perform the steps of the Gauss–Jordan method. Consider the system in Example 1 of Section 2.2

$$-3x + 6y = 12$$
$$2x - 4y = -8$$

Our first step is replace this system with the augmented matrix

$$\begin{bmatrix} -3 & 6 & \bigm| & 12 \\ 2 & -4 & \bigm| & -8 \end{bmatrix}$$

To enter this matrix into the calculator first depress the two keys

$$\boxed{\text{MATRX}} \quad \boxed{\triangleright}$$

Now depress the $\boxed{1}$ key to create the matrix A. The cursor is now flashing over the first number, designating the order of the matrix. Since our matrix has order 2×3, depress $\boxed{2}$ $\boxed{\text{ENTER}}$ and then $\boxed{3}$ $\boxed{\text{ENTER}}$. Now the cursor asks for a_{11}, so depress the keys $\boxed{(-)}$ $\boxed{3}$ $\boxed{\text{ENTER}}$. The cursor now asks for a_{12}. Depress the keys $\boxed{6}$ $\boxed{\text{ENTER}}$, and so forth, until all the entries of the matrix have been inputed. Now depress the keys 2nd CLEAR and return to the home screen. To display your matrix depress 2nd 1 ENTER.

The first step in the Gauss–Jordan method is to divide the first row by -3. To do this on the calculator, depress the keys MATRX 3. On the screen you will see

$$*Row($$

with a flashing cursor after the (symbol. Now depress the following keys

(−) 1 ÷ 3 ALPHA . 2nd 1 ALPHA . 1).

The line should read

$$*Row(-1/3,[A],1).$$

This indicates that we are multiplying the first row of the matrix A by $-1/3$.

Now depress ENTER and the screen should read

$$\begin{bmatrix} 1 & -2 & | & -4 \\ 2 & -4 & | & -8 \end{bmatrix}$$

This answer is stored in ANS, and by pressing the keys 2nd (—) we can access this answer.

The next step in the Gauss–Jordan method is $R_2 - 2R_1 \to R_2$. To do this on the calculator, depress the keys MATRX 4. On the screen you will see

$$*Row + ($$

with a flashing cursor after the (symbol. Now depress the following keys

(−) 2 ALPHA . 2nd (—) ALPHA . 1 ALPHA . 2).

The line should read

$$*Row + (-2,ANS,1,2).$$

This indicates that we are multiplying the first row of the matrix A by -2 and adding the result to the second row.

Now depress ENTER and the screen should read

$$\begin{bmatrix} 1 & -2 & | & -4 \\ 0 & 0 & | & 0 \end{bmatrix}.$$

This is the final step in the Gauss–Jordan method.

Swapping two rows is done in a similar fashion.

Matrix Inverse

Finding the inverse of a square matrix on the calculator is particularly easy. We shall find the inverse of the matrix

$$\begin{bmatrix} 1 & 1 & 2 \\ 2 & 3 & 2 \\ 1 & 1 & 3 \end{bmatrix}$$

found in Example 2 of Section 2.5.

Enter this 3×3 matrix as the A matrix in your calculator. Return to the home screen and simply depress the keys

$$\boxed{\text{2nd}} \ \boxed{1} \ \boxed{x^{-1}} \ \boxed{\text{ENTER}}.$$

The inverse of the matrix will appear on the screen.

Statistical Calculations

The calculator can easily calculate the means and standard deviations requested in Chapter 8. For example, let us calculate the mean found in Example 1 of Section 8.2. This data is

Number Sold in One Quarter	1	2	3	4	5
Frequency of Occurrence	10	15	16	7	2

To begin, first clear the statistical section of all old data by depressing the keys

$$\boxed{\text{2nd}} \ \boxed{\text{MATRX}} \ \boxed{\triangleright} \ \boxed{\triangleright} \ \boxed{2} \ \boxed{\text{ENTER}}.$$

Now to enter the data first depress the keys

$$\boxed{\text{2nd}} \ \boxed{\text{MATRX}} \ \boxed{\triangleright} \ \boxed{\triangleright} \ \boxed{1}.$$

Now enter 1 for x_1, 10 for y_1, and so forth, until all the data is entered. Then depress the keys

$$\boxed{\text{2nd}} \ \boxed{\text{MATRX}} \ \boxed{1} \ \boxed{\text{ENTER}}.$$

The first term $\bar{x} = 2.52$ is the mean.

To find the standard deviation of the data 1.020, .980, 1.030, .970, given for the first assembly line at the beginning of Section 8.5, first clear the statistical section of all data as before. Now enter this data as before, except that you do not change the y_i's which are already set equal to 1. Repeat the previous steps and you will obtain the standard deviation as $\sigma_x = .0254950976$.

C.2 PROGRAMS

► *Graphing*

► *Numerical Calculations*

► *Chaos*

Graphing

We will give several programs that are very helpful in Chapter 11 on Discrete Dynamical Systems. The first program will graph any discrete dynamical system. The first step is to input the appropriate formula into the $Y =$ function. For example, if the discrete dynamical system is the one in Example 1 of Section 11.5

$$P_{n+1} = 1.25P_n(1 - .02P_n),$$

then enter the function

$$Y_1 = 1.25x(1 - .02x).$$

For this particular function we first need to set the range as follows:

- $x_{min} = 0$
- $x_{max} = 25$
- $y_{min} = 0$
- $y_{max} = 10$

The program is as follows:

- :ClrDraw
- :disp "INITIAL VALUE"
- :Input A
- :Disp "STEPS"
- :Input N
- :2 → J
- :PT-On(1,A)
- :A → X
- :Lbl 1
- :Y_1 → B
- :PT-On(J,B)
- :All-Off
- :B → X
- :J + 1 → J
- :If J < N
- :Goto 1

When the program is run, you are first asked for an initial value. Input 2. You are then asked for the number of steps. Input 25. The graph should then appear.

Numerical Calculations

The second program produces the actual numerical values P_n for the solution of a dynamical system, including the change in population $P_n - P_{n-1}$ and the annual rate of increase in population, $(P_n - P_{n-1})/P_{n-1}$. The first step is to input the dynamical system as before. Let us use the same one as in the previous demonstration.

The program is as follows:

- :Disp "INITIAL"
- :Input X
- :Lbl 1
- :$Y_1 \to Z$
- :Disp Z
- :Pause
- :$Z - X \to D$
- :Disp D
- :Pause
- :$(D \div X) \to R$
- :disp R
- :Pause
- :$Z \to X$
- :Goto 1

Chaos

One can also create a program that will produce the data found in Figure 11.39 that is an example of chaos.

First enter into your calculator the function

$$Y_1 = 3.7x(1 - x).$$

Now use the following program which gives the two solutions and the difference between the two.

- :Disp "INITIAL"
- :Input X
- :Disp "SECOND INITIAL"
- :Input U
- :Lbl 1
- :$Y_1 \to Z$
- :Disp Z
- :Pause
- :$X \to Y$
- :$U \to X$
- :$Y_1 \to V$
- :Disp V
- :Pause
- :$Z - V \to D$
- :Disp D
- :Pause
- :$Z \to X$
- :$V \to U$
- :Goto 1

When the program requests "INITIAL", type .2, and when the program requests "SECOND INITIAL", type .2001.

Answers

Section 1.1

1. Yes, No   **3.** Yes, No

5. $\{x | 1 \leq x \leq 15, x$ odd integer$\}$, $\{x | 3 \leq x \leq 21, x$ a multiple of 3$\}$

7. $\{x | 2 \leq x \leq 16, x$ an even integer$\}$, $\{x | x$ an odd positive integer$\}$

9. $\{5, 6, 7, 8, 9, 10, 11\}$, $\{2, 4, 6, 8, 10, 12\}$

11. $\{4, 5, 6, 7, 8, 9, 10\}$, $\{1, 3, 5, 7, 9, 11, 13\}$ **13.** F, T, T, F

15.

```
(————————————————)
-1                3
```

17.

```
(————————————————|
2                5
```

19.

```
(————————————————→
-2
```

21.

```
←————————————————|
                1.5
```

23–33.

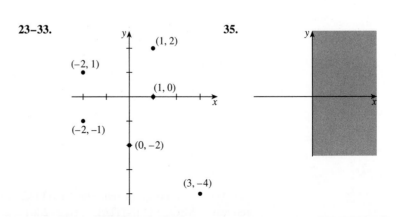

35.

37.

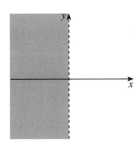

39.

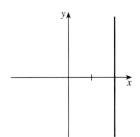

41.

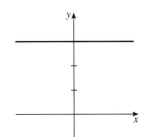

43.

45.

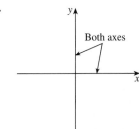

Both axes

47. $\sqrt{5}$

49. $\sqrt{37}$ **51.** $5\sqrt{2}$ **53.** $\sqrt{2}|b - a|$

55.

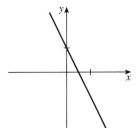

57.

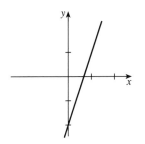

59.

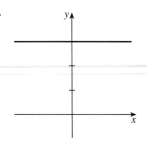

61.

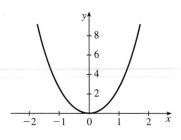

63.

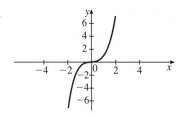

65. 25 miles   **67.** 13 miles

69. 10 days   **71.** $100x + 150y = 3000$

73. $125x + 190y = 800$ **75.** $c = \sqrt{a^2 - b^2}$

Section 1.2

1. 4   **3.** -3 **5.** 0   **7.** Undefined   **9.** $2, \frac{1}{2}, -1$ **11.** $-1, 1, 1$
13. $\frac{1}{2}, -4, 2$ **15.** Undefined, 3, no y-intercept   **17.** 0, no x-intercept, 4
19. $\frac{1}{2}, 1, -\frac{1}{2}$ **21.** $y = -3x + 7$ **23.** $y = -3$ **25.** $y = -x + 2$
27. $x = -2$ **29.** $y = -\frac{b}{a}x + b$ **31.** $y = -\frac{2}{5}x + \frac{11}{5}$ **33.** $y = 0$
35. $x = 1$ **37.** $y = 2x + 5$ **39.** $y = -\frac{5}{3}x + 4$ **41.** $y = -\frac{2}{5}x + 4$
43. No   **45.** $10x + 25y = 235$ **47.** $.08x + .05y = 576$, $160
49. $3x + 4y = 45, 6$ **51.** $m = 89.83$. The length of the twig increases 89.83 times as much as the diameter does.

Section 1.3

1. $C = 3x + 10,000$ **3.** $C = .02x + 1000$ **5.** $R = 5x$ **7.** $R = 0.1x$
9. $P = 2x - 10,000$ **11.** $P = .08x - 1000$ **13.** $x = 2$ **15.** $x = 20$
17. $(1.5, 4.5)$ **19.** $(1, 15)$ **21.** $V = -5000t - 50,000$, $45,000, $25,000
23. $p = -.005x + 22.5$ **25.** (a) $3x + 4y = 120$ (b) $m = -\frac{3}{4}$, making 4 more bookcases requires making 3 fewer desks.
27. (a) $y = 1000 + .05x$, $y = 1500 + .04x$ (b) $50,000

29. $t = \frac{25}{23}$

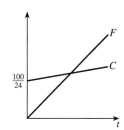

31. $y = 6x + 23$, \$4.5 million.

33. Outside   **35.** (a) Manual (b) Automatic

Volume	Total Costs	
	Manual	Automatic
1,000	\$ 2600	\$8020
10,000	\$17,000	\$8200

Section 1.4

1. $y = .50x + .50$ **3.** $y = 1.1x + 0.1$ **5.** $y = -0.9x + 4.5$
7. $y = -0.7x + 3.4$ **9.** $y = 5x + 5.4$, 35.4 (or rounded to 35)
11. $y = x + 2.5$, \$5.5 million.   **13.** $y = .1006x - 167.12$

Chapter 1 Review

1. $\{x|x$ is an odd positive integer$\}$ **2.** (a) $\{m, i, s, p\}$ (b) $\{-1, 3\}$

3–6.

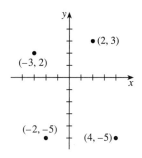

7.

8. $\sqrt{34}$

9.

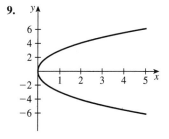

10. $-\frac{7}{3}$ **11.** $\frac{3}{2}$ **12.** Undefined   **13.** $3, -\frac{2}{3}, 2$
14. 0, no x-intercept, -2 **15.** Undefined, 3, no y-intercept
16. $y = -2x + 1$ **17.** $y = -x - 2$ **18.** $y = 3x + 4$
19. $y = \frac{1}{4}x + \frac{31}{4}$ **20.** $C = 6x + 2000$ **21.** $R = 10x$
22. $P = 4x - 2000, 500$ **23.** $x = 150$ **24.** $(1000, 2000)$
25. 150 miles   **26.** 42 minutes   **27.** $p = -\frac{1}{200,000}x + \frac{2}{3}$ **28.** 12,000
29. $4.5x + 5.5y = 15$ **30.** (a) Boston (b) Houston   **31.** $y = \frac{5}{7}x + \frac{1}{7}$

Section 2.1

1. $(2, 5)$ **3.** $(1, -2)$ **5.** $(-1, -1)$ **7.** $(-2, 1)$ **9.** No solution
11. No solution   **13.** No solution   **15.** No solution   **17.** $(1, 2, 0)$
19. $(5, 0, 5)$ **21.** No solution   **23.** $d = 20, q = 5$
25. $n = 10, d = 20, q = 6$ **27.** 10 of A, 6 of B
29. \$700 in first bank, \$300 in second   **31.** 160 of style A and 300 of style B
33. 3 of type A, 5 of type B, 4 of type C   **35.** Since the solution requires a
negative amount (-\$300,000) in low-risk stocks, the client's demands are impossible to
meet.

Section 2.2

1. Yes   **3.** Yes   **5.** No   **7.** $\left.\begin{array}{c} x = 2 \\ y = 3 \end{array}\right\}$, $x = 2, y = 3$

9. $\left.\begin{array}{c} x + 2y = 4 \\ 0 = 0 \end{array}\right\}$, $x = 4 - 2t, y = t, t$ a parameter

11. $\left.\begin{array}{c} x + 2z + 3u = 4 \\ y + 2z + 3u = 5 \end{array}\right\}$, $x = 4 - 2s - 3t, y = 5 - 2s - 3t, z = s, u = t,$
s and t parameters

13. $\left.\begin{array}{c} x + 2z + 4u = 6 \\ y + 2z + 3u = 1 \\ 0 = 0 \end{array}\right\}$, $x = 6 - 2s - 4t, y = 1 - 2s - 3t, z = s, u = t,$
s and t parameters

15. $\left.\begin{array}{c} y + 2z + 2v = 1 \\ u + v = 2 \\ w = 3 \end{array}\right\}$, $x = r, y = 1 - 2s - 2t, z = s, u = 2 - t, v = t, w = 3,$

r and s and t are parameters   **17.** $(1, 2)$ **19.** $(1, 2, 3)$ **21.** $(\frac{3}{2}, \frac{3}{2}, -\frac{1}{2})$
23. $(-1, 1, 1)$ **25.** $(6 - \frac{7}{3}t, 10 - \frac{8}{3}t, t)$ **27.** $(-6 + 3s + 2t, s, t)$
29. $(-7 + 5s - 3t, s, t)$ **31.** $(\frac{3}{2}, -\frac{1}{2} - t, t)$ **33.** No solution
35. $(1, t, t)$ **37.** $(2, 1, 1, 2)$ **39.** $(1, -1, 2, 3)$ **41.** $(1, 3)$
43. No solution   **45.** Solution is $(d - 2b + (2a - c)t, \frac{3b - d}{2} - (\frac{3a - c}{2})t, t)$

47. $(n, d, q) = (-8 + 3t, 44 - 4t, t), 3 \le t \le 11, t$ is an integer
49. 9 of type A and 4 of type B or 3 of type A, 5 of type B, and 4 of type C

51. $900,000 - 2t$ dollars in low risk stocks, t dollars in medium risk stocks, and t dollars in bonds, where $0 \le t \le 450,000$ **53.** 100 of style A, 150 of B, and 50 of C

55. Fired. Solution is $(150, -50 - 2s, s, 500)$ requiring a negative number of Y

57. 3 oranges, 2 cups strawberries, 1 cup blackberries

59. System is $x_1 + x_4 = 700$, $x_1 + x_2 = 600$, $x_2 + x_3 = 800$, $x_3 + x_4 = 900$. Solution is $x_1 = 700 - t$, $x_2 = -100 + t$, $x_3 = 900 - t$, $x_4 = t$, where $100 \le t \le 700$, t is an integer.

Section 2.3

1. 4×2 **3.** 1×2 **5.** 1   **7.** 8   **9.** -2 **11.** 6   **13.** 2

15. $\begin{bmatrix} 5 & 5 & 5 \\ 5 & 5 & 5 \end{bmatrix}$ **17.** $\begin{bmatrix} 1 & 0 & 0 \\ 0 & 1 & 0 \\ 0 & 0 & 1 \end{bmatrix}$ **19.** $\begin{bmatrix} 0 & 0 & 0 \\ 0 & 0 & 0 \\ 0 & 0 & 0 \end{bmatrix}$

21. $x = 3, y = 0, z = 4$ **23.** $\begin{bmatrix} 4 & 5 \\ 1 & 10 \\ -3 & 10 \\ 5 & 6 \end{bmatrix}$ **25.** $[0 \quad 6]$

27. $\begin{bmatrix} 2 & -3 \\ -5 & -2 \\ 3 & -4 \\ -1 & 10 \end{bmatrix}$ **29.** $[-2 \quad -2]$ **31.** $\begin{bmatrix} 2 & 8 \\ 6 & 12 \\ -6 & 14 \\ 6 & -4 \end{bmatrix}$ **33.** $[4 \quad 16]$

35. $\begin{bmatrix} 7 & 6 \\ -1 & 14 \\ -3 & 13 \\ 7 & 14 \end{bmatrix}$ **37.** $[1 \quad 16]$ **39.** Not defined   **41.** $\begin{bmatrix} 0 & 0 \\ 0 & 0 \\ 0 & 0 \\ 0 & 0 \end{bmatrix}$

43. The element in the ith row and jth column of $A + B$ is $a_{ij} + b_{ij}$. But this equals $b_{ij} + a_{ij}$ which is the element in the ith row and jth column of $B + A$.

45. The element in the ith row and jth column of $A + O$ is $a_{ij} + 0$. But this equals a_{ij} which is the element in the ith row and jth column of A.

47. The element in the ith row and jth column of $A - A$ is $a_{ij} - a_{ij}$. But this equals 0, which is the element in the ith row and jth column of O.

49. $X + A = (B - A) + A = B + (-A + A) = B + O = B$

51. $M + A = \begin{bmatrix} 520 & 310 & 100 \\ 365 & 420 & 730 \end{bmatrix}$ **53.** $.9A = \begin{bmatrix} 270 & 180 & 90 \\ 180 & 180 & 360 \end{bmatrix}$

55. $N = \begin{bmatrix} 250 & 300 & 350 & 300 \\ 300 & 200 & 250 & 150 \\ 200 & 240 & 320 & 220 \end{bmatrix}, L = \begin{bmatrix} 300 & 350 & 450 & 400 \\ 320 & 240 & 280 & 250 \\ 250 & 260 & 420 & 280 \end{bmatrix},$

$$N + L = \begin{bmatrix} 550 & 650 & 800 & 700 \\ 620 & 440 & 530 & 400 \\ 450 & 500 & 740 & 500 \end{bmatrix}, L - N = \begin{bmatrix} 50 & 50 & 100 & 100 \\ 20 & 40 & 30 & 100 \\ 50 & 20 & 100 & 60 \end{bmatrix},$$

$$1.2N = \begin{bmatrix} 300 & 360 & 420 & 360 \\ 360 & 240 & 300 & 180 \\ 240 & 288 & 384 & 264 \end{bmatrix}, 0.5(N + L) = \begin{bmatrix} 275 & 325 & 400 & 350 \\ 310 & 220 & 265 & 200 \\ 225 & 250 & 370 & 250 \end{bmatrix}$$

Section 2.4

1. 16   **3.** 23   **5.** Not defined

7. Order of AB is 2×4. Order of BA is undefined.

9. Order of AB is undefined. Order of BA is 5×4

11. Order of AB is 3×5. Order of BA is undefined.

13. Order of AB and BA are 6×6.

15. Order of AB is 20×30. Order of BA is undefined.   **17.** $\begin{bmatrix} 29 \\ 23 \end{bmatrix}$

19. $[43 \quad 29]$ **21.** $\begin{bmatrix} .9 & .8 \\ 3.3 & 2.2 \end{bmatrix}$ **23.** $\begin{bmatrix} 36 \\ 18 \end{bmatrix}$ **25.** $\begin{bmatrix} .8 & 2 \\ 1.3 & 2.8 \end{bmatrix}$

27. $\begin{bmatrix} 3 & 30 & 31 & 14 \\ 8 & 46 & 26 & 26 \end{bmatrix}$ **29.** $[.07 \quad .18 \quad .05]$ **31.** $\begin{bmatrix} 26 \\ 10 \\ 12 \end{bmatrix}$

33. $[-4 \quad 4 \quad 21]$ **35.** $\begin{bmatrix} 5 & 10 & 21 \\ 0 & 4 & 8 \\ 16 & 5 & 15 \end{bmatrix}$ **37.** Not defined   **39.** $\begin{bmatrix} 6 & 10 \\ 15 & 25 \end{bmatrix}$

41. Not defined   **47.** $(A + B)^2 = \begin{bmatrix} 8 & 8 \\ 32 & 40 \end{bmatrix}, A^2 + 2AB + B^2 = \begin{bmatrix} 13 & 13 \\ 32 & 35 \end{bmatrix}$

49. $A(B + C) = AB + AC = \begin{bmatrix} 7 & 6 \\ 10 & 9 \end{bmatrix}, A(BC) = (AB)C = \begin{bmatrix} 1 & 1 \\ 1 & 1 \end{bmatrix}$

53. $A = \begin{bmatrix} 2 & 3 \\ 3 & -5 \end{bmatrix} X = \begin{bmatrix} x_1 \\ x_2 \end{bmatrix}, B = \begin{bmatrix} 5 \\ 7 \end{bmatrix}$

55. $A = \begin{bmatrix} 2 & 5 & 3 \\ 4 & -7 & -2 \\ 5 & -2 & 6 \end{bmatrix}, X = \begin{bmatrix} x_1 \\ x_2 \\ x_3 \end{bmatrix}, B = \begin{bmatrix} 16 \\ 12 \\ 24 \end{bmatrix}$

57. $A = \begin{bmatrix} 2 & -3 & 3 \\ 5 & 6 & -2 \end{bmatrix}, X = \begin{bmatrix} x_1 \\ x_2 \\ x_3 \end{bmatrix}, B = \begin{bmatrix} 3 \\ 1 \end{bmatrix}$

59. $A = [2 \quad 3], X = \begin{bmatrix} x_1 \\ x_2 \end{bmatrix}, B = [7]$

61. $NR = \begin{bmatrix} 27 & 5.3 & 42 \\ 25 & 4.4 & 38 \\ 21 & 3.4 & 35 \end{bmatrix}$, $QC = \begin{bmatrix} 2954 \\ 2714 \\ 2271 \end{bmatrix}$ The number 27 in the first row and first

column indicates that 27 hundred tires are needed in LA. The number 38 in the second row and third column indicates that 38 hundred plugs are needed in NYC, and so forth.

63. (a) $B = \begin{bmatrix} .4 & 0 & 0 \\ .6 & 1 & .5 \\ 0 & 0 & .5 \end{bmatrix}$ (b) $C = \begin{bmatrix} 1 & .2 & 0 \\ 0 & .3 & 0 \\ 0 & .2 & .4 \\ 0 & .3 & .6 \end{bmatrix}$ (c) $W = CZ = C(BY) = (CB)Y =$

$(CB)(AX) = (CBA)X$ (d) $w_1 = 1930$, $w_2 = 1695$, $w_3 = 2550$, $w_4 = 3825$.

Section 2.5

1. Yes   **3.** Yes   **5.** Yes   **7.** Yes   **9.** Yes   **11.** $\begin{bmatrix} 10 & -3 \\ -3 & 1 \end{bmatrix}$

13. $\dfrac{1}{7}\begin{bmatrix} 2 & 1 \\ 3 & 5 \end{bmatrix}$ **15.** No inverse   **17.** No inverse   **19.** $\begin{bmatrix} 1 & -.5 & 0 \\ -1 & 1 & -1 \\ 1 & -.5 & 1 \end{bmatrix}$

21. $\dfrac{1}{3}\begin{bmatrix} 1 & 2 & 4 \\ 1 & -1 & -2 \\ -1 & 1 & -1 \end{bmatrix}$ **23.** $\dfrac{1}{2}\begin{bmatrix} -2 & 0 & 2 \\ 7 & -1 & -3 \\ -3 & 1 & 1 \end{bmatrix}$

25. $\dfrac{1}{2}\begin{bmatrix} 3 & -1 & -1 & -1 \\ 1 & 1 & -1 & -1 \\ -1 & 1 & 1 & -1 \\ -1 & -1 & 1 & 3 \end{bmatrix}$ **27.** $A^{-1} = \dfrac{1}{2}\begin{bmatrix} -1 & 4 \\ -1 & 2 \end{bmatrix}$, $x = -9, y = -7$

29. $A^{-1} = \begin{bmatrix} 3 & -1 \\ -2 & 1 \end{bmatrix}$, $x = 9, y = -3$

31. $A^{-1} = \dfrac{1}{2}\begin{bmatrix} -5 & 4 \\ 3 & -2 \end{bmatrix}$, $x = -7, y = 5$ **33.** $x = 11, y = -3$

35. $x = \frac{11}{7}, y = \frac{27}{7}$ **37.** $x = 3, y = -1, z = -1$

39. $x = 2, y = -\frac{7}{2}, z = \frac{3}{2}$ **41.** $x_1 = -1, x_2 = -1, x_3 = 0, x_4 = 3$

43. $\begin{bmatrix} a & b \\ c & d \end{bmatrix}\dfrac{1}{D}\begin{bmatrix} d & -b \\ -c & a \end{bmatrix} = \dfrac{1}{D}\begin{bmatrix} ad - bc & -ab + ba \\ cd - dc & -bc + da \end{bmatrix} = \begin{bmatrix} 1 & 0 \\ 0 & 1 \end{bmatrix}$

$\dfrac{1}{D}\begin{bmatrix} d & -b \\ -c & a \end{bmatrix}\begin{bmatrix} a & b \\ c & d \end{bmatrix} = \dfrac{1}{D}\begin{bmatrix} da - bc & db - bd \\ -ca + ac & -cb + ad \end{bmatrix} = \begin{bmatrix} 1 & 0 \\ 0 & 1 \end{bmatrix}$

45. (a) \$75,000 in bond fund, \$25,000 in stock fund (b) \$25,000 in bond fund, \$25,000 in stock fund

47. (a) 80 of style A and 40 of style B (b) 10 of style A and 130 of style B.

Section 2.6

1. $\dfrac{1}{41}\begin{bmatrix} 1,100,000,000 \\ 1,800,000,000 \end{bmatrix}$ **3.** $\dfrac{1}{22}\begin{bmatrix} 375,000,000 \\ 437,500,000 \end{bmatrix}$ **5.** $\dfrac{1}{41}\begin{bmatrix} 21,000,000,000 \\ 23,000,000,000 \end{bmatrix}$

7. $\dfrac{1}{48}\begin{bmatrix} 1,550,000,000 \\ 2,700,000,000 \\ 1,600,000,000 \end{bmatrix}$ **9.** $\dfrac{1}{871}\begin{bmatrix} 32,500,000,000 \\ 33,100,000,000 \\ 30,600,000,000 \end{bmatrix}$ **11.** $p_1 = \frac{2}{7}t,\ p_2 = t.$

13. $p_1 = \frac{1}{6}t,\ p_2 = t.$ **15.** $p_1 = \frac{14}{39}t,\ p_2 = \frac{20}{39}t,\ p_3 = t.$

17. $p_1 = \frac{28}{85}t,\ p_2 = \frac{82}{85}t,\ p_3 = t.$

Chapter 2 Review

1. $x = 1,\ y = 2$ **2.** No solution **3.** $x = 1 - 3t,\ y = t$ **4.** $x = 3,\ y = 4$

5. $x = 1.6 - .2t,\ y = .8 + .4t,\ z = t$ **6.** $x = 10 - t,\ y = t,\ z = 0$

7. $x = 1,\ y = 3,\ z = 1$ **8.** $x = -2 + 2t,\ y = 7 - 3t,\ z = t$

9. $\begin{bmatrix} 3 & 9 \\ 6 & 3 \\ -3 & 9 \end{bmatrix}$ **10.** $\begin{bmatrix} -4 & -10 \\ -2 & 0 \\ -6 & 4 \end{bmatrix}$ **11.** $\begin{bmatrix} -1 & -1 \\ 4 & 3 \\ -9 & 13 \end{bmatrix}$ **12.** 14

13. Not defined **14.** Not defined **15.** $\begin{bmatrix} 4 \\ 2 \end{bmatrix}$ **16.** Not defined

17. $\begin{bmatrix} 11 & 3 & 5 \\ 7 & 1 & 0 \\ 7 & 3 & 7 \end{bmatrix}$ **18.** $\begin{bmatrix} 3 & 3 \\ 3 & 16 \end{bmatrix}$ **19.** Not defined **20.** 12

21. Part (a) follows from part (b) by taking $B = A$. To show part (b) notice that

$$AB = \begin{bmatrix} a_{11} & 0 & 0 \\ 0 & a_{22} & 0 \\ 0 & 0 & a_{33} \end{bmatrix}\begin{bmatrix} b_{11} & 0 & 0 \\ 0 & b_{22} & 0 \\ 0 & 0 & b_{33} \end{bmatrix} = \begin{bmatrix} a_{11}b_{11} & 0 & 0 \\ 0 & a_{22}b_{22} & 0 \\ 0 & 0 & a_{33}b_{33} \end{bmatrix}$$

22. $\dfrac{1}{3}\begin{bmatrix} 1 & -1 \\ -2 & 5 \end{bmatrix}$ **23.** Not defined **24.** $-\dfrac{1}{7}\begin{bmatrix} 1 & 3 \\ 3 & 2 \end{bmatrix}$

25. $\dfrac{1}{3}\begin{bmatrix} -12 & 6 & -3 \\ 5 & -2 & 1 \\ -1 & 1 & 1 \end{bmatrix}$ **26.** Does not exist **27.** $\dfrac{1}{3}\begin{bmatrix} 2 & -1 & 0 \\ 6 & 0 & -3 \\ -5 & 1 & 3 \end{bmatrix}$

28. $x = 1,\ y = -4$ **29.** $x = -17,\ y = -16$ **30.** $x = 8,\ y = -3,\ z = 2$

31. $x = 23,\ y = 60,\ z = -53$ **32.** $\dfrac{1}{21}\begin{bmatrix} 550,000,000 \\ 1,150,000,000 \end{bmatrix}$

33. $\dfrac{1}{1597}\begin{bmatrix} 42,000,000,000 \\ 68,300,000,000 \\ 76,800,000,000 \end{bmatrix}$ **34.** $p_1 = \frac{3}{8}t,\ p_2 = t.$

35. $p_1 = \frac{37}{85}t,\ p_2 = \frac{78}{85}t,\ p_3 = t$

36. $1 + t$ three-par holes, $8 - 2t$ four-par holes, t five-par holes, $0 \le t \le 4$, t is an integer

37. 7 standard, 5 deluxe   **38.** 8 of the first and 6 of the second

39. 20 hams and 30 beefs   **40.** 90, 60, 150 liters

41. Plant 1 should produce 100, Plant 2 should produce 200, and Plant 3 should produce 50.

Section 3.1

1.

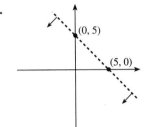

3.

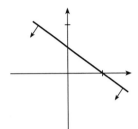

5.

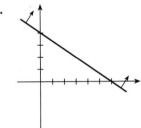

7.

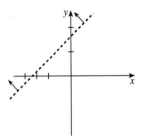

9.

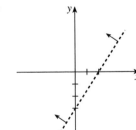

11.

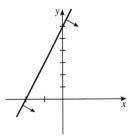

13.

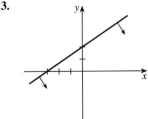

15.

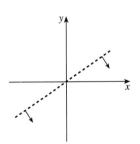

17.

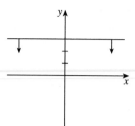

19.

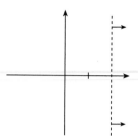

21.

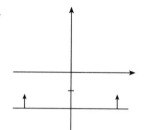

23.

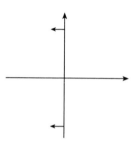

25.

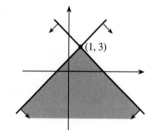

27.

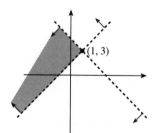

29.

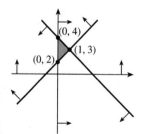

31.

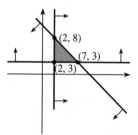

33.

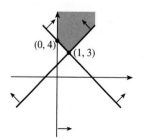

35.

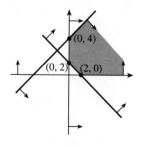

37.

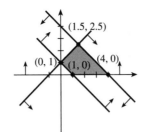

39.

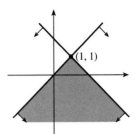

41.

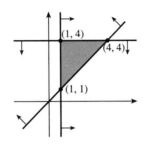

43.

$$x + y \le 200$$
$$x \ge 100$$
$$y \ge 50$$
$$x, y \ge 0$$

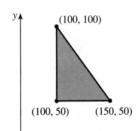

45.

$$x + 2y \le 20$$
$$2x + 2y \le 22$$
$$2x + y \le 15$$
$$x, y \ge 0$$

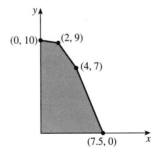

47.
$$x + 3y \leq 45$$
$$10x + 20y \leq 350$$
$$2x + y \leq 55$$
$$x, y \geq 0$$

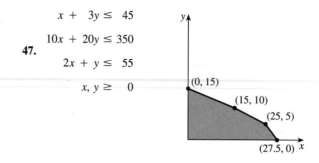

49.
$$x + 4y \leq 800$$
$$x + 2y \leq 500$$
$$2x + y \leq 700$$
$$x, y \geq 0$$

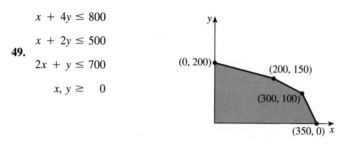

Section 3.2

1. Max = 12, min = 3   **3.** Max = 16, min = 7   **5.** No max, min = 6

7. Max = 1400 at (10, 40)   **9.** Max = 180 at (30, 40)

11. Min = 7 at (3, 4)   **13.** Min = 2 at (1, 0)   **15.** Max = 48 at (8, 4)

17. Min = 20 at (5, 0)   **19.** No solution   **21.** No solution

23. (a) $a \leq \frac{1}{2}$ (b) $\frac{1}{2} \leq a \leq 1$ (c) $1 \leq a \leq 2$ (d) $a \geq 2$

25. (a) A (b) B (c) C (d) D (e) E

27. With the single exception of the point A, starting at any point in the feasible region, we can move a little either in the positive x or positive y direction and still remain in the feasible region and increase z.

29. Min 5000 at $x = 100$, $y = 50$ **31.** Max 530 at (2, 9)

33. Max 115 at (25, 5)   **35.** Max 400 at (300, 100)

37. Min 390,000 at (5, 12) or (3, 15)   **39.** Max 1150 at (200, 150)

41. Max 20 at (5, 15), (6, 14), (7, 13), (8, 12), (9, 11), (10, 10).

Section 3.3

1. Maximize $R = .06x_1 + .08x_2 + .10x_3 + .13x_4$

$$
\begin{aligned}
\text{Subject to} \quad x_1 + x_2 + x_3 + x_4 &\leq 100{,}000 \\
-x_1 \qquad\qquad + x_4 &\leq 0 \\
-x_2 + x_3 \qquad &\leq 0 \\
x_1 \qquad\qquad &\geq 10{,}000 \\
x_1 \qquad\qquad &\leq 40{,}000 \\
x_3 \qquad &\geq 10{,}000 \\
x_1, x_2, x_3, x_4 &\geq 0
\end{aligned}
$$

3. Minimize $C = 20x_1 + 40x_2 + 50x_3 + 20x_4$

Subject to

$$x_1 + x_2 \qquad\qquad\quad \le 500$$
$$x_3 + x_4 \le 100$$
$$x_1 \qquad + x_3 \qquad \ge 150$$
$$x_2 \qquad + x_4 \ge 250$$
$$x_1, x_2, x_3, x_4 \ge 0$$

5. Minimize $\quad C = 4x_1 + 3x_2 + x_3$

Subject to $\quad x_1 \quad + x_2 \quad + x_3 \le 20$
$$30x_1 + 30x_2 + 10x_3 \ge 500$$
$$20x_1 + 20x_2 + 10x_3 \ge 329$$
$$35x_1 + 50x_2 + 20x_3 \ge 425$$
$$x_1, x_2, x_3 \ge 0$$

7. Minimize $\quad C = 20x_1 + 30x_2 + 20x_3$

Subject to $40x_1 + 20x_2 \qquad\quad \ge 4400$
$$30x_1 + 20x_2 + 30x_3 \ge 1800$$
$$10x_1 + 60x_2 + 40x_3 \ge 2800$$
$$x_1, x_2, x_3 \ge 0$$

9. Minimize $\quad C = 40{,}000x_1 + 30{,}000x_2 + 45{,}000x_3$

Subject to $\quad 8000x_1 + 4000x_2 + 4500x_3 \ge 40{,}000$
$$3000x_1 + 6000x_2 + 6500x_3 \ge 24{,}000$$
$$x_1, x_2, x_3 \ge 0$$

11. Minimize $\quad C = 2000x_1 + 3000x_2 + 4000x_3$

Subject to

$$2x_1 + x_2 + x_3 \ge 21$$
$$x_1 + x_2 + 3x_3 \ge 19$$
$$x_1 + 4x_2 + 3x_3 \ge 25$$
$$x_1, x_2, x_3 \ge 0$$

13. Minimize $C = 4x_1 + 2x_2 + 2x_3 + 4x_4$

Subject to

$$x_1 + x_2 \qquad\qquad\quad \le 200 \qquad x_1 \text{ is number from SW to NW}$$
$$x_3 + x_4 \le 600 \qquad x_2 \text{ is number from SE to NW}$$
$$x_1 \qquad + x_3 \qquad \ge 400 \qquad x_3 \text{ is number from SW to NE}$$
$$x_2 \qquad + x_4 \ge 300 \qquad x_4 \text{ is number from SE to NE}$$
$$x_1, x_2, x_3, x_4 \ge 0$$

15. Maximize $\quad R = 50(x_1 + x_3 + x_5) + 65(x_2 + x_4 + x_6)$

Subject to

$$
\begin{aligned}
x_1 + x_2 &\le 200 \\
x_3 + x_4 &\le 30 \\
x_5 + x_6 &\le 60 \\
-.25x_1 + .75x_3 - .25x_5 &\ge 0 \\
-.15x_2 + .85x_4 - .15x_6 &\ge 0 \\
-.10x_2 - .10x_4 + .90x_6 &\ge 0 \\
x_1, x_2, x_3, x_4, x_5, x_6 &\ge 0
\end{aligned}
$$

$x_1 = $ oz of sugar in first candy, $x_2 = $ oz of sugar in second candy, $x_3 = $ oz of nuts in first candy, $x_4 = $ oz of nuts in second candy, $x_5 = $ oz of chocolate in first candy, $x_6 = $ oz of chocolate in second candy.

17. Minimize $\quad C = .50x_1 + .40x_2 + .60x_3 + .75x_4$

Subject to

$$
\begin{aligned}
x_1 + x_2 + x_3 + x_4 &\le 18 \\
x_1 + 2x_2 &\ge 10 \\
100x_1 + 50x_2 + 10x_4 &\ge 800 \\
100x_1 + 200x_3 + 20x_4 &\ge 500 \\
50x_3 + 70x_4 &\ge 200 \\
200x_1 + 150x_2 + 75x_3 + 250x_4 &\le 3000 \\
2000x_2 + 5000x_4 &\le 25000 \\
5x_1 + 20x_2 + 30x_3 + 70x_4 &\le 130 \\
x_1, x_2, x_3, x_4 &\ge 0
\end{aligned}
$$

19. Minimize $z = x_1 + x_2 + x_3 + x_4 + x_5 + x_6 + x_7$

Subject to

$$
\begin{aligned}
x_1 + x_4 + x_5 + x_6 + x_7 &\ge 82 \quad \text{Monday constraint} \\
x_1 + x_2 + x_5 + x_6 + x_7 &\ge 87 \quad \text{Tuesday constraint} \\
x_1 + x_2 + x_3 + x_6 + x_7 &\ge 77 \quad \text{Wednesday constraint} \\
x_1 + x_2 + x_3 + x_4 + x_7 &\ge 73 \quad \text{Thursday constraint} \\
x_1 + x_2 + x_3 + x_4 + x_5 &\ge 75 \quad \text{Friday constraint} \\
x_2 + x_3 + x_4 + x_5 + x_6 &\ge 42 \quad \text{Saturday constraint} \\
x_3 + x_4 + x_5 + x_6 + x_7 &\ge 23 \quad \text{Sunday constraint} \\
x_1, x_2, x_3, x_4, x_5, x_6, x_7 &\ge 0
\end{aligned}
$$

where x_1 is the number of employees who begin their work week on Monday, x_2 the number who begin their work week on Tuesday, and so on.

21. 25 tables and 5 desks. Excess of labor.

23. 300 smallmouth and 100 largemouth. Excess of food I.

25. 200 smallmouth and 150 largemouth. Excess of food III.

27. (a) Labor (b) Wood

Chapter 3 Review

1. 10 sacks of each at a cost of $350
2. 3 of the first and 2 of the second at a cost of $5 per 10 yards
3. 26 chairs and 8 sofas with income $9200
4. (a) Wood in excess by 10 feet (b) 42 chairs and no sofas with income $16,800 and wood in excess by 90 feet and material in excess by 80 square yards
5. 3 of first and 6 of second with profit of $150
6. (a) Labor in excess by 3 hours (b) 6 of the first and 3 of the second with wood in excess by 6 square feet
7. North 2 days and South 3 days with cost $18,000
8. 30 acres in corn and 210 acres in soybeans, with revenue $59,040
9. 12 pounds of the first and 21 pounds of the second with profit of $144

Section 4.1

1. (a)
$$x_1 + x_2 + s_1 = 6$$
$$2x_1 + x_2 + s_2 = 8$$

(b)

x_1	x_2	s_1	s_2	z	
1	1	1	0	0	6
2	1	0	1	0	8
-3	-2	0	0	1	0

(c)

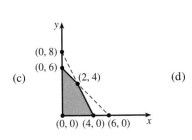

(d)

x_1	x_2	s_1	s_2	(x_1, x_2)	
0	0	6	8	$(0, 0)$	corner
0	6	0	2	$(0, 6)$	corner
0	8	-2	0	$(0, 8)$	not corner
6	0	0	-4	$(6, 0)$	not corner
4	0	2	0	$(4, 0)$	corner
2	4	0	0	$(2, 4)$	corner

3. (a)
$$x_1 + 3x_2 + s_1 = 30$$
$$x_1 + x_2 + s_2 = 12$$
$$4x_1 + x_2 + s_3 = 24$$

(b)

x_1	x_2	s_1	s_2	s_3	z	
1	3	1	0	0	0	30
1	1	0	1	0	0	12
4	1	0	0	1	0	24
-20	-40	0	0	0	1	0

(c)

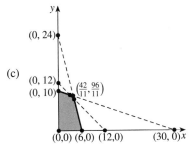

x_1	x_2	s_1	s_2	s_3	(x_1, x_2)	
0	0	30	12	24	(0, 0)	corner
0	10	0	2	14	(0, 10)	corner
0	12	−6	0	12	(0, 12)	not corner
0	24	−42	−12	0	(0, 24)	not corner
30	0	0	−18	−96	(30, 0)	not corner
12	0	18	0	−24	(12, 0)	not corner
6	0	24	6	0	(6, 0)	corner
3	9	0	0	3	(3, 9)	corner
$\frac{42}{11}$	$\frac{96}{11}$	0	$-\frac{6}{11}$	0	$\left(\frac{42}{11}, \frac{96}{11}\right)$	not corner
4	8	2	0	0	(4, 8)	corner

(d)

5.

x_1	x_2	s_1	s_2	z	
1	2	.50	0	0	4
0	−9	−2.5	1	0	0
0	6	2	0	1	16

7.

x_1	x_2	s_1	s_2	z	
0	3.6	1	−.40	0	0
1	.20	0	.20	0	4
0	−1.2	0	.80	1	16

9.

x_1	x_2	s_1	s_2	s_3	z	
1	2	.50	0	0	0	4
0	−9	−2.5	1	0	0	5
0	0	−1.5	0	1	0	3
0	4	1.5	0	0	1	12

11.

x_1	x_2	s_1	s_2	s_3	z	
0	3.6	1	$-.40$	0	0	-2
1	.20	0	.20	0	0	5
0	5.4	0	$-.60$	1	0	0
0	-1.4	0	.60	0	1	15

13.

x_1	x_2	s_1	s_2	s_3	z	
0	0	1	0	$-\frac{2}{3}$	0	-2
0	-9	0	1	$-\frac{5}{3}$	0	0
1	2	0	0	$\frac{1}{3}$	0	5
0	4	0	0	1	1	15

15. Nonbasic: x_1, x_2. Basic: s_1, s_2. $x_1 = x_2 = 0$, $s_1 = 10$, $s_2 = 20$. Corner.
17. Nonbasic: x_1, s_1. Basic: x_2, s_2. $x_1 = s_1 = 0$, $x_2 = 5$, $s_2 = 10$. Corner.
19. Nonbasic: x_2, s_2. Basic: x_1, s_1. $x_2 = s_2 = 0$, $x_1 = 4$, $s_1 = -2$. Not corner.
21. Nonbasic: x_1, s_3. Basic: x_2, s_1, s_2. $x_1 = s_3 = 0$, $x_2 = 2$, $s_1 = 21$, $s_2 = 30$.
Corner.
23. Nonbasic: x_2, s_2. Basic: x_1, s_1, s_3. $x_2 = s_2 = 0$, $x_1 = 2$, $s_1 = 0$, $s_3 = -5$. Not corner.

Section 4.2

1.

x_1	x_2	s_1	s_2	z		
1	2	.5	0	0	4	$x_1 = 4$
0	-9	-2.5	1	0	5	$x_2 = 0$
0	6	2	0	1	16	max $= 16$

3.

x_1	x_2	s_1	s_2	z		
0	1	3	$-.4$	0	1	$x_1 = 2$
1	0	1	.2	0	2	$x_2 = 1$
0	0	4	1.4	1	14	max $= 14$

5.

x_1	x_2	s_1	s_2	z	
$\frac{4}{3}$	0	1	6	0	6
$\frac{1}{3}$	1	0	1	0	1
$\frac{5}{3}$	0	0	3	1	5

$x_1 = 0$
$x_2 = 1$
max $= 5$

7.

x_1	x_2	s_1	s_2	s_3	z	
0	2.5	1	0	22	0	26
0	.5	0	1	5	0	31
1	.5	0	0	4	0	1
0	5	0	0	80	1	40

$x_1 = 1$
$x_2 = 0$
max $= 40$

9.

x_1	x_2	s_1	s_2	s_3	z	
0	5	1	0	0	0	1
0	$\frac{4}{3}$	0	1	0	0	$\frac{1}{3}$
1	1	0	0	0	0	2
0	$\frac{14}{3}$	0	0	0	1	$\frac{26}{3}$

$x_1 = 2$
$x_2 = 0$
max $= \frac{26}{3}$

11.

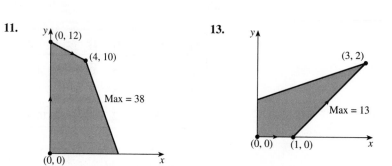

13.

15.

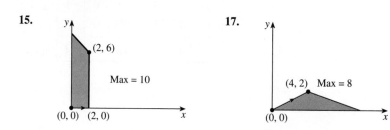

17.

19.

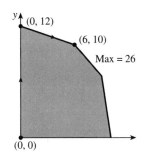

21. Max $= 24$ at $(0, 12, 0)$.

23. Max $= 8$ at $(0, 0, 4)$.   **25.** Max $= 40$ at $(0, 2, 0)$.
27. Max $= 450$ at $(0, 100, 50)$.   **29.** Max $= 8$ at $(0, 1, 2)$.
31. No solution.   **33.** Max $= 260$ at $(0, 4, 2)$.   **35.** Max $= 210$ at $(0, 15, 50)$.
37. Max $= 10,700$ at $(21, 0, 20)$.   **39.** Max $= 164$ at $(20, 5, 16)$.
41. For Exercise 35: s_1 is the number of pounds of excess cashews, s_2 is the number of pounds of excess almonds, and s_3 is the number of pounds of excess peanuts. There is an excess of $s_2 = 3$ pounds of almonds. For Exercise 37: s_1 is the number of feet of excess wood, s_2 is the number of pounds of excess foam rubber, and s_3 is the number of square yards of excess material. There is an excess of $s_1 = 120$ feet of wood. For Exercise 39: s_1 is the number of board-feet of excess wood, s_2 is the number of pounds of excess iron, and s_3 is the number of hours of excess labor. No excess resources.

Section 4.3
1. Minimum 4.5 at $(1.5, 0)$ **3.** Minimum 13 at $(3, 1)$
5. Minimum 70 at $(1, 3)$ and $(3, 2)$ **7.** Minimum 11 at $(4, 1)$
9. Minimum 16 at $(4, 1)$ **11.** Minimum 2 at $(0, 1, 0)$
13. Minimum 472 at $(10, 12, 8)$ **15.** Minimum 3 at $(0, 1, 2)$ **17.** No solution
19. The matrix form of each of these four problems is

$$
\begin{aligned}
\text{Minimize} \quad & w = B^T Y \\
\text{Subject to} \quad & A^T Y \geq C^T \\
& Y \geq 0.
\end{aligned}
$$

and the dual is

$$
\begin{aligned}
\text{Maximize} \quad & z = CX \\
\text{Subject to} \quad & AX \leq B \\
& X \geq 0.
\end{aligned}
$$

For Exercise 7

$$
B = \begin{bmatrix} 2 \\ 3 \end{bmatrix}, \quad
A^T = \begin{bmatrix} 4 & 1 \\ 1 & 1 \\ 1 & 2 \end{bmatrix}, \quad
Y = \begin{bmatrix} y_1 \\ y_2 \end{bmatrix}, \quad
C = \begin{bmatrix} 8 & 5 & 6 \end{bmatrix}
$$

For Exercise 8

$$B = \begin{bmatrix} 20 \\ 30 \end{bmatrix}, \quad A^T = \begin{bmatrix} 3 & 1 \\ 1 & 1 \\ 0 & 1 \end{bmatrix}, \quad Y = \begin{bmatrix} y_1 \\ y_2 \end{bmatrix}, \quad C = \begin{bmatrix} 8 & 4 & 1 \end{bmatrix}$$

For Exercise 9

$$B = \begin{bmatrix} 3 \\ 4 \end{bmatrix}, \quad A^T = \begin{bmatrix} 3 & 2 \\ 1 & 0 \\ 0 & 1 \end{bmatrix}, \quad Y = \begin{bmatrix} y_1 \\ y_2 \end{bmatrix}, \quad C = \begin{bmatrix} 14 & 2 & 1 \end{bmatrix}$$

For Exercise 10

$$B = \begin{bmatrix} 1 \\ 2 \end{bmatrix}, \quad A^T = \begin{bmatrix} -1 & 1 \\ 1 & 1 \\ 0 & 1 \end{bmatrix}, \quad Y = \begin{bmatrix} y_1 \\ y_2 \end{bmatrix}, \quad C = \begin{bmatrix} 0 & 4 & 1 \end{bmatrix}$$

21. Minimum of $4.50 with 4 oranges and 2 cups strawberries

23. First plant 4 days, second plant 2 days. Minimum cost is $230,000.

25. $2250 in first, $8000 in second, none in third

27. 90 bags of type A, 20 of type B, none of type C. Minimum cost is $2400.

29. North mine 8 days, Center 2 days, South 3. Minimum cost is $34,000.

31. 200 pink azaleas and no red from the East, 50 pink and 300 red from the West. Minimum cost is $2650.

33. None from SW to NW, 400 from SW to NE, 200 from SE to NW, 100 from SE to NE. Minimum cost per week is $1600.

35. The primal problem is: Maximize $P = 26x_1 + 24x_2 + 30x_3$ subject to:

$$2x_1 + x_2 + x_3 \le 20$$

$$x_1 + x_2 + 4x_3 \le 30$$

$$x_1 + 3x_2 + 3x_3 \le 40$$

$$x_1, x_2, x_3 \ge 0$$

The dual is: Minimize $z = 20y_1 + 30y_2 + 40y_3$ subject to:

$$2y_1 + y_2 + y_3 \ge 26$$

$$y_1 + y_2 + 3y_3 \ge 24$$

$$y_1 + 4y_2 + 3y_3 \ge 30$$

$$y_1, y_2, y_3 \ge 0$$

We think of y_1, y_2, y_3 as the respective prices of one unit of each of the ingredients. The cost of the ingredients in the first product is $2y_1 + y_2 + y_3$, which must be at least the profit of 26 dollars. This gives the first problem constraint in the dual problem. In the same way we obtain the other two problem constraints. Our cost is $w = 20y_1 + 30y_2 + 40y_3$, which we wish to minimize.

37. The primal problem is: Minimize $\quad C = 75y_1 + 75y_2$

$$\text{Subject to} \quad y_1 + 2y_2 \geq 8$$
$$75y_1 + 60y_2 \geq 420$$
$$y_1, y_2 \geq 0$$

The dual is: Maximize $z = 8x_1 + 420x_2$

$$\text{Subject to} \quad x_1 + 75x_2 \leq 75$$
$$2x_1 + 60x_2 \leq 75$$
$$x_1, x_2 \geq 0$$

We think of x_1 as the price of one gm of fiber (in pure form) and x_2 as one mg of vitamin C (in pure form). If we think of selling fiber and vitamin C separately, then we cannot obtain more for them than their cost combined in oranges and strawberries. This gives the two problem constraints in the dual problem. Our income is $z = 8x_1 + 420x_2$, which we wish to maximize.

39. The primal problem is: Minimize $\quad C = 20y_1 + 30y_2 + 20y_3$

$$\text{Subject to} \quad 40y_1 + 20y_2 \qquad\;\; \geq 4000$$
$$30y_1 + 20y_2 + 30y_3 \geq 2000$$
$$10y_1 + 55y_2 + 40y_3 \geq 2000$$
$$y_1, y_2, y_3 \geq 0$$

The dual is: Maximize $z = 4000x_1 + 2000x_2 + 2000x_3$

$$\text{Subject to} \quad 40x_1 + 30x_2 + 10x_3 \leq 20$$
$$20x_1 + 20x_2 + 55x_3 \leq 30$$
$$30x_2 + 40x_3 \leq 20$$
$$x_1, x_2, x_3 \geq 0$$

We think of x_1 as the price of one pound of nitrogen (in pure form), x_2 as one pound of phosphoric acid (in pure form), and x_3 as one pound of potash (in pure form). If we think of selling nitrogen, phosphoric acid, and potash separately, then we cannot obtain more for them than their cost combined in each of the three types of fertilizers. This gives the three problem constraints in the dual problem. Our income is $z = 4000x_1 + 2000x_2 + 2000x_3$, which we wish to maximize.

41. The primal problem is: Minimize $C = 2000y_1 + 3000y_2 + 4000y_3$

$$\text{Subject to} \quad 2y_1 + y_2 + y_3 \geq 21$$
$$y_1 + y_2 + 3y_3 \geq 19$$
$$y_1 + 4y_2 + 3y_3 \geq 25$$
$$y_1, y_2, y_3 \geq 0$$

The dual is: Maximize $z = 21x_1 + 19x_2 + 25x_3$

$$\text{Subject to} \quad 2x_1 + x_2 + x_3 \leq 2000$$
$$x_1 + x_2 + 4x_3 \leq 3000$$
$$x_1 + 3x_2 + 3x_3 \leq 4000$$
$$x_1, x_2, x_3 \geq 0$$

We think of x_1 as the price of one ton of low-grade ore, x_2 as one ton of medium-grade ore, and x_3 as one ton of low-grade ore. If we think of selling the three grades of ore, then we cannot obtain more for them than their cost combined from each of the three mines. This gives the three problem constraints in the dual problem. Our income is $z = 4000x_1 + 2000x_2 + 2000x_3$, which we wish to maximize.

Section 4.4

  1. Maximum of 6 at (2, 2)   **3.** Maximum of 10 at (2, 4)

  5. Maximum of 16 at (8, 0)   **7.** Maximum of 13 at (2, 3)

  9. Maximum of 1280 at (180, 220, 260) and (620, 0, 40)

11. Minimum of 4.5 at (1.5, 0)   **13.** Minimum of 13 at (3, 1)

15. Minimum of 70 at (1, 3) or (3, 2)   **17.** Minimum of 11 at (4, 1)

19. Minimum of 16 at (4, 1)   **21.** Minimum of 2 at (0, 1, 0)

23. Minimum of 472 at (10, 12, 8)   **25.** Minimum of 3 at (0, 1, 2)

27. Minimum cost of $11,000 with 150 shipped from Stockpile I to Plant A, 150 shipped from Stockpile I to Plant B, none shipped from Stockpile II to Plant A, and 100 shipped from Stockpile II to Plant B.   **29.** Maximum of 8.2 at (30, 30, 40)

31. Minimum cost of $2400 with 90 bags of Type A, 20 bags of Type B, none of Type C

33. The first plant 4 days, the second 2 days with minimum cost of $220,000.

35. The North mine 8 days, the Center 2, and the South 3. $34,000

Section 4.5

  1. (a) The new last column is

$$20 - 2h_2$$

$$30 + h_1 - h_2 - h_3$$

$$25 - h_2 + 3h_3$$

$$40 + 4h_2 + 3h_3$$

(b) $20 - 2h_2 \geq 0, 30 + h_1 - h_2 - h_3 \geq 0, 25 - h_2 + 3h_3 \geq 0$

(c) $(20 - 2h_2, 25 - h_2 + 3h_3)$ (d) $40 + 4h_2 + 3h_3$ (e) 10 (f) (14,16), 46

(g) 0, 4, and 3. The maximum revenue is increased by $0 if the gallons of varnish is increased by 1, by $4 if the board ft of wood is increased by 1, and by $3 if the hours of labor is increased by 1.

  3. (a) The new last column is

$$20 + h_1 + 4h_2 + h_3 + 5h_4$$

$$30 + 3h_1 + 2h_2 + h_4$$

$$25 + 2h_1 + h_2 + 2h_4$$

$$40 + h_1 + 3h_2 + 6h_4$$

$$90 + 2h_1 + 4h_2 + 5h_4$$

(b) $20 + h_1 + 4h_2 + h_3 + 5h_4 \geq 0, 30 + 3h_1 + 2h_2 + h_4 \geq 0$

$25 + 2h_1 + h_2 + 2h_4 \geq 0, 40 + h_1 + 3h_2 + 6h_4 \geq 0$

(c) $(30 + 3h_1 + 2h_2 + h_4, 25 + 2h_1 + h_2 + 2h_4, 40 + h_1 + 3h_2 + 6h_4)$

(d) $90 + 2h_1 + 4h_2 + 5h_4$ (e) 20 (f) (43, 34, 57), 111

(g) 2, 4, 0, and 5. The maximum revenue is increased by $2 if the gal of orange juice is increased by 1, by $4 if the gal of grape juice is increased by 1, by $0 if the gal of grapefruit juice is increased by 1, and by $5 if the gal of cranberry juice is increased by 1.
5. (24, 0) with maximum of 5.   **7.** (0, 15) with maximum of 80.

Chapter 4 Review

1. (500, 0), 500   **2.** (0, 100), 400   **3.** (0, 1), 2   **4.** (35, 0), 70
5. (6, 0), 18   **6.** $(\frac{32}{7}, \frac{25}{7})$, 29   **7.** (7.5, 0), 15   **8.** (27, 2), 60
9. No solution   **10.** (4, 1), 21
11. 100 acres of the first field in wheat, 110 acres of the first field in corn, 100 acres of the second field in wheat, no acres of the second field in corn, $30,900
13. (a) 50 purses, 55 belts, new maximum is $90 (b) 2, 0, 4. The maximum revenue is increased by $2 if the sq ft of leather is increased by 1, by $0 if the oz of brass is increased by 1, by $4 if the hr of labor is increased by 1.

Section 5.1

1. $26.67   **3.** $480   **5.** $2046.67   **7.** $1120.00   **9.** 8.57%
11. 3.5%   **13.** $1980.20   **15.** $5357.14   **17.** 5.11%   **19.** 7.12%
21. $16.67, $983.33   **23.** $500, $4500   **25.** 2.5 years   **27.** 12.5 years.
29. $1666.67, $48,333.33

Section 5.2

1. $1100.00   **3.** (a) $1080.00 (b) $1082.43 (c) $1083.00 (d) $1083.22
(e) $1083.28
5. (a) $21,724.52 (b) $23,769.91 (c) $24,273.39 (d) $24,472.28 (e) $24,523.93
7. (a) $3262.04 (b) $7039.99 (c) $14,974.46 (d) $31,409.42 (e) $93,050.97
(f) $267,863.55
9. (a) $1538.62 (b) $2367.36 (c) $3642.48 (d) $13,267.68
11. (a) 8.16% (b) 8.24% (c) 8.30% (d) 8.32% (e) 8.33%
13. (a) $1784.31 (b) $1664.13 (c) $1655.56
15. (a) $3183.76 (b) $2769.32 (c) $2740.89
17. $31,837.58   **19.** $26,897.08
21. About $3 trillion   **23.** 9.57%   **25.** 81.3 months, 351.6 weeks
27. 128.3 months, 3893 days   **29.** $1104.79 yes   **31.** 6.875%
33. 8.26%, 8.32%, second   **35.** $70,540.50

Section 5.3

1. $17,383.88   **3.** $18,120.60   **5.** $98,925.54   **7.** $15,645.49
9. $2,045,953.85   **11.** $1316.40   **13.** $148.36   **15.** $7.23
17.

$$FV = PMT \frac{(1 + i)^n - 1}{i}$$

$$\frac{iFV}{PMT} + 1 = (1 + i)^n$$

$$\log \left(\frac{iFV}{PMT} + 1 \right) = n \log(1 + i)$$

$$n = \frac{\log \left(\dfrac{iFV}{PMT} + 1 \right)}{\log(1 + i)}$$

19. 34.2 years   **21.** $675,764.89   **23.** $206,349.29
25. $1210.61, $28,209.02, $18,392.71, $1418.71
27. $265.17. Equity: $3316.64, $6944.40, $10,912.47, $15,252.77, $20,000.00
29. $14,836.23. Equity: $129,611.03, $287,529.47, $479.937.77, $714,368.74,
$1,000,000.00. Interest earned is $406,550.80. Interest earned during 18th period is
$7738.86.

Section 5.4

1. $339,758.16   **3.** $286,748.03   **5.** $245,453.69   **7.** $2296.07, $3776.42
9. $2642.37, $5854.22   **11.** $665.30, $139,508.00   **13.** $804.62, $189,663.20
15. $898.83, $61,789.40   **17.** $1014.27, $82,568.60   **19.** $253.63, $2174.24
21. $263.34, $2640.32   **23.** $34.18, $230.48   **25.** $36.15, $301.40
27. $85,812.06, $88.92, $588.80   **29.** $77,412.95, $332.45, $750.54

31.

End of Period	Repayment Made	Interest Charged	Payment Toward Principal	Outstanding Principal
0				10,000
1	2296.07	1000.00	1296.07	8703.93
2	2296.07	870.39	1425.68	7278.25
3	2296.07	727.83	1568.24	5710.01
4	2296.07	571.00	1725.07	3984.94
5	2296.07	398.49	1897.58	2087.36
6	2296.10	208.74	2087.36	0000.00

33.

End of Period	Repayment Made	Interest Charged	Payment Toward Principal	Outstanding Principal
0				10,000
1	2642.37	1500.00	1142.37	8857.63
2	2642.37	1328.64	1313.73	7543.90
3	2642.37	1131.59	1510.78	6033.12
4	2642.37	904.97	1737.40	4295.72
5	2642.37	644.36	1998.01	2297.71
6	2642.37	344.66	2297.71	0000.00

35. (a)

$$PV = PMT \frac{[1 - (1 + i)^{-n}]}{i}$$

$$\frac{PV(i)}{PMT} = 1 - (1 + i)^{-n}$$

$$(1 + i)^{-n} = 1 - \frac{PV(i)}{PMT} = \frac{PMT - PV(i)}{PMT}$$

$$-n \log(1 + i) = \log \frac{PMT - PV(i)}{PMT}$$

$$n \log(1 + i) = \log \frac{PMT}{PMT - PV(i)}$$

$$n = \frac{\log \frac{PMT}{PMT - PV(i)}}{\log(1 + i)}$$

(b) 70 payments.

37. The first. The present value of the first is $67,100.81, which is higher than that of the second, which is $63,817.96

39. The present value of the first option is only $320,882.88. Accept the lump sum.

41. (a) The first saves the most, $24,781.30. (b) The second saves the most, $35,383.35.

43. Your monthly payment in the first case will be $295.24, while using the rebate and the credit union will give payments per month of $292.56. The second is better.

Chapter 5 Review

1. (a) $1469.33 (b) $1485.95 (c) $1489.85 (d) $1491.37 (e) $1491.76

2. (a) $1610.51 (b) $1638.62 (c) $1645.31 (d) $1647.93 (e) $1648.61

3. (a) 9.20% (b) 9.31% (c) 9.38% (d) 9.41% (e) 9.42%

4. (a) 11.30% (b) 11.46% (c) 11.57% (d) 11.61% (e) 11.63%

5. (a) $4631.93 (b) $4505.23 (c) $4496.05

6. (a) $4224.11 (b) $4106.46 (c) $4066.15   **7.** $214,548.21   **8.** $6499.31

9. 9.05%   **10.** $(1 + r)^T = 3$

$$T \log(1 + r) = \log 3$$

$$T = \frac{\log 3}{\log(1 + r)}$$

11. $408,922.62   **12.** $33,111.50; $675,540.00; $14,179.87

13. The present value of the 19 payments of $100,000 each, plus the immediate disbursement of $100,000 is $1,060,359.90.

14. $839.20, $1014.27, $151,760.00, $82,568.60, $95.40, $282.65, $217.03, $642.99

Section 6.1

1. (a) False (b) False.   **3.** (a) True (b) False.

5. (a) False (b) True (c) False (d) True.   **7.** (a) ϕ, {3} (b) ϕ, {3}, {4}, {3, 4}

9. (a)(b)

11. (a) 2 (b) 3 (c) 4 (d) 1

13.

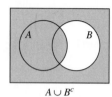

$A \cup B^c$

15.

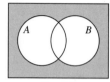

$(A \cup B)^c$

17. (a) I (b) V (c) II (d) VI (e) VIII (f) IV (g) III (h) VII

19.

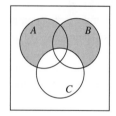

$(A \cup B) \cap C^c$

21.

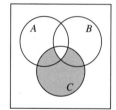

$(A \cap B)^c \cap C$

23.

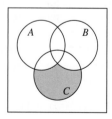

$(A \cup B)^c \cap C$

25. (a) {4, 5, 6} (b) {1, 2, 3, 4, 5, 6, 7, 8}

27. (a) {1, 2, 3}, (b) {9, 10}   **29.** (a) {5, 6} (b) {1, 2, 3, 4, 7, 8, 9, 10}

31. (a) ϕ (b) ϕ

33. (a) People in your state who do not own an auto (b) People in your state who own an auto or a house (c) People in your state who own an auto or do not own a house

35. (a) People in your state who own an auto but not a house (b) People in your state who do not own an auto and do not own a house (c) People in your state who do not own an auto or do not own a house

37. (a) People in your state who own an auto and a house and a piano (b) People in your state who own an auto or a house or a piano (c) People in your state who own both an auto and a house or else own a piano

39. (a) People in your state who do not own both an auto and a house but do own a piano (b) People in your state who do not own an auto, nor a house, nor a piano (c) People in your state who own a piano, but do not own a car or a house

41. (a) $N \cap F$ (b) $N \cap H^c$ **43.** (a) $N \cup S$ (b) $N^c \cap S^c$ or $(N \cup S)^c$

45. (a) $(N \cap H) \cup (S \cap H)$ or $(N \cup S) \cap H$ (b) $N \cap F \cap H^c$

47. (a) $F \cap H \cap (N \cup S)^c$ (b) $F \cap H^c \cap (N \cup S)^c$

Section 6.2

1. 135   **3.** 70   **5.** 60   **7.** 150   **9.** (a) 1100 (b) 750 (c) 100

11. (a) 100 (b) 500   **13.** 110   **15.** 90   **17.** 15   **19.** 7   **21.** 3

23. 56   **27.** (a) 30 (b) 360 (c) 70   **29.** (a) 20 (b) 25 (c) 345   **31.** 30

Section 6.3

1. $5 \cdot 4 \cdot 3 = 60$ **3.** $8 \cdot 7 \cdot 6 \cdot 5 \cdot 4 = 6720$ **5.** $7! = 5040$

7. $9 \cdot 8 \cdot 7 \cdot 6 \cdot 5 \cdot 4 \cdot 3 \cdot 2 = 362{,}880$ **9.** $4 \cdot 30 = 120$

11. $4 \cdot 10 \cdot 5 = 200$ **13.** $10 \cdot 10 \cdot 10 \cdot 10 \cdot 10 \cdot 10 = 1{,}000{,}000$

15. $2 \cdot 10 \cdot 5 \cdot 3 = 300$ **17.** $(26)^3(10)^3 = 17{,}576{,}000$ **19.** 96

21. $5 \cdot 4 \cdot 4 \cdot 3 \cdot 3 \cdot 2 \cdot 2 = 2880$ **23.** $5! = 120$ **25.** $2! \, 4! \, 6! = 34{,}560$

27. $9! 5! = 43{,}545{,}600$ **29.** $3! 4! = 144$ **31.** $12 \cdot 11 \cdot 10 \cdot 9 = 11{,}880$

33. $P(10, 8) = 1{,}814{,}400$ **35.** $(10 \cdot 9 \cdot 8)(8 \cdot 7 \cdot 6) = 241{,}920$

37. $P(11, 4)P(9, 3)P(5, 2) = 79{,}833{,}600$

Section 6.4

1. $(8 \cdot 7 \cdot 6)/(3 \cdot 2 \cdot 1) = 56$ **3.** $(8 \cdot 7 \cdot 6 \cdot 5 \cdot 4)/(5 \cdot 4 \cdot 3 \cdot 2 \cdot 1) = 56$

5. 12   **7.** 35   **9.** 105   **11.** 1   **13.** 1

15.

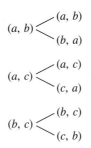

17. 10   **19.** $C(46, 6) = 9{,}366{,}819$ **21.** $C(20, 5) = 15{,}504$

23. $C(20, 4) = 4845$ **25.** $C(40, 5) = 658{,}008$ **27.** $C(11, 3)C(7, 2) = 3465$

29. $C(20, 7)C(6, 3) = 1{,}550{,}400$ **31.** $C(12, 3)P(9, 3) = 110{,}880$

33. $C(11, 4)P(7, 2) = 13{,}860$ **35.** $C(9, 5) = 126$ **37.** $C(8, 5) = 56$

39. $2^8 - C(8, 1) - C(8, 0) = 247$ **41.** $C(12, 3)C(8, 3)C(4, 2)2 = 147{,}840$

43. $C(4, 3) \cdot 48 \cdot 44/2 = 4224$ **45.** $C(4, 2)C(4, 2)C(13, 2) \cdot 44 = 123{,}552$

Section 6.5

1. $a^5 - 5a^4b + 10a^3b^2 - 10a^2b^3 + 5ab^4 - b^5$
3. $32x^5 + 240x^4y + 720x^3y^2 + 1080x^2y^3 + 810xy^4 + 243y^5$
5. $1 - 6x + 15x^2 - 20x^3 + 15x^4 - 6x^5 + x^6$
7. $16 - 32x^2 + 24x^4 - 8x^6 + x^8$
9. $s^{12} + 6s^{10}t^2 + 15s^8t^4 + 20s^6t^6 + 15s^4t^8 + 6s^2t^{10} + t^{12}$
11. $a^{10} - 10a^9b + 45a^8b^2, 45a^2b^8 - 10ab^9 + b^{10}$
13. $x^{11} + 11x^{10}y + 55x^9y^2, 55x^2y^9 + 11xy^{10} + y^{11}$
15. $1 - 12z + 66z^2, 66z^{10} - 12z^{11} + z^{12}$
17. $1 - 12x^3 + 66x^6, 66x^{30} - 12x^{33} + x^{36}$ 19. 210   21. 1260
23. 113,400   25. 5040   27. 1, 7, 21, 35, 35, 21, 7, 1   29. 420
31. 34,650   33. 2520   35. (a) 6 (b) 3 (c) 24 (d) 12 (e) 6
37. (a) 34,650 (b) 3780

Chapter 6 Review

1. (a) Yes (b) No (c) Yes   2. $\{x|x = 5n, n = 1, 2, \ldots, 8\}$
3. $\{0, \sqrt{2}, -\sqrt{2}\}$ 4. $\phi, \{A\}, \{B\}, \{C\}, \{A, B\}, \{A, C\}, \{B, C\}, \{A, B, C\}$

5.

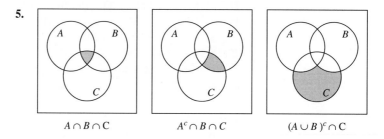

$A \cap B \cap C$ $A^c \cap B \cap C$ $(A \cup B)^c \cap C$

6. $A \cup B = \{1, 2, 3, 4\}, A \cap B = \{2, 3\}, B^c = \{1, 5, 6\}, A \cap B \cap C = \phi,$
$(A \cup B) \cap C = \{4\}, A \cap B^c \cap C = \phi$

7. (a) My current instructors who are less than 6 feet tall. (b) My current instructors who are at least 6 feet tall or are male. (c) My current instructors who are female and weigh at most 180. (d) My current male instructors who are at least 6 feet tall and weigh more than 180 pounds. (e) My current male instructors who are less than 6 feet tall and weigh more than 180 pounds. (f) My current instructors who weigh more than 180 pounds or who are females at least 6 feet tall.

8. (a) M^c (b) $M^c \cap W^c$ (c) $M \cap (H \cup W)$

10.

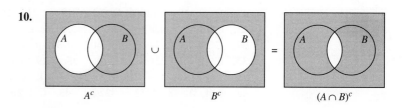

A^c B^c $(A \cap B)^c$

11. 120   12. 10   13. 50   14. (a) 40, (b) 45, (c) 19
15. $P(10, 4) = 10 \cdot 9 \cdot 8 \cdot 7 = 5040, C(10, 4) = 10 \cdot 9 \cdot 8 \cdot 7/(4 \cdot 3 \cdot 2) = 210$

16. 6 **17.** $5!6! = 86,400$ **18.** $P(6, 3)P(5, 3) = 7200$

19. $C(5, 3)C(6, 4) = 150$ **20.** $2[C(6, 6) + C(6, 5) + C(6, 4)] = 44$

21. $C(10, 7) = 120$ **22.** $32 - 80x + 80x^2 - 40x^3 + 10x^4 - x^5$

23. $-36x^7 + 9x^8 - x^9$ **24.** 3150 **25.** 7560 **26.** 369,600

Section 7.1

1. $\{a\}, \{b\}, \{c\}, \{a, b\}, \{a, c\}, \{b, c\}, \{a, b, c\}, \phi$

3. $\{HHH, HHT, HTH, HTT, THH, THT, TTH, TTT\}, \{HHH, HHT, HTH, THH\}$

5. (a) $\{111, 110, 101, 100, 011, 010, 001, 000\}$, (b) $\{111, 110, 101, 011\}$

7. (a) {white, black, red}, (b) {black, red}

9. $\{ABC, ABD, ABF, ACD, ACF, ADF, BCD, BCF, BDF, CDF\}$,

$\{ABC, ABD, ABF, ACD, ACF, ADF\}$

11. (a) $\{0, 1, 2, 3, 4, 5, 6, 7, 8, 9, 10\}$ (b) $E = \{6, 7, 8, 9, 10\}$

(c) $F = \{0, 1, 2, 3, 4\}$ (d) $E \cup F = \{0, 1, 2, 3, 4, 6, 7, 8, 9, 10\}. E \cap F = \phi.$

$E^c = \{0, 1, 2, 3, 4, 5\}. E \cap F^c = \{6, 7, 8, 9, 10\}. E^c \cap F^c = \{5\}$

(e) $E \cup F$ and $E^c \cap F^c. E^c$ and $E \cap F^c. E \cap F^c$ and $E^c \cap F^c.$

13. (a) $\{SSS, SSF, SFS, SFF, FSS, FSF, FFS, FFF\}$ (b) $E = \{SSS, SSF, SFS, FSS\}$

(c) $G = \{SSS, SSF, SFS, SFF\}$ (d) $E \cup G = \{SSS, SSF, SFS, SFF, FSS\}. E \cap G =$

$\{SSS, SSF, SFS\}. G^c = \{FSS, FSF, FFS, FFF\}. E^c \cap G = \{SFF\}. (E \cup G)^c =$

$\{FSF, FFS, FFF\}$ (e) $E \cup G$ and $(E \cup G)^c. E \cap G$ and $G^c. E \cap G$ and $E^c \cap G. G^c$

and $E^c \cap G. E^c \cap G$ and $(E \cup G)^c. E \cap G$ and $(E \cup G)^c.$

15. $E \cap F^c$ **17.** $E^c \cap F^c$ **19.** $E \cap F \cap G^c$

21. $E \cup F \cup G =$ all 26 letters. $E^c \cap F^c \cap G^c = \phi. E \cap F \cap G = \phi. E \cup F^c \cup G$

$= \{a, b, c, d, e, i, o, u\}$

Section 7.2

1. (a) 23 (b) 232 (c) 2317 **3.** .40 **5.** .15 **7.** (a) .12 (b) .80 (c) .08

9. .20 **11.** .56 **13.** A: .125, B: .175, C: .40, D: .20, F: .10

15. .15, .20, .14, .25, .16, .10 **17.** .30 **19.** .72 **21.** .55 **23.** .15

25. Since $f(E) + f(E^c) = f(E \cup E^c) = f(S) = N$,

$$\frac{f(E)}{N} + \frac{f(E^c)}{N} = \frac{N}{N} = 1.$$

For large $N, \dfrac{f(E)}{N} \approx p(E)$ and $\dfrac{f(E^c)}{N} \approx p(E^c)$. Thus $p(E) + p(E^c) = 1$

27. Since $f(S) = f(s_1) + \cdots + f(s_n)$,

$$\frac{f(s_1)}{N} + \cdots + \frac{f(s_n)}{N} = \frac{f(S)}{N} = \frac{N}{N} = 1.$$

For large $N, \dfrac{f(s_1)}{N} \approx p_n$. This implies

$$p_1 + \cdots + p_n = 1.$$

Section 7.3

1. .50, .90 **3.** .18, .63, .62 **5.** .25 **7.** $\frac{1}{6}$ **9.** $\frac{7}{30}$ **11.** .60, .80, 0
13. .70, .70, 0, .30, 1 **15.** .60, .10 **17.** .20, .10 **19.** .20
21. .20, .10, .15 **23.** .85 **25.** (a) $3:17$ (b) $1:3$ **27.** (a) $2:3$ (b) $1:4$
29. $7:5$ **31.** .20 **33.** Since $E \cup E^c = S$ and $E \cap E^c = \phi$,

$$1 = p(S) = p(E \cup E^c) = p(E) + p(E^c).$$

The result then follows.
35. Using the fact that E_1, E_2, and E_3 are pairwise mutually disjoint,

$$\begin{aligned}
p(E_1 \cup E_2 \cup E_3) &= p((E_1 \cup E_2) \cup E_3) \\
&= p(E_1 \cup E_2) + p(E_3) - p((E_1 \cup E_2) \cap E_3) \\
&= p(E_1) + p(E_2) + p(E_3) - p((E_1 \cap E_3) \cup (E_2 \cap E_3)) \\
&= p(E_1) + p(E_2) + p(E_3) - p(\phi \cup \phi) \\
&= p(E_1) + p(E_2) + p(E_3)
\end{aligned}$$

37. Since $E \cap F^c$, $E^c \cap F$, and $E \cap F$ are pairwise mutually disjoint and their union is $E \cup F$, we have

$$\begin{aligned}
p((E \cap F^c) \cup (E^c \cap F)) + p(E \cap F) &= p((E \cap F^c) \cup (E^c \cap F) \cup (E \cap F)) \\
&= p(E \cup F) \\
&= p(E) + p(F) - p(E \cap F)
\end{aligned}$$

The result then follows.
39. Since $E^c \cap F$ and $E \cap F$ are disjoint and their union is F,

$$p(E^c \cap F) + p(E \cap F) = p((E^c \cap F) \cup (E \cap F)) = p(F).$$

The result then follows.
41. .04, .96 **43.** .039, .049, .951 **45.** (a) .009 (b) .001 (c) .006 (d) .974

Section 7.4

1. $\frac{3}{8}, \frac{7}{8}$ **3.** $\frac{1}{12}$ **5.** $\frac{1}{36}, \frac{1}{12}, \frac{5}{36}$ **7.** $\frac{3}{10}, \frac{1}{3}$ **9.** $6 \cdot 7 \cdot 8 / 1330$ **11.** $\frac{20}{323}$
13. $\frac{6}{1326}$ **15.** $\frac{78}{1326}$ **17.** $\frac{208}{1326}$ **19.** $4/2,598,960$ **21.** $624/2,598,960$
23. $10,200/2,598,960$ **25.** $123,552/2,598,960$ **27.** .0833; .2361; .4271; .6181
29. 1 **31.** $\frac{6}{45}$ **33.** $56/2^{10} \approx .0547$ **35.** $\frac{5}{11}$

Section 7.5

1. $\frac{3}{7}, \frac{3}{5}$ **3.** $\frac{2}{3}, \frac{4}{5}$ **5.** 1, 0 **7.** $\frac{2}{3}, \frac{1}{2}$ **9.** $\frac{1}{3}, 1$ **11.** $0, \frac{1}{2}$ **13.** No
15. Yes **17.** No **19.** Yes **21.** $\frac{2}{11}$ **23.** $\frac{1}{7}$
25. (a) $\frac{1}{13}$ (b) $\frac{49}{25 \cdot 33} \approx .059$ **27.** .12, .64, .60 **29.** $\frac{1}{3}, \frac{3}{10}$ **31.** No **33.** .65
35. .72 **37.** First assume that E and F are independent. Then

$$\begin{aligned}
p_2 = p(E \cap F) &= p(E)p(F) \\
&= (p_1 + p_2)(p_2 + p_3)
\end{aligned}$$

$$= p_1 p_2 + p_1 p_3 + p_2^2 + p_2 p_3$$

$$= p_2(p_1 + p_2 + p_3) + p_1 p_3$$

$$= p_2(1 - p_4) + p_1 p_3$$

$$= p_2 - p_2 p_4 + p_1 p_3$$

$$p_2 p_4 = p_1 p_3$$

Now suppose $p_1 p_3 = p_2 p_4$, then

$$p(E)p(F) = (p_1 + p_2)(p_2 + p_3)$$

$$= p_1 p_3 + p_1 p_2 + p_2 p_3 + p_2^2$$

$$= p_2 p_4 + p_1 p_2 + p_2 p_3 + p_2^2$$

$$= p_2[p_4 + p_1 + p_3 + p_2]$$

$$= p_2$$

$$= p(E \cap F)$$

39. If E and F are independent, then $p(E \cap F) = p(E)p(F)$ and

$$p(E^c)p(F^c) = [1 - p(E)][1 - p(F)]$$

$$= 1 - p(E) - p(F) + p(E)p(F)$$

$$= 1 - p(E) - p(F) + p(E \cap F)$$

$$= 1 - p(E \cup F)$$

$$= p((E \cup F)^c)$$

$$= p(E^c \cap F^c)$$

41. Since $E \cap F = \phi$, $p(E \cap F) = 0$. But since $p(E) > 0$ and $p(F) > 0$, $p(E)p(F) > 0$, and therefore $p(E)p(F) > 0 = p(E \cap F)$.

43.

$$P(E^c|F) = \frac{P(E^c \cap F)}{p(F)}$$

$$= \frac{p(F) - p(E \cap F)}{p(F)}$$

$$= 1 - p(E|F)$$

45. $p(E|F) = \dfrac{p(E \cap F)}{p(F)} = \dfrac{p(F)}{p(F)} = 1$

47. $p(E|F) + p(E^c|F) = \dfrac{p(E \cap F)}{p(F)} + \dfrac{p(E^c \cap F)}{p(F)} = \dfrac{p(E \cap F) + p(E^c \cap F)}{p(F)} =$

$\dfrac{p(F)}{p(F)} = 1.$ **49.** .02, .017, $\frac{6}{17}$ **51.** $\frac{8}{26}$ **53.** $\frac{5}{7}, \frac{5}{21}, \frac{1}{21}$ **55.** Yes

57. No   **59.** .000001

Section 7.6

1. $\frac{3}{7}, \frac{57}{93}$ **3.** $\frac{3}{19}, \frac{1}{3}$ **5.** $\frac{4}{11}$ **7.** $\frac{2}{17}$ **9.** (a) $\frac{2}{23}$ (b) $\frac{21}{23}$ **11.** (a) $\frac{1}{56}$ (b) $\frac{5}{28}$

13. $\frac{100}{136} \approx .74$ **15.** $\frac{396}{937} \approx .42, \frac{5}{937} \approx .005$ **17.** $\frac{5}{7}$ **19.** $\frac{10}{19}$ **21.** (a) $\frac{1}{12}$ (b) $\frac{1}{4}$

23. $P(1|N) = \frac{6}{20}, P(2|N) = \frac{4}{20}, P(3|N) = \frac{6}{20}, P(4|N) = \frac{4}{20}$

Section 7.7

1. $5(.2)^4(.8) = .0064$ **3.** $35(.5)^7 \approx .273$ **5.** $70(.25)^4(.75)^4 \approx .087$
7. $20(.5)^6 \approx .3125$ **9.** $35(.1)^4(.9)^3 \approx .00255$ **11.** $10(.1)^3(.9)^2 = .0081$
13. $45(.5)^{10} \approx .0439$ **15.** $56(.5)^{10} \approx .0547$ **17.** $11(.5)^{10} \approx .0107$
19. $7(.6)^6(.4) \approx .1306$ **21.** $7(.6)^6(.4) + (.6)^7 \approx .1586$
23. $(.4)^7 + 7(.6)(.4)^6 + 21(.6)^2(.4)^5 \approx .0962$ **25.** $1 - (.633)^4 \approx .839$
27. $(.839)^{10} \approx .173$ **29.** $6(.085)^2(.915)^2 + 4(.085)^3(.915) + (.085)^4 \approx .03859$
31. $(.03859)^3 \approx .00006$ **33.** $12(.05)(.95)^{11} \approx .341$
35. $(.95)^{12} + 12(.05)(.95)^{11} + 66(.05)^2(.95)^{10} \approx .980$
37. $C(20, 10)(.8)^{10}(.2)^{10} \approx .002$
39. $190(.8)^{18}(.2)^2 + 20(.8)^{19}(.2) + (.8)^{20} \approx .206$ **41.** $190(.05)^2(.95)^{18} \approx .189$
43. $(.95)^{20} + 20(.05)(.95)^{19} + 190(.05)^2(.95)^{18} + 1140(.05)^3(.95)^{17} \approx .984$ -

Chapter 7 Review

1. (a) .016, .248, .628, .088, .016, .004 (b) .892 (c) .264   **2.** (a) $\frac{5}{30}$ (b) $\frac{15}{30}$ (c) $\frac{20}{30}$
3. $p(E \cup F) = .60, p(E \cap F) = 0, p(E^c) = .75$
4. $p(E \cup F) = .55, p(E^c \cap F) = .35, p((E \cup F)^c) = .45$ **5.** $p(b) = \frac{1}{6}$
6. .75   **7.** (a) .09 (b) .91   **8.** (a) .13 (b) .09 (c) .55
9. $\dfrac{5 \cdot 10 \cdot 15}{c(30, 3)} \approx .185, \frac{1}{6}$
10. $C(10, 3)C(5, 4)C(15, 2)/C(30, 9), C(9, 3)C(6, 4)(\frac{10}{30})^3(\frac{5}{30})^4(\frac{15}{30})^2$
11. $p(E|F) = .40, p(E^c|F) = .60, p(F^c|E^c) = \frac{4}{7}$ **12.** Not independent
13. $(.05)^3$ **14.** .254   **15.** (a) $\frac{2}{31}$ (b) $\frac{4}{31}$ **16.** $C(6, 3)(.3)^3(.7)^3$
17. $1140(.55)^{17}(.45)^3 + 190(.55)^{18}(.45)^2 + 20(.55)^{19}(.45) + (.55)^{20} \approx .0049.$ Yes.
18. .50

Section 8.1

1. $\{2, 3, 4, \ldots\}$ infinite discrete   **3.** $[0, 168]$ continuous

5.

A	B	C	D	F
.12	.20	.38	.22	.08

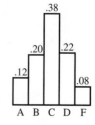

7.

1	2	3	4	5	6	7	8	9	10
.10	.04	.06	0	.08	.20	.20	.22	.06	.04

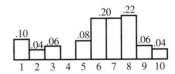

9.

Random variable x	0	1	2	3	4	5
$P(X = x)$	$\frac{6}{36}$	$\frac{10}{36}$	$\frac{8}{36}$	$\frac{6}{36}$	$\frac{4}{36}$	$\frac{2}{36}$

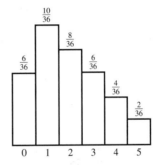

11.

Random variable x	10	7	4	1	-2	-5
$P(X = x)$	$\frac{1}{32}$	$\frac{5}{32}$	$\frac{10}{32}$	$\frac{10}{32}$	$\frac{5}{32}$	$\frac{1}{32}$

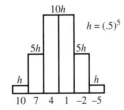

13.

0	1	2
$\frac{21}{45}$	$\frac{21}{45}$	$\frac{3}{45}$

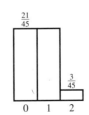

15. $z = .30$ **17.** (a) .15 (b) 0 (c) .40 (d) .23

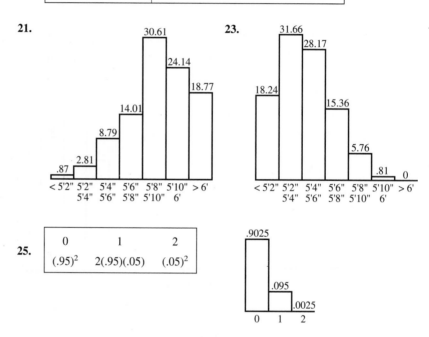

19.

Random variable y	0	1	2	3	4	5
$P(Y = y)$	.20	.15	.05	.35	.15	.10

21.

23.

25.

0	1	2
$(.95)^2$	$2(.95)(.05)$	$(.05)^2$

Section 8.2

1.

0	2	4	6	8	10
.50	.05	.10	.05	.20	.10

$E(X) = 3.4$

3. 0   **5.** .847   **7.** 0, -25% **9.** $-\$.19$ **11.** $-\$1/19$

13. $-\$1/19$ **15.** \$.80 or \$$-.20$ if you account for the cost of the \$1.00 ticket.

17. $\$1/6$ **19.** Yes. $\dfrac{\dfrac{x_1 + \cdots + x_{100}}{100} + \dfrac{x_{101} + \cdots + x_{200}}{100}}{2} = \dfrac{x_1 + \cdots x_{200}}{200}$

21. Yes since

$$\frac{1234 + \cdots + 1333}{100} = \frac{(1234 - 1234) + (1235 - 1234) + \cdots + (1333 - 1234) + 100(1234)}{100}$$

$$= \frac{0 + 1 + \cdots + 99}{100} + 1234$$

23. (a) $E(Y) = (x_1 - c)p_1 + \cdots + (x_n - c)p_n$

$$= [x_1 p_1 + \cdots x_n p_n] - c[p_1 + \cdots + p_n]$$

$$= E(X) - c$$

(b) 13.1 (c) .1   **25.** .09   **27.** .83   **29.** \$1800, \$100   **31.** 2.5

33. \$300, \$255, the first   **35.** 156   **37.** 1.29   **39.** 6.6

Section 8.3

1. (a) $\mu = 3$, var $= 2$, $\sigma = \sqrt{2}$ (b) $\mu = 3$, var $= 1.2$, $\sigma = \sqrt{1.2}$

3. Geddes: $69\frac{1}{3}$, 1.556, 1.247; Schreger: $69\frac{2}{3}$, .222, .471; Alcott: $69\frac{2}{3}$, 2.889, 1.700: Geddes has lowest average, Schreger the most consistent   **5.** $-.30$, 2.01, 1.418

7. 1.5, 3.850, 1.962   **9.** 1, second

11. (a) $\text{Var}(cX) = (cx_1)^2 p_1 + \cdots + (cx_n)^2 p_n - (E(cX))^2$

$$= c^2(x_1^2 p_1 + \cdots + x_n^2 p_n) - (cE(X))^2$$

$$= c^2(x_1^2 p_1 + \cdots + x_n^2 p_n - (E(X))^2)$$

$$= c^2 \text{Var}(X)$$

(b) 3.24 (c) $\text{Var}(\frac{1}{3}X) = .36$ **13.** .96   **15.** $c = 10$

17. Bank: 0, 0, 0. Bond: $-.16$, .034, .185. The bond varies more and carries more risk.

19. U.I. 2.33, U.S.S. 12.834. U.S.S. more risky.

21. A: .83, 1.121, 1.059. B: 1.3, 1.810, 1.345. Salesman B sells the most, but Salesman A is more consistent.

23. Gonorrhea: 789.333, 7531.5, 86.8. Syphilis: 95.17, 560.5, 23.7. The number of gonorrhea cases is varying the most.   **25.** San Diego: 71, 21.5, 4.6. Chicago: 58.75, 375.2, 19.4. Chicago.

27. Marihuana: 23, 52.3, 7.23. Cocaine: 5.67, 6.56, 2.56. Marihuana.

29. Males: 431.8, 13.36, 3.655. Females: 421, 6.000, 2.449. Males.

31. Carbon monoxide: 6.48, .162, .402. Ozone: .124, .0001, .010. Carbon monoxide.

33. Viral hepatitis: .5, .015, $\sqrt{.015}$. Meningitis: .6, .015, $\sqrt{.015}$. They are varying the same.   **35.** $1 - 1/(1.5)^2 = .56$ **37.** .75

Section 8.4

1. .7580   **3.** .3085   **5.** .1151   **7.** .7257   **9.** .1359   **11.** .0919

13. .9772   **15.** .0228   **17.** .383   **19.** .2417   **21.** (a) .6826 (b) .0062

23. 71.165   **25.** .0228   **27.** (a) 15.87% (b) 6.68% (c) 53.28%

29. 116.45 tons   **31.** A 94.2, B 82.9, C 67.1, D 55.8

Section 8.5

1. (a) .1102 (b) .7977   **3.** .0192   **5.** (a) .9901 (b) .9307

7. (a) .9744 (b) .8793   **9.** (a) .1539 (b) .7891   **11.** (a) .9904 (b) .9636

13. .9872

15. Yes, since under the old system the probability of 18 or less defectives is .0166.
17. Yes, since the probability of this response before the change was .0000 to four decimal places.
19. (a) 1139 (b) 113, 906   **21.** (a) 6765 (b) 676,506

Section 8.6

1. $p_3(0) = e^{-3} \approx .0498$ **3.** $p_4(1) = 4e^{-4} \approx .0733$

5. $p_{0.7}(2) = \dfrac{(.7)^2}{2} e^{-0.7} \approx .1217$ **7.** $p_3(0) + p_3(1) + p_3(2) = 8.5e^{-3} \approx .4232$

9. $1 - p_{2.3}(0) \approx .8997$ **11.** $p_{1.3}(1) + p_{1.3}(2) + p_{1.3}(3) \approx .6844$
13. $C(100, 2)(.01)^2(.99)^{98} \approx .1849, p_1(2) \approx .1839$
15. $C(100, 3)(.02)^3(.98)^{97} \approx .1823, p_2(3) \approx .1804$
17. $C(1000, 2)(.0003)^2(.9997)^{998} \approx .03332, p_{.3}(2) \approx .03334$

19.

x	$C(200, x)(.01)^x(.99)^{200-x}$	$\dfrac{2^x}{x!}e^{-2}$
0	.1340	.1353
1	.2707	.2707
2	.2720	.2707
3	.1814	.1804

21. (a) $e^{-2} \approx .1353$ (b) $p_2(0) + p_2(1) + p_2(2) = 5e^{-2} \approx .6767$ (c) .3233
23. (a) $e^{-0.6} \approx .5488$ (b) $p_{0.6}(0) + p_{0.6}(1) + p_{0.6}(2) + p_{0.6}(3) \approx .9966$
25. (a) $e^{-1.5} \approx .2231$ (b) $1 - p_{1.5}(0) - p_{1.5}(1) \approx .4422$ **27.** $13e^{-4} \approx .2381$
29. (a) $5e^{-2} \approx .6767$ (b) $13e^{-4} \approx .2381$ **31.** .0025   **33.** $3.5e^{-2.5} \approx .2873$
35. Mean number of accidents per month is $1.83 = \lambda$.

$100p_{1.83}(0)$	$100p_{1.83}(1)$	$100p_{1.83}(2)$	$100p_{1.83}(3)$	$100p_{1.83}(4)$	$100p_{1.83}(5)$
16	29	27	16	7	3

Since this is very similar to the data, we conclude that the number of accidents occurs randomly.   **37.** $\dfrac{4^{10}}{10!} e^{-4} \approx .0053$

Chapter 8 Review

1.

x	0	1	2	3	4	5
p	.168	.360	.309	.132	.028	.002

2. (a) .20 (b) .35 (c) .30 (d) .70   **3.** 2.97
4. $1.12 or $.12 if the cost of the $1.00 ticket is subtracted.   **5.** $400
6. Wins: $\mu = 28, \sigma^2 = 21.2, \sigma = 4.60$. Loses: $\mu = 12.6, \sigma^2 = 13.44, \sigma = 3.67$

7. 13–29: 3862.4, 501,978.64, 708.5; 30–39: 9338.2, 4,050,740.6, 2012.6. The 30–39 year old group   **8.** 3.709, 1.926   **9.** .96

10. (a) .8078 (b) .1922 (c) .4993   **11.** $c = 2.33$ **12.** .0228   **13.** 38.3%

14. .0107   **15.** .1539   **16.** $\dfrac{5^{10}}{10!}\,e^{-5} \approx .0181$ **17.** $1 - 2.5e^{-1} \approx .0803$

Section 9.1

1. No   **3.** No   **5.** No   **7.** No   **9.** No   **11.** Yes

13. $X_1 = \begin{bmatrix} .5 \\ .5 \end{bmatrix}$, $X_2 = \begin{bmatrix} .45 \\ .55 \end{bmatrix}$, $X_4 = \begin{bmatrix} .4445 \\ .5555 \end{bmatrix}$

15. $X_1 = \begin{bmatrix} .25 \\ .75 \end{bmatrix}$, $X_2 = \begin{bmatrix} .225 \\ .775 \end{bmatrix}$, $X_4 = \begin{bmatrix} .22225 \\ .77775 \end{bmatrix}$

17. $X_1 = \begin{bmatrix} .5 \\ .5 \\ 0 \end{bmatrix}$, $X_2 = \begin{bmatrix} .25 \\ .5 \\ .25 \end{bmatrix}$, $X_4 = \begin{bmatrix} .3125 \\ .3125 \\ .375 \end{bmatrix}$ **19.** $\begin{bmatrix} .7 & .5 \\ .3 & .5 \end{bmatrix}$, .64

21. $\begin{bmatrix} .3 & .8 \\ .7 & .2 \end{bmatrix}$, .65   **23.** .62   **25.** .625   **27.** $\begin{bmatrix} .6 & .1 & .1 \\ .2 & .8 & .2 \\ .2 & .1 & .7 \end{bmatrix}$, .44

29. .28 in first field, .50 in second, .22 in third.

Section 9.2

1. Yes   **3.** No   **5.** Yes   **7.** No   **9.** Yes   **11.** $\begin{bmatrix} \frac{5}{7} \\ \frac{2}{7} \end{bmatrix}$

13. $\begin{bmatrix} .40 \\ .60 \end{bmatrix}$ **15.** $\begin{bmatrix} \frac{4}{7} \\ \frac{3}{7} \end{bmatrix}$ **17.** $\begin{bmatrix} \frac{1}{3} \\ \frac{4}{9} \\ \frac{2}{9} \end{bmatrix}$ **19.** $\begin{bmatrix} \frac{3}{11} \\ \frac{5}{11} \\ \frac{3}{11} \end{bmatrix}$ **21.** $\begin{bmatrix} \frac{5}{8} \\ \frac{3}{8} \end{bmatrix}$ **23.** $\begin{bmatrix} \frac{8}{15} \\ \frac{7}{15} \end{bmatrix}$

25. $\begin{bmatrix} .20 \\ .50 \\ .30 \end{bmatrix}$ **27.** $\begin{bmatrix} .30 \\ .60 \\ .10 \end{bmatrix}$

Section 9.3

1. Yes   **3.** No   **5.** $\begin{bmatrix} 1 & 0 & \frac{1}{3} \\ 0 & 1 & \frac{2}{3} \\ 0 & 0 & 0 \end{bmatrix}$ **7.** $\begin{bmatrix} 1 & 0 & 0 & \frac{1}{6} \\ 0 & 1 & 0 & \frac{1}{3} \\ 0 & 0 & 1 & \frac{1}{2} \\ 0 & 0 & 0 & 0 \end{bmatrix}$

9. $\begin{bmatrix} 1 & 1 & 1 \\ 0 & 0 & 0 \\ 0 & 0 & 0 \end{bmatrix}$ **11.** $\begin{bmatrix} 1 & 0 & \frac{17}{64} & \frac{18}{64} \\ 0 & 1 & \frac{47}{64} & \frac{46}{64} \\ 0 & 0 & 0 & 0 \\ 0 & 0 & 0 & 0 \end{bmatrix}$

13. Expected 4 times in first nonabsorbing state if initially in first nonabsorbing state. Expected 2 times in second nonabsorbing state if initially in first nonabsorbing state. Expected $3\frac{1}{3}$ times in first nonabsorbing state if initially in second nonabsorbing state. Expected $3\frac{1}{3}$ times in second nonabsorbing state if initially in second nonabsorbing state.

15. Expected $\frac{45}{32}$ times in first nonabsorbing state if initially in first nonabsorbing state. Expected $\frac{5}{8}$ times in second nonabsorbing state if initially in first nonabsorbing state. Expected $\frac{5}{16}$ times in first nonabsorbing state if initially in second nonabsorbing state. Expected $\frac{5}{4}$ times in second nonabsorbing state if initially in second nonabsorbing state.

17. 6 if initially in first nonabsorbing state. $\frac{20}{3}$ if initially in second nonabsorbing state.

19. $\frac{65}{32}$ if initially in first nonabsorbing state. $\frac{25}{16}$ if initially in second nonabsorbing state.

21. (a) $\begin{bmatrix} 1 & .25 \\ 0 & .75 \end{bmatrix}$ (b) State 1 is absorbing and one can go from state 2 to state 1.

(c) $\begin{bmatrix} 1 & 1 \\ 0 & 0 \end{bmatrix}$; The second 1 in the first row indicates that the probability is 1 that the task will be learned in the long run.

23. The probability is $\frac{15}{19}$ that the person goes from \$1 to \$0. The probability is $\frac{9}{19}$ that the person goes from \$2 to \$0. The probability is $\frac{4}{19}$ that the person goes from \$1 to \$3. The probability is $\frac{10}{19}$ that the person goes from \$2 to \$3.

25. (a) $\begin{bmatrix} 1 & 0 & \frac{1}{6} & \frac{1}{6} \\ 0 & 1 & 0 & \frac{2}{3} \\ 0 & 0 & \frac{1}{6} & 0 \\ 0 & 0 & \frac{2}{3} & \frac{1}{6} \end{bmatrix}$ (b) The first two states, completion and scrap, are absorbing.

(c) The probability is $\frac{16}{25}$ of going from first manufacturing state to completion. The probability is $\frac{4}{5}$ of going from second manufacturing state to completion. The probability is $\frac{9}{25}$ of going from first manufacturing state to scrap. The probability is $\frac{1}{5}$ of going from second manufacturing state to scrap.

27. Probability of $\frac{5}{6}$ of < 30 becoming paid. Probability of $\frac{2}{3}$ of < 60 becoming paid. Probability of $\frac{1}{6}$ of < 30 becoming bad. Probability of $\frac{1}{3}$ of < 60 becoming bad.

29. Average $\frac{6}{5}$ in manufacturing state 1 if initially in state 1. Average 0 in manufacturing state 1 if initially in state 2. Average $\frac{24}{25}$ in manufacturing state 2 if initially in state 1. Average $\frac{6}{5}$ in manufacturing state 2 if initially in state 2.

31. 3 months for both.   **33.** 4 for room 1 and 3 for room 2.

Chapter 9 Review

1. None   **2.** Regular stochastic   **3.** Regular stochastic

4. Nonabsorbing stochastic   **5.** Nonabsorbing stochastic   **6.** Absorbing stochastic

7. $A^2 = \begin{bmatrix} .55 & .30 \\ .45 & .70 \end{bmatrix}$ $A^4 = \begin{bmatrix} .4375 & .3750 \\ .5625 & .6250 \end{bmatrix}$

The elements in A^2 and A^4 represent the probabilities of going from one state to another in two stages and four stages, respectively.

8. $A^2 = \begin{bmatrix} .19 & .16 & .16 \\ .31 & .34 & .31 \\ .50 & .50 & .53 \end{bmatrix}$ $A^4 = \begin{bmatrix} .1657 & .1648 & .1648 \\ .3193 & .3202 & .3193 \\ .5150 & .5150 & .5159 \end{bmatrix}$

The elements in A^2 and A^4 represent the probabilities of going from one state to another in two stages and four stages, respectively.   **9.** $\begin{bmatrix} .40 \\ .60 \end{bmatrix}$ **10.** $\begin{bmatrix} \frac{16}{97} \\ \frac{31}{97} \\ \frac{50}{97} \end{bmatrix}$

11. $\begin{bmatrix} 1 & 0 & .4 \\ 0 & 1 & .6 \\ 0 & 0 & 0 \end{bmatrix}$ **12.** $\begin{bmatrix} 1 & 1 & 1 \\ 0 & 0 & 0 \\ 0 & 0 & 0 \end{bmatrix}$

13. 40% in the first location and 60% in the second.

14. In the long run $\frac{9}{35}$ in the upper class, $\frac{19}{35}$ in middle class, and $\frac{1}{5}$ in lower class.

15. 50% of the mechanics leave as master mechanics and 50% leave as mechanics.

16. 5   **17.** 7.5

Section 10.1

1. Value is 2. Row picks second row. Column picks first column.

3. Value is -2. Row picks second row. Column picks second column.

5. Value is 5. Row picks third row. Column picks first column.

7. Value is -1. Row picks second row. Column picks second or third column.

9. No   **11.** Value is 0. Row picks third row. Column picks second column.

13. No matter what a is, 1 is a saddle point.

15. $\begin{matrix} p \\ n \\ d \end{matrix} \begin{bmatrix} 0 & 5 & 10 \\ -5 & 0 & 10 \\ -10 & -10 & 0 \end{bmatrix}$; Yes with value 0. Each player plays a penny.

17. $\begin{matrix} p \\ n \\ d \end{matrix} \begin{bmatrix} 1 & -4 & -9 \\ -4 & 5 & -5 \\ -9 & -5 & 10 \end{bmatrix}$; No.

Section 10.2

1. .625   **3.** $-.52$ **5.** 2.1   **7.** 1.56

9. $v = 0, p^T = [.50 \quad .50], q^T = [\frac{2}{3} \quad \frac{1}{3}]$

11. $v = .50, p^T = [.50 \quad .50], q^T = [.75 \quad .25]$

13. $v = .50, p^T = [.75 \quad .25], q^T = [.50 \quad 0 \quad .50]$

15. $v = -.25, p^T = [.75 \quad .25 \quad 0], q^T = [.25 \quad .75]$

17. $v = .50, p^T = [0 \quad .75 \quad .25], q^T = [0 \quad .50 \quad .50]$

19. $v = -\frac{4}{7}, p^T = [\frac{6}{7} \quad \frac{1}{7}]$ **21.** $v = \frac{5}{3}, q^T = [\frac{1}{3} \quad \frac{2}{3}]$

23. $v = \frac{19}{30}$. The first should have a 20% sale $\frac{1}{6}$ of the time and a 30% sale $\frac{5}{6}$ of the time. The second should have a 20% sale $\frac{2}{3}$ of the time and a 30% sale $\frac{1}{3}$ of the time.
25. $v = .50$. Column should advertise one half the time on radio and one half the time in the newspaper. Row should advertise three quarters of the time on TV and one quarter of the time in the newspaper.

Section 10.3

1. $v = -.50, p^T = [\frac{5}{8} \quad \frac{3}{8}], q^T = [.50 \quad .50]$

3. $v = .60, p^T = [\frac{2}{5} \quad \frac{3}{5}], q^T = [\frac{4}{5} \quad \frac{1}{5}]$ **5.** $v = -\frac{1}{3}, p^T = [\frac{1}{3} \quad \frac{2}{3}], q^T = [\frac{1}{3} \quad \frac{2}{3}]$

7. $v = -.25, p^T = [.75 \quad .25], q^T = [.25 \quad .75]$ **9.** $v = -\frac{4}{7}, p^T = [\frac{6}{7} \quad \frac{1}{7}]$.

11. $v = \frac{5}{3}, q^T = [\frac{1}{3} \quad \frac{2}{3}]$ **13.** $v = -.50, p^T = [\frac{5}{8} \quad \frac{3}{8}], q^T = [.50 \quad .50]$

15. $v = .60, p^T = [\frac{2}{5} \quad \frac{3}{5}], q^T = [\frac{4}{5} \quad \frac{1}{5}]$

17. Solving the maximum problem for column (after adding 6 to the payoff matrix) gives the following last line in the final tableau. (Your entry below x_1 may differ.)

x_1	x_2	x_3	s_1	s_2	z	
$\frac{17}{38}$	0	0	$\frac{6}{38}$	$\frac{1}{38}$	1	$\frac{7}{38}$

The value is then $\dfrac{1}{\frac{7}{38}} - 6 = -\frac{4}{7}$. The strategy for column is

$$q^T = \frac{1}{\frac{7}{38}} [0 \quad \tfrac{3}{38} \quad \tfrac{4}{38}] = [0 \quad \tfrac{3}{7} \quad \tfrac{4}{7}]$$

By using the dual theory we can read off the strategy for row by using the numbers under the slack variable and obtain

$$p^T = \frac{1}{\frac{7}{38}} [\tfrac{6}{38} \quad \tfrac{1}{38}] = [\tfrac{6}{7} \quad \tfrac{1}{7}]$$

This agrees with the answer found in Example 9.

19. Solving the minimum problem for row (after adding 3 to the payoff matrix) gives the following last line in the final tableau. (Your entry below y_1 may differ.)

y_1	y_2	y_3	s_1	s_2	z	
$\frac{5}{14}$	0	0	$\frac{1}{14}$	$\frac{2}{14}$	1	$-\frac{3}{14}$

The value is then $\dfrac{1}{\frac{3}{14}} - 3 = \frac{5}{3}$. The strategy for row is

$$p^T = \frac{1}{\frac{3}{14}} [0 \quad \tfrac{2}{14} \quad \tfrac{1}{14}] = [0 \quad \tfrac{2}{3} \quad \tfrac{1}{3}]$$

By using dual theory we can read off the strategy for column by using the numbers under the slack variable and obtain

$$q^T = \frac{1}{\frac{3}{14}} [\tfrac{1}{14} \quad \tfrac{2}{14}] = [\tfrac{1}{3} \quad \tfrac{2}{3}]$$

This agrees with the answer found in Example 11.

21. Each picks even and odd half the time.

23. $v = \frac{19}{30}$. The first should have a 20% sale $\frac{1}{6}$ of the time and a 30% sale $\frac{5}{6}$ of the time. The second should have a 20% sale of $\frac{2}{3}$ of the time and a 30% sale $\frac{1}{3}$ of the time.

25. $v = .50$. Column should advertise $\frac{1}{2}$ the time on radio and $\frac{1}{2}$ the time in the newspaper. Row should advertise $\frac{3}{4}$ of the time on TV and $\frac{1}{4}$ of the time in the newspaper.

Chapter 10 Review

1. $v = 0, p^T = [1 \quad 0], q^T = [0 \quad 0 \quad 1]$

2. $v = -1, p^T = [1, 0], q^T = [0 \quad 1 \quad 0 \quad 0]$

3. $v = \frac{5}{3}, p^T = [0 \quad \frac{2}{3} \quad \frac{1}{3}], q^T = [\frac{2}{3} \quad 0 \quad \frac{1}{3}]$

4. $v = \frac{3}{2}, p^T = [\frac{1}{2} \quad 0 \quad \frac{1}{2}], q^T = [0 \quad \frac{1}{2} \quad \frac{1}{2}]$ **5.** $v = \frac{1}{5}, p^T = [\frac{2}{5} \quad \frac{3}{5}], q^T = [\frac{2}{5} \quad \frac{3}{5}]$

6. $v = \frac{3}{4}, p^T = [\frac{3}{4} \quad \frac{1}{4}], q^T = [\frac{5}{8} \quad \frac{3}{8}]$ **7.** $v = -\frac{1}{4}, p^T = [\frac{3}{4} \quad \frac{1}{4}], q^T = [\frac{1}{4} \quad \frac{3}{4} \quad 0]$

8. $v = \frac{1}{2}, p^T = [\frac{1}{6} \quad \frac{5}{6}], q^T = [\frac{1}{2} \quad \frac{1}{2} \quad 0]$ **9.** $v = -\frac{1}{4}, p^T = [\frac{3}{4} \quad \frac{1}{4}]$

10. $v = \frac{25}{6}, q^T = [\frac{1}{6} \quad \frac{5}{6}]$ **11.** $v = -\frac{1}{4}, q^T = [0 \quad \frac{3}{4} \quad \frac{1}{4}]$

12. $v = \frac{25}{6}, p^T = [0 \quad \frac{5}{6} \quad \frac{1}{6}]$,   **13.** $v = \frac{1}{3}, p^T = [\frac{1}{3} \quad \frac{1}{3} \quad \frac{1}{3}], q^T = [\frac{1}{3} \quad \frac{1}{3} \quad \frac{1}{3}]$

14. If $a \geq b$ or $a \geq c$, then a is a saddle. If $a < b \leq c$, then b is a saddle. If $a < c \leq b$, then c is a saddle.

Section 11.1

1. (a) $106.09 (b) $106.14 (c) $106.17 (d) $106.18 (e) $106.18   **3.** $2158.92

5. $1814.02   **7.** $2457.69   **9.** 10.24 years   **11.** 12.25%   **13.** 6.96%

15. Since $P_N = 2P_0$,

$$P_N = (1 + k)^N P_0$$

$$2P_0 = (1 + k)^N P_0$$

$$2 = (1 + k)^N$$

$$\log 2 = \log(1 + k)^N$$

$$= N \log(1 + k)$$

$$N = \frac{\log 2}{\log(1 + k)}$$

17. .0306   **19.** 23.4 years   **21.** .0120   **23.** 4.14 billion

25. 5 years   **27.** 7.9 years

29. $\frac{1}{2} A_0 = (1 - k)^N A_0$

$$\frac{1}{2} = (1 - k)^N$$

$$\sqrt[N]{\frac{1}{2}} = 1 - k$$

31. 18,508 years old   **33.** 35 days   **35.** 23.78 units

37. $P_{n+1} = .99P_n$, 1.47 million   **39.** $A_{n+1} = .99A_n$, 109.4 milligrams

Section 11.2

1. $2200   **3.** $399,270   **5.** $149,749   **7.** $1068.08
9. $16,555.75   **11.** $1216.13   **13.** $210.67
15. First, since it would take $67,100.81 deposited now to yield annual payments of $10,000 for 10 years. Only $63,817.96 need be deposited now to yield annual payments of $6500 for 20 years.
17. Pick the lump sum of $350,000, since it would take only $320,882.88 to make annual payments of $50,000 for 10 years with the given interest rate.
19. (a) $A_{n+1} = .99A_n + 10$ (b) 453 milligrams
21. (a) $P_{n+1} = (1 - k)P_n + kT$, where T is the total number of employees and k is the constant of proportionality. (b) 1615
23. (a) $T_{n+1} = (1 - k)T_n + kC$, where k is the constant of proportionality. (b) 4.6 minutes

Section 11.3

1. 10. Repeller

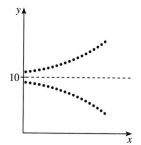

3. 200. Attractor

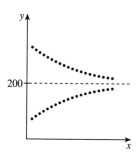

5. 100,000. Attractor

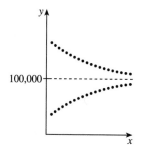

7.

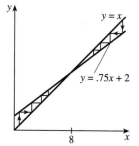

9.

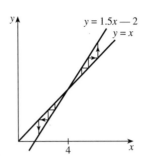

11.

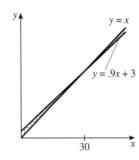

13. $60,000   **15.** $80,000

17. 25 million, less than 25 million, greater than 25 million

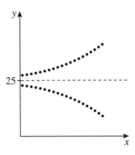

19. For this lake, no matter what the initial amount of pollutants, the total approaches 200 tons in the long-term.

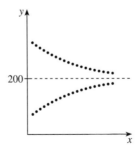

21. Since $a = 1 + k$ and $b = -H$, $\dfrac{b}{1-a} = \dfrac{H}{k}$. Thus

$$P_n = \frac{b}{1-a} + \left(P_0 - \frac{b}{1-a}\right)a^n = \frac{H}{k} + \left(P_0 - \frac{H}{k}\right)(1+k)^n$$

23. 4500 per year

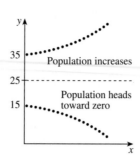

25. When $P_0 = 35$ million tons, the biomass increases without bound. If $P_0 = 15$, the biomass becomes zero in only a few years.

Section 11.4

1. (a) $P_n = 10$. (b) Repeller. (c) Monotonic

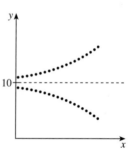

3. (a) $P_n = 200$. (b) Attractor. (c) Monotonic

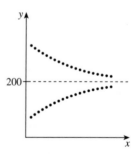

5. (a) $P_n = 50.25$. (b) Attractor. (c) Oscillatory

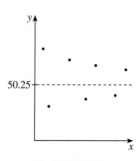

7. (a) $P_n = 49.975$. (b) Repeller. (c) Oscillatory

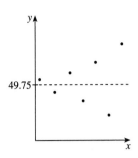

9. (a) $P_n = 50$. (b) Neither. (c) 2-cycle

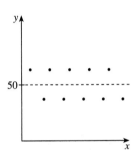

11.

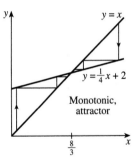

Monotonic, repeller

$y = 3x - 6$

$y = x$

3

13.

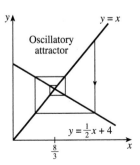

$y = x$

$y = \frac{1}{4}x + 2$

Monotonic, attractor

$\frac{8}{3}$

15.

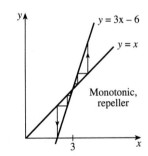

$y = x$

Oscillatory repeller

$y = -2x + 10$

$\frac{10}{3}$

17.

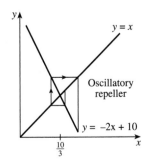

$y = x$

Oscillatory attractor

$y = \frac{1}{2}x + 4$

$\frac{8}{3}$

19.

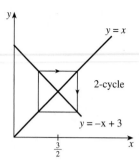

21.

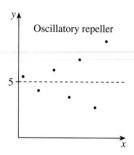

23.

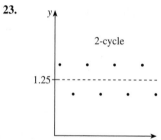

25.

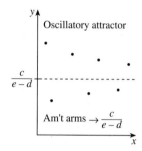

Section 11.5

1.

n	P_n	$P_n - P_{n-1}$	P_n	n	P_n	$P_n - P_{n-1}$	P_n
0	1.000		7.500	8	3.187	.305	5.151
1	1.200	.200	6.563	9	3.476	.289	5.112
2	1.428	.228	6.050	10	3.741	.265	5.083
3	1.683	.255	5.732	11	3.977	.235	5.062
4	1.962	.279	5.522	12	4.180	.202	5.046
5	2.260	.298	5.378	15	4.607	.114	5.019
6	2.570	.310	5.276	20	4.899	.033	5.005
7	2.882	.312	5.204	25	4.975	.008	5.001

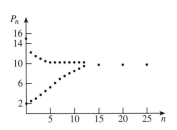

3.

n	P_n	$P_n - P_{n-1}$	P_n	n	P_n	$P_n - P_{n-1}$	P_n
0	2.000		15.000	8	7.791	.686	10.106
1	2.533	.533	12.500	9	8.365	.574	10.070
2	3.164	.631	11.458	10	8.821	.456	10.047
3	3.885	.721	10.901	11	9.167	.347	10.031
4	4.677	.792	10.574	12	9.422	.254	10.021
5	5.507	.830	10.372	15	9.818	.087	10.006
6	6.331	.825	10.243	20	9.975	.012	10.001
7	7.106	.774	10.160	25	9.997	.002	10.000

5.

n	P_n	$P_n - P_{n-1}$	P_n	n	P_n	$P_n - P_{n-1}$	P_n
0	1.000		7.500	8	1.866	.126	5.717
1	1.089	.089	7.083	9	1.996	.130	5.626
2	1.184	.095	6.755	10	2.129	.133	5.548
3	1.284	.100	6.492	20	3.471	.123	5.155
4	1.390	.106	6.277	30	4.387	.065	5.047
5	1.501	.112	6.099	40	4.792	.025	5.014
6	1.618	.117	5.950	50	4.934	.008	5.004
7	1.740	.122	5.824	60	4.979	.003	5.001

7. Each of the equations has the unstable constant solution $P_n = 0$, and the respective stable constant solutions $P_n = 5, 10, 5$.   **9.** 369 million   **11.** 2559 million
13. $P_{n+1} = 1.031 P_n (1 - .0000815 P_n)$ **15.** $P_{n+1} = 1.03 P_n (1 - .0000114 P_n)$

13.

n	P_n	r_n in percent
0	250.0	
1	252.5	.100
2	255.0	.979
3	257.4	.958
4	259.8	.937
5	262.2	.917
6	264.6	.897
7	266.9	.877
8	269.2	.858
9	271.4	.839
10	273.6	.820

15.

n	P_n	r_n in percent
0	853.0	
1	870.1	2.000
2	887.3	1.980
3	904.7	1.960
4	922.2	1.939
5	939.9	1.919
6	957.8	1.898
7	975.7	1.877
8	993.8	1.856
9	1012.1	1.835
10	1030.4	1.814

17.

n	$P_n, H = .1$	$P_n, H = .3$
0	1.000	1.000
1	0.900	0.700
2	0.872	0.568
3	0.861	0.464
4	0.857	0.363
5	0.855	0.248
6	0.854	0.098
7	0.854	$-.132$

In the case $H = .1$, the population heads for about 854,000. In the case $H = .3$, the population becomes zero by the seventh year.

19.

n	P_n	n	P_n
0	1.000	8	3.187
1	1.200	9	3.476
2	1.428	10	3.741
3	1.683	15	4.607
4	1.962	20	4.899
5	2.260	25	4.975
6	2.570	30	4.994
7	2.882		

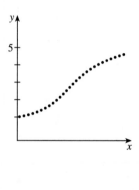

The profits approach $5 million.

21.

n	P_n	n	P_n
0	.100	8	.335
1	.118	9	.379
2	.139	10	.426
3	.163	15	.671
4	.190	20	.854
5	.221	25	.946
6	.255	30	.981
7	.293	35	.994

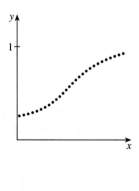

Eventually everyone adopts the innovation.

23.

n	P_n	n	P_n
0	1	7	120
1	2	8	226
2	4	9	401
3	8	10	641
4	16	11	871
5	32	12	983
6	62	13	1000

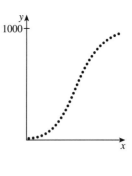

Eventually everyone is infected.

25. 100,000, 4000   **27.** 65 million kilograms, 84.825 million kilograms

Section 11.6

1. Solutions approach .655 (to three decimal places).

3. Solutions approach .664 (to three decimal places).

5. Solutions approach the 2-cycle .608, .722 (to three decimal places).

7. Solutions approach the 2-cycle .633, .699 (to three decimal places).

9. Solutions approach the 2-cycle .452, .842 (to three decimal places).

11. Solutions approach the 2-cycle .442, .848 (to three decimal places).

13. Solutions approach the 4-cycle .512, .879, .373, .823 (to three decimal places).

15. Solutions approach the 4-cycle .467, .861, .413, .839 (to three decimal places).

Exercise 17 Table

n	$P_n, P_0 = .3$	$P_n, P_0 = .301$	n	$P_n, P_0 = .3$	$P_n, P_0 = .301$
0	.777	.778	9	.746	.689
1	.641	.638	10	.702	.793
2	.851	.854	11	.774	.606
3	.468	.460	12	.646	.883
4	.921	.919	13	.846	.382
5	.268	.275	14	.482	.873
6	.726	.738	15	.924	.409
7	.735	.716	16	.260	.895
8	.720	.753			

19. .4794, .8236.   **21.** .558, .765.   **23.** .465, .833.

Chapter 11 Review

1. (a) $1071.23 (b) $1071.86 (c) $1072.29 (d) $1072.46 (e) $1072.50
2. $3306.92   3. 14.2 years   4. 5.95%
5.

$$3P_0 = (1 + k)^N P_0$$

$$3 = (1 + k)^N$$

$$\log 3 = N \log(1 + k)$$

$$N = \frac{\log 3}{\log(1 + k)}$$

6.

$$3P_0 = (1 + k)^N P_0$$

$$3 = (1 + k)^N$$

$$\sqrt[N]{3} = 1 + k$$

7. 1.59%   8. .045   9. 2312 years
10. (a) $V_{n+1} = 1.04V_n$, $V_0 = 100{,}000$ (b) $480,102
11. (a) $P_{n+1} = (.8)P_n$ (b) 9.8   12. $2400   13. $1470.63
14. (a) $738.99 (b) $121,697
15. Take the $500,000 since it only takes $456,427.28 to generate annual payments of $50,000 for 20 years at the given interest rate.
16. $P_{n+1} = 1.02P_n - 4000$, 200,000

17. (a) $6\frac{2}{3}$ (b) repeller (c) monotonic

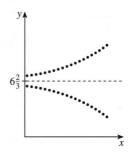

18. (a) 25 (b) attractor (c) monotonic

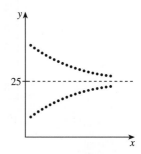

19. (a) 10.5 (b) attractor (c) oscillatory

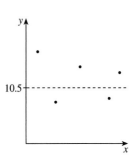

20. (a) 4.76 (b) repeller (c) oscillatory

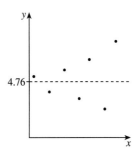

21. (a) 50 (b) neither (c) 2-cycle

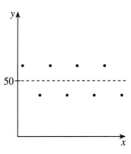

22.

(17)

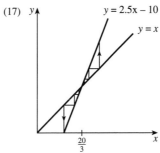

$y = 2.5x - 10$
$y = x$
$\frac{20}{3}$

(18)

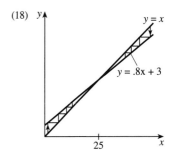

$y = x$
$y = .8x + 3$
25

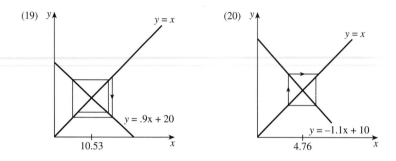

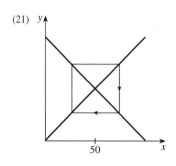

23. $137,142.86   **24.** $83,333.33

25. If initially less than 1000, population will eventually go to zero. If initially 1000, population stays constant. If initially greater than 1000, population will grow.

26. (a) Less than $200,000 (b) equal to $200,000 (c) greater than $200,000.

27. 56.8 million kilograms

Exercise 28 Table

n	P_n	$P_n - P_{n-1}$	P_n	n	P_n	$P_n - P_{n-1}$	P_n
0	.0200		.1500	7	.0905	.0068	.1002
1	.0280	.0080	.1125	8	.0948	.0043	.1001
2	.0381	.0101	.1055	9	.0973	.0025	.1000
3	.0500	.0118	.1026	10	.0986	.0013	.1000
4	.0624	.0125	.1013	11	.0993	.0007	.1000
5	.0741	.0117	.1006	12	.0996	.0004	.1000
6	.0837	.0096	.1003	13	.0998	.0002	.1000

Exercise 29 Table

n	P_n	$P_n - P_{n-1}$	P_n	n	P_n	$P_n - P_{n-1}$	P_n
0	1.600		12.000	7	4.611	.500	8.326
1	1.920	.320	10.500	8	5.010	.488	8.241
2	2.285	.365	9.680	9	5.562	.462	8.179
3	2.693	.408	9.172	10	5.986	.424	8.133
4	3.139	.447	8.836	15	7.370	.182	8.031
5	3.616	.477	8.605	20	7.838	.052	8.007
6	4.112	.495	8.442	25	7.961	.013	8.002

30. In each case the zero solution is not stable and the other constant solution is.

31. 2.43 billion

32. $P_{n+1} = 1.03P_n \left(1 - \dfrac{.03}{1.03(2.43)} P_n \right)$. The population in billions for the next 10 years is: 1.153, 1.171, 1.189, 1.208, 1.226, 1.244, 1.262, 1.280, 1.299, 1.317. The percent change per year is: 1.60, 1.57, 1.55, 1.53, 1.51, 1.48, 1.46, 1.44, 1.42, 1.39.

33. The profits are heading for 12.5 million.

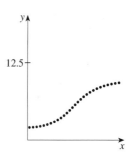

34. 447

35. (a) 157,000 metric tons (b) 3925 metric tons.   **36.** .513, .799

37. .495, .812

Section 12.1

1. Statement   **3.** Statement   **5.** Statement   **7.** Not a statement

9. Statement   **11.** Not a statement   **13.** Statement

15. (a) George Washington was not the third President of the United States. (b) George Washington was the third President of the United States, and Austin is the capital of Texas. (c) George Washington was the third President of the United States, or Austin is the capital of Texas. (d) George Washington was not the third President of the United States, and Austin is the capital of Texas. (e) George Washington was the third President of the United States, or Austin is not the capital of Texas. (f) It is not true that George Washington was the third President of the United States and that Austin is the capital of Texas.

17. (a) George Washington did not own over 100,000 acres of property. The Exxon Valdez was not a luxury liner. (b) George Washington owned over 100,000 acres of property, or the Exxon Valdez was a luxury liner. (c) George Washington owned over 100,000 acres of property, and the Exxon Valdez was a luxury liner.

19. (a) $\sim q$ (b) $p \wedge \sim q$ (c) $p \vee q$ (d) $(\sim p) \vee (\sim q)$

Section 12.2

1.

p	q	$\sim q$	$p \wedge \sim q$
T	T	F	F
T	F	T	T
F	T	F	F
F	F	T	F

3.

p	$\sim p$	$\sim(\sim p)$
T	F	T
F	T	F

5.

p	q	$\sim q$	$p \wedge \sim q$	$(p \wedge \sim q) \vee q$
T	T	F	F	T
T	F	T	T	T
F	T	F	F	T
F	F	T	F	F

7.

p	q	$\sim p$	$p \wedge q$	$\sim p \vee (p \wedge q)$
T	T	F	T	T
T	F	F	F	F
F	T	T	F	T
F	F	T	F	T

9.

p	q	$p \vee q$	$p \wedge q$	$(p \vee q) \wedge (p \wedge q)$
T	T	T	T	T
T	F	T	F	F
F	T	T	F	F
F	F	F	F	F

11.

p	q	$\sim q$	$\sim p$	$p \vee \sim q$	$\sim p \wedge q$	$(p \vee \sim q) \vee (\sim p \wedge q)$
T	T	F	F	T	F	T
T	F	T	F	T	F	T
F	T	F	T	F	T	T
F	F	T	T	T	F	T

Tautology

13.

p	q	r	$p \vee q$	$(p \vee q) \wedge r$
T	T	T	T	T
T	T	F	T	F
T	F	T	T	T
T	F	F	T	F
F	T	T	T	T
F	T	F	T	F
F	F	T	F	F
F	F	F	F	F

15.

p	q	r	$p \wedge q$	$(p \wedge q) \wedge r$	$\sim[(p \wedge q) \wedge r]$
T	T	T	T	T	F
T	T	F	T	F	T
T	F	T	F	F	T
T	F	F	F	F	T
F	T	T	F	F	T
F	T	F	F	F	T
F	F	T	F	F	T
F	F	F	F	F	T

17.

p	q	r	$p \vee q$	$q \wedge r$	$(p \vee q) \vee (q \wedge r)$
T	T	T	T	T	T
T	T	F	T	F	T
T	F	T	T	F	T
T	F	F	T	F	T
F	T	T	T	T	T
F	T	F	T	F	T
F	F	T	F	F	F
F	F	F	F	F	F

19.

p	q	r	$\sim q$	$p \vee \sim q$	$\sim q \wedge r$	$(p \vee \sim q) \vee (\sim q \wedge r)$
T	T	T	F	T	F	T
T	T	F	F	T	F	T
T	F	T	T	T	T	T
T	F	F	T	T	F	T
F	T	T	F	F	F	F
F	T	F	F	F	F	F
F	F	T	T	T	T	T
F	F	F	T	T	F	T

21. (a) True (b) False (c) True (d) True (e) False

23. (a) True (b) False (c) False (d) True

Section 12.3

1. (a) If interest rates drop, then the stock market goes up. (b) Contrapositive: If the stock market does not go up, then interest rates do not drop. $\sim q \rightarrow \sim p$ (c) Inverse: If interest rates do not drop, then the stock market does not go up. $\sim p \rightarrow \sim q$ (d) Converse: If the stock market goes up, then interest rates drop. $q \rightarrow p$

3. (a) If the stock market does not go up, then interest rates do not drop. (b) Contrapositive: If interest rates drop, then the stock market goes up. $p \rightarrow q$ (c) Inverse: If the stock market goes up, then interest rates drop. $q \rightarrow p$ (d) Converse: If interest rates do not drop, then the stock market does not go up. $\sim p \rightarrow \sim q$

5. (a) If interest rates do not drop, then the stock market does not go up. $\sim b \rightarrow \sim q$ (b) Contrapositive: If the stock market goes up, then interest rates drop. $q \rightarrow p$ (c) Inverse: If interest rates drop, then the stock market goes up. $p \rightarrow q$ (d) Converse: If the stock market does not go up, then interest rates do not drop. $\sim q \rightarrow \sim p$

7. $p \rightarrow q$ **9.** $p \rightarrow q$ **11.** $q \leftrightarrow p$ **13.** $\sim q \rightarrow \sim p$ **15.** True

17. False **19.** True **21.** False

23.

p	q	$p \to q$	$(p \to q) \wedge p$
T	T	T	T
T	F	F	F
F	T	T	F
F	F	T	F

25.

p	q	$p \to q$	$q \to p$	$(p \to q) \vee (q \to p)$
T	T	T	T	T
T	F	F	T	T
F	T	T	F	T
F	F	T	T	T

27.

p	q	$p \vee q$	$\sim(p \vee q)$	$\sim p$	$\sim q$	$\sim p \wedge \sim q$	$[\sim(p \vee q)] \leftrightarrow [\sim p \wedge \sim q]$
T	T	T	F	F	F	F	T
T	F	T	F	F	T	F	T
F	T	T	F	T	F	F	T
F	F	F	T	T	T	T	T

29.

p	q	r	$p \to q$	$(p \to q) \to r$
T	T	T	T	T
T	T	F	T	F
T	F	T	F	T
T	F	F	F	T
F	T	T	T	T
F	T	F	T	F
F	F	T	T	T
F	F	F	T	F

31.

p	q	r	$q \wedge r$	$p \vee (q \wedge r)$	$p \vee q$	$p \vee r$	$(p \vee q) \wedge (p \vee r)$	$[p \vee (q \wedge r)] \leftrightarrow [(p \vee q) \wedge (p \vee r)]$
T	T	T	T	T	T	T	T	T
T	T	F	F	T	T	T	T	T
T	F	T	F	T	T	T	T	T
T	F	F	F	T	T	T	T	T
F	T	T	T	T	T	T	T	T
F	T	F	F	F	T	F	F	T
F	F	T	F	F	F	T	F	T
F	F	F	F	F	F	F	F	T

33.

p	q	$p \rightarrow q$	$q \rightarrow p$	$(p \rightarrow q) \leftrightarrow (q \rightarrow p)$
T	T	T	T	T
T	F	F	T	F
F	T	T	F	F
F	F	T	T	T

35.

p	q	r	$p \vee q$	$(p \vee q) \vee r$
T	T	T	T	T
T	T	F	T	T
T	F	T	T	T
T	F	F	T	T
F	T	T	T	T
F	T	F	T	T
F	F	T	F	T
F	F	F	F	F

37.

p	q	$p \wedge q$	$(p \wedge q) \rightarrow p$
T	T	T	T
T	F	F	T
F	T	F	T
F	F	F	T

Tautology

39.

p	q	$p \rightarrow q$	$p \wedge (p \rightarrow q)$	$[p \wedge (p \rightarrow q)] \rightarrow q$
T	T	T	T	T
T	F	F	F	T
F	T	T	F	T
F	F	T	F	T

Tautology

41.

p	q	$p \wedge q$	$q \rightarrow (p \wedge q)$	$p \rightarrow [q \rightarrow (p \wedge q)]$
T	T	T	T	T
T	F	F	T	T
F	T	F	F	T
F	F	F	T	T

Tautology

43.

p	q	r	$p \rightarrow q$	$p \vee r$	$q \vee r$	$(p \vee r) \rightarrow (q \vee r)$	$[p \rightarrow q] \rightarrow [(p \vee r) \rightarrow (q \vee r)]$
T	T	T	T	T	T	T	T
T	T	F	T	T	T	T	T
T	F	T	F	T	T	T	T
T	F	F	F	T	F	F	T
F	T	T	T	T	T	T	T
F	T	F	T	F	T	T	T
F	F	T	T	T	T	T	T
F	F	F	T	F	F	T	T

Tautology

45. Let p be the statement "n^2 is even" and q the statement "n is even." To prove $p \rightarrow q$, we will establish that $\sim q \rightarrow \sim p$. Assume $\sim q$, that is, assume that n is not even. Then n is not divisible by 2 and hence $nn = n^2$ is not divisible by 2. This means that n^2 is not even, that is, $\sim p$ is true.

Section 12.4

1.

p	$p \wedge p$	$(p \wedge p) \rightarrow p$
T	T	T
F	F	T

3.

p	t	$p \vee t$	$(p \vee t) \rightarrow t$
T	T	T	T
F	T	T	T

5.

p	c	$p \wedge c$	$(p \wedge c) \rightarrow c$
T	F	F	T
F	F	F	T

7.

p	q	r	$p \wedge q$	$(p \wedge q) \wedge r$	$q \wedge r$	$p \wedge (q \wedge r)$	$[(p \wedge q) \wedge r] \rightarrow [p \wedge (q \wedge r)]$
T	T	T	T	T	T	T	T
T	T	F	T	F	F	F	T
T	F	T	F	F	F	F	T
T	F	F	F	F	F	F	T
F	T	T	F	F	T	F	T
F	T	F	F	F	F	F	T
F	F	T	F	F	F	F	T
F	F	F	F	F	F	F	T

9.

p	q	$\sim p$	$p \vee q$	$\sim p \wedge (p \vee q)$	$[\sim p \wedge (p \vee q)] \rightarrow q$
T	T	F	T	F	T
T	F	F	T	F	T
F	T	T	T	T	T
F	F	T	F	F	T

11.

p	q	r	$p \leftrightarrow q$	$q \leftrightarrow r$	$(p \leftrightarrow q) \wedge (q \leftrightarrow r)$	$p \leftrightarrow r$	$[(p \leftrightarrow q) \wedge (q \leftrightarrow r)] \rightarrow [p \leftrightarrow r]$
T	T	T	T	T	T	T	T
T	T	F	T	F	F	F	T
T	F	T	F	F	F	T	T
T	F	F	F	T	F	F	T
F	T	T	F	T	F	F	T
F	T	F	F	F	F	T	T
F	F	T	T	F	F	F	T
F	F	F	T	T	T	T	T

13.

p	q	r	$q \vee r$	$p \wedge (q \vee r)$	$p \wedge q$	$p \wedge r$	$(p \wedge q) \vee (p \wedge r)$	$[p \wedge (q \vee r)] \leftrightarrow [(p \wedge q) \vee (p \wedge r)]$
T	T	T	T	T	T	T	T	T
T	T	F	T	T	T	F	T	T
T	F	T	T	T	F	T	T	T
T	F	F	F	F	F	F	F	T
F	T	T	T	F	F	F	F	T
F	T	F	T	F	F	F	F	T
F	F	T	T	F	F	F	F	T
F	F	F	F	F	F	F	F	T

15. $(p \vee q) \vee {\sim}q \Leftrightarrow p \vee (q \vee {\sim}q)$ Law 7b

$\Leftrightarrow p \vee t$ Law 11a

$\Leftrightarrow t$ Law 12a

17. $p \wedge q \Leftrightarrow {\sim}[{\sim}(p \wedge q)]$ Law 1

$\Leftrightarrow {\sim}[{\sim}p \vee {\sim}q]$ Law 9b

19. $({\sim}p \wedge {\sim}q) \wedge {\sim}r \Leftrightarrow [{\sim}(p \vee q)] \wedge {\sim}r$ Law 9a

$\Leftrightarrow {\sim}[(p \vee q) \vee r]$ Law 9a

$\Leftrightarrow {\sim}[p \vee (q \vee r)]$ Law 7b

21. $p \wedge q \Leftrightarrow {\sim}[{\sim}(p \wedge q)]$ Law 1   **23.** $p \rightarrow q \Leftrightarrow {\sim}p \vee q$ Law 3a

$\Leftrightarrow {\sim}[{\sim}p \vee {\sim}q]$ Law 9b

25. $p \vee q \Leftrightarrow {\sim}[{\sim}(p \vee q)]$ Law 1   **27.** $p \rightarrow q \Leftrightarrow {\sim}(p \wedge {\sim}q)$ Law 3b

$\Leftrightarrow {\sim}[{\sim}p \wedge {\sim}q]$ Law 9a

29. Jim does not like Sue and does not like Mary.

31. Sales are not up, or we do not all get raises.

33. ${\sim}(p \rightarrow q) \Leftrightarrow {\sim}[{\sim}(p \wedge {\sim}q)]$ Law 3b

$\Leftrightarrow p \wedge {\sim}q$ Law 1

35. I work hard, and I do not succeed.   **37.** Prices are raised, and sales do not drop.

Section 12.5

1.

p	q	$p \vee q$	${\sim}p$	$(p \vee q) \wedge {\sim}p$	$[(p \vee q) \wedge {\sim}p] \rightarrow q$
T	T	T	F	F	T
T	F	T	F	F	T
F	T	T	T	T	T
F	F	F	T	F	T

Valid

3.

p	q	r	$p \rightarrow q$	$q \rightarrow r$	$(p \rightarrow q) \wedge (q \rightarrow r)$	$p \rightarrow r$	$[(p \rightarrow q) \wedge (q \rightarrow r)] \rightarrow (p \rightarrow r)$
T	T	T	T	T	T	T	T
T	T	F	T	F	F	F	T
T	F	T	F	T	F	T	T
T	F	F	F	T	F	F	T
F	T	T	T	T	T	T	T
F	T	F	T	F	F	T	T
F	F	T	T	T	T	T	T
F	F	F	T	T	T	T	T

Valid

5.

p	q	$p \leftrightarrow q$	$(p \leftrightarrow q) \wedge q$	$[(p \leftrightarrow q) \wedge q] \rightarrow p$
T	T	T	T	T
T	F	F	F	T
F	T	F	F	T
F	F	T	F	T

Valid

7.

p	q	r	$\sim q$	$p \rightarrow \sim q$	$r \rightarrow q$	$(p \rightarrow \sim q) \wedge (r \rightarrow q) \wedge r$	$\sim p$	$[(p \rightarrow \sim q) \wedge (r \rightarrow q) \wedge r] \rightarrow \sim p$
T	T	T	F	F	T	F	F	T
T	T	F	F	F	T	F	F	T
T	F	T	T	T	F	F	F	T
T	F	F	T	T	T	F	F	T
F	T	T	F	T	T	T	T	T
F	T	F	F	T	T	F	T	T
F	F	T	T	T	F	F	T	T
F	F	F	T	T	T	F	T	T

Valid

9.

p	q	r	$p \rightarrow q$	$p \rightarrow r$	$(p \rightarrow q) \wedge (p \rightarrow r)$	$q \wedge r$	$[(p \rightarrow q) \wedge (p \rightarrow r)] \rightarrow (q \wedge r)$
T	T	T	T	T	T	T	T
T	T	F	T	F	F	F	T
T	F	T	F	T	F	F	T
T	F	F	F	F	F	F	T
F	T	T	T	T	T	T	T
F	T	F	T	T	T	F	F
F	F	T	T	T	T	F	F
F	F	F	T	T	T	F	F

Not valid

11. s = "I go out with my sister" $s \vee f \rightarrow \sim d$

f = "I go out with my friend" $\underline{d \qquad}$

d = "I drive my car" $\therefore \sim s$

$((s \vee f) \rightarrow \sim d) \wedge d \Rightarrow \sim(s \vee f)$ Modus tollens

$\Leftrightarrow \sim s \wedge \sim f$ De Morgan

$\Rightarrow \sim s$ Subtraction

Valid.

13. h = "You are happy" $h \leftrightarrow l$

l = "You are healthy" $l \vee s$

s = "You are smart" $\underline{\sim h \qquad}$

$\therefore \sim h \rightarrow \sim s$

In the case that h is false, l false, and s true, $(h \leftrightarrow l) \wedge (l \vee s) \wedge (\sim h)$ is true while $\sim h \rightarrow \sim s$ is false. Thus the argument is not valid.

15. e = "I have an egg for breakfast" $e \wedge j \rightarrow \sim r$

j = "I have orange juice for breakfast" $\sim j \rightarrow \sim g$

r = "I have cereal for breakfast" $\underline{e \wedge g \qquad}$

g = "I have grapefruit juice for breakfast" $\therefore \sim r$

$(e \wedge j \rightarrow \sim r) \wedge (\sim j \rightarrow \sim g) \wedge (e \wedge g) \Leftrightarrow (e \wedge j \rightarrow \sim r) \wedge (\sim j \rightarrow \sim g) \wedge g \wedge e$ Associativity

$\Rightarrow (e \wedge j \rightarrow \sim r) \wedge j \wedge e$ Modus tollens

$\Leftrightarrow (e \wedge j \rightarrow \sim r) \wedge (j \wedge e)$ Associativity

$\Rightarrow \sim r$ Modus ponens

Valid.

17. r = "I clean my room" $r \rightarrow \sim m$

m = "My mother is mad" $\underline{m \qquad}$

$\therefore \sim r$

$(r \rightarrow \sim m) \wedge m \Rightarrow \sim r$ Modus tollens

Valid.

19. b = "Jane eats broccoli" b

s = "Jane eats spinach" s

$$b \rightarrow \sim s$$

$$\therefore s \wedge \sim b$$

$b \wedge s \wedge (b \rightarrow \sim s) \Leftrightarrow s \wedge b \wedge (b \rightarrow \sim s)$ Commutative law

$\Rightarrow s \wedge \sim s$ Modus ponens

$\Leftrightarrow c$ Inverse law

$\Rightarrow c \vee (s \wedge \sim b)$ Addition

$\Leftrightarrow s \wedge \sim b$ Identity law

Valid.

21. b = "I play baseball" $\sim b \rightarrow s$

s = "I play soccer" $\sim s \rightarrow k$

k = "I play basketball" $\therefore \sim s \rightarrow b \vee k$

$(\sim s \rightarrow b) \wedge (\sim s \rightarrow k) \Leftrightarrow (s \vee b) \wedge (s \vee k)$ Implication

$\Leftrightarrow s \vee (b \wedge k)$ Distribution

$\Leftrightarrow \sim s \rightarrow (b \wedge k)$ Implication

Valid.

23. i = "The rate of inflation is increasing" $i \rightarrow g$

g = "The price of gold is increasing" $g \rightarrow \sim b$

b = "The price of bonds is increasing" b

$$\therefore \sim i$$

$(i \rightarrow g) \wedge (g \rightarrow \sim b) \wedge b \Rightarrow (i \rightarrow \sim b) \wedge b$ Hypothetical syllogism

$\Rightarrow \sim i$ Modus tollens

Valid.

25. s = "The price of sugar is rising" $s \rightarrow (d \wedge l)$

d = "The price of candy is rising" $\sim d$

l = "I eat less candy" $\sim l$

$$\therefore \sim s$$

$(s \rightarrow (d \wedge l)) \wedge \sim d \wedge \sim l \Rightarrow (s \rightarrow (d \wedge l)) \wedge (\sim (d \vee l))$ De Morgan

$\Rightarrow \sim s$ Modus tollens

Valid.

27. s = "I go out with my sister" $s \vee f \rightarrow \sim d$

f = "I go out with my friend" d

d = "I drive my car" $\therefore \sim s$

Assume $\sim s$ is false.

Then s is true.

Then $s \lor f$ is true.

If the first hypothesis is false, then we're done.

If the first hypothesis is true, then $\sim d$ is true.

Then the second hypothesis is false.

29. $i = $ "The rate of inflation is increasing" $i \to g$

      $g = $ "The price of gold is increasing" $g \to \sim b$

      $b = $ "The price of bonds is increasing" $\underline{b \qquad\qquad}$

                                                                     $\therefore \sim i$

Assume $\sim i$ is false.

Then i is true.

Suppose each hypothesis is true.

From hypothesis 1, g is then true.

From hypothesis 2, $\sim b$ is then true.

This contradicts hypothesis 3.

Section 12.6

1. $p \lor p \Leftrightarrow p$. Current will flow when p closed.

3.

p	q	$\sim q$	$p \land \sim q$	$\sim p$	$\sim p \land q$	$(p \land \sim q) \lor (\sim p \land q)$
T	T	F	F	F	F	F
T	F	T	T	F	F	T
F	T	F	F	T	T	T
F	F	T	F	T	F	F

Current flows when one switch is open and the other is closed.

5.

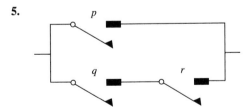

7.

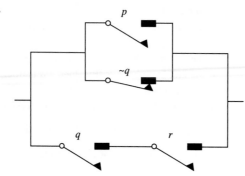

9. $(p \wedge q) \vee p \Leftrightarrow p$ by absorption law.

11. $p \vee (\sim p \wedge \sim q) \vee (p \wedge q) \Leftrightarrow (\sim p \wedge \sim q) \vee p$ Absorption law

$\Leftrightarrow (p \vee \sim p) \wedge (p \vee \sim q)$ Distribution law

$\Leftrightarrow t \wedge (p \vee \sim q)$ Inverse law

$\Rightarrow p \vee \sim q$ Identity law

Current flows if p is closed or q open.

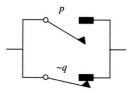

13. $(p \vee q) \wedge (\sim q \vee q) \Leftrightarrow (p \vee q) \wedge t$ Inverse law

$\Leftrightarrow p \vee q$ Identity law

Current flows if either switch is closed.

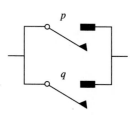

15. $(p \wedge q) \vee (p \wedge r) \Leftrightarrow p \wedge (q \vee r)$ by the distributive law. Current flows if p closed and one of the others closed.

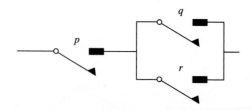

Chapter 12 Review

1. Statement   **2.** Statement   **3.** Not a statement   **4.** Statement

5. (a) Abraham Lincoln was not the shortest President of the United States. (b) Abraham Lincoln was the shortest President of the United States, and Carson City is the capital of Nevada. (c) Abraham Lincoln was the shortest President of the United States, or Carson City is the capital of Nevada. (d) Abraham Lincoln was not the shortest President of the United States, and Carson City is the capital of Nevada. (e) Abraham Lincoln was the shortest President of the United States, or Carson City is not the capital of Nevada. (f) It is not true that Abraham Lincoln was the shortest President of the United States and Carson City is the capital of Nevada.

6. (a) Abraham Lincoln was not the shortest President of the United States. Carson City is not the capital of Nevada. (b) Abraham Lincoln was the shortest President of the United States, or Carson City is the capital of Nevada. (c) Abraham Lincoln was the shortest President of the United States, and Carson City is the capital of Nevada.

7. (a) $\sim q$ (b) $p \vee q$ (c) $p \vee \sim q$ (d) $\sim p \wedge \sim q$

8.

p	q	$\sim p$	$\sim p \wedge q$	$\sim p \vee q$	$(\sim p \wedge q) \wedge (\sim p \vee q)$
T	T	F	F	T	F
T	F	F	F	F	F
F	T	T	T	T	T
F	F	T	F	T	F

9.

p	q	r	$\sim r$	$p \vee q$	$q \wedge \sim r$	$(p \vee q) \vee (q \wedge \sim r)$
T	T	T	F	T	F	T
T	T	F	T	T	T	T
T	F	T	F	T	F	T
T	F	F	T	T	F	T
F	T	T	F	T	F	T
F	T	F	T	T	T	T
F	F	T	F	F	F	F
F	F	F	T	F	F	F

10.

p	q	$\sim q$	$\sim p$	$p \vee \sim q$	$\sim p \vee q$	$(p \vee \sim q) \vee q \vee (\sim p \vee q) \vee \sim q$
T	T	F	F	T	T	T
T	F	T	F	T	F	T
F	T	F	T	F	T	T
F	F	T	T	T	T	T

A tautology

11.

p	q	$\sim p$	$\sim q$	$p \wedge \sim q$	$(p \wedge \sim q) \wedge q$	$\sim p \wedge q$	$[(p \wedge \sim q) \wedge q] \vee [\sim p \wedge q]$
T	T	F	F	F	F	F	F
T	F	F	T	T	F	F	F
F	T	T	F	F	F	T	T
F	F	T	T	F	F	F	F

Not a contradiction

12. (a) True (b) True (c) False (d) True

13. (a) If inflation is increasing, then the price of gold is rising. Contrapositive: $\sim q \to \sim p$. If the price of gold is not rising, then inflation is not increasing. Inverse: $\sim p \to \sim q$. If inflation is not increasing, then the price of gold is not rising. Converse: $q \to p$ If the price of gold is rising, then inflation is increasing. (b) If inflation is increasing, then the price of gold is not rising. Contrapositive: $q \to \sim p$ If the price of gold is rising, then inflation is not increasing. Inverse: $\sim p \to q$. If inflation is not increasing, then the price of gold is rising. Converse: $\sim q \to p$. If the price of gold is not rising, then inflation is increasing. (c) If the price of gold is not rising, then inflation is not increasing. Contrapositive: $p \to q$. If inflation is increasing, then the price of gold is rising. Inverse: $q \to p$. If the price of gold is rising, then inflation is increasing. Converse: $\sim p \to \sim q$. If inflation is not increasing, then the price of gold is not rising.

14.

p	q	$\sim p$	$\sim p \to q$	$\sim(\sim p \to q)$
T	T	F	T	F
T	F	F	T	F
F	T	T	T	F
F	F	T	F	T

15.

p	q	r	$p \wedge q$	$p \wedge q \to r$
T	T	T	T	T
T	T	F	T	F
T	F	T	F	T
T	F	F	F	T
F	T	T	F	T
F	T	F	F	T
F	F	T	F	T
F	F	F	F	T

16.

p	q	$p \to q$	$p \wedge (p \to q)$	$[p \wedge (p \to q)] \to q$
T	T	T	T	T
T	F	F	F	T
F	T	T	F	T
F	F	T	F	T

17.

p	q	r	$q \to r$	$p \to (q \to r)$	$p \wedge q$	$(p \wedge q) \to r$	$[p \to (q \to r)] \to [(p \wedge q) \to r]$
T	T	T	T	T	T	T	T
T	T	F	F	F	T	F	T
T	F	T	T	T	F	T	T
T	F	F	T	T	F	T	T
F	T	T	T	T	F	T	T
F	T	F	F	T	F	T	T
F	F	T	T	T	F	T	T
F	F	F	T	T	F	T	T

18.

p	q	$p \wedge q$	$p \vee (p \wedge q)$	$[p \vee (p \wedge q)] \leftrightarrow p$
T	T	T	T	T
T	F	F	T	T
F	T	F	F	T
F	F	F	F	T

19.

p	q	$p \to q$	$\sim q$	$p \wedge \sim q$	$\sim(p \wedge \sim q)$	$[p \to q] \leftrightarrow [\sim(p \wedge \sim q)]$
T	T	T	F	F	T	T
T	F	F	T	T	F	T
F	T	T	F	F	T	T
F	F	T	T	F	T	T

20. (a) Sue will not major in math and will not major in physics. (b) Sue does not enjoy math or does not enjoy physics.

21.

p	q	r	$\sim q$	$\sim r$	$p \to \sim q$	$\sim r \to p$	$q \wedge (p \to \sim q) \wedge (\sim r \to p) = s$	$s \to r$
T	T	T	F	F	F	T	F	T
T	T	F	F	T	F	T	F	T
T	F	T	T	F	T	T	F	T
T	F	F	T	T	T	T	F	T
F	T	T	F	F	T	T	T	T
F	T	F	F	T	T	F	F	T
F	F	T	T	F	T	T	F	T
F	F	F	T	T	T	F	F	T

Valid

22. s = "It is sunny" $s \to \sim d$

    d = "I study" $\sim s \to \sim b$

    b = "I am at the beach" $\underline{s \wedge b}$

          $\therefore \sim d$

$(s \to \sim d) \wedge (\sim s \to \sim b) \wedge (s \wedge b) \Rightarrow (s \to \sim d) \wedge (\sim s \to \sim b) \wedge s \wedge b$ Associative

          $\Rightarrow (s \to \sim d) \wedge s$ Subtraction

          $\Rightarrow s \wedge (s \to \sim d)$ Commutative

          $\Rightarrow \sim d$ Modus ponens

Valid.

23. $[(p \wedge \sim q) \vee q \vee (\sim p \wedge q)] \wedge \sim q$. Then $[(p \wedge \sim q) \vee q \vee (\sim p \wedge q)] \wedge \sim q$

$\Leftrightarrow [[(p \wedge \sim q) \vee (\sim p \wedge q)] \vee q] \wedge \sim q$ Communitivity, associativity

$\Rightarrow [[(p \wedge \sim q) \vee (\sim p \wedge q)] \vee \sim q] \vee (q \wedge \sim q)$ Distributive law

$\Rightarrow [[(p \wedge \sim q) \vee (\sim p \wedge q)] \vee \sim q] \vee c$ Inverse law

$\Rightarrow [(p \wedge \sim q) \vee (\sim p \wedge q)] \vee \sim q$ Identity law

$\Rightarrow (p \wedge \sim q) \vee [(\sim p \wedge q) \vee \sim q]$ Associative law

$\Rightarrow (p \wedge \sim q) \vee [(\sim p \vee \sim q) \wedge (q \wedge \sim q)]$ Distributive law

$\Rightarrow (p \wedge \sim q) \vee [(\sim p \vee \sim q) \wedge t]$ Inverse

$\Rightarrow (p \wedge \sim q) \vee [\sim p \vee \sim q]$ Identity

$\Rightarrow [(p \wedge \sim q) \wedge \sim p] \vee \sim q$ Associative law

$\Rightarrow [(p \vee \sim p) \wedge (\sim q \vee \sim p)] \vee \sim q$ Distributive law

$\Rightarrow [(t \wedge (\sim q \vee \sim p)] \vee \sim q$ Inverse law

$\Rightarrow [(\sim q \vee \sim p)] \vee \sim q$ Identity law

$\Rightarrow (\sim p \vee \sim q) \vee \sim q$ Communatitive law

$\Rightarrow \sim p \vee (\sim q \vee \sim q)$ Associative law

$\Rightarrow \sim p \vee \sim q$ Indempotent law

Fig.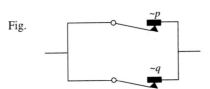

Appendix A.1

1. 10^6 **3.** 10^{-4} **5.** 1000   **7.** 4   **9.** -1 **11.** 2π **13.** 8

15. 81   **17.** $2 \log x + .5 \log y + \log z$ **19.** $.5 \log x + .5 \log y - \log z$

21. $.5 \log x + .5 \log y - .5 \log z$ **23.** $\log x^2 y$ **25.** $\log \dfrac{x^{1/2}}{y^{1/3}}$ **27.** $\log \dfrac{x^3 y}{z^{1/3}}$

29. $.2 \log .3$ **31.** $-.5 \log 3$ **33.** $(1 - \log 2)/3$ **35.** $\pm \sqrt{\log 4}$

37. $10^{-1.5} - 7$ **39.** 2.5

Appendix A.2

1. Yes, 480   **3.** No   **5.** Yes, 52   **7.** Yes, 50, 81,600   **9.** No

11. Yes, 20, 57,700   **13.** 43   **15.** $-.50$ **17.** $d = 1.5, a_1 = 2$

Answers

Appendix A.3

1. (a) $1050.00 (b) $1051.16 (c) $1051.25 (d) $1051.27 (e) $1051.27 (f) $1051.27 (g) $1051.27

3. $1061.84   5. $1648.72   7. $13,463.74   9. 5.65%   11. 6.40%

13. $1652.99   15. $2732.37   17. $27,323.72   19. $22,217.99

21. $561.79, yes   23. 6.77%, 6.72%, first   25. $67,032.01

Index